Special Publication No. 6
of the Society for Geology
Applied to Mineral Deposits

Mineral Deposits within the European Community

Edited by
J. Boissonnas and P. Omenetto

With 221 Figures

Springer-Verlag
Berlin Heidelberg New York
London Paris Tokyo

Dr. JEAN BOISSONNAS
Commission of the European Communities
Directorate-general for Science, Research and Development
Rue de la Loi 200
1049 Brussels, Belgium

Professor Dr. PAOLO OMENETTO
Istituto di Mineralogia e Petrologia
Corso Garibaldi 37
35100 Padova, Italy

ISBN 978-3-642-51860-7 ISBN 978-3-642-51858-4 (eBook)
DOI10.1007/978-3-642-51858-4

Library of Congress Cataloging-in-Publication Data. Mineral deposits within the European Community. (Special publication no. 6 of the Society for Geology Applied to Mineral Deposits) 1. Mines and mineral resources–European Economic Community countries. I. Boissonnas, J. (Jean), 1935-. II. Omenetto, Paolo. III. Series: Special publication ... of the Society for Geology Applied to Mineral Deposits ; no. 6. TN55.M563 1988 553'.094 88-3079

Softcover reprint of the hardcover 1st edition 1988

Typesetting: ASCO Trade Typesetting Ltd., North Point, Hong Kong

2132/3130-543210

Preface

The initial idea for this book arose from a conversation with Professor H.J. Schneider, editor in chief of Mineralium Deposita. He was participating in one of the research projects and suggested that it might be worth while to collect papers on the EC programme in a Special Publication of the SGA. I am pleased to acknowledge the advice he gave me on that occasion and also his help in establishing the first contacts with Springer Verlag.

Special acknowledgements are due to Dr. Ph. Bourdeau of the EC Commission, who was the director responsible for the conception, formulation and overall coordination of the 1978–81 and 1982–85 programmes. Authors and readers alike should be aware that these programmes would not have taken place (and thus by implication no projects could have been funded) without the active support of the national delegations in the Committees advising the Commission on the programmes.

Professor P. Omenetto, acting chairman of SGA during the years 1986–87, has given me constant assistance and encouragement during the successive phases of our editorial work. Despite his many duties, he has even managed to produce a chapter for the book!

Finally, I wish to extend my thanks to my English-speaking colleagues who have helped me on points of style for the preface and introduction, also to my 14-year old son Rémi, a virtuoso on the home-computer and my co-author for the subject index.

Note on the acknowledgments to EC contracts. Contracts with the European Community (EC) are acknowledged at the end of each paper. The letters MSM = (*M*étaux et *S*ubstances *M*inérales, Metals and Mineral Substances) refer to the contracts of the 1982–85 period, whereas the letters MPP = (*M*atières *P*remières *P*rimaires, Primary Raw Materials) designate the projects which were started under the 1978–81 programme. The acronym EEC (European Economic Community) appears in most of the acknowledgement; although widely used in the public, it tends nowadays to be replaced by EC (European Community) in official documents.

Brussels, January 1988 J. Boissonnas

Contents

List of Contributors

You will find the addresses at the beginning of the respective contribution

Introduction

1. Introduction to the Research and Development Programmes of the European Community in the field of Primary Raw Materials

Programme Definition and Objectives. After the oil crisis of 1973–74, serious concern arose within European governments over the short- to medium-term prospects for the supply of mineral raw materials. Although action was taken in some individual countries, it was also thought advisable to promote R&D activities at European Community (EC) level. Discussions between the EC Commission and the Member States eventually led the EC Council of Ministers to adopt, in 1978–79, a series of multiannual research programmes. Of special relevance here is the programme on Primary Raw Materials (1978–81). Also approved were initiatives on Uranium Exploration and Extraction, and on Secondary Raw Materials (recycling).

A further series of programmes was adopted for the period 1982–85. It included the continuation of earlier programmes (but the title Primary Raw Materials was changed to Metals and Mineral Substances) as well as two new initiatives on Substitution and on Wood as a Renewable Raw Material. At present, a third round of programmes is under way for 1986–89.

The programmes on Primary Raw Materials – or Metals and Mineral Substances – are centered on *non-fuel minerals,* excluding iron and building materials. Initially, the main objective was to improve indigenous EC supply. Another aim was to enhance the competitivity of the EC mining sector. Nowadays, in a worldwide situation of over-supply and depressed prices, the latter aim has taken precedence.

In order to meet the first objective, i.e. improving EC indigenous supply, it was decided (1) to encourage the potential for the discovery of deposits in Europe and (2) to optimize methods for the extraction and processing of "domestic" ores. Programmes were accordingly divided into three major research areas: exploration R&D, mining technology and mineral processing.

We are concerned here only with the research area: exploration R&D. EC-funded activities in this field were encouraged out of the belief that there remains a potential for discoveries within the territory of the Community, especially for concealed or buried deposits. Note, however, that in its R&D programmes, the EC does not support routine exploration such as industry would normally undertake. It seeks instead to help to refine concepts, methods and techniques. A significant effort is devoted to economic geology, alongside topics such as geochemical and geophysical prospecting, and remote sensing.

Programme Budgets. While the 1978–81 programme on Primary Raw Materials was allocated 18 million ECU (MECU) in a Community of 9 Member States, this figure rose to only 20 MECU for the 1982–85 period (despite inflation in the meantime), in a Community reinforced by Greece. It remains at 20 MECU for the ongoing 1986–89 programme, although Spain and Portugal have joined the club. Exploration R&D absorbed about half of the budget of the first two programmes but is now down to 20% of the total.

Programme Management. In order to carry out its R&D programmes, the European Commission negotiates contracts with universities, research establishments and private companies in the Member States. Costs are shared, with the EC contributing up to 50%. Preference is given to projects submitted jointly by universities (or research centres) and industry, and also involving international partnership. During the lifetime of any programme, "contact groups" of contractors are set up to ensure that university and industrial scientists investigating related topics meet at regular intervals (at least once a year) to exchange ideas and information on the progress of their research.

2. General Presentation of the Book

In keeping with SGA interest, papers collected in this volume are concerned with the *geology and geochemistry of ores and their host rocks.*

This is *not* a treatise on deposits in the European Community. The areal coverage, although widespread, reflects primarily the response of scientists to calls for proposal, and the subsequent selection of projects.

For reasons explained in section I above, the focus of the book is on EC territory. However, some of the work described was in fact carried out in non-EC countries where research groups had a special interest. This happened particularly in areas where observation was facilitated by excellent conditions of exposure (Oman, for example).

Most of the work described here belongs exclusively to the 1982–85 programme, many projects of which extended well into 1986, if not early into 1987. However, a few projects began under the first programme and were subsequently continued under the second (1982–85). Since Portugal and Spain did not join the Community until 1 January 1986, their scientists could not take part in the programme and therefore the book contains no papers by authors from those two countries.

Clearly, the contents do not represent all that was published as a result of the programme. Some papers have been published already in such journals as Mineralium Deposita or the Transactions of the Institution of Mining and Metallurgy.

The book is divided into three parts:

I. Tungsten (and more or less associated elements such as tin, molybdenum, etc.)
II. Chromite and the platinum-group elements
III. Base metals and other commodities (phosphates, placer minerals) in sediment-hosted deposits.

This threefold division corresponds roughly to three major geological environments: granites, ultrabasic rocks, sediments. In practice, however, a breakdown by commodities has been found preferable, chiefly because tungsten, which is the metal most studied in the book, occurs in metasediments as well as in relation with granites.

Most of the projects have involved the cooperation of scientists from at least two countries. As mentioned above, it should be borne in mind that the programme management had set up a number of "contact groups". There were three of these, corresponding to the subdivisions listed above. Particularly strong links were developed within the Tungsten group, to the extent that a number of projects were carried out as complementary facets of a joint venture.

Part I Tungsten (Tin, Lithium, Molybdenum, Tantalum)

Most types of tungsten deposits occur in Europe: wolframite in quartz veins and in greisens, scheelite in contact skarns, stratabound scheelite in various lithologies. They have received much attention from scientists in recent years, and it is hardly surprising that their study should take up a significant portion of this book. Tungsten, alone or in association with tin (mainly) or lithium, is the subject of 11 out of 13 chapters in Part I and 29 chapters in the whole book.

Geographical coverage ranges from classic provinces, such as SW England, to newly discovered districts (Sicily), with a strong focus on the Pyrenees where research was stimulated by the existence of the important Salau deposit.

In several chapters, there is an emphasis on the search for geochemical signature of fluids, particularly with a view to discriminating between barren and potentially mineralized situations. Whether this approach can effectively be built up into a relatively simple and inexpensive way to select anomalies, as proposed by a number of authors, is a matter for each reader to decide on the basis of his own experience.

Part I also presents papers on molybdenum and tantalum, which reveal a shift towards northern extra-EC countries (Norway, Greenland). In contrast to the granite-related context of most preceding papers, the study on tantalum takes the reader on an excursion to the world of peralkaline syenites.

The first chapter, by Rossi et al., is a comparative study of some granites of Western Europe, both metallogenic (in Brittany and SW England) and barren (in the Pyrenees and Corsica). The aim is to understand the origin of Sn and W, and the behaviour of these metals during magmatic processes.

It is concluded that a high degree of fractional crystallization is a necessary prerequisite for the concentration of metals in a granite, provided, however, that the initial protolith is enriched in those metals. Indeed, a relatively simple set of analyses (REE, Sn, W) is capable of discriminating metallogenic granites.

Shepherd and Miller demonstrate a clear distinction between stanniferous and tungsteniferous ore fluids in the province of SW England. W-bearing fluids are characterized i. a. by enhanced levels of CO_2 and N_2, and a distinctive CO_2-N_2-Ar signature. By contrast, Sn-bearing fluids are depleted in dissolved gas. Those features are sufficiently well developed to warrant their use as an exploration index for granite-related vein/greisen-type tungsten deposits in the area. A comparison with deposits in other areas of Europe, irrespective of age, reveals similar characteristics for the W-bearing fluids.

Guy et al. provide an extensive discussion of stable isotope, major and trace element data on mineralized and barren skarns of the French Pyrenees. For stage I of skarn formation, they demonstrate the dominance of metamorphic waters in barren skarns and of either methamorphic or magmatic waters in mineralized skarns, whereas fluids responsible for stage II hydrosilicate alteration (and for economic concentration of scheelite as well) were dominantly of meteoric origin. The magmatic-hydrothermal model, so overwhelmingly dominant in the literature, is discussed. A potential exploration tool could be developed out of the preposition that $\delta^{18}0$ values of minerals from ore-bearing skarns are lower than those of barren skarns.

Comparing selected granitoids in the eastern Pyrenees and NW Portugal, Salemink and De Jong emphasize the existence of a greater crustal component in the latter magmatic suite. In granitoids associated with large W-skarn deposits, as opposed to barren situations, they recognize a primary zonal distribution of incompatible elements and a specific $\delta^{18}0$ signature. Elevated Cu-Zn contents seem to be good indicators, within the granite bodies, of areas with an increased activity of metallogenic fluids which, in favourable situations, may produce skarn deposits.

In their presentation of Salau, Fonteilles et al. show how ideas on ore controls (lithological, structural and mineralogical) have evolved over the years, and how this has led step by step to better understanding of the deposit, enabling new discoveries. Salau is a difficult deposit, structurally complex; the study is offered as food for thought in the exploration of other, comparable, occurrences.

The contact skarns at Traversella (Piemontese Alps) and Costabonne (E. Pyrenees) occur in very similar environments. According to Dubru et al., 1. fractional crystallization appears to be the major igneous process involved in both cases, although at different degrees of evolution; 2. early metasomatic columns and P, T conditions are similar, and 3. differences in scheelite parageneses of both deposits can be explained by major differences in the activities of components in the ore fluids.

Other aspects of hydrothermal processes are discussed in the next two chapters.

Van Gaans et al. study the spatial variation of the imprints of the different types of hydrothermal alteration in the Sn-W specialized granite of Regoufe, Portugal. They stress the importance of mineralogy or major element chemistry in the response to hydrothermal processes. Data for the study were obtained by Integral Rock Analysis, a fast and newly developed method for the acquisition of large quantities of detailed rock geochemical data, based on direct XRF spectrometry on unpolished rock sections.

Raimbault describes REE spectra in hydrothermal minerals such as apatite. He shows how they can be used to assess genetic relations between granites and W mineralization in the French Massif Central, or to constrain the genesis, chemistry and evolution of mineralizing fluids.

Two chapters describe occurrences of stratabound scheelite in very different settings. De Smedt and Sonnet discuss the formation stages of calc-silicate gneiss (CSG) lenses included in high-grade metasediments of SE France. The CSG were primarily derived by isochemical transformation of a mixed limestone-greywacke protolith and subsequently modified by infiltration metasomatism. The authors argue for a peri-anatectic origin of the W- bearing fluids, an alternative to perigranitic or sedimentary-exhalative models often invoked in similar circumstances.

Who among us, only 5 years ago, knew of NE Sicily as a W-rich district? Scheelite concentrations, some being of unusually high grade, are found at various levels of a sequence of phyllites, black shales and carbonates of the pre-Hercynian basement. The chapter by Omenetto et al. provides a detailed account of the lithological and structural setting, associated polymetallic mineralization, parageneses (note in some cases the association with tourmaline), and draws our attention to possible comparisons with other segments of the circum-mediterranean Hercynian belt, both in Europe and in North Africa.

The Caledonides of SE Ireland contain spodumene pegmatites and some occurrences of scheelite. According to McArdle and Kennan, mineralization is confined to a major shear zone, and only to that part of it which traverses a volcanic sequence and its associated sediments, where these contain coticules and tourmalinites. The shear zone developed synchronously with the emplacement of the Leinster granite nearby.

The last two chapters of this section lead us to more remote areas. Schønwandt reviews an extensive range of porphyry Mo deposits in two provinces of the North Atlantic: the Permian Oslo graben and the Tertiary igneous province of eastern Greenland, both representing intraplate activity. Molybdenite typically occurs in association with syenite-granite complexes, and in that context is spatially related to highly fractionated alkali-rich, high silica intrusives. Striking similarities are noted with Climax (Colorado), both in style of mineralization and alteration, and in the associated granites.

The Motzfeldt Centre, one of the major complexes in the Gardar province of alkaline magmatism, contains zones of pyrochlore enrichment which offer good economic perspectives. The deposit was discovered by the combined processing of geochemical and remote sensing data during an earlier EC-funded project. Tukiainen's paper describes the complex and the mineralization. In the outer unit, extreme in situ differentiation produced a peralkaline residuum rich in volatiles and incompatible elements (Nb, Ta...). Secondary concentration of those elements took place during a subsequent phase of greatly increased hydrothermal activity.

Part II Chromite and Platinum-Group Elements

Chromite and the platinum-group elements (PGE) are critical metals for industrial uses, and possible disruption of supply from external sources (e.g. South Africa) would be felt acutely in the European Community. Accordingly, research on those metals was given some priority in the EC programmes. EC countries offered several exploration opportunities, mostly restricted to the ophiolitic environment and, initially, to chromite: France (New Caledonia), Greece, the United Kingdom (Shetlands). Projects on chromite were carried out in those areas, as well as in Oman, during the 1978–81 programme.

Research in the Shetlands has proved especially rewarding. As a spin-off from the 1978–81 project, a systematic study for platinum-group minerals (PGM) was initiated under the second programme. This very soon resulted in the discovery of unexpected grades of all six PGE, prompting a great deal of academic and industrial interest in ophiolites as a possibly better source of PGE than was previously believed.

This section of the book is comparatively short, partly because projects were limited in number and also because much of the EC-funded research has already been published elsewhere, particularly in the proceedings of the IMM conference on the metallogeny of basic and ultrabasic rocks (Edinburgh, 1985). For a more complete coverage of the EC programme in this field, the reader is referred to the conference volume (Gallagher et al. 1986) namely to the chapters by Christiansen, Dunlop and Fouillac, Johan and Augé, Legendre and Augé, Makovicky et al., Prichard et al.

Part II encompasses a wide range of topics, from field investigations to experimental work in the laboratory.

Roberts and his co-workers provide clues to the structural controls of chromitites in the Vourinos complex. Emplacement fabrics appear to be more widely distributed than in most other ophiolite systems. Detailed studies of two key areas enabled the authors to discriminate between mantle (plastic) and emplacement (brittle to ductile) structures, and to assess their respective roles in each deposit. Since this chapter was written, the approach described has proved fruitful in the siting of successful drill-holes.

Augé and Johan have compared chromitites from Vourinos (Greece), Troodos (Cyprus), Semail (Oman) and New Caledonia. Compositions reflect the diversity of ophiolitic mantle sequences. Differences in the parageneses of interstitial vs. included silicates are indicative of unusual chemical characteristics of the ore-forming systems. Early crystallized PGM (Os, Ir and Ru mineral phases) were found in all the studied ore bodies, except in N. Oman. Their compositions and variations from one deposit to another enable the discussion of the thermodynamic conditions of formation of chromite bodies.

Following on the discovery, recalled above, of unexpected grades of all six PGE in the Shetlands, Prichard and Lord now provide evidence for a Supra Subduction Zone origin of the ophiolite. They report on the occurrence of PGE concentrations in chromite-rich samples at all levels in the ultramafic part of the complex. The discovery of PGM in fresh dunite adjacent to chromite lenses implies that they were concentrated by primary igneous processes and that concentrations are not solely related to alteration zones, as could have been envisaged.

Makovicky and Rose-Hansen have studied the solubility of Pt and Pd in base metal sulphides (pyrrhotite and pentlandite), by means of laboratory syntheses at 900° and 500° C, followed in some cases by prolonged annealing at 300°. Pentlandite concentrates Pd. Solubilities of Pt and Pd in pyrrhotite are appreciable at 900°, but they drop as temperature decreases. Such results can explain natural occurrences of PGE associated with – and exolved from – base metal sulphides.

Part III Base Metals, Phosphates, Placer Minerals

The unifying concept in Part III is that of sediment-hosted deposits. The focus is on lead and zinc in carbonates, particularly in central Ireland, but also in the British Pennines, the Italian Alps, Sardinia, Montagne Noire (France) and southern Spain. Other geological settings and types of mineralization are described in a review paper on the Rhodope massif in northern Greece. Part III also includes investigations on phosphate rock in Belgium and on placer deposits off the coast of N. Greece.

The challenge of exploration for buried carbonate-hosted ore deposits is taken up in the first five chapters, which present multidisciplinary studies together with research on the application of rock geochemistry and related disciplines.

The first two contributions are examples of the integrated use of geological information in the development of exploration strategies. Plant et al. describe an approach based on new metallogenic models and on methods of basin analysis usually applied to hydrocarbon exploration. Image analysis was used to process a wide range of spatially related geological, geochemical and geophysical data from eastern England and prospective areas were identified. Syngenetic/syndiagenetic Irish-style deposits are believed to have been formed by the expulsion of fluids from half-graben basins, which reached the sea floor via listric faults, whereas the proposed model for epigenetic Pennine-style deposits (a fluoritic sub-type of MVT deposits) involves dewatering of shale basins and deposition in fractures in platform limestones. Buried, high heat production granites locally focussed ore fluid flow so that mineral zones are spatially related to their subcrop.

Phillips et al. use combinations of geophysical, structural, lithological and remote sensing data to assess the structural control of mineralization at Silvermines and Navan in central Ireland. Economic deposits formed in dilation zones which developed where Caledonian shear zones, reactivated during the Carboniferous, terminated against buried Caledonian granites or volcanic complexes. In further exploration, it will therefore be essential to identify these shear zones and their terminations against rigid blocks. This purpose can be served by a multidisciplinary approach such as outlined above.

Rock geochemistry is the main topic of the next two chapters, both being on the Lower Carboniferous of central Ireland. Van Oyen and Viaene have studied an area in the vicinity of the Navan mine, evaluating lithogeochemical data for a number of elements in relation to such features as sediment petrography, carbonate diagenesis and mineralogy. They demonstrate the existence of large-scale lithological and lithogeochemical trends, pointing towards mineralization. In presenting the Tynagh case history, Clifford et al. give us the exploration geologist's view. They describe methods of enhancing the anomaly contrast and of identifying vectors towards potential mineralized bodies, and conclude that rock geochemical studies, when coupled with detailed geological control, can provide an effective tool in exploring for base metal sulphides in carbonate environments. This seems especially true in areas of poor exposure such as central Ireland, where most of the material available for study is in the form of drill cores.

The paper by Cazalet et al. describes the results of an investigation into the possible use of light (C1–C5) hydrocarbons as geochemical pathfinders for mineral exploration. Most of the research was carried out in central Ireland. Distinct anomalies in the gas content of rocks have been found in close spatial relationship with ore bodies. Although more work is needed to fully understand the origin of anomalies and to perfect the method, this novel field of research has obvious potential for large scale reconnaissance: like conventional soil- and rock geochemistry, it increases the effective size of exploration targets.

During the Lower Cambrian, carbonate shelf sedimentation prevailed in Sardinia and Montagne Noire (S. France). The primary objective of the next two chapters is to correlate metallogeny with the paleogeographic evolution of those two areas. Summarizing data from SW Sardinia and the southern slope of Montagne Noire, Courjault-Radé and Gandin stress the significance of repeated tensional phases, which ultimately resulted in the collapse of the shelf at the Lower-Middle Cambrian boundary. Episodes of metallogenesis can be clearly related to these phases of instability.

In their study of the northern side of Montagne Noire, Lescuyer et al. provide evidence for still greater instability of the shelf margin. They also describe in some detail a number of deposits, some being stratiform and sedimentary-hydrothermal, others being related to late Hercynian magmatism and fracturation.

Fontboté and Gorzawski have used Sr and stable isotopes to trace the diagenetic evolution of carbonate-hosted Pb-Zn ore deposits in the Triassic of southern Spain and around the San Vicente mine in the Liassic of Peru. In both districts, there is a clear relationship between the depositional environment of host rocks and the occurrence of ore. The model proposed implies that ore was formed during relatively early stages of diagenetic evolution and provides, for those deposits, an alternative to models based on late- to post-diagenetic migration of basinal brines.

The mineralizations of the Paleocarnic Alps (Italy) are bound to a paleo-relief of Devonian limestones. According to Brigo et al., ore concentration took place during a phase of erosion and karstification of the paleo-relief. Initial hydrothermalism, possibly linked to Carboniferous volcanism, is suggested by a study of Ga/Ge ratios. Extensive low temperature remobilization is indicated by REE distribution patterns and other lines of evidence.

Nesbitt et al. describe the geological setting of base metal mineralization in the Rhodope region of NE Greece. All major deposits, both within the Lower Paleozoic (?) basement and the Mesozoic-Cenozoic cover sequences, appear to be of Tertiary age. Throughout the region the bulk of the mineralization is, in general, fracture-controlled. Field and remotely sensed data provide the basis for a metallogenic model in which magma genesis, basin development and fracturation are closely linked. Pb-isotope data are used to discuss the source of metals. This study has obvious implications for defining a strategy of future exploration in the area.

In 1979–1980, a working party of EC experts had listed possible subjects on phosphate rock for future inclusion in an R&D programme, bearing in mind that a critical situation might arise in the EC if external sources of supply were severed. Only one of those subjects was retained into the Exploration area of the 1982–85 programme, but the work proved to reveal an unexpectedly large resource. Robaszynski and Martin describe the Late Cretaceous phosphatic chalk of the Mons basin in Belgium. The thickness of the formation generally exceeds 20 m, even 70 m in places. Admittedly, the grade is only 5 to 10% P_2O_5; however, there seems to exist some scope for benefication.

The text of the 1982–85 programme specified that research could be carried out "in the immediate near-shore areas" of the EC continental shelf. Accordingly, the EC has supported investigations in the Aegean sea. The paper by Perissoratis et al.

reports on occurrences of heavy minerals, some of them quite significant, off the coast of N. Greece (East of Macedonia) and discusses the relevant geochemical data.

Conclusion

From this brief presentation of the contents of each paper, the reader will have formed an idea of the variety of topics addressed and of the results achieved. This book was intended as a contribution to illustrate the vitality of European research in economic geology and related fields. It is much to be hoped that research budgets may not drop to the level where most teams are dispersed and work is discontinued. It is not enough to bring together researchers from several countries for the duration of a 3–4 year contract. Such an effort must be maintained, otherwise its benefits will inevitably decline. European scientists and mining companies must be able to retain their capability to face international competition in the event that exploration activities worldwide begin to recover in the future.

Brussels, January 1988 J. Boissonnas

Part I
Tungsten (Tin, Lithium, Molybdenum, Tantalum)

Metallogenic W and Sn Granites: Genesis and Main Distinguishing Features

PH. ROSSI[1], A. COCHERIE[1], G. MEYER[2], A.M. FOUILLAC[3], and A. AUTRAN[3]

Abstract

Metallogenic granites of Western Europe and unmineralized granites from Corsica and the Pyrenees have been studied geochemically to characterize the metallogenic granites. To obtain reliable results we have (1) verified the cogenetic nature of each of the suites using stable (O, D) and radiogenic (Sr) isotope geochemistry and trace element (e.g., REE) geochemistry and (2) analyzed selected samples for Sn and W using the very sensitive and accurate INAA method.

Thus it has been demonstrated that the preconcentration of metals in a granite is the first condition necessary for mineralization to occur; the second condition is that the magmatic liquid must be highly differentiated, leading to a rather flat REE pattern with a marked negative Eu anomaly; the third condition is related to the collection of the metals by a high temperature, late-magmatic fluid; only restricted amounts of meteoric and/or metamorphic water are involved in the case of the studied targets.

1 Introduction

One of the problems that confronts the mineral exploration geologist is to have available a tool that will discriminate favorable areas for exploration. In the case of W and/or Sn mineralization (except skarns), their constant link with highly silicic leucocratic granite has led to the concept of specialized granites (Stemprok et al., 1977). The relation between granite and mineralization said to be of "acid origin", even though well-known from a spatial viewpoint by most mining geologists, remains insufficiently understood from a genetic viewpoint. It is important to be able to distinguish, among metallogenic granites, between granites that are themselve metal sources and granites that act simply as the medium that sets off

[1] GIS (BRGM — CNRS) 1A, Rue de la Férollerie, 45071 Orleans Cedex 02 (France)
[2] Laboratoire P. SÜE, CEN Saclay, B.P. n° 2, 91190 Gif-Sur-Yvette (France)
[3] BRGM B.P. 6009, 45060 Orleans Cedex (France)

Mineral Deposits within the European Community
(ed. by J. Boissonnas and P. Omenetto)

convecting hydrothermal phenomena because of their thermal effect on the country rock.

First of all there is the simple generally observed fact that no economic W and Sn deposits (except for some skarns) exist in the surroundings of gabbros, diorites, and granodiorites. This indicates that the simple direct thermal effect of a granite related to mineralization is not alone sufficient for the mineralization to occur.

Secondly, assuming that large-scale convective hydrothermal circuits are involved, the key fluids are more likely to be abundant in highly silicic and sodi-potassic granites. These fluids can induce both mechanical and chemical effects:

1. They can induce hydraulic fracturating or reactivation through the stresses caused by the intrusion. The ratio between magmatic and nonmagmatic waters may vary in time and space within and around the pluton during unroofing and cooling.
2. They can collect hygromagmaphile elements and lead to ore deposits.

Although the first point is now accepted, the second is still debated and the origin of elements of economic interest is controversial (Marignac and Weisbrod 1986). Are the Sn and W of magmatic or country rock origin? The aims of this chapter are:

To determine the origin of the metals and their behavior during magmatic processes (post-magmatic processes are not considered here). If metals have a magmatic origin, the knowledge of their behavior and content through these magmatic processes can be a useful tool to discriminate metallogenic granites.

The samples investigated were from Cadomian (late Precambrian) and Variscan metallogenic granites in Western Europe. Their geochemical evolution was compared with that of calc-alkaline, unmineralized granites from Western Corsica and the Pyrenees. Analysis of radiogenic Sr isotopes and of trace elements in each of the plutonic masses studied ensured that the intrusions in which the evolution of metallic-element concentration was studied were indeed cogenetic. This point seems to have been ignored up to now, as the correlations established between Sn/W/Ta and an indicator of the degree of differentiation of the granitic magma (Rb/Sr or DI) were mostly based on samples from different intrusions in the same region, whose cogenetic relationships were far from clear. Without rigorous examination of these conditions, no hypothesis that tries to quantify enrichment processes of certain metallic elements by fractional crystallization from a silicate melt can be considered valid. Although preliminary geochemical studies (Sr isotopes, trace elements, etc.) do not prove that the samples come from a suite of rocks resulting from fractional crystallization of a single granitic magma, they must nevertheless be regarded as at least a necessary condition.

The recent development of sensitive and reliable methods for measuring Sn contents to 1 ppm and W to 0.2 ppm (neutron activation on irradiated powder, Meyer et al., 1985) has shed new light on the geochemical behavior and concentration mechanisms of Sn and W at low grades which enables a criterion to be proposed for discriminating between unmineralized and metallogenic magmatic suites. The approach adopted was thus to study, by comparison between unmineralized and metallogenic cogenetic entities, the behavior of elements such as W, Sn, and Ta (hygromagmaphile elements) in different granites, in order to define their modes of concentration.

Chondrite-normalized REE patterns are also good indicators of the state of differentiation within a granite system and of the type of magmatic association to which it belongs. With such a powerful tool to indicate the degree of maturity of the silicate melt, it was obvious that this information could be combined with data that directly concern the evolution of W and Sn, both in mineralized and unmineralized rocks.

We have also used the geochemistry of the stable isotopes of D and O to attempt to quantify the amount of magmatic fluid involved in the hydrothermal circuits when a batholith is emplaced (e.g., the Fougères and St. Renan batholiths, in the west of France), and to demonstrate the interaction, or lack of interaction, of fluids from outside the magmatic system, which could drain the metallic elements from the host rock towards the granite or vice versa.

2 Geological Setting of the Granite Intrusions Studied

Leucogranites were sampled from the Cadomian batholith centered on Fougères in Brittany, France (Fig. 1, Tables 1, 2). A large part of this batholith is composed of medium- to coarse-grained peraluminous biotite and/or cordierite granodiorite dated at 521 ± 11 Ma (Autran et al., 1983) to 540 ± 10 Ma (Pasteels et al., 1982), and intruded by small bodies of leucogranite dated at 490 ± 14 Ma (Fouillac et al., 1986). The leucogranites show high levels of W and Sn, but only one occurrence has produced economic mineralization, mined at Montbelleux mine. The crustal derivation of these granites is attested by their peraluminous composition, by the presence of numerous metamorphic xenoliths, and by the strontium (Sr_i 0.7093 ± 0.0005 for the granodiorite and 0.715 ± 0.003 for the leucogranite, Fig. 2a) and oxygen ($10.8 < \delta^{18}O < 12.9$) geochemistry. Thus the granodiorites and leucogranites are not cogenetic. Only leucogranites will be considered here.

Variscan mineralized granites were sampled at Dartmoor in Devon (United Kingdom), at St. Renan in Brittany (France) (Fig. 1).

The samples of megacrystic medium- to fine-grained granite from the Dartmoor batholith (Tables 1, 3), northeast of Plymouth, define an isochron (Fig. 2b) giving an age of 276 ± 8 Ma and indicating that the granites are of continental derivation and cogenetic ($Sr_i = 0.7090 \pm 0.0019$, Schneider, pers. commun.). Mineralization of similar age occurs in the Vitifer mine located in the center of the granite body (Shepherd, pers. commun.).

The St. Renan batholith in northwestern Brittany is a medium-to coarse-grained biotite-muscovite granite (Tables 1, 4) extending 20 km E–W and 7 km N–S intruding the Lesneven gneissic basement (Chauris, 1965). Many zones in the roof of the pluton are greisenized, and numerous cassiterite flats have been exploited since Roman times. The various phases of the intrusion do not define an isochron (Fig. 2c). However, a model age of 314 ± 21 Ma has been calculated, giving an initial

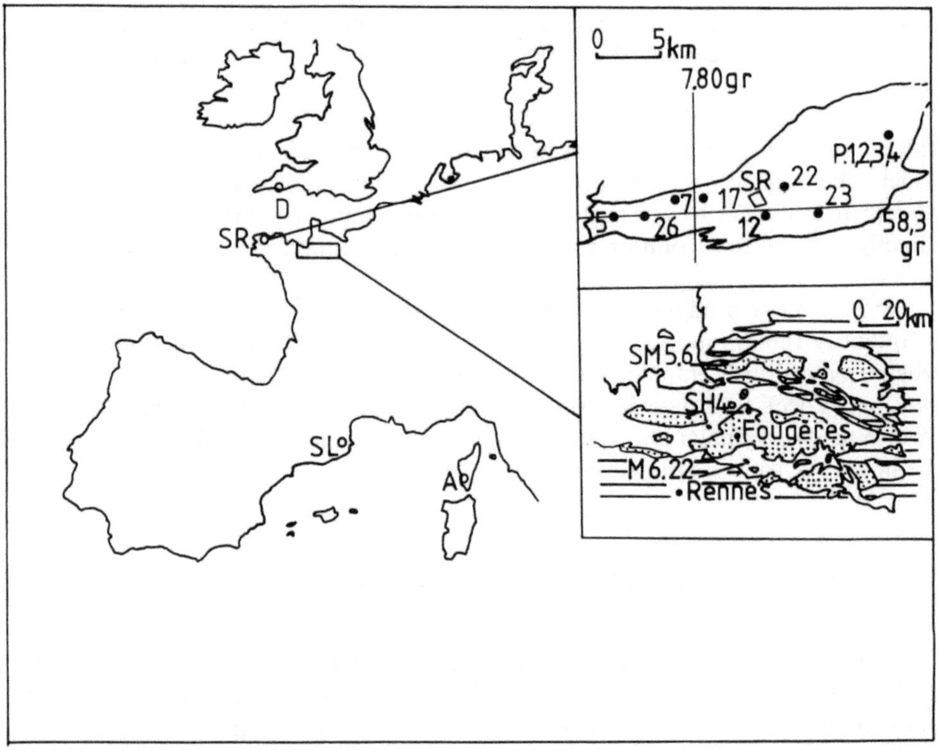

Fig. 1. Location of granite intrusions mentioned in this paper. Mineralized: *D* Dartmoor (Sn–W); *SR* St. Renan (Sn–W). The geological sketchmap of the Cadomian Fougères batholith shows the location of granodiorite and monzogranite (*dotted areas*) and of leucogranites (*black spots*). The schists and graywacke of the Cadomian basement are shown in *white* and the post-granitic deposits as *horizontal lines*. Unmineralized: *SL* St. Laurent (Pyrénées); *A* Ajaccio (Corsica)

strontium ratio of 0.7087 $\pm$ 0.0021, indicating derivation from a continental protolith. In the Penfeunten area, the granite is mainly mineralized in Sn (Charoy 1975), the contact between the Penfeunten "innengranite" unit and the St. Renan unit being marked by a vertical stockscheider more than half a meter thick.

Unmineralized calc-alkaline granites were sampled in Corsica and in the Pyrenees (Table 5). The geology and geochemistry of these Corsican granites were described in detail by Cocherie (1984) and by Rossi (1986). The sequence of main intrusive phases can be summarized as follows: (1) emplacement of a large granodioritic unit (about 300 km^2); (2a) monzogranites intruding the previous unit; (2b) sheets or massifs of basic rocks in spatial association with these two units, emplaced synchronously or slightly later; (3) emplacement of leucocratic monzogranite units extending over several kilometers and controlled by N 50° trending structures. These granites are medium to fine grained and characterized by Fe-biotite and almandine-spessartine garnet as characteristic accessory phase.

The plutonic association of the St. Laurent de Cerdans massif in the Pyrénées was studied by Autran et al. (1970), Fourcade and Allegre (1981) and Cocherie (1984). This plutonic association is composed of a granodioritic to monzogranitic

Table 1. Rb–Sr isotopic data for Fougères, Dartmoor and St. Renan granites. For Rb–Sr data of the Corsican and Pyrenean granites, see Cocherie (1984)

	Rb (ppm)	Sr (ppm)	$^{87}Rb/^{86}Sr$	$^{87}Sr/^{86}Sr$
Fougères leucogranites				
SM 6 B	220.2	60.3	10.66	0.79015
SM 5	199.7	40.5	10.43	0.81436
SH 4	172.6	16.0	31.88	0.93714
M 9	221.9	18.3	36.03	0.96830
Dartmoor granite				
922	368.5	95.3	11.25	0.75276
1	416.1	80.6	15.03	0.76875
S 25	414.3	70.1	17.23	0.77723
35	483.5	79.4	17.74	0.77815
22	433.1	43.7	28.98	0.82087
16	434.3	26.4	48.45	0.90012
St. Renan granite				
SR 23	185.4	182.9	2.94	0.72104
SR 7	241.2	160.1	4.37	0.72963
SR 17	288.1	136.3	6.14	0.73705
SR 11	249.4	106.4	6.81	0.73713
SR 26	271.5	68.7	11.50	0.76173
SR 5	346.1	74.6	13.51	0.76843
SR 10	367.1	43.2	24.87	0.81792

suite cross-cut by a miarolitic leucocratic granite. These later rocks are characterized by red-colored K-feldspar and Fe biotites. Compared to the other Pyrenean granites, the St. Laurent massif is geochemically marked by a more potassic and less aluminous trend.

The calc-alkalic affinity of these granites can be refined by using biotite chemistry (Nachit et al., 1985). Figure 3 shows representative biotite compositions plotted on an Al versus Mg diagram. The dotted line marks the boundary between "calc-alkalic" and "aluminous and potassic" (i.e., muscovite-bearing rocks). It can be seen that the composition of the most ferrous biotites (corresponding to the most evolved melts) converge within the same area. On this diagram, the compositions of biotites of hypersilicic leucocratic granites associated with Sn and W deposits from Europe (this work) are compared to the biotites of granites of the same nature from China (Eocene granites, Western Yunnan, unpubl. data) and Brazil (Precambrian granites of Goias state; Botelho, 1985). This diagram indicates that granitic suites which are characterized by contrasted values of Al/Mg ratio in biotites (i.e., aluminous and potassic suites or calc-alkalic) can both lead to leucocratic melts enriched in W and Sn. This evidence leads us to think that granitic magmas have the capacity to concentrate metals, whatever the nature of the suite may be. It can also be seen that Mg decreases more quickly in mineralized than in unmineralized suites. This feature might serve as a good discriminant factor between these two suites. Nevertheless more data are needed in order to test this tool.

Table 2. Major (% oxide) and trace element (in ppm) concentrations in leucogranites of Fougères (Brittany)

	Leucogranites				Montbelleux mine					
	SM 5	M 9	SH 4	SM 6B	M 5	M 6	M 7	M 12	M13	M 22
SiO_2	74.20	75.20	75.80	74.50	83.68	77.19	83.38	76.32	77.75	63.90
Al_2O_3	14.00	13.85	13.30	14.25	9.14	12.91	8.85	13.18	13.77	19.03
Fe_2O_3	1.17	0.53	0.86	1.25	0.38	0.40	0.65	0.81	0.31	0.22
FeO	*	*	*	*	0.93	1.20	1.20	1.70	0.50	0.72
CaO	0.60	0.18	0.40	0.60	0.29	0.15	0.47	0.20	0.16	3.39
MgO	0.41	0.30	0.43	0.36	<0.20	<0.20	<0.20	<0.20	<0.20	<0.20
Na_2O	3.05	3.10	3.05	3.25	2.08	4.73	0.70	4.35	5.96	8.16
K_2O	4.05	4.70	4.35	4.10	1.78	1.43	2.38	2.03	1.15	1.76
MnO	<0.02	<0.02	0.02	0.02	0.11	0.10	0.12	0.14	0.04	0.18
TiO_2	0.10	0.06	0.09	0.11	<0.02	<0.02	<0.02	<0.02	<0.02	0.02
P_2O_5	0.40	0.17	0.20	0.40	0.12	0.05	0.19	0.05	0.06	0.06
PF	1.05	1.05	0.90	0.85	1.11	1.37	1.53	1.17	0.68	2.92
Total	99.03	99.14	99.44	99.69	99.84	99.75	99.69	100.17	100.60	100.56
U	3.7	4.1	8.7	3.6	15.5	7.7	11.0	11.0	14.9	14.4
Th	3.0	2.4	1.54	1.21	8.6	12.8	8.9	7.5	9.5	12.3
Ta	1.74	1.62	0.98	2.18	6.8	8.8	7.7	4.1	4.8	5.6
Rb	200	222	170	210	375	360	490	230	300	400
Sr	39	16	13	56	31	37	29	29	62	72
Ba	254	157	83	154	13	12	24	27	18	47
W	1.4	5.0	3.1	5.8	539	634	876	4.5	6.0	6.2
Sn	6.8	11.3	5.4	9.9	61	48	224	36	83	101
La	6.9	4.5	2.1	3.8	3.3	2.7	2.7	3.2	1.95	3.8
Ce	13.5	9.1	4.7	7.1	11.0	11.7	10.1	11.1	8.3	14.7
Nd	6.0	3.9	2.5	2.3	5.8	7.4	5.4	5.5	5.2	7.9
Sm	1.6	1.4	0.65	0.90	2.4	3.2	2.2	2.3	2.4	3.7
Eu	0.31	0.14	0.07	0.20	0.11	0.07	0.17	0.04	0.09	0.13
Gd	1.44	1.44	0.79	0.94	2.1	2.48	1.92	1.91	1.75	3.3
Tb	0.28	0.26	0.17	0.18	—	—	—	—	—	—

	Leucogranites				Montbelleux mine					
	SM 5	M 9	SH 4	SM 6B	M 5	M 6	M 7	M 12	M13	M 22
Dy	—	—	—	—	2.6	3.1	2.4	2.6	2.1	3.9
Er	—	—	—	—	1.2	1.4	1.1	1.2	0.90	1.7
Tm	0.17	0.19	0.11	0.15	—	—	—	—	—	—
Yb	1.20	1.10	1.11	1.00	2.0	2.8	1.84	2.2	1.94	3.3
Lu	0.20	0.18	0.20	0.16	0.38	0.54	0.34	0.36	0.31	0.57
Σ REE	36.3	26.5	15.3	19.9	33.3	38.3	30.4	32.8	26.9	46.4
La/Yb	5.8	4.1	1.9	3.8	1.7	0.96	1.5	1.5	1.0	1.2
Eu/Eu*	0.63	0.30	0.30	0.67	0.15	0.08	0.26	0.06	0.14	0.12

Σ REE = total REE content with interpolated values.
Eu* = interpolated value of Eu between Sm and Gd.

Table 3. Major (% oxide) and trace element (in ppm) concentrations in granitic rocks of Dartmoor in Devon (U.K.)

	Coarse grain granites with phenocrists		Coarse grain granites			Fine grain granites			
	1	4	3	21	5	12	13	16	26
SiO_2	71.10	74.60	75.40	75.70	76.50	77.10	77.10	77.40	
Al_2O_3	13.70	14.40	12.80	13.20	13.00	12.44	12.80	12.80	
Fe_2O_3	2.00	1.45	1.60	1.90	1.35	1.05	1.35	1.10	
FeO	1.50	0.78	0.60	0.28	0.74	<0.15	<0.15	<0.15	
CaO	1.20	0.80	0.52	0.47	0.47	0.48	0.28	0.24	
MgO	0.61	0.53	0.32	0.28	0.31	<0.20	<0.20	<0.20	
Na_2O	2.95	3.25	3.00	3.15	2.80	3.35	2.55	3.05	
K_2O	4.80	4.70	4.60	4.35	4.70	4.20	4.95	4.40	
MnO	0.08	0.07	0.06	0.10	0.06	0.06	0.05	0.04	
TiO_2	0.42	0.25	0.17	0.17	0.17	0.04	0.03	0.03	
P_2O_5	0.21	0.17	0.18	0.19	0.19	0.06	<0.05	<0.05	
PF	0.78	0.54	0.50	0.47	0.61	0.84	0.68	0.63	
Total	99.35	101.54	99.75	100.26	100.90	99.97	100.19	100.05	
U	13.4	16.4	31.8	26.1	25.1	10.1	12.3	9.7	—
Th	26.2	13.1	15.3	11.4	11.8	12.7	14.6	14.3	—
Ta	3.26	3.58	5.61	5.61	5.07	6.14	4.00	3.33	2.85
Rb	410	385	620	670	650	830	640	440	620
Sr	75	45	15	17	<10	27	<10	21	15
Ba	265	155	115	85	75	<10	<10	85	11
W	8.1	1.0	15	13	16	28	15	7.3	13
Sn	4.0	7.9	15	13	28	20	19	19	21
La	39.1	21.2	16.1	14.5	14.9	10.7	14.8	9.8	12.5
Ce	81.0	44.8	34.1	30.8	37.0	29.1	30.3	24.2	27.1
Nd	38.7	19.5	17.3	15.7	16.1	12.6	17.8	9.9	17.4
Sm	8.5	4.2	4.2	3.9	4.3	4.2	5.5	3.7	5.4
Eu	0.91	0.50	0.28	0.26	0.27	0.19	0.33	0.37	0.36
Gd	7.1	3.9	3.9	4.0	4.1	4.8	5.5	3.8	6.1

	Coarse grain granites with phenocrists		Coarse grain granites			Fine grain granites			
	1	4	3	21	5	12	13	16	26
Dy	5.8	3.4	3.8	3.9	3.9	6.4	7.2	5.5	8.4
Er	3.1	1.8	1.9	1.8	2.0	4.1	4.4	3.5	4.9
Yb	4.4	2.0	2.1	2.0	2.2	5.5	5.9	5.2	6.7
Lu	0.51	0.31	0.31	0.26	0.29	0.79	0.82	0.77	0.94
Σ REE	201	110	90	83	91	83	99	72	95
La/Yb	11.4	10.6	7.6	7.4	6.9	2.0	2.5	1.9	1.9
Eu/Eu*	0.36	0.38	0.21	0.20	0.20	0.13	0.18	0.30	0.19

Σ REE = total REE content with interpolated values.
Eu* = interpolated value of Eu between Sm and Gd.

Table 4. Major (% oxide) and trace element (in ppm) concentrations in granitic rocks of the St. Renan batholith (Brittany)

	Penfeunten innengranite			St. Renan granite						
	P 1	P 4	P 3	SR 7	SR 17	SR 5	SR 26	SR 11	SR 23	SR 10
SiO_2	75.10	75.28	76.25	71.93	74.21	73.87	74.14	72.46	73.05	74.42
Al_2O_3	14.17	12.95	12.91	14.84	14.96	14.66	14.61	14.35	14.43	15.04
Fe_2O_3	0.94	2.07	1.54	0.37	0.60	0.33	0.31	0.37	0.35	0.10
FeO	0.75	0.86	1.10	1.50	0.68	1.00	0.86	1.90	1.30	0.75
MnO	0.06	0.08	0.09	0.03	0.03	0.04	0.03	0.04	0.03	0.03
MgO	<0.20	<0.20	<0.20	0.44	<0.11	<0.20	<0.20	0.35	0.33	<0.20
CaO	0.50	0.53	0.51	1.08	0.77	0.68	0.62	1.13	1.28	0.22
Na_2O	2.97	2.11	1.23	3.45	3.26	3.23	3.36	3.20	3.37	3.15
K_2O	4.41	3.60	3.97	4.63	4.65	4.85	4.72	5.03	4.76	4.77
TiO_2	0.04	0.04	0.04	0.25	0.10	0.14	0.10	0.41	0.24	0.05
P_2O_5	0.30	0.34	0.30	0.12	0.15	0.18	0.22	0.18	0.09	0.13
P.F.	1.14	1.76	1.99	1.56	1.12	1.15	1.12	0.57	0.81	1.64
Total	100.58	99.82	100.13	100.20	100.64	100.33	100.29	99.99	100.04	100.50
U	—	—	—	—	—	—	—	—	—	—
Th	1.52	1.31	1.18	13.0	6.9	10.7	19.2	33.1	23.0	5.1
Ta	12.4	8.5	9.5	1.39	1.50	1.97	1.54	1.25	0.81	1.46
Rb	550	520	470	241	288	346	272	249	185	367
Sr	23	19	24	160	136	75	69	106	183	43
Ba	12	18	10	624	373	364	351	846	192	133
W	8.3	27	6.4	1.8	5.7	2.3	1.9	1.3	1.4	3.7
Sn	67	50	93	13	13	15	7.5	6.0	5.7	11
La	2.18	2.60	1.62	—	—	26.0	10.40	49.81	34.79	8.39
Ce	4.14	5.75	3.56	50.8	27.8	43.09	21.55	104.15	68.53	17.63
Nd	2.06	2.44	1.69	—	—	18.31	10.32	45.43	28.37	8.11
Sm	0.50	0.64	0.47	—	—	3.64	2.53	8.12	5.22	2.27
Eu	0.05	0.06	0.04	1.05	0.83	0.47	0.49	0.93	0.92	0.33
Gd	0.44	0.55	0.43	—	—	3.26	2.87	4.71	3.66	2.18
Dy	0.53	0.62	0.48	—	—	1.98	2.51	2.80	2.49	2.25

	Penfeunten innengranite			St. Renan granite						
	P 1	P 4	P 3	SR 7	SR 17	SR 5	SR 26	SR 11	SR 23	SR 10
Er	0.25	0.29	0.23	—	—	0.92	1.13	1.44	1.03	1.00
Yb	0.37	0.41	0.33	1.0	1.1	0.75	0.76	1.03	0.96	0.99
Lu	0.06	0.06	0.05	—	—	0.14	0.12	0.12	0.18	0.17
Σ REE	11.3	14.3	9.5	—	—	104	56.2	231	155	46.2
La/Yb	5.9	6.3	4.9	—	—	35	14	48	71	8.5
Eu/Eu*	0.33	0.31	0.28	—	—	0.42	0.56	0.46	0.65	0.46

Σ REE = total REE content with interpolated values.
Eu* = interpolated value of Eu between Sm and Gd.

Table 5. Major (% oxide) and trace element (in ppm) concentrations for the unmineralized calc-alkalic granites from Corsica and Pyrenees

	Corsican granite												Pyrenean Granite			
	CLB 41	CLB 42	CLB 38	CLB 77	CLB 69	CLB 68	CLB 67	CLB 65	CLB 61	CLB 63	CLB 53	CLB 55	ALB 19	ALB 15c	ALB 20	ALB 21d
SIO_2	71.98	73.36	73.74	74.74	75.09	75.23	75.94	76.59	75.98	77.55	76.85	75.56	69.59	74.5	75.33	76.44
Al_2O_3	14.49	14.24	13.51	13.27	13.40	13.27	12.82	16.62	12.48	12.36	12.77	12.63	14.53	13.1	12.48	12.34
Fe_2O_3*	2.77	2.44	2.35	2.25	2.32	2.08	1.39	1.02	1.54	1.29	0.89	1.09	2.74	1.26	1.25	1.13
MnO	0.06	0.06	0.1	0.05	0.07	0.07	0.07	0.06	0.06	0.06	0.08	0.07	0.05	0.04	0.03	0.03
MgO	0.57	0.37	0.38	0.25	0.29	0.24	0.15	0.01	0.03	—	<0.1	0.16	0.36	—	0.02	—
CaO	2.39	1.72	1.79	1.43	1.38	1.44	0.43	0.43	0.63	0.37	0.26	0.52	1.71	0.54	0.53	—
Na_2O	3.69	3.73	3.69	2.77	3.24	3.48	3.55	3.33	3.32	3.09	3.91	4.49	3.3	3.09	3.2	3.45
K_2O	2.99	3.51	3.57	4.95	4.10	4.10	4.42	4.85	4.96	4.61	4.15	4.41	4.12	5.16	5.01	5.11
TiO_2	0.24	0.23	0.21	0.16	0.23	0.23	0.07	0.01	0.06	0.03	<0.01	0.02	0.34	0.13	0.09	0.1
P_2O_5	0.05	0.11	0.06	—	—	—	—	—	—	—	—	0.01	0.07	—	—	—
P.F.	0.43	0.28	0.34	0.28	0.48	0.44	0.64	n.d.	0.35	0.58	0.52	0.49	2.06	0.73	1.41	0.26
Total	99.66	100.05	99.76	100.15	100.60	100.58	99.48	98.92	99.41	99.94	99.43	99.45	98.87	98.55	99.35	98.86
U	0.95	2.3	1.7	4.2	7.94	4.00	4.1	8.3	14.2	8.5	17.4	12.4	7.2	5.8	7.9	10.2
Th	10.1	12.7	13.1	15.3	18.8	16.2	14.3	20.2	32.1	29.0	10.8	15.2	16.9	19.0	25.7	29.3
Ta	0.41	0.96	0.58	1.18	2.28	2.22	2.1	3.7	3.21	3.8	4.93	4.80	1.46	1.47	2.14	3.00
Rb	74	90	136	178	190	206	213	297	303	306	395	352	170	238	249	246
Sr	251	181	130	69	66	68	34	17	17	20	1	23	123	73	38	13
Ba	1186	1414	515	501	451	410	302	128	98	108	37	119	522	432	227	80
W	0.13	0.24	0.24	0.50	0.35	0.89	1.0	0.81	0.33	0.62	2.2	1.4	0.35	0.56	0.22	0.54
Sn	1.3	1.9	0.8	1.3	2.7	5.5	3.2	6.3	7.2	3.9	8.5	8.4	6.0	10.6	6.8	11

	Corsican granite												Pyrenean Granite			
	CLB 41	CLB 42	CLB 38	CLB 77	CLB 69	CLB 68	CLB 67	CLB 65	CLB 61	CLB 63	CLB 53	CLB 55	ALB 19	ALB 15c	ALB 20	ALB 21d
La	42.1	48.8	29.2	34.8	29.2	30.0	19.2	8.86	19.6	18.2	7.85	15.5	32.0	20.4	18.4	12.5
Ce	72.8	86.3	55.2	63.8	55.8	56.9	37.6	21.1	44.1	39.2	19.3	32.2	60.2	39.0	38.5	27.6
Nd	23.5	29.4	21.6	—	—	—	16.1	—	—	22.2	—	14.6	25.6	18.6	20.4	15.4
Sm	3.28	4.77	3.97	4.92	4.76	5.04	3.07	3.85	6.09	6.25	2.77	3.83	5.47	4.36	5.36	4.63
Eu	0.96	0.94	0.55	0.78	0.63	0.70	0.21	0.39	0.33	0.22	<0.07	0.14	0.84	0.59	0.36	0.20
Gd	2.28	2.36	2.64	—	—	—	3.07	—	—	6.75	—	3.64	4.46	4.18	6.0	6.0
Tb	0.31	0.43	0.38	0.61	0.82	0.81	0.61	0.76	1.28	1.30	1.00	0.78	0.85	0.79	1.19	1.24
Tm	—	0.40	0.14	—	—	—	0.46	—	—	1.00	—	0.68	0.43	0.58	0.87	1.08
Yb	0.85	3.07	1.08	2.61	3.33	3.39	3.09	—	6.79	6.62	4.89	4.00	2.90	3.65	5.61	6.88
Lu	0.14	0.56	0.19	—	—	—	0.53	—	—	1.18	—	0.70	0.54	0.66	0.99	1.21
Σ T.R.	157	192	125	154	141	145	93	—	135	119	68	87	149	107	111	96.5
La/Yb	50	15.9	27.0	13.3	8.8	8.9	6.2	—	2.9	2.8	1.6	3.9	11.0	5.6	3.3	1.8
Eu/Eu*	1.08	0.86	0.52	0.54	0.40	0.43	0.21	0.29	0.15	0.10	<0.06	0.12	0.52	0.43	0.19	0.12

Σ REE = total REE content with interpolated values.
Eu* = interpolated value of Eu between Sm and Gd.
Fe_2O_3* = total Fe_2O_3.

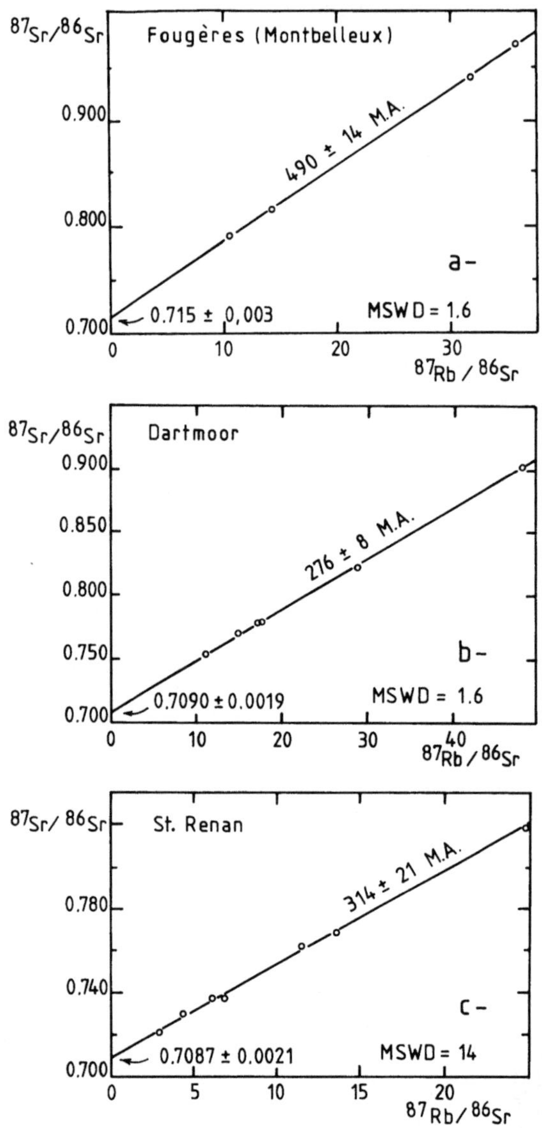

Fig. 2a–c. Rb/Sr isochrons for **a** leucogranites from the Fougères batholith, **b** Dartmoor, **c** St. Renan. Ages and initial ratios were calculated using $^{87}Rb = 1.42\ 10^{-11} y^{-1}$ and an absolute error of 0.0002 for $^{87}Sr/^{86}Sr$ and a relative error of 2% for $^{87}Rb/^{86}Sr$

3 Geochemical Study

3.1 Analytical Methods

Oxygen extraction and isotope analyses were carried out in the BRGM laboratories on samples weighing approximately 10 mg. The standard procedure developed by Clayton and Mayeda (1963) was used. Analysis of standard NBS 28 gave a figure

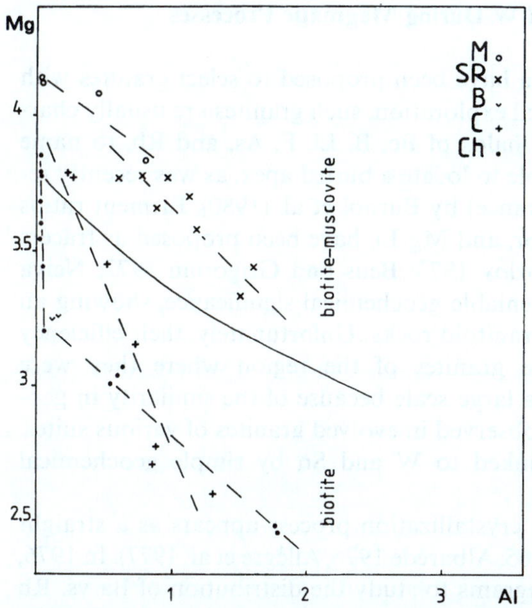

Fig. 3. Composition of biotites from granite samples (mineralized and unmineralized) plotted on an Mg–Al diagram. *M* Montbelleux; *SR* St. Renan; *B* Brazil; *Ch* China; *C* Corsica

of 9.6‰ in relation to SMOW, with an analytical accuracy of ±0.2. Oxygen extraction and isotope measurements were duplicated or triplicated for each sample. D/H extraction and ratio determination were conducted in the BRGM laboratories on samples weighing approximately 150–250 mg. The procedure developed by Friedman (1953) was used. Results are given in relation to SMOW, the analytical accuracy being ±0.02.

Rb–Sr analyses were carried out in the BRGM laboratories. Rubidium and strontium concentrations were determined by standard methods, using a Finnigan-Mat 261 double-collector mass spectrometer. The uncertainties (2 σ level) for $^{87}Rb/^{86}Sr$ ratios are estimated to be about 2%. Replicate $^{87}Sr/^{86}Sr$ determinations indicate an overall precision of ±0.02% (2 σ) on the ratio. These values were used to weigh the points when determining the best-fit line to the data points on an isochron diagram, and to determine the accuracy of fit (MSWD) of the points to this line. The decay constant used is $^{87}Rb = 1.42 \times 10^{-11} y^{-1}$.

Major and rare-earth elements (REE) were analysed by ICP emission spectrometry at the CRPG in Nancy using the method described by Govindaraju et al. (1976). Other trace elements (including Th, Ta, and W) were analysed in the P. Süe laboratory (Saclay, France) by instrumental neutron activation (INAA) using the method described by Chayla et al. (1973). Sn was determined by multiparameter coincidence spectrometry using the INAA method developed by Meyer et al. (1985). These procedures allow the application of low detection thresholds, particularly for W (0.2 ppm) and Sn (0.5 to 1 ppm).

3.2 Geochemical Behavior of Sn and W During Magmatic Processes

Various types of discriminant criteria have been proposed to select granites with metallogenic potential. In geochemical exploration, such granites are usually characterized by distinctive geochemical halos of Be, B, Li, F, As, and Rb, to name but a few. These halos make it possible to locate a buried apex, as was recently reported from Neuf-Jours (Corrèze, France) by Burnol et al. (1980). Element ratios within the rock, such as Rb/Ba, Rb/Sr, and Mg/Li, have been proposed as tracers for Sn–W granites (Tauson and Kozlov 1973; Beus and Grigorian 1977; Neiva 1984). Such indications have an undeniable geochemical significance, showing an advanced state of differentiation of granitoid rocks. Unfortunately, their efficiency to discriminate is restricted to the granites of the region where they were tested, and they cannot be used on a large scale because of the similarity in geochemical trends for these indicators observed in evolved granites of various suites. The elements considered are not linked to W and Sn by simple geochemical mechanisms.

As is well known, the fractional crystallization process appears as a straight line on a log-log diagram (Masuda 1965; Albarède 1976; Allègre et al. 1977). In 1976, Mc Carthy and Hasty used such diagrams to study the distribution of Ba vs. Rb and Sr vs. Rb in granitic melts during crystallization. It has more recently been shown (Cocherie 1986) that a systematic approach can successfully be used for plutonic suites to calculate the D and F parameters of the sequence of fractional crystallization without any assumptions. The same approach has been used for the present study, bearing in mind the following objectives:

1. To check the cogenetic nature of the suites and identify the main magmatic process.
2. To study the geochemical behaviour of economic metals during the magmatic process thus identified.

In Fig. 4, analyses representative of the Corsican monzogranite and leucomonzogranite cogenetic suite (Cocherie et al. 1984; Cocherie 1984), the Dartmoor leucogranite, and the St. Renan leucogranite have been plotted on two log-log diagrams (Ce and Th versus Sn). The correlations observed are compatible with the suites evolving according to a fractional crystallization process, as previously demonstrated in the case of the barren Corsican batholith on the basis of Rb, Sr, Ba and REE analysis (Cocherie 1984).

As is generally the case, the LREE (Ce in particular) lose their hygromagmaphile properties in the more acidic terms of plutonic rock suites because significant amounts of LREE-bearing accessories were precipitated during the later stages of crystallization. Similarly, the Th content increases until it precipitates at a concentration of about 30 ppm. The two diagrams in Fig. 4 show that Sn is clearly hygromagmaphile, whereas LREE and Th represent compatible elements in both mineralized and evolved barren complexes. The Sn enrichment factor during fractional crystallization is apparently greater than one order of magnitude. The crystallization process therefore plays an important part in Sn preconcentration.

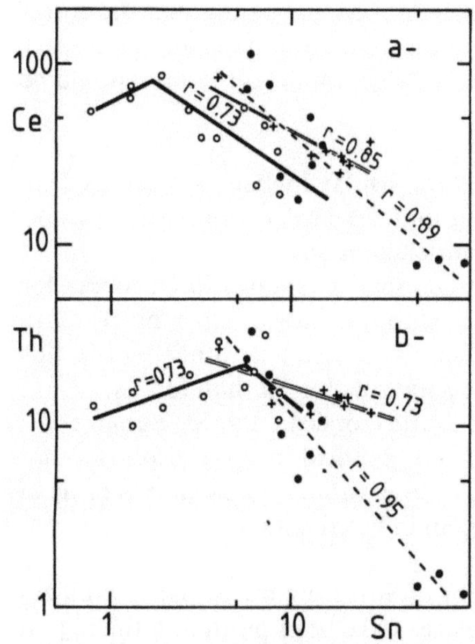

Fig. 4a, b. Logarithmic plots for Ce (**a**) and Th (**b**) versus Sn. At the start of the fractional crystallization process, Ce and Th concentrations increase with the increasing Sn content. At higher degrees of crystallization, the Th and Ce contents decrease whereas Sn remains hygromagmaphile. *Open circle* Corsican granitoids; *black dot* St. Renan leucogranite; *cross* Dartmoor granite

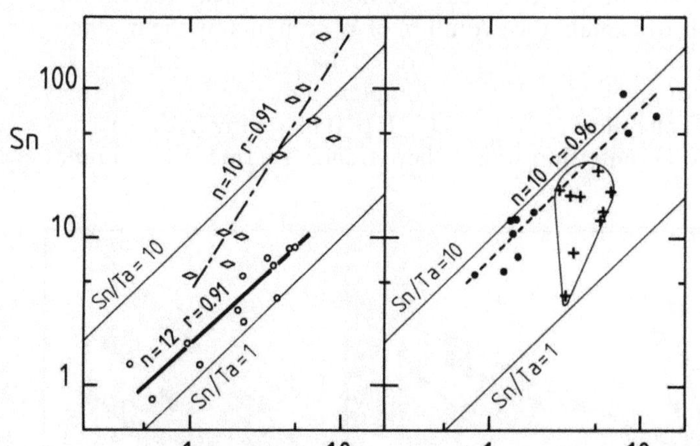

Fig. 5. Logarithmic plots for Sn vs. Ta (ppm). The Corsican protolith displays clearly lower Sn values than the protoliths for the other granitoids. The Fougères-Montbelleux granitoids are marked by *diamonds*. Other symbols as in Fig. 4

In order to characterize more accurately the geochemical behavior of Sn, Sn analyses from Corsican granitoids and three granitoids associated with mineralization (Dartmoor, St. Renan and Fougères-Montbelleux) have been plotted against Ta on a log-log diagram (Fig. 5). Ta has been chosen for its perfect hygromagmaphile behavior, thus Ta appears to be a reliable and sensitive indicator of the differentiation stages.

The representative points fit with linear correlations, indicating that the corresponding suites are cogenetic. Nevertheless, we must point out the surprising scatter of the points representing the Dartmoor granite. Other important points should be emphasized:

1. The granitoids associated with mineralization display higher Sn contents than the barren Corsican granitoids, indicating that the initial protoliths of granitoids associated with mineralization are enriched in Sn.
2. The Sn/Ta ratio increases during the fractional crystallization process for Montbelleux and Dartmoor mineralized granites, indicating a more rapid increase in the Sn content relative to the Ta content, in other words this evolution indicates lower bulk partition coefficients for Sn than for Ta.
3. This Sn/Ta ratio remains constant for both the Corsican and St. Renan complexes. If we suppose that the D_{Ta} values are identical in both it follows that the D_{Sn} values must also be identical. These observations emphasize the role of fractional crystallization in concentrating Sn in granitic melts.

In order to study the geochemical behavior of W during fractional crystallization, the representative points for these complexes have been plotted in the log Sn vs. log W diagram (Fig. 6). All the points for each complex fit a linear correlation, indicating that it is possible to explain the evolution of W contents by the mechanism of fractional crystallization. It can be seen, however, that two trends appear for the leucogranites of the Fougères-Montbelleux complex, indicating two different sets of thermodynamic and chemical conditions (P, T, pH) involving hydrothermal influences during the late differentiation stage. These trends are traced by samples

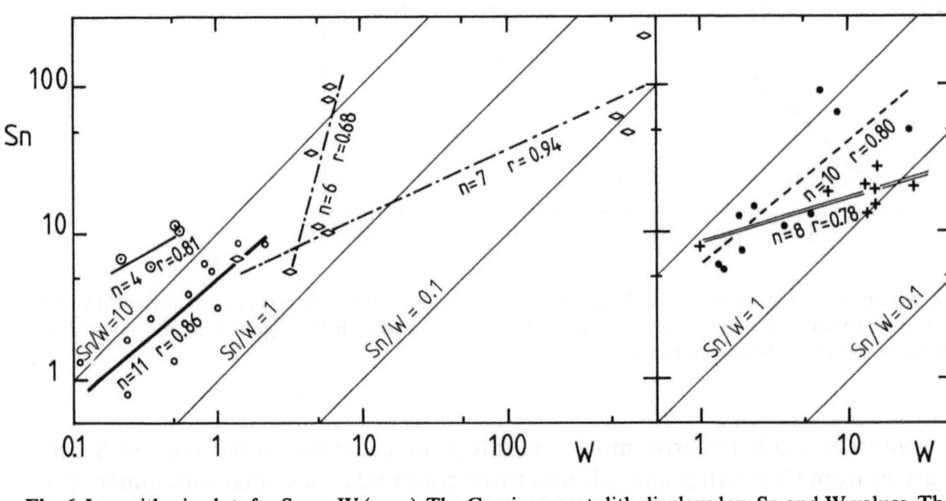

Fig. 6. Logarithmic plots for Sn vs. W (ppm). The Corsican protolith displays low Sn and W values. The Fougères-Montbelleux granitoids display two separate trends. In the general case W appears to have a more hygromagmaphile behavior than Sn both in the mineralized and barren granites (four representative points of the Pyrenean granite of St. Laurent La Junquera are marked by *pointed open circles*). For other symbols see Figs. 4 and 5

from two different areas. The following points should also be noted:

1. As in the log Sn vs. log Ta plot, the slope in the log Sn vs. log W plot is similar for the Corsican and St. Renan granites. The main difference is the level of the initial concentrations in the protolith (nearly one order of magnitude higher for the St. Renan granite). It can be seen that the four representative points of the St. Laurent La Junquera barren granite indicate an end of evolution similar to that for the Corsican granite (cf. Table 5, and Cocherie 1984), though the W content is lower.
2. The two other granites associated with mineralization (Montbelleux and Dartmoor) exhibit a rapid increase of the W/Sn ratios due to a very large increase of W concentration in the case of the Montbelleux trend (r = 0.94). This rapid evolution of W content could be attributed to very low D_W associated with high degrees of fractional crystallization, but, if we compare these trends with the trends of the Corsican and St. Renan granites, which are the result of typical magmatic evolution, it seems more likely that a hydrothermal process is involved.

From the simple Sn versus SiO_2 plot in Fig. 7 it can be seen that all the granite suites associated with mineralized bodies display a somewhat high Sn content for a given SiO_2 concentration (used as a rough index of differentiation) when compared to the unmineralized Corsican suite.

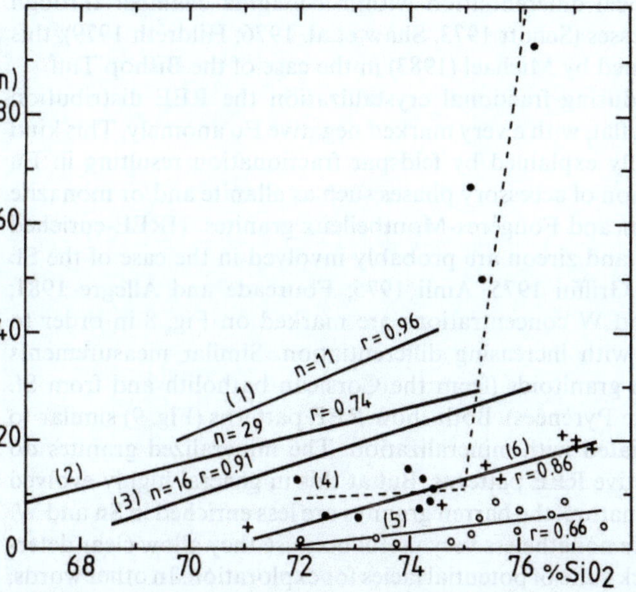

Fig. 7. Sn versus SiO_2 plot for various granites. *(1)* Chanon and St. Silvain (French Massif Central); *(2)* Viseu (Portugal); *(3)* Western Margeride (French Massif Central); *(4)* St. Renan; *(5)* Corsica; *(6)* Dartmoor. *(1)*, *(2)* and *(3)* are from Roger and Derre (1980). The unmineralized Corsican monzogranites show a low initial Sn content indicative of the low initial Sn level in the corresponding protolith

3.3 Rare Earth Elements

REE distribution patterns were studied in order to examine the genetic relationships between the various members of each suite of rocks and to attempt to characterize the mineralized granites.

For chemical compositions that are broadly comparable, various authors (Cocherie 1978; Fourcade and Allègre 1981; Vidal et al. 1982; Miller and Mittlefehldt 1982; Cocherie et al. 1984; Le Bel et al. 1984; and Drysdall et al. 1984, among others) were able to show extreme variations in REE contents. Here, we are dealing with types of granite that were described by Cocherie (1985) as the last stages in magmatic differentiation, and which are also commonly enriched in Sn, W, Nb and Ta, among other elements. The corresponding chondrite-normalized REE patterns show impoverished light REE contents, compared to the other granite types (decreased La/Yb), which are usually accompanied by a very strong negative Eu anomaly (Eu/Eu* ≪ 1) giving seagull-shaped profiles. The most commonly invoked mechanism to explain this evolution of the REE profiles is that of calls upon very advanced fractional crystallization, involving the participation of accessory minerals rich in REE, such as allanite, monazite, zircon, and xenotime. Percolation of fluids enriched in anions (F^-, Cl^-, CO_3^{2-}) is commonly also invoked either as a complementary phenomenon (Raimbault, 1984; Cocherie, 1984) or, as the predominant phenomenon (Taylor and Fryer 1983; Le Bel and Laval 1986), the latter interpretation being based on the experimental work by Flynn and Burnham (1978) and Wendlandt and Harrison (1979). Finally, the possibility should be mentioned of geochemical differentiation within a magma chamber through thermo-gravitational processes (Schott 1973; Shaw et al. 1976; Hildreth 1979); this model, however, is questioned by Michael (1983) in the case of the Bishop Tuff.

Figure 8 shows that during fractional crystallization the REE distribution patterns become somewhat flat, with a very marked negative Eu anomaly. This kind of pattern can be essentially explained by feldspar fractionation resulting in Eu depletion and in fractionation of accessory phases such as allanite and/or monazite in the case of the Dartmoor and Fougères-Montbelleux granites. HREE-enriched minerals such as xenotime and zircon are probably involved in the case of the St. Renan granite (Amli and Griffin 1975; Amli 1975; Fourcade and Allegre 1981; Cocherie 1984). The Sn and W concentrations are marked on Fig. 8 in order to monitor increased values with increasing differentiation. Similar measurements were made for two barren granitoids (from the Corsican batholith and from St. Laurent La Junquera in the Pyrénées). Both show REE patterns (Fig. 9) similar to those for granitoids associated with mineralization. The mineralized granites do not, therefore, have distinctive REE patterns. But at this in general highly evolved stage, for the same REE signature, the barren granites are less enriched in Sn and W.

Rare-earth elements are nonetheless very useful because they allow clear determination within a given rock suite of potential facies for exploration. In other words, if a granite displays an REE distribution pattern typified by limited LREE and HREE fractionation and a marked Eu anomaly, and is characterized by low Sn and W values (<10–15 ppm and <1–2 ppm respectively), it cannot host economic concentrations of such metals.

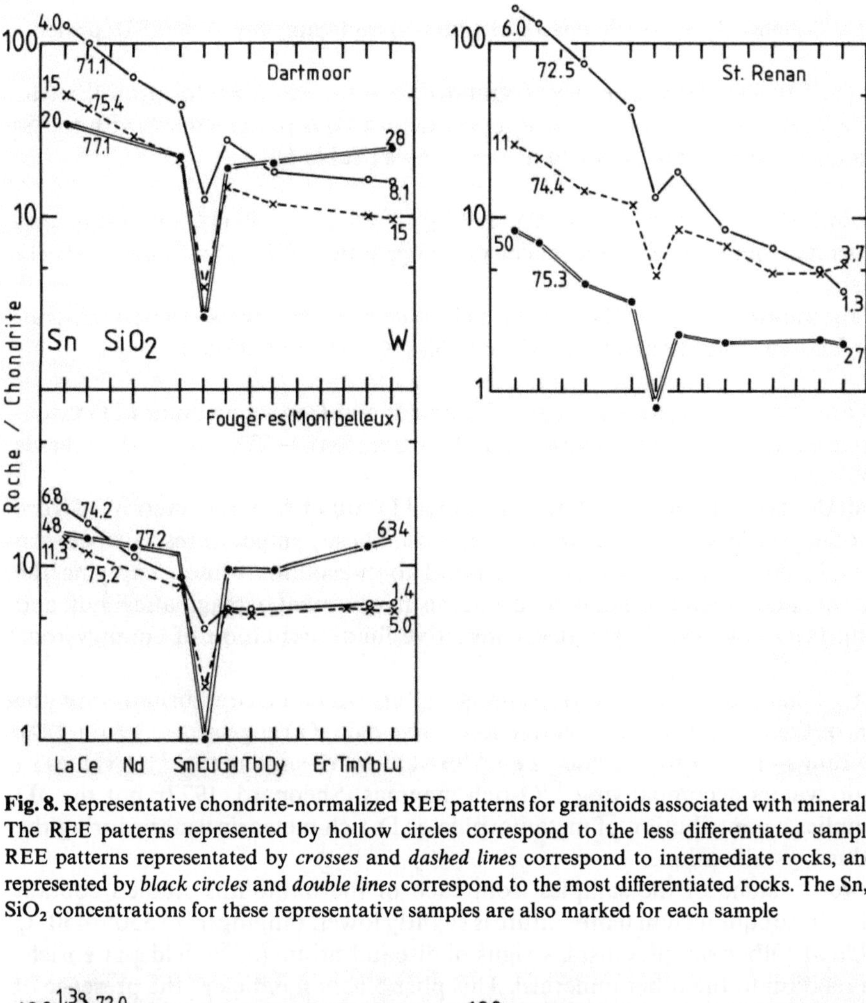

Fig. 8. Representative chondrite-normalized REE patterns for granitoids associated with mineralization. The REE patterns represented by hollow circles correspond to the less differentiated samples. The REE patterns representated by *crosses* and *dashed lines* correspond to intermediate rocks, and those represented by *black circles* and *double lines* correspond to the most differentiated rocks. The Sn, W and SiO_2 concentrations for these representative samples are also marked for each sample

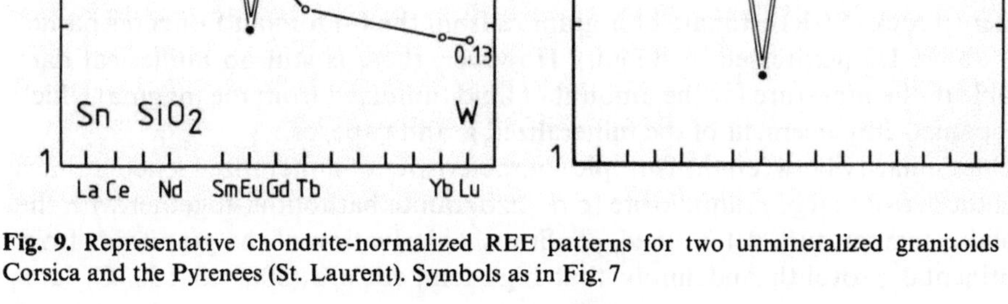

Fig. 9. Representative chondrite-normalized REE patterns for two unmineralized granitoids from Corsica and the Pyrenees (St. Laurent). Symbols as in Fig. 7

3.4 $^{18}O/^{16}O$ and D/H Geochemistry of Fluids Accompanying Mineral Deposits

With regard to the derivation of the granitoïds from a continental protolith, the whole rock geochemistry of stable isotopes (O and D) is in agreement with the Sr geochemistry. The figures obtained are as follows (Table 6):

The Fougères Cadomian batholith: $10.8 \leqq \delta^{18}O_{WR} \leqq 12.9$ (14 analyses), and the associated mineralized Montbelleux leucogranite: $10.2 \leqq \delta^{18}O_{WR} \leqq 12.0$ (11 analyses).
St. Renan granite: $9.4 \leqq \delta^{18}O_{WR} \leqq 12.0$ (12 analyses), and the associated mineralized Penfeunten "innengranite": $10.3 \leqq \delta^{18}O_{WR} \leqq 10.6$ (3 analyses).

The ∂D are very homogeneous in both Cadomian and Hercynian units: ∂D muscovite varies from -52 up to -40‰ and ∂D biotite from -67 up to -53‰ versus SMOW.

In all the cases studied, high-temperature (HT) equilibria were determined from the isotopic composition of mineral separates. These temperatures fall between 500 °C and 600 °C (Table 6) and correspond to subsolidus values, implying the existence of late-magmatic phenomena and the involvement of magmatic fluids, and excluding the existence of extended convective fluid circulation of country rock origin.

The calculated isotopic compositions of the water in equilibrium with the minerals of these granitoïds at 500 to 600°C have a $\delta^{18}O$ range of $+7.4$ to $+12$‰ and δD range of -35 to -20‰. The $\delta^{18}O$ values are similar to $\delta^{18}O$ values of magmatic waters accompanying ^{18}O-rich magmas (Sheppard, 1977); but the δD range implies a contribution of metamorphic or D-rich meteoric waters during the subsolidus crystallization.

However, some of the samples from each of the batholiths studied contain minerals whose equilibrium temperature is slightly lower, ranging from 350°–450 °C (see Table 6). Other samples display signs of disequilibrium ($\delta^{18}O$ feldspar enrichment in relation to the other minerals). This phenomenon indicates the presence of a fluid phase rich in ^{18}O, which re-equilibrated with the feldspar during the cooling process. This fluid phase was of limited volume, so that the isotopic disequilibrium of the feldspars is slight. In the case of the Montbelleux W-granite and of the St. Renan granite, the participation of a component of metamorphic origin has been identified (Fouillac et al. 1986). A meteoric origin is excluded as the emplacement of the batholith occurred during a glacial climatic period. Exchange with such meteoric water would have lowered the δD values of the minerals. The proportion of water to rock (W/R) estimated for granites from the Cadomian Fougères batholith is about 1/3 (expressed in weight). However, there is still no numerical data available in the literature for the amount of fluid (unmixed from the magma) which accompanied emplacement of the mineralized granite apexes.

The similarity between the isotopic characteristics of mineralized leucogranites and of those from large granodiorite to monzogranite batholiths, together with the subsolidus temperature determined, confirm the derivation of the granitoids from a continental protolith and imply that high-temperature (500° to 550 °C) late-

Table 6. Stable isotope data on whole rocks and separated minerals

Samples		^{18}O							D			
		Qz	K-Spar and Plaio.	Musc.	Biot.	Chl.	Plag.	Ilm	Musc.	Biot.	Chl.	T°–C
Cadomian batholite[a]			$10.8 < {}^{18}O_{W.R.} < 12.9$ (4 an)									
Biotite cordierite granodiorite	(FOU 14)	12.7	11.6	10.4	7.35				−42	−55		510°
Biotite granodiorite	(FOU 5)	11.65	10.35		6.7					−67		540°
Biotite granodiorite	(FOU 22)	11.00	11.1 (K.F)		7.1		11.3			−88		635°[b]
Leucogranite	(SHI)	12.0	11.4	9.6					−42			580°[b]
Leucogranite	(SMGA)	13.9	11.1			5.7				−65		370°
Leucogranite	(SM8)	13.1	12.3	10.2	7.5				−43	−59		485°[b]
Montbelleux[a] (Mineralized granite)			$10.2 < {}^{18}O_{W.R.} < 12.4$ (11 an)									
W leucogranite	(M 6)	11.7	10.55	9.0					−48			540°
Greisen	(M 7)	11.45		18.5					−47			515°[b]
Albitite	(M 13)	11.3	11.4	8.3					−47			510°[b]
Albitite	(M 22)	13.15	11.0	9.1					−46			510°
Saint-Renan batholith			$9.4 < {}^{10}O_{W.R.} < 12.0$ (8 an)									
Biotite monzogranite	(SR 11)	11.5	9.1	8.0	4.9				−40	−53		455°
Biotite monzogranite	(SR 23)	11.4	9.8	6.5						−55		550°
Biotite monzogranite	(SR 26)	13.1	11.2	8.8	6.5				−52	−65		435°
Penfeunten innengranite												
Muscovite granite	(P 1)	11.2	10.4	8.5					−48			535°
Muscovite granite	(P 2)	11.6		8.4					−44			490°[b]
Muscovite granite	(P 3)	11.7	9.7	8.5					−45			490°

[a] Localization on Fouillac et al. (1986).
[b] Isotopic temperatures obtained from (Qz-Mica) fractionation only.

magmatic phenomena predominated during deposition of the W–Sn mineralization. A limited amount of water external to the granitic system has been identified, such water being of meteoric and/or metamorphic origin.

4 Conclusions

This extensive geochemical study of varied granitic bodies leads to the following observations of geochemical and metallogenic interest:

1. The hygromagmaphile behavior of Sn is clearly demonstrated both in the case of the barren granite from the western Corsica batholith and in the cases of the granites associated with mineralization. The same general geochemical behavior is seen for W, but its concentration in the magmas is more easily disturbed by interaction with fluids.
2. Nevertheless, because of their hygromagmaphile properties fractional crystallization is an important process for the concentration of these metals.
3. A general consequence of this geochemical process is the particular shape of REE patterns for the highly differentiated bodies, i.e., rather flat REE patterns with pronounced negative Eu anomalies.
4. While the extraction of metals from magma leading to economic concentration is due to postmagmatic phenomena, which play an important role in the deposition of minerals in veins, etc., a high degree of fractional crystallization is a necessary precondition for the concentration of metals in a granite. It is also necessary, however, that the initial protolith has rather high W and Sn contents. For example, a granite with an SiO_2 content of 72% and less than 10 ppm of Sn will be considered barren. The answer to Lehmann's question (1982) about the metallogeny of tin: "Magmatic differentiation versus geochemical heritage?" is that both phenomena are effective and their synergy necessary.
5. We must also take into consideration the role of the fluids in the collection of the metals. This study indicates that they are partly of magmatic origin, that limited amounts of meteoric and/or metamorphic water were involved in the studied targets, and that they act mainly at subsolidus temperatures (500°–550 °C).
6. Thus a relatively few sophisticated exploration tool based on REE, Sn, and/or W determinations, can discriminate zones of metallogenic granite. An accurate and precise analytical method is of course necessary.

Acknowledgements. This study was finanically supported under EEC (*Contract MSM-031-F*) to the project Deposits of Tungsten and Associated Metals in Western Europe and a grant from the Ministère de la Recherche et de la Technologie, France, contract n° 83 EO 976 to the project: Test d'un nouvel outil de prospection stratégique des granitoïdes à Sn — W et éléments connexes. This paper was presented at the European meeting on Tungsten Deposits at Toulouse, France, on May 12–14, 1986 and has benefited from discussion with the participants. The authors thank L. Le Bel who promoted this study, F. Schneider for unpublished data, and J. Boissonnas and M. Cuney for constructive review.

References

Albarede F (1976) Some trace element relationships among liquid and solid phases in the course of the fractional crystallization of magmas. Geochim. Cosmochim Acta 40:667–673

Allegre CJ, Treuil M, Minster JF, Minster JB, Albarede F (1977) Part I. Fractional crystallization processes in volcanic suites. Contrib Mineral Petrol 60:57–75

Amli R (1975) Mineralogy and rare earth geochemistry of apatite and xenotime from the Gloserheia granite pegmatite, Froland, southern Norway. Am Miner 60:607–620

Amli R, Griffin W (1975) Microprobe analysis of REE minerals using empirical correction factors. Am Miner 60:599–606

Autran A, Fonteilles M, Guitard G (1970) Relations entre les intrusions de granitoïdes, l'anatexie et le métamorphisme régional considérés principalement du point de vue du rôle de l'eau: cas de la chaîne hercynienne des Pyrénées orientales. Bull Soc Géol Fr 12:673–731

Autran A, Beurrier M, Calvez JY, Cocherie A, Fouillac AM, Rossi Ph (1983) Caractérisation des granitoïdes du batholite mancellien, implications métallogéniques. C.R. final ATP CNRS. Résumé in: Principaux résultats Scient et Techn du BRGM. p. 57

Beus AA, Grigorian SV (1977) The use of geochemical specialization of rocks in geochemical exploration. In AA Levinson (Editor), Geochemical exploration methods for mineral deposits, Chap 4. Applied Publishing Ltd., Willmette, Ill. p. 53–71

Botelho N (1985) Caracterizaçao e evoluçao das micas do granito estanifero da Pedra Branca (Goias) em relaçao aos processos mineralizadores tardi posmagmaticos. Revista Bras Geoc 15 (3):259–268

Bottinga Y, Javoy M (1975) Oxygen isotope partitioning among the minerals in igneous and metamorphic rocks. Rev Geophys Space Phys 13:401–418

Burnol L, Peronne Y, De Vaucorbeil H (1980) La coupole cachée de leucogranite de Neuf Jours (Corrèze) et les minéralisations en tungstène associées. Chron Rech Min 455:93–116

Charoy B (1975) Les phénomènes de greisenisation dans le district minéralisé de Penfeunten (Massif de St. Renan, Massif Armoricain). Aspects pétrologiques, géochimiques et caractérisation génétique. Bull BRGM II 5:363–383

Chauris L (1965) Les minéralisations pneumatolitiques du massif Armoricain. Mem BRGM 31:217 p

Chayla B, Jaffrezic H, Joron JL (1973) Analyse par activation dans les neutrons épithermiques. Application à la détermination d'éléments en traces dans les roches. C R Acad Sci, Paris, 277, série D:273–275

Clayton RN, Mayeda TK (1963) The use of Bromine Pentafluoride in the extraction of oxygen from oxides and silicates for isotopic analysis. Geochim Cosmochim Acta Vol 27:42–43

Cocherie A (1978) Géochimie des terres rares dans les granitoïdes. Thèse 3ème cycle, Université de Rennes I, unpublished, 207 p

Cocherie A (1984) Interaction manteau-croûte: son rôle dans la genèse d'associations plutoniques calco-alcalines, contraintes géochimiques (éléments en traces et isotopes Sr et O). Thèse d'état, Université de Rennes I, BRGM (1985), doc 90:246 p

Cocherie A (1985) REE geochemistry of mineralized granites compared with highly acidic unmineralized granites. Commission of the European communities Dublin, May 1986, 4 p

Cocherie A (1986) — Systematic use of trace element distribution patterns in log-log diagrams for plutonic suite. Geochim Cosmochim Acta 50:2517–2522

Cocherie A, Rossi Ph, Le Bel L (1984) The variscan calc-alkalic plutonism of western Corsica: mineralogy and major and trace element geochemistry. Phys Earth Planet Inter 35:145–178

Drysdall AR, Jackson NJ, Ramsay CR, Douch CJ, Hackett D (1984) Rare element mineralization related to precambrian alkali granites in the Arabian shield. Econ Geol 79:1366–1377

Flynn RT, Burnham CW (1978) An experimental determination of rare earth partition coefficients between a chloride-containing vapor phase and silicate melts. Geochim Cosmochim Acta 42:685–701

Fouillac AM, Cocherie A, Rossi Ph, Calvez JY, Autran A (1986) Etude géochimique du batholite mancellien (Massif Armoricain). Rapport BRGM 86 DT 037 MGA. 15 p unpubl.

Fourcade S (1981) Géochimie des granitoïdes. Thèse d'Etat. Paris VII, 189 p

Fourcade S, Allegre CJ (1981) Trace element behavior in granite genesis. A case study. The calc-alkaline plutonic association from the Querigut complex (Pyrénées, France). Contrib Mineral Petrol 76:177–195

Friedman I (1953) Deuterium content of natural waters and other substances. Geochim Cosmochim Acta 4:89–103

Govindaraju J, Mevelle G, Chouard C (1976) Automated optical emission spectrochemical bulk analysis of silicate rocks with microwave plasma excitation. Analyt Chem 48:1325–1331

Hildreth W (1979) The Bishop Tuff: evidence for the origin of compositional zonation in silicic magma chambers. Geol Soc Amer Sp paper 180:43–75

Le Bel L, Laval M (1986) Felsic plutonism in the Al Amar — Idsas area, Kingdom of Saudi Arabia. J Afr Earth Sci 4:87–98

Le Bel L, Li Yi-Dou, Sheng Ji-Fou (1984) Granitic evolution of the Xihuashan-Dangping (Jiangxi, China) tungsten-bearing system. Tschermaks Min Petr Mitt 33:149–167

Lehmann B (1982) Metallogeny of tin: magmatic differenciation versus geochemical heritage. Econ Geol 77:50–59

Marignac Ch, Weisbrod A (1986) L'origine des gisements stannowolframifères: une approche critique. Colloque Européen "Gisements de tungstène". Toulouse 12–14 mai 24–25

Masuda A (1965) Geochemical constants for rubidium and strontium in basic rocks. Nature 205:555–558

Mc Carthy TS, Hasty RA (1976) Trace element distribution patterns and their relationship to the crystallization of granitic melts. Geochim Cosmochim Acta 40:1351–1358

Meyer G, Jaffrezic H, Treuil M (1985) Analyse instrumentale de l'étain par activation neutronique et spectrométrie de coïncidence. Geostandards Newsl 9:19–82

Michael PJ (1983) Chemical differentiation of the Bishop Tuff and other high-silica magmas through crystallization processes. Geology 11:31–34

Miller CF, Mittlefehldt DW (1982) Light rare earth element depletion in felsic magmas. Geology 10:129–133

Nachit H, Razafimahefa N, Stussi JM, Carron JP (1985) Composition chimique des biotites et typologie magmatique des granitoïdes. CR Acad Sci Paris, 301, Série II, n° 11:813–818

Neiva AMR (1984) Geochemistry of tin-bearing granitic rocks. Chem Geol 43:241–256

Pasteels P, Dore F (1982) Age of the Vire-Carolle granite. In: Numerical dating in stratigraphy. ODIN GS (ed) WILEY New York

Raimbault L (1984) Géologie, pétrographie et géochimie des granites et minéralisations associées de la région de Meymac (Haute-Corrèze, France). Thèse doc Ingénieur, Paris, 482 p

Roger G, Derre C (1980) Processus géochimiques de concentration liés à l'évolution de magmas granitiques. Rapport ATP CNRS, 129 p

Rossi Ph (1986) Organisation et genèse d'un grand batholite orogénique: le batholite calco-alcalin de la Corse. Thèse d'Etat Toulouse. Doc. BRGM n° 107, 292 p

Rossi Ph, Cocherie A, Fouillac AM, Calvez JY (1985) Géochimie des éléments en traces et des isotopes (Sr, O) des granites de Panasqueira (Portugal). Rapport périodique CCE 1985

Schott J (1973) Contribution à l'étude de la thermodiffusion dans les milieux poreux. Application aux possibilités de concentrations naturelles. Thèse d'état, Toulouse, 198 p

Shaw HR, Smith RL, Hildreth W (1976) Thermogravitational mechanisms for chemical variations in zoned magma chambers: Geol Soc Am Abstracts with programs 8:1102

Sheppard SMF (1977) The identification of the origin of ore-forming solutions by the use of stable isotopes. In: Volcanic processes in ore genesis. Inst Min Metal Geol Sc London, pp 25–41

Stemprok M, Burnol L, Tischendorf G (1977) Metallization associated with acid magmatism, Geological Survey of Czecoslovakia, Prague 2:41–96

Tauson LV, Kozlov VD (1973) Distributions functions and ratios as estimators of ore bearing potential of granites. In: Jone M.J. (ed), Geochemical exploration 1972. Inst Min Metal, London: 37–44

Taylor RP, Fryer BJ (1983) Rare earth element lithogeochemistry of granitoïd mineral deposits. CIM Bull 76:74–84

Vidal Ph, Cocherie A, Le Fort P (1982) Geochemical investigations of the origin of the Manaslu leucogranite (Himalaya, Nepal). Geochim. Comoschim Acta 46:2279–2292

Fluid Inclusion Volatiles as a Guide to Tungsten Deposits, Southwest England: Application to Other Sn–W Provinces in Western Europe

T.J. SHEPHERD and M.F. MILLER[1]

Abstract

Analysis of fluid inclusion volatiles in vein quartz, by gas mass spectrometry, demonstrates a clear distinction between stanniferous and tungsteniferous ore fluids in the Hercynian metallogenic Sn–W province of SW England. The tungsteniferous fluids are characterized by enhanced levels of CO_2 and N_2 (>0.5 mol%), a distinctive CO_2–N_2–Ar signature and strong CO_2–N_2 covariance. By contrast, the stanniferous fluids are depleted in dissolved gas; thus confirming the widely recognized 'W–CO_2' association. These features are sufficiently well developed to warrant their use as an exploration index for granite-related vein/greisen-type tungsten deposits in this region. Comparison with tungsten deposits in NW England, Eire, France and Portugal reveals similarly high levels of dissolved gas and an inter-correlation between CO_2, CH_4 and N_2 which are independent of the age of mineralization. This suggests that such features are fundamental characteristics of tungsteniferous fluids and allows the exploration criteria established for SW England to be applied to other Sn–W provinces in Western Europe. However, no specific signature or volatile content is universally applicable, and in keeping with good exploration practice it is recommended that routine orientation studies be carried out in each district.

1 Introduction

The association of CO_2-rich fluids with endo- and exogranitic tungsten deposits has been recognized for some time and the relationship has come to be accepted as a chemical characteristic of tungsteniferous fluids (Higgins 1980; Shepherd et al. 1976; Naumov and Ivanova 1971). Indeed, the ubiquitous occurrence of CO_2 has led some workers to advocate the transport of tungsten by carbonate complexes in hydrothermal solutions (Higgins 1980). Since the solubility of CO_2 in aqueous solution is a function of pressure, temperature and salinity, the amount of CO_2 carried by the fluids can be quite variable. Deposits formed at high temperatures

[1] British Geological Survey, 64–78 Grays Inn Road, London WCIX 8NG, UK

Mineral Deposits within the European Community
(ed. by J. Boissonnas and P. Omenetto)

and pressures show relatively higher levels of dissolved CO_2 (Ivanova 1974) than those formed at shallower depths (Kelly and Turneaure 1970; Landis and Rye 1974). Where wolframite or scheelite are minor phases with respect to other ore minerals, the fluid inclusion evidence is often equivocal. Some of this uncertainty can be attributed to the difficulty in assigning different generations of quartz (in which the inclusions occur) to the deposition of specific ore minerals. There remains, however, the fundamental problem concerning the actual chemical relationship between CO_2 and W, and from an exploration point of view, whether or not tungsteniferous fluids can be differentiated from stanniferous fluids according to their CO_2 content.

CO_2-rich fluids are by no means unique to tungsten and occur in association with U, Au, Mo, Hg, Bi and Sb deposits (Brill 1982; Cuney 1978; Leroy 1978; Rice et al. 1985; Smith et al. 1984). Furthermore, CO_2 is the secondmost abundant crustal volatile and is an important component of metamorphic fluids (Touret 1977; Weisbrod et al. 1976). To develop an exploration strategy for tungsten based on the observed fluid inclusion "W–CO_2" signature requires both a knowledge of the variation in CO_2 content of the tungsteniferous fluids and additional criteria for discriminating between different CO_2-rich hydrothermal and metamorphic fluids. Also, it is important to know why CO_2 appears to enhance W transport and to what extent it is responsible for ore deposition. This chapter addresses the first set of criteria and in doing so may provide an insight into the second.

The usefulness of fluid inclusions in mineral exploration has been much discussed and remains an area of controversy. Their major contribution is in providing a better understanding of the physico-chemical characteristics of ore environments and thus improving the geological models used in exploration. Porphyry coppers, Kuroko deposits and vein-type uranium deposits are good examples in which inclusion studies of this nature have been successfully applied (Leroy 1978; Nash 1976; Roedder 1971; Takenouchi 1980). Cases in which inclusions have been used to detect blind ore deposits are regrettably few, though the work of Smith and Kesler (1985) on Archean gold deposits points the way forward. The analysis of inclusion volatiles is still in its infancy and as yet has only been applied systematically to gold deposits in Archean Greenstone belts and epithermal precious metal deposits of Mesozoic-Cenozoic age (Hedenquist and Henley 1985).

The work reported here describes a detailed fluid inclusion volatile investigation supported by thermometric analysis of several tin and tungsten deposits in the Hercynian granite belt of SW England to establish the following points:

1. Can the W–CO_2 signature, as applied to SW England, be quantified for use as a fluid inclusion exploration index for granite-related tungsten deposits?
2. In areas of Sn–W mineralization, is it possible to discriminate between quartz veins which host Sn and those which are more favourable for W, in the absence of ore minerals?
3. Are there other inclusion volatiles associated with CO_2-rich tungsteniferous fluids which strengthen the discrimination?
4. Can the scientific rationale developed for SW England be applied to other Sn–W metallogenic provinces in Western Europe (Hercynian and Caledonian)?

2 Carbon Dioxide-Bearing Inclusions: Their Recognition and Chemical Analysis

For the majority of tungsten deposits, the evidence for CO_2 is provided by the occurrence of inclusions containing liquid CO_2. Where conditions favour the trapping of a homogeneous H_2O-CO_2 fluid, the inclusions are typically three phase with constant phase ratios (aqueous liquid, liquid CO_2, gaseous CO_2). To an experienced researcher, the recognition of liquid CO_2 presents no problem and it is a simple matter to calculate the mol% CO_2 in solution (Shepherd et al. 1985b). However, fluids in the system H_2O-CO_2-NaCl (Bowers and Helgeson 1983) exhibit liquid immiscibility over a wide range of pressures, temperatures and salinities, and hence heterogeneous H_2O-CO_2 mixtures are probably very common. The trapping of such mixtures is more complex than for a homogeneous fluid and gives rise to coexisting inclusions with variable H_2O/CO_2 ratios and salinities. Depending upon the P–T regime, these may range from liquid CO_2 inclusions with trace water, to aqueous inclusions with trace CO_2. While the non-volatile salts (NaCl, $FeCl_2$, $CaCl_2$ etc.) partition preferentially into the aqueous phase, the volatiles (CH_4, N_2 etc.) are enriched in the immiscible gas-rich phase. From published descriptions of tungsten deposits, heterogeneous H_2O-CO_2 fluid inclusion assemblages are typical, implying either heterogeneous trapping of two immiscible fluids or several pulses of fluid with different compositions. Both cases may be the result of fluid immiscibility. As documented by Pichavant and Ramboz (1982) and Ramboz et al. (1982), detailed analysis is required in order to prove that immiscibility has occurred. Thus, neither the relative abundances of H_2O-rich and CO_2-rich inclusions in a sample, nor the corresponding calculated liquid CO_2 densities, can be used to estimate the total concentration of CO_2 in solution. Moreover, for inclusions containing less than 2 mol% CO_2, the amount of CO_2 present may be insufficient to produce a visible condensed phase at room temperature.

It is evident, therefore, that our knowledge of the CO_2 content of tungsteniferous fluids for different geological environments is at best cursory and does not provide a satisfactory exploration data base. Fluid inclusions provide, in theory at least, the closest approach to determining the nature and composition of the ore-bearing fluid. In practice, however, access to this information is difficult to obtain. Whilst notable success has been achieved by the application of laser Raman spectroscopy to the quantitative analysis of volatiles in individual inclusions (Ramboz et al. 1985), the technique is not at present widely available and, furthermore, is not ideally suited to gaseous poor inclusion fluids (<1 mol% gas) or very small inclusions (<2 μm). Our approach has been based on the judicious use of bulk techniques, with fluid release by heating of the sample. Quantitative analysis was then performed by cryogenic separation in conjunction with volumetric analysis and mass spectrometry. As described below, the reproducibility of replicate CO_2 analyses obtained by this procedure, together with the chemical characterization and analysis of other inclusion volatiles, provided a reliable basis for the recognition of tungsteniferous fluids.

3 Analytical Methods

For the thermometric measurements, samples were analyzed using a Linkam TH600 programmable heating-freezing stage (Shepherd 1980).

For the volatile studies, the inclusion fluids were released under vacuum ($<10^{-4}$ torr) by thermal decrepitation of 0.5 g samples of quartz (grain size 0.5–1 mm) at 550 °C and analyzed using a purpose-built automated extraction line linked to a VG Micromass quadrupole gas mass spectrometer. Particular attention was given to the preparation of samples for volatile analysis to avoid contamination from sources other than the inclusion fluids. Low temperature oxygen plasma discharge, followed by hot 6 M HCl and then several rinses of deionized water, was found to be adequate for removing trace organic material, carbonates, sulphides and iron oxide impurities. Final selection was made by hand-picking individual grains under a low power microscope. To ensure the mobility of water vapour in the extraction line, appropriate sections were maintained at a temperature of ca. 80 °C, using heated tapes. By cryogenic trapping at −196 °C, H_2O and CO_2 (together with trace organic components $>C_1$ and sulphur-containing species) were isolated from the 'non-condensable' components: CH_4, N_2 and noble gases, together with any H_2 and CO that may be present (the origin of which is discussed below). By measuring the pressure of the non-condensable gas fraction using a capacitance manometer (range 10^{-3} to 10 torr), the total number of moles of gas in this fraction could be calculated. The mass spectrometer was then used to quantitatively identify the various individual gas species. Substituting an n-pentane/liquid nitrogen slush bath (−129 °C) for the liquid nitrogen trap, the CO_2 was thus released and its pressure measured in a standard volume. (Traces of H_2S would also be expected to be present in this fraction. However, partly because of the nature of the samples used in this investigation, and partly because of possible interference effects associated with the use of a heated, all-metal extraction line system, no H_2S was detected.) For the determination of H_2O, a reduction furnace containing zinc at ca. 400 °C was used to convert the water vapour to H_2, the pressure of which was subsequently recorded. Values for the volumes of CO_2 and H_2O released from replicate quantities of host mineral provided a valuable check of sub-sample homogeneity. Detection limits of the experimental system were as follows: total non-condensables (without quantifying the component gases), ~0.1 nl; for speciation of this fraction, 0.5 nl. The minimum volume (STP) of CO_2 that could be detected was also ~0.1 nl. In the case of water, the minimum quantity that could be accurately measured by the volumetric procedure was ca. 50 μg. An aspect of the non-condensable gas analysis which requires further consideration concerns the observation that in many instances, H_2 was detected (albeit at low levels), as were traces of CO. Discussion of the origin of these components is presented later. However, it may be stated here that their true concentrations in the fluid at room temperature cannot be readily determined by the use of bulk extraction techniques. Thermodynamic calculations show that the equilibrium concentrations of these components are often substantially lower than the experimentally detected levels obtained by bulk-release methods, implying that trace organic contamination of the host matrix is probably a contributory factor.

In view of the CO_2 enrichment of tungsteniferous fluids, stable carbon isotope analysis of this component was initiated. At present, this investigation is incomplete and only preliminary findings are reported here. The experimental procedure adopted for the release and isolation of CO_2 from the quartz was similar to that described above. Sufficient quartz to give a minimum of 50 μl (STP) CO_2 was used; this requirement being imposed by the sensitivity of the VG Micromass 903 mass spectrometer used for the analyses. Data reproducibility, using five replicate analyses, was found to be within 1‰, which was considered acceptable for bulk extraction of fluid inclusion material.

Six samples were also selected for detailed argon isotopic analysis using a VG 1200 gas mass spectrometer operating in the static mode. Argon was released from the inclusions directly using a radio-frequency induction heater, and passed into a high vacuum, noble gas purification line. Standard isotope dilution techniques, utilizing an enriched ^{38}Ar spike, were employed to determined the isotopic ratios 40/38, 38/36 and 40/36. Taking a value of 299.76 for the atmospheric Ar 40/36 ratio, the measured ratios were then normalized to give the proportion of radiogenic to atmospheric ^{40}Ar in the sample.

4 Tin-Tungsten Deposits, Southwest England

4.1 Metallogenic Setting

The Cornubian metallogenic province, the name given to the extensive area of polymetallic mineralization associated with the Hercynian granites of SW England, occupies an area of 3800 km^2 and is regarded as a classic example of hydrothermal mineralization associated with acid magmatism (see Fig. 1). The granites outcrop as five major masses (Dartmoor, Bodmin, St. Austell, Carnmenellis, Lands End) and several smaller intrusives (Godolphin, Cligga Head, St. Michael's Mount, Kit Hill, Hingston Down and Hemerdon) emplaced into a low grade regionally metamorphosed Devonian-Carboniferous volcano-sedimentary sequence at the close of the Hercynian orogeny. Apart from certain lithium-rich varieties, they show only minor variations in bulk mineralogy and chemistry, and are typically coarse-grained, porphyritic, two mica, peraluminous, biotite-rich 'S-type' leucogranites (Stone and Exley 1985). Geochemically they display high $\delta^{18}O$ values, high K : Na ratios, high normative corundum, low Fe_2O_3 : FeO ratios, a deficiency in magnetite, hornblende and sphene, high LREE/HREE ratios and characteristic negative Eu anomalies (Darbyshire and Shepherd 1985). Closely associated with the granites are swarms of granite-porphyry dykes, known locally as elvans, which show an intimate temporal link with mineralization.

Opinion regarding the genesis of the ores is divided, but on certain aspects there is general agreement:

1. The distribution of Sn–W–Cu ores is spatially related to granite cupolas or inferred granite ridges at shallow depth.

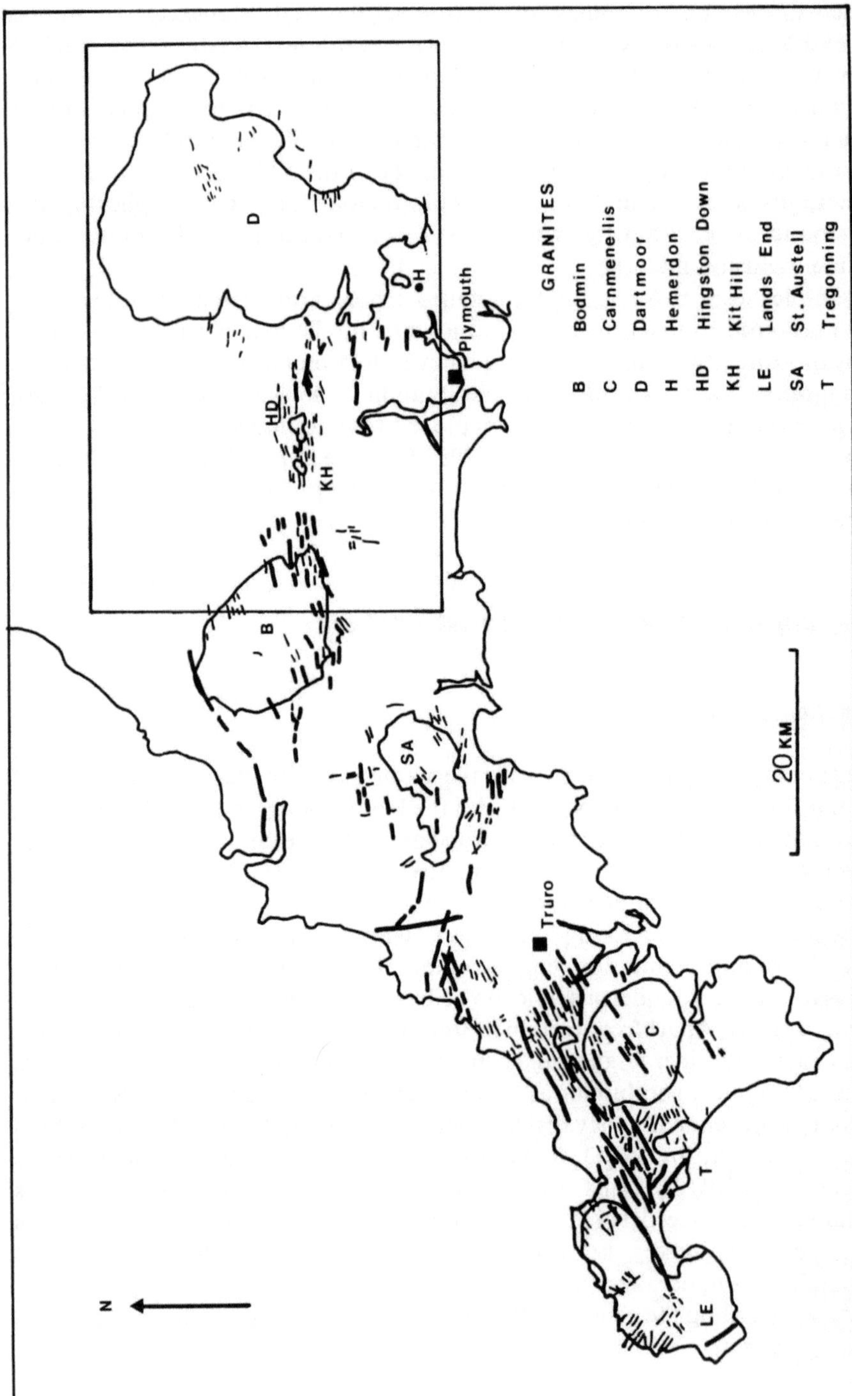

Fig. 1. Cornubian metallogenic province showing the relative distribution of mineral veins (*light lines*) and granite porphyry dykes (*heavy lines*) with respect to the Hercynian granites. *Boxed area* encloses the Bodmin Moor-Dartmoor study area

2. Within individual centres of mineralization there is a broad zonation of ore minerals.
3. Mineralization was synchronous with granite magmatism but continued for some considerable time after granite emplacement.
4. Post-magmatic hydrothermal mineralization/alteration shows a complex and repetitive history, with much overlapping of stages.

Except for minor occurrences of cassiterite-wolframite in pegmatitic vugs of the lithium mica/topaz granites (Manning 1985), the tin-bearing skarns of the Geevor and Levant Mines and the magnetite skarns of eastern Dartmoor, little metalliferous mineralization is associated with the magmatic phase. Of greater importance are events during the post-magmatic phase. Fracture controlled tourmalinization and greisenization are found in all granites; the greisens forming selvages to quartz and quartz-tourmaline veins which carry significant amounts of wolframite and cassiterite. These frequently form sheeted vein complexes centered on small granite cupolas and constitute potential large tonnage, low grade W–Sn deposits. Currently, the most important economic mineralization is associated with late post-magmatic, polymetallic (Sn–W–Fe–As–Cu–Pb–Zn), E–W trending quartz fissure veins, breccia pipes and irregular metasomatic replacement bodies which form complex systems in and around the granite. The paragenetic succession is chronologically and structurally complex due to the superimposition of several mineralizing pulses, and displays all the features of polyascendent zonation. Correlation between individual veins, even within the same district, is difficult. The final metalliferous mineralizing event of economic significance is represented by N–S trending 'cross-course' veins which carry Pb–Zn–Ag sulphides in association with quartz, fluorite, barite and carbonates. Their exact age is unknown but they were emplaced during a different tectonic regime from that of the preceding stage.

Since the early ideas of mineral zonation based on the outward migration of ore fluids from the granites, giving rise to concentric ore zones controlled by sequential deposition of ore minerals ($Sn \rightarrow Cu \rightarrow Zn + Pb \rightarrow Fe + Sb$) along a falling temperature gradient, the emanative centre theory of Dines (1956) has remained unchallenged. Each centre is thought to mark the exit point in the consolidated granite carapace of magmatic-hydrothermal fluids, and was proposed to account for the irregular distribution of Sn and Cu. Recent stable isotope studies (Jackson et al. 1982; Shepherd et al. 1985b) have shown that meteoric water was a major component of the ore fluids and various modified models have been proposed which invoke leaching of metals from the surrounding sedimentary-volcanic rocks.

Nowhere in these models is adequate explanation given for the distribution of tungsten ores. It is loosely assumed that the genetic models proposed for Sn are also valid for W because of their common association. With few exceptions, the major tungsten occurrences/deposits of SW England are not associated with significant quantities of Sn. Furthermore, where cassiterite and wolframite occur in the same vein they often occupy different structures (Dines 1956). Thus, there is sufficient evidence to suggest that the formation of tungsten ores is independent of the deposition of tin ores and may be related to fundamental differences in the chemistry of the ore fluids or environment of deposition.

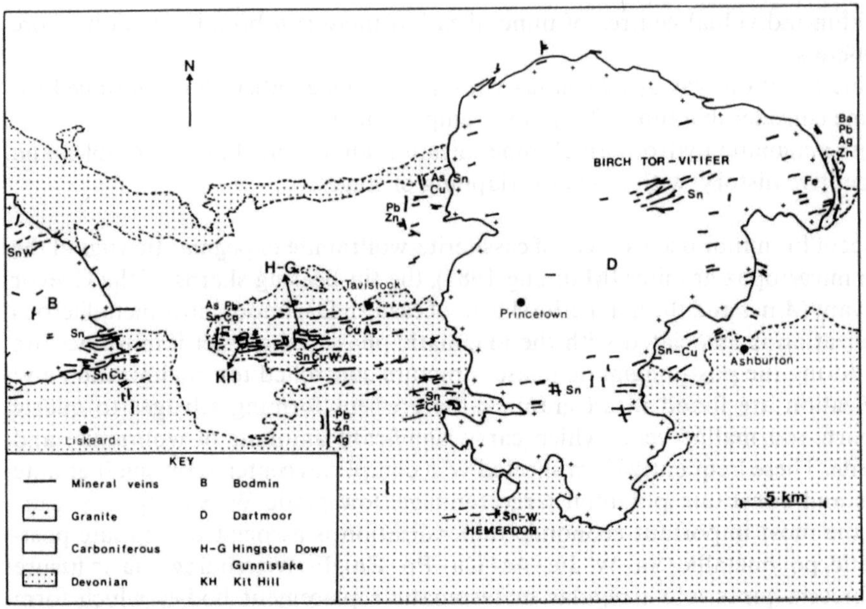

Fig. 2. Simplified geological map of the Bodmin Moor-Dartmoor study area, eastern sector of the Cornubian metallogenic province

4.2 Geology and Mineralization of the Dartmoor-Bodmin Study Area

4.2.1 Geology

The area investigated includes the Dartmoor Granite, the eastern part of the Bodmin Granite and the intervening tract of Palaeozoic rocks (Fig. 2). The Palaeozoics, known locally as Killas, range in age from Middle Devonian to Upper Carboniferous and comprise a tightly folded sequence of limestone, slates and mudstones with inter-bedded tuffs and lavas. Throughout the area the rocks are disrupted by a series of large-scale allochthonous thrust sheets which in places have superimposed older strata on younger. Regional metamorphism is high anchizone/low greenschist facies. Vitrinite reflectance values (1.8–5.7% R) for maturated organic material, which occurs throughout the Devonian-Carboniferous sedimentary succession are in good agreement with reported clay mineral crystallinity values. Close to the granites, the sediments are converted to hornfels and a broad zone of thermal spotting can be recognized.

Between the two main granites (ca. 280–290 Ma) there are three smaller cupolas suggesting the presence of the buried batholith at shallow depth. These granites, namely Kit Hill, Hingston Down and Gunnislake, are similar in composition to the larger masses. East-west trending quartz-porphyry dykes ('elvans') show sharply cross-cutting relationships with both the granites and sediments.

Late magmatic boron metasomatism of the granites is widespread and tourmaline is a very abundant accessory mineral.

4.2.2 Tin-Tungsten Mineralization

Three principal styles of mineralization are present:

1. Sheeted vein complexes, typified by the Hemerdon tungsten deposit, south Dartmoor.
2. Tin-bearing quartz-tourmaline veins of the Birch Tor-Vitifer area, central Dartmoor.
3. East-west trending polymetallic fissure veins with Sn, W, Cu, As and Zn occurring in the area between Bodmin Moor and Dartmoor.

a) Hemerdon Tungsten Deposit. The Hemerdon orebody is located in the northern part of a dykelike granite complex in the extreme south of Dartmoor. Mineralization is represented by three sets of closely spaced quartz veins (ENE–WSW, NE–SW, NNE–SSW) carrying wolframite, hematite, feldspar, minor arsenopyrite and cassiterite (a sulphur-deficient assemblage). Hematite is invariably late-stage and is associated with intense dissolution and cavitation of the wolframite crystals. Veins belonging to the ENE–WSW set commonly have greisen borders. Though sometimes referred to as a W–Sn orebody, the ore zone carries only 250 ppm Sn; 30–50% of which occurs in the greisen borders of the ENE–WSW quartz veins.

b) Birch Tor-Vitifer Tin Veins. Cassiterite-bearing quartz-tourmaline veins are widespread in the Birch Tor-Vitifer area of the Dartmoor Granite, though the deposits are generally small. The veins are near-vertical fracture fillings and show evidence of repeated re-opening during their formation. Deposition of massive iron-rich tourmaline with relatively little quartz was followed by an assemblage of quartz-tourmaline-cassiterite and then by quartz-specular hematite.

c) E–W Trending Polymetallic Fissure Veins. In contrast to the simple, low-sulphide mineralogy of intra-granitic veins, the sedimentary envelope is host to a variety of polymetallic sulphide orebodies. Around the three small cupolas of Kit Hill, Hingston Down and Gunnislake, veins carrying wolframite and cassiterite proximal to the granites pass into complex sulphide veins further out. A paragenesis is recognized with quartz and tourmaline preceding the formation of arsenopyrite, which overlaps with and is succeeded by cassiterite and wolframite. A phase of hematitic wall rock alteration is followed by the development of chlorite and chalcopyrite with associated minor sphalerite, stannite, pyrrhotite, galena, pyrite and late-stage carbonates and fluorite. [For a more detailed description of the geology and mineralization the reader is referred to Beer and Scrivener (1982); Bull (1982); Scrivener (1982); Shepherd et al. (1985a) and Darbyshire and Shepherd (1985).]

4.3 Fluid Inclusion Petrography and Thermometric Results

Four types of aqueous inclusions describe the three main styles of tin/tungsten mineralization (Shepherd et al. 1985a).

Type 1: Two-phase liquid + vapour (L + V) inclusions where L > V.
Type 2: Two-phase vapour-rich, liquid + vapour (L + V) inclusions where V > L.
Type 3: Three-phase liquid + vapour + solid (L + V + S) where the solid phase is halite.
Type 4: Multiphase (L + V + S) inclusions containing halite and up to six other daughter minerals.

Though all four types of inclusion are present at Hemerdon and Birch Tor, types 1 and 2 predominate and, in both cases, are intimately associated. For the polymetallic fissure veins the inclusions are represented almost entirely by type 1 (99%). Inclusions containing liquid CO_2 are *conspicuously absent in all three environments* and the presence of CO_2 is only detected by the development at sub-ambient temperatures of gas hydrates (i.e. clathrates) in type 1 and 2 inclusions. Though thermometric analysis does not distinguish between clathrates of CO_2 and CH_4 if the hydrate dissociation temperature is lower than 10 °C, the inclusion volatile data confirm only a minor contribution of methane.

Considering the thermometric data for Hemerdon, there is a strong bimodal salinity distribution for inclusions which homogenize above 400 °C. A low salinity fluid (< 10 wt% NaCl) characterized by CO_2-bearing type 1 and 2 inclusions, many of which homogenize into the vapour state or show critical phenomena, contrasts markedly with a high salinity fluid (+29 wt% NaCl) represented by type 3 and 4 inclusions. A possible explanation for the T_H-salinity distribution consistent with the geological setting is the unmixing of a moderately saline, magmatic fluid to give a high density, high salinity component and a low density, CO_2-enriched, low salinity component. Whilst acknowledging the problem of verifying fluid immiscibility, several geological aspects support the idea. Firstly, the hydrothermal system at Hemerdon is very localized and the diagnostic fluid components are not present elsewhere within a 6 km radius of the deposit. Secondly, the quartz veins show no evidence of multiple stages of growth normally commensurate with several periods of extended fluid circulation. Thirdly, the chronology of inclusions is extremely complex and characterized by multiple generations of chemically varied pseudosecondaries consistent with rapid changes in fluid chemistry. These features lead us to conclude that the quartz-wolframite veins were deposited from a 'single pass' hydrothermal system linked to a relatively near-surface, magmatic-hydrothermal source undergoing rapid decompression. A change in confining pressures from near-lithostatic to hydrostatic would not only facilitate propagation of hydraulic fractures but also lead to fluid unmixing. The process is envisaged as having taken place at depth, the two components described above ascending as an immiscible mixture. Calculations by Bowers and Helgeson (1983) indicate that any fluid in the system H_2O-CO_2-NaCl containing 10–30 wt% NaCl at 500°–600 °C would unmix to give the observed salinites if the pressure fell below 100–200 MPa. Moreover,

Wilson and Eugster (1984) suggest that temperatures of 400 °C or more, together with relatively low fO_2 and pH, are needed for significant transport of tin (as $SnCl^+$) in supercritical solutions. For a system undergoing fluid unmixing it follows therefore that Sn would preferentially concentrate in the high salinity component.

By inference, in the absence of experimental data to the contrary and the general antipathetic relationship between high salinity fluids and wolframite ores noted elsewhere (Higgins 1980), we propose that W remains in the CO_2-rich, low salinity component. Under such conditions, the formation of a tin or tungsten deposit would depend upon the concentration of available metal in the magmatic aqueous differentiate, the relative amounts and effective physical separation of each fluid component, and their rate of dispersal by entrainment in convecting groundwater systems peripheral to the thermal centre. Though speculative, this hypothesis provides a simple explanation for the frequent spatial and temporal separations between Sn and W ores associated with acid magmatism, and is worthy of further study and experimental verification.

At Birch Tor, as confirmed by volatile analysis, CO_2 is virtually absent. Likewise, CO_2 levels were very low during the deposition of cassiterite in the E–W fissure veins and type 1 inclusions show no evidence of clathrate formation. However, type 1 inclusions associated with quartz/wolframite ores show significant clathrate development, and their salinity and mode of homogenization is comparable to the CO_2-rich type 1 and 2 inclusions developed at Hemerdon.

Thus, thermometric data for the quartz/wolframite ores suggest a comparatively high level of dissolved CO_2 in the tungsteniferous fluids, even though the diagnostic liquid CO_2 inclusions are absent. Data for cogenetic quartz/cassiterite ores indicate a relative CO_2 depletion, thereby confirming a W–CO_2 signature for tungsten deposits in the eastern sector of the SW England metallogenic province.

4.4 Variation in Inclusion Volatiles (CO_2–CH_4–N_2–Ar)

Volatile analyses for the three main styles of Sn–W mineralization are shown in Table 1 and confirm the predominantly aqueous nature of the ore fluids. The principal dissolved gases, in order of abundance, are CO_2, N_2 and CH_4, together with minor quantities of Ar.

For the Birch Tor Sn deposits and E–W trending polymetallic fissure veins carrying Sn or Cu–Sn assemblages, the concentration of dissolved gas is always very low (<0.2 mol%). This contrasts sharply with the marked CO_2 enrichment of the tungsteniferous fluids at Hemerdon, with values of up to 1.5 mol% CO_2. Furthermore, fluids associated with the deposition of quartz-wolframite ores in the E–W veins contain appreciably more CO_2 than those depositing temporally-related quartz-cassiterite ores. Both aspects are illustrated in Fig. 3. Also evident is the prominent linear covariance between CO_2 and N_2, as is the clear distinction between tungsteniferous and stanniferous fluids throughout the region. CH_4 shows a similar but weaker correlation with N_2, implying a basic C–N_2 covariance, the corresponding CO_2/CH_4 ratio controlled perhaps by the oxidation state of the ore fluids.

Table 1. Results of inclusion volatile analysis for quartz samples from tin and tungsten deposits, SW England

Sample No.	H_2O μmol g^{-1}	H_2O mol%	CO_2 mol%	N_2 mol%	CH_4 mol%	H_2 mol%
A. Hemerdon tungsten deposit (Hercynian)						
HEM 79-10	85.5	98.61	1.18	0.18	0.04	0.04
HEM 97-23	85.5	99.48	0.41	0.05	0.02	0.04
HEM 79-26	71.0	99.10	0.66	0.15	0.05	0.05
HEM 79-28	67.5	98.79	0.89	0.20	0.06	0.06
HEM 79-30	66.8	98.51	1.16	0.24	0.03	0.06
HEM 79-42	20.5	98.22	1.25	0.22	0.05	0.27
HEM 79-51	81.5	98.44	1.30	0.19	0.02	0.04
HEM 80-1	74.3	98.44	1.35	0.19	0.01	0.02
HEM 80-5	63.8	98.29	1.40	0.23	0.03	0.05
HEM 80-6	79.2	99.34	0.51	0.08	0.04	0.03
HEM 80-15	68.3	98.51	1.19	0.18	0.07	0.05
HEM 80-25	75.1	98.26	1.40	0.21	0.08	0.05
HEM 80-37	61.1	98.75	0.90	0.21	0.06	0.08
HEM 80-39	59.7	98.80	0.97	0.12	0.08	0.03
HEM 80-42	73.0	98.10	1.46	0.24	0.13	0.07
HEM 80-47	57.9	98.92	0.90	0.11	0.01	0.02
B. Polymetallic east-west fissure veins (Hercynian)						
SW 83-25b	108.0	99.52	0.28	0.09	0.02	0.10
SW 83-25c	90.4	99.32	0.47	0.09	0.02	0.01
SW 83-26a	131.0	99.55	0.29	0.06	0.01	0.08
SW 83-27a	166.0	99.52	0.41	0.04	0.01	0.02
SW 83-28b	62.1	97.99	1.40	0.14	0.39	0.08
SW 83-29b	94.7	99.08	0.78	0.07	0.03	0.05
SW 83-30b	140.0	99.55	0.35	0.05	0.01	0.04
SW 83-32b	72.7	98.74	0.97	0.13	0.07	0.08
SW 83-34c	67.7	97.48	2.26	0.15	0.02	0.09
SW 83-34d	133.0	99.60	0.26	0.04	0.01	0.09
SW 83-37a	133.0	99.74	0.19	0.04	0.01	0.03
SW 83-37b	137.0	99.45	0.35	0.08	0.02	0.10
SW 84-14	37.2	99.71	0.17	0.05	0	0.02
SW 84-15*	87.1	99.29	0.45	0.12	0.01	0.05
SW 84-16*	50.5	98.67	0.91	0.13	0.04	0.09
SW 84-17	240.0	99.87	0.10	0.02	0	0
SW 84-18*	66.4	99.21	0.57	0.13	0.04	0.02
SW 84-20*	94.9	98.58	1.14	0.16	0.03	0.04
SW 84-22	103.0	99.54	0.27	0.04	0.02	0.07
SW 84-24	64.4	99.11	0.50	0.07	0.01	0.12
SW 84-25*	48.8	98.78	0.74	0.23	0.09	0.05
SW 84-27*	65.7	98.80	0.72	0.26	0.07	0.07
C. Birch Tor-Vitifer tin veins (Hercynian)						
SW 81-13	88.2	99.72	0.20	0.05	0	0.03
SW 81-14	114.0	99.87	0.09	0.03	0	0.02
SW 82-13	119.0	99.85	0.09	0.03	0	0.03
SW 82-14	106.0	99.83	0.10	0.04	0	0.03
SW 82-15	146.0	99.86	0.09	0.03	0	0.01
SW 82-19	58.6	99.63	0.15	0.12	0	0.10

* Denotes veins carrying quartz-wolframite ores.

Fig. 3. CO_2–N_2 covariance plot of inclusion volatiles for tin and tungsten vein-type deposits, SW England and Carrock Fell, NW England. ● Quartz-wolframite ores (Hemerdon and other deposits, SW England); ■ quartz-wolframite ores (Carrock Fell, NW England); ○ quartz-cassiterite ores (SW England)

A purely empirical parameter which proved to be of value in distinguishing tungsteniferous from stanniferous fluids was the Ar/N_2 mole ratio; this was noticeably lower in the former cases (<.001 to 0.006 for tungsteniferous fluids; 0.006 to 0.011 for stanniferous fluids). Whereas Ar in fluid inclusion may be derived from a variety of sources (including *in situ* decay of ^{40}K; trapping of ancient atmospheric gas; leaching from surrounding crustal rocks prior to fluid entrapment) this simplistic index was found to be universally applicable to W deposits in the region. It might be expected that for Hemerdon, where the inclusion assemblages are strongly magmatic in character, the Ar would be primarily radiogenic; some stripped from K-bearing minerals during anatexis, the remainder being generated *in situ* by radioactive decay of ^{40}K. However, argon isotope analysis (Table 2) demonstrated that this is not the case; contrary to expectation, atmospherically-derived ^{40}Ar predominates. Since the Ar in the tungsteniferous fluids has a dominant atmospheric component, suggesting the involvement of meteoric groundwaters, it might be expected that, to a first approximation, the level of nitrogen also derived from dissolved ancient atmosphere would result in an Ar/N_2 ratio similar to present-day air (0.0119). This interpretation necessarily neglects the relative solubilities of the two components under P–T and salinity conditions appropriate to those of ore deposition. Even so, the observed N_2 concentration in the fluid greatly exceeds that

Table 2. Results of isotopic analysis of argon released from fluid inclusions for selected quartz samples, Western European tungsten deposits (Prefixes: SW and HEM = SW England; CF = Carrock Fell; Mont = Montredon

Sample No.	Sample wt. (g)	Atmospheric ^{40}Ar (nl)	Atmospheric ^{40}Ar % total	Radiogenic ^{40}Ar (nl g^{-1})
SW 84-16	0.2463	0.4774 ± 0.0067	87.73 ± 1.40	0.271025 ± 0.027629 (10.19%)
CF 76-7	2.0703	0.9230 ± 0.0095	37.65 ± 1.03	0.738321 ± 0.008710 (1.18%)
Mont	1.0341	0.8391 ± 0.0092	69.48 ± 1.10	0.356409 ± 0.011184 (3.14%)
CF 77-77A	4.4084	4.8938 ± 0.0491	64.00 ± 1.00	0.624559 ± 0.012877 (2.06%)
HEM 80-1	5.7336	14.5365 ± 0.1486	91.56 ± 1.02	0.233647 ± 0.029588 (12.66%)
CF 77-39B	2.9974	1.5182 ± 0.0157	55.96 ± 1.03	0.398690 ± 0.006595 (1.65%)

attributable to a purely atmospheric origin implying an additional nitrogen source, possibly of deep-seated origin, but more probably derived from ammonium-rich crustal rocks in the region. [NB. Subsequent to this investigation, experimental $^{40}Ar-^{39}Ar$ studies of the inclusion fluids show that the argon is made up in part of 'excess ^{40}Ar', leached from the surrounding crustal rocks, and in part dissolved ancient atmospheric argon. The ability to quantify these components allows use to be made of the minor radiogenic ^{40}Ar component for dating the mineralization (Kelly et al. 1986).]

4.5 Discussion

Because of the difficulty in providing an *in situ* vein sample grid and the problem of sampling multiple generations of quartz in the E–W fissure veins, rigorous interpretation of the spatial variation in volatile chemistry for the study area is impracticable. However, for veins with a recorded tungsten production, the samples are characterized by relatively high levels of $(CO_2 + N_2) > 0.5$ mol%. Geographically, these veins are within or close to the Kit Hill and Hingston Down granite cupolas in accordance with the known pattern of mineral zonation. East of Hingston Down there are no reports of significant wolframite until one approaches the Hemerdon area. Thus, samples with high $(CO_2 + N_2)$ values from vein structures southeast of the Hingston Down cupola are distinctly anomalous. Therefore, in view of the noted W–CO_2 signature, it is suggested these samples indicate an involvement of ore fluids with a capacity for transporting tungsten. Since there is no geological evidence for sub-outcropping shallow intrusives, the data suggest a possible high level expression of tungsten mineralization at depth. As recognized for other geochemical exploration discriminants, absolute values for volatiles cannot be used to quantify the extent or degree of tungsten deposition but merely indicate the greater tungsten potential for this zone compared to other vein structures east and north of the Kit Hill-Hingston Down axis. A more detailed discussion of CO_2–N_2 covariance is given below.

5 Comparative Study of Selected Western European Deposits

5.1 Description Deposits

Based on the results for SW England, a comparative study was undertaken of other magmatic-hydrothermal tungsten deposits in Western Europe. This work was carried out in close collaboration with ore research groups at the Ecole Nationale Supérieure des Mines de Saint Etienne, France, and the Université de Louvain, Belgium. From their extensive sample collections, suitable material was selected for analysis. Two styles of tungsten mineralization were investigated: quartz veins with wolframite (±scheelite) and scheelite skarn deposits.

Carrock Fell, UK. The Carrock Fell deposit, located in the northern Lake District, is associated with a Caledonian granite which intrudes Ordovician shales and volcanics. The mineralization comprises quartz-wolframite-sulphide veins developed across the granite-country rock contact. Mineralogically, the veins are similar to those at Panasqueira, Portugal (Bussink 1984) and display many of the same paragenetic features. An Rb–Sr age of ca. 395 Ma for the mineralization is in good agreement with the known age of granite emplacement (Shepherd and Waters 1984; Shepherd and Darbyshire 1981).

Ballinglen, Eire. A tungsten prospect on the eastern flank of the Caledonian Leinster Granite. Mineralization, comprising quartz veinlets with sulphides and minor scheelite, is closely associated with a microgranite dyke swarm which cuts volcano-sedimentary rocks of Cambrian to Upper Ordovician age. Because the dykes are presumed part of the main intrusive complex, the mineralization is considered to be of the same age (Steiger and Bowden 1982).

Montredon, France. An important deposit in the southern Massif Central associated with a 530-Ma granite-orthogneiss dome which intrudes schists of Cambrian age. The mineralization has many features in common with Hemerdon and consists of a sheeted quartz vein complex carrying wolframite and sulphides in the upper part of the intrusive and extending outwards into the enclosing schists. Though spatially coincident with the exposed orthogneiss, it is thought that the mineralization could be much younger and related to a deeper granite of possible Hercynian age (Beziat et al. 1980).

Neuf Jours, France. A tungsten propsect in the Haute Correze district of the Massif Central. The mineralization comprises quartz-wolframite-sulphide veins developed within the greisenized apical portion of a leucogranite. The granite intrudes biotite-silliminite gneisses of the Sornac-Saint Germain Lavolps Series (Raimbault 1984).

Justes, Portugal. The Justes (or Cumieira) deposit is part of the Hercynian metallotect of northern Portugal and consists of disseminated wolframite and cassiterite, contained within a partially-greisenized, pipelike, intragranitic breccia zone. The host intrusive belongs to the younger, late-Hercynian granite series and intrudes schists of presumed Cambrian age (Garcia 1986).

Salau, France. One of Western Europe's most important tungsten deposits and an excellent example of scheelite skarn mineralization. Located in the axial zone of the French Pyrenees, the mineralization is associated with calc-silicate skarns, developed Devonian limestones and shales at the contact with a Hercynian granodiorite stock. The orebodies occur as a series of massive pyrrhotite-scheelite lenses within the skarn units and represent the re-working and concentration of subeconomic scheelite in the skarns by later hydrothermal processes (Derre et al. 1980; Fonteilles et al., this Vol.).

5.2 Variation in Inclusion Volatiles (CO_2–CH_4–N_2–Ar)

Analytical data for the above deposits are summarized in Tables 2 and 3. From an inspection of the data, it is evident that there are certain characteristics which transcend differences in age, geological setting and style of mineralization:

1. Tungsteniferous fluids are enriched in CO_2 and N_2, and locally CH_4.
2. CO_2 and CH_4 show a positive linear correlation with N_2.
3. For those deposits studied, atmospheric Ar constitutes a significant proportion of the total Ar budget indicating a major involvement of meteoric water.

Taking the CO_2–N_2 covariance diagram for SW England, addition of data for Carrock Fell reveals an extension of the linear trend to higher absolute CO_2 and N_2 concentrations (Fig. 3). Accepting the slight scatter, the distribution of sample points demonstrates a high degree of statistical correlation and proves that the relationship between CO_2 and N_2 is not restricted to Hercynian deposits, but applies equally to Caledonian deposits. Thus, tungsteniferous ore fluids of differing age may have a common source and/or are generated under similar conditions. Including data for the other European deposits (Fig. 4) produces a more complex pattern but one which shows two distinct CO_2–N_2 trends. Whilst Neuf Jours agrees with the main trend, samples from Justes, Montredon and the Veronique orebody (Salau) define an N_2-depleted trend. Note that the unmineralized skarn samples from Salau and Costabonne (i.e. material presumed to be the protore for the pyrrhotite-scheelite orebodies) are poor in dissolved gases. Though less well defined, the bivariate plot for CH_4 and N_2 shows a similar pattern, implying a moderate degree of inter-correlation between CO_2, N_2 and CH_4.

During the investigation, a small number of samples yielded anomalously high levels of CO and H_2. Though thermodynamic calculations predict that these two species should be present under equilibrium conditions in a C–H–O fluid, realistic estimates of the P–V–T conditions prevailing during fluid entrapment suggest that, for these samples, the equilibrium concentrations of CO and H_2 should be well below experimental detection. Furthermore, although there is a linear correlation

Table 3. Results of inclusion volatile analysis for quartz samples from other Western European tungsten deposits

Sample No.	H_2O μmol g^{-1}	H_2O mol%	CO_2 mol%	N_2 mol%	CH_4 mol%	H_2 mol%
A. Carrock Fell tungsten vein deposit, NW England, UK (Caledonian)						
CF 76-7	107.0	97.43	2.03	0.36	0.13	0.03
CF 76-25	90.9	97.85	1.83	0.25	0.04	0.02
CF 77-6	51.5	97.63	1.91	0.31	0.10	0.02
CF 77-39A	95.4	98.20	1.29	0.32	0.16	0.02
CF 77-77A	79.8	97.68	1.86	0.31	0.10	0.02
CF 77-77B	85.3	98.50	1.29	0.16	0.02	0.02
CF 77-98	94.7	97.17	2.26	0.40	0.14	0.02
B. Ballinglen tungsten vein deposit, Eire (Caledonian)						
BG1	52.2	97.64	0.57	0.77	0.86	0.10
BG5	34.8	97.25	0.90	0.78	0.91	0.11
BG6	21.6	97.18	0.89	0.78	0.95	0.09
C. Neuf Jours tungsten vein deposit, France						
205A	78.5	98.51	1.08	0.22	0.06	0.13
205C	59.2	97.66	1.62	0.26	0.21	0.25
319A	38.0	97.61	1.31	0.30	0.19	0.58
S1.4	93.0	95.54	3.70	0.51	0.15	0.09
S1.24	47.2	98.67	0.72	0.14	0.06	0.40
S5.6	91.0	98.38	1.18	0.18	0.06	0.19
D. Montredon tungsten vein deposit, France						
Mont.	107.0	95.92	3.61	0.11	0.30	0.04
E. Justes-Vila Real tungsten vein deposits, Portugal						
DG 431	68.0	96.56	3.14	0.07	0.19	0.01
DG 599	55.3	95.36	4.13	0.25	0.18	0.07
DG 601	34.3	97.24	1.94	0.07	0.58	0.05
DG 694	57.7	99.25	0.64	0.06	0.01	0.02
DG 695A	37.1	98.40	1.44	0.10	0.01	0.03
DG 853	49.1	97.60	2.08	0.07	0.19	0.02
DG 861	55.4	97.90	1.52	0.03	0.44	0.04
F. Costabonne tungsten skarn deposit, France						
CB9	51.0	99.64	0.22	0.04	0.01	0.06
JF7	46.3	99.39	0.46	0.03	0.01	0.05
JF30	80.9	99.25	0.47	0.01	0.20	0.03
1102.5	86.2	99.03	0.78	0.04	0.01	0.09
G. Salau tungsten skarn deposit, France						
230/3V	73.6	99.51	0.11	0.12	0.20	0.20
230/25	27.4	87.17	11.88	0.41	0.42	0.04

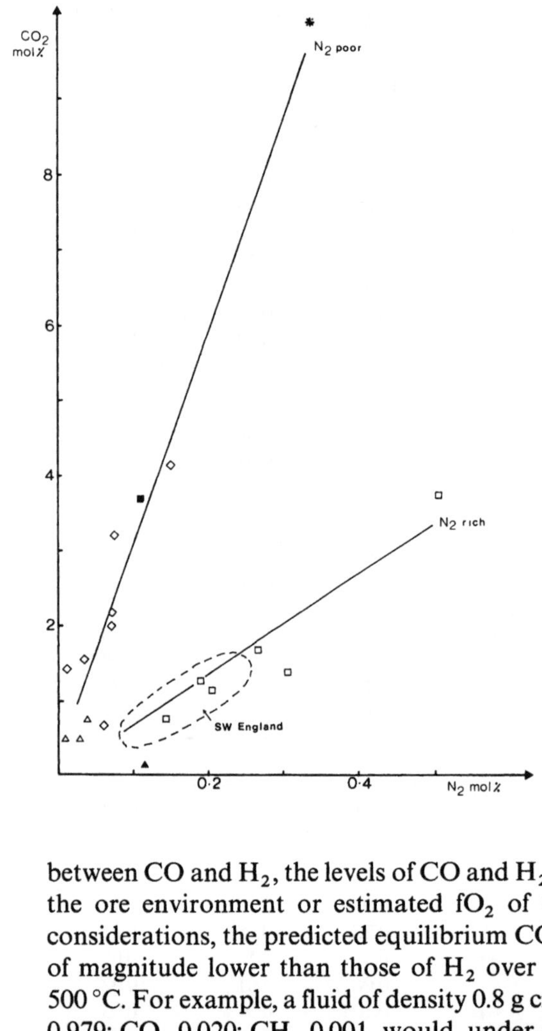

Fig. 4. CO_2–N_2 covariance plot of inclusion volatiles for tin and tungsten deposits in France and Portugal. Vein-type deposits: □ Neuf Jours; ■ Montredon; ◇ Justes, Vila Real. Skarn-type deposits: ▲ Salau; ∗ Salau (Veronique orebody); △ Costabonne

between CO and H_2, the levels of CO and H_2 released are not consistent with either the ore environment or estimated fO_2 of the ore fluids. From thermodynamic considerations, the predicted equilibrium CO concentrations should be two orders of magnitude lower than those of H_2 over the P–T range 0.3–3.2 kb and 300°–500 °C. For example, a fluid of density 0.8 g cm^{-3} having initial mole fractions: H_2O 0.979; CO_2 0.020; CH_4 0.001, would, under equilibrium conditions, give rise to an H_2 mole fraction of $\sim 3.4 \times 10^{-5}$ at 300 °C (total pressure $\sim$760 bar), rising to 5.5×10^{-4} at 500 °C (total pressure 3.50 Kbar). The corresponding CO mole fractions are $\sim 2.3 \times 10^{-7}$ and 6.9×10^{-6} respectively. Calculations were performed using a modified Redlich-Kwong equation of state (Holloway 1981) to determine total pressure and fugacity coefficients.

These observations lead us to conclude that high CO and H_2 levels are most likely the result of pyrolysis of organic matter; the source of which has not been identified. Several authors (Aleksandrova et al. 1980; Naumov et al. 1976) have previously reported that pyrolysis of organic material (from micropores and fractures in the host mineral) frequently occurs when releasing inclusion fluids by thermal means. Piperov and Penchev (1973) have suggested that reduction of H_2O by Fe^{2+} may be a contributory source of the observed levels of H_2, together with the oxidation of CH_4 by H_2O at high temperatures. The latter process would also

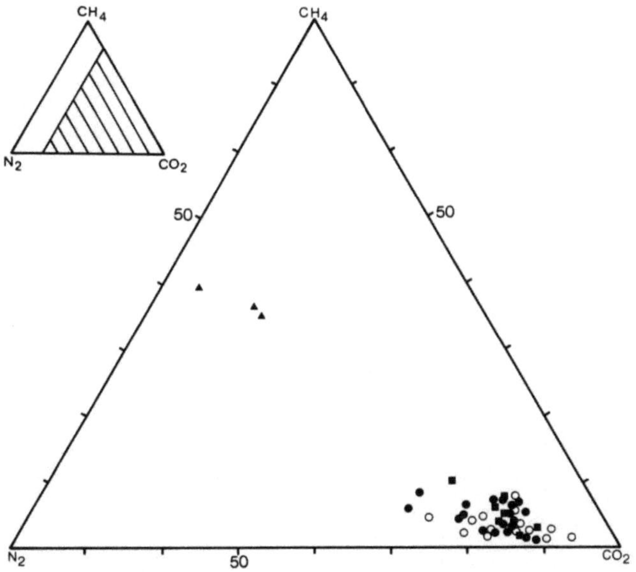

Fig. 5. CO_2–CH_4–N_2 covariance plot of inclusion volatiles for tin and tungsten vein-type deposits, SW England, Carrock Fell, NW England and Ballinglen, Eire. ● Quartz-wolframite ores (Hermerdon and other deposits, SW England); ■ quarter-wolframite ores (Carrock Fell, NW England); ○ quarter-cassiterite ores (SW England); ▲ quarter-scheelite ores (Eire)

give rise to CO production, whereas pyrolysis of organic matter in the absence of excess H_2O generally yields CH_4 (Evans and Felbeck 1983). Furthermore, the quartz lattice itself (Gougel 1963), together with the reduction of H_2O by Si· and Si–O· surface radicals (Kita et al. 1982), are additional potential sources of H_2.

5.3 Discussion

To evaluate the CO_2–CH_4–N_2 covariance in more detail, the mole values given in Tables 1 and 3 have been normalized and the data replotted on CO_2–CH_4–N_2 triangular diagrams. Data for SW England and Carrock Fell plot within a very well-defined field, which also encloses the field for genetically-related stanniferous fluids (Fig. 5). This is not altogether surprising since fluids derived from the same source might be expected to retain or display an inheritance signature. The compositional overlap also substantiates the model proposed for SW England, whereby Sn-rich and W-rich fluids are derived from a parent fluid by fluid immiscibility (see above). Figure 6 shows the respective fields for other deposits and further emphasizes the distinction between N_2-rich and N_2-depleted deposits as shown in Fig. 4. Including data for Panasqueira (Bussink 1984) produces a third array, which projects towards the anomalous field for Ballinglen and a late-stage post-ore vein for Salau. However, all three arrays originate in the CO_2-rich corner of the diagram. To account for this pattern, several factors need to be considered:

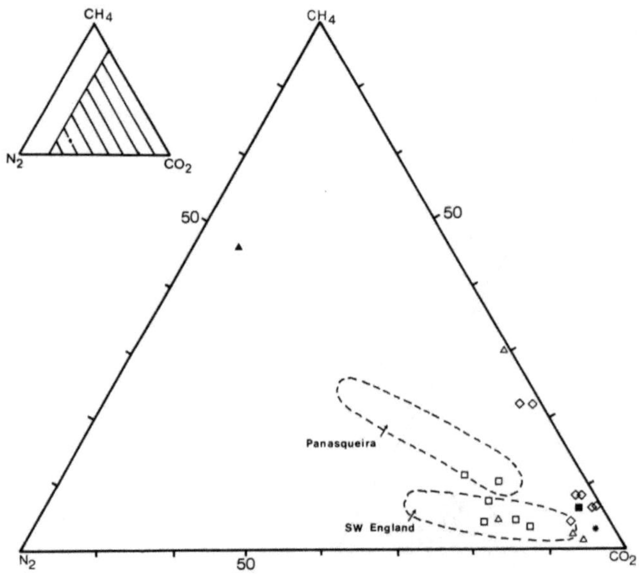

Fig. 6. CO_2–CH_4–N_2covariance plot of inclusion volatiles for tin and tungsten deposits in France and Portugal. Vein-type deposits: □ Neuf Jours; ■ Montredon; ◇ Justes, Vila Real. Skarn-type deposits: ▲ Salau; * Salau (Veronique orebody); △ Costabonne. Data for Panasqueira from Bussink (1984)

1. Nature of the source region with respect to initial volatile composition.
2. Extent of fluid-graphite equilibrium and its control on fO_2.
3. Fluid mixing.

According to Ramboz et al. (1985), changes in CO_2 and CH_4 can often be explained by graphite buffering. In the presence of graphite, fluids above 400°C correspond to the system H_2O–CO_2–CH_4. Below 400°C, graphite becomes deactivated and fluid compositions are best described by the more reduced system H_2O–CH_4. Therefore, in the presence of graphite, changes in fO_2, though controlling CO_2/CH_4 ratios, will tend to produce trends parallel to the CO_2–CH_4 plane, at constant N_2. This may be true for Justes, Montredon and Salau (Veronique orebody), but it is certainly not the case for the other deposits. The arrays shown in Fig. 6 reflect systematic changes in the $N_2/(CO_2 + CH_4)$ ratio. Because of the strong CO_2–N_2 correlation SW England, Carrock Fell and Neuf Jours, and the predominance of CO_2 as the main non-aqueous volatile, the greater proportion of N_2 released from these samples must be derived from the higher temperature, CO_2-rich inclusion assemblages. This suggests that the observed variation relates either to fundamental changes in the magmatic source region or an increasing addition of nitrogen from outside the system.

Carbon isotope studies of the CO_2 component (presently in progress) indicate that, for SW England, the $\delta^{13}C$ value is typically between −9 and −10‰ (PDB). Such values result in general from mixing of an isotopically-light carbon component (either carbonate-derived or of deep-seated origin), with a minor, sedimentary 'organic' component ($\delta^{13}C < -20‰$) already mixed in the granitic melt (Hoefs 1980). Mantle values, as estimated from carbonatites and kimberlites, are typically -4 to $-7‰$ (Deines and Gold 1973). In the case of Carrock Fell, $\delta^{13}C$ values range from -11.1 to $-13.3‰$, indicative of a greater sedimentary component.

An important question still to be resolved is the mechanism whereby CO_2 (and possibly N_2) derived from a sedimentary source become incorporated in a magmatic-hydrothermal system. One scenario would be the assimilation of pelitic material during the anatexis of 'S'-type granite magmas. Another possible mechanism is the devolatilization of sedimentary rocks in the thermal metamorphic aureole of a magma chamber and mixing of the volatiles released with magmatic aqueous fluids. Both models assume that the tungsten is released to the fluid phase from the silicate melt. However, a third mechanism can be envisaged whereby tungsten is remobilized from strata-bound enrichments adjacent to the granites. As yet, no specific scheelite deposit has been linked to the genesis of specific vein or skarn deposits. Whichever applies, it is clear that tungsten is transported to the site of deposition in CO_2-rich, low salinity fluids.

6 Conclusions

As stated in the introduction, a principal aim of the investigation was to compare and contrast the chemistry of the fluid inclusion volatiles associated with granite-related Sn and W mineralization, to define new exploration criteria for tungsten deposits. In particular, it was hoped to quantify and improve the empirical W–CO_2 signature and test its potential as an exploration index. For the Bodmin-Dartmoor study area, volatile analyses demonstrate a marked enrichment in CO_2 and N_2 for tungsteniferous fluids, combined with very low Ar/N_2 ratios. By comparison, stanniferous fluids are noticeably depleted in non-aqueous volatiles. The pattern for Hemerdon is repeated for other Hercynian tungsten deposits in SW England, suggesting that such features are common to the Cornubian metallogenic province. Complementary thermometric data confirm the existence of CO_2-rich fluids. Unlike many tungsten deposits described in the literature, liquid CO_2 inclusions are absent and the presence of CO_2 is only revealed by the development of gas hydrates at low temperatures or by volatile analysis. Hence, optical examination does not provide a reliable method for discriminating between tungsteniferous and stanniferous quartz veins. Occasionally, high ($CO_2 + N_2$) values are recorded for unrelated styles of mineralization, but the fluids have Ar/N_2 ratios close to atmospheric. The Ar/N_2 ratio thereby offers a useful secondary discriminant. We believe the CO_2–N_2–Ar signature established for SW England constitutes a practical exploration index for granite-related tungsten deposits in areas of poor exposure, or where sample material is limited (e.g. drill cores).

Analysis of other tungsten deposits in Western Europe (Caledonian and Hercynian) reveals a similar enrichment in CO_2 and N_2, and suggests that a high dissolved gas content (>0.5 mol%) and strong CO_2–N_2 covariance are primary characteristics of tungsteniferous ore fluids, irrespective of age. Locally, CH_4 is an important constituent. Though not yet proven, it would also appear that the fluids involved in the formation of high-grade ore zones within scheelite skarn deposits (i.e. Salau) have a volatile signature similar to that of quartz-wolframite vein deposits. In contrast, fluids responsible for the main stage barren skarns are very low in dissolved gases. Substantial data for tin deposits is lacking, but preliminary volatile analyses for quartz-cassiterite ores from the Panasqueira region support the proposed model.

No specific signature or volatile content can be applied to all tungsten deposits, but within individual mineral provinces there is a common pattern. As a general guide, hydrothermal quartz veins containing inclusions with enhanced levels of CO_2 and N_2 may be regarded as potential tungsten targets. However, in keeping with good exploration practice, orientation studies should be carried out in each region to establish the local threshold of dissolved gas and CO_2–N_2–CN_4 pattern.

Acknowledgements. We are grateful to Amax Exploration (UK) Inc. for access to borehole cores from their Hemerdon property and for valuable discussions with their staff. The authors especially wish to acknowledge the collaboration, assistance and advice of B. Guy, D. Garcia, M. Perrin and colleagues (Ecole Nationale Supérieure des Mines de Saint Etienne) and J. Verkaeren, Ph. Sonnet, G. van Marcke de Lummen and colleagues (Universite Catholique de Louvain) in providing information and samples for tungsten deposits in France and Portugal, and for their enthusiastic support for the research. Our thanks are extended to J. Dubessy (CREGU, Nancy) for providing a constructive review of the manuscript and helpful suggestions for improvement. Financial support for this work was provided in part by the EEC (*Contract No. MSM-102-UK*). This work is published with the approval of the Director of the British Geological Survey.

References

Aleksandrova ES, Bannikova LA, Sushchevskaya TM (1980) Errors in gas analysis by thermal decrepitation of inclusions. Geokhimiya 11:1710–1716; Geochem Int 17:66–71

Beer KE, Scrivener RC (1982) Metalliferous mineralization. In: Geology of Devon, Durrance EM and Manning DAC (eds), Univ Exeter, pp 117–147

Beziat P, Prouhet JP, Tollon F (1980) Le district de Montredon-Labessonnié (Tarn) W, Sn, F. XXVI Geological Congress, Paris. Gisements Francais, Fascicule E7

Bowers TS, Helgeson HC (1983) Calculation of the thermodynamic and geochemical consequences of nonideal mixing in the system H_2O–CO_2–NaCl on phase relations in geologic systems: equation of state for H_2O–CO_2–NaCl fluids at high pressures and temperatures. Geochim Cosmochim Acta 47:1247–1275

Brill H (1982) Fluid inclusion study of Sn–W–Au, Sb– and Pb–Zn mineralizations from Brioude-Massiac district (French Massif Central). Tschermaks Mine Pet Mitt 30:1–16

Bussink RW (1984) Geochemistry of the Panasqueira tungsten-tin deposit, Portugal. Geol Ultraiectina 33:1–170

Cuney M (1978) Geologic environment, mineralogy and fluid inclusions of the Bois Noirs–Limouzat uranium vein, Forez, France. Econ Geol 73:1567–1610

Darbyshire DPF, Shepherd TJ (1985) Chronology of granite magmatism and associated mineralization, SW England. J Geol Soc (Lond) 142:1159–1178

Deines P, Gold DP (1973) The isotopic composition of carbonatite and kimberlite carbonates and their bearing on the isotopic composition of deep-seated carbon. Geochim Cosmochim Acta 37:1709–1733

Derré C, Fonteilles M, Nansot LY (1980) Le gisement de scheelite de Salau, Ariège–Pyrénées. XXVI Geological Congress, Paris. Gisements Francais. Fascicule E9

Dines HG (1956) The metalliferous mining region of south-west England. Memoir, Geological Survey of Great Britain

Evans RJ, Felbeck GT Jr (1983) High temperature simulation of petroleum formation. I. The pyrolysis of Green River shale. Org Geochem 4:135–144

Garcia D Le pipe bréchique à W–Sn de Cumieira, Vila Real, Nord Portugal. Comm du Service Geol du Portugal (in press)

Gougel R (1963) Die chemische Zusammensetzungen der in den Mineralien einiger Granite und ihrer Pegmatite eingeschlossenen Gase und Flüssigkeiten. Geochim Cosmochim Acta 27:155–181

Hedenquist JW, Henley RW (1985) The importance of CO_2 on freezing point measurements of fluid inclusions: evidence for active geothermal systems and implications for epithermal ore deposition. Econ Geol 80:1379–1406

Higgins NC (1980) Fluid inclusion evidence for the transport of tungsten by carbonate complexes in hydrothermal solutions. Can J Earth Sci 17:823–830

Hoefs J (1980) Stable isotope geochemistry. Springer, Berlin Heidelberg New York, 208 pp

Holloway JR (1981) Compositions and volumes of supercritical fluids in the Earth's crust. In: Hollister LS, Crawford ML (eds) Mineralogical Association of Canada, short course in fluid inclusions: applications to petrology, Chapter 2. pp 13–28

Ivanova GF (1974) Geochemical and physico-chemical conditions of tungsten migration and deposition. In: Stemprok M (ed) Metallization associated with acid magmatism. Ustredni Ustav Geologicky, Prague, pp 267–269

Jackson NJ, Halliday AN, Sheppard SMF, Mitchell, JG (1982) Hydrothermal activity in the St. Just mining district, Cornwall, England. In: Evans AM (ed) Metallization associated with acid magmatism. Wiley, New York, pp 393–399

Kelly S, Turner G, Butterfield AW, Shepherd TJ (1986) The source and significance of argon isotopes in fluid inclusions from areas of mineralization. Earth Planet Sci Lett 79:303–318

Kelly WC, Turneaure FS (1970) Mineralogy, paragenesis and geothermometry of the tin and tungsten deposits of the Eastern Andes, Bolivia. Econ Geol 65:609–680

Kita I, Matsuo S, Wakita H (1982) H_2 generation by reaction between H_2O and crushed rock: an experimental study on H_2 degassing from the active fault zone. J Geophys Res 87:10789–10795

Landis GP, Rye RO (1974) Geologic, fluid inclusion, and stable isotope studies of the Pasto Bueno tungsten-base metal ore deposit, northern Peru. Econ Geol 69:1025–1059

Leroy J (1978) The Margnac and Fanay uranium deposits of the La Crouzille district (Western Massif Central, France): geologic and fluid inclusion studies. Econ Geol 73:1611–1634

Manning DAC (1985) Comparison of influence of magmatic water on form of granite-hosted Sn–W deposits and associated tourmalinization from Thailand and southwest England. In: High heat production (HHP) granites, hydrothermal circulation and ore genesis. Special conference issue. Inst Mining and Metallurgy, St. Austell, England, pp 203–212

Nash JT (1976) Fluid inclusion petrology: data from porphyry copper deposits and applications to exploration. US Geol Surv Prof Pap 907D:16

Naumov GB, Mironova OF, Naumov VB (1976) Carbon compounds in fluid inclusions in hydrothermal quartz. Geochem Int 13:164–171

Naumov VB, Ivanova GF (1971) The pressure and temperature conditions for formation of wolframite deposits. Geochem Int 8:381–394

Pichavant M, Ramboz C (1982) Fluid immiscibility in natural processes: use and misuse of fluid inclusion data. I. Phase equilibria analysis — a theoretical and geometrical approach. Chem Geol 37:1–28

Piperov NB, Penchev NP (1973) A study on gas inclusions in minerals. Analysis of the gases from micro-inclusions in allanite. Geochim Cosmochim Acta 37:2075–2097

Raimbault L (1984) Géologie, pétrographie et géochimie des granites et minéralisations associées de la région de Meymac (Haute Corrèze, France). Thèse de Docteur Ingénieur, Ecole Nationale Supérieure des Mines de Paris

Ramboz C, Pichavant M, Weisbrod A (1982) Fluid immiscibility in natural processes: use and misuse of fluid inclusion data. II. Interpretation of fluid inclusion data in terms of immiscibility. Chem Geol 37:29–48

Ramboz C, Schnapper D, Dubessy J (1985) The P–V–T–X–fO_2 evolution of H_2O–CO_2–CH_4-bearing fluid in a wolframite vein: reconstruction from fluid inclusion studies. Geochim Cosmochim Acta 49:205–219

Rice CM, Harmon RS, Shepherd TJ (1985) Centraly City, Colorado: the upper part of an alkaline porphyry molybdenum system. Econ Geol 80:1769–1796

Roedder E (1971) Fluid inclusion studies on the porphyry-type ore deposits at Bingham, Utah, Butte, Montana and Climax, Colorado. Econ Geol 66:98–1200

Shepherd TJ (1980) Temperature-programmable heating-freezing stage for microthermometric analysis of fluid inclusions. Econ Geol 75:1244–1247

Shepherd TJ, Darbyshire DPF (1981) Fluid inclusion Rb–Sr isochrons for dating mineral deposits. Nature 290:578–579

Shepherd TJ, Waters P (1984) Fluid inclusion gas studies, Carrock Fell tungsten deposit, England: implications for regional exploration. Miner Dep 19:304–314.

Shepherd TJ, Beckinsale RD, Rundle CC, Durham J (1976) Genesis of Carrock Fell tungsten deposits, Cumbria: fluid inclusion and isotopic study. Trans Inst Mine Metall (Sect B: Appl Earth Sci) 85:B63–B73

Shepherd TJ, Miller MF, Scrivener RC, Darbyshire DPF (1985a) Hydrothermal fluid evolution in relation to mineralization in southwest England with special reference to the Dartmoor-Bodmin area. In: High heat production (HHP) granites, hydrothermal circulation and ore genesis. Special conference issue. Inst Mining and Metallurgy, St. Austell, England, pp 345–364

Shepherd TJ, Rankin AH, Alderton DHM (1985b) A practical guide to fluid inclusion studies. Blackie, Glasgow

Smith TJ, Kesler SE (1985) Relation of fluid inclusion geochemistry to wall rock alteration and litho-geochemical zonation at the Hollinger-McIntyre gold deposit, Timmins, Canada. CIM Bull 78:35–46

Smith TJ, Cloke PL, Kesler SE (1984) Geochemistry of fluid inclusions from the McIntyre-Hollinger gold deposit, Timmins, Ontario, Canada. Econ Geol 79:1265–1285

Steiger A, Bowden A (1982) Tungsten mineralisation in south-east Leinster, Ireland. In: Brown AG (ed) Mineral exploration in Ireland: progress and developments 1971–1981. Wexford Conference 1981 Irish Assoc Econ Geol pp 108–114

Stone M, Exley CS (1985) High heat production granites of southwest England and their associated mineralization: a review. In: High heat production (HHP) granites, hydrothermal circulation and ore genesis. Special conference issue. Inst Mining and Metallurgy, St. Austell, England, pp 571–593

Takenouchi S (1980) Fluid inclusions studies in Japan and their application to mineral exploration. In: 4th Joint Meeting MMIJ–AIME, Tokyo. Tech Session A–1:1–16

Touret J (1977) The significance of fluid inclusions in metamorphic rocks. In: Fraser DG (ed) Thermodynamics in geology Reidel, Dordrecht, pp 203–227

Weisbrod A, Poty B, Touret J (1976) Les inclusions fluides en géochimie-pétrologie: tendances actuelles. Bull Soc Fr Minér Cristallogr 99:140–152

Wilson GA, Eugster HP (1984) Cassiterite solubility and tin-chloride speciation in supercritical solutions. In: Abstacts of the Annual Meeting of the Geological Society of America and Associated Societies, Reno, Nevada, USA, Geol Soc Am (Abstract No 34613)

Geochemical and Isotope (H, C, O, S) Studies of Barren and Tungsten-Bearing Skarns of the French Pyrenees*

B. GUY[1], S.M.F. SHEPPARD[2], A.M. FOUILLAC[3], R. LE GUYADER[1,4], P. TOULHOAT[1], and M. FONTEILLES[1]

Abstract

Both scheelite-bearing and barren skarns occur in Lower Palaeozoic carbonate formations in contact zones of Hercynian granites in the Pyrenees. Chemical analyses of the mineralized Costabonne skarns which developed in marbles show that W, U, Fe, Mn, Zn, S and less clearly Ta and Nb were introduced and elements such as Ti, Al, Zr, Hf and the REE were typically immobile.

D/H and $^{18}O/^{16}O$ analyses of minerals and fluid inclusions from the early Stage I (garnet, pyroxene, ±scheelite) indicate the dominance of metamorphic waters in barren skarns and either metamorphic or magmatic waters in mineralized skarns. Comparable minerals are typically 1‰ or more enriched in ^{18}O in barren skarns relative to mineralized ones. Fluids responsible for Stage II hydrosilicate-sulphide alteration (amphibole, chlorite, calcite, quartz, sulphides ± scheelite) were dominantly of meteoric origin whose δD value may have evolved during skarn development. $^{13}C/^{12}C$ and $^{18}O/^{16}O$ systematics of calcite are interpreted in terms of a hydrothermal decarbonation model, although a magmatic carbon contribution cannot be excluded for some calcites. $^{34}S/^{32}S$ analyses of minerals, with $\delta^{34}S_{\Sigma S} \sim +2 \pm 1‰$ for the mineralizing fluids, do not discriminate between a magmatic or country rock source for the sulphur. In mineralized skarns, tungsten concentrations increase with increase in the importance of Stage II development.

The difference in O-isotope composition of comparable minerals from mineralized and barren skarns is a potential prospecting tool. More data are required to test this possibility and its applicability to non-Pyrenean skarns.

1 Introduction

In the French and Spanish Pyrenees (Fig. 1), skarns occur quite commonly near Hercynian granitoids where the Palaeozoic metasedimentary country rocks include

---

* CRPG Contribution No. 696

[1] Departement Géologie UA CNRS No. 384 "Métallogénie et Pétrologie", Ecole des Mines, 158 Cours Fauriel, 42023 Saint-Etienne Cedex 2, France

[2] Centre de Recherches Pétrographiques et Géochimiques, (CNRS) B.P. 20, 54501 Vandoeuvre-les-Nancy Cedex, France

[3] BRGM-D.T. MGA Section Isotopes, BP 6009 45060 Orléans Cedex 2, France

[4] Laboratoire P. Süe, CEN Saclay, BP n02, 91191 Gif-Sur-Yvette Cedex, France

Mineral Deposits within the European Community
(ed. by J. Boissonnas and P. Omenetto)

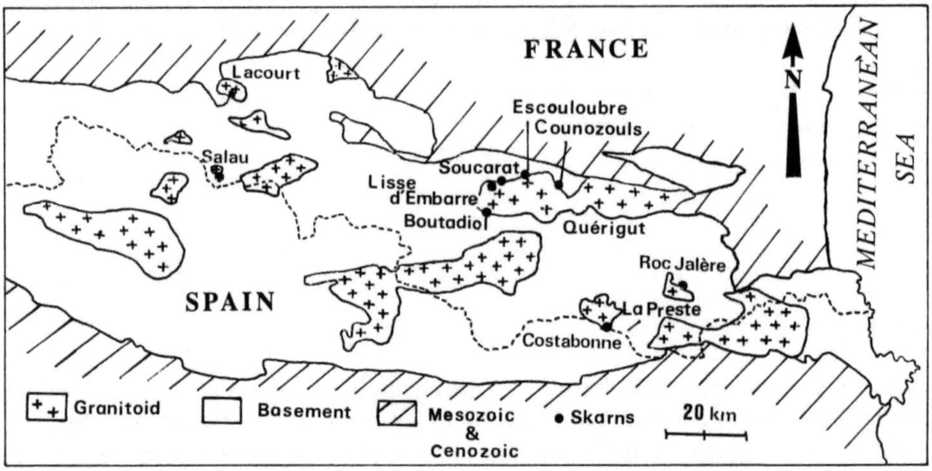

Fig. 1. Simplified geological map of the Central and Eastern Pyrenees showing the location of the principal granitoid massifs and the barren and mineralized skarns discussed in the text

limestones and carbonate-bearing rocks (e.g. Autran 1980). Many of these skarns contain tungsten as scheelite and some were being exploited until very recently (e.g. Anglade mine at Salau). In this chapter 'mineralized' skarns will refer to those with scheelite and base metal sulphides in addition to the usual calc-silicate assemblages, whilst in 'barren' skarns, such minerals do not occur except for minor pyrite or pyrrhotite. In both types of skarns evidence is present for metasomatism and hydrothermal phenomena such as veins, replacement phenomena, etc. Mineralized skarns include Salau, Costabonne, Roc Jalère and Lisse d'Embarre (Fig. 1). Except for Salau and Costabonne, these are all small skarns with very minor scheelite mineralization. Barren skarns include Lacourt, Soucarat and other Quérigut skarns, except Lisse d'Embarre, such as Counozouls, Escouloubre, etc. The small Boutadiol skarn will be considered here as unmineralized because scheelite has not been observed, although base metal sulphides are present. Because criteria are being sought to help to distinguish between mineralized and barren skarns and both mineralized and sterile skarns or calc-silicate hornfelses can be closely associated in space, an (M) or (B) will follow the skarn name to indicate the general presence or absence respectively of scheelite at that locality. An (M), however, does not imply that the specific sample analyzed contains scheelite.

This chapter presents a summary of some of our geochemical results — H-, C-, O- and S-isotope, major and trace element data — on a number of these mineralized and barren skarns. The possible consequences of these results for the prospection of tungsten-bearing skarns is then considered. Although the location and shape of orebodies is structurally controlled, structural aspects of skarn formation are not considered here.

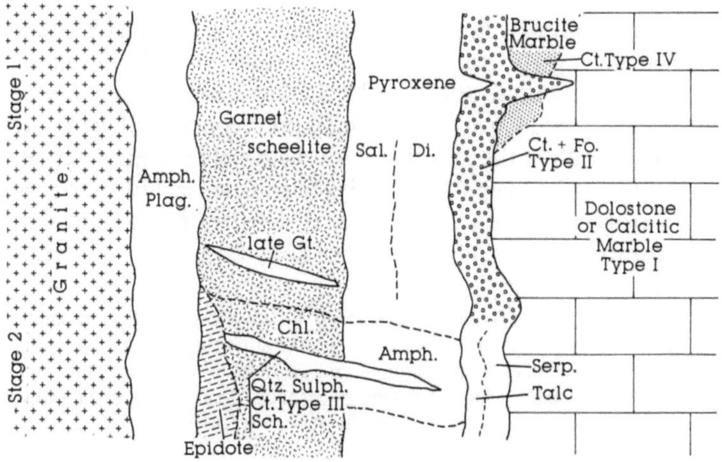

Fig. 2. Schematic summary of zoning and mineral paragenesis of the Costabonne skarn developed between granite (*left*) and dolomitic marble (*right*). In the early Stage I (*top*) a forsterite-calcite (Type II) or magnesian skarn forms in dolomites at possibly T > 500 °C for $X_{CO_2} = 0.05$ and 2 kb total pressure. The magnesian skarn with early diopside and andraditic garnet is followed by salite, spessartine-grandite and scheelite (the different garnet generations are not distinguished). In the later Stage II (*bottom*) Mn-rich grossularite develops in the garnet zone, quartz veins with actinolite or chlorite, calcite (Type III), scheelite and sulphides, and retrograde alteration at the magnesian front to serpentine and talc. Abbreviations: *Amph* amphibole; *Chl* chlorite; *Ct* calcite; *Di* diopside; *Fo* forsterite; *Gt* garnet; *Plag* plagioclase; *Qtz* quartz; *Sal* salite; *Sch* scheelite; *Serp* serpentine; *Sulph.* Sulphides (After Guy 1979)

2 Samples and Their Geological Setting

Rock and mineral samples were selected for analysis from the unaltered and altered country rocks and through the different zones of the skarn to the contact zone which is often but not necessarily, against granitoid (Fig. 2). Many skarns developed at the expense of relatively pure calcitic or dolomitic marbles. For example, at Costabonne the marbles are dominantly dolomitic, whilst at Salau calcitic marbles, with or without graphite, are abundant. Skarns also form in metamorphosed shales, sandstones, tuffs, etc. where these are interlayered with limestones and other carbonate-bearing formations (e.g. Costabonne, Salau, etc.). This study principally considers skarns which developed at the expense of marbles. Skarns formed in pelitic rocks at Costabonne have been discussed by van Marcke de Lummen and Verkaeren (1986).

Figure 2 presents a schematic summary of the zoning and paragenesis of the Costabonne skarns. Four petrographically different types of carbonate were studied. Carbonates derived from the pre-existing limestones were analyzed from a few tens of metres within the marbles to the contact zone with the skarns (Type I). Type II calcites are those which form part of a skarn zone (e.g. calcite-forsterite assemblage which formed at the expense of dolostones at Costabonne, Fig. 2). Calcite of Type III was precipitated either during carbonation reactions in the massive garnet or pyroxene skarns during the hydrosilicate stage (formation of amphibole, chlorite, etc.) or as calcite-bearing skarn veinlets in skarn or marble. Type IV calcites occur in brucite marbles formed during the decarbonation of dolostones followed by hydration reactions, as at Costabonne (Guy 1979; Dubru 1986). Some minor alteration calcite, which is a post-skarn formation, is not considered here. At Soucarat (B) the skarns occur both in the marbles, with the development of garnet, and in the schists, with development of epidote. Analyzed silicates include both the Stage I primary skarn minerals (garnet, pyroxene) and later hydrosilicate alteration minerals (amphibole, serpentine, chlorite) of the earlier anhydrous skarn minerals (Stage II).

The nature of the granitoids in the Pyrenees is quite variable. Skarns are associated with calcalkaline granitoids at Quérigut and Salau (Leterrier 1972; Soler 1980). The granite at Costabonne has also been considered to be calcalkaline (Autran 1980; Soler 1980; Salemink and De Jong, this Vol.) but Stussi and de la Roche (1984) defined it as alumino-potassic. The isotope evidence for the dominantly crustal origin of such Hercynian granitoids in the Pyrenees has been reviewed by Sheppard (1986b).

3 Major and Trace Elements

From over 200 chemical analyses on Pyrenean skarns only a small selection can be presented here. The most complete study is on the Costabonne skarn. Table 1 gives a representative set of major and trace element analyses of the principal rock types from the unaltered schist and marbles and different zones of the skarns. Only general trends are discussed here.

Comparing the chemistry of the marbles with that of the associated skarns (cf. 1200A, SCF 32 and CF 26 with CF 16), the concentrations of nearly all elements increase during the development of the skarn with the notable exceptions of Ca, Mg, Sr and CO_2. Except for these four elements, the initial relatively pure marbles are strongly depleted in nearly all other elements. In contrast the contents of a much wider range of elements are significant in the schists and granites (Table 1). The garnet zone of the schist-skarn is enriched in Fe, Mn, Mg, Ca, Zn and Sb but depleted in Si, Na, K, Rb, Cs, Ba, Th and Co (cf. CF 5 and R 20). The content of a number of other elements such as Ti, Al, Zr, Hf, P, Ta, Sc and sometimes the REE either remain essentially constant or there is not a systematic increase or decrease in their content. However, in the biotite, amphibole and pyroxene zones between the schists and garnet zone, the contents of 23 trace elements are very similar to those of the schist, except for Rb, Ba, Zn and W (van Marcke de Lummen and Verkaeren 1986). There is ~ 2 ppm W throughout the schist-skarn compared with 11 ppm W in the analyzed schist. However, the W content of the schist should probably be interpreted with caution because (1) its value is six times that of average shale and sandstone (Turekian and Wedepohl 1961), and (2) too few analyses exist on schists from Costabonne to know whether such values are typical for the non-mineralized schists.

Taking the data on skarns in both marbles and schists, it has been established that the following elements were transported by a fluid phase into or out of the skarnified rock: Ca, Fe, Mn, Mg, Rb, Cs, Sr, Ba, W and Zn. In particular, W is added during the first stage of skarn development with garnet or pyroxene and during the hydrosilicate-sulphide stage of the metasomatism of marbles, but it is

Table 1. Major and trace element analyses of selected rocks from Costabonne[a]

	CF 5 sch	R 20 banded sk gt	CF 16 gt sk	CF 21 gt px sk	R 17 px gt sk	SCF 233b salite px sk	SCF 232 diop sk	1200 C cc fo sk	1200 A dolost sk contact	SCF 32 dolost 100 m	CF 26 cc marble
SiO_2	67.5	43.3	38.6	40.7	40.1	50.8	54.2	16.1	0.0	0.1	0.3
TiO_2	0.81	0.51	—	—	—	—	—	—	0.01	0.01	—
Al_2O_3	14.7	10.1	4.9	5.9	4.9	0.2	0.04	—	—	—	—
Fe_2O_3	5.80	9.67	24.0	17.61	18.02	10.92	4.08	0.56	1.45	0.46	0.11
MnO	0.08	3.32	1.51	1.93	1.87	3.81	1.12	0.13	0.28	0.20	0.08
MgO	2.27	3.53	0.92	2.25	2.19	9.74	16.10	16.70	20.17	19.82	1.50
CaO	2.75	27.23	28.87	28.98	27.24	24.08	23.97	35.54	30.67	32.06	55.00
Na_2O	2.11	—	—	—	0.20	0.22	—	—	0.01	—	0.09
K_2O	1.93	—	—	0.04	0.05	—	—	—	—	—	0.01
P_2O_5	0.15	0.14	—	0.02	—	0.01	0.05	—	0.03	—	0.01
P.F.	1.26	2.51	0.84	2.06	3.25	0.59	0.72	32.40	47.31	48.07	43.66
Total	100.52	100.27	99.86	99.58	98.00	100.40	100.41	101.47	99.90	100.67	100.74
U	2.5	3.2	17	5.9	5.9	4.9	0.45	0.40	0.07	0.04	0.07
Th	17	4.8	0.22	0.09	0.03	0.15	0.02	0.02	0.01	0.02	0.02
Zr	314	160	46	30	20	25	—	—	—	—	—
Hf	8.2	4.9	0.15	0.15	0.10	0.10	—	—	0.30	—	0.02
Ta	1.6	1.1	0.35	0.6	0.19	0.01	—	—	—	—	0.02
W	11	1.4	68	53	130	46	470	2	0.7	—	0.2
Sb	0.02	0.12	0.5	0.02	0.08	0.1	0.06	0.03	0.04	0.02	0.01
Rb	135	1.8	2	2	5.4	0.5	1	0.4	—	0.2	0.45
Cs	10.9	0.16	0.98	1.55	2.7	0.03	0.01	0.02	0.01	0.04	0.07
Sr	356	—	13	15	24	11	12	164	88	60	164
Ba	218	9	40	30	12	20	—	—	5	2	2
Sc	13	21	0.13	0.1	0.17	0.06	0.01	0.08	0.08	0.02	0.01
Cr	68	120	110	76	110	48	17	0.8	3.5	2	1.5
Co	15	9.6	17	6.5	6.5	27.5	5.5	0.55	0.5	0.2	0.08
Ni	27	27	3.5	8	8.9	18	1	3	2	0.65	0.2

Table 1 (*continued*)

	CF 5 sch	R 20 banded sk gt	CF 16 gt sk	CF 21 gt px sk	R 17 px gt sk	SCF 233b salite px sk	SCF 232 diop sk	1200 C cc fo sk	1200 A dolost sk contact	SCF 32 dolost 100 m	CF 26 cc marble
Zn	111	690	78	412	325	1000	280	6	10	10	2
La	42	20	1	0.7	0.66	0.34	0.45	1.5	0.82	0.44	1.94
Ce	77	37	9.5	4.7	5.3	2.1	1	4.4	2.6	0.64	2.7
Sm	6.6	4.26	1.5	1.2	1	0.05	0.08	0.5	0.3	0.06	0.3
Eu	1.72	0.61	0.76	0.75	0.52	0.03	0.01	0.22	0.05	0.02	0.08
Tb	0.92	0.9	0.1	0.04	0.11	0.01	0.01	0.06	0.04	0.08	0.04
Yb	3.75	3.02	0.2	0.07	0.03	0.1	0.05	0.12	0.1	0.02	0.1
Lu	0.4	0.48	0.01	—	0.01	—	—	0.02	0.01	—	0.01

[a] Major elements in wt % (XRF), trace elements in ppm (neutron activation) at Ecole des Mines, Saint-Etienne and Laboratoire P. Süe, Saclay; Sr by XRF at Univ. de Lyon, Nb by XRF at Ecole des Mines, St. Etienne. Abbreviations: sch = schist; sk = skarn; gt = garnet; px = pyroxene; diop = diopside; cc = calcite; fo = forsterite; dolost = dolostone. The horizontal bar (—) indicates that the element is below detection limit.

possibly removed (or remains inert?) during the skarnification of schists. During the initial development of the calc-silicate skarn zones from marbles, CO_2 was lost on a massive scale through decarbonation reactions. Later on during the evolution of the skarn, however, carbonation reactions occurred with the formation of calcite as part of the hydrosilicate alteration stage of the earlier massive skarn (Fig. 2). These data (Table 1) are principally for Stage I samples. Nevertheless, they may not represent only Stage I processes because Stage I is rarely completely devoid of Stage II alteration at Costabonne.

Studies on the barren Soucarat skarn (Toulhoat 1982) have shown that Fe and Mn were not added to the skarn. In this example, Al and Si were the principal elements added to the marble bands and these elements were probably derived from the adjacent schists. Conversely, Ca was lost from the marble to the schists.

Sulphur was not one of the elements analyzed for in Table 1. From thin section and field observations S was only systematically added to the mineralized skarns and the Boutadiol skarn which does not contain scheelite. At Salau sulphide first appears in the assemblage calcite-quartz-pyrrhotite-scheelite without accompanying alteration of the Stage I silicates. Sulphur was also added with the later Stage II hydrosilicate-sulphide development of the skarn when primary minerals were altered to hydrous phases.

4 Stable Isotope Geochemistry

Hydrogen, carbon, oxygen and sulphur isotope analyses have been carried out on a wide range of minerals and/or fluid inclusions from several Pyrenean skarns. The data are presented as δ values defined as follows:

$$\delta x = \left[\frac{R_x - R_{std}}{R_{std}}\right] \times 1000, \qquad \text{in per mil (‰)}$$

where R = (D/H), ($^{13}C/^{12}C$), ($^{18}O/^{16}O$) or ($^{34}S/^{32}S$), x is the sample and std the standard, SMOW for H and O, PDB for C and CDT for S. Well-established analytical techniques were used on all samples. Table 2 lists a selection of these data together with a brief description of the samples.

4.1 C- and O-Isotope Compositions of Carbonates

The $^{13}C/^{12}C$ and $^{18}O/^{16}O$ ratios of the four types of carbonate defined above were analyzed. These data are plotted in Fig. 3 with reference to (1) the 'limestone box' which represents the isotope composition of most diagenetically altered limestones (e.g. Keith and Weber 1964), and (2) the 'primary magmatic carbon box' or PMC which includes the isotope compositions of most primary igneous carbonates (and diamonds) (Taylor et al. 1967; Sheppard and Dawson 1975; Deines 1980).

The four types of carbonate are only shown for Costabonne because of the large number of data points on the figure. Except principally for Lacourt, the data for other skarns closely follow the Costabonne pattern. Type I carbonates have C-Isotope compositions essentially indistinguishable from limestone values except for the ^{13}C-depleted samples which invariably come from the contact zone with skarn. O-isotope compositions may be similar to limestone values or depleted in ^{18}O. The most ^{18}O-depleted samples ($\delta^{18}O < \sim +14$) often come from the contact zone with skarn. Type II calcites with $-3 > \delta^{13}C > -8$ are markedly depleted in ^{13}C relative to limestone values despite the fact that these carbonates were produced during replacement reactions of limestone. Type III calcites with $-3 > \delta^{13}C > -10$ overlap the Type II field and can be even more ^{13}C-depleted than Type II. Calcites

Table 2. H-, C-, O- and S-isotope analyses of selected samples from Pyrenean skarn deposits

Sample No.	and description[a]	Mineral	δD	δ^{13}C	δ^{18}O	δ^{34}S
A. Costabonne (M)						
CB 12-1	Dolostone	Dolomite		−0.4	23.4	
CB 12-2	Ct–Fo skarn	Calcite		−6.5	12.3	
		Serpentine	−105		7.5	
DV 5	Dolostone, contact with skarn (DV 2)	Dolomite		−4.0	12.6	
DV 2	Skarn	Calcite		−6.3	10.8	
CB 33-1	Dolostone, contact with skarn (CB 33-2)	Dolomite		−2.7	13.0	
CB 33-2	Ct–Fo veinlet	Calcite		−6.5	9.6	
CB 47	Gt–Px skarn	Calcite		−8.8	9.1	
		Pyrite				3.4
		Sphalerite				3.6
CB 49	Gt–Px skarn	Calcite		−9.0	8.4	
		Pyrite				2.7
CB 39	Late Ct vein	Calcite		−8.3	9.4	
		Pyrite				5.4
CB 1	Px skarn	Pyrite				3.6
		Sphalerite				2.6
NT 6	Ct–Fo skarn	Serpentine	−109		5.3	
C 40	Px skarn	Actinolite	−93		7.8	
1053	Px skarn	Phlogopite	−113		7.6	
C 41	Gt skarn	Chlorite	−108		5.8	
2033 SNI	Gt skarn	Garnet			6.6	
CB 450	Early quartz with Gt	Quartz (FI)	−43			
Fk GO II[b]	Late quartz in endo skarn	Quartz (FI)	−25			
NTI	Brucite marble	Calcite		−7.0	10.4	
B. Salau (M)						
SAC 1	Marble far from skarn	Calcite		+2.3	25.4	
SAC 2	Graphite marble	Calcite		−1.3	11.9	
M 1000	Px skarn	Calcite		−7.0	11.0	
SAC 6	Scheelite-poor	Calcite		−5.1	13.5	
SA 311 PS	Ore	Pyrrhotite				1.0
		Chalcopyrite				1.4
SA 522 PS	Ore	Pyrrhotite				1.5
		Chalcopyrite				0.8
S 1619	Gdr, little alt.	Biotite	−47			
79-6	Gdr, 'fresh'	Biotite	−73			
79-19	Greisen	Muscovite	−38		10.6	
79-13	Gt skarn, late	Garnet			10.6	
79-2	Qtz–Ct–Chl.Po vein	Chlorite	−48		4.3	
79-5	Late vein	Amphibole	−79		6.8	
C. Lisse d'Embarre (M)						
78-148	Endoskarn	Pyroxene			7.7	
		Hornblende	−66		8.6	
78-153	Border of granite	Hornblende	−64		9.0	
		Calcite		−4.6	12.2	
78-156	Endoskarn	Biotite	−47		7.6	
78-161	Gt skarn	Garnet			8.6	
78-159	Gt–Px skarn	Calcite		−2.1	12.8	

Table 2 (*continued*)

Sample No.	and description[a]	Mineral	δD	$\delta^{13}C$	$\delta^{18}O$	$\delta^{34}S$
D. Boutadiol (B)						
78-168	Magnetite skarn	Fe Pargasite	−119		8.5	
U 14	Fe skarn	Pyrrhotite				−1.2
E. Lacourt (B)						
LA 1′	Px skarn	Calcite		−14.1	13.9	
LA 5	Qtz–Ct vein	Calcite		−4.7	15.3	
LA 12	Px skarn	Wollastonite			10.2	
		Calcite		−9.4	13.7	
LA 13 A	Gt hornfels	Calcite		−16.0	14.2	
		Pyrite				−1.7
LA 14	Gt skarn	Garnet			11.5	
LA17	Hornfels	Calcite		−16.1	14.0	
		Pyrite				−6.5
	Vein in LA 17	Pyrite				5.1
F. Soucarat (B)						
P 3A	Marble	Calcite		−2.2	17.2	
P 3B	Gt skarn	Calcite		−8.1	14.1	
SO1	Gt–Px skarn	Calcite		−11.9	13.0	
78-121	Gt–Px skarn	Wollastonite			9.7	
78-164	Px skarn	Pyroxene			10.7	
		Calcite		+2.8	15.3	

[a] Abbreviations: Chl = chlorite; Ct = calcite; Fo = forsterite; Gdr = granodiorite; Gt = garnet; Po = pyrrhotite; Px = pyroxene; Qtz = quartz.
[b] F. Kalaydjian, analyst.

from the brucite marbles (Type IV), although less strongly depleted in ^{18}O relative to limestones, are similarly ^{13}C-depleted to Type II and III skarn calcites.

The inverted L- or hyperbolic-shaped field for Type I, II and III Costabonne (M) carbonates is followed in general form by all Pyrenean skarns (Fig. 3). The field for Salau (M) calcites is indistinguishable from that of Costabonne. On the other hand, calcites from non-mineralized skarns such as Soucarat (B) and other barren Quérigut skarns, and Lacourt (B) are typically less strongly depleted in ^{18}O than those with mineralization. Similarly, the C- and O-isotope compositions of most other analyzed skarns in the world follow the general pattern of Costabonne (see Fig. 6 in Valley 1986).

Two fundamentally different interpretations are possible for the L-shaped field displayed by most skarns in the $\delta^{13}C$–$\delta^{18}O$ diagram (Fig. 3: (1) a *magmatic-hydrothermal* model where the compositions of the carbonates represent various mixtures between the initial limestone value and magmatic carbon and oxygen, the PMC pole, and (2) a *hydrothermal decarbonation* model where the C-isotope composition of the carbonates reflects either directly or indirectly differing degrees of decarbonation of limestone carbonate, and the O-isotope composition is largely controlled by the hydrothermal fluid and the temperature. In both models the variation in $\delta^{18}O$ of Type I carbonates, where decarbonation reactions were generally unimportant, is most readily explained as due to exchange of the carbonate with an externally derived aqueous-rich hydrothermal fluid. The O-isotope composition of the most ^{18}O-depleted carbonates was controlled by that of the fluid phase and the temperature of exchange. For example, a calcite with $\delta^{18}O = +12$ is in equilibrium with H_2O at +6.4 at 300°C and +9.5 at 450°C (O'Neil et al. 1969). The lack of C-isotope variations of Type I carbonates from the limestone values, except for a few of the more ^{18}O-depleted carbonates, which are in contact with skarn, implies that

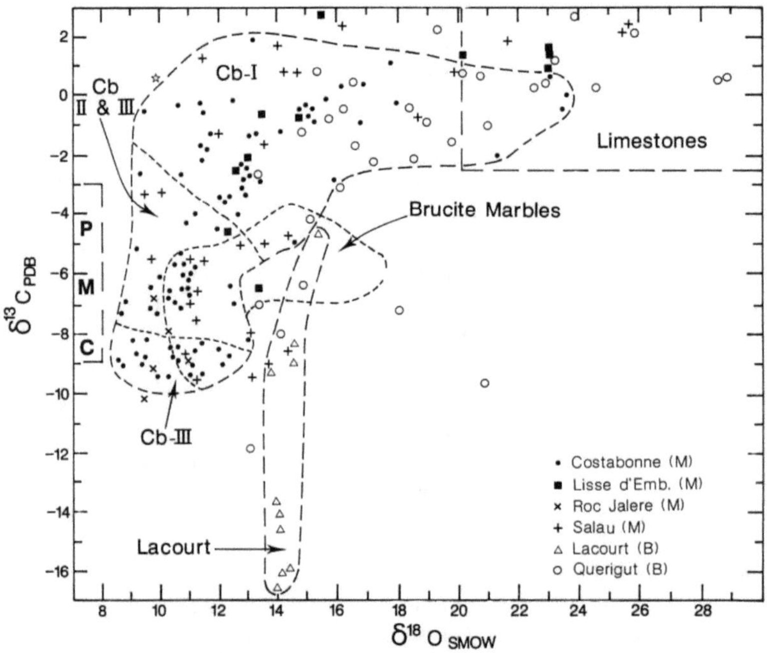

Fig. 3. Plot of $\delta^{13}C$ versus $\delta^{18}O$ for carbonates from Pyrenean skarns and marbles. The *boxes* for limestones and primary magmatic carbon (*PMC*) are given for reference. Fields labelled *Cb-I*, *Cb-II* and *Cb-III* are for Type I, Type II and Type III calcites respectively, from Costabonne. The field labelled *brucite marbles* is based in part on additional data (not plotted) from Dubru (1986; A.M. Fouillac, analyst). Note that *open symbols* are for barren skarns

the carbon content of the fluid phase was very low. It therefore had essentially no buffering capacity on the marble carbonates except within a few centimetres of skarn.

The *magmatic-hydrothermal* model interprets the C- and O-isotope compositions of Type II and III calcites as largely being controlled by a hydrothermal fluid whose isotope composition was determined by the contribution of carbon and oxygen derived (1) from and/or equilibrated with magma, and (2) from the marbles. The large range of $\delta^{13}C$ values thus simply reflects the equally large range of magmatic C to limestone C ratios because CO_2-carbonate C-isotope fractionation factors are essentially independent of temperature for $T > 300\,°C$. This interpretation has been advocated by most workers in the field whether for mineralized or barren skarns (e.g. Taylor and O'Neil 1977; Guy 1980; Bowman et al. 1985a,b; Brown et al. 1985; Valley 1986). It is largely based on the similarity of C- and O-isotope compositions of the magmatic carbonate field (PMC on Fig. 3) with the skarn carbonates. Such a mixing model will not generally generate calcites whose isotope compositions plot on a straight line between the PMC and limestone poles. The C/O atomic ratio in the 'magmatic' fluid is unlikely to be 1/3 (see fluid inclusion data below).

Bowman et al. (1985b) have also argued that mass-balance calculations indicate that the $\delta^{13}C$ and $\delta^{18}O$ values of such skarn calcites cannot be derived from simple decarbonation of pre-existing limestone or marble. Note, however, that C and O are not necessarily coupled, particularly as the CO_2 content of the aqueous fluid is low (see below); their argument may therefore not be valid. The typically 1 to 5‰ or more enrichment in ^{18}O of skarn calcites relative to PMC is explained by O-isotope exchange with more ^{18}O-rich igneous or metasedimentary minerals and/or the submagmatic temperatures of exchange. The magmatic-hydrothermal model, however, cannot readily account for skarn calcites with $\delta^{13}C < -10$ (Fig. 3), as is observed at Lacourt (B) and Soucarat (B).

The *hydrothermal decarbonation* model interpretes the $\delta^{13}C$ values of Type II and III calcites in terms of a Rayleigh distillation or continuous CO_2 fluid removal process from limestone carbonate and the $\delta^{18}O$ values by a hydrothermal exchange process with an externally derived fluid (see Valley 1986 for review). Carbon and oxygen are therefore not coupled because the mass of oxygen in the CO_2 is relatively minor compared to the quantity of H_2O oxygen. Fluid inclusion data and mineralogical arguments indicate that X_{CO2} is usually <0.1 (e.g Taylor and O'Neil 1977), in support of this argument. The calculations here assume $\delta^{13}C = 0$ for the initial carbonate and a CO_2-calcite fractionation of 2.6 which is essentially independent of temperature between 300° and 700 °C (Bottinga 1968).

Application of the Rayleigh model to Type II calcites requires that either they represent the carbonate residues or they are in exchange equilibrium with residues after 65 to 95% of the original mass of carbon has been distilled off from the system. In situ decarbonation alone is not sufficient, because only 50% of the carbonate C is liberated as CO_2 during the production of forsterite + calcite from dolomite + quartz. Calcite II C has therefore exchanged with a ^{13}C-depleted fluid phase whose C-isotope composition was externally controlled. This fluid must have come from a part of the system which had undergone more complete decarbonation. Similarly, Type III calcites represent the residues after 65 to 98% decarbonation. Because Type III carbonates were introduced into the mineral assemblage during carbonation reactions or as vein material, this carbon was transported to the site of precipitation from an external source region which had undergone major (65%) to essentially complete decarbonation (98%). The massive garnet and pyroxene zones in another part of the system could be suitable source rocks for such carbon, because they represent completely decarbonated metasomatized marbles. This interpretation combined with the observed range of $\delta^{13}C$ values requires that the first 65% or so of the limestone C involved in decarbonation reactions is not reused to form Type III calcites. The large range of $\delta^{13}C$ values for Type III calcites is just what one would expect in a system where the C-isotope composition of the fluid is controlled by a source region that has undergone moderate to complete decarbonation. The rarity of preserved carbonate rocks which have undergone about 70 to 90% decarbonation in the skarn system is not in contradiction with this interpretation.

This Rayleigh model implies that calcites with $\delta^{13}C$ at −9‰ can only represent a few percent ($< \sim 5\%$) of the total quantity of initial carbonate involved in the decarbonation processes. Although quantitative data are lacking, at least in a qualitative sense, the abundance of calcite in a given skarn system increases as $\delta^{13}C$ increases from −9 towards 0 in support of this model. Thus, the density of sample points on Fig. 3 does not reflect the relative importance by mass of the different types of carbonate. More precise modal data on the different calcite types are needed to rigorously test this proposition.

If graphite participated in some of these reactions — graphite is consumed during skarn-forming processes at Lacourt (B) — then the necessary degree of decarbonation is diminished by an extent dependent on the graphite to carbonate carbon ratio, or the final $\delta^{13}C$ values of the calcites are more strongly ^{13}C-depleted (e.g. −16‰ at Lacourt). The Lacourt data (Fig. 3) demonstrate that at least in some skarn systems the role of graphite can be clearly recognized.

The brucite marbles (Type IV calcite), which occur locally within the skarn-bearing metasedimentary section, have $\delta^{13}C$ values similar to Type II and III skarn calcites but have $\delta^{18}O$ values up to 3 or 4‰ higher (Fig. 3). They probably did not form by simple decarbonation of dolomitic marbles (Guy 1979; Dubru 1986). O-isotope and possibly also C-isotope exchange with the infiltrating hydrothermal fluid is implied by these data, as for Type II calcites.

The O-isotope composition of the Type II and III carbonates was dominantly controlled by exchange with the H_2O-rich fluid whose $^{18}O/^{16}O$ ratio was controlled by the source of the fluid, fluid-rock interaction and therefore the fluid to exchangeable mineral ratio, and the temperature of exchange. Taking a temperature of 450°C a calcite with $\delta^{18}O = +10$ is in equilibrium with water at +7.5‰ (O'Neil et al. 1969). Although this composition is indistinguishable from primary magmatic water values, formation waters, metamorphic waters or exchanged meteoric waters can also have such values (see Sheppard 1986a for review). At a lower temperature such as 300 °C the water is at +4.4‰ and is depleted in ^{18}O relative to magmatic water.

Based on the C- and O-isotope data alone, a choice between the magmatic-hydrothermal and hydrothermal-decarbonation model is not that evident despite the fact that most such studies of skarns have strongly favoured the magmatic-hydrothermal interpretation. The analysis of the data has shown that the hydrothermal-decarbonation model can satisfactorily explain both the general pattern of the $\delta^{13}C$–$\delta^{18}O$ data and the relative importance of the $\delta^{13}C$ values. A combined magmatic hydrothermal-decarbonation model is also possible. These results will be discussed further below after the presentation of the other isotope results.

4.2 H- and O-Isotope Compositions of Minerals

Minerals analyzed for both their D/H and $^{18}O/^{16}O$ ratios are plotted on a δD–$\delta^{18}O$ diagram (Fig. 4) with the primary magmatic water box, metamorphic water field and meteoric waterline as references. Other hydrous minerals and fluid inclusions from quartz samples which were only analyzed for δD are also included in this figure. Table 2 gives some additional O-isotope data on predominantly Stage I anhydrous skarn minerals such as garnet, pyroxene and wollastonite.

The $\delta^{18}O$ values of silicate minerals range from about +4 to +12‰. Silicate minerals from barren skarns are possibly slightly more enriched in ^{18}O (> ~1‰) relative to comparable minerals from mineralized skarns (Fig. 5); a similar observation was made above for calcite. More data on comparable silicate minerals are needed to test this possibility. The δD values on minerals range from −120 to −30‰, a relatively large variation. In particular, essentially the complete range is observed at Costabonne. The five Costabonne samples which are all markedly D-depleted come from mineralized skarns in dolomitic marbles. In contrast, the non-mineralized skarns associated with calcareous schists (van Marcke de Lummen 1983; van Marcke de Lummen and Verkaeren 1986) are at the D-rich end of the δD range. Their δD values are, however, indistinguishable from the mineralized skarns at Salau and Lisse d'Embarre.

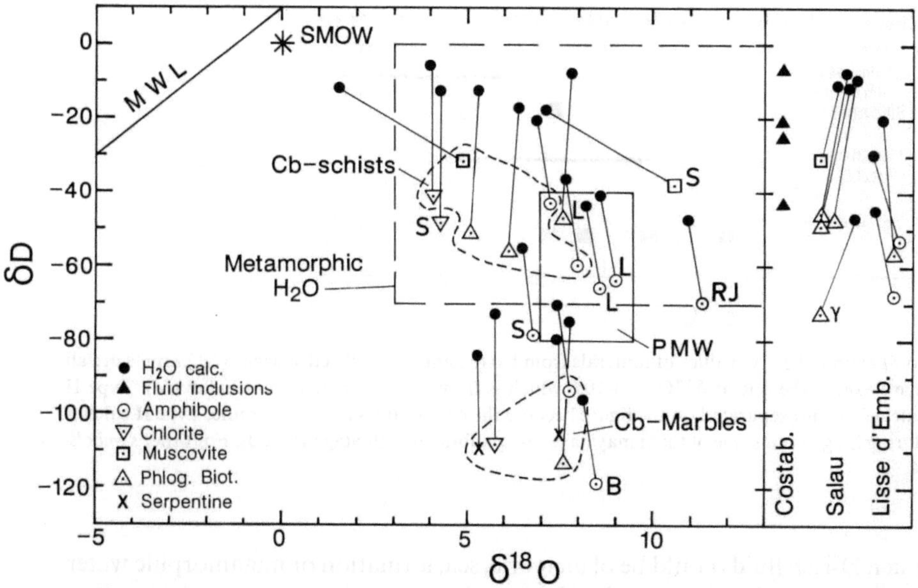

Fig. 4. Plot of δD versus $\delta^{18}O$ for hydrous minerals, fluid inclusions and calculated isotope composition of waters from Pyrenean skarns. The primary magmatic water field (*PMW*), metamorphic water field and meteoric water line (*MWL*) are given for reference. The fields labelled *Cb-marbles* and *Cb-schists* are for skarns developed in marbles and schists from Costabonne. The mineral data in the Cb-schists field are from van Marcke de Lummen (1983; A.M. Fouillac, analyst). *Lines* connect calculated isotope composition of water with mineral data; see text for discussion. Abbreviations: *B* Boutadiol; *L* Lacourt; *RJ* Roc Jalère; *S* Salau; γ granite

4.3 H- and O-Isotope Compositions of Fluids

The isotope composition of the hydrothermal fluid can be either measured directly on fluid inclusions (Table 2) or calculated from the chemical and isotope composition of the mineral by applying the relevant mineral-H_2O isotope fractionation factor at the temperature of formation (e.g. Taylor 1974, 1979; Sheppard 1977a, 1986a). Both approaches are used here. For the mineral data, the H- and O-isotope composition of H_2O in equilibrium with the mineral has been calculated for 300 °C using the experimentally determined or estimated fractionation factors from Taylor (1974), Suzuoki and Esptein (1976), Friedman and O'Neil (1977) and Graham et al. (1984, 1987). Although there are a number of uncertainties in these calculations (composition effects, fractionation factor, etc.), the relative isotope compositions which are calculated will not change very much if the temperature is varied from 250° to 400 °C. For example, at 400 °C δD H_2O would be depleted by about 5‰ and $\delta^{18}O$ H_2O would be within about ±1‰ of the 300 °C value.

At Costabonne the three fluid inclusion samples with δD between −6 and −25‰ are for skarns which developed in the granite. These directly measured values are indistinguishable from the calculated H_2O values for the skarns associated with the schists. The most D-depleted fluid inclusion sample (−43‰) comes from a quartz-bearing garnet skarn which is considered to be early (Stage I).

Inspection of Fig. 4 emphasizes the contrast in δD H_2O values between the schist ($\delta D \sim -10$‰) and marble ($\delta D \sim -80$‰) skarns at Costabonne. Thus, at least two different fluids were involved during the evolution of the Costabonne system. Comparable D-depleted fluids have also been observed at Boutadiol. The hydrothermal fluids at ~ -10‰ are also typical of skarns at Lisse d'Embarre and

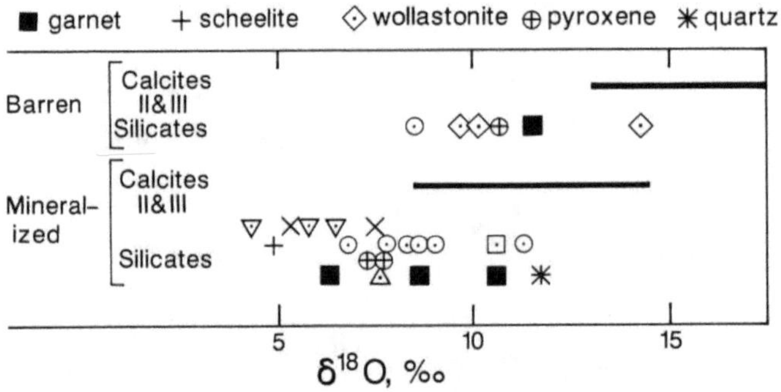

Fig. 5. Comparison of $\delta^{18}O$ values of minerals from barren and mineralized skarns. The garnets are all from Stage I except the one at $\delta^{18}O = +10.6$ which is from a late garnet vein at Salau (M). Type II and III calcites are not separated on the figure because the $\delta^{18}O$ values are comparable; Type II calcites formed during Stage I but some of them may have re-equilibrated with Stage II fluids. For other *symbols*, see Fig. 4

Salau. Such D-rich fluids could be of meteoric, sea, formation or metamorphic water origin (e.g. Sheppard 1986a). A significant primary magmatic water component, however, can be excluded from this hydrosilicate stage (Stage II) of the skarn evolution.

Certain samples from Costabonne, Salau, Lisse d'Embarre and Roc Jalère indicate that fluids with $\delta D \sim -35$ to -60‰ were also involved. Some of these samples probably formed early in the skarn development and are consistent with a magmatic or mixed magmatic-meteoric origin (e.g. CB450, 78-148, 79-6, in Table 2). The sample 79-5, however, is an amphibole which comes from a late vein with quartz-calcite-pyrrhotite. This sample implies that either magmatic waters were available during Stage II, but after the main development of the meteoric-hydrothermal system, or that non-magmatic waters, which isotopically were very similar to magmatic water values, were present during the evolution of the system.

Interpretation of the isotope composition of Hercynian hydrothermal waters in terms of their origin is hampered because of our rudimentary knowledge of the isotope composition of meteoric waters and how they varied in space and time at this epoch. In this discussion attention is focussed on the δD values because these are particularly sensitive indicators of the origin of the fluid unless water to rock ratios are very high (e.g. Sheppard 1986a).

Similarly, D-rich Hercynian hydrothermal fluids at ~ -10‰ are known from Cornwall, England and Bohemia (Sheppard 1977b; Jackson et al. 1982), Panasqueira and other tungsten deposits of Portugal (Turpin et al. 1981; C.R.P.G. unpublished data) and the Trois Seigneurs massif in the Pyrenees (Wickham and Taylor 1985). These have been interpreted to be of meteoric or seawater origin. More D-depleted Hercynian waters of inferred meteoric origin are at ~ -50‰ for the St. Sylvestre massif in the Massif Central (Turpin and Sheppard 1983), at ~ -50 to -70‰ in the late Hercynian veins ($\sim$280 Ma) in the La Lauzière massif, Western Alps (Negga et al. 1986) and at -43 to -55‰ (Kelly and Rye 1979) or -125‰ (Campbell et al. 1984) for Panasqueira, Portugal.

For the Pyrenean skarns, the D-rich hydrothermal fluids at −10‰ are interpreted here to be of probable meteoric origin although a metamorphic origin cannot be excluded. The $\delta^{18}O$ values are consistent with this interpretation and imply that water/rock ratios were small to moderate (i.e. the $\delta^{18}O$ of H_2O was controlled in part by rock oxygen compositions through mineral-H_2O exchange reactions). Today, such D-rich meteoric waters are characteristic of regions which are not only at low latitude but also at low altitudes where mean annual air temperatures of the meteoric recharge areas are relatively high (see Sheppard 1986a for review). By analogy, it is inferred that altitudes were around sea level in the Pyrenees, at least during this part of the Hercynian. These meteoric-hydrothermal fluids were therefore associated with the main hydrosilicate-sulphide development and mineralization stage at Salau (Stage II), Lisse d'Embarre and part of the Costabonne system.

The more D-depleted hydrothermal fluids at −40‰ to −55‰ or even those at −80‰ could contain a more or less important magmatic or metamorphic water component. It is assumed that the primary magmatic water box satisfactorily represents the isotope composition of magmatic waters associated with Pyrenean magmatism. Although none of our samples can be used to unambiguously characterize the isotope composition of magmatic waters in these skarn deposits, data on fresh granodiorites and monzogranites from Quérigut (Javoy and Fourcade 1980) indicate that waters in equilibrium with these magmas would plot within the primary magmatic water box. However, the even more D-depleted water associated with the pargasite from Boutadiol, the nature and late appearance of some of the minerals (e.g. serpentine and chlorite at Costabonne), and the evidence for D-depleted fluids at Panasqueira (~ −125‰) suggest that a non-magmatic source is probable at least for some of these samples. Possibilities include a meteoric origin, or perhaps an organic water contribution as was proposed by Sheppard (1986a) for the D-depleted waters at Panasqueira. Methane and nitrogen of probable organic origin are major constituents of the fluids associated with the early sulphides at Salau (J. Dubessy, pers. comm. 1987).

A meteoric origin at Costabonne implies that the isotope composition of local meteoric waters evolved rapidly (?) during the development of the skarn assemblages and that the more D-depleted waters developed later after uplift either of the Costabonne region or along the principal trajectory of the precipitating air masses. A change of δD of meteoric waters from −10 to −80‰ today implies a decrease in the mean annual air temperature from about 16° to 3 °C (Dansgaard 1964); this would be equivalent to about 1400 m of uplift. Isotopic exchange between the hydrous minerals of the skarns developed in the marbles and recent relatively D-depleted meteoric waters is not considered to be probable because four different types of minerals are involved with different kinetic properties and grain sizes. It would also have to be a local phenomenon that did not affect the majority of the samples.

For Stage I, reconnaissance fluid inclusion data show that the associated fluids were dominantly saline and aqueous (Toulhoat 1982). The D/H ratio of these solutions, however, has not yet been well characterized because of sampling problems i.e. low abundance of early fluid inclusions and sometimes the presence of secondary inclusions. The δD values of these earlier fluids could be in the range −40 to −55‰.

Although the $\delta^{18}O$ data on Stage I minerals are rather limited except for Type II calcites, they are used to calculate $\delta^{18}O$ values of waters in equilibrium with them at temperatures of 400° to 600 °C (Friedman and O'Neil 1977; Matthews et al. 1983). For the barren Boutadiol, Lacourt and Soucarat skarns, $\delta^{18}O$ H_2O are in the range +11 to +16‰, whilst for the mineralized Lisse d'Embarre and Roc Jalène skarns, $\delta^{18}O$ H_2O are between +7.5 and 10.5‰. More ^{18}O-enriched waters at about +12‰ are associated with the late garnet from Salau. Water values that are higher than about +10‰ are typical of metamorphic waters and not primary magmatic waters (Sheppard 1986a). Values between 10 and 7‰ could be dominantly of magmatic or metamorphic origin, or even highly evolved meteoric waters. These data therefore imply that barren skarns were characterized by relatively high ^{18}O metamorphic waters. Unfortunately, the data on the mineralized skarns alone cannot be used to define their origin. If they are also of metamorphic origin, then their lower ^{18}O values imply that these waters have been influenced more strongly by exchange with silicate-rich rocks such as schists, metavolcanics and granitoids that are less ^{18}O-rich than limestones.

The O-isotope composition of Stage II fluids are also often more ^{18}O-depleted in mineralized skarns than in barren skarns, assuming comparable temperatures. This is most clearly seen for Type III calcites (Fig. 5), but it also applies to many of the hydrous silicates. This difference could be a result of higher water to rock ratios in the mineralized skarns and/or exchange with limestone oxygen was less dominant in the mineralized skarns than in the barren ones. Thus, for mineralized skarns silicate rocks may have played a more dominating role in determining the isotope geochemistry of the fluids during both Stages I and II relative to barren skarns.

4.4 S-Isotope Compositions of Sulphides

The principal sulphide minerals, which were introduced after Stage I and during the hydrosilicate alteration stage of the skarns, are pyrite (e.g. Costabonne) or pyrrhotite (e.g. Salau) and these may be associated with sphalerite and more minor chalcopyrite, molybdenite, bismuthinite, arsenopyrite, etc. The $^{34}S/^{32}S$ ratios of the principal sulphides are presented in Fig. 6. The total range of $\delta^{34}S$ values is quite small for the mineralized skarns of Costabonne, Roc Jalère and Salau (0 to +6‰) and even smaller for a given mineral. For the barren skarns of Lacourt and those about the Quérigut granitoids (Boutadiol, Balbonne, Puyvaladar and Counozouls), the $\delta^{34}S$ values are, however, more variable with $-13 < \delta^{34}S < +6$. A pyrrhotite sample from the relatively sulphide-rich but scheelite-absent Boutadiol skarn has $\delta^{34}S$ of −1.2‰. Although the data on barren skarns are still quite limited, the $^{34}S/^{32}S$ ratios can apparently be used to distinguish mineralized from barren skarns.

For sulphide minerals which are in equilibrium with each other, the sequence of concentrating ^{34}S is chalcopyrite < pyrrhotite ~ sphalerite < pyrite (e.g. Ohmoto and Rye 1979). About half of the co-existing pyrite-sphalerite pairs from Costabonne could be in isotope equilibrium at temperatures of about 300 °C; the others are in disequilibrium. At Salau the pyrrhotite-chalcopyrite pairs are in disequilibrium.

At Costabonne there are two distinct groups of $\delta^{34}S$ values for pyrite. The group at ~ +5.7‰ is for clear skarns were diopside + forsterite was altered to tremolite + talc + calcite + pyrite; the group at ~ +3.7‰ is for dark skarns where pyroxene + garnet was altered to amphibole + calcite + quartz ± scheelite ± pyrite ± sphalerite.

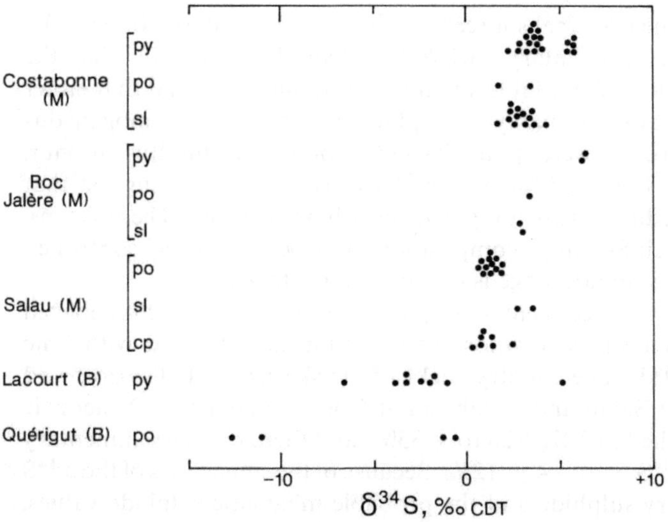

Fig. 6. Compilation of $\delta^{34}S$ values of sulphides from Pyrenean skarns. The Quérigut data are for four small barren skarns: Counozouls, Puyvalador, Boutadiol and Balbonne (in sequence of increasing $\delta^{34}S$). Abbreviations: *py* pyrite; *po* pyrrhotite; *sl* sphalerite; *cp* chalcopyrite

4.5 S-Isotope Composition of Fluids

The $^{34}S/^{32}S$ ratio of the fluid can be estimated from the mineral analysis and knowledge of the temperature, pH, fO_2 and composition of the fluid (e.g. Ohmoto 1986). The fO_2 conditions for the skarns are low, being close to the pyrite-pyrrhotite boundary. The value of the pH is probably variable from near-neutral to slightly basic because of the stability of muscovite, in the endoskarns, and calcite. Reduced sulphur species are therefore overwhelmingly dominant in the fluid. At 300 °C, aqueous H_2S is the principal species under the estimated fO_2-pH conditions and the $\delta^{34}S$ value of the total sulphur in solution is about $+2 \pm 1$‰ for the mineralized skarns. Under the proposed fO_2-pH conditions, a small increase in pH will cause an increase in the $\delta^{34}S$ value of precipitating pyrite, all other things being equal; the observed variations in $\delta^{34}S$ pyrite at Costabonne (Fig. 6) can be explained if the pH is a function of the host rock mineralogy.

A $\delta^{34}S$ value of the fluid which is close to zero is often interpreted to be of magmatic or deep-seated origin (e.g. Ohmoto 1986; Guy 1980; Kelly and Rye 1979; Taylor and O'Neil 1977). However, such an interpretation merits closer attention. Ohmoto (1986) has emphasized that both mantle-derived and crustal-derived magmas may assimilate crustal sulphur during their emplacement. Magmatic sulphur, therefore, does not necessary have $\delta^{34}S \sim 0$. In addition, isotope studies have shown that Hercynian dioritic to granitic magmatism was formed during crustal melting processes (e.g. Sheppard 1986b), implying equally the role of crustal sulphur in any accompanying magmatic-hydrothermal system. Preliminary S-isotope analyses on Pyrenean granitic rocks range from -12 to $+7$‰ (A. Autran, pers. comm.). Magmatic sulphur might thus be the source of the sulphur in the skarns.

Alternatively, could the principal sources of sulphur have been hydrothermally leached sulphur from the sedimentary rocks? This hypothesis requires that the average isotope composition of leachable sulphur was about +2‰ and that either the sedimentary sulphur, most probably as sulphide, was relatively homogeneous or the hydrothermal processes were quite efficient at homogenizing this sulphur. The range of $\delta^{34}S$ values observed for barren skarns (Fig. 6) would suggest that the $^{34}S/^{32}S$ ratios of the sediments are not particularly homogeneous. These skarns, however, are small and their S-isotope compositions may be very locally controlled because the hydrosilicate-sulphide stage is of minor importance.

The isotope composition of sedimentary sulphides is typically 35 ± 12‰ depleted in ^{34}S relative to seawater sulphate whose composition has changed with time (e.g. Schidlowski et al. 1983). The country rocks of the skarns are Palaeozoic and in particular Devonian at Salau and Cambrian at Costabonne. Early Palaeozoic seawaters were very enriched in ^{34}S (+28 to +35‰) and therefore the sedimentary sulphides typically had $\delta^{34}S = \sim -4 \pm 12$‰. Because of the similarities of the $\delta^{34}S$ values of such sedimentary sulphide and the probable magmatic sulphide values, the existing S-isotope data cannot be used to distinguish between these two sulphur sources.

5 Discussion

Tungsten-bearing skarns result from a complex sequence of reactions in impure carbonate sediments which occur adjacent to or near granodioritic or quartz monzonitic plutonic intrusions. To improve our powers of prospecting for mineralized skarns, it is crucial to clearly define the relative roles of the intrusion and its country rocks. However, an understanding of which key elements are required to be associated in the right sequence to form a mineralized skarn has been elusive for two principal reasons. Firstly, some of the later skarn events are superimposed upon earlier phases of the skarn-forming processes, partially or completely modifying the geochemical signatures of preceding stages. Secondly, many of the geochemical techniques which potentially can yield information on whether the source of elements such as hydrogen, carbon, sulphur and the metals is from the magma (or igneous rock) or its country rocks require a major data base for their interpretation. This is usually too incomplete in the area of study to enable one to arrive at conclusions with some confidence.

Skarn formation in the Pyrenees is preceded by the emplacement of the pluton and the development of contact metamorphism. In calcareous sediments calc-silicate hornfels form as a result of mineral reactions on a local scale. There is little evidence at this stage that there was any significant introduction of elements. The granitoid rocks near the Quérigut skarns of Boutadiol (B), Counozouls (B), Escouloubre (B), Lisse d'Embarre (M) and Soucarat (B) and at Salau are of calcalkaline association, whilst the Costabonne granite is alumino-potassic.

The skarns developed at or near the plutonic contacts initially with the formation of anhydrous calc-silicate assemblages in calcitic metasediments (e.g. Salau) and magnesian silicates in dolomitic formations (e.g. Costabonne, Fig. 2). In the barren skarns, evidence for a metamorphic water source is quite convincing for some Stage I assemblages. For the mineralized skarns, the isotopic data cannot discriminate between the relative importance of metamorphic versus magmatic waters. Primary scheelite was introduced during the development of Stage I. At Costabonne, it is preferentially associated with the second generation of grossularitic garnet rather than with the earlier andraditic garnet or ferrosalite, whilst at Salau, it is associated with hedenbergite; primary grade in both deposits is similar at 0.1 to 0.4% WO_3 (Guy 1979; Dubru et al., this Vol.).

Is the role of magmatic waters in mineralized skarns one of the key differences between mineralized and barren skarns? Alternatively, are metamorphic waters involved in both skarn types: in mineralized skarn schists or igneous rocks, which are more ^{18}O-depleted than limestones, may indeed have played a more dominant role in controlling the isotopic composition of the fluid? Although a choice does not seem possible at present, the metamorphic water model merits further study. The latter model, for example, can explain the possibly rather passive role of the granite because (1) contact hornfelses involving little or no introduction of elements were developed *before* Stage I skarn formation, and (2) evidence for either a magma with a high $H_2O + CO_2$ content or unmixing and liberation of a magmatic-hydrothermal fluid is rarely very evident in the Pyrenees; this is in sharp contrast with porphyry copper systems. The magmatic-hydrothermal interpretation of skarns implies that the granitoid magma which was the source of these fluids released them at a level beneath that of skarn formation; the responsible granitoid, therefore, cannot generally be observed (see Fonteilles et al., this Vol.).

The Stage II fluids with $\delta D \sim -10‰$ as at Salau, Lisse d'Embarre and the Costabonne schist-skarns have a well-defined meteoric signature. The Boutadiol and Costabonne marble skarns are also considered to be formed from meteoric-hydrothermal fluids but with $\delta D \sim -80‰$. This interpretation implies that at Costabonne there were two major periods of hydrosilicate-sulphide alteration. The marble skarns are probably later and may have formed after considerable uplift of the region. Based on the H- and O-isotope data, and in particular the δD values at $-80‰$, a magmatic or even possibly metamorphic origin for these fluids cannot be discounted.

The $\delta^{13}C$ and $\delta^{18}O$ values of the Type II and III calcites associated with the hydrothermal waters at $\delta D \sim -10$ and $\sim -80‰$ are indistinguishable. It was argued above that the $\delta^{13}C$ values of all of these calcites, whether from mineralized or barren skarns, can be explained by a decarbonation model. A magmatic carbon source cannot so readily explain the large variations of $\delta^{13}C$ from -16 to -3 in all of the calcites.

Similarly, it has been shown above that the S-isotope data cannot discriminate between a magmatic and sedimentary metamorphic source. Although the different elements are not necessarily coupled, the combined isotope data for mineralized and barren Stage II skarns can be accounted for by a meteoric-hydrothermal model. The deep circulating fluids which may be restricted essentially to the country rocks as at Costabonne or may involve both the country rocks and intrusion as at Salau could leach both their sulphide sulphur and carbonate carbon from deeper levels of the metasedimentary pile. This interpretation also implies that the source of many of the metals for Stage II could be in the metasediments rather that in the associated granitoids. The granitoids nevertheless play the crucial role of providing the necessary energy to drive the meteoric-hydrothermal convection system.

In Stage II, scheelite is associated with quartz + calcite + sulphides ± amphibole with typical grades of about 1 to 2% WO_3 at both Costabonne and Salau (Guy 1979). Tungsten may therefore be strongly enriched during Stage II processes. If this W is dominantly derived from the reworking of Stage I W, then the volume of Stage I scheelite-bearing assemblages involved must be substantially larger than the volume of Stage II. Alternatively, W must be added to the system. At Salau, for example, the volume of Stage I is very much larger than that of Stage II; also Stage II is only locally well developed. At Costabonne, Stage II is more pervasive and less intensely developed. Stage I grossular and its scheelite were not altered during the hydrosilicate-sulphide alteration stage. The source or sources of W are therefore still unresolved.

6 Conclusions

This study of barren and tungsten-bearing skarns in the Pyrenees has developed a metamorphic and/or magmatic water source for Stage I and meteoric water source for Stage II interpretation of the combined isotope data. In the barren skarns, metamorphic waters were dominant during Stage I. Although the isotope data on Stage I mineralized skarns cannot discriminate between a dominantly magmatic or metamorphic source of the fluids, the O-isotope compositions of these fluids were typically 1‰ or more depleted in ^{18}O relative to the fluids in barren skarns, assuming comparable temperatures of formation. The possible role of magmatic waters in skarns may be one of the key necessities for it to be mineralized. Alternatively, the metamorphic water interpretation for mineralized skarns implies that carbonate rocks have probably played a less influencial role relative to some lower ^{18}O metasilicate assemblages in controlling the isotope composition of Stage I fluids than in barren skarns. These metasilicate rocks may also have been a source of W because such rocks typically have higher W contents than limestones (Turekian and Wedepohl 1961).

Stage II hydrothermal fluids are dominantly of meteoric origin. The C- and O-isotope systematics of Type II and III calcites can most readily be explained by a hydrothermal-decarbonation process followed by precipitation to give Type III calcite and exchange to give Type II calcite. Leaching of sulphur from the dominantly sedimentary section permeated by the meteoric-hydrothermal system can account for the S-isotope systematics. A significant magmatic contribution of carbon and sulphur is possible in some samples, but it is not demanded by the data.

The following elements were added or subtracted to the system during skarnification: Ca, Fe, Mn, Mg, Rb, Cs, Sr, Zn and S. In the mineralized skarns W and U were also added. Other elements and notably Ti, Al, Zr, Hf and the REE were typically immobile.

The importance of Stage II mineralization is related to the degree of development of the hydrosilicate-sulphide alteration processes. These are very well but locally developed at Salau where there is an important mine (Fonteilles et al., this Vol.). At Costabonne, Stage II is widely developed but its intensity is quite variable, there is no mine. In the barren skarns Stage II is never more than weakly developed.

It is possible that Stages I and II were formed under different pressure regimes. Fluid pressures during Stage I could have been close to lithostatic, consistent with a metamorphic or magmatic aqueous fluid source, whilst Stage II pressures were close to hydrostatic to enable meteoric waters to enter the system. This implies that magma emplacement, uplift and erosional history could have played a key role in the development of fracture permeability.

Most of the isotope data summarized here are directly comparable with those presented, for example, by Taylor and O'Neil (1977) on the Osgood Mountains, Nevada mineralized skarns. They interpreted their data in terms of a magmatic source for Stage I and a mixed magmatic–meteoric source for Stage II. However, it has been shown that a single interpretation of much of the data is not evident and that the isotope composition of meteoric waters may not be constant during the evolution of the total skarn system. This latter point is reasonable although rarely emphasized in the literature; the skarn deposits did not form initially in a near-surface environment and important erosion and uplift may have occurred during their evolution.

Despite the overwhelming dominance of the magmatic-hydrothermal model for tungsten-bearing skarns in the literature (Einaudi et al. 1981 and references therein; Bowman et al. 1985a,b; Brown et al. 1985; Salemink and De Jong, this Vol.) it is hoped that this study has shown that (1) the interpretation of some of the geochemical data on these deposits is not unequivocal, (2) the metamorphic-hydrothermal followed by meteoric-hydrothermal model of mineralized skarns merits further examination, and (3) there is a necessity to examine as much of the system as possible, i.e. igneous rocks and country rocks and how they may vary with depth. Additionally, the magmatic water proponents must present more forceful arguments and facts in support of their interpretation.

Additional O-isotope data on silicate minerals are needed to verify the proposition that the $\delta^{18}O$ values of minerals from mineralized skarns are typically lower than those from barren skarns. This could be developed and could aid the prospection of tungsten-bearing skarns. Data should also be sought on non-Pyrenean skarns to determine whether the proposition can be applied to mineralized skarns in general.

Acknowledgements. The major part of this study was supported by the Commission of European Communities (*Contracts No. MPP-080-F and MSM-040-F*) to whom we are grateful. The authors have benefitted from discussions with members of the Geology Department of the Ecole des Mines de Saint-Etienne, the Applied Geology Department of the Université Catholique de Louvain and the C.R.P.G., and in particular with M. Dubru, D. Garcia, L. Raimbault, G. van Marcke and J. Verkaeren. L. Nansot is thanked specially for welcoming us to the Salau Mine. Trace element analyses by neutron activation were performed in the laboratory of P. Sue through a collaboration with M. Treuil. The technical assistance of P. Coget, Ch. Lehmann, A. Legros, L. Monnier and L. Santangelo are appreciated.

References

Autran A (1980) Le cadre paléozoïque et orogénique hercynien de la formation des gîtes de skarns à tungstène des Pyrénées. Mém BRGM 99:193–204

Bottinga Y (1968) Calculation of fractionation factors for carbon and oxygen isotopic exchange in the system calcite–carbon dioxide–water. J Phys Chem 72:800–808

Bowman JR, Covert JJ, Clark AH, Mathieson GA (1985a) The Cantung E zone scheelite orebody, Tungsten, NW Territories: O, H, and C isotope studies. Econ Geol 80:1872–1985

Bowman JR, O'Neil JR, Essene EJ (1985b) Contact skarn formation at Elkhorn, Montana. II: Origin and evolution of C–O–H skarn fluids. Am J Sci 285:621–660

Brown PE, Bowman JR, Kelly WC (1985) Petrologic and stable isotope constraints on the source and evolution of skarn-forming fluids at Pine Creek, California. Econ Geol 80:72–95

Campbell A, Rye D, Petersen U (1984) A hydrogen and oxygen isotope study of the San Cristobal mine, Peru: implications of the role of water to rock ratio for the genesis of wolframite deposits. Econ Geol 79:1818–1832

Dansgaard W (1964) Stable isotopes in precipitation. Tellus 16:436–468

Deines P (1980) The carbon isotopic composition of diamonds: relationship to diamond shape, color, occurrence and vapor composition. Geochim Cosmochim Acta 44:943–962

Dubru M (1986) Pétrologie et géochimie des marbres à brucite et des borates associés au gisement de tungstène de Costabonne (Pyrénées Orientales, France). Thèse Doctorat Université Catholique de Louvain, 443 pp

Einaudi MT, Meinert LD, Newberry RJ (1981) Skarn deposits. Econ Geol 75:377–397

Friedman I, O'Neil JR (1977) Compilation of stable isotope fractionation factors of geochemical interest. In: Fleischei M (ed) Data of Geochemistry, 6th edn. US Gov Printing Office, Washington DC

Graham CM, Harmon RS, Sheppard SMF (1984) Experimental hydrogen isotope studies. I. Systematics of hydrogen isotope exchange between amphibole and water. Am Miner 69:128–138

Graham CM, Viglino JA, Harmon RS (1987) An experimental study of hydrogen isotope exchange between aluminous chlorite and water. Am Miner (in press)

Guy B (1979) Pétrologie et géochimie isotopique (S, C, O) des skarns à scheelite de Costabonne (Pyrénées Orientales, France). Thèse Ing Doct Ecole des Mines, Paris, 270 pp

Guy B (1980) Géochimie isotopique du soufre, du carbone et de l'oxygène des skarns des Pyrénées. Mém BRGM 99:282–292

Jackson NJ, Halliday AN, Sheppard SMF, Mitchell JG (1982) Hydrothermal activity in the St. Just mining district, Cornwall, England. In: Evans AM (ed) Metallization association with acid magmatism. Wiley, New York, pp 137–179

Javoy M, Fourcade S (1980) La géochimie des isotopes de l'oxygène et de l'hydrogène dans le complexe éruptif du Quérigut (Ariège). 8e Réunion Annuelle des Sciences de la Terre, Marseille, 196

Keith ML, Weber JN (1964) Carbon and oxygen isotopic composition of selected limestones and fossils. Geochim Cosmochim Acta 28:1787–1816

Kelly WC, Rye RO (1979) Geologic, fluid inclusion, and stable isotope studies of the tin-tungsten deposits of Panasqueria, Portugal. Econ Geol 74:1721–1822

Leterrier J (1972) Etude pétrographique et géochimique du massif granitique de Quérigut (Ariège). Sci Terre, Nancy, Mém 23:320

Marcke de Lummen G van (1983) Pétrologie et géochimie des skarnoides du site tungstifère de Costabonne (Pyrénées Orientales). Thèse Doct, Univ Cath, Louvain, 293 pp, annexes 150 p

Marcke de Lummen G van, Verkaeren J (1986) Physicochemical study of skarn formation in pelitic rock, Costabonne peak area, eastern Pyrenees, France. Contrib Miner Pet 93:77–88

Matthews A, Goldsmith JR, Clayton RN (1983) Oxygen isotope fractionations involving pyroxenes: the calibration of mineral-pair geothermometers. Geochim Cosmochim Acta 47:631–644

Negga HS, Sheppard SMF, Rosenbaum JM, Cuney M (1986) Late Hercynian U-vein mineralization in the Alps: fluid inclusion and C, O, H isotope evidence for mixing between two externally derived fluids. Contrib Miner Pet 93:179–186

Ohmoto H (1986) Stable isotope geochemistry of ore deposits. In: Valley JW, Taylor HP Jr, O'Neil JR (eds) Reviews in Mineralogy, vol 16. Stable isotopes in high temperature geological processes. Miner Soc Am, Washington DC, pp 491–559

Ohmoto H, Rye RO (1979) Isotopes of sulfur and carbon. In: Barnes HL (ed) Geochemistry of hydrothermal ore deposits, 2nd edn. Wiley, New York, pp 509–567

O'Neil JR, Clayton RN, Mayeda T (1969) Oxygen isotope fractionation in divalent metal carbonates. J Chem Phys 51:5547–5558

Schidlowski M, Hayes JM, Kaplan IR (1983) Isotopic inferences of ancient biochemistries: carbon, sulfur, hydrogen, and nitrogen. In: Schopf JW (ed) Earth's earliest biosphere, its origin and evolution. Princeton University Press, Princeton, New Jersey, pp 149–186

Sheppard SMF (1977a) The identification of the origin of ore-forming solutions by the use of stable isotopes. In: Volcanic processes in ore genesis. Inst of Mining and Metallurgy and Geological Society of London, pp 25–41

Sheppard SMF (1977b) The Cornubian batholith, SW England: D/H and $^{18}O/^{16}O$ studies of kaolinite and other alteration minerals. J Geol Soc 133:573–591

Sheppard SMF (1986a) Characterization and isotopic variations in natural waters. In: Valley JW, Taylor HP Jr, O'Neil JR (eds) Reviews in mineralogy, vol 16. Stable isotopes in high temperature geological processes. Miner Soc Am, Washington DC, pp 165–183

Sheppard SMF (1986b) Igneous rocks III: isotopic case studies of magmatism in Africa, Eurasia and Oceanic Islands. In: Valley JW, Taylor HP Jr, O'Neil JR (eds) Reviews in Mineralogy, vol 16. Stable isotopes in high temperature geological processes. Miner Soc Am, Washington DC, pp 319–371

Sheppard SMF, Dawson JB (1975) Hydrogen, carbon and oxygen isotope studies of megacryst and matrix minerals from Lesothan and South African kimberlites. In: Ahrens LH et al. (eds) Physics and chemistry of the earth, vol 9. Pergamon, New York, pp 747–763

Soler P (1980) Etude pétrologique du gisement de Salau et de son enveloppe immédiate. Mém BRGM 99:217–229

Stussi JM, La Roche H de (1984) Le magmatisme orogénique granitique de la chaîne varisque française. Typologie chimique et répartition spatiale. CR Acad Sci, Paris 298(2):43–48

Suzuoki T, Esptein S (1976) Hydrogen isotope fractionation between OH-bearing minerals and water. Geochim Cosmochim Acta 40:1229–1240

Taylor BE, O'Neil JR (1977) Stable isotope studies of metasomatic Ca–Fe–Al–Si skarns and associated metamorphic and igneous rocks, Osgood Mountains, Nevada. Contrib Miner Pet 63:1–49

Taylor HP Jr (1974) The application of oxygen and hydrogen isotope studies to problems of hydrothermal alteration and ore deposition. Econ Geol 69:843–883

Taylor HP Jr (1979) Oxygen and hydrogen isotope relationships in hydrothermal mineral deposits. In: Barnes HL (ed) Geochemistry of hydrothermal ore deposits, 2nd edn. Wiley, New York, pp 236–277

Taylor HP Jr, Frechen J, Degens ET (1967) Oxygen and carbon isotope studies of carbonatites from the Laacher See District, West Germany and the Alno District, Sweden. Geochim Cosmochim Acta 31:407–430

Toulhoat P (1982) Pétrographie et géochimie des isotopes stables (H, O, C, S) des skarns du Quérigut (Pyrénées). Thèse 3e Cycle, Univ Paris VI, 268 pp

Turekian KK, Wedepohl KH (1961) Distribution of the elements in some major units of the earth's crust, Bull Geol Soc Am 72:175–192

Turpin L, Sheppard SMF (1983) Stable isotope studies on Saint Sylvestre granitic complex and associated uranium mineralizations (Limousin, Massif Central, France). Terra Cognita 3:176

Turpin L, Ramboz C, Sheppard SMF (1981) Chemical and isotopic evolution of the fluids in the Sn–W deposit, Panasqueira, Portugal. Terra Cognita, Special Issue, p 42

Valley JW (1986) Stable isotope geochemistry of metamorphic rocks. In: Valley JW, Taylor HP Jr, O'Neil JR (3ds) Reviews in mineralogy, vol 16. Stable isotopes in high temperature geological processes. Miner Soc Am Washington DC pp 445–489

Wickham SM, Taylor HP Jr (1985) Stable isotopic evidence for large-scale seawater infiltration in a regional metamorphic terrane; the Trois Seigneurs Massif, Pyrenees, France. Contrib Miner Pet 91:122–137

Petrochemical and $^{18}O/^{16}O$ Characteristics of 'W-Skarn Associated' and 'W-Barren' Granitoids in the (E-)Pyrenees and NW Portugal

J. SALEMINK and A.F.M. DE JONG[1]

Abstract

Petrochemical and $^{18}O/^{16}O$ results are presented concerning the occurrence of tungsten skarn deposits associated with Hercynian granitoids in the (E-)Pyrenees and NW Portugal. The investigation comprises a comparison of the bulk-rock, major trace element and oxygen isotope behaviour of granitoids in the 'I-type' Pyrenean and 'S-type' Portuguese magmatic suites. In more detail the magmatic and post-magmatic element distributions and $^{18}O/^{16}O$ characteristics are compared of the W-skarn associated granites of Costabonne and Covas, the (almost) W-barren granite of Batère and the Arga granite which is associated with Sn–(Ta–W) aplo-pegmatitic mineralizations.

The evidence presented does not point to a significant magmatic differentiation between 'W-skarn associated' granitoids and granitoids without W-skarn formations. In the investigated areas in the (E-)Pyrenees and NW Portugal, however, the more important W-skarn deposits of Costabonne and Covas (and also Salau ?) appear to be preferentially related to calcalkaline biotite granites-granodiorites with a primary zonal distribution of the refractory elements, and with $\delta^{18}O_{WR}(SMOW) = 8.5–9.5‰$.

In contrast to the (almost) W-barren situations at Batère and Arga, the geochemical data from the W-skarn associated granitoids of Costabonne and Covas reveal a statistical correlation between copper and zinc, and comparable Symap contour plots showing elevated Cu–Zn dispersions pointing towards the skarn deposits. The elevated Cu–Zn dispersions seem to be good indicators of areas with an increased activity of the dominantly magma-derived metasomatic fluids which, in favourable country rocks (marbles or calc-silicate rocks), may produce contact metasomatic skarn deposits with ore-grade tungsten contents.

1 Introduction

In the Pyrenees a number of granitoid plutons of Hercynian age is associated with contact metasomatic tungsten skarns (e.g. Costabonne, Salau), while others are

[1] Institute of Earth Sciences, State University of Utrecht P.O. Box 80.021, 3508 TA Utrecht, The Netherlands

Mineral Deposits within the European Community
(ed. by J. Boissonnas and P. Omenetto)

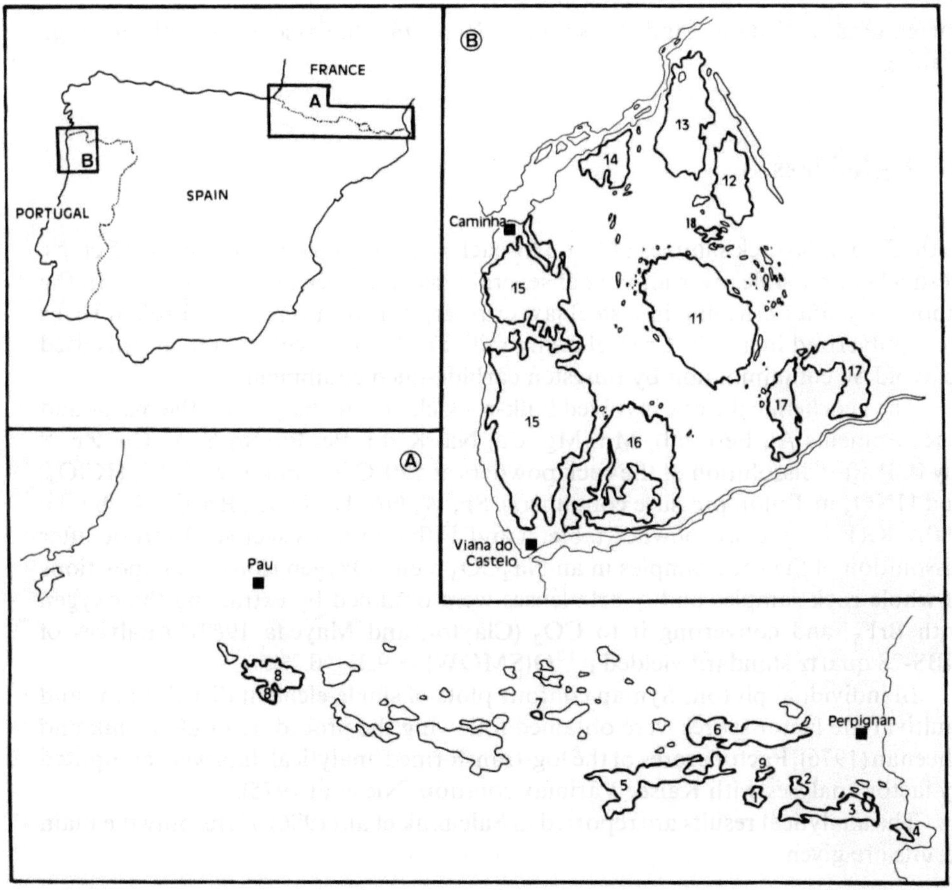

Fig. 1. Investigated areas with plutons. *1* Costabonne; *2* Batère; *3* St. Laurent-La Junqueira; *4* Roda; *5* Mt. Louis; *6* Querigut; *7* Salau; *8* Cauteret-Panticosa; *9* Castell (anatect. grt.); *11* Arga; *12* Covas; *13* Cerveira; *14* Lanhelas; *15* Ancora; *16* Perre; *17* St. Ovidio

'barren', that is they have not produced significant W-concentrations (e.g. Batère) (see the reviews by Autran et al. 1970, 1980). In northern Portugal Hercynian age granitoids are found of which a few are associated with W-skarns (e.g. Covas), while most ar barren or have produced Sn–(Ta–W) aplo-pegmatitic or quartz vein mineralizations (e.g. Arga) (see Schermerhorn 1982).

In this investigation a comparison is made between the petrochemical evolution of the (E-)Pyrenean and NW-Portuguese magmatic suites (Fig. 1). The purpose is to investigate the petrochemical similarities and differences between the two magmatic provinces and, possibly, to recognize a characteristic development of 'W-skarn associated' granitoids.

In more detail the research comprises the magmatic and post-magmatic, hydrothermal element distributions and $^{18}O/^{16}O$ characteristics in and around the W-skarn associated granites of Costabonne and Covas, the (almost) W-barren granite of Batère and the Sn–(Ta–W) pegmaplite associated with the Arga granite.

2 Techniques

In the field bulk-rock samples of 1–1.5 kg each were collected by taking 10–15 chips, fresh where possible, over an area of several tens of m^2 per sample location. In the laboratory, after crushing in a steel jaw crusher, split portions of the broken rocks were pulverized in an Ni–Cr steel swing mill. Fe–Ni–Cr steel equipment was used to avoid W contamination by tungsten carbide-lined equipment.

The geochemical work involved bulk-rock chemical analyses on the major and trace elements Al, Fe(total), Mn, Mg, Ca, Na, K, Li, Be, Ba, Sr, Y, V, Cu, Zn, S (by ICP after dissolution of the rock powders at 120°C in a mixture of HF, $HClO_4$, and HNO_3 in Teflon pressure containers), Sn, W, Nb, Ta, Ti, Zr, Rb, Cs, Sr, Ba, U, P (by XRF on pressed powder tablets) and F (by an ion-selective electrode after dissolution of the rock samples in an Na_2CO_3 melt). Oxygen isotope compositions of whole rock samples and quartz lenses were obtained by extracting the oxygen with BrF_5, and converting it to CO_2 (Clayton and Mayeda 1963). Analyses of NBS-28 quartz standard yielded $\delta^{18}O(SMOW) = 9.3 \pm 0.2‰$.

In individual plutons Symap contour plots of single-element distributions and multivariate factor scores were obtained following the procedure of Dougenik and Sheenan (1976). Factor scores of the log-transformed analytical data were computed by factor analyses with Kaiser Varimax rotation (Nie et al 1975).

The analytical results are reported in Salemink et al. (1986). Here, only the main results are given.

3 The (E-)Pyrenean and (NW-)Portuguese Magmatic Suites

In Fig. 2a, b, c, d variation diagrams are given in which respectively the MgO, Rb, Cu and W contents of the investigated (E-)Pyrenean and NW-Portuguese granitoid samples are plotted against their (somewhat modified) Kuno's Solidification Index:

$$S.I.^* = \frac{100\ MgO}{MgO + FeO^T + Na_2O + K_2O},$$

where FeO^T = total Fe calculated as FeO (instead of FeO + Fe_2O_3). In Fig. 3 $\delta^{18}O_{WR}(SMOW)$ values of whole rock granitoid samples from the investigated plutons are plotted against their Solidification Index, S.I.* (see also Appendix).

The variation diagrams of Figs. 2a, b and 3 illustrate that the granitoid plutons from the (E-)Pyrenees and NW Portugal exhibit a regional difference in rubidium content, oxygen isotope composition and acidity. In general, the analytical results show that the more aluminous, alkaline-calcalkaline 2-mica granites from the NW-Portuguese magmatic suite are more acid, higher in Rb, Li, Cs, Be, Sn, Nb, P,

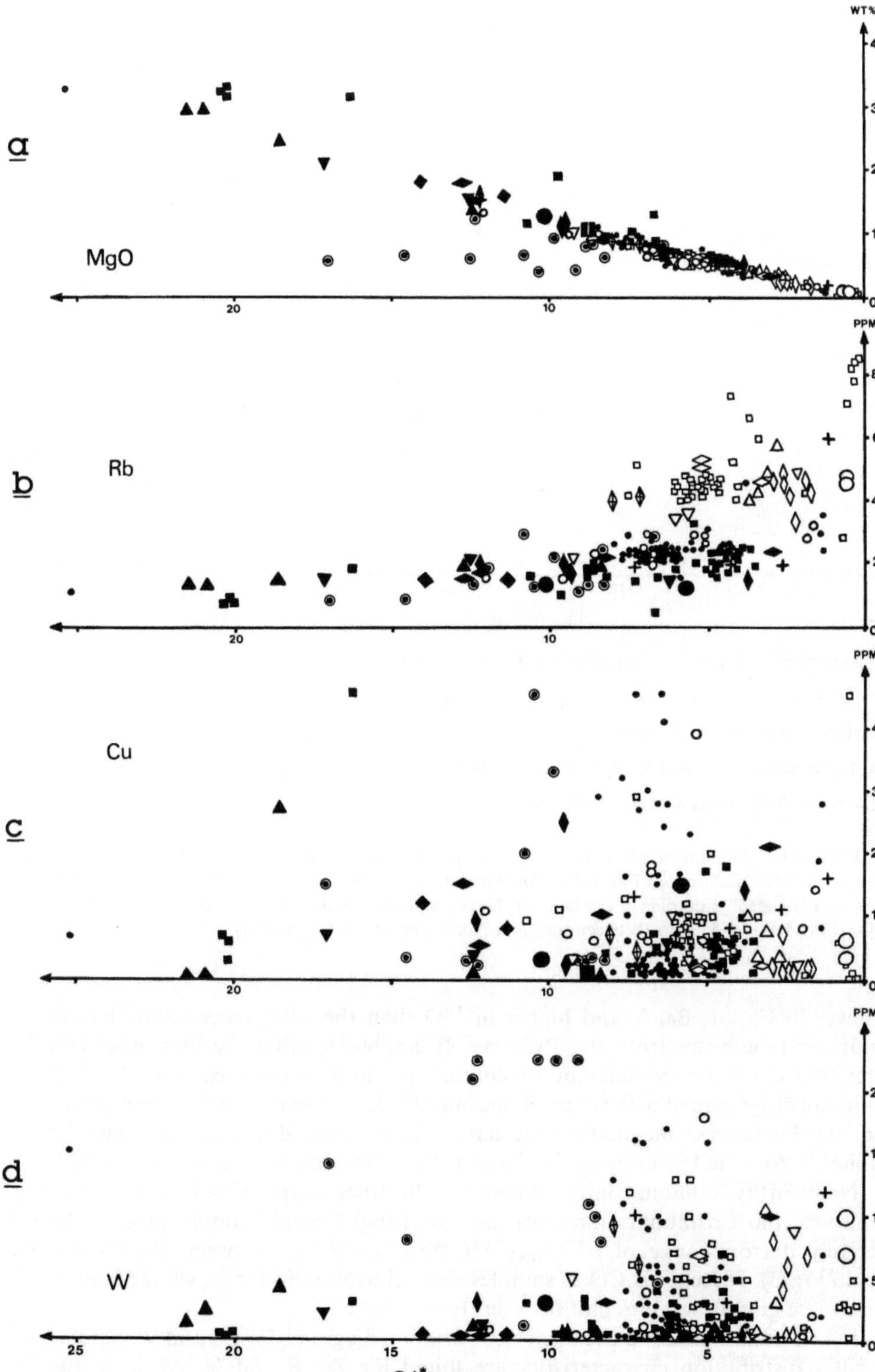

Fig. 2a–d. Variation diagrams of MgO, Rb, Cu and W versus the Solidification Index, S.I.* = 100 MgO/(MgO + FeO^T + Na_2O + K_2O) (see text). *Filled symbols* represent Pyrenean granitoids; *open symbols* are NW-Portuguese granitoids (for denotation of individual symbols see Fig. 3)

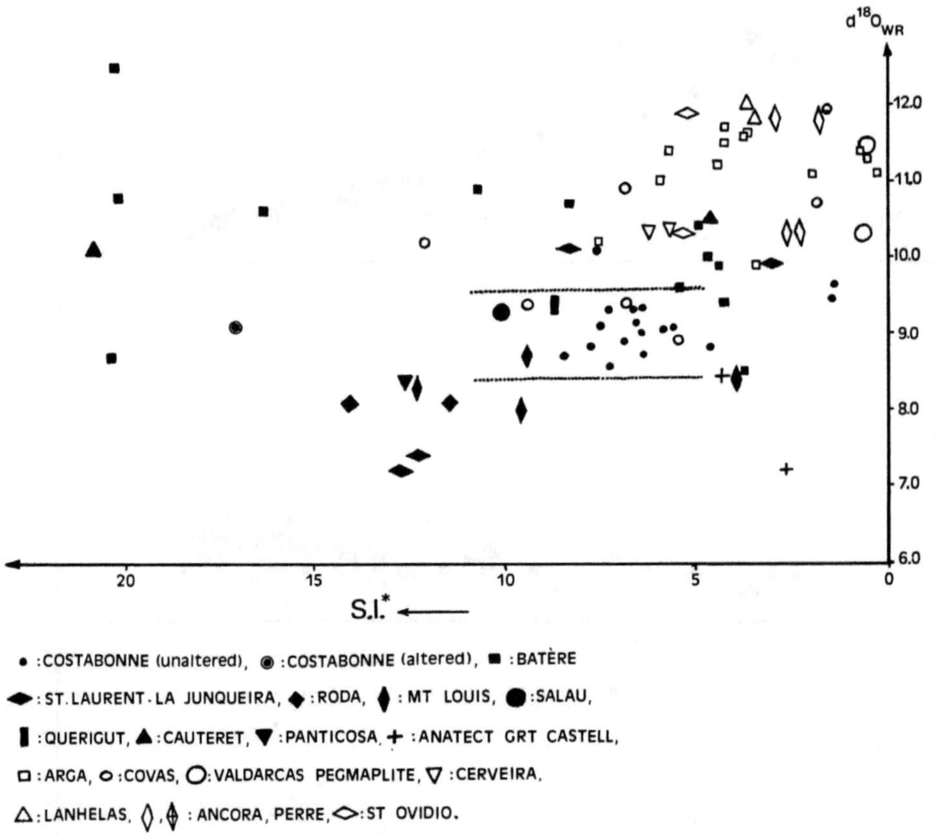

Fig. 3. Variation diagram of whole rock $^{18}O/^{16}O$ compositions of Pyrenean and NW-Portuguese granitoids vs. the Solidification Index, S.I.*. In contrast to most other granitoids, the samples from the 'W-skarn associated' granitoids of Costabonne, Covas, and also Salau, plot in the range S.I.* = 5–10; $\delta^{18}O_{WR}$(SMOW) = 8.5–9.5‰ (between the *dashed lines*) (see text and Appendix)

F, lower in Ca, Sr, Ba, Y and higher in ^{18}O than the more calcalkaline biotite granites-granodiorites from the Pyrenees. It emphasizes that the two magmatic suites were derived from different 'protoliths'. The lithogeochemical and $^{18}O/^{16}O$ results are in agreement with the available initial $^{87}Sr/^{86}Sr$ evidence that the granites from the Portuguese magmatic suite have a larger crustal component than the granitoids from the Pyrenees (cf. e.g. Pinto 1983; Vitrac-Michard and Allegre 1975).

Noteworthy is that in contrast to most of the other samples from the Pyrenean granitoids, the Costabonne granites and the (single) Salau sample plot in the relatively narrow range of $\delta^{18}O_{WR}$(SMOW) = 8.5–9.5‰ (between the dashed lines in Fig. 3). Three of six Covas samples also fall within this range, while all other Portuguese granite samples plot between 10 and 12‰.

Figure 2c and d show quite dispersed variation diagrams for Cu and W. Similar irregular distribution characteristics are found for Zn, S and U. As such, the variation diagrams of these typically hydrothermally mobilized elements differ

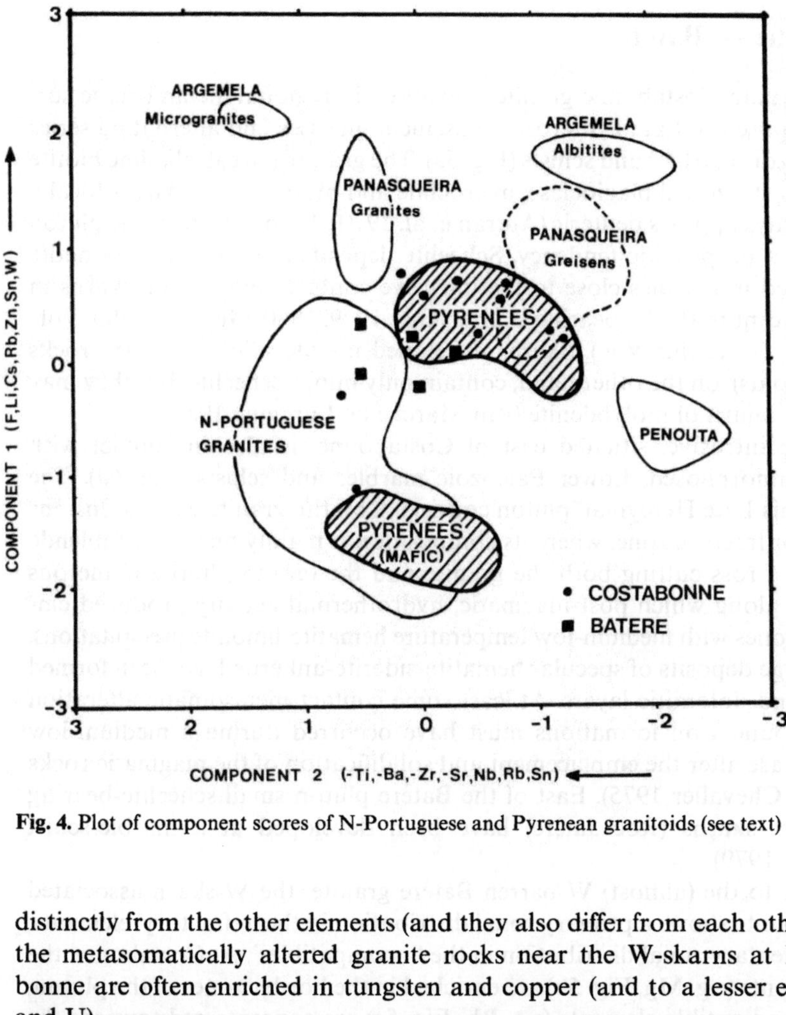

Fig. 4. Plot of component scores of N-Portuguese and Pyrenean granitoids (see text)

distinctly from the other elements (and they also differ from each other). Especially the metasomatically altered granite rocks near the W-skarns at Pic de Costabonne are often enriched in tungsten and copper (and to a lesser extent in Zn, S and U).

As also indicated by Fig. 2c and d copper and tungsten are only slightly concentrated by increasing magmatic differentiation (decreasing Solidification Index, S.I.*). The unaltered samples from the W-skarn associated granites of Costabonne, Salau, and also Covas, are not specifically enriched in W or Cu, or in any of the other major trace elements analyzed. The behaviour of tungsten, moreover, points to a dominantly metasomatic-hydrothermal mobilization and concentration of this element, instead of to a distinct lithogeochemical specialization of the magma.

This is confirmed by the component score plot of Fig. 4. The Costabonne granite and the other investigated Pyrenean granitoids plot near the barren megacrist granodiorites and biotite granites from the Portuguese Hercynian belt, and not with such specialized Sn–W–Nb–Ta mineralized granites like Panasqueira and Argemela, which are specifically enriched in Rb, Li, Cs, Be, F, P, Sn and W (Oosterom et al. 1982, 1984; Bussink 1984).

4 Costabonne — Batère

The Late Hercynian Costabonne granite is exposed in regional metamorphic surroundings composed of Late Proterozoic basement gneisses and alternating series of Lower Paleozoic marbles and schists (Fig. 5a). The granite is a calcalkaline biotite granite with quartz, zoned plagioclase, microcline and biotite. Muscovite is locally present and always appears deuteric (Autran et al. 1970). In the centre of the pluton the granite has a porphyritic tendency. Scheelite deposits are formed in garnetite skarns developed in marbles close to the intrusive contact, and in endoskarns in the granite adjacent to the W-bearing skarns (Guy 1979, 1980; Guy et al., this Vol., see also Dubru et al., this Vol.). Skarns developed in calc-silicate country rocks (marls or skarnoids), on the other hand, contain only minor scheelite, but they may have notable amounts of molybdenite (van Marcke de Lummen 1983).

The Batère intrusive, situated east of Costabonne, is also in contact with regionally metamorphosed, Lower Paleozoic marbles and schists (Fig. 6a). The main body of this Late Hercynian pluton consists of biotite granite with extensions north of a major fracture zone, where its composition is mainly biotite-hornblende (grano-)diorite. Cross-cutting both the granite and the (grano-)diorite numerous fractures occur along which post-magmatic, hydrothermal activity produced cm-wide bleached zones with medium-low temperature hematite-limonite precipitations. In the north large deposits of specular hematite-siderite-ankerite have been formed in limestones and dolomitic layers. At least some contact metasomatic alteration of the strata-bound iron formations must have occurred during a medium-low temperature phase after the emplacement and solidification of the magmatic rocks (Guitard 1973; Chevalier 1975). East of the Batère pluton small scheelite-bearing pyroxene-garnet skarns (Roc Jalère) have been developed at a marble-schist boundary (Guy 1979).

In contrast to the (almost) W-barren Batère granite, the W-skarn associated Costabonne granite shows a primary zonal distribution in the 'refractory' elements, and a complementary zonal distribution in the 'incompatibles'. At Costabonne the refractory elements (e.g. Mg, Fig. 5b) are enriched in the border zones of the pluton, whereas the incompatible elements (e.g. Rb, Fig. 5c) are concentrated towards the centre of the pluton, suggesting that magmatic differentiation occurred during the emplacement and (gradual) crystallization of the Costabonne pluton. In the Batère intrusive there is a sharp contrast in refractories and incompatibles between the hornblende-biotite (grano-) diorite in the north and the main body biotite granite, but there is only a weak primary magmatic differentiation variation in the granite body (Fig. 6b, c).

Also in contrast to the Batère situation, the geochemical data from the Costabonne pluton reveal a significant statistical correlation between Cu and Zn, and fairly similar Symap contour plots. At Costabonne elevated contents of both Cu and Zn point towards the (chalcopyrite-sphalerite containing) W-skarns at Costabonne, and towards the south, near La Preste, where also marbles are exposed close to the intrusive contact, possibly with (hidden ?) skarn deposits (Fig. 5d, e). At Batère Cu and Zn have no statistical correlation and their Symap contour plots are unrelated (Fig. 6e, d).

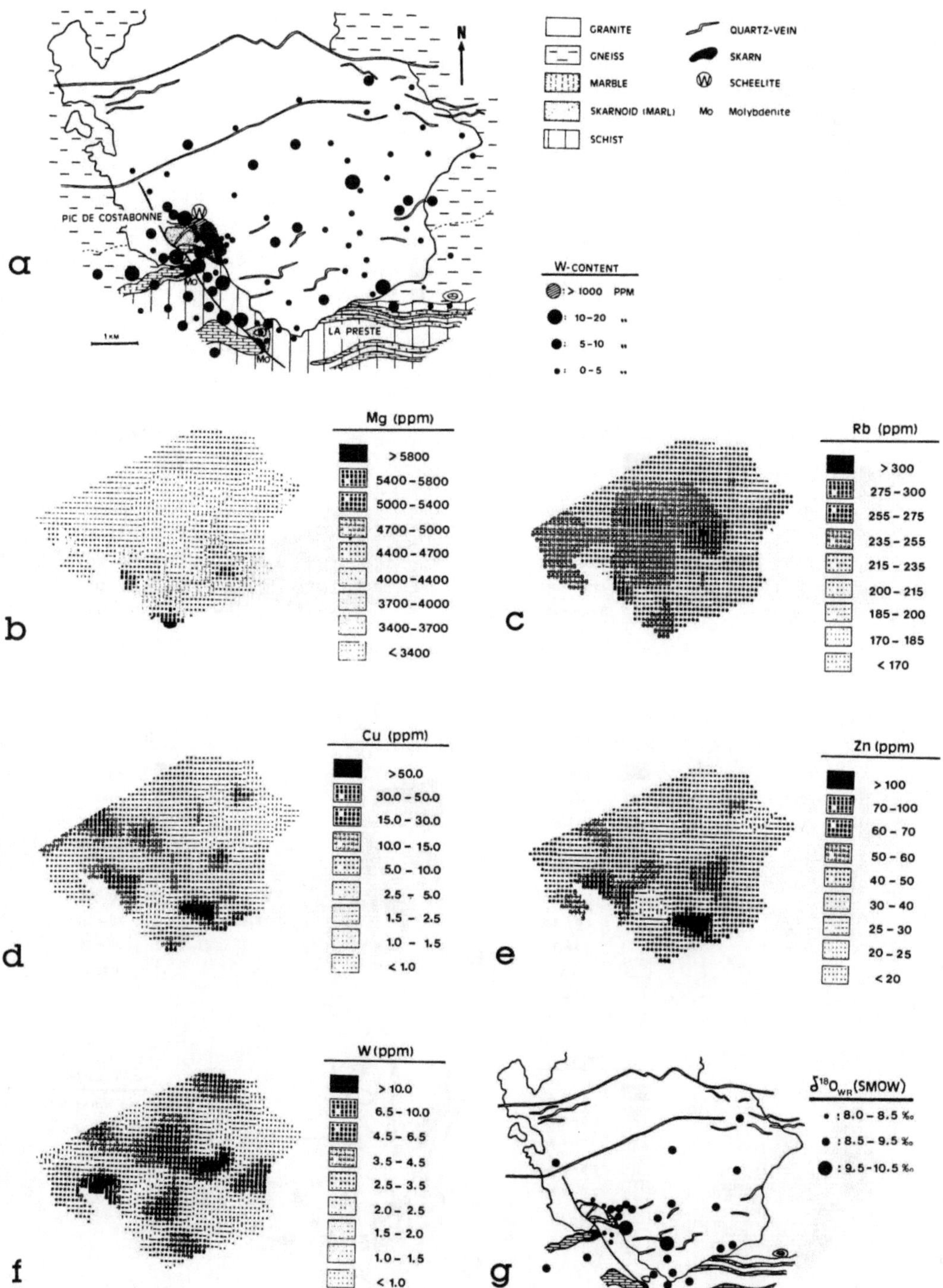

Fig. 5a–g. Geologic setting with sample locations and W contents **(a)**, single-element Symap contour plots of Mg **(b)**, Rb **(c)**, Cu **(d)**, Zn **(e)**, W **(f)**, and $\delta^{18}O_{WR}$ values **(g)** for the Costabonne pluton

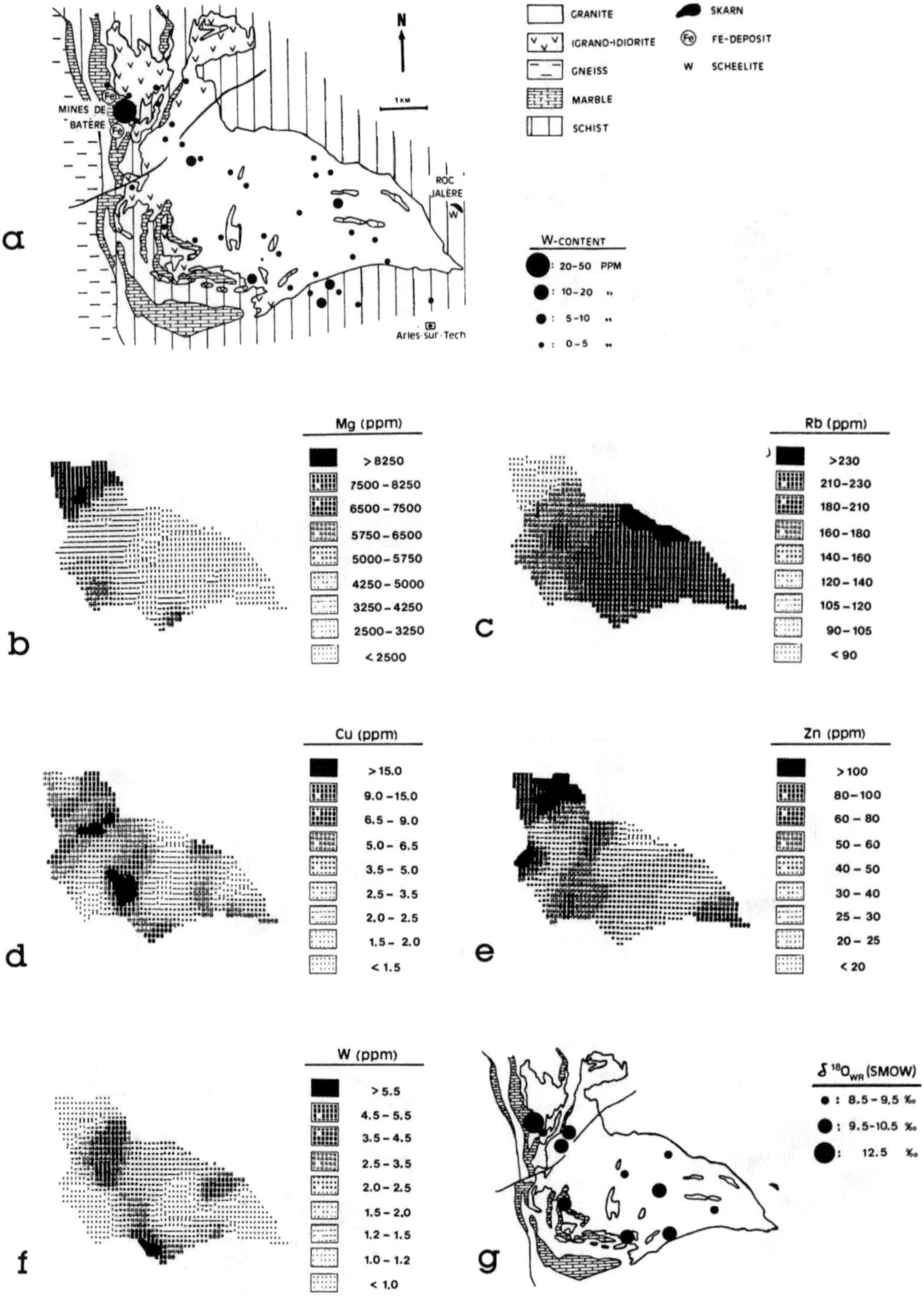

Fig. 6a–g. Geologic setting with sample locations and W contents **(a)**, single-element Symap contour plots of Mg **(b)**, Rb **(c)**, Cu **(d)**, Zn **(e)**, W **(f)**, and $\delta^{18}O_{WR}$ values **(g)** for the Batère intrusive

Although tungsten distribution patterns must be considered with care due to the X-ray fluorescence detection limit of 2–3 ppm, the W-contour plot for Costabonne shows two significant features, indicated by W-contents well above the detection limit (5 ppm) (Fig. 5f). There is a slight tungsten enrichment in the more differentiated, central parts of the pluton, and there is a strong W increase in the metasomatically affected granite rocks near the W-skarns at Pic de Costabonne. At Batère W contents are uniformly low and the W distribution pattern is undecisive (Fig. 6f).

$^{18}O/^{16}O$ compositions of whole rock samples from the Costabonne granite are fairly uniform: $\delta^{18}O_{WR}$(SMOW) = 8.5–9.5‰ (Fig. 5g). The uniform distribution of this typically magmatic $^{18}O/^{16}O$ composition (Taylor 1968) does not point to a significant contribution of externally derived, low $\delta^{18}O$-meteoric fluids during the post-magmatic cooling stages of the intrusive. Moreover, the somewhat increased $^{18}O/^{16}O$ composition of the Costabonne granite relative to many of the other Pyrenean granitoids (Fig. 3) could point to a moderate amount of magmatic assimilation of the isotopically heavier country rocks into the Costabonne magma. A moderate amount of hybridization of the Costabonne granite magma is also suggested by its somewhat increased initial $^{87}Sr/^{86}Sr$ ratio relative to most of the other Pyrenean granitoids (Vitrac-Michard and Allegre 1975).

At Batère, on the other hand, the western and central parts of the intrusive are systematically higher in $\delta^{18}O_{WR}$ than the east, from $\delta^{18}O_{WR}$ = 10–11‰ in the west to 8.5–9.5‰ in the southeast (Fig. 6g). The highest $\delta^{18}O_{WR}$ values at Batère are found in clearly hydrothermally altered granites, indicating that the high ^{18}O content in the Batère granite relative to the other Pyrenean plutons (Fig. 3) probably results from post-magmatic, hydrothermal alterations at lower temperatures.

In the mica schists in the contact metamorphic aureole surrounding the Costabonne granite, $\delta^{18}O_{WR}$ values of whole rock samples are higher than the granite data, but also fairly constant ($\delta^{18}O_{WR}$ = 9.0–11.2‰). Within the contact metamorphic aureole $^{18}O/^{16}O$ compositions of quartz lenses, however, decrease systematically with decreasing distance from the intrusive contact (see Table 1). This $\delta^{18}O$ decrease in the quartz lenses can be explained by a temperature-dependent $^{18}O/^{16}O$ fractionation between the quartz lenses and a more or less constant $\delta^{18}O$-metamorphic fluid in a closed system. The results indicate that the schists were only affected by contact metamorphic temperature rises, and not by extensive hydrothermal alterations. Also, there is no chemical evidence for the existence of large-scale hydrothermal convection cells. Apparently, the contact metamorphosed schistous country rocks were largely impermeable during the post-magmatic cooling stages of the intrusive system.

C–S–O–H isotope studies on the contact metasomatic skarn deposits at Costabonne, formed mainly in the marbles and marls, show that the scheelite-forming metasomatic fluids must have equilibrated with the magma (Autran et al. 1980; see also van Marcke de Lummen. 1986). Skarn minerals, formed at subsequent stages during the skarn formation, indicate a gradual decrease of the $^{18}O/^{16}O$ composition of the metasomatic solutions with decreasing temperature (see Table 2 and Fig. 7). The gradual $\delta^{18}O_{H_2O}$ decrease can be explained by rather small water/rock ratios and a continuous re-equilibration of the dominantly magma-

Table 1. $^{18}O/^{16}O$ compositions of bulk schist samples and autochtonous quartz lenses in the contact metamorphic aureole around the Costabonne pluton (see text)

Sample		Distance (m)	$\delta^{18}O_{WR}$(Schist)	$\delta^{18}O$(Quartz)
Py034	Bi-hornfels	5	9.1	
031	Bi-hornfels	30		11.1
035	Bi-hornfels	50	9.8	
036	Bi-mu-schist	150	9.4	
249	Bi-mu-schist	500		13.7
250	Bi-mu-schist	500		11.3
041	Chl-bi-mu-schist	500	9.1	12.8
032	Chl-mu-schist	700		12.0
005	Chl-mu-schist	800	11.2	13.1
071	Chl-mu-schist	900		12.6
251	Chl-mu-schist	1000		13.3
255	Chl-mu-schist	1300		12.3
018	Chl-mu-schist	1400		13.4
002A	Chl-mu-schist	1500	10.0	14.1
003	Chl-mu-schist	1900	9.5	14.8
131	Chl-mu-schist	2000	10.6	14.4

Table 2. $^{18}O/^{16}O$ compositions of skarn minerals from Costabonne, formed in subsequent stages of the metasomatic evolution at the measured or estimated formation temperature T(meas) or T(est.), and the $^{18}O/^{16}O$ composition of the co-existing H_2O fluid (the fractionation curves used were taken from the compilation of Friedman and O'Neil 1977)

Sample No.	qtz	mt	hm	cpx[a]	T(meas)	T(est.)	$\delta^{18}O_{H_2O}$	Minerals[a]
Py040B	11.8			7.4	550		9.5	px–qz
040C	11.5					590–500	9.2–8.7	ad–qz
077-1	13.1					550–500	10.8–10.3	ad–qz
216-2	11.3					550	9.0	ad–qz
400	11.8					550	9.5	ad–qz
203B2	11.0					550–500	8.7–8.2	ad–qz
007-2	12.2					500–450	9.4–8.6	ep–qz
010-1	9.3					500–450	6.5–5.7	ep–qz
203B1	11.4					500–450	8.6–7.8	ep–qz
213	10.4	−3.2			400		5.8	qz–mt/hm
216-1	10.2		−1.7		440		6.4	qz–hm
040D	12.2					250	3.2	qz–lim

[a] *Abbreviations*: *cpx* clinopyroxene; *ad* andradite garnet; *ep* epidote; *mt* magnetite; *hm* hematite; *lim* limonite; *qz* quartz.

derived fluids with the wall rocks through which they pass, at gradually decreasing temperatures (Salemink et al. 1983; Salemink 1985).

The isotope and chemical evidence suggests that the skarn formation at Costabonne took place in a largely closed fluid flow system, whereby the dominantly magma-derived metasomatic solutions were transported from the (crystallizing) magma towards the permeable marble 'vents'. The areas with elevated Cu–Zn

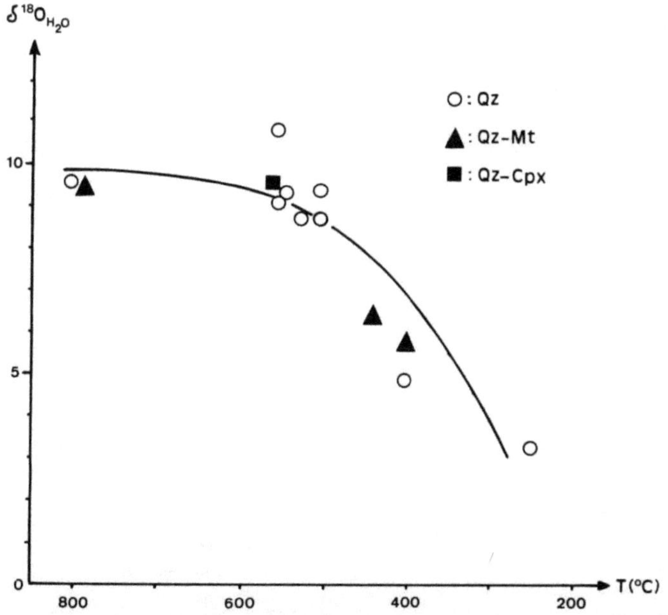

Fig. 7. Plot of $\delta^{18}O(H_2O$ fluid) against the (measured or estimated) formation temperature of subsequent skarn mineralizations from Costabonne (see text and Table 2)

(and W) contents in the Costabonne granite, concentrated near the skarn deposits, probably reflect regions with an increased activity of the metasomatic solutions that transported copper, zinc, and also tungsten, towards the (chalcopyrite-sphalerite-scheelite containing) skarn deposits.

5 Arga — Covas

The biotite-muscovite Arga granite intruded a series of folded, Silurian metasedimentary schists, quartzites, acid metavolcanics and calcsilicate rocks after the main phase of the Hercynian regional folding and metamorphism (Dias and Boullier 1985) (Fig. 8a). The granite is a coarse, middle-grained, 2-mica granite with a porphyritic tendency; in the granite, post-magmatic alterations caused muscovitization and chloritization (Dias 1984). In the area surrounding the Arga pluton there are numerous pegmaplites and hydrothermal quartz veins, some of which are mineralized in tin with some tungsten and minor tantalum. East of Arga, at Cabraçao, a sizable pegmatite with cassiterite and tantalite-columbite with wolframite was formerly exploited. The deposit seems to be related to the alkali-calcalkali parent magma of Serra de Arga (Cotelo Neiva 1954). West of Arga Sn-bearing pegmatites occur in swarms in and just outside the contact zone of the granite.

The smaller Covas granitoid outcrop north of Serra de Arga is exposed as a coarse-grained, biotite-muscovite granite along its western margin, and as a

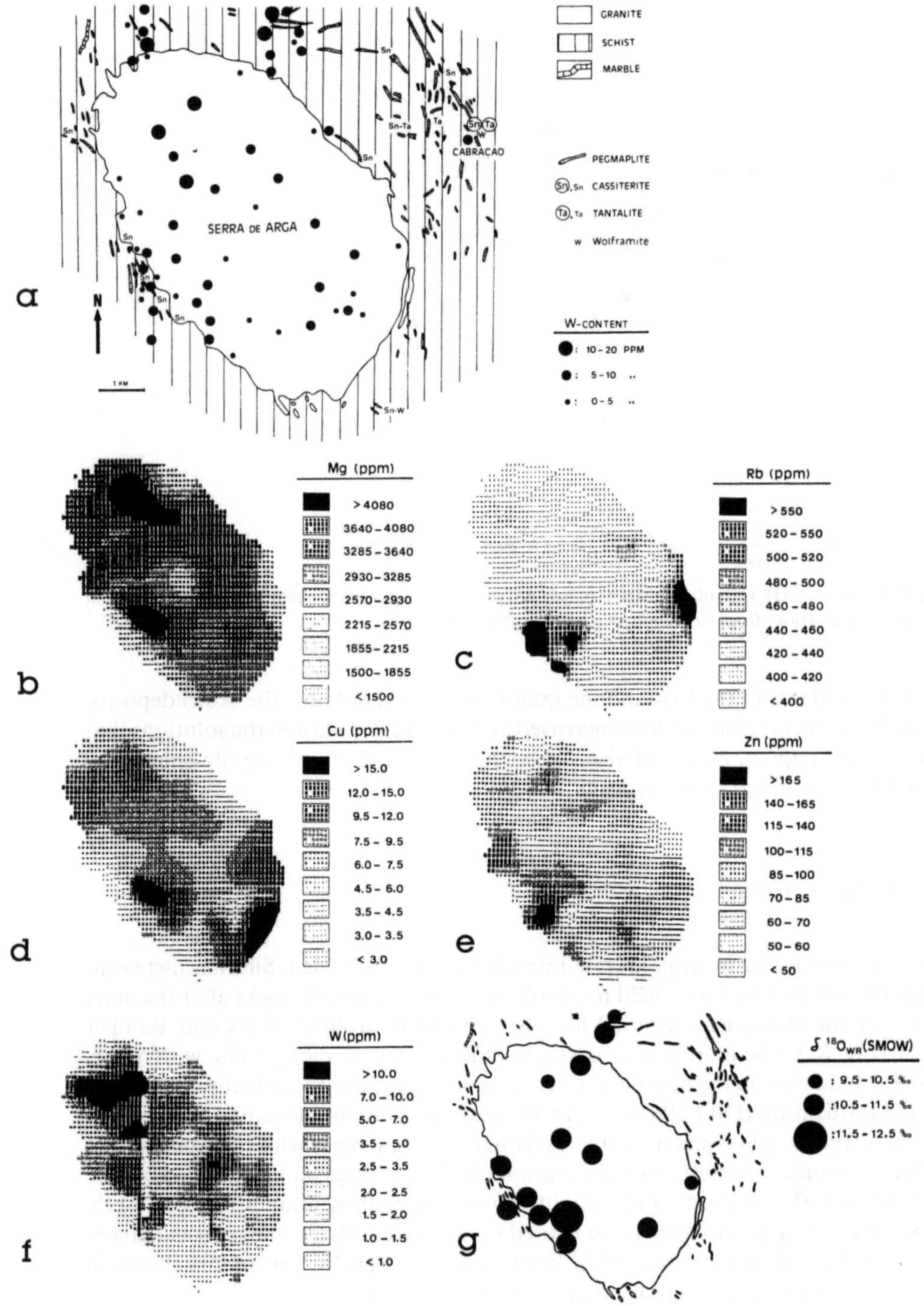

Fig. 8a–g. Geologic setting with sample locations and W contents **(a)**, single-element Symap contour plots of Mg **(b)**, Rb **(c)**, Cu **(d)**, Zn **(e)**, W **(f)**, and $\delta^{18}O_{WR}$ values **(g)** for the Arga pluton

medium-grained, biotite-rich variety in its central-eastern reaches (Fig. 9a; after Texeira and Torre de Assuncao 1961). Southwest of the Covas pluton the scheelite-bearing Valdarcas metasomatic pyroxene-garnet-vesuvianite skarn deposits are developed in a metasedimentary horizon of impure marbles situated on top of a massive quartzite layer (Coelho et al., in press). Near the skarns there are many barren aplitic pegmatites (Conde et al. 1971). In this study only a limited number of samples from the Covas plutonic body was analyzed. A more detailed petrological investigation of the Covas pluton and the nearby Valdarcas scheelite skarn deposits is carried out by the university of Porto (J. Coelho) in close cooperation with the Ecole des Mines de St. Etienne (D. Garcia).

Compared to the surrounding S-type 2-mica granites, the Covas granitoid rocks are more calcalkaline and more like the Pyrenean granitoids, with higher Ca, Sr, and lower Rb, Li, Cs, Be, Sn, F, (Nb, P), and also lower W (see Fig. 2). The Symap contour plots of the refractory and alkali elements in the Covas intrusive reveal a complicated magmatic evolution. Most refractory elements show a concentration in the centre of the pluton (e.g. Mg, Fig. 9b), while rubidium shows the opposite trend with low Rb contents in the core of the body and higher Rb concentrations near the intrusive margins (Fig. 9c). Other elements, such as Fe and Al, increase to the southwest.

The major and trace element geochemistry of the Arga granite compares well with that of the surrounding 2-mica granites (see Fig. 2). In the Arga granite the refractory and alkali elements are fairly uniformly distributed, and the intrusive only shows a weak dispersion in its primary magmatic setting Fig. 8b, c). In the Arga pluton, however, there are two distinct areas with strong enrichments in the 'hygromagmatophile' elements Rb, Li, Sn, P, F, (Nb, Be, Cs). One concentration is located in the east and points towards the Cabraçao Sn–Ta(–W) pegmatite deposit. The other is situated in the southwest, near the swarm of pegmatites with small Sn mineralizations (see Fig. 8c).

Although insufficient samples were collected for a reliable statistical treatment, preliminary results for the Covas situation reveal a relevant statistical correlation between Cu and Zn, and comparable Symap contour plots with elevated Cu–Zn dispersions pointing towards the Valdarcas scheelite skarn deposits (apart from a high Cu–Zn anomaly in the north due to one anomalous sample (Fig. 9d, e)). At Arga Cu and Zn have no statistical correlation and their Symap contour plots are unrelated (Figs. 8d, e).

The W content in the Covas granite rocks is invariably low (0–4 ppm, with one incidental exception of 18 ppm), and its distribution pattern is insignificant (Fig. 9f). At Arga W contents are relatively high (10–20 ppm), but the tungsten distribution pattern (Fig. 8f) gives no indication for a W enrichment towards the Sn–(Ta–W) mineralizations.

Whole rock $^{18}O/^{16}O$ compositions of the Arga granite are fairly uniformly $\delta^{18}O_{WR} = 11.0–11.5‰$ (Fig. 8g). They fall clearly within the range of 'S-type' granites which have a large crustal component. There is no evidence for large-scale, post-magmatic alterations by externally derived, low $\delta^{18}O$-meteoric waters.

Whole rock $^{18}O/^{16}O$ compositions of granite samples from the central-eastern parts of the Covas intrusive body are uniformly low: $\delta^{18}O_{WR} = 8.9–9.4‰$ (Fig. 9g).

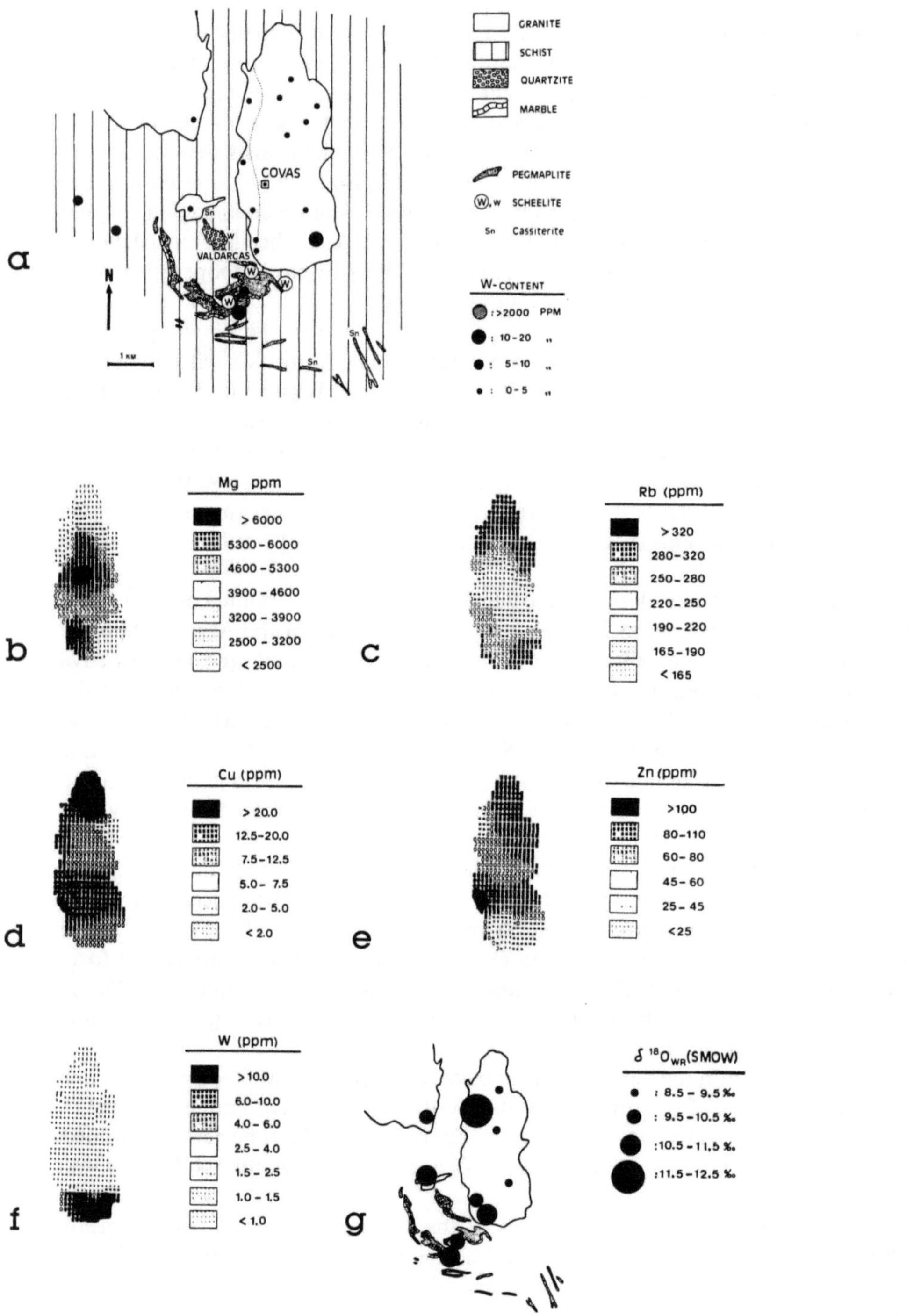

Fig. 9a–g. Geologic setting with sample locations and W contents **(a)**, single-element Symap contour plots of Mg **(b)**, Rb **(c)**, Cu **(d)**, Zn **(e)**, W **(f)**, and $\delta^{18}O_{WR}$ values **(g)** for the Covas granitoid outcrop

In addition to its calcalkaline character and mafic tendency, these low $\delta^{18}O$ compositions point to a magma source of rather deep-seated origin for the central-eastern section of the Covas granitoid outcrop. The more leucocratic, coarse-grained biotite-muscovite granite near the western intrusive margin has $\delta^{18}O_{WR} = 10–12‰$, and it probably constitutes a separately intruded S-type magma (D. Garcia, pers. comm.).

6 Conclusion

The lithogeochemical data and petrological arguments presented do not point to a distinct magmatic differantiation between 'tungsten skarn associated' granitoids and granitoids without tungsten skarn formations. In the investigated areas in the (E-)Pyrenees and NW Portugal, however, the more important W-skarn deposits of Costabonne and Covas (and Salau ?) appear to be preferentially related to calcalkaline biotite granites-granodiorites (Solidification Index, S.I.* = 5–10) with a primary magmatic, zonal distribution of the refractory elements, and with $\delta^{18}O_{WR}(SMOW) = 8.5–9.5‰$. Possibly an initially rather deep-seated, calcalkaline parent magma, after a moderate amount of crustal assimilation, together with a gradual magmatic differentiation during emplacement and crystallization and an outflow of dominantly magma-derived metasomatic solutions, creates a favourable condition for the formation of W-skarn deposits.

In contrast to the (almost) tungsten barren situations at Batère and Arga, the geochemical data from the W-skarn associated granitoids of Costabonne and (less obviously) Covas show a statistical correlation between Cu and Zn, and comparable trends in the Cu and Zn Symap contour plots with elevated Cu–Zn dispersions pointing towards the skarn deposits. The elevated dispersions of copper and zinc could be good indicators of areas with an increased activity of the metasomatic fluids which, in favourable country rocks (marbles or calc-silicate rocks), may produce contact metasomatic skarn deposits with ore-grade tungsten contents.

Appendix

Whole rock $^{18}O/^{16}O$ compositions and Solidification Indices, S.I.*, of (E-)Pyrenean and NW Portuguese granitoids (see Fig. 3) [the chemical analyses are given in Salemink et al. (1986); the locations of the granitoids are given in Fig. 1].

Sample		S.I.*	$\delta^{18}O_{WR}$(‰)	Sample		S.I.*	$\delta^{18}O_{WR}$(‰)
COSTABONNE				*Costabonne (cont.)*			
Py011	Bi-granite	6.62	9.3	025	Bi-granite	6.36	8.7
012	Bi-granite	6.91	8.9	026	Bi-granite	6.71	8.7; 8.9
016	Bi-granite	7.55	10.1	027	Bi-granite	7.19	8.6
024	Leucogranite	17.05	9.1	033	Bi-granite	8.42	8.7

Sample		S.I.*	$\delta^{18}O_{WR}$ (‰)
Costabonne (*cont.*)			
046	Bi-granite	6.40	9.2; 8.8
047	Mu-granite	1.46	9.5
049	Bi-granite	5.83	8.9; 9.2
060	Bi-granite	5.09	8.3
060B	Bi-granite	6.41	9.3
093	Bi-granite	7.25	9.3
095	Bi-granite	4.57	9.0; 8.8
099	Bi-granite	7.48	9.3; 9.0
118	Bi-granite	6.57	9.3; 8.9
121	Bi-granite	4.55	9.3; 9.0
BATÈRE			
Py073	Hbl–bi–grdt.	20.17	12.5
074A	Hbl–bi–grdt.	20.34	8.7
079	Hbl–bi–grdt.	20.19	10.2
074B	Altered grdt.	16.30	10.6
Py076	Bi-granite	8.24	10.7
081	Bi-granite	4.41	9.9
084	Bi-granite	4.27	9.4
085	Bi-granite	4.96	10.4
090	Altered grt.	10.70	10.9
104	Bi-granite	3.69	8.5
Py106	Bi-granite	5.41	9.6
111	Bi-granite	4.68	10.0
ROSAS			
Py042	Hbl–bi–grdt.	11.36	8.1
043	Hbl–bi–grdt.	14.07	8.1
LA JUNQUEIRA			
Py044	Hbl–bi–grdt.	3.01	9.9
045	Hbl–bi–grdt.	8.28	10.1
134	Hbl–bi–grdt.	12.74	7.2
135	Hbl–bi–grdt.	12.23	7.4
MT. LOUIS			
Py139	Hbl–bi–grdt.	9.54	8.0
140	Hbl–bi–grdt.	3.81	8.4
141	Hbl–bi–grdt.	9.46	8.7
142	Hbl–bi–grdt.	12.30	8.3
SALAU			
P1	Bi-granite	10.13	9.3
QUERIGUT			
P2	Bi-granite	8.67	9.4
CAUTERET			
P4B	Monzonite	4.68	10.5
P6A	Granodiorite	20.89	10.1

Sample		S.I.*	$\delta^{18}O_{WR}$ (‰)
PANTICOSA			
P7	Monzonite	12.39	8.4
P9	Diorite	35.83	7.5
CASTELL			
Py137	Bi-mu-granite	4.24	8.4
138	Bi-mu-granite	2.59	7.2
COVAS			
Po533	2Bi-mu-granite	6.86	10.9
534	Bi-mu-granite	12.02	10.2
562	Bi-granite	9.40	9.4
565	Bi-mu-granite	5.44	8.9
566	Bi-mu-granite	1.63	11.9
568	Bi-mu-granite	6.78	9.4
VALDARCAS			
Po535	Pegmaplite	0.59	10.3
536B	Pegmaplite	0.72	11.4
ARGA			
Po504	Bi-mu-granite	4.21	11.5
505	Bi-mu-granite	7.51	10.2
510	Pegm. grt.	3.44	9.9
518	Bi-mu-granite	5.66	11.4
525	Bi-mu-granite	3.73	11.6
526	Bi-mu-granite	4.38	11.2
527	Bi-mu-granite	4.23	11.7
528	Pegmaplite	0.23	11.1
530	Bi-mu-granite	1.96	11.1
543	Bi-mu-granite	4.70	11.6
546	Bi-mu-granite	5.88	11.0
LANHELAS			
Po557	Bi-mu-granite	3.74	12.0
558	Bi-mu-granite	3.44	11.7
CERVEIRA			
Po559	Bi-mu-granite	5.62	10.3
560	Bi-mu-granite	6.11	10.3
ST. OVIDIO			
Po517	Bi-mu-granite	5.27	11.9
616	Bi-mu-granite	5.26	10.3
ANCORA			
Po571	Bi-mu-granite	2.23	12.1
572	Bi-mu-granite	2.58	11.8
602	Bi-mu-granite	1.74	11.8
PERRE			
Po610	Bi-mu-granite	2.93	11.8

Acknowledgements. Part of this investigation was financed by the EEC, *contract No. MSM-074-NL.* B. Guy of the Ecole Supérieure des Mines de St. Etienne, France, and J. Verkaeren, M. Dubru and G. van Marcke de Lummen of the Université Catholique de Louvain, Belgium, are thanked for their introduction and cooperation in the Costabonne work. D. Garcia of the Ecole des Mines de St. Etienne is acknowledged for his communications on the Covas area. G. van Beusekom and S.P. Vriend are gratefully thanked for their guidance in the statistical analyses and construction of the Symap contour plots.

References

Autran A, Fonteilles M, Guitard G (1970) Relations entre les intrusions de granitoides, l'anatexie et le métamorphisme régional, considerées principalement du point de vue de rôle de l'eau: cas de la chaîne hercynienne des Pyrénées Orientales. Bull Soc Geol Fr (7)12:673–731

Autran A, Derre C, Fonteilles M, Guy B, Soler P, Toulhoat P (1980) Minéralisations liées aux granitoides, 2eme partie: la genèse des skarns a tungstène dans les Pyrénées. Mem BRGM 99:190–322

Bussink RW (1984) Geochemistry of the Panasqueira tungsten-tin deposit, Portugal. Geol Ultraiectina 33:170

Chevalier P (1975) Le gîsement de siderite de Batère (Pyrenées Orientales, France). Bull BRGM 5:385–406

Clayton RN, Mayeda TK (1963) The use of bromine pentafluoride in the extraction of oxygen from oxides and silicates for isotopic analysis. Geochim Cosmochim Acta 27:43–52

Coelho J, Garcia D, Fonteilles M (1985) Les skarns à scheelite de Covas (Minho, Nord Portugal): Pétrographie et minéralogie des paragenèses primaires. Comun Serv Geol Port 71:123–138

Conde LN, Pereira V, Ribeira A, Thadeu C (1971) Jazigos hypogenicos de estanho e wolframio, Vol 7. I Congresso Hispano-Luso-Americano de Geologia Economica, Madrid-Lisboa, Livro-Guia Excursao, 81 pp

Cotelo Neiva JM (1954) Pegmatitos com cassiterite e tantalite-columbite de Cabraçao (Ponte de Lima — Serra de Arga). Mem Not Univ Coimbra 36:61

Dias G (1984) Granitos hercinicos sintectonicos do area de Ponte de Lima (Norte de Portugal) — Evoluçao geoquimica. Mem Not Univ Coimbra 98:9–33

Dias G, Boullier AM (1985) Evolution tectonique, métamorphique et plutonique d'un secteur de la chaîne hercynienne ibérique (Ponte de Lima, Nord du Portugal). Bull Soc Geol Fr (8)1:423–434

Dougenik JA, Sheenan DE (1976) Symap user's reference manual. Harvard Univ, Massachusetts, 170 pp

Friedman I, O'Neil JR (1977) Compilation of stable isotope fractionation factors of geochemical interest. In: Data Geochem, 6th edn. Geol Surv Prof Pap 440 KK

Guitard G (1973) Sur la genèse des gîsements métasomatiques de talc et de chlorite magnesienne des Pyrenées et sur les rélations entre le talc et le magnesite. In: Raguin E (ed) Les roches plutoniques dans leurs rapports avec les gîtes mineraux. Coll Sci intern Masson, Paris, pp 369–395

Guy B (1979) Pétrologie et géochimie isotopique (S, C, O) des skarns de Costabonne. Thèse Ing Doct, Ecole des Mines de Paris, 240 pp

Guy B (1980) Etude géologique et pétrologique du gîsement de Costabonne, In: Minéralisations liées aux granitoides. Mem BRGM 99:236–250

Marcke de Lummen G van (1983) Pétrologie et géochemie des skarnoids de sîte tungstifère de Costabonne (Pyrenées Orientales). Thèse Doct, Univ Louvain, 293 pp

Marcke de Lummen G van (1986) Oxygen and hydrogen isotope evidence for influx of magmatic water in the formation of W-, Mo- and Sn-bearing skarns in pelitic rocks (Costabonne, France and Land's End, England). In: Proceedings of the VII IAGOD meeting, Sweden, Aug 18–22, 1986

Nie NH, Hull CH, Jenkins JG, Steinbrenner K, Bent DH (1975) Statistical package for the social sciences, 2nd edn. McGraw-Hill, New York, 675 pp

Oosterom MG, Vriend SP, Bussink RW, Moura ML (1982) Geochemistry of tin, tungsten and related elements such as tantalum and niobium. Final report EEC Environment and Raw Materials Research Programmes, contract 025-79-MPP NL, 58 pp

Oosterom MG, Bussink RW, Vriend SP (1984) Lithogeochemical studies of aureoles around the Panasqueira tin-tungsten deposit, Portugal. Miner Dep 19:283–288

Pinto MS (1983) Geochronology of Portuguese granitoids: a contribution. Stud Geol Salmanticensia 18:277–306

Salemink J (1985) Skarn and ore formation at Seriphos, Greece, as a consequence of granodiorite intrusion. Geol Ultraiectina 40:231 pp

Salemink J, Schuiling RD, Jong AFM de, Anten P (1983) Quantification of skarn and ore formations at Seriphos, Greece. Final report EEC Environment and Raw Materials Research Programmes, contract MPP-142-NL, 74 pp

Salemink J, Jong AFM de, Oosterom MG (1986) Lithogeochemical parameters and models in the search for tungsten mineralization. Final report EEC Environment and Raw Materials Research Programmes, contract MSM-074-NL, 74 pp

Schermerhorn LJG (1982) Framework and evolution of Hercynian mineralization in the Iberian Meseta. Comun Serv Geol Port 68:91–140

Taylor HP (1968) The oxygen isotope geochemistry of igneous rocks. Contrib Miner Pet 19:1–71

Texeira C, Torre de Assuncao C (1961) Carta geologica de Portugal na escala 1/50,000. Noticia explicative de folha 1-C, Caminha. Serv Geol Portugal, Lisboa, 41 pp

Vitrac-Michard A, Allegre CJ (1975) A study of the formation and history of a piece of continental crust by $^{87}Sr/^{86}Sr$ method: the case of the French oriental Pyrenees. Contrib Miner Pet 51:205–212

Ore Controls for the Salau Scheelite Deposit (Ariege, France): Evolution of Ideas and Present State of Knowledge

M. FONTEILLES[1,2], L. NANSOT[4], P. SOLER[2,3], and A. ZAHM[2,5]

Abstract

In the Salau skarn-type scheelite deposit, ore controls are of three kinds: lithologic, structural and mineralogical. Lithologic controls include the proximity of a granodioritic narrow apex, and a special banded formation called 'barrégiennes', composed of thin alternating beds of limestone and shales or sandstone. Structural controls include the folding pattern of barrégiennes, early blastomylonite and two types of faults: f1, the drainage channel of fluids responsible for the main stage of mineralization; f2, a series of reverse faults with individual displacements of 100-m magnitude which delimit the main orebodies. The general pattern of the deposit as a whole is an aureole of orebodies in barrégiennes in or near an f1 fault, around an apophysis of the granite. In the conclusion of this chapter, the predicted (already partly confirmed by drilling) location of new orebodies is deduced.

From a mineralogical standpoint, the Salau skarns can be described as having been developed in a two-stage process: the first stage produced clearly zoned skarn with grossularite and hedenbergite (always on the marble side) and low-grade scheelite; the second stage was cross-cutting in relation to f1 fractures, with almandine-spessartite rich garnet often in direct contact with the marble, and (somewhat later) an abundance of pyrrhotite and high-grade scheelite. The fluids of the first stage come from granite, whereas almandine-spessartite rich garnet is an indicator of the action of a fluid in equilibrium with country rocks (including limestone), and is the best mineralogical guide to high-grade scheelite ore in the Salau mine. The mechanism of precipitation of rich ore is inferred to be an inter-reaction in the later stage of two fluids of contrasting origins (one originating in the granite, the other one circulating in the country rocks). It is shown that the various types of chemical changes observed in rocks (through exchange or precipitation) can be explained by this model.

[1] Laboratoire de Géologie Appliquée, Univ. Paris VI, 4 Place Jussieu, 75252 Paris Cedex 05, France
[2] UA 384 CNRS 'Pétrologie et Métallogénie', 4 Place Jussieu, 75252 Paris Cedex 05, France
[3] ORSTOM 213 rue Lafayette, 75010 Paris, France, Lab. Géochimie Comp. et Syst., Univ. Paris VI, and Lab. CNRS-CEA Pierre Süe (CEN Saclay), 91191 Gif Sur Yvette Cedex, France
[4] Société Minière d'Anglade (SMA), Salau, 09140 Seix, France
[5] Laboratoire de Géologie. Ecole des Mines de Saint-Etienne, 158 Cours Fauriel, 42023 St. Etienne Cedex, France

Mineral Deposits within the European Community
(ed. by J. Boissonnas and P. Omenetto)

In the Appendix an outline of the case history is given, and explained by the successive development of new ideas induced both by the development of ore extraction and by the progress of the scientific analysis of structural setting, mineralogy and petrology.

1 Introduction

The basic method of mining geology is still that of reasoning by analogy and extrapolation, traditionally known as the method of ore controls. Even when geophysical and geochemical methods for prospecting and test drilling are used systematically and extensively, this cannot be done efficiently without some previous idea of the favourable areas, nor can they be interpreted without some basic ideas on the geology. Sufficient knowledge and understanding of the controls can save much time and money. Faulty knowledge may lead to abandonment of deposits because their extensions could not be found in the form of non-outcropping orebodies separated from previously known bodies. It can be assumed that this was the case for a fair number of mines until the 1950s. When metal prices are low, it may be wise to look for extensions of high-grade deposits, which represent large savings on investments and hold promise of finding ore of the same quality as in the bodies first discovered, rather than prospecting other occurrences. However, present methods are based on the recognition of types, although effective, they are not sufficiently fine and selective to guarantee that the potential deposits found will be of an economic grade under present conditions.

Naturally, the method of ore controls no longer amounts to simple reasoning by analogy. There is a fast-growing body of ideas and methods for analyzing geological reality in order to understand how the ore has been deposited, deformed, displaced, etc. Actually, these new methods only serve to provide new ore controls which are more subtle and more precise, but which still hardly ever have an absolute value. However proud we may be of progress in this field, it must be recognized that these methods do not, in general, enable us to definitely determine where we shall find ore, or of what grade it will be. There are at least three reasons for this:

1. Our knowledge of the processes is usually incomplete and our methods of analysis imperfect. Thus, the validity of our extrapolations beyond the ranges observed is not guaranteed.
2. Some processes are of a random nature, for example fracturing phenomena whose frequency is known but only in a statistical sense, arrangement into channels on a more or less horizontal surface (peneplain, delta), etc.
3. Geological phenomena of the most varied kind may be superimposed and occur locally, upsetting a 'law' which was of necessity derived from a limited subset of these phenomena.

As a mine develops, mineralization controls become more accurate. We come to know the limits of their validity, which may be of two kinds:

1. The discovery of other orebodies may often show that a law which appeared general, is not, and that there are other orebodies which are differently located or organized which thus enables other laws to be established (other controls independent of those already known which may be applied in parallel).
2. The systematic study of orebodies located or arranged according to a particular control often produces conditions limiting or specifying their range of application, sometimes throwing the initial formulation into doubt and enabling another to be proposed.

The aim of this chapter is to show how the available ore controls have evolved over time as a mine developed, using as an example the scheelite skarn deposit at Salau (Ariège). This case is interesting for various reasons: firstly, because some of us were able to follow the evolution of ideas over more than 20 years since the discovery of the deposit; and secondly, because this was a difficult case, with rich ore but without continuity or consistent grades; and finally, because in the present state of knowledge, the mine reserves are limited. Under such conditions, it was advisable for us to combine all our ideas on the structure of the deposit and methods of finding extensions. Thus, certain characteristics of the structure, which had not been clear at previous stages, seemed important enough to justify new research in the mine. Actually, it has led to the discovery of a new body of high-grade ore, the dimension of which is still unknown because the mine was closed just at that time due to the present very low price of tungsten and other financial problems. However, it will probably reopen in the future as it is clearly one of the highest grade tungsten deposits known at present and the ore reserves might in fact still be large. We believe that the understanding so acquired of a rich, but very complex, scheelite skarn deposit could also be useful for guiding research on several analogous occurrences presently being studied in France and elsewhere in western Europe.

2 Geological Controls: Exploration for Extensions of the Deposit

2.1 Scheelite Mineralizations

The first scheelite mineralizations of any magnitude found in the Pyrenees were at Costabonne. The grades and reserves of this deposit finally proved to be inadequate after fairly detailed investigation, but attempts were made to use the experience gained to look for other scheelite deposits. Controls were defined by analogy with Costabonne. A search was made for skarn-type deposits developed in Pre-Hercynian limestone or dolomite strata in the neighbourhood of the Hercynian granitic intrusions.

Investigations were made near, but not necessarily at, the contact, since the southern skarn at Costabonne developed on a schist-dolomite contact, while the northern skarn developed on a granite-dolomite contact (Dubru et al., this Vol.)

The Salau deposit was discovered by BRGM (Bureau de Recherhes Géologiques et Minières) in 1960 during these prospections.

NB. *Salau ores are classified hereafter into four types (0, 1, 2, 3), according to their relative size or economic importance. Thus, type 3 ore is the most important, whereas type 0 is very subordinate.*

2.2 Lithological and Petrological Controls: Type 3 Ore

Following the discovery of the mineralizations in the Bois d'Anglade at Salau (Fig. 1), and the first gallery works carried out by BRGM at level 1430 as well as the systematic prospecting of the la Fourque granite contacts, which were carried out simultaneously, it was determined (Autran et al. 1980) that:

1. The first major orebodies (*type 3 ore*) were developed in banded iron-rich skarns derived by metasomatosis of Devonian limestones (known locally as barrégiennes), which are characterized by alternations every 1–10 cm corresponding to varying proportions of sandstone- or shale-type materials and limestone.
2. In these barrégiennes, the skarns were developed at the granite contact. Embayments are structures favouring this development, and in general, contacts where the granite actually cuts across strata were more favourable than subparallel contacts. In particular, these resulted in the development of much thicker skarns (normal contact thickness multiplied by a factor of about 10).

The pure or graphitic limestones overlaying the barrégiennes in the Devonian series at Salau were considered unfavourable, as the skarns developing at contacts between granite and limestones of this type or in veins in it are generally very thin (10 cm to 1 m maximum thickness). Contacts of granite and Devonian limestones are very complex in the upper part of the Ravin de la Fourque (Fig. 5), with apophyses, embayments, etc. These structures are generally considered to favour skarn-type mineralizations. However, at Salau such mineralizations are of secondary or very limited importance. The main explanation for this, following the line of argument developed above, was found in the fact that a fair proportion of these limestones are pure limestones and not barrégiennes.

Another peculiarity of the Salau deposit, as far as it was known at this stage, was the systematic association, without exception, of economic grade scheelite with pyrrhotite. The highest grades often appeared in the limestone contact zone (marble line). Dispersed low-grade scheelite was observed in skarns which had not been invaded by pyrrhotite, but WO_3 contents were not above 0.3–0.4%. The grades associated with pyrrhotite were always from 1% to several percent (type 3 ore), even though massive pyrrhotite could also be barren. All efforts to distinguish barren pyrrhotite from pyrrhotite associated with scheelite, using mineralogical or chemical data or considering the silicate gangue, were in vain. All indications are that the pyrrhotite involved is the same, formed at the same stage of mineralization. The Costabonne occurrence, in which there is no pyrrhotite (and very little pyrite) was, on the whole, poor compared with the Salau deposit, with grades of 0.3–0.4%.

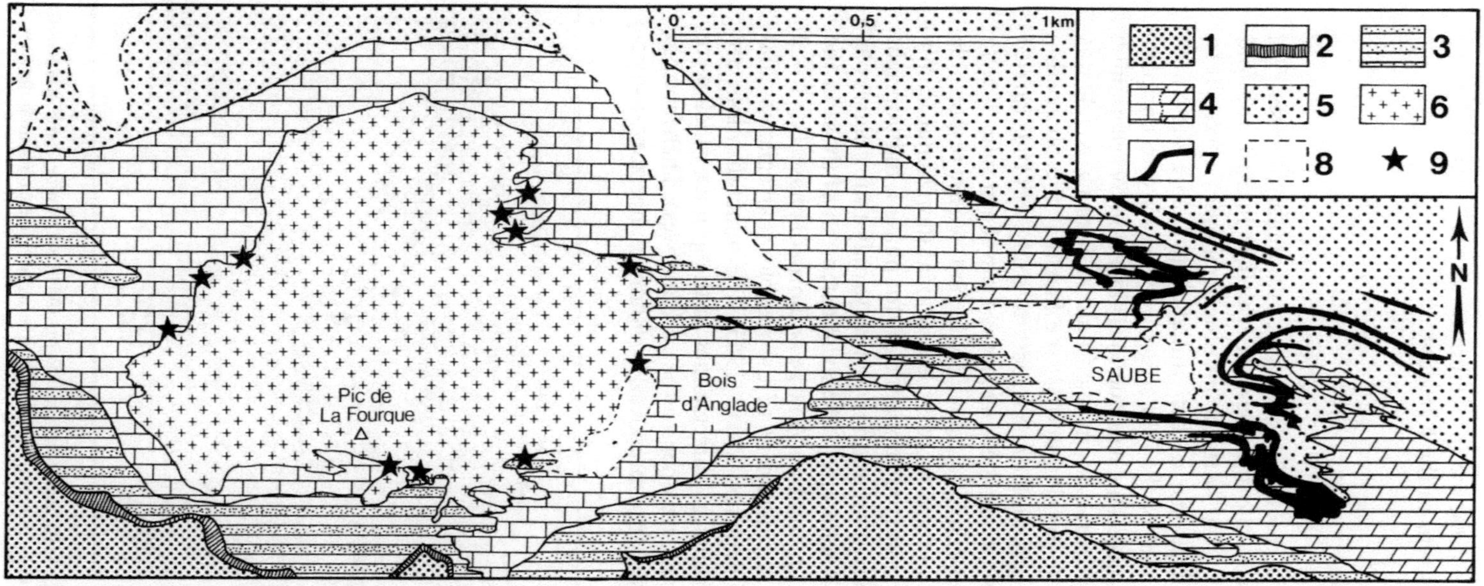

Fig. 1. Geological surroundings and distribution of scheelite occurrences around the la Fourque granitic apex (Derré 1983 and Nansot, unpublished SMA documents). *1* Shale-sandstone series (Ordovician). Salau series (Devonian): *2* crinoidal limestone; *3* Barrégiennes; *4* Salau limestone and dolomite; *5* bluish shale; *6* la Fourque granodioritic apex; *7* microgranite or microgranodiorite sills; *8* quaternary cover; *9* Scheelite occurrences

This gave rise to the idea that the presence of pyrrhotite was an essential factor influencing the value of scheelite skarns in the Pyrenees.

2.3 Controls by the Fold Structure

Since at this stage of investigations, most of the ore and, presumably, the reserves were contained in the barrégiennes, the thickness of which is limited, the fold structure of these and their intersections with the granitic apophyses appeared to be the chief controls for mineralizations in their initial state before they were disturbed by faulting. The broad lines of the fold structure were hard to decipher in the sector where the mine is located. This was due to (1) invasion by the granite; (2) disturbances induced by the mineralization itself and by faulting which was intense at the southern edge of the granite; and (3) the fact that the observable outcrops were limited and barely accessible. The structural study was thus carried out on the eastern side of the valley facing the mine, away from the granite. Kaelin (1982) recognized three main phases of folding, of which the first two are responsible for the overall structural pattern and are more or less coaxial (horizontal E–W), although of contrasting styles: horizontal recumbent folds for phase 1, tight upright folds for phase 2. Phase 3 has a different orientation and plays only a subsidiary role.

The structures unraveled by Kaelin are in the axial extension of those in the mine, of which they provide an understanding. A large antiform of phase 2 or 1 + 2 (Fig. 2) thus explains the overall arrangement of the marbles, barrégiennes and shales as mapped, for example in detail by Derré (1983). The mineralizations are largely distributed in the embayments of barrégiennes in the granite which were folded in stage 2. These embayments are open to the east and west. Their bottom is defined by the intersection of this large, practically cylindrical, horizontal struc-

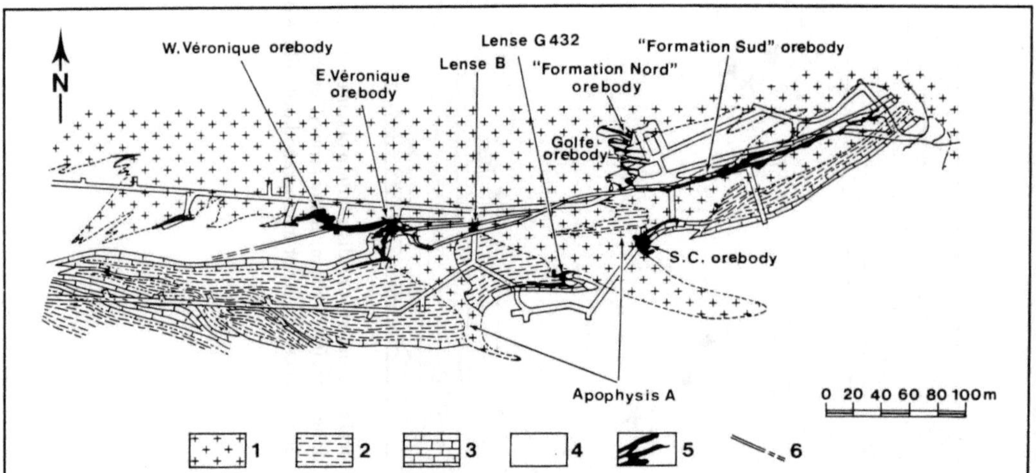

Fig. 2. Plan of anticlinal structure at level 1430 (according to Derré, 1983). *1* Granite; *2* metashales; *3* marble and limestone; *4* calcic hornfels (Barrégiennes); *5* skarns and mineral bodies; *6* main faults

ture with a major granite apophysis oriented to the southeast, apophysis A. Part of this cylindrical structure, which opens towards the east, corresponds to the Bois d'Anglade embayment and to the Quer de l'Aigle occurrences. The part opening towards the west makes up the NW part of the 'Véronique' orebody.

Here, a feature observed long ago should be pointed out: the parallel arrangement of the granite and subhorizontal ridges at the bottom of the Bois d'Anglade, clearly visible on the block diagram shown in Soler (1977).

Small-scale boundary shapes with a horizontal axis have been found above 1486, where they form the upward closure of the Véronique orebody, and at around 1150 (according to borehole surveys) where they presently participate in the downward closure of the Véronique orebody which has already been prospected. The sections drawn up by Nansot for this part of the deposit, all based on borehole surveys (Fig. 3), show that the ore-bearing limestone layer is delimited by two roughly parallel, cylindrical, horizontal axis surfaces, the one forming the roof consisting of shales, the other, which forms the floor, consisting of granite. All these observations appear to show that locally the granite-limestone boundary is inherited from a pregranitic shale-limestone boundary. This granite structure, which forms a horizontal roof or arch above Véronique, inherited from the pre-existing shale structure, probably explains, through its action of blocking the fluids in their upward movement, the rich mineralization of the upper part of Véronique.

2.4 Controls by Faulting[1]: Type 1 and 2 Ores

The existence of many silicified crush zones oriented EW in the granite and their importance in the morphology of the exposed granite have long been known. Recognition of the role of these faults in the delimitation of the orebody may be attributed to Soler (1977), who showed that a major fault of this family separates the NW and SE parts of the Véronique orebody which was discovered shortly before. He showed, step by step, that the 'Formation Sud' which forms the southern edge of the Bois d'Anglade embayment (the first orebody discovered at Salau), could also be regarded as a fault linked to the Véronique fault, but with a different filling.

Kaelin (1982) made a systematic distinction. He identified two principal stages of fracturing which he called faults f1 and f2.

2.4.1 Faults f1

The faults f1 have a strike of 80 to 90 and a northerly dip of 70° to 80°. However, they appear as faults only in silicate rocks, granites and shales. In the limestone they disappear as the limestone flows and adapts to the movements of the blocks of

[1] Note that Ledru and Autran (1987) have a completely different kinematic and dynamic conception of the development of the Salau structure from ours. It did not appear fruitful to discuss this approach in this chapter, since Ledru's work, which appeared recently, did not play any part in the history of the identification of controls actually used in developing the deposit. The interested reader should refer to the original article.

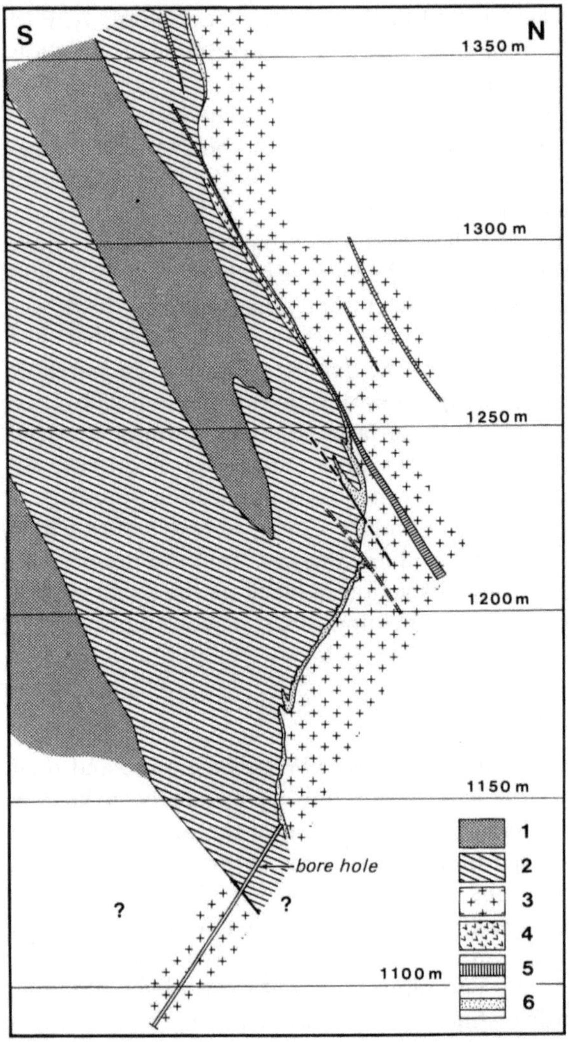

Fig. 3. N–S section at Véronique (Nansot, unpublished SMA documents). *1* Metashales; *2* marble; *3* granite; *4* breccia, including ore, granite and marble; *5* faults f2; *6* skarns and mineralized pyrrhotite

silicate rock with complex disharmonic folds which are emphasized by the graphitic banding.

The high-grade association of pyrrhotite and richly mineralized scheelite has mostly developed in the skarns along this type of fault, usually removing all apparent traces of crushing or deformation at a sample scale. Among the characteristics of this formation, the presence of approximately 1-cm spots of quartz or quartz + clinopyroxene or amphibole and of lenses of graphitic limestone some tens of centimetres in size, contorted and disrupted in a complicated way, should be noted. These spots and lenses are surrounded by massive pyrrhotite (±scheelite).

This type of 'filling' of the f1 faults is identical with the mineralization in the Formation Sud of the Bois d'Anglade embayment, which also has approximately the same strike. This is what we call *type 2 ore*.

The two known occurrences of wolframite in the quartz, where the Bois d'Anglade Formation Sud pinches out in the granite (Fonteilles and Machairas 1968; Soler 1977), could be part of this mineralization in the fault f1 developed under intragranitic conditions without influence from the calcic environment.

2.4.2 Faults f2

The faults f2 have a strike angle of 80 to 140 (mean about 120) and a northerly dip of 40° to 80°. They are filled with quartz accompanied by scheelite in large crystals, which are infrequent throughout and are of no economic interest. Unlike f1 they often develop remarkable slickensides. They clearly intersect with the f1 faults, at the same time limiting the workable, mineralized panels. In fact, these are not single faults but a series of regularly spaced parallel faults whose displacements are all of the same type. According to Kaelin (1982), the movement of these faults is reverse and senestral (although the extent of this movement had in no case been determined at the time he was studying them).

With the exploitation of the Véronique orebody, a new type of ore was distinguished, consisting of crushed quartz, sometimes presenting a mottled appearance, rich in scheelite (with associated arsenopyrite and pyrrhotite) distributed along a subvertical, nearly 5-m-thick mylonitic zone f0, striking 120 (Fig. 4). This mylonitic zone is interrupted to the north by an f2 fault. Although of economic grade (0.7–2.5% WO_3, average 1%), this type of ore is still small in volume at the present state of knowledge of the deposit. This zone, whose horizontal extension is small (less than 50 m) becomes poorer to the east and has not been followed at all levels. It disappears locally into the massive pyrrhotite (+scheelite), which fills the faults f1, in particular at levels 1470 and 1486. This ore thus appears to be the oldest of economic significance developed in the deposit. In support of the notion that this ore was developed in a comparatively early stage, it should be noted that this mineralized mylonite is folded by the third stage of folding defined by Kaelin (1982). This ore is designated *type 1*.

2.5 Type 2 Ore in 'Véronique Southeast' and Estimation of Displacements Along Faults f2

The upper part of the 'Véronique Southeast' orebody is extremely dislocated by faulting, and thus far its characteristics are not well understood up to the present stage of development. At the deeper levels, which have now been reached with boreholes and galleries, it has escaped the action of the late faults f2 and its characteristics are clearly apparent. Between levels 1150 and 1250 it appears as a several metres thick orebody with a fairly constant horizontal extension of about 50 m in an easterly direction, along the contact between the granite and the graphitic limestone. The ore is of type 2 and is very characteristic, with contorted decimetric lenses of graphitic limestone surrounded by massive pyrrhotite containing scheelite. WO_3 contents are fairly constant, of the order of 1.5 to 2.0%. At its extremity, very

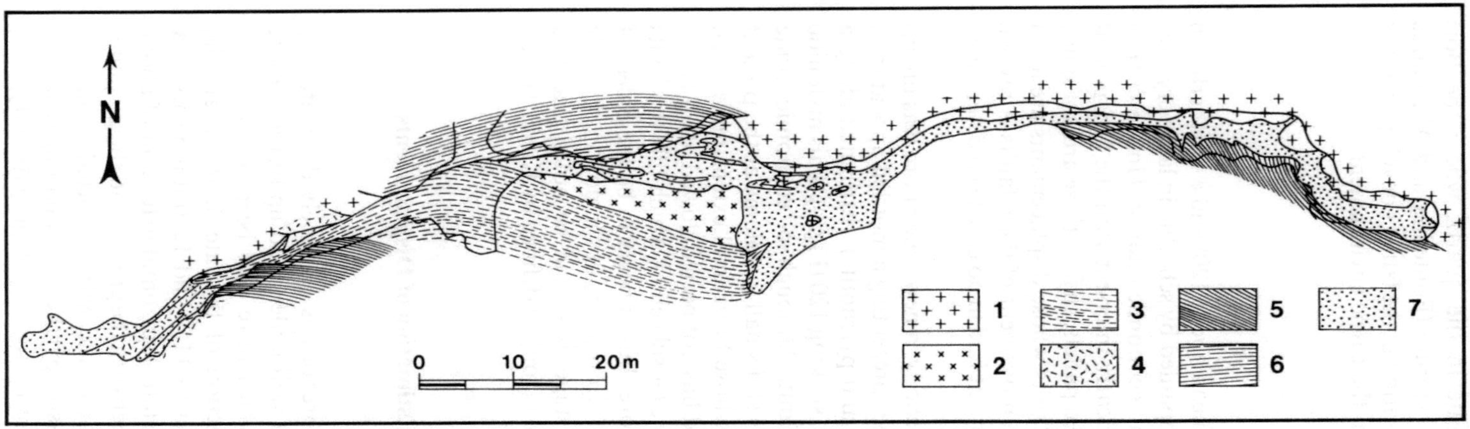

Fig. 4. Location of type 1 ore (site plan DV 1470, Kaelin, 1982). *1* Granite; *2* albitized granite; *3* silicified and mineralized mylonites (f0 and type 1 ore); *4* marble; *5* calcic hornfels; *6* skarns developed at the expense of calcic hornfels; *7* massive pyrrhotite (partly mineralized)

locally, skarn residues are observed in the pyrrhotite mass, then the orebody pinches out in apophysis A. To the west it is disrupted and loses continuity, and contents often fall below the cut-off grade.

Further up, boreholes show that grades are low and mineralization is thin between levels 1250 and 1350. Nansot's sections (Fig. 3) show that this drop in grade is a result of shearing and drawing-out of the orebody along the largest f2 fault known in the mine. Observations in galleries show that it is in fact a type 2 ore, completely dislocated in lenses and overturned blocks.

Above level 1350, the Véronique Southeast orebody again develops in the same way with virtually the same grades as below 1250, and this can be followed upwards continuously to about 1480, despite a fairly dislocated character, in detail. The shift between 1250 and 1350 gives the extent of the translation movement associated with the main plane f2. This is indeed a reverse fault with a throw of about 100 m, but with negligible horizontal displacement, which contradicts the initial impression derived from the striations.

This conclusion is very important as the results available at present from boreholes show that the 'Véronique West' column stops suddenly at about 1150, where the marbles disappear between granite at the footwall and to the north, and shales to the south. Comparisons between the various N–S sections drawn up from the borehole results suggest that the northern boundary of the shales is aligned in an approximately EW plane, which can be interpreted as an f2 plane. It may be expected that movement along this plane is of the same type as that along the main fault f2, with an amplitude of the same order. This shows us where to drill to find the extension of Véronique Southeast. This drilling was recently completed before the mine was closed, and *new high-grade ore was in fact found in this way.*

To summarize, the Véronique Southeast mineralized column is now known to be over more than 300 m in height without major variations in horizontal dimension or grades. A considerable extension downwards may be anticipated, with grades and dimensions of the same order. This column is not vertical; its projection on a vertical east-west plane has a westward pitch of about 70°, with the granite apophysis A forming its eastward boundary. The column is also not flat and, irrespective of the throws of the reverse faults f2, it is located on a cylindrical surface with a rather variable SSW dip, which is greater in the upper part than in the lower part, where it is ca. 45°. The axis of this cylindrical surface, like that of the folds of the shaley roof and in the limestone, dips to the east at about 15° to 20°.

2.6 Overall Layout of Known Orebodies Around Apophysis A. Role and Significance of Faults f1 and f2 and Type 2 Ore

With the possible exception of zones V and VI, which are minor orebodies located in the upper part of the deposit (around 1620 and 1750 respectively, Fig. 5), on a projection on an E–W vertical plane, all the known orebodies at Salau are situated in an aureole around a parabolically-shaped barren area formed by apophysis A.

Broadly speaking, the orebodies appear to be distributed in space over a cylindrical surface with the axis striking 80 and dipping 15° to the east, close to the

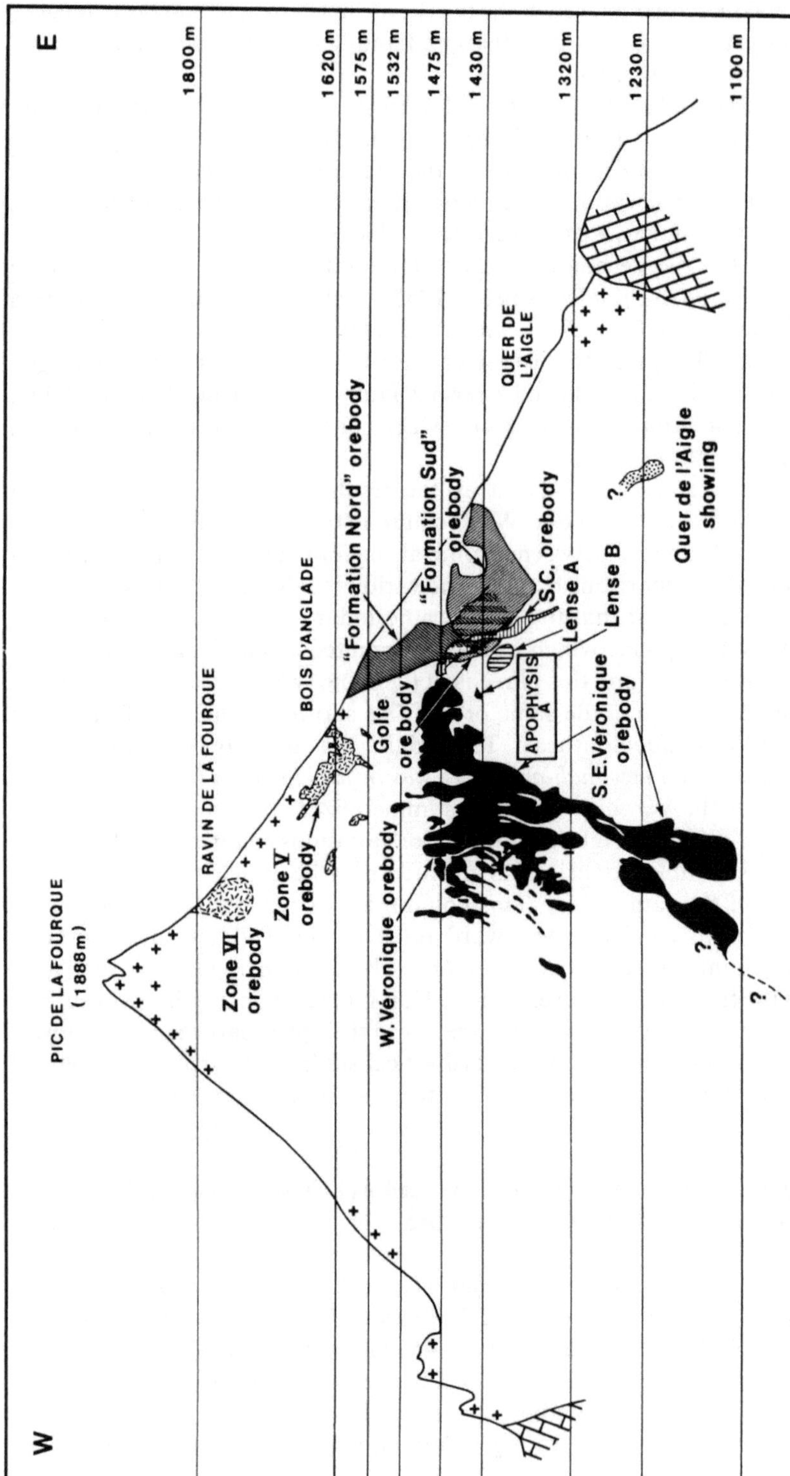

Fig. 5. Overall distribution of orebodies around granitic apophysis A (projection on an EW vertical plane, Nansot, unpublished SMA documents)

intersection of this surface and the granite apophysis A. More precisely, this surface and this intersection are outlined by the type 2 orebodies Véronique Southeast and Formation Sud in the Bois d'Anglade. Therefore, we interpret this distribution as being linked to a fault f1 intersecting with the granite apophysis A. This fault dies away in the limestones which flow long the granite blocks.

Type 2 ore has thus only developed in the immediate neighbourhood of the points where this fault emerges from the granite, i.e. in an aureole around apophysis A. According to Kaelin (1982), fault f1 was thus the channel through which the fluids responsible for the mineralization arrived. The observed granite only played a passive role and remained mostly barren. The source was lower along f1.

In view of this, the type 2 ore developed in the skarns (always very limited in extent) formed at the contact between the granite and the graphitic limestone. This small volume of skarn (porous medium) was the route which the solutions had to take, and is a 'throttling' [Pélissonier (1965) used the word '*étranglement*'], which explains the richness of this mineralization and the almost complete transformation of these skarns into massive pyrrhotite and scheelite ore. Thus type 3 ore appears to be the result of an 'oil-stain' extension of the mineralization where the fluid was able to reach and penetrate the Barrégiennes, which are clearly much more favourable to transformation and circulation. This oil-stain development, by percolation, explains the ragged aureole formed by the type 3 orebodies around the most proximal and regular type 2 orebodies. Zones V and VI could be slightly more remote extensions of these type 3, discontinuous mineralizations.

The Formation Sud of the Bois d'Anglade embayment is interrupted downwards by f2. The throw of this main f2 being of the order of 100 m, the extension of the Formation Sud, if any, should be found about 100 m below the bottom of the embayment. In view of the folding axes observed in the lower part of Véronique, such mineralizations could be the continuous westward extension of the known Quer de l'Aigle occurrences and would represent the equivalent offset downwards by a fault f2 of the zone where the Bois d'Anglade embayment opens towards the east, becoming progressively poorer in grade. The Quer de l'Aigle and its westward extension would thus complete the circular arrangement of mineralized formations around the apophysis A of the granite.

2.7 Probability of Extension

Fault 1, the channel for the mineralizing fluids, with its strike of 80°, diverges progressively from the la Fourque granite towards the west. This may explain why no new economic orebody was found on the extending gallery 1430 westwards and, if our line of reasoning is correct, leaves little chance of finding any in the southwestern part of the granite contacts at any level.

On the other hand, the two known groups of occurrences to the northeast and northwest on either side of the granite (Fig. 1) could correspond to the intersection of another f1 plane with the edges of the granite. If this is so, investigations should be carried out to determine whether this other f1 plane develops orebodies analogous to Véronique Southeast and the Formation Sud in the Bois d'Anglade, at greater depth.

3 Mineralogical Controls (Pyrrhotite, Garnet II): Their Application to the Assessment of Occurrences of Scheelite Skarns

The skarns and mineralization developed during a whole series of hydrothermal events: additions and renewals by successive solutions with different characteristics. Each of these episodes is characterized by the development of a particular, more or less specific mineralogy.

If there is a distinctive mineralogy associated with the stage at which the economic mineralization is developed, the idea naturally arises to consider the presence of this mineralogy as a favourable sign and to systematically examine whether it appears in the scheelite occurrences whose value is to be determined. In some cases these mineralogical properties have even been taken as a condition sine qua non for the presence of economic ore.

This proposal was made for the pyrrhotite which so clearly accompanies most of the scheelite at Salau. In our opinion, this concept is erroneous and the application of such a criterion could well result in some of the richest deposits being abandoned. As an example we could mention the large scheelite skarn deposit at Shizuyan (China), where pyrrhotite is either absent or present in very minor quantities.

In fact, a comparative study to determine whether the supposed mineralogical criterion occurred with sufficient frequency in the known deposits throughout the world had been omitted.

3.1 Controls in the "Golfe" Ore Body

In the initial history of the Salau mine, the skarns discovered were mainly exoskarns, differing principally in the nature of the substrate rock. On pure limestone (whether or not graphitic), in contact with granite or in veins, skarns are characterized by three monomineral zones: an internal grossular garnet zone, a hedenbergite zone (by far the largest) and an external zone of white calcite separating the silicate skarn from the graphitic marble. A small proportion of the garnet probably developed in an endoskarn. The development of a plagioclase-clinopyroxene ± quartz paragenesis on the external edge of the granite in contact with the skarns may also be linked with endoskarn development.

On the barrégiennes, the development of two transformation zones yielding banded skarns may be observed. The light-coloured external zone is characterized, among other features, by a salite-type pyroxene, the darker internal zone by hedenbergite. The other minerals are grossular, etc. The modifications of the granite are apparently the same as those in the former case. WO_3 contents in the unmodified skarns are of the order of 0.3–0.4%.

After the development of the skarns proper, two successive episodes are noted:

1. A stage in which the garnet and the adjacent granite are epidotized with transformation of clinopyroxene into hornblende ± calcite ± quartz. This stage corresponds to a transformation of the skarns without any marked additions (except perhaps for iron in certain cases);
2. A stage in which a paragenesis of the 'propylitic' type is usually developed. We have called the resulting rock RHO (Rock of Hydrothermal Origin, as opposed to skarns as such) which, depending on the substrate rock, consists of:
   a) On the granite: albite (transparent), chlorite, sometimes biotite, muscovite, quartz ± epidote (type 0 ore).
   b) On the skarns: actinolite, epidote, calcite, quartz (type 2 and 3 ores).

The presence of chalcopyrite, bismuth, often massive magnesian tourmaline (dravite) and sometimes apatite implies metasomatic additions. The scheelite content is extremely variable and often high. This episode is the one at which mineable ore is developed.

3.2 Garnet II Stage in "Véronique"

This simple model was greatly complicated after the discovery of the Véronique orebody. A new stage of development of typical skarns characterized in particular by garnet II rich in almandine-spessartine components appears in this orebody after the epidote stage and before the RHO stage. Among the minerals associated with the garnet, a blue-green hornblende and, rather unusually, a scapolite may be observed. This garnet is also accompanied by masses of black quartz.

The main characteristics of this garnet stage are the following:

1. These skarns have always developed on silicate rocks, either granite or epidotized granite, or else first-stage (exo)skarns. At this stage there is no development of skarns at the expense of the carbonated country rock;
2. They may consist of veins intersecting the primary hedenbergite, but these veins grade into indistinct recrystallization veins in the calcite where they emerge from the primary skarn;
3. Regardless of the type of involved silicate rock (granite or hedenbergite skarn, for example), the garnet tends to develop at the contact with the pre-existing silicate rock and the limestone (always on the silicate rock side) and not as an internal zone which is the usual position of skarn garnets.

The residual skarns observed in the type 2 ore masses generally display very well this stage of garnet II.

3.3 Direction of Fluid Movement

The first-stage skarns developed mostly as exoskarns. The fluids were in equilibrium with the minerals present in the granite, and were very aggressive towards the carbonate environment (destruction of graphite by oxidation, conversion of carbonate into ferrous or aluminous silicates).

The epidote stage, in view of its oxidizing nature (which is thus foreign to graphitic marble) could be the lower temperature outcome of this episode, the source of the fluid being essentially the same.

For the second-stage skarns, the RHO and the associated ore, the reverse is true, the fluid was in equilibrium with the external carbonate medium and reacted with the primary silicates; note in particular the high CO_2 fugacity (development of carbonates and scapolites) and the reducing character of this fluid, which transforms more or less pistacite-rich epidote into grossular-almandine garnet.

We conclude that the fluid responsible for the formation of the initial stages of the skarns (and perhaps for the epidotization) came from the granite, whereas the fluid responsible for garnet I and the RHO came from the limestone country rock. The isotope data (C, O, H) available at present (Guy 1979; Toulhoat 1978) agree with this interpretation. In both cases it is a question of the immediate origin of the fluid and not its ultimate origin.

3.4 Economic Significance of the Stages

This is an important result, because the scheelite contents associated with the first-stage skarn are still low (0.3–0.4%, not recoverable at present) and the mineable ore is linked with late transformations associated with fluids 'originating' in the country rock. Note that this has no implications with regards to the quantities of tungsten added at this stage. It is even possible that the recoverable mineralizations could result from simple, late reconcentration phenomena. At all events, it seems that the appearance of such solutions was indispensable for the skarns to be of an economic grade. The pyrrhotite, which was from the start regarded as a mineralogical control, is indeed deposited by these solutions. But it is an unreliable indicator in other areas. The presence of pyrrhotite is related to other factors, in particular sulphur fugacity, which appear to be independent from those which favour transport and deposition of tungsten.

On the other hand, type II garnet, rich in almandine-spessartine, may be a good indicator that a solution of the type which interests us here was present.

The idea of assigning an important role to garnet II as an indicator of the proximity of rich mineralizations is the result of repeated observation (Shimazaki 1977; Brown et al. 1985) of the occurrence of this mineral in economic deposits. Investigation of the presence or absence of this type of garnet in known deposits should be systematized.

4 Speculation on the Part Played by Granite and on the Origin of the Fluids (based on the consideration of the controls discussed above)

It is perhaps surprising not to find the composition of the granite among the ore controls proposed here. This is because we do not know whether the granite observed in the outcrop and in the mine played any part in the process other than a purely passive one.

It should be recalled that this granite may be classified in the common calc-alkaline Pyrenean categories of granite, but is more basic (granodiorite) than average, with many even more basic xenoliths (diorite). It is distinguished from the normal trend by a slightly higher iron content at the same stage of evolution (Soler 1977); and it shares this characteristic with the Batère granite which appears to be associated with a small, richly mineralized skarn at Roc Jalère (Salemink and De Jong, this Vol.). Note that granites, as little evolved as those of Salau, are not, in general, very favourable to the development of tungsten mineralizations, a fact which may be correlated with the low tungsten content of magmas at this stage of evolution, which in turn probably results in low W contents in fluids issuing from these magmas. However, the Costabonne granite does not show this relatively high iron content. It is also more evolved and accompanied at its margins by small bodies of white granite (Le Guyader 1982; Dubru et al., this Vol.). The notion that granite is the source of the fluids responsible for the mineralization is a classic one. At Salau, with regards to the rich mineralization associated with massive pyrrhotite (RHO stage), the presence of abundant tourmaline and occasional pockets of apatite may be considered as an indication in favour of this idea, the more so since irregular occurrences of fibrous tourmaline (dravite) have developed locally in the granite, together with small, blind, greisenized veins close to the mineralized zones. Note also that two occurrences of wolframite have been described, occurring as residues in the scheelite (Fonteilles and Machairas 1968; Soler 1977). All these observations suggest that the mineralization is related to granites which are more evolved than those observed in the mine. Moreover, the relationship which has recently been shown between rich mineralization and a subvertical f1 fault, together with the lack of change with increasing depth in the type 2 ore down the 350 m of the Véronique Southeast column, does not speak in favour of a nearby source of tungsten. At all events, the source is not the adjacent granite. Note that these observations give rise to hopes that the economic mineralizations may extend downwards. The fluid was channelled by fault f1. We should note that such channelling of the fluid, which for the most part explains the distribution of the ore, is only conceivable if the source is very localized, which contradicts the hypothesis which is sometimes advanced that common fluids circulate in the country rock.

It is not impossible that a very evolved, light-coloured granite comparable to the Costabonne granite was active at depth and fed fault f1 with mineralizing fluids. The apparent evolution of the fluids in the light of the above discussion may be summarized as follows:

At the early stage, the fact that the skarns are chiefly exoskarns and the evidence of important additions (Fe, Si, W, etc.) imply a source irrelevant to the country rock, possibly a granitic one.

The garnet II stage is clearly related to a fluid originating in the country rock, in equilibrium with the marble and graphite. There is nothing to suggest that any tungsten was added at this stage. The RHO may be related, as the early stage, to a fluid derived from a deep granitic source.

Two difficulties remain with regard to such a representation:

1. For the garnet II stage, a fairly massive fringe of black quartz often developed at the edge of this garnet on the marble side.
2. For the RHO stage, there was progressive but intense enrichment in scheelite on the edges of the 'skarns' (in a broad sense) up to the contact with marble, where the grade drops sharply to zero ('marble line'), a phenomenon which is present in most scheelite skarn deposits. Contradicting the proposed representation, this phenomenon could suggest that the tungsten-bearing fluid was very low in silica which circulated in the marble and that the scheelite was precipitated by a reaction with the silica in the skarns or the granite. Such silica-poor fluids could transport aluminium (Pascal 1984) and could thus be responsible for the formation of garnet II by replacement of hedenbergite on the edge of the skarns. However, this contradicts the successive nature of the development of the garnet II and the RHO.

One way of resolving these contradictions would be to assume that the precipitation of the scheelite arises from the meeting of two fluids with different sources and compositions, one in equilibrium with the carbonate environment, the other with the granitic silicate environment (the latter being able to transport tungsten). The garnet II stage would only represent a negative fluctuation in the supply of granitic fluid, during which the external fluid temporarily penetrated the silicate environment near its edges. The black quartz deposit would then represent a positive fluctuation, when the fluid originating from the granitic medium again tended to penetrate the marble. Knowledge of the isotope composition of the oxygen and the hydrogen in the fluid inclusions in this black quartz and in the quartz associated with the RHO seems essential at this stage.

The marble line could be explained by a very large scheelite precipitation where the two fluids came into contact. The random distribution of the tungsten contents at a metric scale and in samples would then be explained by fluctuations in the path and derivation of these two flows of fluids.

Appendix

We have collated in three stables the successive geological controls used for the ore at Salau. In the second and third tables (position with respect to the granite and structural controls), the comments will, we hope, make possible a better understanding of how the ideas evolved and how the emphasis placed on certain controls changed as knowledge of the deposit increased.

A. Type of control: lithology (*nature of substrate*)		Underlying ideas and comments
General		
1st idea	Marble or dolomite	
At Salau		
2nd idea (1965)	Main part played by barrégiennes (banded limestone with narrow layers of shale or sandstone)	Type 3 ore ('Golfe', 'S.C', 'Véronique NW' orebodies)

3rd idea (1976)	Undetermined nature of the Formation Sud and Véronique SE substrate (massive pyrrhotite with contorted lenses of graphic limestone)	Type 2 ore
4th idea (1980)	Mineralized mylonite south of Véronique	Type 1 ore
5th idea (1984)	Development of rich ore at the expense of more or less skarnified granite (first discovered in the upper part of Véronique NW); sometimes biotitization	Type 0 ore
6th idea (1986)	Development of type 2 ore, at the expense of the pure ± graphitic limestones, by superimposition on previous skarnification	Type 2 ore is associated with and equivalent to type 0 ore on a different substrate rock

B. Second type of control: position with respect to the granite		Underlying ideas and comments
General		
1st idea	Granite-limestone contact	1st interpretation: the granite immediately in contact is the source of fluids
at Costabonne		
2nd idea	This is not the only interesting contact; there are also skarn veins, and above all skarn bodies, in other contacts between marble and silicate rocks other than granite, for example shale strata	2nd interpretation: the source is not the adjacent granite but a deeper part of the granite intrusion (interpretation 1 is false) 3rd interpretation: the limestone silicate rock contacts are the main sites of fluid circulations
at Salau		
3rd idea (1965)	The skarns and the mineralization (at the granite contact) were mostly developed in the barrégiennes (type 3 ore) but to a very limited extent in the pure ± graphitic marble	Interpretation 3: the fluid is mostly guided by the pelitic beds in the marble
4th idea	There are mineralized mylonites of recoverable grades over 50 m horizontally southwards from the mineralized body	
5th idea (1979)	The rich (pyrrhotite) ore developed where a particular family of faults (f1) cut across the barrégiennes	5th interpretation: circulation of fluids along the veins; the barrégiennes as traps; abandonment of interpretation 3
6th idea (1986)	A rather special type of rich pyrrhotite ore (type 2) may develop on the pure ± graphitic previously skarnified limestones (the 3rd idea above again loses importance)	

7th idea	This type 2 ore is arranged in a narrow aureole around granite apohysis A on a surface f1; type 3 ore in a ragged aureole round the former	Type 3 ore is explained by 'leakage anomalies', which explain its random nature; the channel through which the solutions arrived is a single, well-defined fault f1; they have a deep origin

C. Third type of control: structural control		Underlying ideas and comments
General		
1st idea	Granitic apices of very small dimensions are the most favourable; diameter of la Fourque granite, 1.2 km	1. Essentially subvertical, monoclinal structure of the series
2nd idea (1965)	Major part played by semi-enclosed structures (embayments, roof pendants, etc.) The Bois d'Anglade embayment and the 'S.C': two embayment structures (open towards the east)	2. Faults play only a very subsidiary role
(1977)	Véronique Northwest: an embayment open to the west; the main orebodies are blow 1500 m, where the granite forms the roof of the limestone	3. Granite as caprock
3rd idea (1977)	The large fault in Véronique: a fault delimiting a northern and a southern compartment, the relationship between which is not known The large Véronique fault continues eastwards through the Formation Sud of the Bois d'Anglade embayment	4. Abandoning 2; abandoning 3 except around the 1486 m level The deposit is divided into panels by the fault 5. What is the continuation of the Véronique North and Bois d'Anglade embayment on the other side of the fault? Perhaps Véronique South and 'S.C' respectively
4th idea (1979)	Part played by fold structures; importance of early folds with EW subhorizontal axis and vertical plane; existence of a large limestone anticline with a shale core south of Véronique The deposit developed in the embayments was created at the intersection of apophysis A with this anticline; the structure discovered towards the east (Saubé area) is a good model of the folded structure around the mine The large Véronique fault is a reverse fault and the position of the granite forming the roof of the ore in Véronique is to a large extent due to the displacement of this fault, regularly spaced	Abandoning 1 Hypothesis 5 is false; how to determine the throw of f2?
5th idea (1978–80)	Discovery of the mylonitic zone f0, an early structure with scheelite mineralization but poor in pyrrhotite	

6th idea	Recognition in Véronique of an f1 fault invaded and cemented by mineralization rich in scheelite and pyrrhotite	
	This f1 fault in the silicate rocks degenerates into a flow zone in the marble; it predates f2, which interrupts and displaces the orebodies associated with f1	
7th idea	The ore SE of the la Fourque granite has developed mainly where an f1 fault with strike 80 emerges from the granite apophysis A and disappears in the limestone	This explains the absence of ore SW of the granite
	The ore is arranged in two columns, one west of apophysis A (Véronique SE), the other to the east (Formation Sud of Bois d'Anglade); the displacement caused by the reverse fault f2 may be assessed at about 100 m following the line of greatest slope, from examination of the Véronique	The solution formation is not the extension of the large f2 fault in Véronique but of f1 displaced and deformed by f2
	SE orebody between 1350, 1250; the downward closure of the deposit at level 1150, shown in borehole surveys is due to displacement of the Véronique SE orebody by a replica of f2	The slightly mineralized body at the Quer de l'Aigle may represent an extension of the Bois d'Anglade embayment south of the main f2 fault

Acknowledgements. Part of this research was supported under EEC (*Contracts No. MPP-080-F*) *and MSM-040-F*. The authors wish to thank the Société Minière d'Anglade, for permitting access to the Salau mine and publication of the present contribution, Jean Boissonnas for helpful discussions and an anonymous translator from the EC Commission for the translation from French.

References

Autran A, Derre C, Fonteilles M, Guy B, Soler P, Thoulhoat P (1980) Genèse des skarns à tungstène dans les Pyrénées. In: Z. Johan (ed) Minéralisations liées aux granitoïdes part 2. Mém BRGM 99:193–319

Brown PE, Bowman JR, Kelly WC (1985) Petrologic and stable isotope constraints on the source and evolution of skarn-forming fluids at Pine Creek, California. Econ Geol 80:72–95

Derre C (1983) La province à Sn–W ouest-européenne. Histoire de divers types de gisements du Massif Central, des Pyrénées et du Portugal. Distribution des gisements. Thèse d'Etat, Université Paris VI

Fonteilles M, Machairas G (1968) Eléments d'une description pétrographique et métallogénique du gisement de scheelite de Salau (Ariège) 2ème série. Bull. BRGM 3:62–85

Guy B (1979) Pétrologie et géochimie isotopique (S, C, O) des skarns de Costabonne. Thèse Doct Ing Ecole des Mines de Paris

Le Guyader R (1982) Elements traces dans les skarns à scheelike et les roches associées à Costabonne (Pyrénées Orientales-France). Thèse 3ème cycle, Université Paris VI

Kaelin JL (1982) Analyse structurale du gisement de scheelite de Salau (Ariège-France). Thèse Doct Ing Ecole des Mines de Saint-Etienne

Ledru P Autran A (1987) Relationship between fluid circulation, ore deposition and shear zones: New evidence from the Salau scheelite deposit (French Pyrenees), Econ Geol 82:224–229
Pascal ML (1984) Nature et propriétés des espèces en solution dans le système K_2O–Na_2O–SiO_2–Al_2O_3–H_2O–HCl: contribution expérimentale. Thèse d'Etat, Université Paris VI
Pelissonnier H (1965) Le problème de la concentration naturelle des substances minérales. Ann Mines 12:889–924
Shimazaki H (1977) Grossular-spessartime-almandine garnets from some japanese scheelite skarns. Can Miner 15:74–80
Soler P (1977) Pétrographie, thermochimie et métallogénie du gisement de scheelite de Salau (Pyrénées Ariégeoises-France). Thèse Doct Ing Ecole Mines Paris
Toulhoat P (1978) Pétrographie et géochimie des isotopes stables (D/H, $^{18}O/^{16}O$, $^{13}C/^{12}C$, $^{34}S/^{32}S$) des skarns du Quérigut — comparaison avec les skarns à scheelite des Pyrénées. Thèse 3ème cycle, Université Paris VI

Distribution of Scheelite in Magnesian Skarns at Traversella (Piemontese Alps, Italy) and Costabonne (Eastern Pyrenees, France): Nature of the Associated Magmatism and Influence of Fluid Composition

M. DUBRU, J. VANDER AUWERA, G. van MARCKE de LUMMEN, and J. VERKAEREN[1]

Abstract

The Costabonne and Traversella scheelite-bearing skarns are compared on the basis of associated igneous rock composition, early metasomatic columns, parageneses of the scheelite and differences in the fluid composition. Both series of intrusive rocks are of calc-alkaline affinity and are the result of a process of fractional crystallization. These rocks are of granitic composition at Costabonne and of dioritic composition at Traversella. The early metasomatic columns are similar: dolomite/forsterite + calcite/pyroxene/garnet, but the silicates are richer in Fe and Mn at Costabonne. At Traversella, Fe is introduced during a later hydrothermal stage (magnetite). Two generations of scheelite are found at Costabonne, one associated with the garnet and a second in association with amphibole, quartz and calcite as an alteration product of pyroxenes. At Traversella, scheelite is introduced during the hydrothermal stage by replacement of calcite in the outer zones. These differences in scheelite distribution may be explained by differences in fluid compositions under similar pressure and temperature conditions. At Costabonne, activities of Ca and Si were higher and activities of Mg and W were lower than at Traversella.

1 Introduction

In this chapter, we will compare the tungsten-bearing skarns of Traversella and Costabonne. Both deposits are located in very similar geological environments, i.e. they are developed on dolomitic marbles at the contact with late tectonic calc-alkaline intrusive bodies. According to the classification proposed by Zharikov (1970) and Einaudi and Burt (1982) they may be considered as magnesian skarns since they contain magnesian silicates (forsterite, diopside, etc.) in the outer zones of the metasomatic columns.

[1] Laboratoire de Minéralogie et de Géologie Appliquée, Université de Louvain, (UCL), 3, Place L. Pasteur, B-1348 Louvain-la-Neuve, Belgique

Mineral Deposits within the European Community
(ed. by J. Boissonnas and P. Omenetto)

In the first section, we analyze the process of magmatic differentiation giving rise to the associated intrusive rocks and show that they are linked by fractional crystallization.

In the second and the third sections we present a paragenetic analysis of the early metasomatic columns and the subsequent hydrothermal stages. Theoretical considerations on metasomatic processes (e.g. Korzhinskii 1970) have emphasized the critical importance of chemical parameters imposed by the fluid during the various stages of skarn formation. In this chapter, we show that, notwithstanding similarities between both deposits, important differences in scheelite distribution do exist because of specific differences in some chemical parameters of the fluids.

2 Geological Setting

2.1 Traversella

The Traversella intrusion was emplaced 33 million years ago during the Oligocene (Hunziker 1974) in the Sesia-Lanzo thrust sheet, internal Alps (Fig. 1). According to Müller (1912) and Kennedy (1931), the country rocks consist mainly of gneisses and eclogitic micaschists interbedded with lenses of eclogites and dolomitic limestones. These formations were metamorphosed in the blue-schist facies during the cretaceous (Compagnoni et al. 1977).

The marbles are exposed west and north of the intrusion. They are partly replaced by magnetite and scheelite-bearing skarns. Copper and, later, iron have been mined since the Roman period. The Traversella mine closed down in 1969 and the area has been recently re-investigated for tungsten.

2.2 Costabonne

The Costabonne skarn complex is located (Fig. 1 in Guy et al., this Vol.) in the southern part of the Canigou massif (Pyrenees axial zone) which consists of a Precambrian gneissic basement covered by a Palaeozoic metamorphic series (Fig. 2). Both series were folded and metamorphosed during the Hercynian orogeny up to the sillimanite grade (Guitard 1970) and were intruded by several late Hercynian granitic stocks, including the Costabonne granite (Autran et al. 1970). Most of the gneisses, lying in the central part of the Canigou massif, are orthogneisses derived from Precambrian calc-alkaline granites. The lower part of the Palaeozoic metamorphic series (Canaveilles Formation) of Precambrian to early Ordovician age contains essentially micaschists with thin sandstone and quartzite layers and thick limestone and dolomite beds.

The Costabonne granite intruded almost entirely the gneisses, only its southern margin is now in contact with the lower part of the Canaveilles Formation.

Fig. 1. Simplified geological map of the Traversella area (After Müller 1912). *1* Traversella diorite; *2* porphyrite; *3* lamprophyre; *4* marbles; *5* micaschists; *6* moraine.

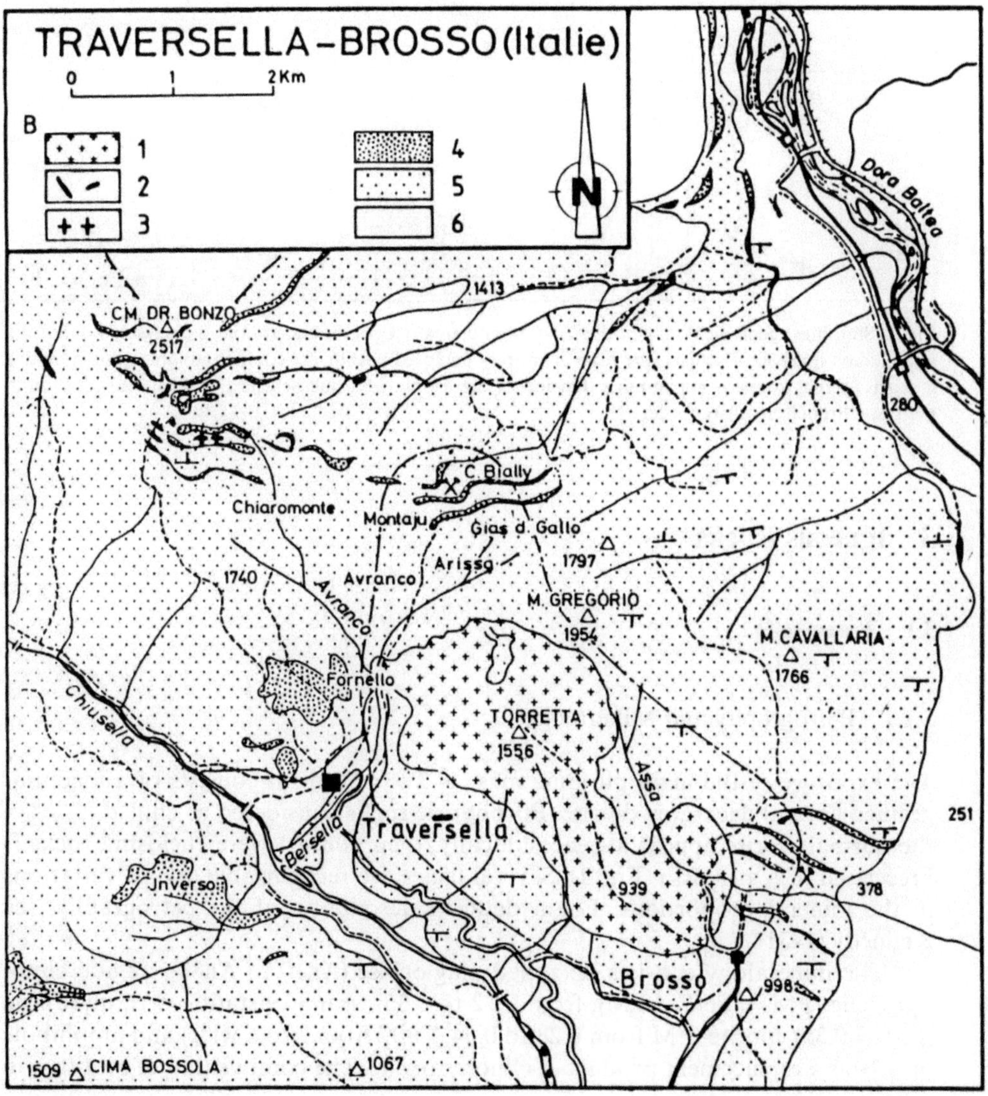

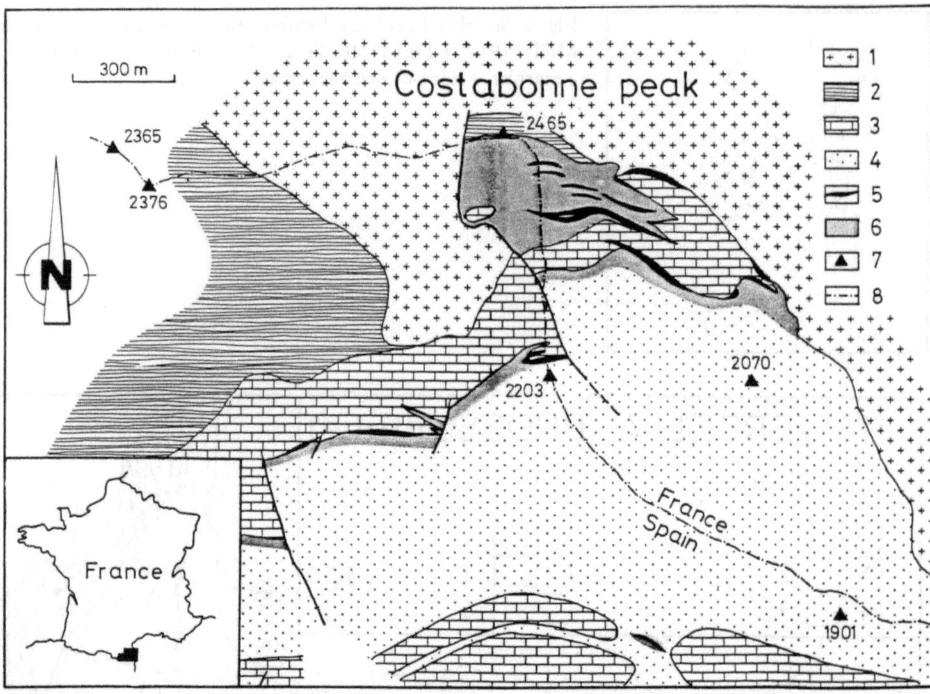

Fig. 2. Simplified geological map of the Costabonne area. *1* Costabonne granite; *2* gneisses; *3* dolomitic and calcitic marbles; *4* schists (the map is entirely enclosed within the contact metamorphic aureole); *5* skarn (garnet zone); *6* maximum extension of metasomatized micaschists; *7* peaks with altitude in metres; *8* French/Spanish border

3 Igneous Rocks

3.1 Traversella

3.1.1 Petrography and Mineralogy

The exposure of the main plutonic mass is 2 × 4.5 km. It consists of a medium-grained (2 mm) dark-grey diorite. Porphyritic rocks considered as chilled facies of the normal diorite appear at places: locally, their contact were enclosing rocks is brecciated (e.g. at Arissa, Fig. 1). Several facies are recognizable due to variations in the mineral proportions. For a detailed description of the petrography, see Kennedy (1931).

The mineralogy consists of: zoned plagioclase (An10 to An80), augitic clinopyroxene (FM = Fe/Fe + Mg from 0.12 to 0.35), scarce enstatitic orthopyroxene (FM = 0.35), biotite (FM from 0.28 to 0.54, TiO2 from 3 to 6 wt%) and amphibole mostly as a replacement product of clinopyroxene. The composition of biotite is in agreement with that associated with amphibole or pyroxene in magmatic rocks after Nockolds (1947).

Some samples show cumulative textures and consist mostly of clinopyroxene and biotite with minor amounts of magnetite, ilmenite and plagioclase.

Leucocratic aplite dykes (1- to 50-cm-thick), sometimes containing garnet, are found cutting across the diorite in marginal areas. Graphic textures are common. The only occurrence of pegmatite (grain size up to 3 cm) forms a vein of about 50 cm in width.

3.1.2 Whole Rock Compositions

The composition of the main intrusive body corresponds to that of monzonite to diorite, whereas the leucocratic dykes are of granitic composition (Table 1). The diorites are of calc-alkaline to calcic affinity, using the Peacock Suite Index (1931); however, they are rich in potassium. The nature of the Traversella igneous rocks was discussed by Kennedy (1931). He reported that they are monzonites and diorites and correspond to the potassic series intermediate between the pure 'diorite series' and 'syenite series' found in the peri-Adriatic province. The rocks show a remarkable differentiation trend from basic to intermediate terms between 55 and 66% SiO_2 (Figs. 3 and 4). The few samples with cumulative textures are poorer in silica. The granitic rocks appear to lie on the same trend but since they are located very close to the Al_2O_3 pole in Fig. 3, conclusions are questionable inasmuch as there is an important composition gap between these rocks and the diorites (Figs. 3 and 4).

A model of magmatic differentiation by fractional crystallization and cumulate separation has recently been tested using major and trace elements (van Marcke de Lummen and Dubru, in preparation). The bi-logarithmic diagram U:Th (Fig. 5) shows a straight line passing through the origin for the diorites and the cumulative

Table 1. Composition of the Traversella Igneous Rocks[a]

	1 C	2 C	3 D	4 D	5 D	6 D	7 AD	8 AD
SiO2	44.57	48.55	56.10	60.70	62.87	66.54	73.20	76.05
TiO2	2.08	1.34	0.97	0.82	0.65	0.57	0.35	0.27
Al203	13.59	14.31	17.75	16.96	15.89	15.55	14.28	12.75
Fe203	11.72	12.41	8.30	5.23	5.37	4.34	1.88	1.42
MnO	0.15	0.20	0.16	0.11	0.08	0.07	0.05	0.05
MgO	13.66	6.99	3.94	2.86	2.08	1.68	0.60	0.20
CaO	11.15	9.51	7.48	6.59	4.32	3.79	2.45	0.90
Na20	0.65	0.68	1.41	1.75	1.42	1.96	1.08	1.03
K20	1.70	2.29	2.81	3.23	3.76	3.01	4.12	6.05
P205	0.11	0.63	0.43	0.42	0.41	0.41	0.14	0.08
L.I.	0.63	0.78	0.42	0.71	0.83	0.51	0.26	0.27
Total	100.01	97.69	99.77	99.38	97.68	98.44	98.42	99.07

[a] Abbreviations: *C* cumulatic rocks; *D* diorite; *AD* aplite dyke. Total iron as Fe_2O_3.

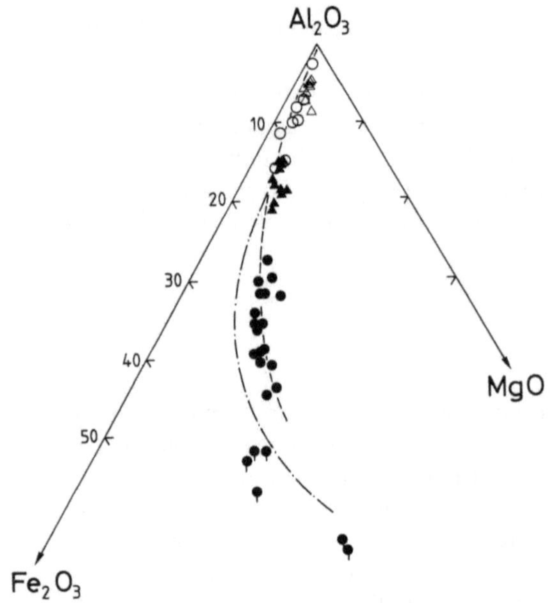

Fig. 3. Al_2O_3:Fe_2O_3:MgO diagram (weight proportions). Composition of the Traversella and Costabonne intrusive rocks (X-ray fluorescence analyses). Traversella: diorite (*filled circles*), granitic dykes (*open circles*) and cumulates (*filled circles with vertical bars*). Costabonne: granite (*filled triangles*) and leucogranites (*open triangles*); data from Le Guyader (1982). *Dashed line*: differentiation trend of the Traversella diorite. *Dashed–dotted line*: differentiation trend of calc-alkaline series (After Besson and Fonteilles 1974)

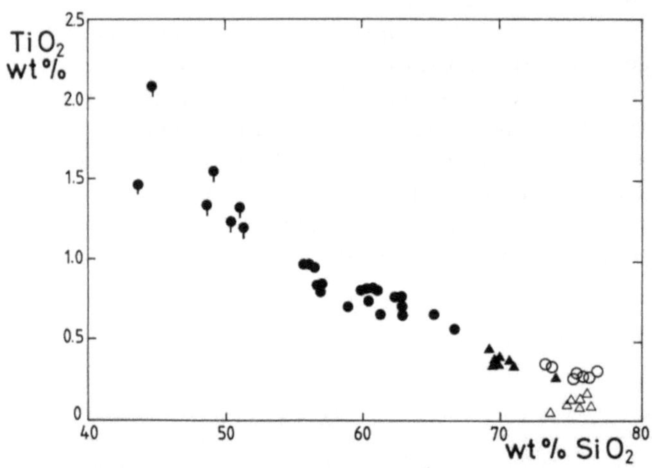

Fig. 4. Plot of TiO_2 versus SiO_2 (w + %). Same *symbols* as in Fig. 3

rocks, which indicates that a process of fractional crystallization took place (Treuil and Joron 1975). The distribution of the REE (Fig. 6) shows the same genetic relationships for the diorites. As shown in Figs. 5 and 6, the composition of the aplite dykes does not fit the composition trend of the diorite. This suggests that they are not co-magmatic with the rest of the pluton.

Petrographic and chemical criteria such as the presence of basic xenoliths, the existence of contemporaneous andesitic volcanism in the area (Venturelli et al. 1984), the occurrence of hornblende, biotite and titanite, as well as interstitial or

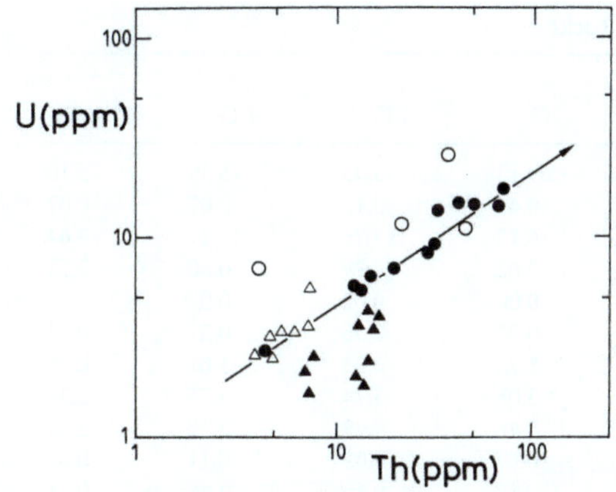

Fig. 5. Bi-log U:Th diagram (ppm). Same *symbols* as in Fig. 3

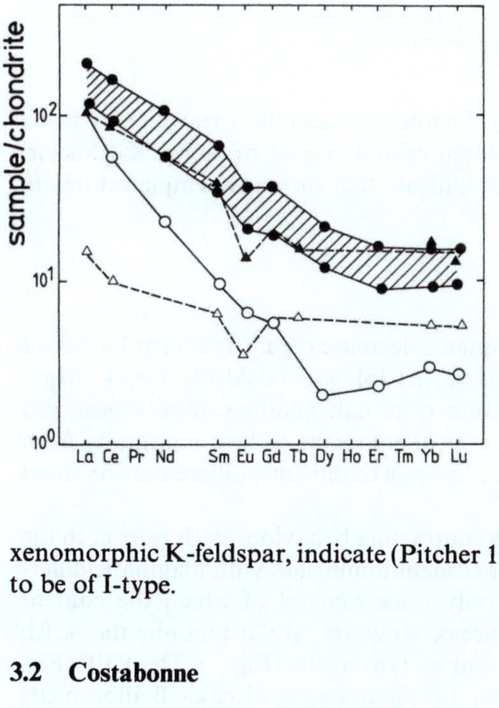

Fig. 6. REE distribution patterns of Traversella and Costabonne intrusive rocks. Same *symbols* as in Fig. 3. *Ruled area*: composition field of the Traversella diorite. NA analyses from Le Guyader (1982) for Costabonne and ICP (CRPG, Nancy) analyses for Traversella (this study)

xenomorphic K-feldspar, indicate (Pitcher 1983) that the Traversella pluton is likely to be of I-type.

3.2 Costabonne

3.2.1 Petrography and Mineralogy

The Costabonne granite forms a roughly circular mass about 10 km in diameter. Its contacts with the country rocks are perfectly sharp. The main mass consists of monzogranite with a granular texture (grain size up to 4 mm). The mineralogy consists of: quartz, zoned plagioclase (An05 to An 45), perthitic K-feldspar, enclosing the other minerals, biotite (containing inclusions of accessory minerals), deuteric muscovite, apatite, zircon, monazite, ilmenite and magnetite.

Table 2. Composition of the Costabonne Igneous Rocks[a]

	1 G	2 G	3 G	4 G	5 LG	6 LG	7 LG
SiO2	70.70	69.77	69.82	69.13	75.03	75.59	75.10
TiO2	0.37	0.37	0.38	0.43	0.12	0.07	0.07
Al203	15.31	14.06	15.29	15.12	13.76	14.17	13.64
Fe203	2.55	2.29	3.01	2.62	0.79	0.50	0.53
MnO	0.05	0.09	0.02	0.06	0.02	0.07	—
MgO	0.99	0.88	0.92	0.91	0.53	0.21	0.22
CaO	0.36	1.58	1.75	1.72	0.34	1.61	0.26
Na20	3.22	3.68	2.65	3.08	3.74	5.75	3.39
K20	5.49	4.56	4.08	5.48	4.99	3.35	5.51
P205	0.16	0.15	0.16	0.19	0.08	0.11	0.07
L.I.	1.23	1.16	1.16	0.83	0.66	0.46	0.54
Total	100.30	98.46	98.13	99.64	100.12	101.83	99.87

[a] Abbreviations: *LG* leucogranites: *G* granites. Total iron as Fe_2O_3.

A number of dykes (up to 1-m-thick) of white granites (alaskites) cut across the country rocks. They exhibit pegmatitic textures and consist of albite, K-feldspar, muscovite and quartz. Field relationships indicate that they were emplaced before the skarns.

3.2.2 Whole Rock Compositions

With increasing SiO_2, the major oxide contents decrease regularly except for Na_2O and K_2O which remain nearly constant (Fig. 4, Table 2). The Al_2O_3:Fe_2O_3:MgO diagram (Fig. 3) confirms that the granite is of calc-alkaline affinity (see also Salemink and De Jong, this Vol.). Moreover, there seems to be a continuity from the granites to the alaskites which appear to belong to the same differentiation trend (Figs. 3 and 4).

The trace elements U and Th show a contrasting behaviour with respect to the igneous rocks of Traversella, in that their content diminishes with magmatic differentiation (Fig. 5). Rb seems to be the only trace element of which the content increases during differentiation in both areas. However, at Costabonne the K/Rb ratio decreases, whereas it remains constant at Traversella (Fig. 7). The REE (Fig. 6) show normal distribution patterns for the final stages of calc-alkaline melts evolving through a fractional crystallization process (Cocherie 1984).

4 Description of Early Skarns

4.1 Location of the Skarn Bodies

At Traversella (Fig. 8) as well as at Costabonne (Fig. 9), the skarns are located either on the contacts between the igneous rock and the country rocks or on

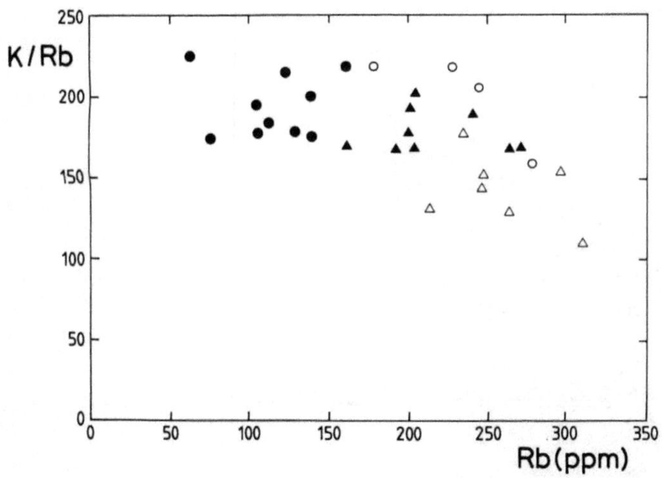

Fig. 7. Plot of K/Rb ratio versus Rb (ppm). Same *symbols* as in Fig. 3

the contacts between dolomite and metapelites or even inside the metapelites themselves.

At Traversella the main skarn bodies are located between a thick dolomite bed and the metapelites which dip towards the pluton. Four main masses of mineralized skarns, roughly parallel to the western contact of the diorite, form the main minable zone of Traversella, extending in a north-south direction for about 1.2 km. Each mass of skarn consists of a sheetlike body a few tens of metres in length, a few hundreds of metres in height and up to a few tens of metres in width.

At Costabonne, the country rocks are roughly parallel to the granite contact. Skarn bodies form elongated, continuous masses of variable width (up to 30 m), extending for a few hundreds of metres in a horizontal direction as well as in vertical extension.

4.2 Early Skarn Zonation

In skarns, a zonation pattern (metasomatic column) is clearly recognizable: dolomite/calcite + forsterite/diopside/salite/garnet at Costabonne (Guy 1980; see Fig. 2 in Guy et al., this Vol.), and dolomite/calcite + forsterite/pyroxene/garnet at Traversella (Vander Auwera 1985).

Other columns exist (e.g. van Marcke de Lummen and Verkaeren 1986; Dubru 1986), but this goes beyond the scope of this work.

Temperature and pressure conditions were estimated to be 500°–600 °C and 1.0–1.5 kb resspectively, at Traversella (Vander Auwera, in prep.) and 550°–600° and 1.5–2.0 kb at Costabonne (Guy 1980; van Marcke de Lummen and Verkaeren 1986; Dubru 1986).

Regardless of the total width of the skarn bodies, the thickness of the inner zones at Costabonne (salite/garnet) is always considerably greater than the outer

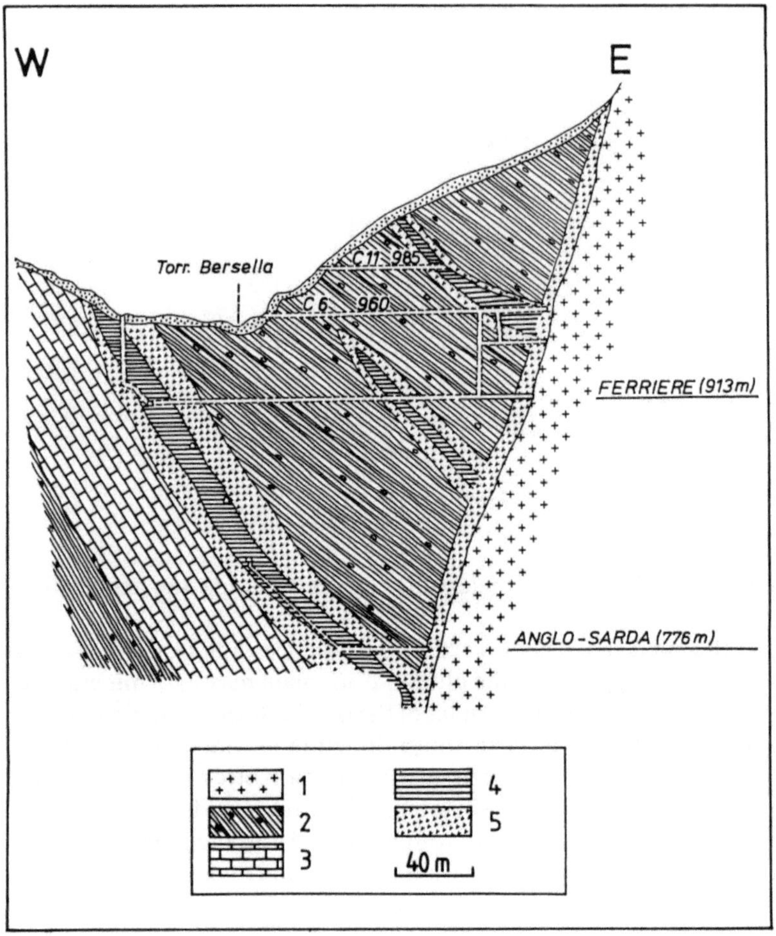

Fig. 8. Schematic E–W cross-section of the Traversella deposit (FIAT, unpublished data). *1* Diorite; *2* micaschists; *3* marbles; *4* magnetite and scheelite mineralization; *5* skarns

ones (calcite + forsterite/diopside), which appear mainly as veinlets cross-cutting the dolomites. As Traversella, on the contrary, the garnet-pyroxene zone seems to be of limited extension relative to the thickness of the calcite-forsterite and pyroxenite zones (Zuchetti 1966).

4.3 Composition of Silicate Phases (Fig. 10)

The forsterite is rich in Mg (olivine) in both deposits: Fe < 15% (atomic proportions). The Mn content reaches 1% (at. prop.) at Costabonne and is lower than 0.5% (at. prop.) at Traversella.

The pyroxene is mainly diopsidic at Traversella (Fe < 20%, at. prop.). At Costabonne, the pyroxene compositions show a wider range: up to 55% Fe (at.

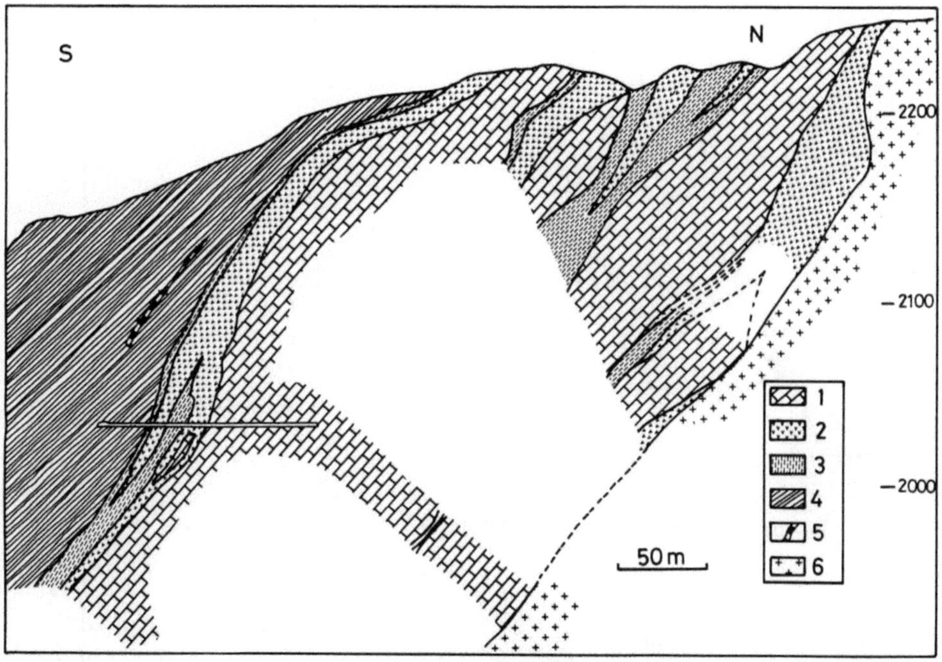

Fig. 9. S–N cross-section of Costabonne (After Guitard and Laffitte 1958). *1* Dolomitic marbles; *2* skarns; *3* micaschists, partly metasomatized; *4* micaschists; *5* leucogranite dykes; *6* granite; *7* quartz veins

prop.), but a gap is found between 25 and 35% Fe. The more diopsidic compositions are restricted to the so-called diopside zone, whereas the iron-rich compositions are found in the salite zone.

At Costabonne, two generations of garnet are present. The first one (garnet S1) is rich in andradite (>90%) and poor in almandine and spessartite. The second one (garnet G2), of later formation, shows a wide range of compositions from 10 to 85% andradite and is rich in almandine and spessartite (<15%). At Traversella, only one generation of garnet has been observed. In includes both an andradite-rich (>90%) garnet and a grossular-rich garnet (>60%). The occurrence of a grossular-rich garnet seems to be due to the presence of spinel in the replaced rock dolomite.

5 Hydrothermal Alteration and Scheelite-Bearing Parageneses

5.1 Traversella

The bulk of the scheelite occurs in association with the alteration parageneses in the calcite + forsterite zone. Three main steps can be distinguished in the alteration process of this zone (Fig. 11).

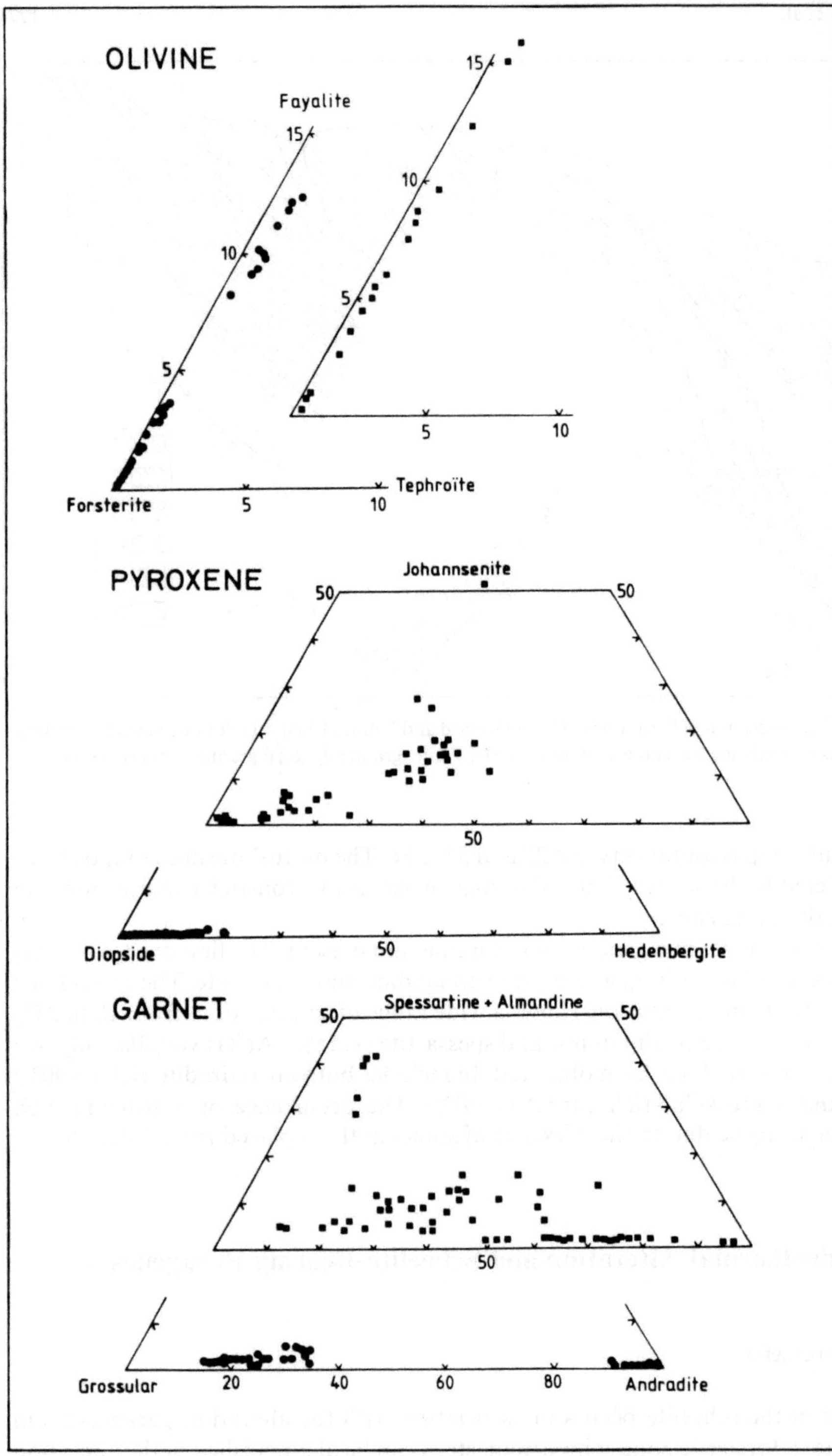

Fig. 10. Composition of the silicates of the Traversella and Costabonne early skarn zoning (microprobe analyses). *Circles*: Traversella; *squares*: Costabonne (data from Guy 1980 and Le Guyader 1982)

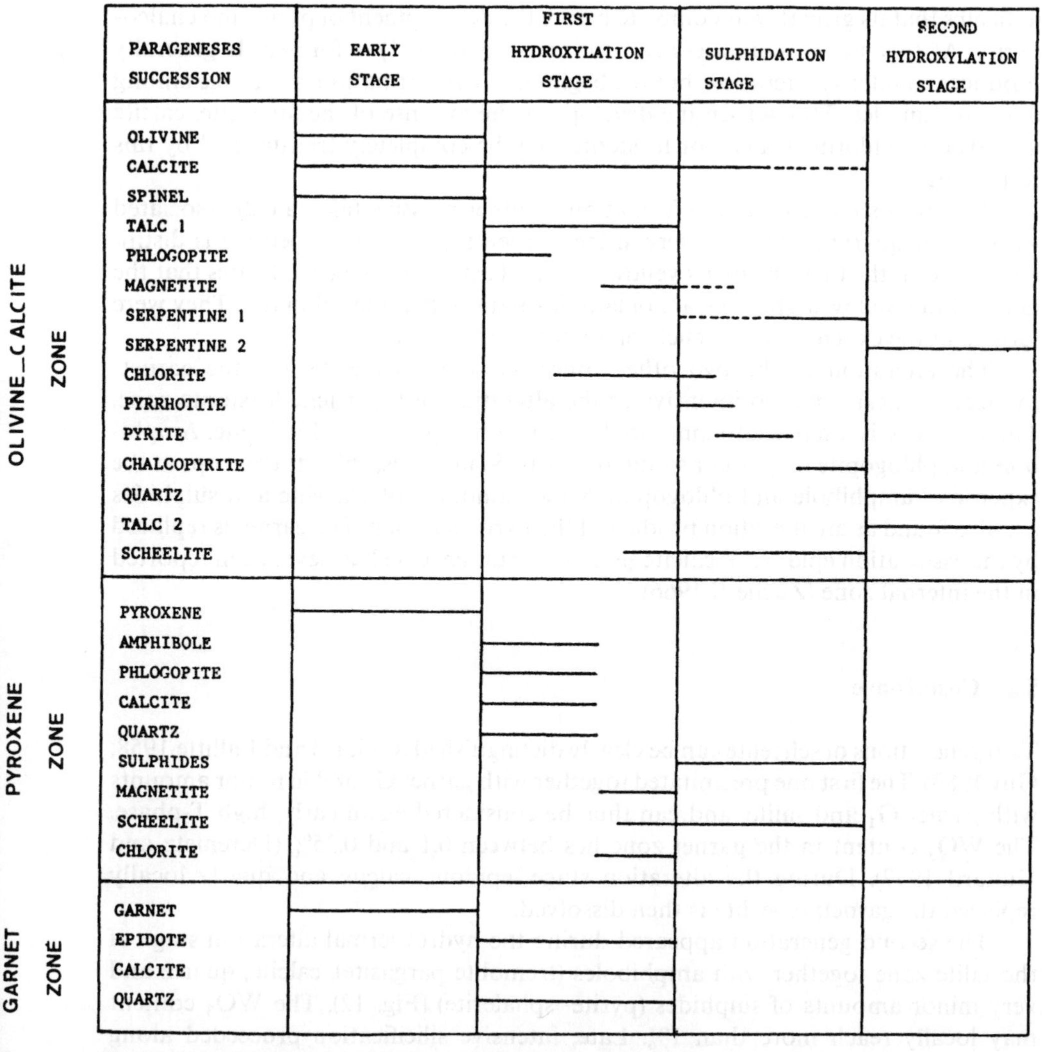

Fig. 11. Scheelite and alteration mineral parageneses at Traversella

During the first step, talc (talc 1) or phlogopite develops at the expense of the forsterite, but this alteration is not complete and relicts of forsterite are still present. Magnetite replaces either dolomite (magnetite fringe bordering the calcite + for sterite veins) or forsterite. The precipitation of scheelite at the expense of the calcite seems to begin after the development of the magnetite. But these two minerals together are stable. During this step, the phlogopite, when it occurs, may be transformed into chlorite.

The second step is characterized by the development of the sulphides. They appear in the following sequence: pyrrhotite, pyrite, chalcopyrite. They replace either the magnetite or the calcite. Scheelite is still stable, but textural evidence indicates that its growth was complete before the development of pyrite and chalcopyrite. At this step, a colourless serpentine (serpentine 1) is formed. It generally produces an intense alteration, but the beginning of its formation is variable among different samples. This serpentine develops at the expense of the forsterite, calcite and also the chlorite. Calcite or forsterite may be completely transformed by this serpentine.

The third step includes the formation of either massive talc (talc 2) associated with minor quartz or massive serpentine (serpentine 2). This serpentine is distinguished from the first one by its yellow colour. Textural evidence indicates that the talc and this yellow serpentine are of later formation than the sulphides. They were formed at the expense of the relicts of forsterite and calcite.

The alteration of the two other zones (i.e. the pyroxenite and the garnet-pyroxene zones) is not so intensive as the alteration of the calcite-forsterite zone. The pyroxene is altered into amphibole + calcite + quartz or phlogopite. Amphibole and phlogopite may occur simultaneously. Sometimes, chlorite develops at the expense of amphibole and phlogopite. Minor amounts of scheelite and sulphides are also found as an alteration product of the pyroxene zone. The garnet is replaced by the association epidote ± calcite ± quartz, but scheelite has never been reported in the internal zone (Zuchetti 1966).

5.2 Costabonne

Two generations of scheelite can be clearly distinguished (Guitard and Laffitte 1958; Guy 1980). The first one precipitated together with garnet G_2 and in minor amounts with garnet G_1 and salite, and can thus be considered as an early, high T-phase. The WO_3 content in the garnet zone lies between 0.1 and 0.35% (Deremetz and Guitard 1957). During the alteration stage, epidote, calcite and quartz locally replaced the garnet; scheelite is then dissolved.

The second generation appeared during the hydrothermal alteration stage of the salite zone together with amphiboles (tremolite-pargasite), calcite, quartz and very minor amounts of sulphides (pyrite–sphalerite) (Fig. 12). The WO_3 content may locally reach more than 1%. Late, intensive silicification proceeded along fractures together with calcitization of the silicates.

In the calcite–forsterite zone, the olivine is extensively transformed to serpentine and talc, but no scheelite is found.

6 Fluid Composition and Scheelite Precipitation

As demonstrated above, in both deposits, the minable scheelite concentrations are closely related to the secondary hydrothermal alteration stage. But, although the early skarn zonings developed at Costabonne and Traversella are similar, the scheelite parageneses are quite different. This is due to the result of dissimilar ore-forming conditions.

	PARAGENESES SUCCESSION	EARLY STAGE	HYDROXYLATION STAGE
OLIVINE-CALCITE ZONE	OLIVINE	——————	
	CALCITE	——————	——
	SPINEL	——————	
	SERPENTINE		——————
	TALC		——————
PYROXENE ZONE	DIOPSIDE	——————	
	SALITE	——————	
	AMPHIBOLE		——————
	CALCITE		——————
	SCHEELITE		——————
	QUARTZ		——————
GARNET ZONE	GARNET	——————	
	EPIDOTE		——————
	CALCITE		——————
	QUARTZ		——————
	SCHEELITE	———	

Fig. 12. Scheelite and alteration mineral parageneses at Costabonne

At Traversella, the main stage of scheelite crystallization (step 2) is thought to have been taken place at 350°–400 °C and 1.0 to 1.5 kb (Vander Auwera, in prep.). The scheelite precipitation at Costabonne probably took place in the same temperature range under a pressure of 1.5 to 1.7 kb (Guy 1980; Varenne 1983). Because the hydroxylation reactions, which are involved here, are not sensitive to such a small pressure variation, pressure and temperature are not likely to be the determining factors for the difference in scheelite distribution. These determining factors are probably a_{SiO_2}, a_{H_2O}, a_{WO_3} which are perfectly mobile constituents (Korzhinskii 1959), but probably also $a_{Ca^{2+}}$, because it has a principal influence on the precipitation of scheelite, although it is an inert component.

The presence of serpentine (mostly antigorite in both deposits) implies a very low X_{CO_2} (Trommsdorff and Evans 1977). Figure 13A and B shows that under the conditions 450 °C, 1 kb and $X_{CO_2} < 0.05$, diopside is no longer stable and is replaced by tremolite, calcite and quartz. But if X_{CO_2} tends to O, diopside may be stable in the same range of activity of Si (a_{Si}) and Ca (a_{Ca}) as antigorite and tremolite. The small alteration degree of the diopside at Traversella may thus suggest that the X_{CO_2} was lower than at Costabonne. The associations tremolite + calcite + quartz + scheelite and serpentine (antigorite) + scheelite appear under different activity conditions (Fig. 13A, B, C): a_{Ca} and a_{Si} were higher and a_{Mg} lower by approximately a

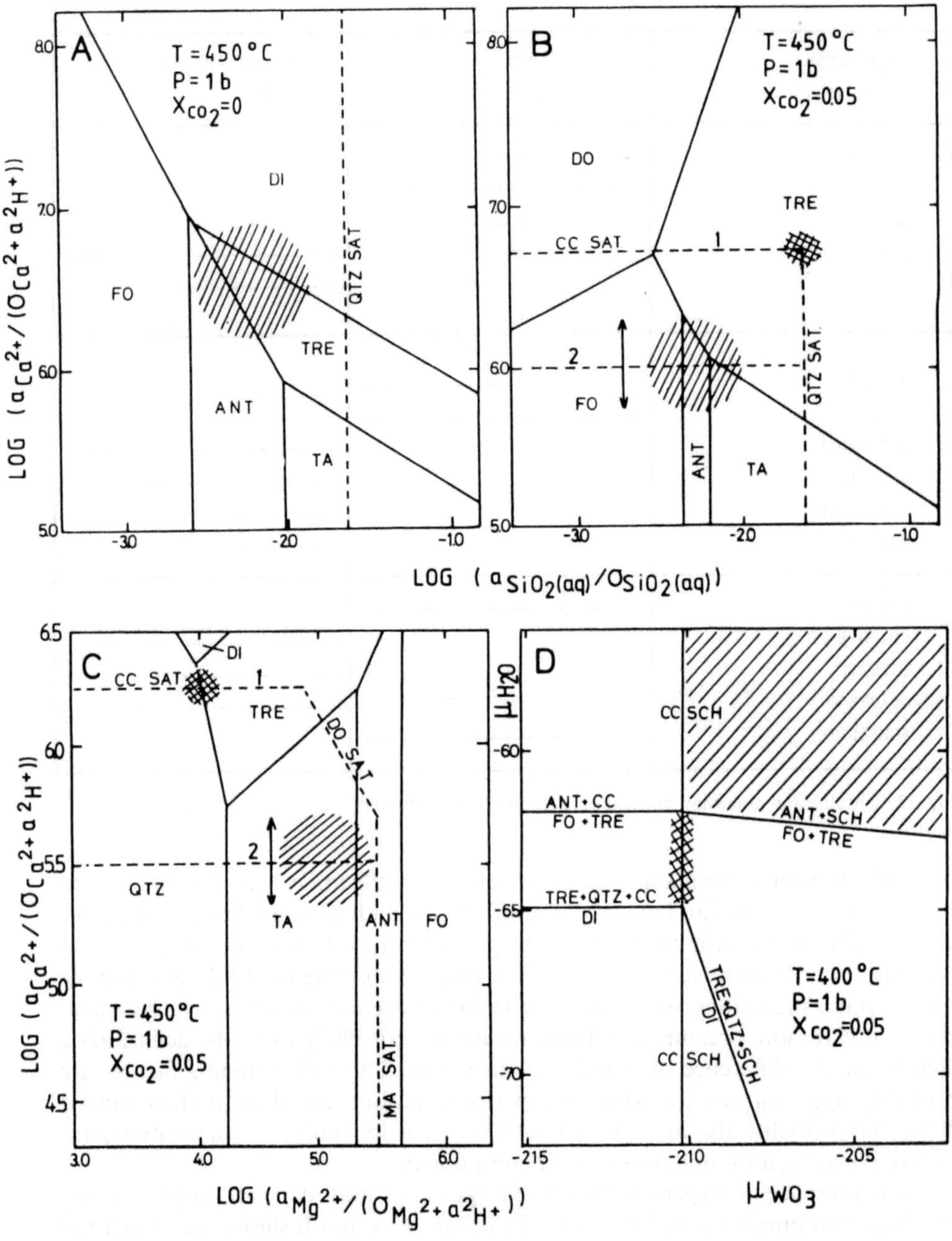

Fig. 13A–D. Phase relations in the system CaO–MgO–SiO_2–CO_2–HCl at 1 kb ($P_f = PH_2O + PCO_2$). **A** Si–Ca activity diagram at 450 °C and $X_{CO_2} = 0$; **B** Si–Ca activity diagram at 450 °C and $X_{CO_2} = 0.05$; **C** Mg–Ca activity diagram at 450 °C and $X_{CO_2} = 0.05$; **D** WO_3–H_2O chemical potential diagram at 400 °C and $X_{CO_2} = 0.05$ (HCl absent). *Dashed lines* represent the quartz, calcite or magnesite saturation levels in the fluid. Scheelite saturation levels at Costabonne (*1*) and Traversella (*2*). The *hachured* and *cross-hachured* areas correspond to the estimated genetic conditions for Traversella and Costabonne deposits respectively. *Di* Diopside; *Fo* forsterite; *Ant* antigorite; *Tre* tremolite; *Ta* talc; *Do* dolomite; *Cc* calcite; *Ma* magnesite; *Qtz* quartz; *Sch* scheelite

factor of 10 at Costabonne compared to Traversella. Due to this lower a_{Ca} at Traversella, the presence of scheelite in indicative of higher a_{WO_3} (Fig. 13B). Furthermore, Fig. 13D indicates that both a_{H_2O} and a_{WO_3} were probably higher at Traversella. These differences in fluid compositions may, in turn, be linked to the differences in the degree of evolution of associated intrusive rocks.

7 Conclusions

1. Both Costabonne and Traversella skarn deposits present a set of intrusive rocks of calc-alkaline affinity, linked by a process of fractional crystallization. Igneous compositions have, however, different degrees of evolution. At Costabonne, the intrusive rocks are monzogranites, whereas at Traversella they are monzonites–diorites (except for small granite dykes).

2. Early skarn development is similar in both deposits, in that the columns present a similar succession of metasomatic zones. At Costabonne, the silicate phases are richer in Fe and Mn. Introduction of large quantities of Fe (magnetite) at Traversella occurred during the later hydrothermal alteration stage. At Costabonne, sulphides are rare, whereas they are very abundant at Traversella.

3. Scheelite is present in minor amounts in association with the early formed silicates at Costabonne. The bulk of the scheelite appeared at a later stage during the alteration of the pyroxene zone. At Traversella, precipitation of scheelite also took place at a later stage as an alteration product of calcite in the calcite–forsterite zone.

These two examples show that there is not a unique location of scheelite in magnesian skarns. Given the identity of the replaced rocks (dolomitic marbles), of the early metasomatic columns and of the P, T conditions, differences of a_{Ca}, a_{Si}, a_{Mg}, a_{WO_3} and X_{CO_2} may account for the uneven distribution of scheelite.

Acknowledgements. Microprobe analyses have been carried out by J. Wautier (CAMST, UCL–FNRS). The authors are indebted to G. Meulemans, R. Paques and M.N. Hoet for technical assistance and to J. Naud for analytical support. G. Martinotti and M. Zerbato (GEOMINERARIA ITALIANA srl) are gratefully acknowledged for geological advice and for providing unpublished data. This work was financially supported by EEC (*Contract No. MSM-127-B*), SPPS (*Contract No. MP/CE/13*) and FNRS grants.

References

Autran A, Fonteilles M, Guitard G (1970) Relations entre les intrusions de granitoïdes, l'anatexie et le métamorphisme régional, considérées principalement du point de vue du rôle de l'eau: cas de la chaîne hercynienne des Pyrenées-Orientales. Bull Soc Geol Fr 7, 12, (4):673–731

Auwera J Vander (1985) Iron and tungsten skarns at Traversella (Italian Alps). Fortschr Mine Bd 63 (1): 244

Besson M, Fonteilles M (1974) Relations entre les comportements contrastés de l'alumine et du fer dans la differentiation des séries tholéitiques et calco-alcaline. Bull Soc Fr Miner Crist 97:443–449

Cocherie A (1984) Interaction manteau-croûte: son rôle dans la genèse d'associations plutoniques calco-alcalines, contraintes géochimiques (éléments en traces et isotopes du strontium et de l'oxygène). Documents BRGM 90

Compagnoni R, Dal Piaz GV, Hunziker JC, Gasso G, Lombardo B, Williams PF (1977) The Sesia-Lanzo zone, a slice of continental crust with alpine high pressure-low temperature assemblages in the western Italian Alps. Rend Soc It Mine Pet 33:281–334

Dubru M (1986) Pétrologie et géochimie du marbre à brucite et des borates associés au gisement de tungstène du Pic de Costabonne (P.O. France). Thèse, Université de Louvain, Louvain-la-Neuve

Einaudi MT, Burt DM (1982) Introduction, terminology, classification and composition of skarn deposits. Econ Geol 77:745–754

Guitard G (1970) Le métamorphisme hercynien mésozonal et les gneiss oeillés du massif du Canigou (Pyrenées Orientales). Mém BRGM 63

Guitard G, Laffitte P (1958) Les calcaires métamorphiques et les skarns du pic de Costabonne (Pyrénées Orientales). Sci Terre 6:59–131

Guy B (1980) Etude géologique et pétrologique du gisement de Costabonne. In: Minéralisations liées aux granitoïdes. II. La genèse des skarns à tungstène dans les Pyrenées. Mém BRGM 99

Guyader R Le (1982) Elements-traces dans les skarns à scheelite et les roches associées à Costabonne (Pyrenées orientales, France). Thèse, Université de Paris VI, Paris

Hunziker JC (1974) Rb–Sr and K–Ar age determination and the alpine tectonic history of western Alps. Mém Inst Geol Mine Univ Padoue 31:1–55

Kennedy WQ (1931) The igneous rocks, pyrometasomatism and ore deposition at Traversella, Piedmont, Italy. Schweiz Mine Pet Mitt 11:77–138

Korzhinskii DS (1959) Physicochemical basis of the analysis of mineral paragenesis. Consultants bureau, New York, 142 pp

Korzhinskii DS (1970) The theory of metasomatic zoning. Clarendon, Oxford

Marcke de Lummen G van, Verkaeren J (1986) Physico-chemical study of skarn formation in pelitic rocks, Costabonne peak area, eastern Pyrenees, France. Contrib Mine Pet 93:77–88

Müller F (1912) Die Erzlagerstätten von Traversella im Piedmont, Italien. Prakt Geol Bd 20:209–240

Nockolds SR (1947) The relation between chemical composition and paragenesis in the biotite micas of igneous rocks. Am J Sci 245:401–420

Pitcher WS (1983) Granite type and tectonic environment. In: Hsü KJ (ed) Mountain building processes. Academic Press, London, pp 19–40

Treuil M, Joron JL (1975) Utilisation des éléments hygromagmatophiles pour la simplification de la modelisation quantitative des processes magmatiques. Exemples de l'Afar et de la dorsale océanique. Rend Soc It Mine Pet 31:125–174

Trommsdorff V, Evans BW (1977) Antigorite–ophicarbonates: contact metamorphism in Valmolenco, Italy. Contrib Mine Pet 62:301–312

Varenne JL (1983) Etude des inclusions fluides dans la scheelite de l'indice tungstifère de Costabonne. Travail d'option. ENSM Saint-Etienne, 85 pp

Venturelli G, Thorpe RS, Dal Piaz GV, Moro A del, Potts PJ (1984) Petrogenesis of calc-alkaline, shoshonitic and associated ultrapotassic oligocene volcanic rocks from the northwestern Alps, Italy. Contrib Mine Pet 86:209–220

Zharikov VA (1970) Skarns, Part 1. Int Geol Rev 12:541–559

Zuchetti S (1966) Studi sul giacimento di Traversella (Torino). I corpi mineralizzati a scheelite. Symposium internationale sui giacimenti minerari delle alpi. Trente 3:929–960

Assessment of Mineralogical Influences on the Element Mobility in the W–Sn Enriched Granite of Regoufe and Its Derivatives (Portugal) by Means of XRF Analysis of Unpolished Rock Sections

P.F.M. Van Gaans, S.P. Vriend, R.P.E. Poorter, and J.B.H. Jansen[1]

Abstract

The relation of rock chemistry, mineralogy and geochemical processes was studied in the hydrothermally altered, W–Sn specialized granite of Regoufe, northern Portugal. To this end unpolished rock sections sawn from small drill cores were directly analyzed by X-ray fluorescence spectrometry, which is the basic aspect of an approach called Integral Rock Analysis (IRA). Chemical variation within the granite and its derivatives is mainly due to pervasive (auto)metasomatic activity. The effect of chemical weathering on the rock chemistry of this denuded granite is negligible. With the aid of factor analysis the imprints of albitization, muscovitization, apatitization and mineralization are traced throughout the granite and associated dyke system. Alteration generally increases from west to east. In the NE area the separate effect of disseminated wolframite mineralization, apart from the common W–Sn quartz-vein association, and late sericitization are recognized. The importance of mineralogy or major element chemistry in the response to the hydrothermal processes is typically evidenced by the element associations of Sr and of Ti and Zr. Sr preferentially substitutes for K in feldspar and mica in the western region, whereas it mainly substitutes for Ca in phosphates in the most altered eastern zone. As expected, Ti and Zr are closely related to biotite in the western part and appear to remain concentrated in the biotite alteration products. Depletion of Ti and Zr by leaching of refractory minerals is linked to Na_2O, P and metal enrichment through albitization, apatitization and mineralization. The IRA approach offers a rapid method for the acquisition of large quantities of detailed rock geochemical data. An additional advantage is that data are indicative of mineralogy. Selected rock sections were investigated microscopically and by electron-probe microanalysis, resulting, among others, in the discovery of the trace minerals columbotantalite and scorodite in the granite.

[1] Department of Geochemistry and Experimental Petrology, Institute of Earth Sciences, University of Utrecht, P.O. Box 20081, 3508 TA Utrecht, The Netherlands

Mineral Deposits within the European Community
(ed. by J. Boissonnas and P. Omenetto)

1 Introduction

Not all W–Sn enriched granites have affiliated ore deposits. Understanding of the crystal-chemical parameters that lead to dispersion and concentration of elements may improve the discrimination between barren and productive granitoids (Stemprok 1979). The main rock-forming processes are reflected by rock chemistry. The composition of a sample of a particular granite may reflect the composition of the primary magma, metasomatic and post-magmatic processes, metamorphism, and in the case of surface samples, the effects of weathering (Govett and Nichol 1979). The impact of processes, however, is also determined by the mineralogical nature of the host rock itself. Especially trace element behaviour strongly depends upon the interaction of magma or fluid with existing host minerals (Mellinger 1984). The integration or rock chemistry, mineralogy and mineral chemistry is therefore a prerequisite to a full understanding of the history of the rock. Results of conventional chemical analysis and electron-probe microanalysis are often difficult to interrelate, owing to differences in scale and nature of the object studied. Integral Rock Analysis (IRA) is an attempt to fill this gap. This approach was applied to the granite of Regoufe, Portugal, an example of a 'specialized' granite (Tischendorf 1977), to gain insight in the interaction of mineralogy with metasomatic processes.

2 Geology

The granite of Regoufe, located about 200 km north of Lisbon, is a hydrothermally altered, W–Sn specialized granite of Hercynian age (280 ± 9 Ma, Pinto 1985). The geology of the region was studied by Sluijk (1963). The granite, with a surface expression of about 6 km^2, discordantly intruded the metasedimentary Beira Schist Formation. The contact is steep in the east and dips away at a low angle in the west. A petrological sketch map is given in Fig. 1. Vriend et al. (1985) distinguished two major rock types on the basis of mineralogical variation. A porphyritic two-mica (P2M) granite with tourmaline grades towards the east into a muscovite albite (MA) granite rich in arsenopyrite, with virtually no K-feldspar megacrysts, biotite or tourmaline (Fig. 1). The eastern part is the most altered. Granitic and aplitic rocks are exposed in rows of small outcrops W and SW of the granite and form a few ring-shaped dykes. Several tungsten-bearing quartz veins in and around the granite have been mined in the past. Post-magmatic processes, including mineralization, are related to various trace element trends within the granite (Vriend et al. 1985; Voncken et al. 1986). Schist roof pendants occur at the higher altitudes. Study of the chemical variation and the interaction of rock type and hydrothermal fluids at the direct contact of a schist inclusion (van Gaans et al. 1986a) revealed enrichment of the inclusion in W, Ta, Nb, Sn, Rb, Cs and K and depletion in P, Ca, Sr, Na, Ti and Zr, relative to normal schist concentrations, which is in accordance with the mineralization trend within

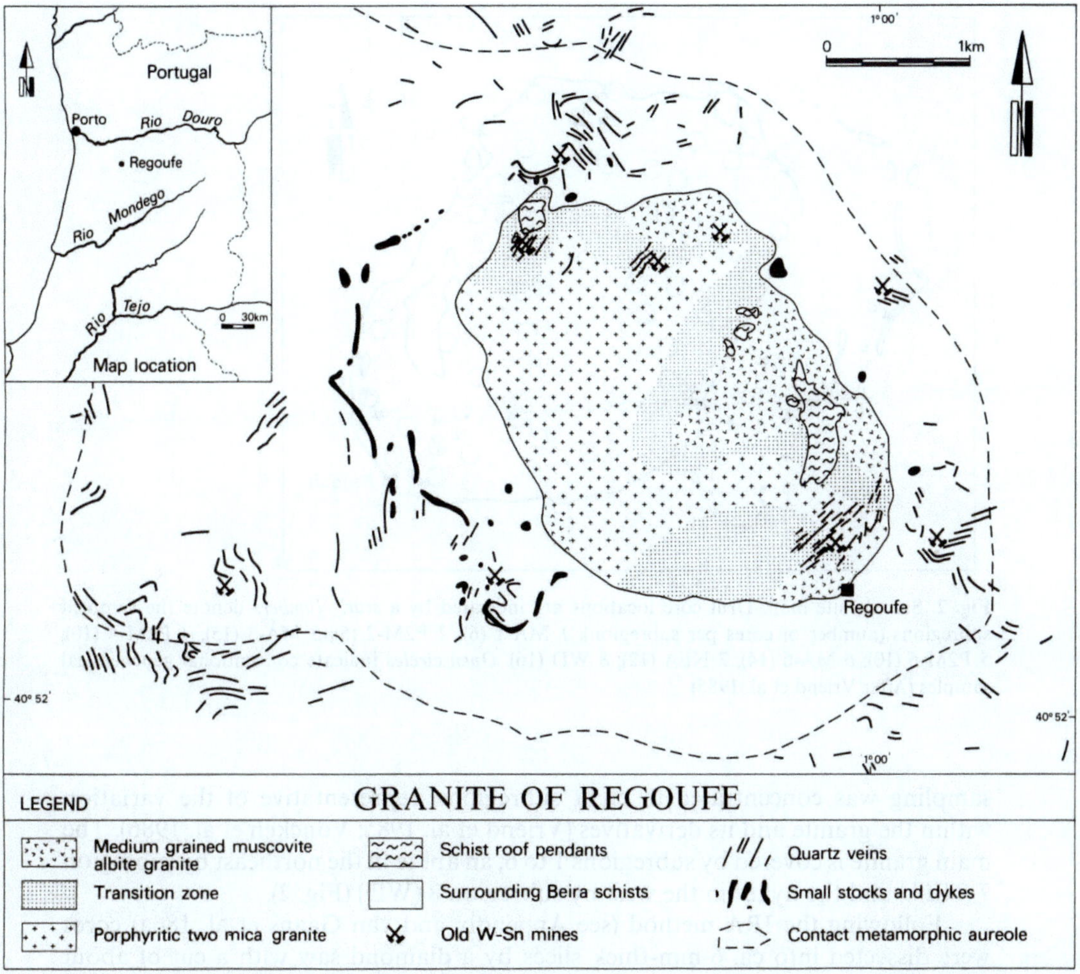

Fig. 1. Petrological sketch map of the Regoufe granite (After Vriend et al. 1985)

the granite itself (Vriend et al. 1985). The schist inclusion appeared to be a favourable deposition site for the ore-related elements. The effect of albitization and greisenization in the schist decreased steeply with increasing distance from the granite contact.

3 Sampling, Analytical and Statistical Techniques

Some 90 small drill cores, with a diameter of 24 mm and a length varying between 5 and 20 cm, were collected in the Regoufe granite and its associated dyke system. Detailed study in a number of subregions of the granite was thought to result in a better understanding of the rock-forming processes (Voncken et al. 1986). Therefore,

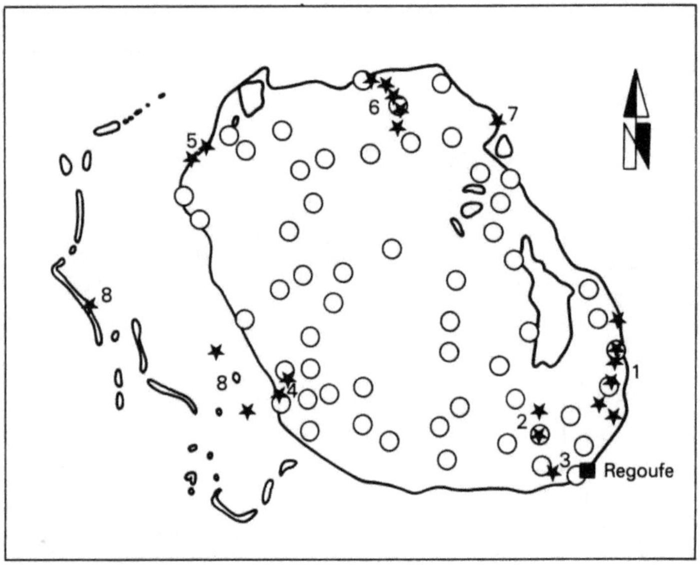

Fig. 2. Sample site map. Drill core locations are indicated by a *star*. *Numbers* denote the sampled subregions (number of cores per subregion); *1* MA-1 (6); *2* P2M-2 (5); *3* MA-3 (15); *4* P2M-4 (10); *5* P2M-5 (10); *6* MA-6 (14); *7* NEA (12); *8* WD (16). *Open circles* indicate conventional geochemical samples (After Vriend et al. 1985)

sampling was concentrated in eight subregions, representative of the variation within the granite and its derivatives (Vriend et al. 1985; Voncken et al. 1986). The main granite is covered by subregions 1 to 6, an aplite in the northeast by subregion 7 (NEA) and the dykes in the west by subregion 8 (WD) (Fig. 2).

Following the IRA method (see Appendix and van Gaans et al. 186a) cores were dissected into ca. 6-mm-thick slices by a diamond saw with a cut of about 3 mm. The terms core, slice and section are illustrated in Fig. 3. Ten to 15 sections per core provided sufficient data for the study of the granite on the various spatial scales. Calculations showed that 10 to 15 sections are sufficient to adequately estimate (trace) element contents, including those occurring in small discrete particles, and to give a reasonable chance of detecting trace minerals (Grassia 1986).

Some 1000 rock sections were analyzed by XRF for SiO_2, Na_2O, K_2O, CaO, Cs, Sn, Ti, P, Ta, Nb, W, Rb, Sr and Zr using a Philips PW 1400 with automatic sample changer. The exposed part of the sections was 22 mm. Pressed powder tablets of artificial and international natural granite standards were included for calibration. As the focus of this study is inter-element correlations and not absolute tenors, for reasons of expediency, no further matrix corrections were applied.

Results of 55 conventionally analyzed samples of Regoufe granite were used for comparative purposes (Vriend et al. 1985; Vriend unpubl. data). The samples are composites of ten chips collected over an area of 500 m^2. Major elements were analyzed wet chemically (Shapiro 1967) and XRF analyses of pressed powder briquettes were made for trace elements. IRA concentration levels for most elements

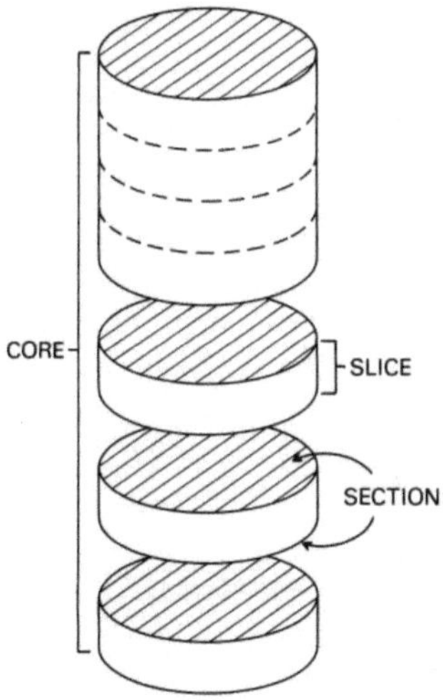

Fig. 3. Illustration of the terms core, slice and section

are, considering the spatially different sample coverage (Fig. 2), in agreement with the conventionally obtained results (Table 1). Significant differences in the concentration level between IRA and conventional results are due to uncorrected mineralogical matrix effects (de Jongh 1970; van Gaans et al. 1986b) which do not greatly influence correlation coefficients (van Gaans et al. 1986a). For all elements the displayed chemical variation among subregions within the granite is in accordance with the distribution maps of conventional results (Vriend et al. 1985, 'Vriend, unpubl. data). Thus, the IRA data set has an internally consistent, relative basis.

Data were interpreted with the aid of SPSS and BMDP statistical software (Nie et al. 1975; Hull and Nie 1981; Dixon 1981). The element association patterns were studied by means of component analysis, a type of factor analysis (Le Maitre 1982; Jöreskog et al. 1976). The number of factors was chosen with application of the Gutman criterion, i.e. only unrotated factors with eigenvalues greater than 1.0 are retained. Pearson correlations, used as a measure of association, are adversely affected by the presence of outliers and by skewness of the frequency distributions. Histograms showed that 3.7% of the analyzed sections significantly deviated from the main population for one or more elements. These outliers were removed from the data set. A selection of rock sections with outlying or extreme results was polished for electron-probe microanalysis (EPMA). For the main population a transformation of the general form $x' = \ln(x\text{-alpha})$, with alpha adjustment to obtain minimum skewness for the distributions (Miesch 1981; Selinus 1983), was applied to all variables.

Table 1. Results of Integral Rock Analysis (IRA); averages for the various subregions of the granite and its derivatives.[a]

Subregion	Western dikes (WD) 8	Porphyritic two-mica granite (P2M) 2	4	5	mean P2M	Miscovite albite granite (MA) 1	3	6	mean MA	NE Aplite (NEA) 7	Mean main granite 1–6	Vriend et al. (1985)
W	13	15	13	14	14	16	15	15	15	23	15	16
Sn	54	56	39	55	47	66	58	54	58	85	54	54
Nb	27	18	19	20	19	31	33	33	33	41	29	37
Ta	32	28	24	23	25	32	38	35	36	37	33	16
Ti	170	309	322	361	331	153	90	102	108	82	187	354
Zr	24	31	30	36	32	20	14	17	16	15	21	35
Cs	46	55	29	31	34	48	32	28	34	57	34	49
Rb	635	662	532	513	557	727	739	661	708	963	663	683
K_2O	4.09	4.90	4.32	4.39	4.44	3.99	3.91	4.00	3.96	3.74	4.13	4.09
SiO_2	76.8	74.2	74.3	74.7	74.4	73.4	75.1	74.2	74.4	77.0	74.4	73.1
Sr	46	38	38	41	39	35	50	43	44	64	42	36
CaO	0.27	0.11	0.20	0.22	0.19	0.24	0.40	0.25	0.32	0.24	0.27	0.27
P	1649	1386	1301	1246	1301	2062	2305	2036	2163	2164	1855	1958
Na_2O	4.39	2.90	3.46	3.27	3.31	3.68	4.42	4.03	4.13	2.60	3.83	3.53
N	247	47	132	68	247	100	200	175	475	144	722	55

[a] The selection of elements measured reflects the major rock-forming processes. IRA data are on an internally consistent, relative basis. The conventional results for the main granite (Vriend et al. 1985; Vriend unpubl. data; see also Fig. 2) are added for comparison. Oxides are in wt%, elements in ppm. N is the number of rock sections or samples.

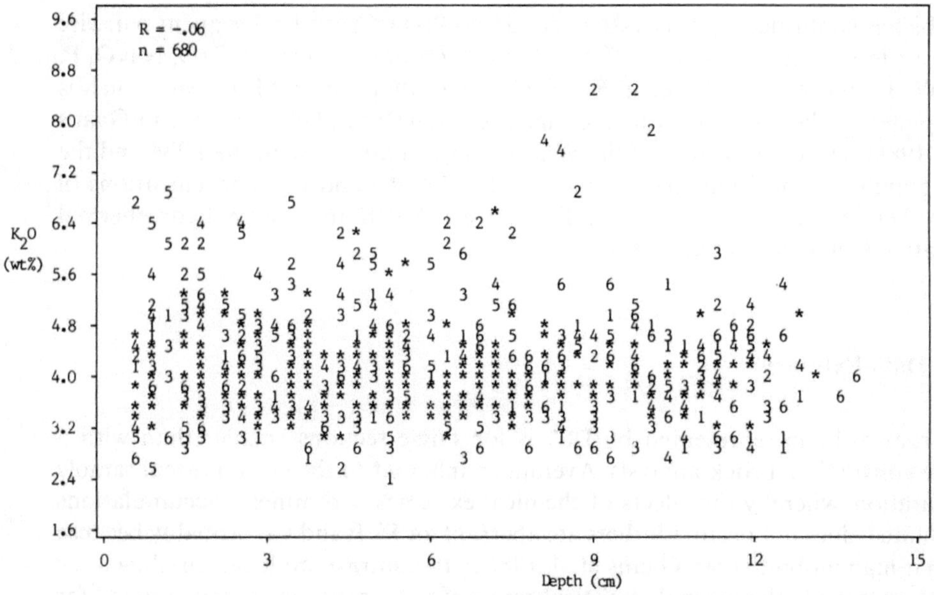

Fig. 4. Variation of K_2O with depth as an inverse measure of weathering for the main granite. The *numbers* in the plot indicate subregions, an *asterisk* is used for coinciding points

4 Lithogeochemistry

4.1 Weathering

The Regoufe granite has a slightly weathered appearance, while generally a soil cover is lacking. Of the major oxides, K_2O is most sensitive to the weathering of granite (Chessworth 1979). No trend of K_2O content with depth is evident for the Regoufe granite (Fig. 4). Also, none of the other analyzed elements exhibit an obvious increase or decrease in concentration with depth. Clearly, erosion proceeds more rapidly than chemical weathering in the Regoufe environment. In the present study on (trace) element behaviour the influence of weathering on chemistry need not be considered and the collected cores can be used integrally. An additional conclusion is that conventional samples taken at the immediate surface also have no significant overprint of chemical weathering.

4.2 Spatial Element Distributions

The granite subregions 2, 4 and 5, located in the P2M granite and in the transition zone, are chemically distinct from subregions 1, 3 and 6 in the MA granite (Table

1), which is confirmed by Analysis of Variance (ANOVA) and subsequent pairwise contrast tests (van Gaans et al. 1985). The MA granite is higher in CaO, Na_2O, P, Rb, W, Ta and Nb and lower in K_2O, Ti and Zr than the P2M granite, which is interpreted as the result of hydrothermal alteration (Vriend et al. 1985; van Gaans et al. 1985). The composition of the WD is intermediate between the P2M and the MA granite compositions. High tenors of Sn, Cs, Rb and low concentrations of K_2O, Na_2O, Ti and Zr in the NEA are attributed to intense hydrothermal alteration (van Gaans et al. 1985).

4.3 Data Extremes

The rock volume represented by IRA is for single sections smaller than with a single conventional rock analysis. Averaging, inherent to the conventional sample preparation, whereby the effects of chemical extremes and mineral accumulations are diluted, does not occur. Outliers are aberrant for Si, P and Cs, probably because of their high mobility (van Gaans et al. 1986a). In contrast, no local enrichment or depletion outside the normal statistical range of concentrations is encountered for Sn and Zr, suggesting no secondary entrapment within the granite. The dominating mineralogical/structural factor causing an extreme can be determined, and interesting phenomena may be discovered in rock sections falling in the tails of the frequency distributions. Trace minerals that were unknown in the granitic rocks of Regoufe were discovered by EPMA of selected sections. Columbo-tantalite grains $[(Nb_{1.38}Ta_{0.57}Ti_{0.07}Fe_{0.55}Mn_{0.44})O_6]$ were identified in a section of MA-6, which was highest in Ta and Nb (Plate 1). Scorodite $[(Fe,Al)(As,P,Bi)O_4 \cdot 2H_2O]$ was detected in the section highest in P_2O_5 (5.3%) of the NEA. A phosphorus outlier of 3.0% P_2O_5 from the eastern margin of the Regoufe granite (MA-1) (versus a maximum of 0.71% in conventional samples) was microscopically identified as an apatite bearing quartz veinlet, thus clearly showing the mobility of P.

4.4 Element Association Patterns Within the Main Granite

R-mode Component Analysis was applied to the analytical results of single sections, to the means per core and for comparison also to the conventionally analyzed samples of the main granite, to study the interrelation of the chemical elements. The Varimax-rotated factor-loading matrices for sections (F_{sect}), cores (F_{core}) and conventional bulk samples (F_{bulk}) are listed in Table 2. Ta was excluded from the factor model for sections because it showed no significant correlations with the other elements, due to the large relative random error for analytical values near the detection limit.

The mineralogical influence on the element association patterns is emphasized by the extracted factors for rock sections. CaO, Sr and P form an apatite factor

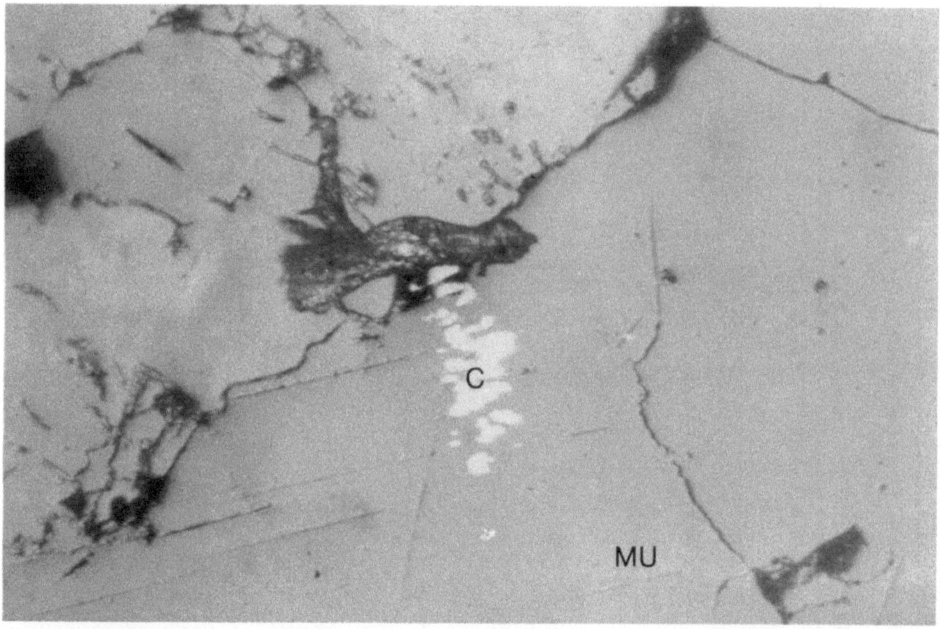

Plate 1. Example of columbo-tantalite (*C*) embedded in muscovite (*MU*). Incident light, width of photograph ca. 0.2 mm

$F3_{sect}$. Apatite occurs as primary magmatic inclusions and as a late hydrothermal mineral (Sluijk 1963; this study). This factor is therefore an expression of apatitization. The combination of CaO with Sr is present for all three types of samples in $F3_{sect}$, $F3_{core}$ and $F3_{bulk}$. K_2O, Rb and negative SiO_2 in $F4_{sect}$ may reflect muscovitization or the closure effect (Le Maitre 1982) between K-feldspar (muscovite, biotite) and quartz. The positive association of Rb with K for sections contrasts with their inverse association in $F1_{core}$ and $F1_{bulk}$, which reflect the observed increase of Rb and decrease of K with alteration (Table 1). Evidently, the small-scale mineralogical association dominates for rock sections, whereas the larger scale alteration trend dominates for means per core and bulk samples. For similar reasons, the negative association of Ti and Zr with Nb, Ta and Rb in $F1_{core}$ and $F1_{bulk}$ is obliterated for sections where the chemical or spatial relation of all these elements to biotite takes its effects. $F1_{sect}$ has high positive loadings for Ti and Zr and negative loadings for Na_2O and also P; it links the leaching of refractory minerals to processes like albitization and apatitization. Its mirror image forms part of $F1_{core}$ and the almost identical $F1_{bulk}$, which have a more pronounced apatite component. The loadings of Ta, Nb and Rb, and for bulk samples also Sn, on $F1_{core}$ and $F1_{bulk}$ relate leaching of refractories to mineralization. $F2_{sect}$ with high positive loadings for Nb, W, Cs, Sn and Rb obviously describes a mineralization process. $F2_{core}$ is similar to $F2_{sect}$ with only a moderate Nb contribution. $F2_{bulk}$ is a simple W–Sn mineralization factor; here, the relation of mineralization to the other processes is expressed more explicitly in $F1_{bulk}$.

Table 2. Varimax-rotated factor-loading matrices for the Regoufe main granite (only loadings of over 0.4 are given)

A. IRA data of sections					B. IRA mean data per core				C. Conventional data (after Vriend et al., 1985)				
	$F1_{sect}$	$F2_{sect}$	$F3_{sect}$	$F4_{sect}$		$F1_{core}$	$F2_{core}$	$F3_{core}$		$F1_{bulk}$	$F2_{bulk}$	$F3_{bulk}$	$F4_{bulk}$
W		0.67			W		0.70		W		0.91		
Sn		0.88			Sn		0.85		Sn	0.63	0.51		
Nb		0.79			Nb	0.69	0.55		Nb	0.87			
					Ta	0.69			Ta	0.82			
Ti	0.86				Ti	−0.92			Ti	−0.94			
Zr	0.91				Zr	−0.92			Zr	−0.93			
Cs		0.73			Cs		0.82		Cs				0.83
Rb		0.67		0.61	Rb	0.47	0.71		Rb	0.86			
K_2O				0.88	K_2O	−0.70			K_2O	−0.63			0.48
SiO_2				−0.75	SiO_2			0.75	SiO_2				−0.58
Sr			0.76		Sr			0.80	Sr			0.91	
CaO			0.82		CaO	0.64		0.47	CaO			0.78	
P	−0.44		0.73		P	0.82			P	0.88			
Na_2O	−0.63				Na_2O	0.84			Na_2O	0.71	−0.43		

Table 3. Varimax-rotated factor-loading matrices for the sampled subregions (IRA data of sections; only loadings of over 0.4 are given)[a]

WD

	F_{min}	F_{biot}	F_{ap}	F_{mu}
W	0.74			
Sn	0.72			
Nb	0.67			
Ti		0.86		
Zr		0.91		
Cs		0.68		0.48
Rb				0.88
K_2O				0.90
SiO_2				−0.72
Sr			0.66	
CaO			0.85	
P			0.89	
Na_2O			0.70	

P2M-2

	F_{min}	F_{ap}	F_{mu}
W	0.61		
Sn	0.96		
Nb			0.95
Ti	0.91		
Zr	0.52		
Cs		0.84	
Rb			0.88
K_2O			0.92
SiO_2			−0.84
Sr			0.87
CaO		0.81	
P		0.88	
Na_2O		0.87	

P2M-4

	F_{min}	$F_{biot-alb}$	F_{ap}	F_{mu}
W	0.72			
Sn	0.90			
Nb	0.92			
Ti	0.63	0.57		
Zr		0.86		
Cs	0.66	0.50		
Rb	0.46			0.82
K_2O				0.90
SiO_2		0.48		−0.56
Sr				0.68
CaO			0.94	
P			0.93	
Na_2O		−0.43	0.63	

P2M-5

	F_{min}	F_{biot}	F_{ap}	F_{mu}
W	0.71			
Sn	0.81			
Nb	0.78			
Ti	0.64	0.69		
Zr		0.90		
Cs	0.62	0.56		
Rb				0.87
K_2O				0.91
SiO_2				−0.81
Sr	−0.53			0.66
CaO			0.92	
P			0.88	
Na_2O			0.87	

MA-1

	F_{min-mu}	F_{biot}	F_{ap}	F_{alb}
W	0.71			
Sn	0.84			
Nb	0.77			
Ti	0.42	0.85		
Zr		0.90		
Cs	0.58	0.66		
Rb	0.94			
K_2O	0.74			
SiO_2	−0.64			0.49
Sr				0.59
CaO			0.92	
P			0.91	
Na_2O				−0.86

MA-3

	F_{min}	F_{ap}	F_{mu-alb}
W	0.59		
Sn	0.85		
Nb	0.78		
Ti	0.89		
Zr	0.56		
Cs	0.72		0.49
Rb	0.54		0.74
K_2O			0.85
SiO_2			−0.86
Sr		0.80	
CaO		0.93	
P		0.92	
Na_2O	−0.55		0.44

MA-6

	F_{min1}	F_{ap}	F_{ser}	F_{min2}	F_{alb}
W	0.49			0.64	
Sn	0.90				
Nb				0.87	
Ti	0.86				
Zr	0.58		−0.53		
Cs	0.65		0.67		
Rb			0.94		
K_2O			0.89		
SiO_2					−0.69
Sr		0.60			
CaO		0.96			
P		0.93			
Na_2O					0.74

NEA

	$F_{min1-alb}$	F_{ap}	F_{ser}	F_{min2}	F_{alb}
W				0.75	
Sn	0.79			0.50	
Nb				0.75	
Ti	0.92				
Zr	0.50		−0.64		
Cs			0.69		
Rb			0.97		
K_2O			0.90		
SiO_2					−0.88
Sr		0.64	0.42		
CaO		0.86			
P		0.91			
Na_2O	−0.52				0.69

[a] The tentative geochemical interpretation of the factors is indicated by the suffixes: *min* mineralization; *min1* Sn(,W) quartz-vein type mineralization; *min2* disseminated W mineralization; *biot* biotite; *ap* apatite/apatitization; *mu* muscovitization; *alb* albitization; *ser* late sericitization.

4.5 Element Association Patterns Within the Subregions

Component Analysis was applied to rock section data for the subregions (Table 3). A sufficient number of analyses (Howarth and Sinding-Larsen 1983) per subregion is available (Table 1).

A mineralization factor F_{min} is present for all subregions, commonly composed of Sn, W, Nb, Cs, Rb and mostly Ti. For the WD this association is more or less divided over a simple mineralization factor and the biotite factor. For MA-6 and NEA an Nb–W factor F_{min2}, with only a minor Sn contribution in NEA, is found separate from the major Sn-related factor F_{min1}. The two factors probably express different mineralization types. In the NE part small disseminated wolframite mineralizations occur within the granite (this study) which is in contrast to the common quartz-vein, W–Sn type mineralization.

Zr and Ti, which are depleted in the Regoufe granite (Vriend et al. 1985), invariably combine in F_{biot}. Except for the NEA, they load on the same factors as Cs, which is highly enriched. Zr–Ti depletion may be related to albitization or apatitization (see above). The association with the incompatible element Cs probably stems from the close association of Ti and Zr with biotite or its alteration products (van Gaans et al., in prep.). In the three-factor models (P2M-2 and MA-3) Ti, Zr and Cs load positively on F_{min}, indicating that the ore elements have some spatial relation to (altered) biotite. In the factor models for the MA-6 and NEA, Zr loads on two different factors. Outside the Zr–Ti association on F_{min}, an alkali factor F_{ser} with a negative Zr loading is present. This alkali metal versus Zr factor points towards the leaching of Zr with late sericitization which is largely restricted to the NE part of the Regoufe area.

SiO_2 inversely associates either with Na_2O in F_{alb} (MA, P2M-4, NEA) or with K_2O, Rb and Sr or Cs in F_{mu} (P2M-2, P2M-4, MA-1, WD), in MA-3 with both in one factor F_{alb-mu}. This may reflect albitization and muscovitization respectively, or the predominance of either albite or K-feldspar as the most abundant mineral next to quartz. Although the Na_2O content of the NEA is relatively low (Table 1), the relative importance of albite over K-feldspar, as observed in thin section, is confirmed by F_{alb} in the component analysis.

K_2O and Rb generally load on the same factor, either F_{mu} or F_{ser}, in the WD, the MA granite and the NEA together with Cs, in the P2M granite and the NEA with Sr. In F_{mu} they oppose SiO_2, in F_{ser} (MA-6, NEA) Zr.

CaO and P consistently form an apatite factor F_{ap}, in the P2M granite together with Na_2O, in the MA granite and the northeastern aplite commonly together with Sr (MA-3, MA-6, NEA) and in the western dykes (WD) with both.

The Sr affinities thus vary among subregions. Sr associates with K_2O and Rb and inversely with SiO_2 for the P2M granite (F_{mu}). In the factor model for P2M-5 Sr also loads negatively on F_{min}. Sr loads on F_{ap} in the WD and the NEA and in part of the MA granite (MA-3 and MA-6), in MA-1 Sr and SiO_2 together oppose Na_2O in F_{alb}. In the NEA Sr also loads on F_{ser}. The above indicates that Sr acts mainly as a substitute for K in the P2M granite and mainly as a substitute for Ca in the MA granite and the aplites and dykes. In the NEA both substitutions appear to be of importance. This ambiguous behaviour of Sr explains why the conventional general study of Vriend et al. (1985) failed to explain the Sr variance.

5 Discussion and Conclusions

Owing to the small rock volume, mineralogically determined element associations greatly affect the association patterns for rock sections. Larger scale processes and features are stressed if mean concentrations per core are considered. Their association patterns therefore more closely resemble the patterns obtained on conventional samples. The expression of metasomatic and ore-forming processes varies with the scale of observation. Therefore, the resolution of rock chemical data is different for each scale. Mineralogical and local phenomena may go unnoticed if only bulk chemistry is considered. Integration of the association patterns of all scales, including results of conventional chemical analysis and microscope techniques, elucidates the interaction of the various rock-forming phenomena.

The influence of weathering on rock chemistry of the Regoufe granite is negligible for major and trace elements, provided that the physical rock structure is retained. Erosion is obviously faster than chemical weathering in this type of environment.

The effects of the hydrothermal processes on rock chemistry are recognized throughout the Regoufe granite and its derivatives. Albitization, W–Sn quartz-vein type mineralization, muscovitization and apatitization and in the NE, also disseminated W-mineralization and late sericitization, are distinguished. Although the exact compositions and relative importance of the various factors differ among subregions, association patterns of the individual subregions, based on rock sections, basically describe the same features and processes as identified for the whole main granite. The post-magmatic processes obviously not only induced chemical variation *among*, but also *within* subregions. Therefore, despite local variation, the detailed results presented here indicate the pervasive nature of the processes. The decrease in the extent of alteration from the roof zone towards the deeper parts of the granite is apparent from the gradually changing concentrations and by the shifts in the element association patterns, whereby the distinct patterns for the NE are noteworthy.

The element correlations clearly illustrate the impact of the mineralogical composition of the granite in hydrothermal alteration. The interaction of mineralogy with metasomatic processes can be identified and evaluated e.g., in an apatite factor, a biotite influence for Ti and Zr, mica, feldspar and quartz components. Sr illustrates both the importance of mineralogy or major element chemistry and the effect of varying alteration. It preferentially substitutes for K, in feldspar or mica, in the P2M granite and for Ca, in phosphate minerals, in the WD and the MA granite. In the NEA both substitutions are of importance.

Facilitated by the selection based on rock chemistry (IRA) of sections, interesting features, like an apatite-bearing quartz veinlet, and some trace minerals were identified. Columbo-tantalite was detected in the MA granite, scorodite was found in the NEA.

Appendix: The IRA Approach

Flat rock sections cut from drill-cores are analyzed by XRF spectrometry as they are, without further sample preparation, yielding a total chemical analysis for each section. At the cost of some loss in accuracy and precision much is gained by: a better resolution of within-sample inhomogeneities, a more direct relation between rock chemistry and mineralogy, the availability of the sections for later investigation, e.g. ore microscopy and electron-probe microanalysis (EPMA) and effectiveness in time and costs. The IRA approach is described in detail by van Gaans et al. (1986a).

Evaluation of the IRA method (van Gaans et al. 1986b, c) showed that precision for repeat analysis of the same rock section at the 95% confidence level is normally within 8%. Precision of XRF analysis of pressed powder tablets is only 25 to 50% better. Accuracy of the method mainly depends on the type of matrix correction applied. Insufficient matrix correction usually causes approximately linear systematic deviations which can be empirically corrected. However, for exploration purposes and for the study of geochemical processes, relative figures are more important than absolute values (Levinson 1974; Fletcher 1981) and in these instances elaborate matrix corrections are rarely needed. Within a geochemical setting of volcanic rocks with associated sulphide deposits (van Gaans et al. 1986c), correlations between IRA and conventional results were better than 0.94 (van Gaans et al. 1986c).

In the same study a considerable reduction of 60% in analyst time was achieved compared to conventional routine XRF analysis of trace and major elements. Instrument time is generally ten times longer using a sequential spectrometer. However, the consequent ten-fold increase of data allows the study of rock chemistry on 'conventional' as well as on smaller, within-sample, down to mineral scales. Moreover, sample throughput is fully automated and if necessary can be greatly increased with a simultaneous spectrometer. For subsequent study by more expensive and more labour-intensive techniques, such as microscopy, X-ray diffraction, electron-probe microanalysis, scanning electron microscopy or neutron activation-induced beta-autoradiography, sections of interest can be selected on the basis of the IRA data, providing a more (cost-)efficient use of these methods.

Acknowledgements. This study was partly financed by the European Communities (*Contract No. MSM-073-NL*). R.D. Schuiling was one of the initiators of the project and is thanked for his critical comments. The Netherlands Organisation for the Advancement of Pure Research (ZWO) financially supported the electron microprobe facilities. H. Dols and H. van Veen collected the Regoufe drill cores and assisted with the sample preparation and XRF analysis. J. van der Wal supervised the use of the XRF equipment. J. de Groot made polished and thin sections. A. Trappenburg and J. van Bergenhenegouwen prepared the drawings.

References

Chessworth W (1979) The major element geochemistry and the mineralogical evolution of granitic rocks during weathering. Phys Chem Earth 11:305–313

Dixon WJ (ed) (1981) BMDP Statistical software 1981. University of California Press, Berkeley, Calif

Fletcher WK (1981) Analytical methods in geochemical prospecting. Elsevier Scientific Publishing Company, Amsterdam, 255 pp. (Handbook of Exploration Geochemistry Govett, GJS ed) vol 1)

Gaans PFM van, Vriend SP, Schuiling RD (1985) Integral rock analysis; a new approach to lithogeochemical exploration. Application: the granite of Regoufe. Report to the Commission of the European Communities, contract MSM-073-NL.

Gaans PFM van, Vriend SP, Schuiling RD (1986a) Integral rock analysis; a new approach in lithogeochemical exploration with use of X-ray fluorescence spectrometry. Geol Mijnbouw 65:205–213

Gaans PFM van, Vriend SP, Wal J van der, Schuiling RD (1986b) Integral rock analysis; a new approach to lithogeochemical exploration. Application: carboniferous sediments of a coal exploration drilling, Limburg, the Netherlands. Report to the Commission of the European Communities, contract MSM-073-NL

Gaans PFM van, Vriend SP, Meyer HC, Finlow-Bates T, Wal J van der, Schuiling RD (1986c) Integral rock analysis; a new approach to lithogeochemical exploration: exploration for volcanogenic massive sulphides, the Pyrite Belt, Huelva, Spain. A pilot study based on Ti–Zr chemistry. Report to the Commission of the European Communities, contact MSM-073-NL

Gaans PFM van, Vriend SP, Poorter RPE, Jansen JBH (in prep) Changing element association patterns with hydrothermal processes in the W–Sn enriched Regoufe granite and its derivatives, Portugal

Govett GJS, Nichol I (1979) Lithogeochemistry in mineral exploration. In: Hood PJ (ed) Geophysics and geochemistry in the search of metallic ores. Geol Surv Can Ec Geol Rep 31:339–362

Grassia A (1986) "Discovery" sampling in geological research Part 2. Math Geol 18:323–328

Howarth RJ, Sinding-Larsen R (1983) Multivariate analysis. In: Howarth RJ (ed) Statistics and data analysis in geochemical prospecting. Elsevier Scientific, Amsterdam pp 207–286 (Handbook of Exploration Geochemistry, vol 2)

Hull CH, Nie NH (1981) SPSS update 7–9. McGraw-Hill, New York

Jöreskog KG, Klovan JE, Reyment RA (1976) Geological factor analysis. Elsevier Scientific, Amsterdam (Methods in Geomathematics 1)

Jongh WK de (1970) Heterogeneity effects in X-ray fluorescence analysis. Philips Analytical Eequipment Bulletin

Levinson AA (1974) Introduction to exploration geochemistry. Wilmette, Illinois, USA, 612 pp

Maitre RW le (1982) Numerical petrology. Statistical interpretation of geochemical data. Elsevier Scientific, Amsterdam

Mellinger M (1984) The application of correspondence analysis to the study of lithogeochemical data: general strategy and usefulness of various data-coding schemes. J Geochem Expl 21:455–469

Miesch AT (1981) Estimation of the geochemical threshold and its statistical significance. J Geochem Expl 16:49–76

Nie NH, Hull CH, Jenkins JG, Steinbrenner K, Bent DH (1975) Statistical Package for the Social Sciences, 2nd edition. McGraw-Hill, New York

Pinto MS (1985) Carboniferous granitoids of Portugal: some geochemical and geochronological aspects. In: Lemos de Sousa MJ, Wagner RH (ed) Papers on the Carboniferous of the Iberian Peninsula. Sedimentology, Stratigraphy, Paleontology, Tectonics and Geochronology pp 15–33

Selinus O (1983) Factor and discriminant analysis to lithogeochemical prospecting in an area of central Sweden. J Geochem Expl 19:619–642

Shapiro L (1967) Rapid analysis of rocks and minerals by a single-solution method. US Geol Surv Prof Pap 575B:187–191

Sluijk D (1963) Geology and tin-tungsten deposits of the Regoufe area, northern Portugal. Thesis, University of Amsterdam

Stemprok M (1979) Mineralized granites and their origin. Episodes 3:20–24

Tischendorf G (1977) Geochemical and petrographic characteristics of silicic magmatic rocks associated with rare-element mineralization. In: Stemprok M, Burnol L, Tischendorf G (eds) MAWAM (Metallization associated with acid magmatism), vol 2. Geologic Survey, Prague; Stuttgart, Schweizerbart, pp 41–96

Voncken JHL, Vriend SP, Kocken JWM, Jansen JBH (1986) Determination of beryllium and its distribution in rocks of the Sn–W granite of Regoufe, N-Portugal. Chem Geol 56:93–103

Vriend SP, Oosterom MG, Bussink RW, Jansen JBH (1985) Trace element behaviour in the W–Sn granite of Regoufe, Portugal. J Geochem Expl 23:12–25

The Recording of Fluid Phases Through REE Contents in Hydrothermal Minerals. A Case Study: Apatites from the Meymac Tungsten District (French Massif Central)

L. RAIMBAULT[1]

Abstract

The use of REE spectra in hydrothermal minerals is described and a method for the quantification of some geochemical parameters of mineralizing fluids (chondrite-normalized La to Yb ratio, Eu anomaly) is proposed.

This method is used in order to describe and elucidate the complex relations between various granites and related tungsten mineralizations in the Meymac area (French Massif Central). Perspectives for a use in geochemical exploration are also presented.

1 Introduction

The use of rare earth elements (REE) as geochemical tracers of hydrothermal phenomena is subject to several difficulties related partly to the insufficient knowledge of REE behaviour under hydrothermal conditions and partly to the lack of a remnant of the main phase, i.e. the fluid phase, that therefore cannot be analyzed. In contrast to the field of igneous petrology, where accumulation of both experimental and analytical data on natural systems, over the last decades, allows precise quantitative modellization of magmatic systems, studies of hydrothermal systems remain scarce in spite of recent developments. The role of REE tracer of fluorite has been known for a few years (e.g. Marchand et al. 1976; Grappin et al. 1979); since that time some other minerals have been used for this purpose, e.g. scheelite (Cottrant 1981; Vinogradova et al. 1982) and apatite (Knutson et al. 1985). However, most of these studies remain qualitative (see e.g. Muecke and Clarke 1981, where calculations are applied to magmatic evolution only). Furthermore, REE geochemistry in hydrothermal solutions appears to be more complicated than under magmatic conditions. In the latter case, the composition of the medium is somewhat buffered by the limited possible variations of the major oxides; in the former one,

[1] Laboratoire de Géologie, UA CNRS no. 384, Ecole des Mines, 158, cours Fauriel, F — 42023 Saint-Etienne Cedex 2, France; and Groupe des Sciences de la Terre, Laboratoire Pierre Süe, CEN Saclay — BP no. 2, F — 91191 Gif sur Yvette Cedex, France

Mineral Deposits within the European Community
(ed. by J. Boissonnas and P. Omenetto)

ligands play an important role in REE fractionation (e.g. Möller 1983; Flynn and Burnham 1978). The concentration of these elements, as well as their nature (F^-, Cl^-, CO_2^{2-}, PO_4^{3-} and other ions), may be highly variable in hydrothermal fluids.

Finally, experimental data are too scarce (Marchand 1976; Flynn and Burnham 1978) to enable complete comprehension of the hydrothermal behaviour of REE.

In this chapter we will present a short review of available data on REE partitioning in hydrothermal systems, and show how it is possible to take into account the fluid phase without hypothetical considerations of fluid/magma interaction.

2 REE Under Hydrothermal Conditions

2.1 Experimental Data

At present, two sets of experimental data are available: $D^{REE}_{fluorite/fluid}$ (fluorite/fluid REE partition coefficients; Marchand 1976) at 120 °C in a chlorine-bearing environment, and $D^{REE}_{melt/vapour}$ (Flynn and Burnham 1978) at 800 °C. The latter data indicate that under magmatic conditions, Cl and F are efficient complexing agents, while carbon dioxide is not. Furthermore, they show a strong correlation between $D^{REE}_{melt/vapour}$ and chloride molality. However, neither the global shape of the D vs atomic number curve, nor the Eu anomaly, vary with m_{Cl}, so that it is possible to deduce the global form of REE spectra without precise knowledge of the fluid chemistry. Ratios such as Eu/Eu* or $(La/Yb)_n$ (La to Yb ratio, normalized to chondritic values) have intrinsic values independent of the relative concentrations of ligands (but not from their nature!), so that the distribution coefficient:

$$K^{La/Yb} = (La/Yb)_X/(La/Yb)_{fluid},$$

(where X is a hydrothermal mineral or a melt) can be used, excluding the measure of absolute D values.

2.2 Constraints Deduced from Studies on Natural Systems

Using such experimental data, as well as analytical data on natural vein-mineral pairs thought to be in equilibrium, a set of values proportional to REE partition coefficients have been calculated for apatite, scheelite and wolframite (Raimbault 1985). The $K^{La/Yb}_{X/fluid}$, deduced from these results, is shown in Table 1. The influence of temperature cannot yet be evaluated due to the rarity of experimentally calibrated work. According to Raimbault (1985), such an influence seems to be negligible as a first approximation in the temperature range observed (approximately 200° to 500 °C). K values for scheelite have been omitted in Table 1, since some new data give a value close to that of apatite (0.2), in contrast to the previously obtained value of 3.3. However, more work is needed for an accurate determination.

Table 1. Values of $K^{La/Yb}_{X/fluid}$ for some minerals and melts

	$(La/Yb)_X/(La/Yb)_{fluid}$	
X	Experimental	Natural systems
Fluorite	0.10[a]	
Scheelite	0.2 to 0.44[b]	?
Apatite		0.19
Wolframite		0.012
Granitic melt, 4 kb	0.55(8)[c]	
1.25 kb	0.44[c]	

[a] Marchand (1976), 120 °C.
[b] Raimbault and Baumer (unpublished data), 600 °C, 1 kb.
[c] Flynn and Burnham (1978), 800 °C.

2.3 Use of REE in Hydrothermal Minerals as Geochemical Tracers

Some analyses of hydrothermal minerals within single-zoned crystals may show considerable variations. Knutson et al. (1985), for example, report REE contents of a zoned apatite from the Panasqueira Sn–W deposit, varying from 1 to 30 in relative values! However, even in this extreme case, the global shape of the spectrum does not change. Such variations can be related to local crystallization conditions, while the REE spectrum is inherited from the fluid phase in the entire vein. This supports the claim that hydrothermal phases with a high mineral/fluid partition coefficient, such as apatite, fluorite and scheelite, are 'recording minerals' of fluid-REE geochemistry (Grappin et al. 1979). For a closed system, their own resulting spectrum is directly proportional to that of the initial solution, whereas in an open system, the ratio of the two spectra is proportional to the partition coefficient, which allows us an indirect measurement of fluid REE contents. Particularly, comparisons of fluid spectra, obtained from different sources (hydrothermal minerals, granites considered as representative of melts), enable identification of these fluids and their source.

The Eu anomaly (Eu/Eu*) is strongly dependent upon redox conditions controlling the Eu^{2+}/Eu^{3+} ratio. A method for the evaluation of dominant species in solution has been recently proposed (Guion et al. 1985). When the trivalent state predominates, Eu can be treated in the same way as other trivalent REE and the anomaly in the mineral is inherited from the solution. However, in any case the oxidation level of Eu must be carefully discussed for detailed utilization of Eu anomalies.

3 Apatites from the Meymac Tungsten District

3.1 Results

We present here a short discussion of apatite analyses from the Meymac tungsten district (French Central Massif: see Fig. 1). Samples have been taken over a wide

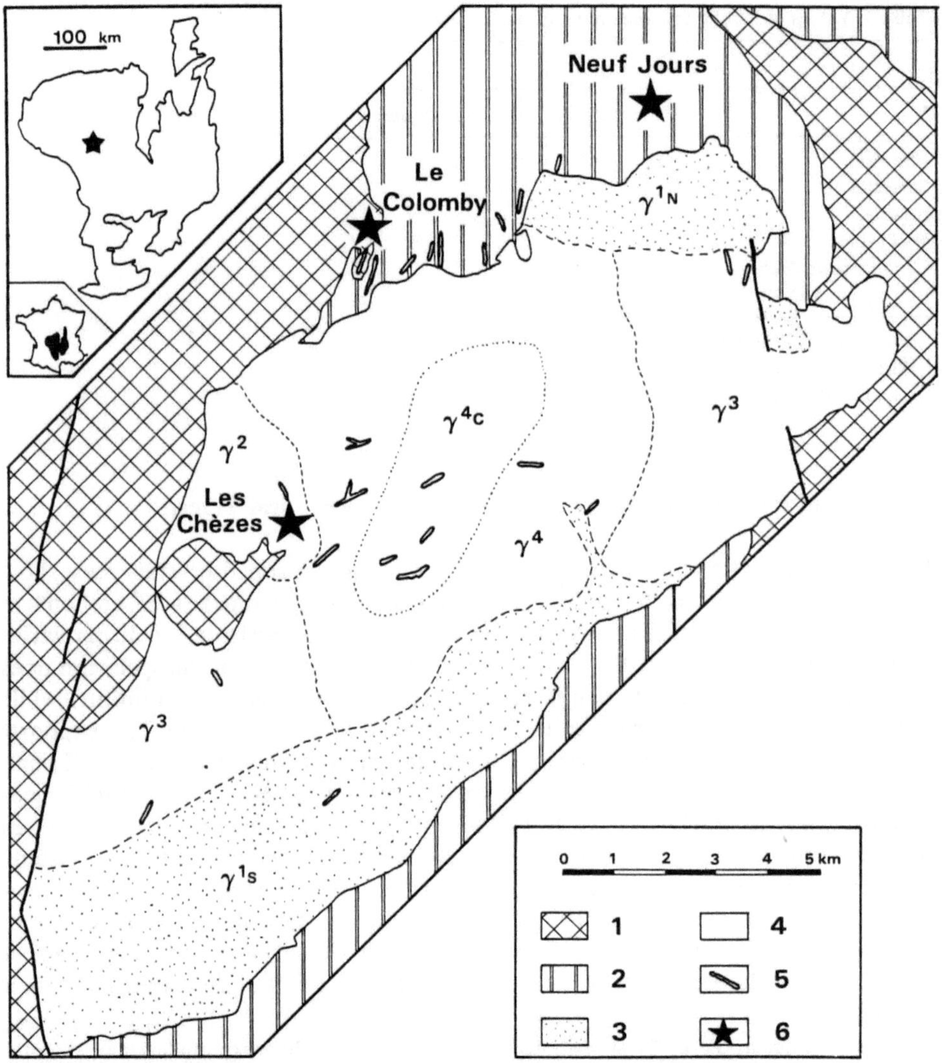

Fig. 1. Geological map of the Meymac tungsten district. *1* Surrounding granites; *2* metamorphic rocks; *3, 4* Meymac granite, with indication of facies; *5* aplitic rocks; *6* main mineralizations

area (about 100 km^2) and this allows us to illustrate the possibilities of the method for purposes of mineral reconnaissance.

A complete description of the geological setting is given by Raimbault (1984). The porphyritic Meymac granite intrudes catazonal metamorphic rocks and some aplites cut both the granite and the country rocks. At Neuf Jours (NJ) a leucogranitic cupola representing a late stage of the Meymac granite is greisenized and poorly mineralized in wolframite in some quartz veins. Other hydrothermal phenomena also occur in this area, the most important being those of Les Chèzes (scheelite-bearing greisen pipes 10 km SW of NJ) and Le Colomby (cassiterite- and wolframite-bearing quartz veins, 5 km W of NJ).

Table 2. Apatite analyses

	1 S3–4	1 S1–23	1 S1–24	1 S1–29	2 202	2 S5–12	3 S5–5w	3 S5–5g	3 S5–6d	3 S5–6w
La	182	231	217	178	204	179	27.2	8.8	33	66
Ce	370	524	532	445	404	492	38	13	48	108
Nd	201	270	298	335	252	324	17	6	20	54
Sm	46	94	94	75	60	75	3.8	0.90	3.2	8.8
Eu	13.0	21.1	16.8	19.6	33.1	34.7	22.2	16.5	16.9	32.0
Gd	—	—	—	—	—	—	—	—	—	—
Tb	15.0	18.2	20.2	15.4	12.3	13.2	0.90	0.42	0.75	2.3
Dy	—	—	—	—	—	76	—	—	—	—
Ho	29	30	22	15	19	19	—	—	—	—
Tm	—	—	—	—	—	9.0	—	—	—	—
Yb	67	47	55	43	52	41	4.0	2.3	4.5	5.8
Lu	10.9	7.8	8.3	6.1	8.1	6.4	0.68	0.31	1.10	1.03
	4a 321T	4a 321QR	4b 448A	4b 449B	5a 333D	5a CHS23g	5b 536	6a CHS23o	6b 631C	7 S5–13
La	113	172	151	367	398	788	684	38	480	23
Ce	318	431	364	848	1022	1730	742	108	1066	98
Nd	261	251	286	372	860	1490	880	39	520	32
Sm	106	95	90	120	229	317	195	18.1	133	6.7
Eu	6.3	7.0	6.3	6.6	15.4	34.6	25.8	4.0	32.8	4.7
Gd	—	—	—	—	—	—	215	—	—	—
Tb	26	30	29	21	33	38	28	3.3	15.3	1.83
Dy	—	—	—	—	—	—	193	—	77	—
Ho	40	55	49	37	49	65	31	3.0	23	5.5
Tm	—	—	—	—	—	—	14.6	—	9.4	—
Yb	105	181	138	88	109	101	106	8.7	46	64
Lu	16.0	29	21.2	13.8	18.0	15.3	16.1	1.24	6.9	15.3

Analyses are given in Table 2. Numbers at the top indicate the origin of apatite crystals:(1) NJ cupola greisens;(2) NJ quartz-tourmaline veins at some distance from the cupola (400 m);(3) NJ apatite-wolframite quartz veins and greisenized granite veins in gneisses at some distance from the cupola (500 m);(4a) aplite intrusive in the Meymac granite and(4b) quartz veins from Le Colomby;(5a) Meymac porphyritic granite and(5b) an arsenopyrite-bearing quartz vein found as an isolated sample 2 km N of Le Colomby;(6a) modified granite under Les Chèzes greisens and(6b) greisen in Vedrenne granite, found as an isolated sample 2 km W of Les Chèzes;(7) NJ sulphide veinlet. Analyses show well-defined groups on an REE vs. Eu/Eu* diagram (Fig. 2, where REE indicates the sum of rare earths, including interpolated elements), corresponding to a sample origin. Figure 3 presents the fields of major apatite types.

3.2 Interpretation

Type 1 apatites have a narrow variation range: $(La/Yb)_n = 1.75$ to 3.14; Eu/Eu* = 0.50 to 0.75; REE = 1250 to 1620 ppm, indicating uniformity of fluid during grei-

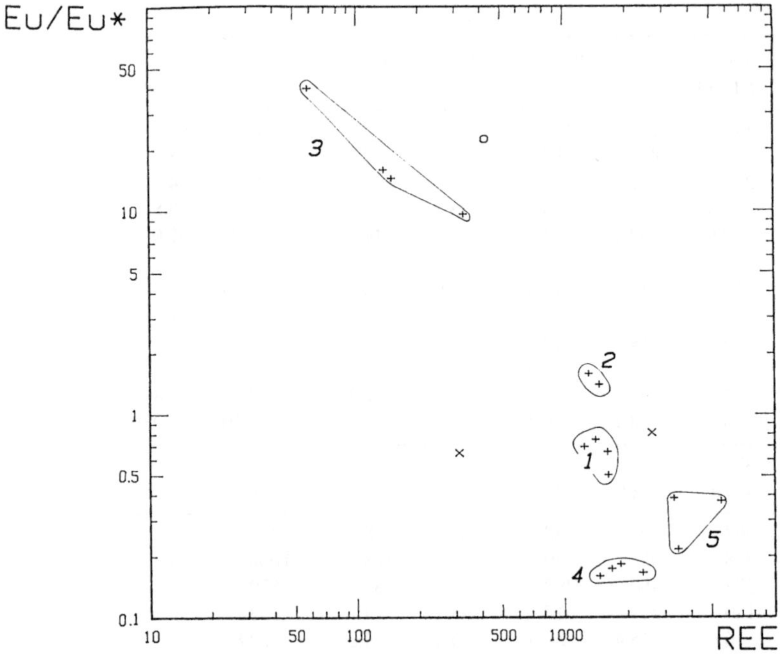

Fig. 2. Eu/Eu* vs. REE plot for Meymac apatites. *Numbers* refer to apatite types (see text); type 6 apatites are represented by *x* and a Panasqueira apatite (Knutson et al. 1985) by *o*

senization. Assuming a trivalent state of Eu, we can derive some parameters for this fluid: $(La/Yb)_n = 9.2$ to 16.5 and Eu/Eu* = 0.50 to 0.75, which is almost identical to the values calculated (using the experimentally determined vapour/melt partition coefficients of Flynn and Burnham) for the fluid associated to the leucogranite melt from the cupola (9.9 and 0.70 respectively). This granite appears also as the source of the mineralizing fluid under oxidizing conditions.

Type 2 apatite spectra (Fig. 3) are similar to those of type 1, except for their Eu anomaly: fluids become enriched in Eu, so that an increasing distance from the source produces an increase of Eu/Eu* up to 1.4–1.6. The other parameters remain remarkably constant.

In contrast to other groups, *type 3* apatites show a wide range of $(La/Yb)_n$, Eu/Eu* and REE values: 2.45 to 7.30, 9.6 to 40 and 58 to 327 respectively, and these large variations are correlated (La/Yb and REE) or anticorrelated (Eu/Eu* with others). No granitic magma can be related to such a fluid, the main characteristics of which are its low REE content and a very pronounced Eu anomaly. It is interesting to note that similar features have been reported from Panasqueira apatites (Knutson et al. 1985): REE from 30 to 400x chondritic and Eu/Eu* from about 8 to 30 (see Figs. 2 and 3 where one analysis has been plotted for comparison). This similarity can be considered a positive tool for mineral exploration, although the manner in which such a fluid is produced is not yet understood.

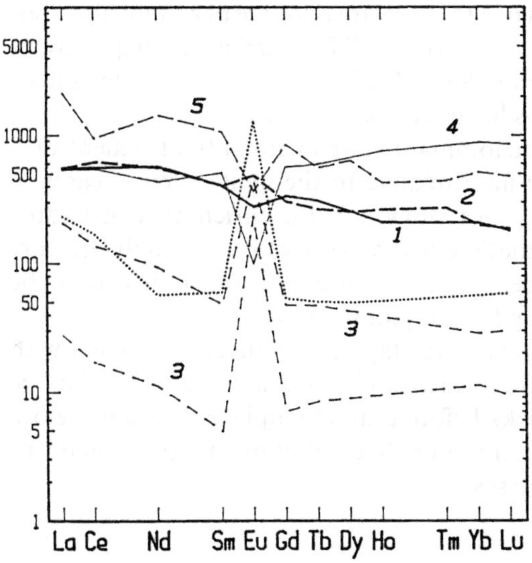

Fig. 3. REE spectra of some selected apatites. *Numbers* refer to apatite types (see text); *dotted line* represents a Panasqueira apatite (Knutson et al. 1985)

The *type 4* analyses group both magmatic and hydrothermal apatites, but their spectra are indistinguishable, proving clearly that Le Colomby fluids originate from aplitic magmas. Mineralization is not of economic importance. Eu anomalies are similar for hydrothermal and magmatic apatites and aplites, with Eu/Eu* values of 0.17, 0.16–0.18 and 0.22. Furthermore, oxygen fugacities in aplitic melt and hydrothermal fluid were high enough to cause predominance of Eu^{3+}.

Type 5 analyses also represent magmatic granitic apatites and hydrothermal apatite (see Table 2, 5b). Dispersion within this group is more important, due to granite evolution from one facies to another. However, the hydrothermal apatite is sufficiently similar to that of the granite, allowing one to identify the granite as the source for the hydrothermal solution. No known mineralization is associated with this occurrence.

Type 6 apatites have quite different REE contents (REE = 311 to 2620) and similar shapes but significantly different slopes for their spectra ($(La/Yb)_n$ = 2.83 to 6.70). For the first one, a granitic melt identical to the NJ leucogranite cannot be excluded as the source of fluid; however, taking into account the two apatites, another evolved granite, known from drilling under the Les Chèzes deposit, can best explain the calculated fluid characteristics ($(La/Yb)_n$ = 14.9 to 35, Eu/Eu* = 0.66 to 0.83, calculated from apatite analyses, as compared to values of 17.6 and 0.64 calculated from granite analyses). Actually, more data are needed for a better assignment of fluid source.

3.3 Constraints on Magmatic Fluids

This study gives as by-products some constraints on the genesis and chemistry of fluids associated with granitic melts:

1. Almost all the granite types known in the area are a source of a fluid; however, there is no evidence that these fluids are important LREE carriers as suggested by several authors (e.g. Muecke and Clarke 1981; Vidal et al. 1982). Apatites under magmatic and hydrothermal conditions have identical spectra.

2. Similarly, in several cases the predominant state of Eu is the trivalent one, so that no Eu anomaly appears in the fluid relative to the melt. Even in cases of more reducing conditions, apatite does not show any Eu enrichment in hydrothermal environments relative to magmatic ones. A positive Eu anomaly appears only at some distance from the source, so that such a phenomenon cannot be interpreted as an indication of unmixing from a granitic melt.

3. Although there are many successive fluid stages, only those associated with the more evolved granitic melts are ore-bearing. This strongly suggests that W (and Sn) are not present in country rocks before granite emplacement and before mobilization during granite cooling, but it is more likely that magmatic enrichment is a necessary step in ore-forming processes.

4 Conclusions and Perspectives

This method gives a useful framework for the study of hydrothermal solutions, especially when ore-source granites can be identified: in this case the genetic relations can be assessed from analyses on both hydrothermal and magmatic phases. In other cases, one can deduce important constraints on fluid evolution, for example interaction with surrounding rocks, resulting in the appearance of a positive Eu anomaly with increasing distance from the source. In the case of Neuf Jours-like ore deposits, analyses of hydrothermal apatites collected in the field (for example as isolated hand specimens) without detailed knowledge of the geological environment, can give useful indications on the vicinity of the source of this hydrothermal event. Moreover, the presence of apatites with similar REE contents in different deposits (such as Panasqueira and Neuf Jours) indicates the existence of reproducible processes in ore deposition. Although such processes are not yet understood, the discovery of such REE spectra in hydrothermal apatites would represent in the future a promising prospecting tool for mineral reconnaissance.

Further experimental work is still needed for a more complete interpretation of REE spectra in hydrothermal minerals. Some minerals must be investigated, and the roles of temperature, pressure and of the contents in complexing agents must be experimentally calibrated, in order to provide means to measure some physico-chemical parameters of ore-bearing fluids.

Acknowledgements. This study has been supported under *EEC* (*Contract MSM-040 F*). Sampling was possible due to the assistance of BRGM and SNEA geologists. Fruitful discussions with my colleagues were greatly appreciated, especially with JL Joron, who directed the analytical work, and B. Guy, who reviewed the first draft of this paper. L. Santangelo typed the manuscript.

References

Cottrant JF (1981) Cristallographie et géochimie des terres rares dans la scheelite. Application à quelques gisements français. Thesis, PM Curie University, Paris

Flynn RT, Burnham CW (1978) An experimental determination of rare earth partition coefficients between a chloride containing vapor phase and silicate melts. Geochim Cosmochim Acta 42:685–701

Grappin C, Treuil M, Yaman S, Touray JC (1979) Le spectre des terres rares de la fluorine en tant que marqueur des propriétés de dépôt et des interactions entre solutions minéralisantes et roches sources. Miner Dep 14:297–309

Guion JL, Touray JC, Joron JL, Tollon F (1985) Détermination de l'état de valence prédominant de l'europium en solutions hydrothermales à partir des spectres de terres rares de couples scheelite-feldspath. Application au district de Montredon (Tarn). Bull Minér 108:851–853

Knutson C, Peacor DR, Kelly WC (1985) Luminescence, color and fission track zoning in apatite crystals of the Panasqueira tin-tungsten deposit, Beira-Baxa, Portugal. Am Min 70:829–837

Marchand L (1976) Contribution à l'étude de la distribution des lanthanides dans la fluorine. Etude expérimentale et application au gîte de Maine (Saône et Loire, France). Thesis, Orléans University

Marchand L, Joseph D, Touray JC, Treuil M (1976) Critères d'analyse géochimique des gisements de fluorine basés sur l'étude de la distribution des lanthanides. Miner Dep 11:357–379

Möller P (1983) Lanthanoids as a geochemical probe and problems in lanthanoid geochemistry. Distribution and behaviour of lanthanoids in non-magmatic phases. In: Sinha SP (ed) Systematics and the properties of the lanthanides. Reidel, Dordrecht, pp 561–616

Muecke GK, Clarke DB (1981) Geochemical evolution of the South Mountain Batholith, Nova Scotia: rare-earth-element evidence. Can Miner 19:133–145

Raimbault L (1984) Géologie, pétrographie et géochimie des granites et minéralisations associées de la région de Meymac (Haute Corrèze, France). Thesis, ENS Mines, Paris

Raimbault L (1985) Utilisation des spectres de terres rares des minéraux hydrothermaux (apatite, fluorine, scheelite, wolframite) pour la caractérisation des fluides minéralisateurs et l'identification des magmas sources et des processus évolutifs. Bull Minér 108:737–744

Vidal P, Cocherie A, Le Fort P (1982) Geochemical investigations of the origin of the Manaslu leucogranite (Himalaya, Nepal). Geochim Cosmochim Acta 46:2279–2292

Vinogradova LG, Barabanov VF, Gordukalov AI, Sukharzhevskii SM (1982) Contents and composition of REE in scheelites (in Russian). Zap Vsesojuz Miner Obshch 111:98–109

Genesis of Scheelite-Bearing Calcsilicate Gneisses in the Tanneron Massif (Var, France)

A. DE SMEDT[1,2] and P. SONNET[1]

Abstract

The Tanneron gneissic massif (Var department, SE France) contains calcsilicate gneiss lenses, some with scheelite mineralization. Mineralized and barren lenses have been sampled in the abandoned La Favière tungsten mine (10 km W of Cannes). They are made up of concentric zones, namely a muscovite-free gneiss outer shell, a barren pyroxene gneiss zone, a mineralized pyroxene band and in places a marble core. Bulk rock and microprobe analyses have been made on each zone. The bulk rock geochemistry of the barren zones can be explained by isochemical transformation of a mixed limestone-greywacke protolith. On the other hand, because of their low Ti content, the scheelite-bearing zones are interpreted as metasomatized marble. Field observations and pyroxene, plagioclase and biotite compositions suggest that *both* mineralized and barren zones, at least partly, formed by infiltration metasomatism as products of a metasomatic column established at the marble-gneiss contact. Metasomatism is distinct from and post-dated the metamorphic formation of barren calcsilicate gneiss. It involved fluids with high W, F and Cl activity.

1 Introduction

Calcsilicate gneiss (CSG) lenses are widespread in the Tanneron Massif, SE France (10 km W of Cannes). Some of them have tungsten mineralization. They were mined from 1982 to 1986 in the La Favière scheelite deposit which during these 4 years yielded 150 000 tonnes at 1 wt% WO_3. The workings are now inaccessible and the surrounding exposure is poor.

In this chapter, a metasomatic origin will be advanced for the formation of the mineralized CSG. The source of the fluids is thought to have been of peri-anatectic nature (Sonnet et al. 1985). This model differs from the peri-granitic or sedimentary-

[1] Laboratoire de Minéralogie et de Géologie Appliquée, Université Catholique de Louvain, 3, Place L. Pasteur, Bâtiment Mercator, 1348 Louvain-la-Neuve, Belgium
[2] Aspirante du Fonds National Belge pour la Recherche Scientifique

Mineral Deposits within the European Community
(ed. by J. Boissonnas and P. Omenetto)

exhalative models invoked for similar scheelite-bearing CSG studied elsewhere (see for instance, Newberry 1983; Casquet and Tornos 1984; Plimer 1980; Boyer and Routhier 1974; Skaarup 1974).

2 Geological Setting

The Tanneron Massif is situated in the crystalline part of Provence (Fig. 1). It comprises a para- and orthogneissic basement (Precambrian) and a paragneissic cover (early Palaeozoic) (Crevola 1977). The cover series is geochemically related to a shale-greywacke sequence and includes a thick acid volcanic intercalation at the top (Sonnet et al. 1985).

Both basement and cover were affected by the same regional metamorphism, presumably polyphased, of Caledonian age. Conditions during the last metamorphic event were of the amphibolite facies and are characterized by the presence of garnet almandite and sillimanite. The geothermometers anorthite = grossularite + sillimanite + quartz (Ghent 1976; Ghent et al. 1979) and garnet-biotite (Thompson 1976; Ferry and Spear 1978) indicate fairly constant conditions (somewhere between 550° and 650 °C at 3 to 6 kb) in the massif, but with a systematic 50 °C increase where migmatites are present. Migmatitic features, located in the basement and at the base of the cover, consist of small bodies (centimetric to decimetric) of poorly foliated granite with intrusive to progressive (schlieren) contacts. In addition, veins of

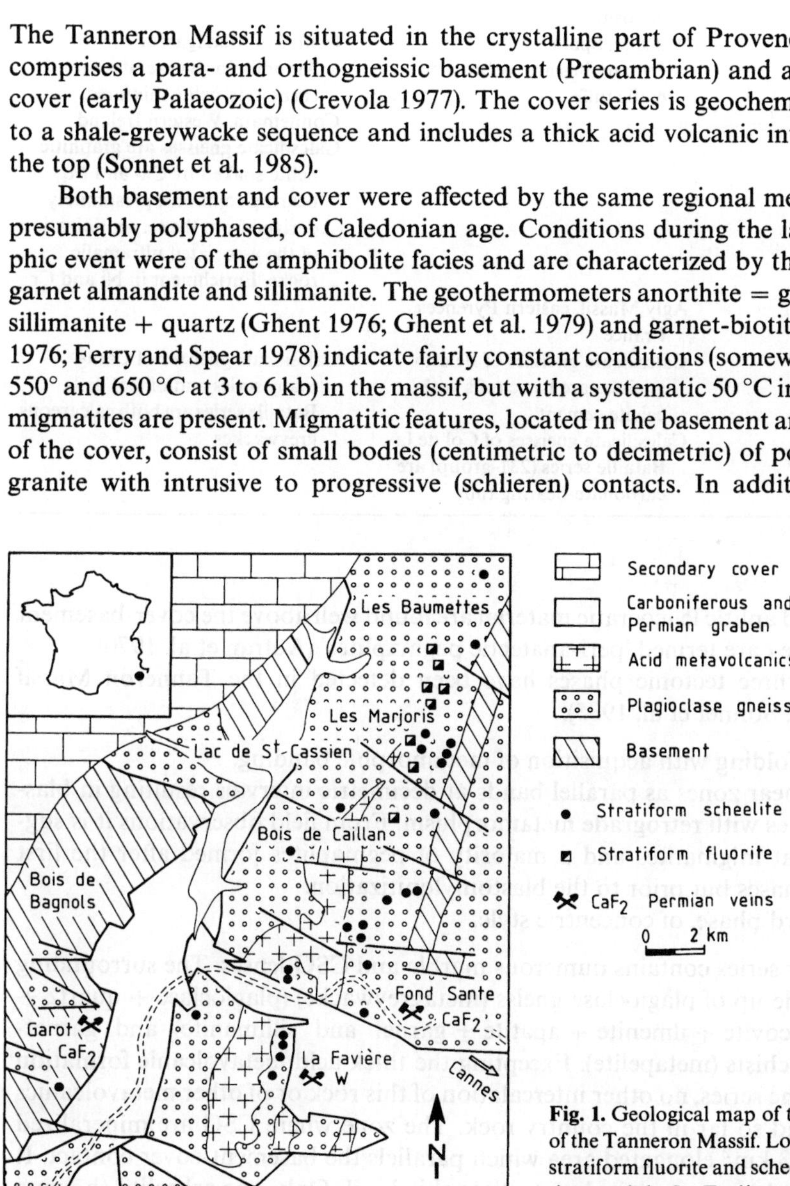

Fig. 1. Geological map of the western part of the Tanneron Massif. Localization of the stratiform fluorite and scheelite mineralizations and the La Favière scheelite mine

Table 1. Description and location of CSG (metavolcanites or metasediments) given in the literature

References	Metavolcanites	Metasediments
P.C. van de Kamp 1968	SE Ontario, Grenville Province, Canada Carbonate-silicate gneisses and pyroxene-scapolite gneisses are andesite and dacite tuffs mixed with limestone upon deposition	
B. Moine 1978	Aracena, Spain Pyroxenites are carbonated mafic tuffs	Centre of Madagascar Amphibolo-pyroxenites are carbonate-pelite mixtures.
B.E. Leake 1980		Connemara, Western Ireland Calcsilicate gneisses are graphitic pelite soaked by Ca- and Mg-bearing solutions, presumably derived from the serpentinization of the associated ultramafic rocks. Enrichment in Ni and Cr
M. Fonteilles 1981	Agly Massif, eastern Pyrenees, France Calcsilicate gneisses of Belesta gneisses are basic tuffs with calcite cement Calcsilicate gneisses of Col de la Bataille series (231-group) are carbonate-bearing tuffs	Calcsilicate gneisses of Caramany series and of Col de la Bataille series are both calcareous greywackes

pegmatitic and aplitic leucocratic material are found well above the cover-basement boundary. They are termed 'peri-anatectic pegmaplites' (Autran et al. 1970).

At least three tectonic phases have been detected in the Tanneron Massif (Crevola 1977; Sonnet et al. 1985):

1. Isoclinal folding with acquisition of metamorphic banding.
2. Ductile shear zones as parallel bands at decimetric intervals resulting in blastomylonites with retrograde metamorphism. From field observations it is suggested that migmatites and a majority of pegmaplites formed after the first folding phases but prior to the blastomylonitization.
3. A late third phase, of concentric style.

The cover series contains numerous marble and CSG lenses. The surrounding rocks are made up of plagioclase gneiss (metagreywacke) (plagioclase + quartz + biotite + muscovite + ilmenite + apatite ± garnet and sillimanite) and garnet-bearing micaschists (metapelite). Excepting the thick acid metavolcanic formation at the top of the series, no other intercalation of this rock or of other metavolcanics has been found so far in the country rock. The zone where CSG are mineralized forms a 11×3 km^2 elongated area which parallels the basement-cover contact. It is not restricted to any particular stratigraphic level. Only one scheelite showing has been found so far in the basement. It occurs in a problematic CSG lense enclosed in an orthogneiss (at Garot, Fig. 1). Stratiform fluorite scheelite-free deposits are found in the vicinity of several mineralized CSG lenses (Crevola and Sonnet 1984). Although they are not mineralized with scheelite, they are thought, on tectonic evidence, to have formed at the same time as the W-bearing CSG lenses.

3 Description of the Calcsilicate Gneiss Lenses

3.1 Zoning of the Lenses

The same zonal pattern is found in mineralized and barren CSG lenses. It consists of a definite sequence of zones (Fig. 2).

3.1.1 Muscovite-Free Gneiss

The CSG lenses are surrounded by a decimetric shell of fine-grained (500μ) *muscovite-free, biotite gneiss* which makes the transition with the enclosing plagioclase gneiss.

3.1.2 Amphibole Gneiss

At the outer edge of the CSG lenses, there is a zone of fine-grained *amphibole gneiss* (amphibole replaces biotite). This zone is never thick and may be absent locally. Therefore, no analyses have been performed on this zone.

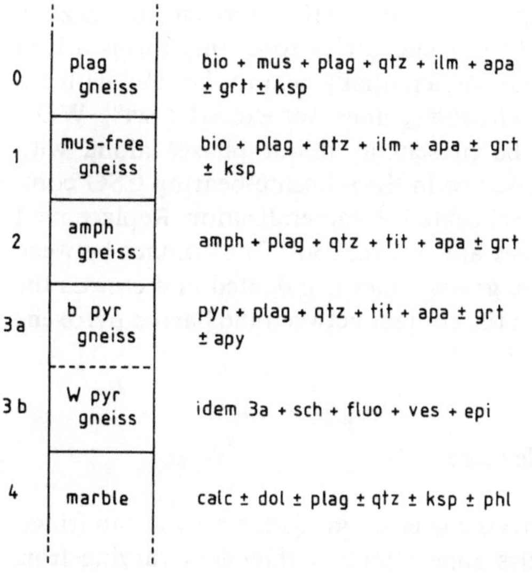

Fig. 2. The zone sequence in CSG lenses. The minerals present in each zone are listed in order of decreasing abundance: *amph* Amphibole; *apa* apatite; *apy* arsenopyrite; *bio* biotite; *calc* calcite; *dol* dolomite; *epi* epidote; *fluo* fluorite; *grt* garnet; *ilm* ilmenite; *ksp* K-feldspar; *mus* muscovite; *phl* phlogopite; *plag* plagioclase; *pyr* pyroxene; *qtz* quartz; *sch* scheelite; *tit* titanite; *ves* vesuvianite. *b pyr gneiss* barren pyroxene gneiss; *W pyr gneiss* mineralized pyroxene gneiss

3.1.3 Pyroxene Gneiss

The main body of the lenses is made up of *pyroxene gneiss* (pyroxene replaces amphibole). Where the lense is barren, the pyroxene gneiss is homogeneously fine-grained (500μ) and foliated. Where it is mineralized, in addition to the *barren pyroxene gneiss*, an inner zone of *scheelite-bearing pyroxene gneiss* exists, coarse-grained (3 mm) and in most cases lacking foliation. Apart from the presence of scheelite, the mineralogical associations are the same in both rocks (however, see Sect. 3.2.1).

3.1.4 Marble

Coarse-grained (3 mm), white marble forms the core of the lenses. In addition to disseminated dolomite grains in variable proportions, accessory minerals include quartz, K-feldspar, plagioclase and phlogopite. No scheelite has been observed in the marble.

3.2 Definition of Two Types of Mineralized Lenses

Mineralized CSG lenses are divided into two types depending on the presence or absence of a marble core. This factor controls the relative size of the fine- and coarse-grained pyroxene zones, the abundance of scheelite, and, to some extent, the mineral paragenesis.

3.2.1 Type 1: Lenses with a Marble Core

Where marble is present in the core of the lenses, the barren pyroxene gneiss zone is well developed; while the mineralized pyroxene gneiss zone only forms a thin band (Fig. 3a). In this band, scheelite occurs preferentially near the contact with the marble. The tungsten content is low and usually does not exceed 1 wt% WO_3. Vesuvianite, epidote and fluorite can be present as minor phases along with scheelite. However, fluorite is extremely scarce in the scheelite-bearing CSG compared with the nearby stratiform fluorite scheelite-free mineralization. Replacement structures at the boundary between zones are conspicuous. The contact between the mineralized and the barren pyroxene gneiss zones is indented and crosses the foliation. These features also characterize the contact between the barren pyroxene gneiss and the muscovite-free gneiss zones.

3.2.2 Type 2: Lenses Without a Marble Core

Where marble is absent, the barren pyroxene gneiss zone forms only a thin fringe, whereas the mineralized pyroxene gneiss zone attains a thickness varying from

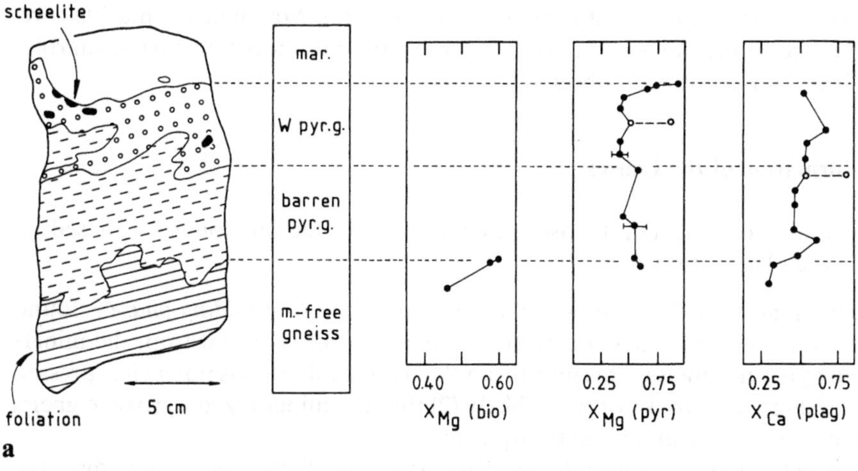

Fig. 3a, b. Example of chemical features of zones of typical samples from type 1 **(a)** and type 2 **(b)** zoned mineralized lenses. Plot of biotite, pyroxene and plagioclase composition across the samples. The *horizontal scales* give the Mg/Mg + Fe (X_{Mg}) ratios of biotite and pyroxene, and the Ca/Ca + Na (X_{Ca}) ratio in plagioclase. *Solid circles* represent individual analyses or the average of several point analyses within a small area. *Horizontal bars* connected through some points show the total range of compositional variation within the area. *Broken lines* between two *open circles* represent the zonation of a single grain. *mar.* Marble; *W pyr. g.* mineralized pyroxene gneiss; in **a**, *barren pyr. g.* barren pyroxene gneiss and *m.-free gneiss* muscovite-free gneiss, the amphibole gneiss zone is absent; **b**, *m.f.g.* muscovite-free gneiss and the *triangles* barren pyroxene gneiss + undifferentiated amphibole gneiss

several decimetres up to 2 m (Fig. 3b). Replacement features between zones are barely noticeable. The tungsten content of the mineralized pyroxene gneiss is high (between 5 and 9 wt% WO_3). Scheelite is coarse-grained (1 cm) and evenly distributed.

Generally, most barren CSG lenses in the Tanneron Massif lack a marble core. At La Favière mine, barren marble-free lenses are common but barren marble-bearing lenses have not been observed.

3.3 Deformation of the Lenses

The chronology of tungsten deposition relative to the tectonic phases is based on the following observations:

1. The metamorphic foliation which accompanies *the first folding phase* is found in the various lithologies constituting the lenses but is absent from the mineralized pyroxene gneiss. This may mean that (1) the foliation is not well expressed in coarse-grained rocks or more likely (2) that the mineralized pyroxene gneiss formed after the peak of metamorphism.
2. The effects of the *second folding phase* occur in all the rocks, including the mineralized pyroxene gneiss. Where the pyroxene gneiss is only slightly affected by phase 2, scheelite occurs as randomly oriented poeciloblasts. In contrast, in strongly deformed pyroxene gneiss, it appears as streaks with aspect ratios of more than 10:1. This linear fabric parallels minor sheath folds. Deformation of the mineralized pyroxene gneiss is relatively uncommon. The relative widths of the successive zones of the CSG lenses have presumably been modified by the second fold phase. Where marble with interbedded CSG occurs, intrafolial folds are observed rather than boudinage. This indicates that the marble was deformed in preference to other lithologies. As marble acted as 'strain-softening' rock, replacement features and zone widths were presumably better preserved in type 1 marble-bearing lenses than in type 2 marble-free lenses.

4 Major and Trace Element Geochemistry

Analyses of La Favière samples are given in Tables 2 and 3. Analyzed metavolcanics in Table 4 come from the massive acid intercalation which is the only metavolcanic recorded so far in the surroundings of the deposit. In the ($CaO:SiO_2$) diagram (Fig. 4), a solid line has been drawn between the plagioclase gneisses and the marbles for which the compositions A and B respectively have been chosen as representative. The line depicts the compositions of rocks resulting from sedimentary mixing of greywacke (plagioclase gneiss A) with limestone (marble B). Due to metamorphic decarbonatization, the compositions of the mixed rocks are displaced away from the mixing line along path lines radiating from the axis origin (= CO_2 pole). The broken line denotes the totally decarbonated mixed rocks. Mineralized pyroxene gneisses fit well with this model as they are enclosed between the solid and broken lines. For barren pyroxene gneiss, a more siliceous greywacke or a metavolcanic rock can be chosen as possible end member for the mix; compared with mineralized pyroxene gneiss (all above the 80% line), barren pyroxene gneiss compositions correspond to a slightly greywacke-richer initial mix. The position of some barren

Table 2a. Major elements (%) and Cr (ppm) analyses of mineralized CSG lenses of type 1 (La Favière Mine)[a]

	25M	300MP	300BP	300MFG
SiO_2	3.32	54.48	59.15	66.87
TiO_2	0.07	0.39	0.73	0.77
Al_2O_3	0.89	15.58	13.59	12.95
Fe_2O_3	0.74	6.13	5.41	4.98
MnO	0.05	0.16	0.15	0.08
MgO	1.75	2.90	2.83	2.38
CaO	51.00	14.07	12.30	4.44
Na_2O	tr	0.01	1.09	1.11
K_2O	0.15	1.74	0.68	2.54
P_2O_5	0.31	0.26	0.20	0.20
WO_3	0	0.20	0	0
LOI	41.47	2.79	2.71	1.82
Total	99.75	98.71	98.84	98.14
Cr	7	42	89	67

	124/1MP	124/2MP	124BP	124MFG
SiO_2	40.09	44.75	58.14	61.11
TiO_2	0.20	0.27	0.76	0.76
Al_2O_3	12.99	16.27	13.68	16.06
Fe_2O_3	5.94	6.58	6.26	7.08
MnO	0.20	0.20	0.28	0.10
MgO	3.45	3.10	3.08	3.01
CaO	26.52	19.68	13.66	4.34
Na_2O	0.22	0.23	0.64	1.29
K_2O	0.97	1.21	0.55	2.95
P_2O_5	0.41	0.35	0.24	0.24
WO_3	1.28	0.03	0	0
LOI	4.82	4.76	2.71	1.20
Total	97.09	97.43	100.00	98.14
Cr	38	35	82	90

[a] The number of the analysis refers to the CSG lense and the letter to the zone in the lense. (*M* marble; *MP* mineralized pyroxene gneiss; *BP* barren pyroxene gneiss; *MFG* muscovite-free gneiss); total iron as Fe_2O_3; *LOI* Loss on ignition.

pyroxene gneisses cannot be explained by a mere mixing model. Muscovite-free gneisses and barren pyroxene gneisses are closely grouped and coincide with the compositional field of the enclosing plagioclase gneisses. From this, one may ascribe a greywacke protolith to the muscovite-free gneiss.

The same conclusions also apply to the (Al_2O_3:CaO:MgO + FeO) diagram (Fig. 5). The mineralized and barren pyroxene gneisses are nearly identical and plot along a line connecting the domains of the marbles and the plagioclase gneisses. Metavolcanics can no longer be considered as a possible end member for the mix.

In the (TiO_2:Al_2O_3) and (TiO_2:Cr) diagrams (Figs. 6 and 7), the plagioclase gneiss, the muscovite-free gneiss and the barren pyroxene gneiss all plot in the same compositional field. Thie confirms the mixing model for the barren pyroxene gneiss

Table 2b. Major elements (%) and Cr (ppm) analyses of mineralized CSG lenses of type 2 (La Favière Mine) (for explanation, see Table 2a)

	27MP	27BP	53/1MP	53/2MP	53BP
SiO_2	54.39	64.24	55.07	53.38	59.03
TiO_2	0.14	0.73	0.18	0.15	0.69
Al_2O_3	13.34	12.80	15.75	14.78	14.05
Fe_2O_3	3.91	5.10	3.95	4.51	5.50
MnO	0.14	0.13	0.12	0.14	0.15
MgO	2.31	2.70	2.29	2.70	2.28
CaO	12.40	10.43	12.01	12.73	10.96
Na_2O	0.77	0.51	0.80	0.82	1.66
K_2O	0.62	0.38	0.57	0.53	0.58
P_2O_5	0.35	0.24	0.30	0.32	0.33
WO_3	7.42	0	5.44	9.03	0
LOI	4.65	1.04	1.78	2.15	2.54
Total	100.44	98.30	98.26	101.24	97.77
Cr	59	57	57	66	60

	53MFG	79MP	79MFG	130MP	130/2BP
SiO_2	62.55	49.75	59.37	51.89	63.43
TiO_2	0.77	0.24	0.76	0.13	0.84
Al_2O_3	15.46	16.04	16.77	14.91	12.17
Fe_2O_3	5.85	4.97	6.66	5.21	4.73
MnO	0.10	0.16	0.11	0.13	0.11
MgO	2.51	3.17	3.07	3.08	2.34
CaO	3.96	13.75	4.17	13.43	10.09
Na_2O	1.44	0.48	1.91	0.46	0.33
K_2O	2.17	0.88	2.62	0.63	0.55
P_2O_5	0.30	0.41	0.26	0.32	0.23
WO_3	0	7.10	0	8.79	0
LOI	2.07	2.64	1.66	2.14	3.17
Total	97.18	99.59	97.36	101.12	97.99
Cr	79	62	79	67	71

and the greywacke protolith for the muscovite-free gneiss. However, these diagrams do not support the mixing model for the mineralized pyroxene gneiss. This rock can either be interpreted in Fig. 6 as (1) calcium enriched, acid metavolcanics (calcareous tuffs) because they both have low Ti and high Al contents; or in Fig. 7, as (2) metasomatized marble because of the low Ti content they have in common. In this second interpretation, Al and Cr must have been introduced into the marble metasomatically.

5 Mineral Composition

Microprobe analyses of biotite, pyroxene and plagioclase grains were performed along profiles across the successive zones in selected mineralized CSG lenses of types

Table 3. Major elements (%) and Cr (ppm) analyses of barren CSG lenses from La Favière Mine (samples 43) or from elsewhere in the Tanneron Massif (samples 19, 28 and 57); total iron as Fe_2O_3; *LOI* Loss on ignition

	43BP	43MFG	19BP	28/1BP	28/2BP	57BP
SiO_2	55.02	62.59	61.54	70.37	56.24	60.86
TiO_2	0.39	0.80	0.57	0.59	0.93	0.69
Al_2O_3	17.14	15.76	12.36	9.65	16.52	14.18
Fe_2O_3	5.64	6.07	5.20	5.26	7.57	3.82
MnO	0.15	0.09	0.12	0.54	0.27	0.28
MgO	3.34	3.16	3.44	1.63	3.44	1.46
CaO	12.59	4.20	13.96	8.91	11.91	8.44
Na_2O	0.41	1.37	0.01	0.01	0.02	0.36
K_2O	2.11	2.36	0.62	0.63	1.13	2.20
P_2O_5	0.25	0.27	0.20	0.22	0.41	0.19
WO_3	0	0	0	0	0	0
LOI	2.12	0.60	0.53	0.95	0.97	6.03
Total	99.16	97.27	98.55	98.76	99.41	98.53
Cr	46	67	84	107	37	—

1 and 2 from La Favière. Results are presented for one representative lense of each type (Fig. 3a and b). Data concerning barren CSG, cited in the text for comparison, is derived from samples from La Favière and elsewhere in the Tanneron Massif.

5.1 Composition of Biotite

Biotite in muscovite-free gneiss has the same $X_{Mg} = 0.5$ (X_{Mg} = Mg/Mg + Fe) value in type 1 lenses as in type 2 lenses. Close to the contact with the barren pyroxene gneiss, there is a sharp increase in X_{Mg} from 0.5 to 0.7.

Biotites of mineralized CSG lenses show higher Si, Mg, F and Cl contents than biotites of barren CSG lenses and other biotite-bearing rocks from the massif (plagioclase gneisses, orthogneisses, migmatites) (Figs. 8 and 9).

5.2 Composition of Pyroxene

The compositions of the pyroxenes show that they belong to the diopside-hedenbergite series.

In type 1 lenses, the pyroxene composition maintains a constant Hed 50 value across the barren and the mineralized pyroxene gneiss bands. Close to the marble contact, a very sharp decrease in iron is occasionally observed. Zoned pyroxene (Fig. 10) with a diopsidic core occurs in places. A striking feature is that the increase in iron towards the edge of the crystal is steplike.

In type 2 lenses, the pyroxene composition is constant (Hed 50).

The pyroxenes of barren CSG lenses have the same characteristics as these of type 2 lenses.

Table 4. Major elements (%) and Cr (ppm) analyses of silico-aluminous rocks (*L* acid metavolcanics; *PG* plagioclase gneiss); total iron as Fe_2O_3; *LOI* Loss on ignition

	L3	L4	L5	L6	L7
SiO_2	68.95	73.15	71.64	75.22	73.92
TiO_2	0.37	0.27	0.42	0.07	0.17
Al_2O_3	15.27	13.47	14.08	13.49	13.38
Fe_2O_3	2.76	2.65	3.27	1.08	1.67
MnO	0.05	0.03	0.05	0.03	0.03
MgO	1.37	0.82	1.46	0.09	0.57
CaO	0.44	0.10	0.15	0.40	0.89
Na_2O	4.40	2.71	1.53	2.60	3.09
K_2O	4.26	5.62	5.62	5.56	4.61
P_2O_5	0.19	0.05	0.11	0.10	0.05
WO_3	0	0	0	0	0
LOI	1.11	1.62	1.30	1.58	1.38
Total	99.17	100.49	99.63	100.22	99.76
Cr	—	7	—	—	—

	L8	L9	L10	18PG
SiO_2	76.40	75.70	72.50	64.43
TiO_2	0.01	0.04	0.20	0.81
Al_2O_3	12.50	13.40	14.25	14.28
Fe_2O_3	1.23	0.75	1.83	6.08
MnO	0.01	0.03	0.05	0.10
MgO	0.01	0.13	0.40	3.15
CaO	0.60	0.32	0.60	1.30
Na_2O	3.00	3.14	2.91	2.92
K_2O	4.45	5.12	4.91	3.49
P_2O_5	0.01	0.01	0.03	0.19
WO_3	0	0	0	0
LOI	0.87	0.80	1.35	1.45
Total	99.09	99.44	99.03	98.20
Cr	—	—	11	—

5.3 Composition of Plagioclase

In type 1 lenses, the Ca content of plagioclase in the muscovite-free gneiss has a constant value of approximately An30 like that of the enclosing plagioclase gneiss. In the barren pyroxene gneiss, it can either increase continuously from An30 to An50 or jump directly to An50. In the mineralized pyroxene gneiss, the Ca content of plagioclase is constant at An50. Zoned plagioclase grains are found with a Ca-rich core (An85) and an An50 margin.

In type 2 lenses, the Ca content is similar in both the muscovite-free gneiss and the pyroxene gneiss (between An35 and An50).

In barren CSG lenses, the plagioclase composition is An30 in the plagioclase- and muscovite-free gneiss and jumps to An95 in the pyroxene gneiss. Such com-

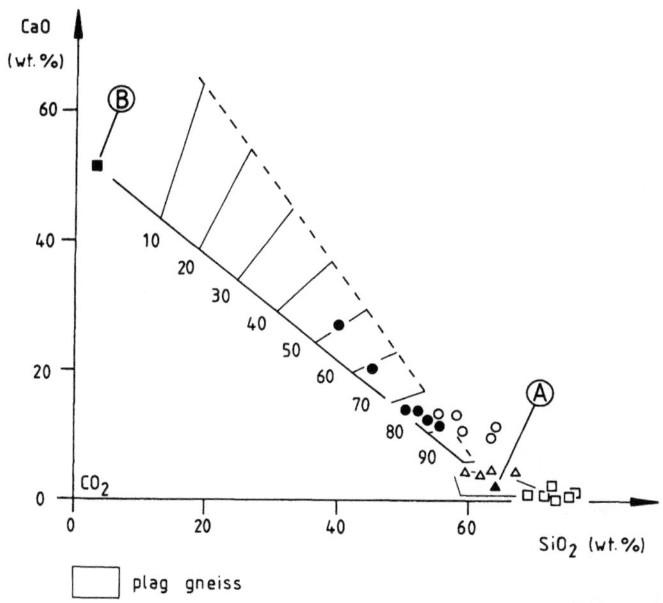

Fig. 4. ($CaO:SiO_2$) wt% diagram for La Favière samples. *Open circles*: barren CSG; *solid circles*: mineralized CSG; *solid square*: marble; *open squares*: acid metavolcanics; *solid triangle*: plagioclase gneiss; *open triangles*: muscovite-free gneiss. The *solid line* denotes the mixing line before decarbonatization between *A* (plagioclase gneiss 18PG, Table 4) and *B* (marble 25M, Table 2a); the *broken line* denotes the mixing line after decarbonatization; the proportions of plagioclase gneiss in the mix are given in wt%

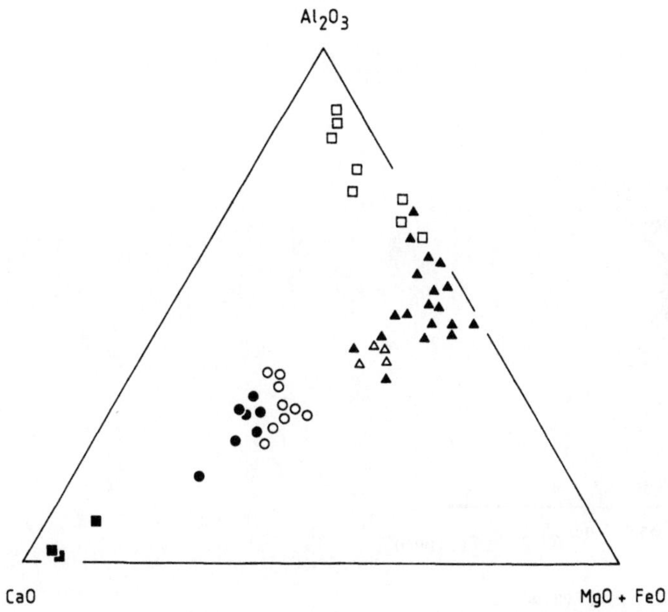

Fig. 5. ($Al_2O_3:CaO:MgO + FeO$) mol. prop. diagram for La Favière samples; *symbols* as in Fig. 4

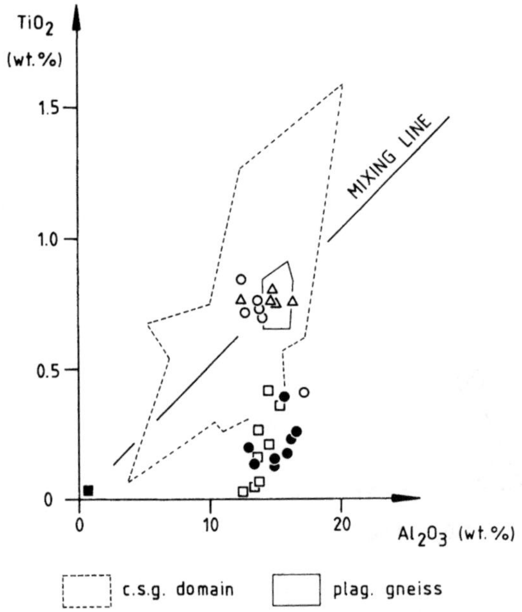

Fig. 6. (TiO_2:Al_2O_3) wt% diagram for La Favière samples. *Symbols* as in Fig. 4; the *broken line* is the contour of several CSG compositional domains taken from the literature (see Table 1)

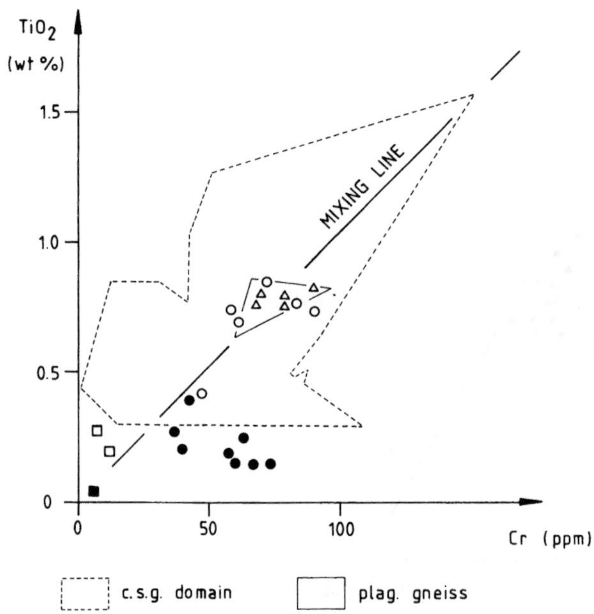

Fig. 7. (TiO_2 wt%:Cr ppm) diagram; *symbols* as in Fig. 4

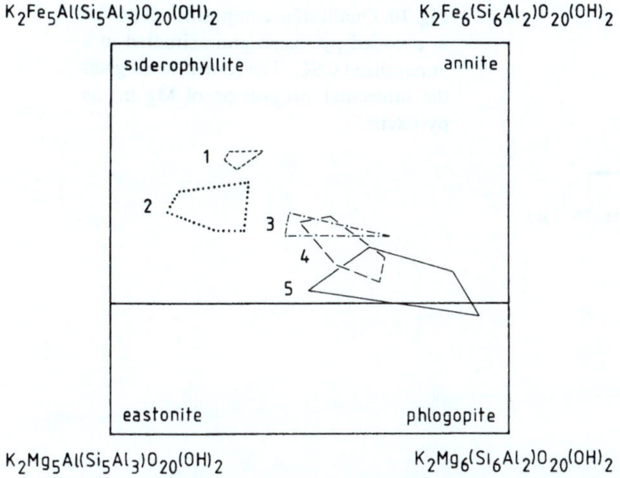

Fig. 8. Plot of biotite analyses according to their tetrahedral and octaedral content in the phlogopite, annite, eastonite and siderophyllite system. *Domain 1*: migmatites and pegmatites; *2* plagioclase gneisses; *3* a muscovite-free gneiss zone of a barren CSG lens; *4* a muscovite-free gneiss zone of a mineralized CSG lense (type 1); *5* a muscovite-free gneiss zone of a mineralized lense (type 2)

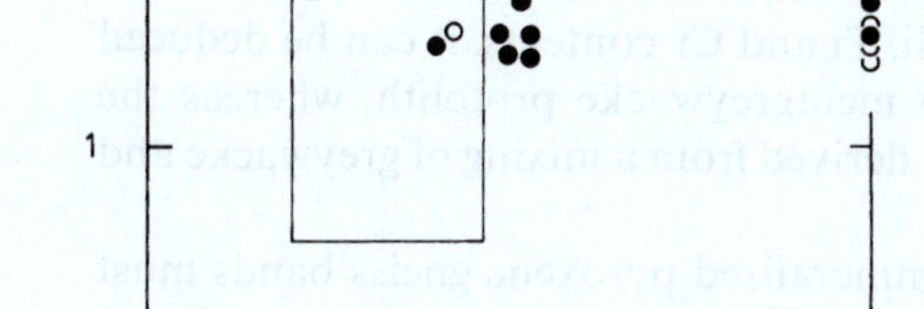

Fig. 9. [IV(Cl):IV(F)] diagram for biotites from several migmatites (*squares*), plagioclase gneisses (*open triangles*), a muscovite-free gneiss zone of a barren CSG lense (*solid triangles*) and a muscovite-free gneiss zone of a mineralized CSG lense (type 1: *solid circles*; type 2: *open circles*). Composition field for igneous (*1*) and skarn (*2*) biotites are taken from van Marcke de Lummen and Verkaeren (1985). The intercept values for Cl and F are a measure of the relative Cl and F enrichment in biotite with correction of the Fe–F and the Mg–Cl avoidance (see Munoz 1984); relative F enrichment increases towards the bottom and relative Cl enrichment increases towards the left

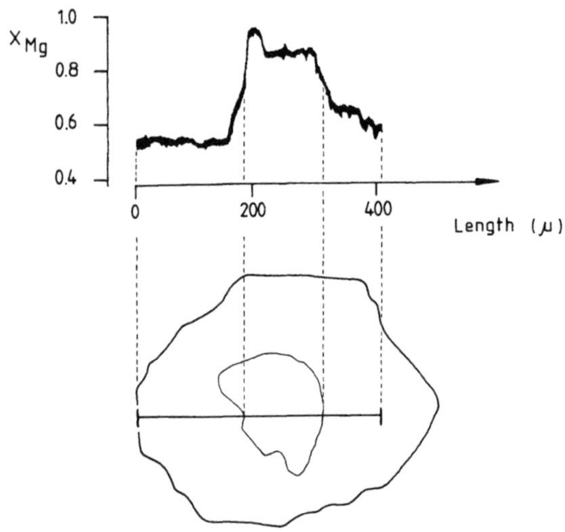

Fig. 10. Qualitative compositional profile in a zoned pyroxene grain situated in a mineralized CSG. The *vertical scale* gives the molecular proportion of Mg in the pyroxene

positions (bytownite to anorthite) are the most commonly reported plagioclase compositions in CSG formed in the amphibolite facies (Misch 1964; Thompson 1975; Moine 1978).

6 Discussion

The regional context of scheelite-bearing CSG in the Tanneron Massif is very similar to that of W mineralizations elsewhere such as NE Brazil (Goni and Picot 1965; Ebert 1970), Colorado, USA (Tweto 1960) and Bindal, Norway (Skaarup 1974).

The barren and mineralized CSG lenses are enclosed in plagioclase gneiss (derived from a former mixed shale-greywacke sequence) and are locally associated with marble. From their Ca, Mg, Fe, Al, Si, Ti and Cr contents, it can be deduced that the muscovite-free gneiss zone has a metagreywacke protolith, whereas the barren pyroxene gneiss zone was probably derived from a mixing of greywacke and limestone.

On the basis of their Ti content, the mineralized pyroxene gneiss bands must be interpreted either as metasomatically Al- and Cr-enriched, impure limestones or as Ti-poor and Cr-rich metavolcanics. Ti, Al and, to a lesser extent, Cr are usually thought to be inert during metasomatism (for instance, in skarns, Fonteilles 1978; Thompson 1975; in gneisses, Moine 1978). The second interpretation is thus more likely geochemically as it involves no Al-metasomatic exchange. However, the first interpretation is prefered because no metavolcanics occur in the wall rock of the mineralized CSG lenses. Furthermore, the massive acid volcanics at the top of the series have a composition which cannot fully account for the composition of the protolith for the mineralized lenses.

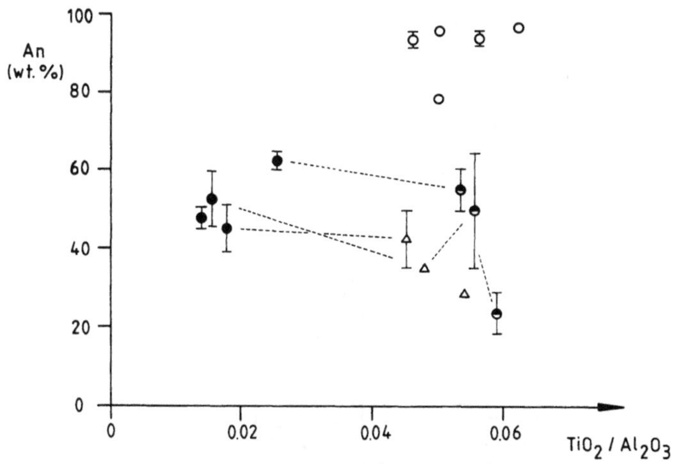

Fig. 11. (An mol%:TiO_2/Al_2O_3 wt% ratio) diagram. The *horizontal scale* gives the TiO_2/Al_2O_3 whole rock ratio and the *vertical scale*, the molecular proportion of anorthite in the plagioclase which was obtained by microprobe in the same sample. *Solid circles*: Mineralized pyroxene gneiss; *half-filled circles*: barren pyroxene gneiss from a mineralized CSG lense; *open circles*: pyroxene gneiss from a barren CSG lense; *open triangles*: muscovite-free gneiss; for each point, the symbol represents the average of several plagioclase grains in the sample and the *vertical bar* through the symbol shows the range of compositional variation; the *broken lines* connect samples from the same lense

The following features also support a metasomatic origin for the mineralized pyroxene gneiss zone and suggest that the other zones of the mineralized lenses (i.e. the muscovite-free gneiss and the barren pyroxene gneiss) had a similar origin.

1. Figures 6 and 7 depict the variation field for several barren CSG compositional domains from the literature (see Table 1). It is noteworthy that the mineralized pyroxene gneisses plot outside that field.
2. The Ca content of plagioclases of mineralized lenses (An50) is much lower than the constant An95 of barren lenses. In Fig. 11, the plagioclase composition is plotted against the whole rock TiO_2/Al_2O_3 ratio for the different bands constituting the barren and mineralized CSG lenses. The constant An content across the bands in mineralized CSG lenses suggest that the plagioclase composition was determined neither by the sediment composition (= TiO_2/Al_2O_3 ratio) nor by the local diffusion gradient but rather by a percolating fluid.
3. The high F, Cl, Si and Mg contents of the biotites in mineralized lenses suggest that these rocks were percolated by fluids with a high activity in F, Cl, Si and Mg.
4. As shown previously (Sonnet et al. 1985), the muscovite-free gneiss and the pyroxene gneiss of mineralized CSG lenses are enriched in Be and Ta, suggesting the influence of fluids in equilibrium with a highly differentiated rock (end of crystallization or beginning of fusion).
5. Replacement features observed in type 1 lenses suggest that the mineralized pyroxene gneiss zone formed, at least partly, in the barren pyroxene gneiss zone,

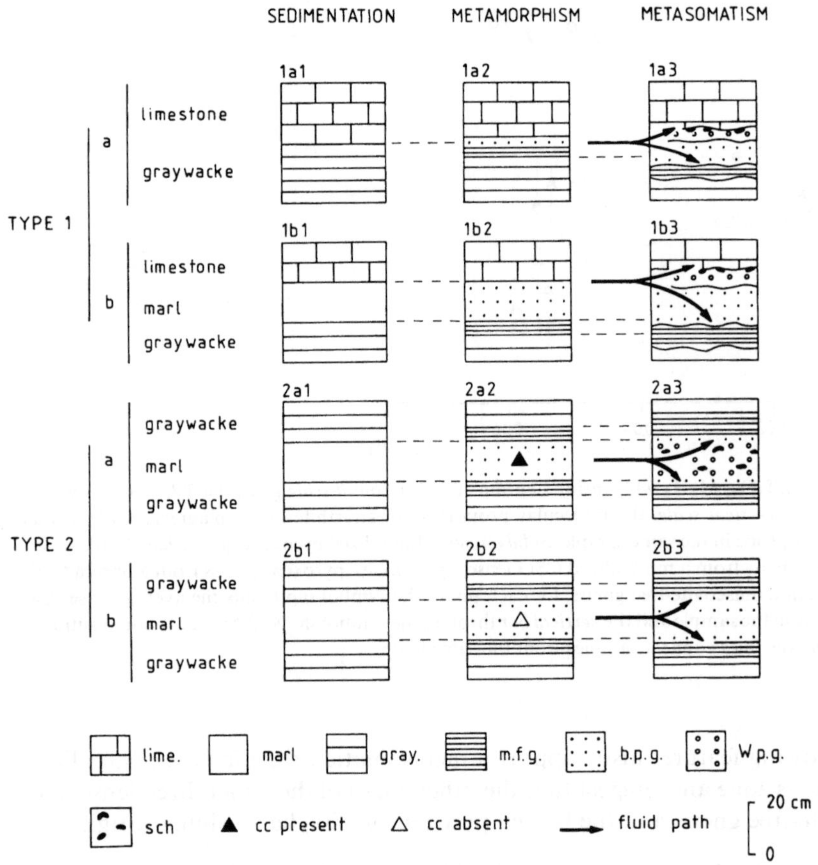

Fig. 12. Possible effects of metamorphism and metasomatism on carbonated lenses with different structures occurring in greywackes. *lime.* Limestone; *gray.* greywacke; *m.f.g.* muscovite-free gneiss; *b.p.g.* barren pyroxene gneiss; *W p.g.* mineralized pyroxene gneiss; *sch* scheelite; *cc* calcite

and that the barren pyroxene gneiss zone formed in turn in the muscovite-free gneiss.

6. In mineralized lenses, some pyroxene and plagioclase grains are distinctly zoned. The crystal cores (more anorthitic or diopsidic) are likely to be relicts of the minerals which previously existed in the replaced rock.

Accordingly, the following steps in the formation of the W-bearing CSG lenses are envisaged:

1. Deposition of limestone (case 1a1 or 1b1 in Fig. 12) or marl (2a1 or 2b1) in a greywacke sequence. The limestone-greywacke boundary may have been sharp (1a1) or may involve a marly transition zone (1b1).
2. Metamorphism with crystallization of An90 plagioclase. A muscovite-free gneiss and a pyroxene gneiss zone are formed by diffusion between incompatible

rocks (1a2) or by partial (2a2) or total decarbonatization of the marl (1b2 and 2b2).

3. Influx of a fluid carrying W, Al, Cr, Be, Ta, F, Cl channelled by the marble-plagioclase gneiss or marble-CSG contact and percolating outwards into the marble and the plagioclase gneiss. As a result, a barren pyroxene gneiss zone developed in the plagioclase gneiss (1a3) or in already formed metamorphic pyroxene gneiss (1b3, 2a3, 2b3). For type 1 lenses, a mineralized pyroxene gneiss zone developed in the marble (1a3, 1b3). The crystallization of scheelite in the pyroxene gneiss formed in the marble, rather than in the pyroxene gneiss formed in the plagioclase gneiss, is similar to that which is observed in skarns. Whenever skarns (transformed marble) occur alongside skarnoids (transformed silico-aluminous rocks), scheelite mostly forms in the former (Newberry 1983; van Marcke de Lummen and Verkaeren 1986). For type 2 lenses, it is thought that scheelite crystallized in rocks which, due to incomplete metamorphic decarbonization, were still rich enough in calcite (2a3).

7 Conclusions

In the Tanneron Massif, scheelite-bearing CSG lenses occur along with barren lenses and marble beds in gneisses belonging to a former shale-greywacke sequence. The formation of the mineralized lenses post-dates the metamorphic foliation.

Plagioclase, biotite and pyroxene compositions indicate the imprint of a percolation metasomatism following the metamorphic formation of CSG. Field observations also support the model of a metasomatic column established at the marble-gneiss contact. Geochemical data suggest that the barren pyroxene gneiss originated, before metasomatism, from metamorphic CSG or muscovite-free gneiss, whereas the mineralized pyroxene gneiss represents metasomatized marbles. The source of the W-rich fluids is thought to be of peri-anatectic origin.

The scheelite-bearing CSG of the Tanneron Massif are, at first sight, very similar to the barren ones. Detailed mineralogy and bulk geochemistry are necessary in order to demonstrate the metasomatic origin of the former. This convergence between mineralized and barren CSG may be explained by the distal character of the deposit relatively to the source of fluids. After sufficient interaction with the environment, hydrothermal fluids presumably acquired chemical characteristics and produced results similar to those of the fluids in equilibrium with the plagioclase gneiss.

Acknowledgements. The authors wish to thank G. Crevola for geological advice and assistance during field work, M. Lequertier and Entreprises Gagneraud et Fils for allowing access to the La Favière Mine, and J. Verkaeren, G. van Marcke de Lummen, M. Demange, B. Moine, M. Fonteilles and A. Lees for kindly reading the manuscript and offering a number of valuable suggestions. Field work and analyses were financed by the EEC (*Contracts No. MPP-143-B* and *MSM-127-B*) and the Service de la Programmation de la Politique Scientifique (*Contract No. MP/CE/13*). Microprobe analyses were carried out at the Centre d'Analyse par Microsonde pour les Sciences de la Terre, financially supported by the Fonds National Belge pour la Recherche Scientifique.

References

Autran A, Fonteilles M, Guitard G (1970) Relations entre les intrusions de granitoïdes, l'anatexie et le métamorphisme régional, considérées principalement du point de vue du rôle de l'eau: cas de la chaîne hercynienne des Pyrénées Orientales. Bull Soc Géol Fr 12(7):673–731

Boyer F, Routhier P (1974) Extension régionale de couches à schéelite dans la couverture métamorphique de la zone axiale de la Montagne Noire (Hérault, France). C R Acad Sci Paris 279 D:1829–1832

Casquet C, Tornos F (1984) El skarn de W–Sn del Carro del Diablo (Sistema Central Espanol). Bol Geol Min CXV-1:58–79

Crevola G (1977) Etude pétrographique et structurale de la partie orientale du massif de Tanneron (Provence cristalline). Thèse de 3ème cycle, Université de Nice, 355 p, unpublished

Crevola G, Sonnet P (1984) Une minéralisation fluorée stratiforme, à déformations varisques dans le massif métamorphique du Tanneron (Provence cristalline). C R Acad Sci Paris 299(12):805–809

Ebert H (1970) The Precambrian geology of the "Borborema" belt (State of Paraiba and Rio Grande do Norte; northeastern Brazil) and the origin of its mineral provinces. Geol Rundsch 59:1292–1326

Ferry JM, Spear FS (1978) Calibration of the partitioning of Fe and Mg between biotite and garnet. Contrib Miner Pet 66:113–117

Fonteilles M (1978) Les mécanismes de la métasomatose. Bull Minér 101:166–194

Fonteilles M (1981) Anatexis of a metagraywacke series in Agly massif, eastern Pyrenees, France. J Fac Sci Univ Tokyo Sec II Geol Mineral Geogr Geophys 20(3):181–240

Ghent ED (1976) Plagioclase-garnet-Al_2SiO_5-quartz: a potential geobarometer-geothermometer. Am Miner 61:710–714

Ghent ED, Robbins DB, Stout MZ (1979) Geothermometry, geobarometry and fluid compositions of metamorphosed calcsilicates and pelites, Mica Creek, British Columbia. Am Miner 64:874–885

Goni J, Picot P (1965) Certaines particularités minéralogiques des tactites à schéelite du NE du Brésil. Bull Soc Fr Minér Crist 88:11–16

Kamp PC van de (1968) Geochemistry and origin of metasediments in the Haliburton–Madoc south-eastern Ontario. Can J Earth Sci 5(6):1337–1372

Leake BE (1980) Some metasomatic calc-magnesian silicate rocks from Connemara, western Ireland: mineralogical control of rock composition. Am Miner 65:26–36

Marcke de Lummen G van, Verkaeren J (1985) Mineralogical observations and genetic considerations relating to skarn formation at Botallack, Cornwall, England. In: High Heat Production (HHP) granite, hydrothermal circulation and ore genesis. Inst Miner Metall, London, pp 535–547

Marcke de Lummen G van, Verkaeren J (1986) Physicochemical study of skarn formation in pelitic rocks, Costabonne peak area, eastern Pyrenees, France. Contrib Miner Pet 93:77–88

Misch P (1964) Stable association wollastonite-anorthite and other calcsilicate assemblages in amphibolite-facies crystalline schists of Nange Parbat, northwest Himalayas. Beitr Miner Pet 10:315–356

Moine B (1978) Héritage sédimentaire ou volcanosédimentaire et échanges de matière dans la formation des gneiss calcomagnésiens. Bull Soc Géol Fr 20(1):11–20

Munoz JL (1984) F–OH and Cl–OH exchange in micas with applications to hydrothermal ore deposits. In: Bailey SW (ed) Micas, Reviews in Mineralogy. Miner Soc Am 13:469–494

Newberry RJ (1983) Tungsten-bearing skarns of the Sierra Nevada. I. The Pine Creek Mine. California. Econ Geol 77:823–844

Plimer IR (1980) Exhalative Sn and W deposits associated with mafic volcanism as precursors to Sn and W deposits associated with granites. Miner Dep 15:275–289

Skaarup P (1974) Stratabound scheelite mineralization in skarns and gneisses from the Bindal area, northern Norway. Miner Dep 9:299–308

Sonnet P, Verkaeren J, Crevola G (1985) Scheelite-bearing calcsilicate gneisses in the Provence crystalline basement (Var, France). Bull Miner, 108:197–210

Thompson AB (1975) Calcsilicate diffusion zones between marbles and pelitic schists. J. Pet 16:314–346

Thompson AB (1976) Mineral sections in pelitic rocks. II. Calculation of some P–T–X (Fe–Mg) phase relations. Am J Sci 276:425–454

Tweto O (1960) Scheelite in the Precambrian gneisses of Colorado. Econ Geol 55:1406–1428.

Scheelite-Bearing Metalliferous Sequences of the Peloritani Mountains, Northeastern Sicily (with some Remarks on Tungsten Metallogenesis in the Calabrian-Peloritan Arc)

P. OMENETTO[1], V. MEGGIOLARO[1], P. SPAGNA[1], L. BRIGO[2], P. FERLA[3], and J.L. GUION[4]

Abstract

The Peloritani Mountains (Northeastern Sicily) are part of a segment of the Hercynian chain recognized in the Calabrian-Peloritan Arc, geotectonically defined by a pre-Hercynian crystalline basement (Aspromonte Nappe + Mandanici Unit) overthrust and overturned on its Paleozoic (Cambro-Ordovician up to Devonian-Carboniferous) volcanosedimentary cover. Tungsten (scheelite) and associated polymetallic stratabound ores, more or less intensely affected by the pre-Hercynian and Hercynian tectonometamorphic events, are confined to the pre-Hercynian basement. In particular, the most significant scheelite-tourmaline (arsenopyrite) and scheelite-carbonate-quartz (albite) mineralizations could be considered as tungsten (B-, As-, silica, Na-)-bearing exhalites within a peculiar volcanosedimentary uppermost unit of "Cambrian" age (roughly corresponding to the present low-metamorphic Alpine Mandanici tectonic unit). The Mandanici Unit stratabound ores, essentially outcropping in Sicily, show a metal assemblage [tungsten-fluorine-Pb-Zn (Ag, As) with subordinate antimony and copper] largely inherited by the skarns and hydrothermal veins associated with the late Hercynian S-type granitoids, widespread in Calabria. On the whole, the geological and metallogenic framework of the "Calabro-Peloritan Hercynian Range" sensu Ferla et al. (1982–83) displays interesting analogies and predictable correlation possibilities with those of other segments of the circummediterranean Hercynian Chain.

Introduction

Tungsten metallogenesis in the metamorphic rock belt of the Peloritani Mountains (Northeastern Sicily) is a geologically significant element, recently defined during

[1] Istituto di Mineralogia e Petrologia, Università di Padova, Corso Garibaldi 37, I-35100 Padova, Italy

[2] Istituto di Mineralogia, Università di Ferrara, Corso Ercole I d'Este 32, I-44100 Ferrara, Italy

[3] Istituto di Mineralogia, Petrografia e Geochimica Università di Palermo, Via Archirafi 36, I-90123 Palermo, Italy

[4] Laboratoire de Minéralogie, Université Catholique de Louvain, Place Pasteur 3, 1348 Louvain-la-Neuve, Belgium

Mineral Deposits within the European Community
(ed. by J. Boissonnas and P. Omenetto)

the research programs carried out by ENTE MINERARIO SICILIANO (EMS) in cooperation with the EEC. At the outset, the reliability of this exploration target was founded on the positive results of the investigation for scheelite performed by some of the present authors in the Calabrian Arc on SNIA BPD-Mining Division prospects (Aspromonte, Serre and Longobucco regions). Owing to the predictable connections (partly confirmed by preliminary observations) between Sicilian and Calabrian high-grade metamorphic units of the Calabrian-Peloritan Arc, no special hopes were attached to the low-grade units (and in particular to the Mandanici Unit, mostly outcropping in Sicily). On the contrary, the prospecting results indicated a totally capsized situation, with the most interesting discoveries concentrated only within the Mandanici Unit. At the present time, mining reconnaissance workings on Sicilian tungsten (geochemical and geophysical exploration, drilling, mineral processing tests) are in progress.

1 Peloritani Mountains: Lithostratigraphic Setting and Mineralization

The Peloritani Mountains in Northeastern Sicily are part of a metamorphic/magmatic rock belt comprised between the Appennine Range and the Sicilian Maghrebids. As briefly summarized in Fig. 1, the scientific debate concerning the geotectonic interpretation of this belt (so-called Calabrian-Peloritan Arc) involves the existence of a nappe-structured Africa-verging Hercynian orogenic segment, only weakly affected by Alpine tectonics and developed in the external (central and southern) sector of the Arc.

According to one of the present authors (Ferla 1974a, 1974b, 1982–83; Ferla et al. 1982–83; Censi and Ferla 1982–83) the Peloritani Mountains crystalline buildup pertains integrally to the "Calabro-Peloritan Hercynian Range" (Fig. 1). It is composed of two separate lithostratigraphic elements: (1) a pre-Hercynian basement (Mandanici Unit and Aspromonte Nappe); (2) a Paleozoic volcano-sedimentary cover sequence (South Peloritan Complex according to Ferla 1982–83).

1.1 Paleozoic Volcanosedimentary Sequence

This sequence (and its own Meso-Cenozoic cover) crops out in the form of several tectonic wedges, due to Alpine imbrication at the front Peloritan Arc/Sicilian

Fig. 1. Distribution of the most significant scheelite occurrences in the Calabrian-Peloritan Arc: tectonic sketch map according to Ferla et al. 1982–83 (model A by Lorenzoni, Zanettin Lorenzoni, Ferla and collaborators, bibliography 1978–1983, synoptic sketch map in Ferla et al. 1982–83). Different tectono-stratigraphic attributions (model B by Bonardi, Giunta and collaborators, bibliography 1976–1984, synoptic sketch map in Bonardi et al. 1982, with later modifications) are indicated in brackets in the map. Basically, according to A the CPA is formed by juxtaposed Alpine *l.s.* and Hercynian ranges; according to B, it is a composite belt of Alpine l.s.age. A brief comment to legend: the CPA is a metamorphic/magmatic rock belt comprised between the Appennine Range *(5)* and the Sicilian Maghrebids *(6)*. The northern sector is dominated by the Alpine Chain *(1)* Europe-vergent and (foll. A)

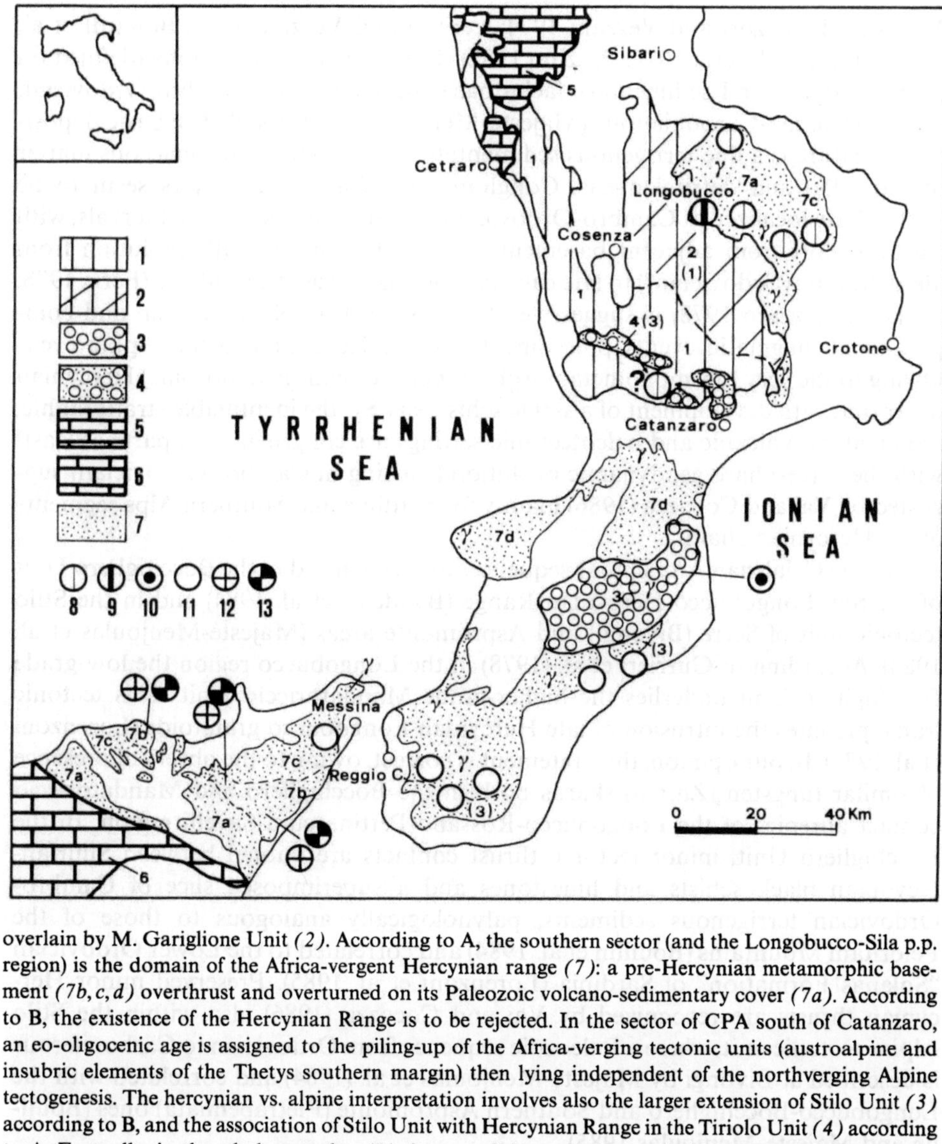

overlain by M. Gariglione Unit *(2)*. According to A, the southern sector (and the Longobucco-Sila p.p. region) is the domain of the Africa-vergent Hercynian range *(7)*: a pre-Hercynian metamorphic basement *(7b, c, d)* overthrust and overturned on its Paleozoic volcano-sedimentary cover *(7a)*. According to B, the existence of the Hercynian Range is to be rejected. In the sector of CPA south of Catanzaro, an eo-oligocenic age is assigned to the piling-up of the Africa-verging tectonic units (austroalpine and insubric elements of the Thetys southern margin) then lying independent of the northverging Alpine tectogenesis. The hercynian vs. alpine interpretation involves also the larger extension of Stilo Unit *(3)* according to B, and the association of Stilo Unit with Hercynian Range in the Tiriolo Unit *(4)* according to A. Factually, in the whole complex *(7)* the age of the main metamorphic and magmatic events is unanimously considered as pre-alpine l.s.: late Hercynian granitoids (γ) are particularly widespread in the Calabrian sector of the range. *Scheelite ores*: *8* skarns of Longobucco region, in the contact aureole between late Hercynian granodiorites and Bocchigliero (Mandatoriccio p.p.) units. W–Cu–Zn (garnet, fluorite, hedenbergite, idocrase; minor amphibole, epidote, scapolite); *9* vein skarn/greisen and hydrothermal vein stockworks with Zn, Pb, fluorite and minor scheelite in the granodiorites of Longobucco region; *10* disseminated Mo–W "porphyry-type" ores in the granodiorites and microgranites of Bivongi area (Stilo Unit); *11* "La Favière type" l.s. ores (skarn and skarnoid facies) with probable composite influence of volcanic (mafic), perianatectic, contact/granitic, and retrograde metamorphic environments, in the medium-high grade (p.p. metagranitic) sequences of Aspromonte Nappe *(7c)* in Southern Aspromonte and Northeastern Peloritani Mountains; *12* stratabound tourmaline + scheelite ores in the Mandanici Unit (*7b*, SE- and NW-Peloritani Mts.-cf. text); *13* stratabound Mangiaracina-type high grade scheelite ores in the Mandanici Unit (*7b*, SE- and NW-Peloritani Mts.-cf. text).

Maghrebids (Atzori and Vezzani 1974; Lentini and Vezzani 1975; Bonardi et al. 1976; Atzori and Ferla 1979; Pezzino 1982). The sequence stratigraphy (defined on paleontological and sedimentary facies basis) develops from Cambro-Ordovician up to Devonian-Carboniferous (Majesté-Menjoulas et al. 1986). The basic depositional feature is a fine terrigenous sedimentation very rich in carbonaceous matter (up to 10%) and detrital micas. Conglomeratic channel-like layers seem to be particularly localized at Cambro-Ordovician and Silurian-Devonian intervals, with clastic supply from adjacent basement sources. Volcanism, with evolution from alkalibasaltic (Ordovician?) to late calcalkaline basic to acidic products (Ferla 1978; Ferla and Azzaro 1978) is suggestive of the association of extensional and compressional magmatic events preceding the main Hercynian metamorphic cycle. Owing to the very low monometamorphic (anchi-to semi-metamorphic) Hercynian overprint, with development of a single schistosity S_1, the identifiable stratigraphic, sedimentary, volcanic and paleotectonic settings are congruent (pro parte at least) with the three-phase geodynamic evolution (starting in Cambro-Ordovician) suggested by Vai and Cocozza (1986) mainly for Sardinia and Southern Alps segments of the Hercynian chain.

In the Calabrian Arc, similar sequences are recognized in the Bocchigliero Unit of the Sila-Longobucco Hercynian Range (Baudelot et al. 1984) and in the Stilo tectonic unit of Serre (Bivongi) and Aspromonte areas (Majesté-Menjoulas et al. 1984). According to Gurrieri et al. (1978) in the Longobucco region the low-grade Bocchigliero Unit underlies the higher-grade Mandatoriccio Unit. This tectonic frame predates the intrusion of late Hercynian Longobucco granitoids (Lorenzoni et al. 1978). In our opinion, this statement is correct, owing to the observed presence of similar tungsten (Zn, Cu) skarns both in the Bocchigliero and Mandatoriccio contact aureoles of the Longobucco-Rossano-Pettinascura-Savelli region. In the Bocchigliero Unit, minor tectonic thrust contacts are quoted between Silurian-Devonian black schists and limestones and a superimposed slice of Cambro-Ordovician terrigenous sediments, palynologically analogous to those of the Peloritani Mountains (Bouillin et al. 1984) and correlated to the Lower Ordovician "Solanas Formation" of Sardinia (Lorenzoni et al. 1985). Preserved minor Hercynian thrusts are recognized by Vai and Cocozza (1986) also within the Stilo Alpine tectonic unit, where a Paleozoic sequence from Ordovician to Carboniferous is described at Bivongi by Majesté-Menjoulas et al. (1984) and correlated with the Longobucco-Bocchigliero and Southern Aspromonte (Pietrapennata) ones (Bouillin and Majesté-Menjoulas 1985).

In the Paleozoic volcanosedimentary sequence of Peloritani Mountains mineralizations are scarce and poorly known (Ferla 1982–83). Three types of ore occurrences are recognizable: (a) copper-pyrite (Zn, Pb) Kieslager, apparently hosted in black schists of Silurian-Devonian interval. Similar Kieslager are reported by Vighi (1949) and Bonardi et al. (1982) in connection with Silurian "ampelitic" schists and underlying "Ordovician" metabasites in the Bivongi area; (b) scattered self-sealing veinlets (not extending to the overlying Meso-Cenozoic cover) of prevailing Pb–Sb–Cu(Ag) sulphosalts in phyllites (Triscari and Saccà 1984); (c) Hg-anomalies (cinnabar in alluvium: Brondi and Ferretti 1966). Noteworthy is the ubiquitous *absence* of scheelite showings and alluvial anomalies.

1.2 Mandanici Unit

The Mandanici Unit is distinctive for the Peloritani Mountains geotectonic domain. In the Calabrian Arc only the small outcrops of the Cardeto window are assigned, on purely geological grounds, to this unit. A further discriminating parameter to be considered is the high concentration of both polymetallic (Ferla 1982–83) and tungsten stratabound ores (Brigo and Omenetto 1982–83; Omenetto 1984; Omenetto et al. 1986a, 1986b) with a typical metal assemblage (lead-zinc-silver; antimony (copper); fluorine-tungsten) apparently lacking in Calabria.

In consequence of assumed Hercynian thrusting and of the strong mechanical effects of Alpine imbrication, no exposures are known of the primary transition between the Mandanici lower-grade "basal" phyllites and the originally overlying Paleozoic volcanosedimentary sequence. Indirect evidence is provided by the clastic heritage in the detritic horizons of the Paleozoic sequence: (1) detrital micas with characteristics of crystal lattice and composition comparable with those of phyllite-forming micas of Mandanici Unit; (2) fragments of ilmenite-bearing phyllites, sulfide-bearing quartzites, chert in conglomeratic layers; (3) clastic heavy minerals such as ilmenite and tourmaline in metasandstone intercalations. According to Majesté-Menjoulas et al. (1986) a positive correlation (Acritarchs, volcanism) is predictable between Peloritan and Grande Kabylie (Algeria) Cambro-Ordovician sequences. In Grande Kabylie it is interesting to note that the conglomeratic-arenaceous base of the Djebel Aïssa Mimoun series (Upper Cambrian-Caradoc), unconformably overlying the Djebel Beloua phyllitic series, includes abundant reworked fragments and detrital micas from phyllite source (Baudelot and Géry 1979). A similar situation is recorded for Petite Kabylie (Baudelot et al. 1981). The lithological characters (Baudelot et al. 1981; Bossiére and Raymond 1972) of Grande-Petite Kabylie phyllitic sequences (graphitic and arenaceous phyllites with marble and porphyroid intercalations, grading to micaschists injected by pegmatite, quartz veins and amphibolites) are fairly similar to those peculiar to the Mandanici Unit sequence, with perhaps the exception of acidic metavolcanics.

Because of the complicated and, as a rule, poorly mapped tectonic pattern involving the boundary between the Mandanici Unit and the overlying Aspromonte Nappe (detailed field studies on this subject are in progress), the more reliable markers to distinguish the top-bottom limits of the Mandanici Unit are lithologies, metamorphic grade, and ores.

1.2.1 Lithostratigraphy and Metamorphism

The overturned lithostratigraphic column of the Mandanici Unit is essentially formed by epimetamorphic rocks. Common lithologies are phyllites (mainly carbonaceous and/or quartzose; also iron carbonate-, chlorite + albite-, chloritoid-, and garnet-bearing) with marble/calcschist intercalations, quartzites, metabasites (chlorite schists), grading upwards to micaschists and amphibolites of Aspromonte Nappe affinity. The metamorphic evolution (also common to the overlying Aspromonte Nappe micaschists and gneisses) is characterized by the following events:

1. A pre-Hercynian deformation phase F_1, with development of planar foliation S_1 (characterized by the syntectonic crystallization of ilmenite in phyllites and of kyanite in the micaschists and gneisses). This phase, tentatively considered as "Caledonian" epoch by Ferla (1974a, 1982–83) and Ferla et al. (1982–83), is also registered in the Grande and Petite Kabylie "phyllitic" basement (Bossiére and Raymond 1972, Baudelot and Géry 1979, Baudelot et al. 1981) where it is regarded as eo-Caledonian or older (late Cadomian ?) in age.
2. A main Hercynian deformation phase F_2, with development of strong foliation S_2, roughly parallel (on metric scale) to the original bedding S_0. This foliation is recognizable in the anchimetamorphic to greenschist facies lithologies of Mandanici Unit and in amphibolite facies sequences of Aspromonte Nappe.
3. A regional, mainly cataclastic deformation phase (F_3) with retrograde metamorphism.
4. A late Hercynian, regional thermal metamorphism, affecting in particular the northern sector of the Peloritani Mountains and associated with the granitic intrusions of Capo Rasocolmo and Capo d'Orlando.

1.2.2 Volcanism

In the less metamorphic dark-grey phyllite zone and in the basal part of the green phyllite zone (Fig. 2), evidence of synsedimentary volcanic activity is observable in the form of metabasite and chlorite schist layers, ranging in thickness from one to several tens of meters, with large areal diffusion and often showing gradual transition to the associated ancient sediments. The mafic lithologies are referable to hyaloclastitic or tuffitic basaltic products progressively mingled with pelitic material. The paragonite content of the paragonite marble (Ferla and Lucido 1973) possibly derives from metamorphism of submarine alteration products of volcanic glass. These processes appear to be coupled with palagonitic/spilitic alteration, nowadays recognizable in the metabasite assemblages (actinolite-albite-epidote-chlorite-sphene-calcite) grading to pure albite + chlorite zones. The pre-metamorphic alteration and sediment contamination produced chemical variability (Table 1) on a very small (outcrop) scale. However, the contents of titanium oxide (up to 4%) and of less "mobile" trace elements indicate a transitional character for this volcanic activity, in connection with a (local ?) extensional geodynamic regime.

1.2.3 Ore Distribution

The practical congruence between bedding S_0 and the dominant foliation S_2, as well as the preservation of the F_2 deformation characteristics within the whole regional extension of the Mandanici Unit, justify as feasible the comparison between the single lithostratigraphic sequences. The most reliable correlation elements are represented by three marble marker levels (Fig. 2: upper marble, paragonite marble, lower marble; according to Ferla 1982–83: Marbles I, II, III) and by the recognized

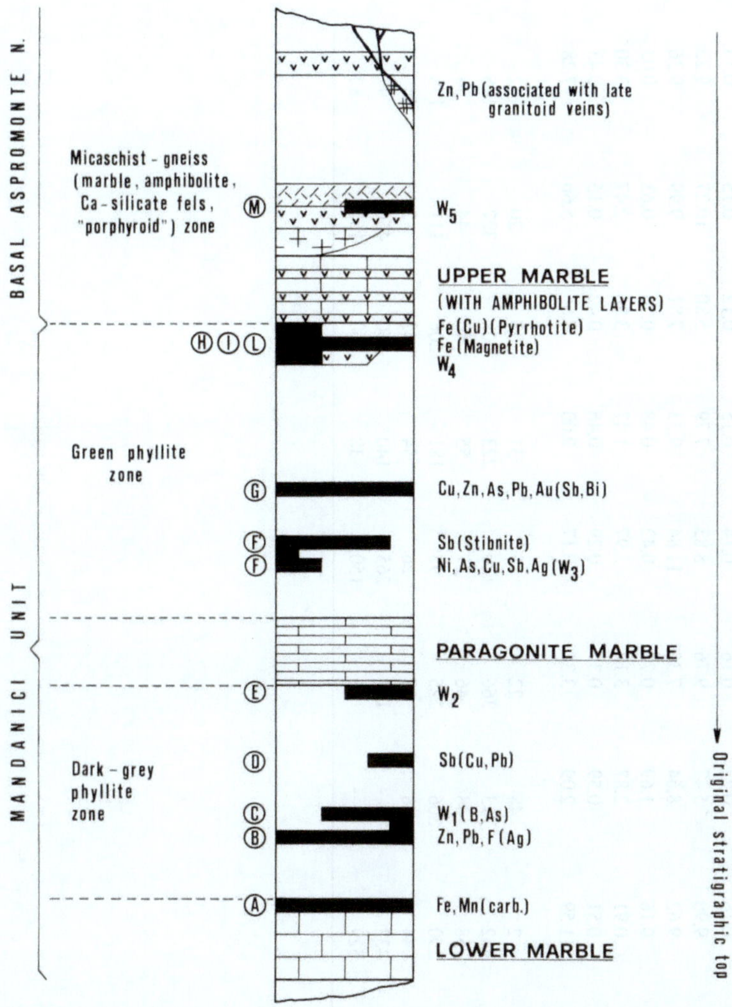

Fig. 2. Comprehensive metallogenic sequence of the pre-Hercynian basement (Mandanici Unit and Aspromonte Nappe)

spatial persistence of a number of stratabound ore-bearing horizons (Figs. 2, 3). Two metalliferous phyllitic zones are distinguishable, each one sandwiched between two marble levels:

1. A lower *dark-grey phyllite zone*, comprised between the lower marble and paragonite marble and characterized by noticeable thickness variations and lateral/vertical sedimentary facies changes. This zone includes the following mineralized levels (Fig. 3):

a) at the top of the lower marble: ubiquitous, thick "crusts" of (mostly oxidized) primary mm-rhythmic Fe–Mn(Mg) carbonates, with traces of galena, pyrite and

Table 1. Selected chemical analyses (XRF-spectrometry) of metavolcanic layers in the dark-grey phyllite zone and in the basal green phyllite zone of Mandanici Unit

	VACCO 1[a]	VACCO 2[a]	VACCO 3[a]	VACCO 4[a]	CAL 83[a]	RFM 42I[b]	RMS 20[b]	RFC 2A[b]	RFG 8[b]	RFM 45A[c]	RTM 9[c]
SiO_2 (wt.%)	40.35	31.04	39.15	36.18	43.66	45.59	39.17	40.05	50.16	45.55	42.75
TiO_2	2.28	2.18	2.18	3.12	3.69	2.42	1.35	2.83	1.13	1.19	3.11
Al_2O_3	15.13	14.93	12.46	13.49	14.38	13.07	14.28	13.24	13.68	13.78	12.12
Fe_2O_3	15.69	12.78	14.29	14.28	16.69	14.03	9.49	15.14	7.56	10.10	12.19
MnO	0.14	0.18	0.17	0.16	0.22	0.16	0.14	0.15	0.12	0.15	0.13
MgO	11.54	10.80	9.19	9.90	5.47	9.76	8.83	7.79	5.80	10.27	9.23
CaO	5.15	10.52	9.80	9.62	8.64	7.35	11.84	9.33	7.81	9.99	9.36
K_2O	0.14	0.12	0.09	0.16	1.69	0.75	0.42	0.48	0.76	0.63	0.03
Na_2O	1.03	5.72	2.41	0.91	2.87	3.12	3.97	1.47	3.86	2.47	3.30
P_2O_5	0.23	0.25	0.31	0.51	0.59	0.37	0.29	0.46	0.42	0.15	0.43
P.C.	8.43	11.44	10.02	11.59	2.09	3.31	10.17	9.00	8.65	5.66	7.28
Rb (ppm)	2	513	2	2	48	22	15	31	33	20	2
Sr	153	346	404	432	713	366	120	223	208	327	185
Y	38	54	42	58	48	46	27	55	40	44	34
Zr	273	231	199	290	306	225	93	181	206	111	178
Nb	43	42	34	44	48	28	26	54	42	2	46
Cr	404	513	473	432	—	152	265	140	150	456	97
Ni	382	582	548	529	—	59	150	38	60	155	82

[a] Metabasites above paragonite marble.
[b] Metabasites between lower and paragonite marbles.
[c] Metabasites below lower marble.

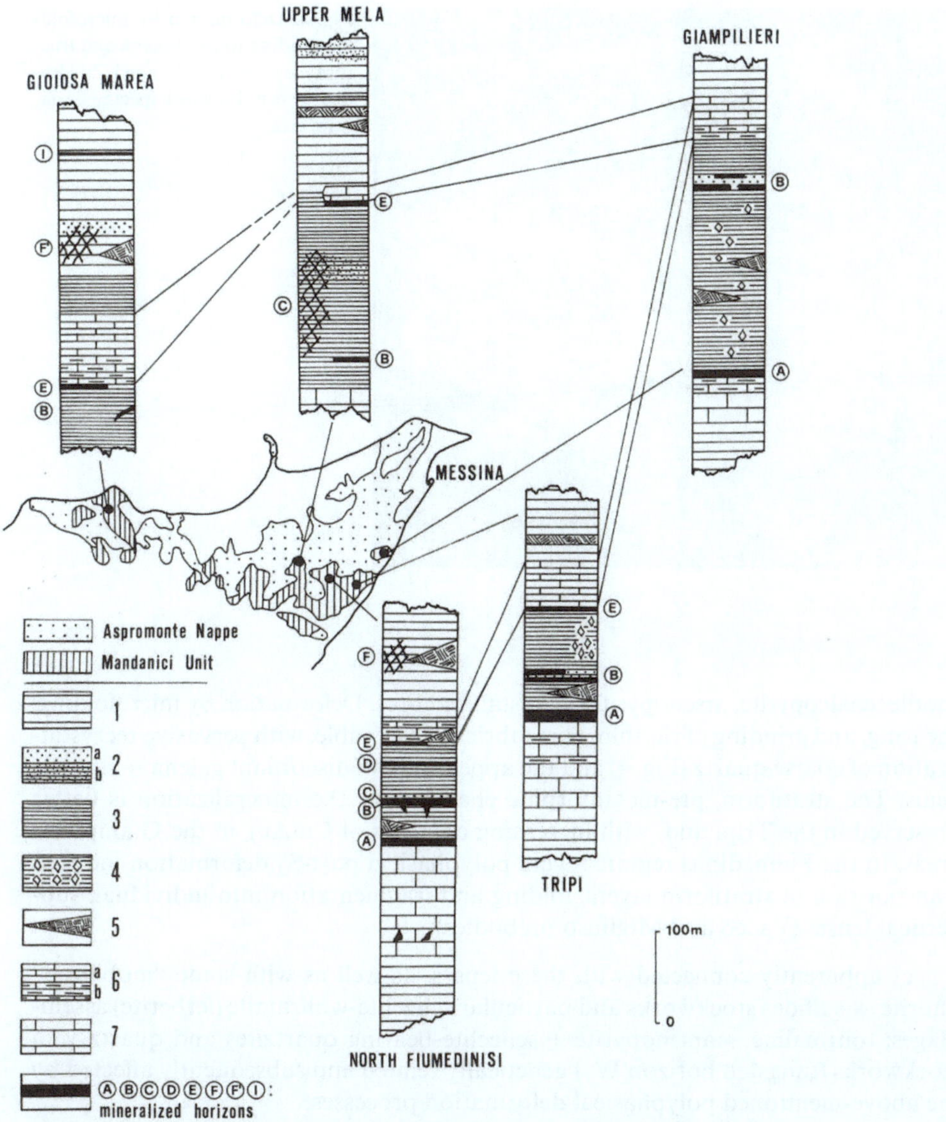

Fig. 3. Comprehensive lithostratigraphic columns of the Mandanici Unit with indication of the main mineralized horizons. *1* Phyllites; *2*: *a* quartzites, *b* quartzitic phyllites; *3* graphitic phyllites; *4* iron carbonate phyllites; *5* metabasites; *6*: *a* paragonite marble, *b* calcschists; *7* marbles. Mineralized horizons: cf. Fig. 2

barite. In the basal section of the marble, quartz ± scheelite veinlets and karst cavities filled with Cu–Sb–sulfosalts and Fe–Mn oxides + barite are observable;

b) in medium-thick (⩾50 m) series of graphitic phyllites and quartzites, with minor volcanogenic intercalations: greenish (chlorite-bearing) and gray quartzites, rhythmically alternating with fine-banded, rich sphalerite + fluorite (galena, pyr-

Fig. 4. Deformation by microfolding grading to axial plane and fracture cleavage of a banded Zn–Pb-quartz ore. Polished specimen, natural size

rhotite, chalcopyrite, arsenopyrite) recrystallized ores. Deformation by microfolding, shearing, and grinding of rhythmic ore fabric is observable, with pervasive recrystallization of coarse quartz (Fig. 4) and the appearance of discordant galena + fluorite veins. The stratiform, pre-metamorphic character of the mineralization is better preserved in the Tripi and, with increasing contents of Cu(Au), in the Giampilieri areas. In the Fiumedinisi region, severe polyphasical post-S_2 deformation induced transposition of stratiform layers, folding and fragmentation into individual, subvertical lenses (Vacco and Migliuso orebodies);

c) apparently connected with these lenses, as well as with some "mobilized" fluorite + sulfides stockworks and particular scheelite-wolframite (ferberite) assemblages: tourmaline + arsenopyrite ± scheelite-bearing quartzites and quartz vein stockworks (tungsten horizon W_1) genetically related and subsequently affected by the above-mentioned polyphasical deformation processes;

d) within thick ($\geqslant$100 m) sequences of phyllites, quartz-phyllites, quartzites (graphitic p.p.) grading downwards to mineralized quartz-siderite calcschist facies and to iron carbonate phyllites: small Sb (Cu, Pb) sulphosalt occurrences;

e) in sequences of reduced ($\leqslant$20 m) thickness ("black schists", graphitic and ankeritic phyllites) immediately below the paragonite marble: tungsten ores (horizon W_2) with scheelite contents up to 50%. Ore paragenesis is monotonous (quartz, carbonate(Fe), Na-plagioclase, apatite with traces of pyrite, graphite) but the textural features are quite variable, because of intense deformation/shearing endured by the primary ore when coming (as in the North Fiumedinisi region) directly into tectonic contact with the overthrust Aspromonte Nappe.

2. An upper *green phyllite zone* (Fig. 2) found between the paragonite marble and the upper marble and characterized by increasing (upwards) metamorphic grade and the abundance of volcanogenic supply. As far as the type of volcanism is concerned, we must distinguish between the bottom (transitional) and top (tholeiitic) sections of the green phyllitic zone. Mineralized horizons:

f–f′) in the sequence developed above or laterally to paragonite marbles, particularly widespread (with significance of an actual stratigraphic maker) are strongly albitized transitional metabasites. In association late hydrothermal vein-like bodies exist, mineralized with chlorite (p.p. iron-rich), albite, carbonates, sulfides, only locally showing a week (Alpine ?) foliation. Confined to these sequences are a number of vein systems and vein stockworks outcropping (and partly mined in the past) in the areas of North Fiumedinisi, South Fiumedinisi and Gioiosa Marea (Fig. 3). A first type of mineralization is represented by brecciated phyllitic bodies cemented by quartz, sericite, arsenopyrite (pseudomorphically replaced by chlorite) grading to Cu–Ni–As–Sb–Ag assemblages (chalcopyrite, gersdorffite, tetrahedrite, polybasite, bournonite) with traces of gold (0.75 ppm Au) in the North Fiumedinisi area (Vacco, Giampaolo, Intera). Similar ores are found in the South Fiumedinisi region (Triscari 1985) at Zilli (Sb-bearing gersdorffite, Fe-tetrahedrite, traces of chalcopyrite with quartz, siderite, ankerite gangue) and considered to be connected with the major vein system of the old S. Carlo mine, where the filling-up of ac-tension joints in the green phyllites is provided by Cu–Sb–Ag(Bi, Pb, Zn) ores (Donati et al. 1978: tetrahedrite, chalcopyrite, bournonite, stromeyerite and traces of Pb–Zn sulfides, arsenopyrite, boulangerite, jamesonite, bismuthinite) with small amounts of scheelite (W_3 in Fig. 2) in chalcedony-like quartz matrix (Triscari and Saccà 1982). A second ore type is characterized by the association stibnite-pyrite (marcasite) with minor amounts of sphalerite (galena) and traces of Ag and Au, being mined in the past at the Montagnareale (Patti) mine near Gioiosa Marea. Antimony ores are apparently linked, in the form of kidneys formed by fine vein-stockworks, to a black quartzite horizon;

g) intimately mixed with graphitic phyllites, chlorite schists, amphibole schists are stratiform Kieslager, mainly developed in the Bafia area (Val Pomia and Val Carbone mines, now worked out). The paragenesis consists of chalcopyrite (with inclusions of star-like sphalerite), arsenopyrite, pyrite, sphalerite, galena, native gold, pyrrhotite, tetrahedrite, boulangerite, tetradymite, native silver. "Gangue" minerals are the essential components of the host rock: chlorite, amphibole, ilmenite, rutile, quartz, calcite (Omenetto 1972);

h–i) in the overlying iron-rich metasediments (chloritoid-bearing phyllites, ± garnetiferous) below the upper marble: stratiform lenses of magnetite and/or pyrrhotite (with traces of Cu, Zn sulfides). Magnetite deposits are widespread both in western and eastern Peloritani Mountains.

1.3 Aspromonte Nappe

According to Ferla (1982–83) the observable tectonic contacts between the Aspromonte Nappe and the underlying Mandanici Unit (also involving part of the tungsten horizon W_2 in the Fiumedinisi region) are to be considered Hercynian (with Alpine echos) minor thrusts, internal to a major geotectonic buildup with normal lithostratigraphic Mandanici Unit/Aspromonte Nappe transition (Fig. 2). The boundary is marked by the upper marble level, grading to a thick sequence of high-medium grade metamorphic graywackes. The marble-bound section of the paragneissic sequence is particularly characterized by a "variegated" alternance including, in addition to marbles and Ca-silicate fels, pelitic and mafic/acidic volcanic intercalations (kyanite-staurolite micaschists, amphibolites and less frequent augengneisses). Minor foliated and unfoliated granitic, aplitic, and pegmatitic vein-like bodies could be related to respectively ancient (eo-Caledonian/late-Cadomian ?) and younger (Hercynian) late-orogenic magmatic activity, this last genetically related to the deepest effects (crustal anatexis) of the late-Hercynian regional thermal metamorphic event, with poor evidence in the Northern Peloritani Mountains (small granitic intrusions of Capo Rasocolmo–Capo d'Orlando) and, on the contrary, impressively widespread in Calabria.

With the exception of some hydrothermal base metal vein occurrences, observed in connection with granitoids veins and (Ferla 1982–83) close to the Capo Rasocolmo—Capo d'Orlando intrusions, the metallogenic interest of the Aspromonte Nappe in the Peloritani Mountains appears to be restricted to the above-mentioned variegated alternances around the upper marble horizon. Here, monomineralic tungsten ores (horizons $H-W_4$ and $M-W_5$ in Fig. 2) are represented by scheelite-bearing ($WO_3 \leqslant 1\%$) amphibolites (where scheelite is a component of a peculiar metamorphic assemblage: tremolite-ferroactinolite, white mica, epidote, chlorite, apatite, altered plagioclase and abundant sphene). According to Ghezzo (1967), these amphibolites derive from retrogressive metamorphism of high-grade horn-blende-plagioclase amphibolites (very frequent and practically scheelite-free) under tectonic influence (along thrust contacts ?), thus evoking the problem (not yet adequately approached) of tungsten source and concentration mechanism. Chemically (Ferla and Azzaro 1978), the amphibolites correspond to island arc tholeiites (with minor associated metarhyolites), indicating orogenic volcanic activity. Additional tungsten-bearing lithologies (again with scheelite as unique metalliferous component) are Ca-silicate fels (garnet, epidote, pyroxene, apatite, plagioclase, sphene, wollastonite quartz, carbonate), pegmatitic-aplitic gneisses, rare marbles and quartzites. These scheelite occurrences of the Aspromonte Nappe in Sicily, although interesting, have been poorly explored and studied up till now. They could be included in the same group of tungsten (with traces of Fe, Zn, Cu, Bi) ores discovered in the Aspromonte Nappe of Calabria (Fig. 1) and defined as more complex when compared with "La Favière type" ores (peri-anatectic sensu Sonnet et al. 1985) because of the probable composite influence of closely spaced syngenetic volcanic (mafic), perianatectic, contact/granitic, and retrograde metamorphic environments (lithologically reflected in the "variegated alternance sequences"). The additional fact is the presence of amphibolites as possible "preconcentration" loci for tungsten. According to Marignac and Weisbrod (1986), metamorphism (regional or confined to intrusions) creates effective conditions for tungsten mobilization with formation of Ca-silicate quartzites or skarnoids: including, in our opinion, as mobilization agent also the metamorphic environment developing peri-anatectic features.

The preceding remarks explain the potential interest (for tungsten, possibly for other metals) of those medium-high metamorphic grade lithological packages including the components of the "variegated sequences". In fact, this type of tungsten ore shows world-wide diffusion, in different geotectonic domains with ages ranging from pre-Cambrian to Early-Middle Paleozoic (Arribas 1979a, Höll 1980, Salim et al. 1980, Barnes 1983, Brigo and Omenetto 1983, Aissa 1986, Delakowitz and Hack 1986, Raith 1986, Pertold and Suk 1986). As tungsten concentration essentially results from the matching effects of different physicochemical processes, not ubiquitously and not equally intense, the economic interest of the related orebodies is normally limited.

2 Depositional Features of Tungsten Ores in the Mandanici Unit

As previously indicated, the most significant tungsten occurrences in the Mandanici Unit are hosted in the dark-grey phyllite zone found between the lower marble and the paragonite marble (Figs. 2, 3). The sedimentation environment (platform/back-platform basin with restricted circulation) is defined by shallow-water carbonates (Censi and Ferla 1982–83) associated with pelitic material rich in organic carbon (shales) and silica (lithic sandstones); with intermittent volcanogenic supply. Synsedimentary tectonic instability in the basin is suggested by thickness and sedimentary facies changes.

2.1 Pre-Metamorphic Stratiform Mineralization

Very scarce examples are observable in the Fiumedinisi area, and are not indicated in the column of Fig. 3. The mineralization consists of abundant microporphyroblastic ($\phi < 0.1$ mm) scheelite ($WO_3 \geqslant 1\%$) lining the planar foliation of sericite-chlorite quartzites, with subordinate plagioclase and K-feldspar (accessories apatite, zircon, pyrite, rutile and green tourmaline). Cross-cutting the foliation are some self-sealing scheelite veinlets. These tungsten-bearing quartzites are apparently linked to the top section (below the paragonite marble) of thick (100 m) ankerite-rich carbonaceous phyllite sequences. The top section includes several layers of pyrite- and rutile-bearing quartzites, showing foliation-parallel abundant green tourmaline (B_2O_3 3000 ppm) and traces of arsenopyrite, sphalerite, tetrahedrite, and possibly stibnite. Upwards, the quartzite layers grade to graphitic schists and paragonite marble.

2.2 Scheelite-Tourmaline Mineralization

The scheelite-tourmaline mineralization is particularly well exposed in the North Fiumedinisi area, close to the main Pb–Zn–F(Ag) stratabound ores. In this area the structural pattern essentially derives from a number of post-S_2 (the main

Hercynian foliation according to Ferla 1982–83) deformation events. The S_2 slaty cleavage is involved in a N 120°–130°/E 10° microfolding, locally grading to axial plane foliation and fracture cleavage (S_3). The formation of fissure systems (N 60° and N 90°) follows, predating larger-scale folding phases (N 115°, N 150° and N 45°) and late (Alpine) block faulting.

The scheelite mineralization (WO_3 0.17–11.1%, average 2.5%) is linked to a 5-m-thick horizon of quartz phyllites, capping the Vacco-Migliuso polymetallic bodies. The quartz phyllites are lined by a complex network of tourmaline-bearing quartz veins, up to 10 cm thick, including scheelite crystals and traces of albite, sericite, pyrite, and arsenopyrite. Laterally to the veins, the phyllites are tourmalinized over distances of several decimeters. Tourmalinization develops along the more or less deformed foliation planes. Minor quantities of scheelite appear in zones of superimposed pervasive silicification. Quartz-vein filling and tourmalinization are related to a cataclastic phase later than N 120°–130° microfolding, and predate the N 150°/N 45° deformation phases affecting both phyllites and quartz veins.

In order to clarify the relationships between conformable and discordant tourmalinized structures, a mineralogical study was performed by the present authors in collaboration with Y. Fuchs (Fuchs et al., in preparation) on three groups of tourmaline crystals: (1) foliation-parallel, (2) at the wall and (3) within the quartz veins. The results are briefly summarized here. Tourmalines are optically identical, yellow to amber in color. Composition is very similar, resulting appreciably alkali-deficient (with lack of proton deficiencies). According to the data compiled by Foit and Rosenberg (1977) and Fuchs (pers. commun.), tourmaline crystal growth occurred from a single fluid phase, at relatively high temperature ($\geqslant 350\,°C$) and low pH.

2.3 Scheelite-Carbonate-Quartz Mineralization

This mineralization invariably lies at the transition of graphitic schists and black shales to paragonite marble (E-horizon in Figs. 2, 3), apparently in areas of strongly reduced thickness ($\leqslant 20$ m) of the dark-gray phyllite zone. The single orebodies are lens-shaped, normally less than 1 m thick, and of limited extent although very rich in scheelite (up to 50%). In the low-metamorphic outcrops of Tripi-South Fiumedinisi area, far from tectonic contact with the Aspromonte Nappe, paragonite marbles and calcschists are extensively enriched in a quartz + albite (An_{0-2}) $\pm$ apatite paragenesis, also forming individual plagioclase quartzite layers at the marble base. Scheelite (with minimum contents of 10%) is essentially connected with quartz-albite concentrations, also co-existing in the marble with pyrite-graphite seams grading to stylolitic joints. Higher grade scheelite + carbonate ores are localized within polyphasical breccias in the marble, with later silicification. In the North Fiumedinisi area, close to the tectonic contact with the Aspromonte Nappe, the same type of ores are involved in a tectonized shear zone at the transition between silicified and albitized black shales and blastomylonitic paragonite marble *boudins*. Scheelite ores show strong evidence of shearing and tectonitic compaction. In the surround-

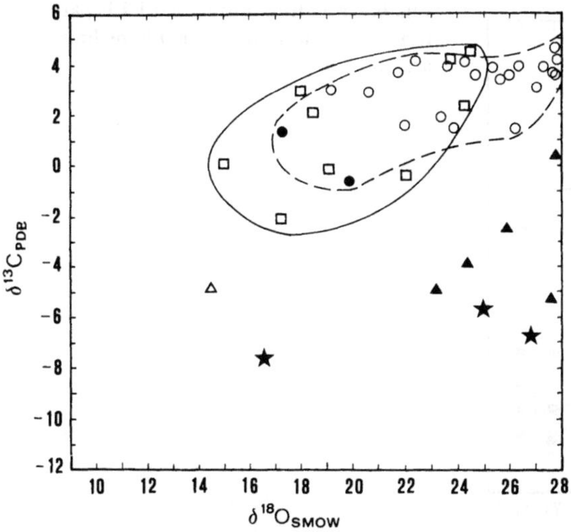

Fig. 5. Oxygen and carbon isotope composition of carbonates from scheelite-carbonate-quartz mineralization. *Stars* calcite of mineralized facies; *black circles* paragonite marbles near mineralization; *open triangles* vein calcite cutting mineralization; *open circles* paragonite marbles (after Censi and Ferla 1982–83); *open squares* lower marbles (*ibidem*); *black triangles* graphitic phyllites (*ibidem*)

ings very frequent are layers of deeply albitized metabasites (cf. Sect. 2.2.2). In the Gioiosa Marea area (T. Zappardino) paragenetically similar ores (undeformed) are characterized by the very coarse-grained (up to some centimeters in ϕ) porphyroblastic growth of scheelite in a roughly parallel, mm- to cm-spaced fabric.

Geochemical studies on this type of tungsten ore are in progress. Preliminary data are available concerning oxygen and carbon isotope composition of carbonates pertaining to the host rock (paragonite marble), to ore facies and to late barren cross-cutting veins. With respect to those of barren marble (Fig. 5) ore facies calcites show increasing $\delta^{18}O$ (from +18 to +25) and decreasing $\delta^{13}C$ (from 0 to −6), provisionally indicating fluids peculiar to metamorphic domain, with some different fluid mixing for late barren vein calcite. The scheelite REE-patterns of carbonate-quartz-and tourmaline-bearing mineralizations are shown in Fig. 6. The only possible remarks concern the distinctly higher contents of scheelite from tourmaline-rich ores, and the slightly positive Eu-anomaly (higher oxygen fugacity ?) for scheelite from carbonate + quartz ores close to paragonite marble.

3 Discussion and Concluding Remarks

Insufficient geochemical data, necessary to correctly explain the mechanism of tungsten ore deposition in the Peloritani Mountains geological domain, make any genetic statement on the matter only provisional. However, the previously exposed geological and metallogenic facts are certainly exploitable for a brief discussion.

A first problem concerning the main tungsten ore types is their "stratabound" character. Scheelite-tourmaline mineralization of North Fiumedinisi area shows general epigenetic features. In addition, we must point out that the vertical extent

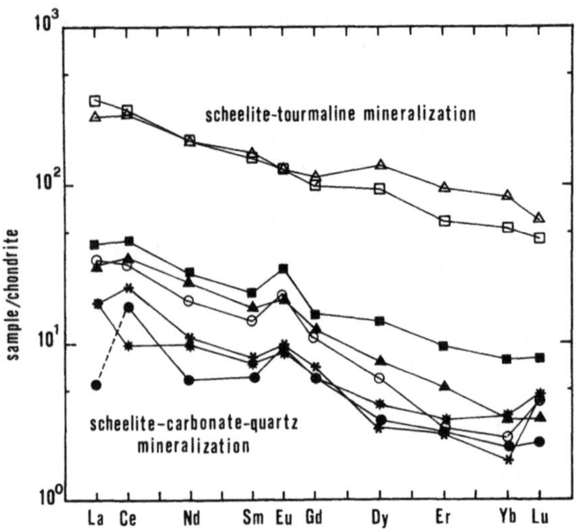

Fig. 6. Chondrite-normalized REE patterns for scheelites of the Fiumedinisi area ores

(Fig. 3) of the tourmalinization process is considerably greater in the Upper Mela area, comparatively more influenced by the peripherical effects of the late Hercynian regional thermal metamorphism, reaching its climax in the Gioiosa Marea area (characterized by the static crystallization of andalusite, sillimanite and K-feldspar in the Mandanici Unit phyllites and by the pegmatitic intrusions of Capo Calavà). As previously indicated, the Gioiosa Marea (T. Zappardino) scheelite-carbonate-quartz ores exhibit "giant" porphyroblastic texture. Significant also is the "hypothermal" paragenesis of some strictly associated scheelite-bearing veins, including coarse-grained arsenopyrite, iron-rich sphalerite with dominant pyrrhotite inclusions and some peculiar intergrowths of galena + pyrrhotite + native bismuth. Also in this case the presently observable mineralized features bear the signature of the high temperature metamorphic recrystallization and mobilization effects. Coupling the few geochemical data given in Sections 2.2. and 2.3., it is tempting to conclude as to the positive role of regional thermal metamorphism in modelling the last face of the ores. On the other hand, the lateral lithostratigraphic persistence of both scheelite + tourmaline and scheelite + carbonate + quartz "horizons" remains confirmed from Ionian to Tyrrhenian slopes of Peloritani Mountains (Fig. 3). In the first case, probable boron and tungsten "protoconcentrations" (not ubiqitously revealed by metamorphic mobilization, as the scarce evidence of pre-metamorphic scheelite-and tourmaline-bearing lithologies seems to demonstrate) are to be ascribed to exhalative activity in connection with mafic volcanism (a genetic statement quite common to the voluminous literature: Cunningham et al. 1973, Plimer 1980, Barnes 1983, Appel 1985, Brodtkorb (de) et al. 1985, Arribas Rosado 1986, Delakowitz and Hack 1986, Raith 1986, etc.). In the second case, the real problem is to demonstrate the "primary" or "secondary" character of this unusually rich tungsten ore. In fact, to our knowledge, no examples exist of strata-bound mineralizations with comparable scheelite grades at the orebody scale, with perhaps the exception of some tungsten-bearing silica sinters defined in parts of ancient deposits (Höll 1986) or found in very recent to present-day epithermal environments (Henley et al. 1986).

Looking at the Calabrian Arc, it is interesting to note that the mineralized skarns and hydrothermal veins, genetically related to the abundant, S-type late Hercynian granitoids (intruded in sequences of the Paleozoic volcano-sedimentary cover, as in the Longobucco region) clearly reflect the bulk paragenesis of the pre-Ordovician (seemingly Cambrian) Mandanici Unit stratabound ores, inherited in the form of typical associations: scheelite-fluorite-Zn(Cu) skarns and Pb–Zn-fluorite-barite (scheelite) veins, where fluorite (economically exploited) appears as the most discriminating element (cf. Fig. 1).

By taking into account the considerable problems related to the reconstruction of the geodynamic evolution of the circummediterranean Hercynian Chain by geological (stratigraphic, tectonometamorphic, magmatic) correlation between the different segments (Vai 1979, Schmidt and Soellner 1982, Matte 1986, Vai and Cocozza 1986), it is stimulating to confirm the precise metallogenic correlation possibilities between segments such as Calabrian-Peloritan Arc and Montagne Noire (similar Cambrian sequences and global metal spectrum (W, F, B, Pb, Zn), presence of tourmaline-scheelite and scheelite-black schist ores, common F-bearing scheelite skarns linked to Hercynian granites; cf. Boyer and Routhier 1958, 1974, Béziat and Tollon 1979, Demange 1983, Tollon et al. 1984, Béziat et al. 1986, Caleffi et al. 1986, Safa et al. 1987). With the support of analogous patterns from the Hercynian belts of Spain (Arribas 1979a, 1979b; Arribas-Rosado 1986), these correlation attempts could surely be extended to other segments of the circummediterranean Hercynian Chain, such as the Grande and Petite Kabylie basements, where Touahri and Fuchs (1986) describe, at the lithostratigraphic level of Mandanici Unit, comparable Zn–Pb-fluorite-chalcopyrite-stibnite-tetrahedrite ores (with minor amounts of silver and traces of gold) without (?) tungsten; Sardinia (Sulcis region); etc. (Omenetto et al., in preparation).

Acknowledgements. The authors wish to thank Rudi Höll and Christian Marignac for helpful discussions, Philippe Sonnet and Yves Fuchs for the unvaluable collaboration. Research performed in the framework of the EMS/EEC (*Contract N. MSM-061-I*).

References

Aissa DE (1986) Les gisements et indices métallifères de l'Edough (Algérie). In: Petrascheck-Jankovic (ed) Geotectonic evolution and metallogeny of the Mediterranean area and western Asia, Öst Ak Wiss 8:235–249

Appel PWU (1985) Strata-bound tourmaline in the Archean Malene supracrustals, West Greenland. Can J Earth Sci 22:1485–1491

Arribas A (1979a) Les gisements de tungstène de la zone de Morille (Province de Salamanca, Espagne). Chr Rech Min 450:27–34

Arribas A (1979b) Le gisement de tungstène de Barruecopardo. Chr Rech Min 450:42–49

Arribas-Rosado A (1986) The significance of tourmaline in stratabound tungsten deposits in Spain. In: Tungsten Meeting Toulouse May 1986, Extended Abstracts p 33

Atzori P, Ferla P (1979) Caratteristiche del metamorfismo ercinico sulle successioni sedimentarie e magmatiche del basamento paleozoico delle unità inferiori dei M. Peloritani. Mem Soc Geol It 20:447–452

Atzori P, Vezzani L (1974) Lineamenti petrografico-strutturali della catena peloritana. Geol Rom 13:21–27

Barnes RG (1983) Stratiform and stratabound tungsten mineralisation in the Broken Hill Block, N.S.W. J Geol Soc Australia 30:225–239

Baudelot S, Géry B (1979) Découverte d'Acritarches du Cambrien supérieur et du Trémadoc dans le massif ancien de Grande Kabylie (Algérie). CR Acad Sci Paris 288 D:1513–1516

Baudelot S, Bouillin JP, Coiffait Ph (1981) Découverte d'Ordovicien inférieur daté par Acritarches dans l'Ouest de la Petite Kabylie (Algérie): conséquences structurales. CR Ac Sci Paris 293 (II):611–614.

Baudelot S, Bouillin JP, Majesté-Menjoulas C (1984) Données récentes sur le Paléozoique du massif de la Sila (Calabre, Italie). In: Sassi and Julivert (eds) IGCP N.5 Newsletter 6:9–10

Béziat P, Tollon F (1979) Le district à W–Sn–F de Montredon-Labessonié (Tarn, France). Chr Rech Min 450:13–18

Béziat P, Boudrie M, Guion JL, Tollon F (1986) Le district a W (Sn–F) de Montredon-Labessonnié (Tarn). Présence de concentrations stratiformes et filoniennes. In: Tungsten Meeting Toulouse May 1986, Extended Astracts p 65

Bonardi G, Giunta G, Liguori V, Perrone V, Russo M, Zuppetta A (1976) Schema geologico dei Monti Peloritani. Boll Soc Geol It 95:49–74

Bonardi G, De Vivo B, Giunta G, Lima A, Perrone V, Zuppetta A (1982) Mineralizzazioni dell'arco calabro-peloritano. Ipotesi genetiche e quadro evolutivo. Boll Soc Geol It 101:141–155

Bossière G, Raymond D (1972) Sur l'âge probablement anté-Llanvirn du métamorphisme des schistes satinés au Nord de la Grande Kabylie (Algérie). CR Acad Sc Paris 275 D: 641–644

Bouillin JP, Majesté-Menjoulas C (1985) Stratigrafia e struttura del Paleozoico calabrese. In: Meeting Siena Evoluzione Paleozoico Italiano, Note Brevi Riassunti p 102

Bouillin JP, Baudelot S, Majesté-Menjoulas C (1984) Mise en évidence du Cambro-Ordovicien en Calabre centrale (Italie). Affinités paléogéographiques et conséquences structurales. C R Acad Sci Paris 298 (2/3):89–92.

Boyer F, Routhier P (1958) Observations sur deux niveaux minéralisés dans le Paléozoique inférieur des monts du Minervois (Montagne Noire, Aude). Bull Soc geol France 6:257–266

Boyer F, Routhier P (1974) Extension régionale de couches à scheelite dans la couverture métamorphique de la zone axiale en Montagne Noire (Hérault, France). CR Acad Sc Paris 279 D:1829–1832

Brigo L, Omenetto P (1982–83) Inquadramento geotettonico e litostratigrafico delle mineralizzazioni a scheelite dell'arco calabro-peloritano. Rend Soc It Miner Petrol 38 (3):1480–1481

Brigo L. Omenetto P (1983) Scheelite-bearing occurrences in the Italian Alps: geotectonic and lithostratigraphic setting. In: Schneider HJ (ed) Mineral deposits of the Alps and of Alpine Epoch in Europe. Springer, Berlin Heidelberg New York, pp 41–50

Brodtkorb MK (de), Pezzutti N, Brodtkorb A, Schidlowski M (1985) Tourmaline schists and their relationship to Precambrian scheelite deposits from the San Luis Province, Argentina. In: German K (ed) Geochemical aspects of ore formation in recent and fossil sedimentary environments. Monograph Series Mineral Deposits N.25, Borntraeger, pp 151–160

Brondi A, Ferretti O (1966) Studio mineralogico delle alluvioni dei Monti Peloritani. Riv Min Sic 100/102:3–28

Caleffi D, Moine B, Béziat P, Tollon F (1986) Géochimie et minéralogie des schistes noirs dans l'environnement des minéralisations stratiformes en scheelite du district de Montredon (Montagne Noire). In: Tungsten Meeting Toulouse May 1986, Extended Abstracts pp 69–70

Censi P, Ferla P (1982–83) I marmi dei Monti Peloritani. Composizione isotopica dell'ossigeno e del carbonio e ricostruzione degli ambienti formazionali. Rend SIMP 38(3):1101–1117

Cunningham WB, Höll R, Taupitz KC (1973) Two new tungsten-bearing horizons in the Older Precambrium of Rhodesia. Mineral Deposita 8:200–203

Delakowitz B, Hack M (1986) Tungsten deposits in the Sierras Pampeanas, Province of San Luis, Argentina. A. "Sierra del Morro-Oeste", B. "Pampa del Tamboreo". In: Tungsten Meeting Toulouse May 1986, Extended Abstracts pp 41–43

Demange M (1983) La Montagne Noire: Geotraverse A_1 et coupes stratigraphiques de corrélations (S.C.F.). In: Sassi-Szederkényi (eds) IGCP 5 Newsletter 5:267–277

Donati G, Stagno F, Triscari M (1978) Ricerche sulle manifestazioni metallifere dei M. Peloritani. III) Giacimenti delle contrade S. Carlo, Montagne e viciniori presso Fiumedinisi (Messina). Atti Acc. Peloritana Pericolanti Messina, Cl. Sc. Mat. Fis. Nat. 56:177–238

Ferla P (1974a) Aspetti petrogenetici e strutturali del polimetamorfismo dei Monti Peloritani (Sicilia). Per Miner 43:517–590

Ferla P (1974b) Contributo alla conoscenza della natura del basamento pre-ercinico nei M. Peloritani (Sicilia). Le metapsammiti devoniane contenenti clasti filladici associate a filladi a paragonite del settore di Floresta. Miner Petr Acta 20:63–77

Ferla P (1978) Natura e significato geodinamico del vulcanismo preercinico presente nelle filladi e nei semiscisti dei Monti Peloritani (Sicilia). Rend SIMP 34(1):55–74

Ferla P (1982–83) Inquadramento geologico-petrografico delle mineralizzazioni metallifere nei Monti Peloritani (Sicilia). Rend Soc It Miner Petr 38(3):1075–1091

Ferla P, Azzaro E (1978) Natura del vulcanismo precedente al metamorfismo ercinico nelle unità inferiori dei M. Peloritani orientali. Boll Soc Geol It 97:767–774

Ferla P, Lucido G (1973) Paragenesi a paragonite-cloritoide e paragonite-calcite nelle rocce di basso grado metamorfico del settore di Gioiosa Vecchia (Peloritani occidentali). Per Miner 42:381–437

Ferla P, Lorenzoni S, Zanettin Lorenzoni E (1982–83) Geological constitution and evolution of the Calabro-Peloritan Hercynian Range. Rend Soc It Min Petrol 38(3):951–962

Foit FF Jr, Rosenberg PE (1977) Coupled substitutions in the tourmaline group. Contr Min Petr 62:109–127

Ghezzo C (1967) Notizie petrografiche e geologiche sui terreni metamorfici rilevati nella tavoletta II S.O. del Foglio Castroreale (Monti Peloritani). Rend Soc Min It 23:47–58

Gurrieri S, Lorenzoni S, Zanettin Lorenzoni E (1978) L'unità pre-alpina di Bocchigliero (Sila-Calabria). Boll Soc Geol It 97:711–716

Henley RW, Hedenquist JW, Roberts PJ (eds) Guide to the active epithermal (geothermal) systems and precious metal deposits of New Zealand. Monograph Series on Mineral Deposits n. 26, Borntraeger 1986.

Höll R (1980) Tungsten in Central Europe. 26th IGC Paris Abstracts Volume 3(13):948

Höll R (1986) Nature and origin of the tungsten deposit Felbertal/Austria. In: Tungsten Meeting Toulouse May 1986, Extended Abstracts pp 7–8

Lentini F, Vezzani L (1975) Le unità meso-cenozoiche della copertura sedimentaria del basamento cristallino peloritano (Sicilia nord-orientale). Boll Soc Geol It 94:537–554

Lorenzoni S, Messina A, Russo S, Stagno F, Zanettin Lorenzoni E (1978) Le magmatiti dell'unità di Longobucco (Sila-Calabria). Boll Soc Geol It 97:727–738

Lorenzoni S, Minzoni N, Zanettin Lorenzoni E (1985) La "Formazione di Solanas" nella Sila e nell'Aspromonte (Calabria). In: Meeting Siena Evoluzione Paleozoico Italiano, Note Brevi Riassunti p 47

Majesté-Menjoulas C, Bouillin JP, Cygan C (1984) La série de Bivongi, type de succession paléozoïque (Ordovicien a Carbonifère) de Calabre méridionale. C R Acad Sci Paris 299(2/6):249–252

Majesté-Menjoulas C, Bouillin JP, Cygan C, Fournier-Vinas C (1986) Les formations paléozoïques (Cambrien à Carbonifère) des Monts Péloritains (Sicile). Premierès datations par Acritarches et Conodontes. C R Ac Sci Paris 303(2/14):1315–1320

Marignac Ch, Weisbrod A (1986) L'origine des gisements stanno-wolframifères: une approche critique. In: Tungsten Meeting Toulouse May 1986, Extended Abstracts pp 24–25

Matte Ph (1986) La chaine varisque parmi les chaines paléozoiques périatlantiques, modèle d'évolution et position des grands blocs continentaux au Permo-Carbonifère. Bull Soc géol France (8) 2/1:9–24

Omenetto P (1972) Le manifestazioni metallifere di Bafia nei Monti Peloritani (Sicilia). Mem Ist Geol Min Univ Padova 29:1–16

Omenetto P (1984) La scheelite nei terreni cristallini della Sicilia nord-orientale. EMS Informazioni 1(9):10–13

Omenetto P, Meggiolaro V, Spagna P, Brigo L, Ferla P, Guion JL (1986a) Lithostratigraphic, tectonic and geochemical features of tungsten-bearing metalliferous sequences of Southeastern Peloritani Mountains (NE-Sicily). In: Tungsten Meeting Toulouse May 1986, Extended Abstracts pp 34–35

Omenetto P, Meggiolaro V, Spagna P, Brigo L, Ferla P, Guion JL (1986b) Prevariscan-Variscan scheelite ores in Calabrian-Peloritan Arc: tungsten/base metal association in Southeastern Peloritani Mountains (NE-Sicily). In: IGCP 5 Final Meeting Sardinia May 1986, Extended Abstracts pp 62–64

Pertold Z, Suk M (1986) Precambrian of the Bohemian Massif and its metallogeny. In: Pouba Z (ed) Proc. Conf. Metallogeny Precambrian. Geol Surv Prague, pp 9–22

Pezzino A (1982) Confronti petrografici e strutturali tra i basamenti metamorfici delle unità inferiori dei M. Peloritani (Sicilia). Per Miner 51:35–50

Plimer IR (1980) Exhalative Sn and W deposits associated with mafic volcanism as precursors to Sn and W deposits associated with granites. Mineral Deposita 15:275–289

Raith J (1986) Scheelite occurrences in the crystalline area east of the Hohe Tauern, Eastern Alps, Austria. In: Tungsten Meeting Toulouse May 1986, Extended Abstracts p 32

Safa P, Couilloud D, Tessier B, Moine B (1987) Le gisement de tungstène de Fumade (Tarn, France): gitologie, caractères pétrographiques et géochimiques des skarns et des granites. Chron rech min 487:39–51

Salim J, Legrand JM, Dardenne MA (1980) Contrôles géologiques et genèse des minéralisations de tungstène de la province scheelitifère du NE du Brésil. 26th IGC Paris Abstracts Volume 3(13):1004

Schmidt K, Soellner F (1982) Proposals for the geodynamic interpretation of the "Caledonian event" in Central and Southern Europe, with special reference to the Eastern Alps. In: Sassi and Varga (ed) IGCP 5 Newsletter 4, pp 82–93

Sonnet P, Verkaeren J, Crevola G (1985) Scheelite bearing calc-silicate gneisses in the Provence crystalline basement (Var, France). Bull Mineral 108:377–390

Tollon F, Moine B, Prouhet JP (1984) Tungstène lié aux schistes noirs et milieux confinés. Greco 52 — SGA: Concentrations métalliques et milieux confinés, Meeting Paris

Touahri B, Fuchs Y (1986) Les minéralisations de plomb, zinc, cuivre du domaine alpin en Algérie. In: Petrascheck-Jankovic (ed) Geotectonic Evolution and Metallogeny of the Mediterranean Area and Western Asia, Öst Ak Wiss 8:217–233

Triscari M (1985) A first occurrence of gersdorffite in the Peloritani Mts. (Sicily N.E.). Rendic SIMP 40:289–294

Triscari M, Saccà C (1982) Scheelite at the "S. Carlo" Sb–Cu–Ag-Mine, Fiumedinisi, Messina, Italy. Miner Petrogr Acta 26:159–168

Triscari M, Saccà C (1984) Contributo alla conoscenza dei minerali metalliferi dei Monti Peloritani (Sicilia N.E.) VI) La Boulangerite. Rend SIMP 39(1):145–154

Vai GB (1979) Tracing the Hercynian structural zones across "Neo-Europa": an introduction. Mem Soc Geol It 20:39–45

Vai GB, Cocozza T (1986) Tentative schematic zonation of the Hercynian chain in Italy. Bull Soc géol France (8) 2/1:95–114

Vighi L (1949) Mineralizzazioni a solfuri misti e ad ossidi di ferro nel bordo orientale della Serra S. Bruno, in Calabria. Mem e Note Ist Geol Appl Napoli 2:37–74

Controls on the Occurrence and Distribution of Tungsten and Lithium Deposits on the Southeast Margin of the Leinster Granite, Ireland

P. McArdle[1] and P.S. Kennan[2]

Abstract

In the Caledonides of SE Ireland there are significant tungsten and lithium deposits. The wall rocks of both deposits are volcano-sedimentary rocks of Lower Palaeozoic age which occur on the SE flank of the Leinster Granite batholith. Tungsten and lithium occur in geographically separated areas, but both areas are traversed by the East Carlow Deformation Zone, an important shear zone. Tungsten mineralization consists of scheelite hosted in quartz veins or disseminated in their immediate vicinity, but significant scheelite occurrences are confined to microgranite sheets which follow the local lithostratigraphic contacts. Lithium is present in a series of spodumene pegmatites which are characteristically associated with metavolcanic rocks at some distance from the Leinster Granite.

Tungsten and lithium have had a multistage history of evolution. They are regarded as having been present at an early stage in the wall rocks, probably as a result of seafloor hydrothermal activity. Subsequently additional orogenic mechanisms, such as shearing and granite intrusion, effected a remobilization and concentration of these elements which gave rise to the deposits as now observed.

1 Introduction

Lithium and tungsten mineralization in Ireland is confined to sequences of Lower Palaeozoic rocks which were deformed during the Caledonian orogeny (Williams and McArdle 1978). Lithium deposits are known only from the Caledonides of SE Ireland, whereas tungsten deposits occur more widely (McArdle et al. in press). This paper will discuss both types of mineral deposits in SE Ireland. The lithium mineralization contains minor amounts of tantalum and tin, while the tungsten deposits contain minor quantities of tin (Steiger and von Knorring 1974, Steiger and Bowden 1982).

[1] Geological Survey of Ireland, Beggars Bush, Haddington Road, Dublin 4
[2] Dept. of Geology, University College, Stillorgan Road, Dublin 4

Mineral Deposits within the European Community
(ed. by J. Boissonnas and P. Omenetto)

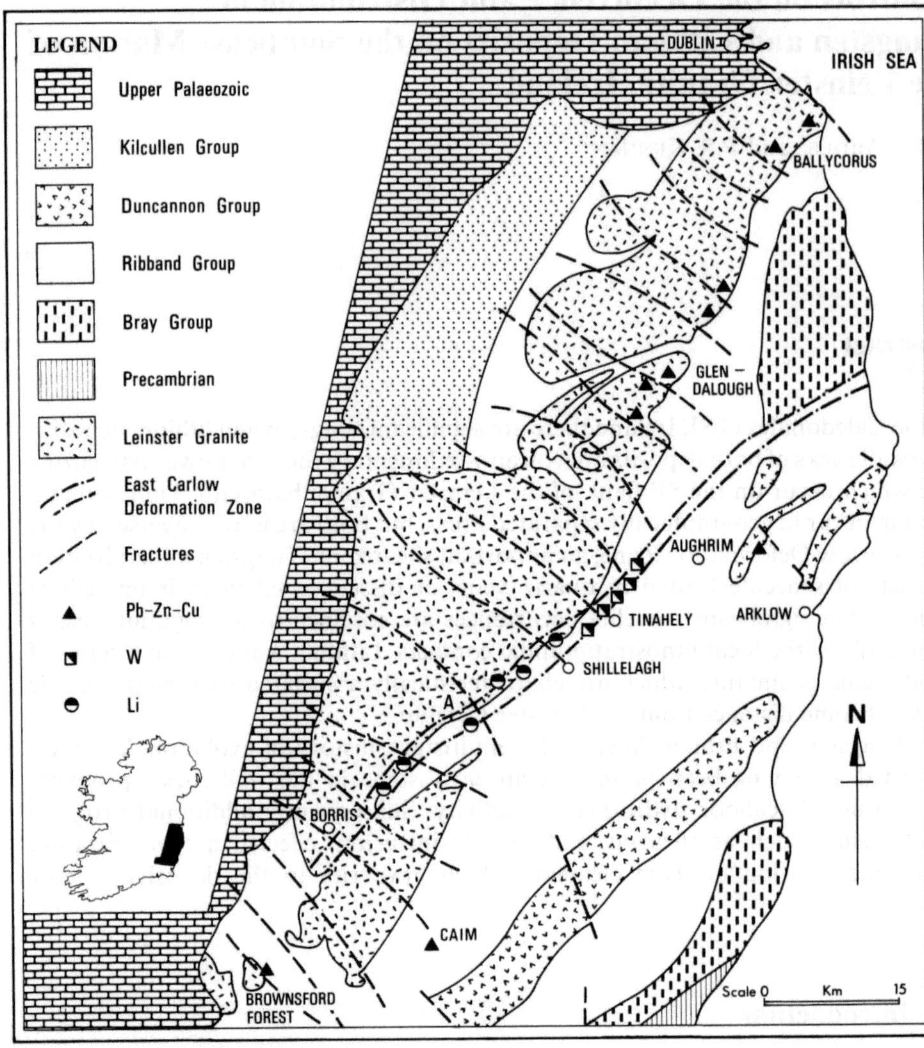

Fig. 1. Metal deposits, lithostratigraphy and structural features in the vicinity of the Leinster Granite in SE Ireland. (After Kennan et al. 1986). *A* Aclare House

Lithium and tungsten deposits are spatially related to the SE margin of the major Leinster Granite batholith which was emplaced in Caledonian times into a series of Cambrian to Silurian volcanic and sedimentary rocks. Lead-zinc mineralization is also related to the same margin of this batholith (Fig. 1). It has been shown that, although this latter mineralization is hosted in veins within the granite, the source of the lead and zinc probably lies in the aureole rocks and that a complex sequence of events gave rise to the concentration of the metals in the granite (Williams and Kennan 1983). The purpose of this paper is to extend the application of these concepts to the lithium and tungsten deposits. It will be argued that the original source of both lithium and tungsten lay in the sedimentary and volcanic rocks of the district and that they were remobilized by later tectonic, plutonic and metamorphic events.

2 Stratigraphy of the Host Rocks

The Lower Palaeozoic rocks of SE Ireland have been divided into four lithostratigraphic groups (Brück et al. 1979 and Fig. 1). The Bray Group (lower to middle Cambrian) comprises greywackes and quartz arenites, the Ribband Group (middle Cambrian to Llandeilo) consists of distal turbidites with local volcanic rocks, the Duncannon Group (largely Caradoc) is composed of volcanic rocks with minor sediments and the Kilcullen Group (lower Ordovician to Wenlock) is composed of a greywacke sequence. The mineralization described in this paper is hosted in, or spatially associated with, rocks of the Ribband Group.

A succession of three formations belonging to the Ribband Group has been described from the district containing the mineralization (McArdle 1981, 1984). The Ballybeg Pelitic Formation, forming the base of the Ribband Group, is 2000–2500 m thick and consists of thinly bedded siltstones, with associated coticule, tourmalinite, dacitic tuff, quartzite and greywacke. The lithologies all formed in a deep water marine environment and the dominant siltstones are regarded as distal turbidites. Within the metamorphic aureole of the Leinster Granite these siltstones are represented by staurolite-biotite-andalusite schists. The uppermost part of the Formation is characterized by significant amounts of tourmalinite, dacitic tuff and spectacular andalusite-quartz segregations and this is the stratigraphic level where the locally important lithium pegmatites occur.

The overlying Kilcarry Volcanic Formation is composed of andesitic lavas, with minor dacitic tuff bands and is 500 m thick. The Formation, occurring only within the Leinster Granite aureole, is represented by hornblende schists. The Monaughrim Semi-Pelite Formation, youngest formation of the Ribband Group, consists of more than 250 m of impure shelf sandstone with some distinctive calcareous horizons. Also confined to the Leinster Granite aureole, these rocks now consist of garnet-biotite semi-pelitic schists and hornfelses.

3 Structure

The Caledonide rocks in this part of SE Ireland have suffered a polyphase tectonic evolution during the late Silurian-early Devonian (McArdle 1984). The overall structural pattern as now displayed was determined during the initial two phases (D1 and D2). During the first (D1) phase the rocks developed a slaty cleavage and associated minor structures under low grade metamorphic conditions. The cleavage dips steeply SE and is parallel to the boundaries of lithostratigraphic contacts and the margin of the Leinster Granite.

The effects of the second phase (D2) were more localized, being characterized by narrow shear zones. One of these is the East Carlow Deformation Zone (McArdle and Kennedy 1985). It occurs along the margin of the Tullow Lowlands Unit of the Leinster Granite and is a dip-slip ductile zone which formed during D2 emplacement of the pluton; indeed the marginal granite was foliated at this time. The zone is a linear feature in which D1 structures are modified and new D2 structures (foliation, microfolds) developed under metamorphic conditions ranging between amphibolite and greenschist facies. Its continuation can be inferred in a northeasterly direction as far as the coastline (Fig. 1). It forms the locus for the lithium and tungsten mineralization described here (McArdle et al. in press and Fig. 1). A suite of granitic and appinitic and lamprophyric intrusions were emplaced along this zone.

Later deformation phases, D3 and D4, had a less significant impact on the structural development of the area. They generally consist of mesoscopic fold structures, but also include late stage large-scale faults.

4 The Leinster Granite

The Leinster Granite (Fig. 1) is composed of five dome-shaped units which were synchronously intruded during D2 deformation in the early Devonian (Brindley 1973; Brück and O'Connor 1977). All the units are exposed close to roof level and roof pendants are prominent in central parts. The main intrusion mechanisms were sheet injection followed by inflation to form domes.

The granite has a uniform composition over large areas. Brück and O'Connor (1977), following Brindley (1954), detailed the main granite types as follows. Type 1 granite is a fine grained quartz diorite which is locally gradational in grain size and composition to Type 2 granite. Type 2 granite forms the main bulk of the pluton. It is a coarse grained two-mica granite: Equigranular Type 2 and Porphyritic Microcline Type 2 varieties can be distinguished in the field based on textural criteria. Type 3 and Type 4 granites are local muscovite-rich granite varieties which share many petrographic features with Type 2 granites (Brück and O'Connor 1977). They are regarded as late stage hydrothermally altered variants of Type 2 granites, mainly distinguished by muscovite and tourmaline (Brück and O'Connor 1977).

5 Tungsten Mineralization

Scheelite mineralization in the Tinahely area (Fig. 1) is associated with a swarm of narrow, parallel-sided sheets of Type 1 granite. Significantly scheelite is hosted in quartz veinlets in these sheets and, in addition, minor amounts of scheelite are present in the country rock sediments. The granite sheets are emplaced along the NE-trending slaty cleavage in rocks belonging to the Ballybeg Pelitic Formation. Coticule and tourmalinite beds are locally present and Steiger and Bowden (1982) recorded sericitic laminae rich in tourmaline. Pyrrhotite and arsenopyrite occur disseminated or locally concentrated along bedding laminae, while the former in places follows the clevage and is deformed by it. The sheets which are mineralized are concentrated within a zone which is 1 km wide and follows the schistosity for 10 km. This zone forms part of the East Carlow Deformation Zone (McArdle and Kennedy 1985).

While mineralized granite sheets are restricted to the vicinity of the East Carlow Deformation Zone, similar sheets without mineralization are present beyond the confines of the Zone. Chevron-style D4 folds have been intensively developed in both granite and country rocks. Lithological variations in the Type 1 granite largely reflect silicification and chloritization as a result of retrograde metomorphism.

Significant scheelite mineralization is confined within, or adjacent to, thin quartz veinlets. Scheelite in the granite sheets occurs as grains within the veinlets or else form fine disseminations. Pyrrhotite may be intergrown with scheelite and the bleached zones are characterized by arsenopyrite crystals. The mineralization is characterized by fluorite, arsenopyrite, pyrrhotite, and scheelite-bearing assemblages. Minor mineral phases include sphalerite, chalcopyrite, galena, stannite, molybdenite, cassiterite, rutile, sphene, tourmaline (schorlite-dravite), apatite, cosalite, native bismuth, proustite and mackinawite (Steiger and Bowden 1982). The arsenic: tungsten ratio is variable but it averages about 7:1.

There is direct evidence linking tungsten to the development of the East Carlow Deformation Zone; sediment-hosted scheelite was recrystallized into quartz segregations which were formed during the D1 or D2 deformation phase. Metamorphic processes accompanying the formation of the East Carlow Deformation Zone subsequently remobilized the tungsten and associated metals and redeposited them in the granite sheets. For example, there is geochemical evidence to suggest an enrichment-depletion pattern in the distribution of associated arsenic values (McArdle et al. in press).

6 Lithium Mineralization

An important development of lithium-bearing pegmatites (Steiger and von Knorring 1974) is present along the flank of the Leinster Granite between Borris and Shillelagh (Fig. 1). No mine production has taken place to date.

Six significant if sub-economic lithium pegmatites have been confirmed to date, largely by diamond drilling (Steiger and von Knorring 1974, McArdle 1984), including the important occurrence at Aclare House (Fig. 2). These pegmatites, as well as seven other reported occurrences, are confined to the vicinity of the Kilcarry Volcanic Formation. Four of the pegmatites contain tourmalinite in their wall rocks. These tourmalinites do not contain anomalous levels of lithium, although high trace levels do occur in other interbanded metasediments (McArdle 1984). The wall rocks also contain disseminated sulphides (pyrrhotite, chalcopyrite, arsenopyrite and pyrite) which are confined to lithological laminae and are stratabound at the top of the Ballybeg Pelitic Formation (McArdle 1984). The pegmatities all occur

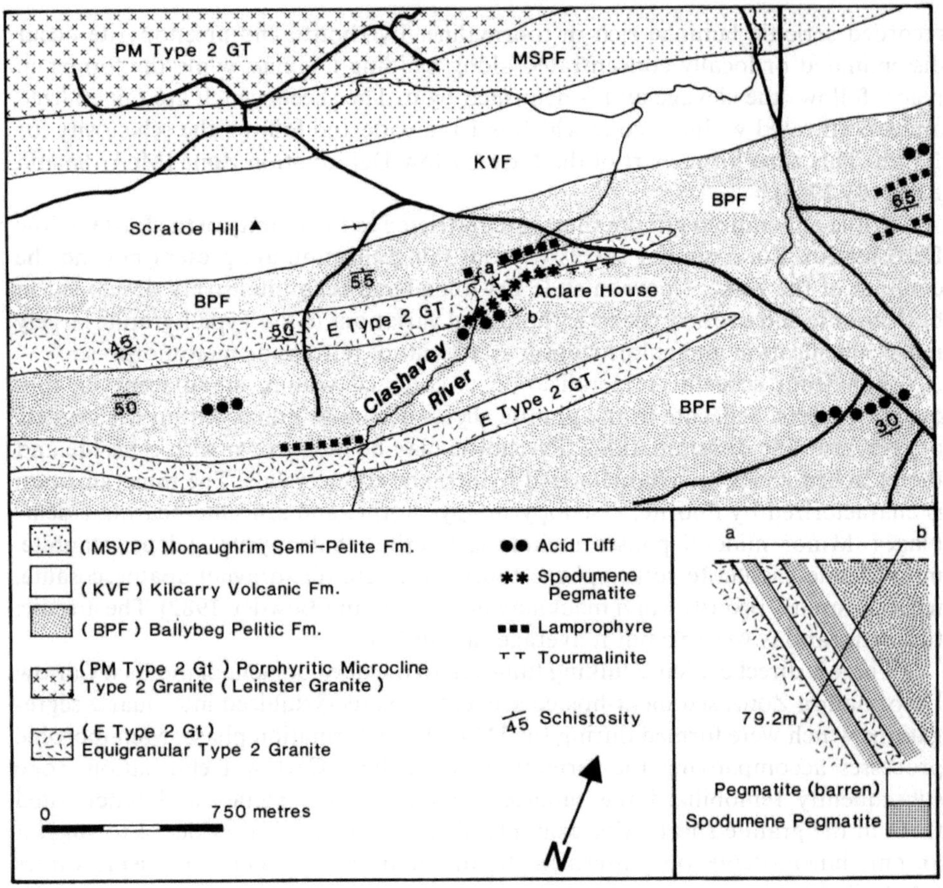

Fig. 2. Geological setting of the Aclare House spodumene pegmatite. (After McArdle 1984). The area is located on Fig. 1 (letter *A*)

within the East Carlow Deformation Zone. On a local scale, individual pegmatites are related to granitic sheets conformably lying in the aureole schists or else occur on the margin of the Leinster Granite; none are known within the Leinster Granite further than 1 km from its margin.

The pegmatites are not internally zoned, but they are composite. They form simple tabular sheets which are generally steeply dipping. Spodumene-bearing sections as narrow as 50 mm are interbanded with other lithologies. Spodumene, showing a good planar fabric, occurs in a matrix of quartz-albite-muscovite. The spodumene is white, contains 0.2% ferric oxide and forms crystals up to 0.3 m long. A bulk sample was found to contain 24% spodumene, equivalent to about 1.5% lithia, 35% quartz, 25% albite, 6% microcline, 9% muscovite and 1% accessories (Steiger and von Knorring 1974).

Tin and niobium-tantalum minerals are constant accessories in these pegmatites in the form of cassiterite and columbite-tantalite. Additional accessories include manganapatite, spessartine, beryl, bertrandite, lithium mica, lithiophilite, tourmaline, and pyrite. One pegmatite (Stranakelly) is greisened in part and contains lithian mica, clevelandite and quartz. Mica-rich areas contain 0.2% by weight of tantalum minerals — uraniferous microlite and manganotantalite (Steiger and von Knorring 1974).

These pegmatites belong to the composite or unzoned lithium pegmatites which are widely recognized as a special category of pegmatites, distinctive from zoned lithium pegmatites, in many parts of the world (for example, Norton 1973; Mulligan 1965; Ginzberg and Lugovskoi 1977 and Gyongyossy and Spooner 1979). Such pegmatites commonly form steeply dipping bodies with an overall regular sheet-like geometry, uniform compositions and textures and higher lithium grades than those of the zoned bodies.

7 Discussion

In summary, the controls on the occurrence and distribution of tungsten and lithium mineralization are threefold, stratigraphic, tectonic and plutonic. The mineralization is confined to a major shear zone, but only to that part which traverses a volcanic formation and its associated sediments — lithologies which, as indicated below, are regarded as favourable for the occurrence of mineralization. The shear zone developed synchronously with granite emplacement, immediately preceding pegmatite development (McArdle and Kennedy 1985). But the following reasons suggest that the lithium pegmatites were not primarily controlled by the emplacement of the Leinster Granite itself:

- they are confined to particular lithologies within the Ribband Group and within the East Carlow Deformation Zone;
- the more important lithium pegmatites lie at a distance from the batholith;
- there are no regional zonation patterns related to distance from the batholith.
- they do not occur in the batholith at a distance greater than 1 km from its margin;
- the known occurrences are essentially concentrated on one flank of the batholith;

Although our present knowledge suggests that the Caledonian-Appalachian Orogen is unevenly endowed with lithium pegmatites, they do recur at various points along the length of the orogen. In the SW of the Appalachians the Carolina Piedmont contains the Kings Mountain Belt, the most important mining centre of pegmatitic lithium in the world (Kesler 1976). The Cherryville Pluton and its surrounding country rocks contain an abundance of pegmatites but the lithium pegmatites occur at some distance from the pluton and are confined to the vicinity of the Kings Mountain Shear Zone (Kesler 1976; Horton 1981). Further NE, near Penobscot Bay in Maine, the Peg Claims spodumene pegmatites form steeply dipping discordant tabular sheets in the country rocks between the Waldboro Granite to the W and the more distant Clark Island Granite to the E (Sundelius 1963). In the New Ross area of Nova Scotia lesser examples of lithium pegmatites are associated with Devonian granites that intrude the quartzites and slates of the Meguma Group (Mulligan 1965; Flanagan 1978).

A common feature of lithium and tantalum pegmatites, in both the Caledonian-Appalachian orogen and elsewhere, is that they tend to occur at a distance from granite plutons. In the Yellowknife-Beaulieu district, North-west Territories, several hundred lithium pegmatites have been identified and they are restricted to the metamorphic aureoles surrounding granite plutons (Mulligan 1965, Lasmanis 1978). In the USSR, Ginzberg and Lugovskoi (1977) describe several lithium pegmatite districts which have a regional association with granite plutons but where individual lithium pegmatites occur at distances of 0.5 to 2.0 km from plutonic contacts. A similar spatial relationship between lithium-bearing pegmatites and granites has been described from Galicia, Spain (Hensen 1967). In the Masuku area of southern Zambia the regional zonation of tin-tantalum-niobium pegmatites cannot be accounted for simply by distance from granitic sources (Legg and Namateba 1982).

8 The Coticule-Tourmalinite Association

In SE Ireland the spatial association between the lithium pegmatites and the coticule/tourmalinite bearing Lower Palaeozoic metasediments (Ballybeg Pelitic Formation) is clear. These metasediments are considered to be the source of lithium for the pegmatites. The main tourmalinites immediately underlie the Kilcarry Volcanic Formation and some others occur either within or immediately above it. It is suggested that they are volcanogenic exhalative sediments and that an unusual range of elements, including boron, lithium and base metals, accumulated either in these tourmalinites or in their associated sediments. Accordingly it is suggested that the lithium in the spodumene pegmatites originally had an exhalative origin and developed under the influence of a hydrothermal system related to volcanism. It seems that the onset of volcanism may have given rise to conditions under which metal enrichment took place on the Iapetus seafloor as a result of hydrothermal activity (Kennan et al. 1986), and that lithium was one of those metals (e.g. Humphris and Thompson 1978, Seyfried et al. 1984).

The Tinahely tungsten mineralization is hosted in that part of the Ribband Group which contains coticule. The coticule and the associated tourmalinite represent metamorphosed chemical sediments which were originally cherty rocks (Kennan and Kennedy 1983). Similar rocks have been described from other strata-bound tungsten deposits (e.g. Holl et al. 1972; Cunningham et al. 1973; Clarke 1983). These metamorphosed manganese-rich ironstones are chemical sediments which were deposited in relatively deep water conditions by means of either sedimentary or volcanogenic processes (Roy 1981; Hutchison 1983).

On a regional scale a consistent W to E sequence in the development of volcanic rocks, tourmalinite and coticule (Fig. 3) suggests the possible occurrence of lithium and tungsten on the NW side of the granite. Lithium appears to be associated preferentially with tourmalinite, and tungsten and base metals with coticule.

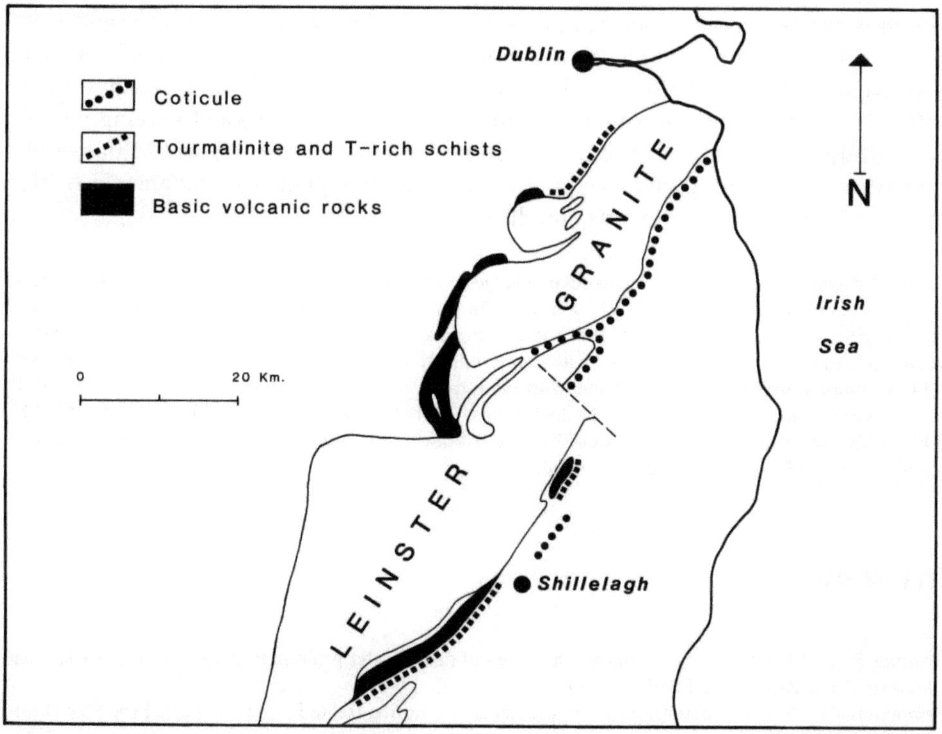

Fig. 3. Lithological relationships of coticule and tourmalinite adjacent to the Leinster Granite

Kennan and Kennedy (1983) have suggested that the coticule/tourmalinite pair of very distinctive lithologies is a feature of the margin of Iapetus Ocean in that this pair (a) is demonstrably continuous over large distances and (b) may therefore have a value in very long range correlation within that ocean. Spodumene pegmatites at Peg Claims lie adjacent to areas where tourmalinite has been recognised as a guide to ore (Slack 1982). The spodumene ores in the Kings Mountain Belt also occur in close proximity to a major belt of coticule-bearing metasediments (Horton 1981, Plate 1). In the light of the orogen-long lateral continuity of the coticule the various pegmatite occurrences are likely to share many mineralogical and chemical features. This is what Stewart (1978) recognised and what lead to his suggestion that various North American lithium-rich pegmatites represented something approaching eutectic compositions.

Stewart (1978) noted (a) that, where regional compositional zoning of pegmatites adjacent to related granite bodies occurs, lithium-rich pegmatites tend to lie farthest from the granite and rarely within it; (b) that such pegmatites constitute only a small proportion of the total number; (c) that the similarity of modes measured for different occurrences were such as to suggest the involvement of eutectic composition in each case; and (d) that as a result, many lithium-rich pegmatites have almost the same composition. These observations are consistent with the origin of pegmatite magma by anatexis of lithium-rich sediments. Lithium-rich granites associated in space and time with lithium-rich pegmatites might well derive from the same source as a consequence of a greater degree of anatexis. However, as Stewart (1978, p. 979) emphasised, it is extremely unlikely that, even with perfect fractionation, pegmatites derived from such a granite magma would ever attain the lithium concentrations encountered in the typical lithium-rich pegmatite. Hence the preference for the generation of pegmatite magmas directly by limited anatexis of sediments enriched in lithium.

Acknowledgements. The authors wish to acknowledge Ennex plc and Irish Base Metals Ltd. for providing access to data and drill core. Mr. J.A. Clifford (Ennex plc) assisted our study in several ways and Mr. J. Kennedy (University College Dublin) drafted the diagrams. The research was included in *Contract MSM 110 EIR* of the EEC Primary Raw Materials Research Programme and Dr. J. Boissonnas (EEC Commission) is thanked for his help and encouragement. This paper was considerably improved by comments and suggestions from J. Boissonnas, Prof. P. Omenetto, Mr. J.A. Clifford, Mr. M. Whitworth and the anonymous reviewers. The contribution of PMcA is published with the permission of the Director, Geological Survey of Ireland.

References

Brindley JC (1954) The geology of the northern end of the Leinster granite: Part 1 — Internal structural features. Proc Roy Ir Acad 56B: 159–90

Brindley JC (1973) The structural setting of the Leinster Granite, Ireland — a review. Sci Proc Roy Dubl Soc 5A: 27–36

Brûck PM, O'Connor PJ (1977) The Leinster batholith: geology and geochemistry of the Northern Units. Geol Surv Ire Bull 2: 107–41

Brûck PM, Colthurst JRJ, Feely M, Gardiner PRR, Penny SR, Reeves TJ, Shannon PM, Smith DG, Vanguestaine M (1979) Southeast Ireland: Lower Palaeozoic stratigraphy and depositional history. In: Harris AL, Holland CH, Leake BE (Eds) The Caledonides of the British Isles — reviewed. Geol Soc Lond Spec Publ No 8, pp 533–44

Clarke MCG (1983) Current Chinese thinking on the South China tungsten province. Trans Instn Min Metall (Sect B: Appl earth Sci) 92: 10–5

Cunningham WB, Holl R, Taupitz KC (1973) Two new tungsten bearing horizons in the older Precambrian of Rhodesia. Mineral Deposita 8: 200–3

Flanagan JT (1978) Lithium deposits and potential of Quebec and Atlantic Provinces, Canada. Energy 3: 391–8

Ginzberg AI, Lugovskoi GP (1977) Deposits of lithium. In: Smirnov VI (ed) Ore deposits of the USSR, Vol 3. Pitman Publishing, London, pp 295–311

Gyongyossy ZD, Spooner ETC (1979) Lithium: an energy element. Minerals Sci Engng 11: 155–76

Hensen BJ (1967) Mineralogy and petrography of some tin, lithium and beryllium bearing albite-pegmatites near Doade, Galicia, Spain. Leidse Geol Meded 39: 249–59

Holl R, Maucher A, Westenberger H (1972) Synsedimentary-diagenetic ore fabrics in the strata- and time-bound scheelite deposits of Kleinarltal and Felbertal in the Eastern Alps. Mineral Deposita 7: 217–26

Horton JW (1981) Shear zone between the Inner Piedmont and Kings Mountain belts in the Carolinas. Geology 9: 28–33

Humphris SE, Thompson G (1978) Trace element mobility during hydrothermal alteration of oceanic basalts. Geochimica et Cosmochimica Acta 42: 127–36

Hutchison CS (1983) Economic deposits and their tectonic setting. The Macmillan Press Ltd, London, 365 p

Kennan PS, Kennedy MJ (1983) Coticules — a key to correlation along the Appalachian-Caledonian orogen. In: Schenk PE (ed) Regional trends in the geology of the Appalachian-Caledonian-Hercynian-Mauritanide orogen. D Riedel Publishing Co, Dordrecht, pp 355–61

Kennan PS, McArdle P, Williams FM, Doyle EM (1986) A review of metal deposits associated with the Leinster Granite and a model for their genesis. In: Andrew CJ, Crowe RW, Finlay S, Pennell W, Pyne JF (eds) Geology and genesis of mineral deposits in Ireland. Ir Assoc Econ Geol, Dublin, pp 201–10

Kesler TL (1976) Occurrence, development and long-range outlook of lithium-pegmatite ore in the Carolinas. In: Vine JD (ed) Lithium resources and requirements by the year 2000. US Geol Surv Prof Paper 1005, pp 45–50

Lasmanis R (1978) Lithium resources in the Yellowknife area, Northwest Territories, Canada. Energy 3:399–407

Legg CA, Namateba C (1982) Regional zoning of tin-tantalum-niobium pegmatites in the Masuku area, Southern Province, Zambia. In: Evans AM (ed). Metallization associated with acid magmatism. John Wiley and Sons Ltd, 181–90

McArdle P (1981) The country rocks flanking the Leinster Granite between Aughrim and Ballymurphy. Geol Surv Ire Bull 3:85–95

McArdle P (1984) The geological setting of mineralization along the southeast margin of the Tullow Lowlands Unit of the Leinster Granite. Unpubl PhD Thesis, Nat Univ Ire, 330 p

McArdle P, Kennedy MJ (1985) The East Carlow Deformation Zone and its regional implications. Geol Surv Ire Bull 3:237–55

McArdle P, Williams FM, Doyle EM, Kennan PS, Moore-Lewy G, Gallagher V, O'Connor PJ (in press) The metallogenic relationships of mineralization along the southeast margin of the Leinster Granite. Geol Surv Ire Report Series RS86/2, 57 p

Mulligan R (1965) Geology of Canadian lithium deposits. Geol Surv Canada Econ Geol Rept No 21. 131 p

Norton JJ (1973) Lithium, cesium and rubidium — the rare alkali metals. In: Brobst DA, Pratt WP (eds) United States mineral resources. US Geol Surv Prof Paper 820, pp 365–78

Roy S (1981) Manganese deposits. Academic Press, London, 458 p

Seyfried WE JR, Janecky DR, Mottl MO (1984) Alteration of the oceanic crust: implications for geochemical cycles of lithium and boron. Geochimica et Cosmochimica Acta 48:557–69

Slack JF (1982) Tourmaline in Appalachian-Caledonian massive sulphide deposits and its exploration significance. Trans Instn Min Metall (Sect B: Appl earth Sci) 91:81–9

Steiger R, Bowden A (1982) Tungsten mineralization in southeast Leinster, Ireland. In: Brown AG (ed) Mineral exploration in Ireland: Progress and developments 1971–1981. Ir Assoc Econ Geol, Dublin, pp 108–14

Steiger R, Von Knorring O (1974) A lithium pegmatite belt in Ireland. J Earth Sci (Leeds) 8:433–43

Stewart DB (1978) Petrogenesis of lithium-rich pegmatites. Amer Miner 63:970–80

Sundelius HW (1963) The Peg Claims spodumene pegmatites, Maine. Econ Geol 58:84–106

Williams CE, McArdle P (1978) Ireland. In: Bowie SHU, Kvalheim A, Haslam HW (eds) Mineral Deposits of Europe vol 1, Northwest Europe. The Instn Min Metall and the Mineral Soc, pp 319–45

Williams FM, Kennan PS (1983) Stable isotope studies of sulphide mineralization on the Leinster granite margin and some observations on its relationship to coticule and tourmalinite rocks in the aureole. Mineral Deposita 18:335–47

Geology and Geotectonic Setting of Cratonic Porphyry Molybdenum Deposits in the North Atlantic Region

H.K. SCHØNWANDT[1]

Abstract

Porphyry molybdenum deposits in the North Atlantic area occur in two provinces: the Permian Oslo Graben and the Tertiary Igneous Province of East Greenland. Both provinces represent intraplate igneous activity. The Oslo Graben is a failed rift, whereas the East Greenland province is a line of intrusive centres along a passive continental margin. The deposits are associated with syenite-granite complexes, but they are spatially related to alkali-rich, high-silica granites. The deposits show striking similarities to the granite-type porphyry Mo deposits in Colorado. A great variety of Mo mineralizations occur in the two North Atlantic provinces, however, only a subvolcanic environment seems to favour generation of porphyry-type deposits.

1 Introduction

Porphyry molybdenum deposits represent the most important source of molybdenum. The majority of these deposits occur in North America where porphyry molybdenum deposits have been classified according to their magmatic affinity into granitic and quartz-monzonitic types (White et al. 1981; Mutschler et al. 1981). Westra and Keith (1981) used a geochemical classification, whereas Sillitoe (1980) proposed a classification based on a geotectonic setting. The large, granite-type porphyry Mo deposits of Colorado (Climax and Henderson) have been linked with a relatively atectonic stage of transition from steepening subduction to back-arc rifting. Questa (New Mexico), Mt. Emmons (Colorado) and probably Pine Grove in Utah have been ascribed to a subsequent stage of crustal extension and bimodal magmatism. All of these deposits are related to the terminal stage of a Wilson cycle.

Porphyry molybdenum deposits in the North Atlantic region are also associated with alkali-rich, high-silica granites (Table 1) but, contrary to the North American occurrences, they are connected with continental rifting during the initial stages of

[1] Department of Geology, University of Aarhus, DK-8000 Aarhus C, Denmark
Present address: The Geological Survey of Greenland, Øster Voldgade 10, DK-1350 Copenhagen K, Denmark

Mineral Deposits within the European Community
(ed. by J. Boissonnas and P. Omenetto)

Table 1. Chemical composition (wt%) of Mo-related granites in East Greenland and in the Oslo Graben[a]

	1	2	3	4	5	6
SiO_2	76.12	76.31	76.25	75.50	75.10	76.80
TiO_2	0.09	0.07	0.09	0.28	0.22	0.15
Al_2O_3	11.75	12.00	11.83	12.20	12.50	11.50
Fe_2O_3	0.13	0.05	0.01	1.50	1.40	1.60
FeO	0.97	0.72	0.72	0.00	0.40	0.27
MnO	0.03	0.05	0.05	0.00	0.05	0.02
MgO	0.11	0.09	0.09	0.14	0.06	0.06
CaO	0.54	0.61	0.88	0.28	0.26	0.11
Na_2O	3.17	3.41	3.63	3.40	4.60	4.60
K_2O	6.19	5.91	5.57	5.00	4.40	4.50
P_2O_5	0.00	0.00	0.00	0.30	0.00	0.00
H_2O	0.84	0.70	0.81	—	—	—
Total	99.94	99.92	99.93	98.60	98.99	99.61

[a] Columns: 1, aplite granite (Malmbjerget); 2, perthite granite (Malmbjerget); 3, quartz-feldspar porphyry (Malmbjerget); 4, aplogranite (Glitrevann caldera); 5, alkali granite (Nordli); 6, quartz-feldspar porphyry (Nordli).

a Wilson cycle. They occur in two metallogenetic provinces: the Permian Oslo Graben in Norway and the Tertiary Igneous Province in East Greenland.

The purpose of this chapter is to emphasize that although the reasons for the rifting and the geotectonic features of the provinces are different, the metallogeny shows striking similarities. It is also emphasized that the North Atlantic porphyry molybdenum occurrences resemble those of Colorado in spite of their different geotectonic settings and the fact that the North Atlantic occurrences are related to granite-syenite complexes, whereas the North American ones only seem to be related to granites.

2 Geotectonic Setting

The Permian Oslo Graben and the Tertiary province in East Greenland both represent intraplate igneous activity, but the geotectonic features of the two provinces are different. The Oslo Graben is a failed rift with a crustal extension at the present level of erosion of 4 km (Ramberg 1976). The Tertiary Igneous Province of East Greenland appears to represent a line of intrusive centres along a half-graben fault structure.

2.1 Oslo Graben

The Oslo Region is the geographical term for an area of approximately 11 000 km^2 around Oslo in southern Norway (Dons 1978; Ramberg and Spjeldnaes 1978). The region is surrounded by a Precambrian basement and is delimited by two dominant fracture lines (NNW and NNE) (Ramberg 1976). The Oslo Graben refers specifically to a subsided crustal block about 220 km in length (Fig. 1). The subsided crustal

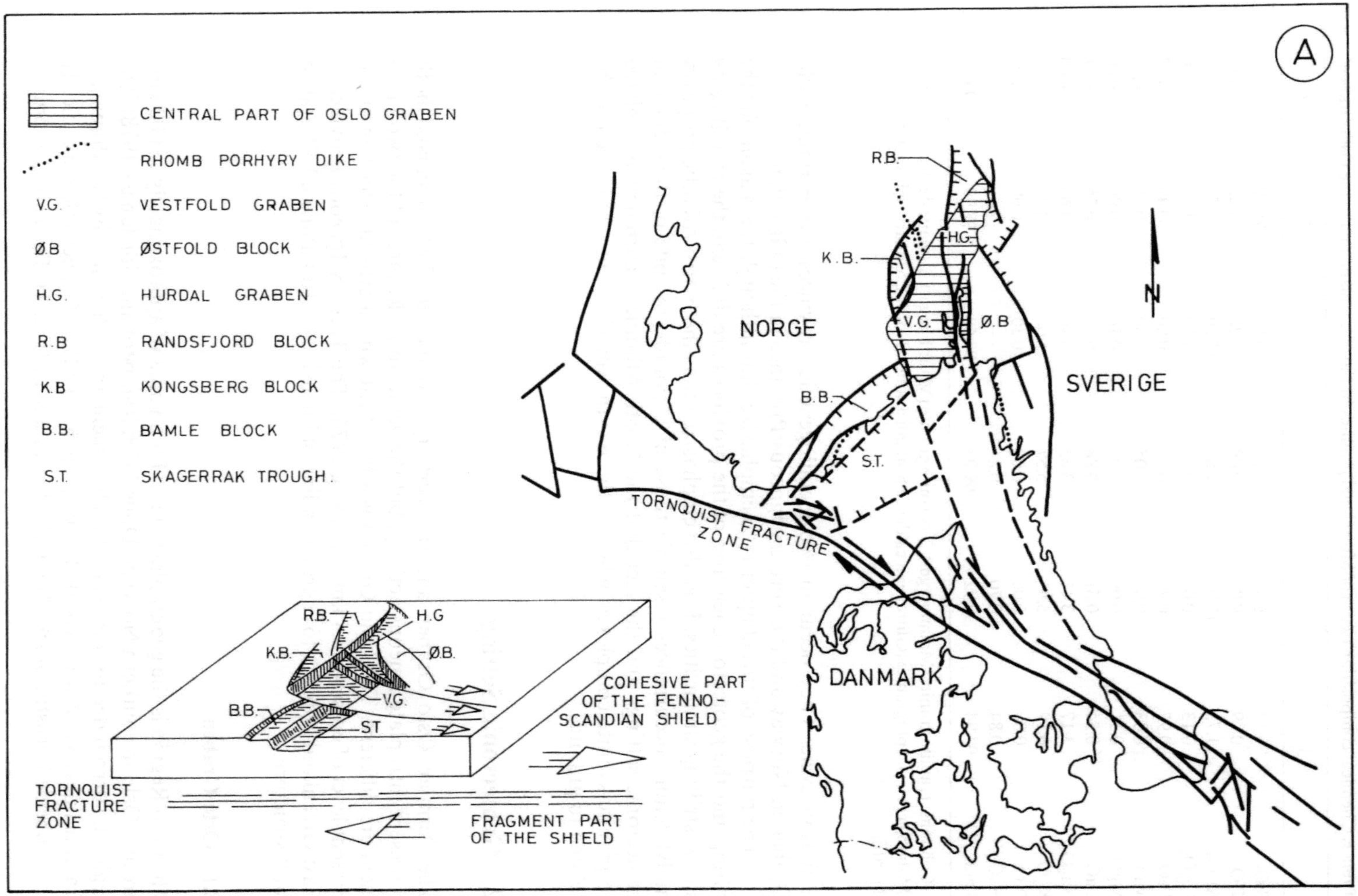

Fig. 1A. A tentative model for the formation of the Oslo Rift as the result of offsets along N–NNW trending secondary faults related to the Tornquist fracture zone. The Glitrevann, Sande and Ramnes calderas are located in the eastern part of the Vestfold Graben

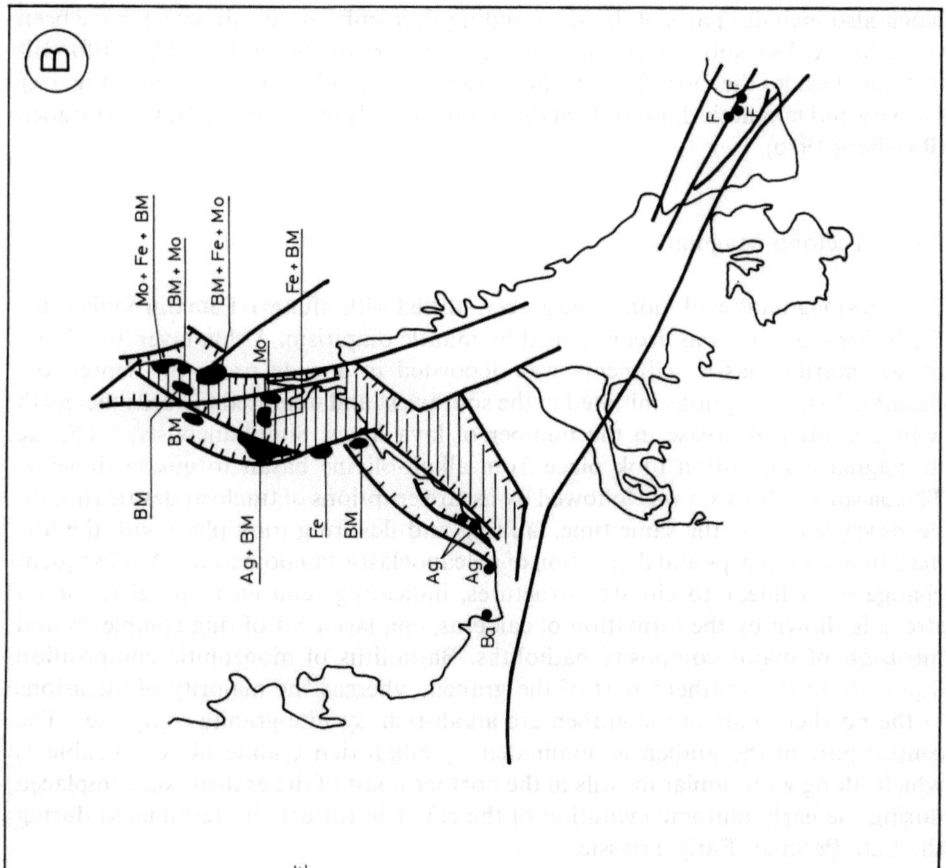

Fig. 1B. Location of main areas of Permian mineralization.

block also includes parts of the surrounding Precambrian terrain which have been subjected to less subsidence than the central part of the graben. The Oslo Rift includes, besides the Oslo Graben, the Skagerrak Trough of the North Sea (Fig. 1A). Gravity and magnetic data confirm this southwesterly extension of the Oslo Graben (Ramberg 1976).

2.1.1 Tectono-Magmatism

Tectono-magmatic evolution of the graben started with rifting in Late Carboniferous/Early Permian and was accompanied by mantle diapirism. A thin layer (0–120 m) of non-marine clastic sediments was deposited in a wide proto-rift depression. Basaltic fissure eruptions initiated in the southwest and migrated towards the north with a gradual decrease in the number of lava flows. Simultaneously, a change in magma composition took place from alkali-olivine basalt to quartz tholeiite. The basaltic volcanism was followed by fissure eruptions of trachyandesitic rhomb-porphyry lavas. At the same time, faulting and flexuring took place with the formation of fault scarps and deposition of volcanoclastic fanglomerates. A subsequent change from linear to circular structures, indicating reduced regional tensional stress, is shown by the formation of calderas, emplacement of ring complexes and intrusion of major composite batholiths. Batholiths of monzonitic composition dominate in the southern part of the graben, whereas the majority of intrusions in the northern part of the graben are alkali-rich, syenite-granite complexes. The central part of the graben is dominated by alkali-rich granite plutons (Table 1) which, along with similar massifs in the northern part of the graben, were emplaced during the early plutonic evolution of the rift. The rift activity terminated during the Late Permian/Early Triassic.

The Oslo Rift can be viewed as a pull-apart basin complex which developed along secondary faults related to the Tornquist lineament (Fig. 1A). This fracture zone is a first-order structural feature which, according to Arthaud and Matte (1977), forms the northern boundary of a mega-shear zone with right-lateral movement between two plates: a northern plate including the Canadian-Fennoscandian shield and a southern plate that includes the African shield. Fracturing in this mega-shear zone began around 300 Ma, but the main movement took place during the Stephanian and later reactivation occurred during the Permian. However, regional evidence suggests that the Tornquist lineament has been a fundamental crustal plate boundary since the Precambrian (Watson 1976; Pegrum 1984).

Right-lateral offsets along the Tornquist fracture zone, combined with changes in the direction of the zone northwest of the Danish coast, created the N–NNW-trending secondary faults (Fig. 1A). The offsets along the secondary faults were partly compensated by the opening of the Oslo Rift along pre-existing NNE- and NNW-trending fractures in the Precambrian basement. As movements have taken place along the Tornquist fracture zone since the Precambrian, the occurrences of geological events along the Oslo Graben lineament can find part of their explanation linked to these movements (Ramberg 1976; Ramberg and Spjeldnaes 1978; Touret 1970; Bylund and Patchett 1977).

Burke and Dewey (1973) and Ramberg (1976) envisaged the rift as being generated by a mantle plume. N–S-trending shear movements were proposed by Ramberg and Larsen (1978) and Schønwandt and Petersen (1983) as being responsible for the rift formation, whereas Ziegler (1978) suggested that the Oslo Graben represents a pull-apart feature at the presumed termination of the Tornquist lineament.

2.2 Tertiary Province of East Greenland

Extensive igneous activity took place in the Kangerdlugssuaq and Mesters Vig areas of East Greenland in lower Tertiary times (Fig. 2). This activity was part of

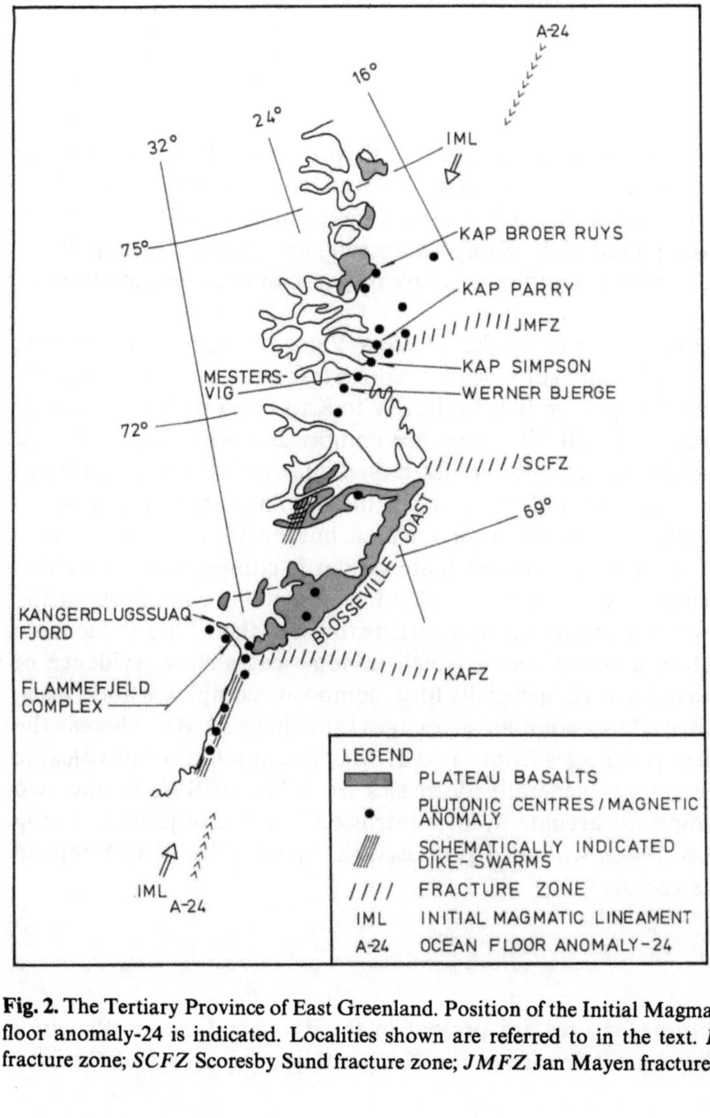

Fig. 2. The Tertiary Province of East Greenland. Position of the Initial Magmatic Lineament and ocean floor anomaly-24 is indicated. Localities shown are referred to in the text. *KAFZ* Kangerdlugssuaq fracture zone; *SCFZ* Scoresby Sund fracture zone; *JMFZ* Jan Mayen fracture zone

Table 2. Tectono-magmatic evolution of the Kangerdlugssuaq area, East Greenland (after Nielsen and Brooks 1981; Nielsen, 1987; and Brooks 1979)

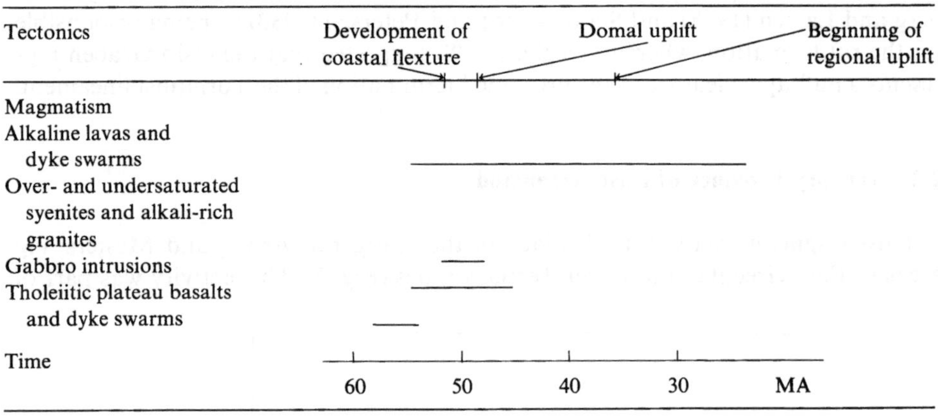

the evolution of the North Atlantic plateau basalt province which includes the Scottish-Hebridean Province, the Faeroe Islands, West Greenland and Baffin Island (Noe-Nygaard 1974). Brooks and Nielsen (1982) have summarized the Tertiary tectono-magmatic development of the Kangerdlugssuaq area (Table 2) and pointed out that the igneous province resembles in many respects an oceanic accretionary plate margin.

The Tertiary igneous province of the Mesters Vig area forms a prominent NE–SW-trending line of plutonic-subvolcanic centres traceable for approximately 125 km from the Werner Bjerge Complex in the SW to Kap Parry in the NE (Fig. 2) (Haller 1971; Noe-Nygaard 1976). The province comprises a wide range of rock types including mafic and ultramafic lithologies, monzonites, alkali syenites, granites and nepheline syenites. Syenites and granites are the dominant rock types of the intrusive centres. No plateau basalts occur in the area, but the plutonic/subvolcanic centres cut a sequence of tholeiitic sills which are similar in composition to plateau basalts which occur elsewhere in East Greenland. The igneous complexes in the Mesters Vig area show sharp cross-cutting contacts towards their Late Palaeozoic to Mesozoic sedimentary country rocks, which in some cases show evidence of partial uplift. The intrusive centres generally form composite complexes and young from NE to SW. The Kap Parry complex gives ages from 40 to 34 Ma, whereas the Werner Bjerge Complex yields ages from 31 to 21 Ma. The amount of subvolcanic rocks in the intrusive centres generally decreases from NE to SW. In the two northwesternmost complexes arcuate syenite intrusions were interpreted as ring dykes by Schaub (1938, 1942), who also proposed that part of the Kap Simpson Complex is a caldera structure.

2.2.1 Tectonics

Two prominent structural features are present in the Tertiary Igneous Province of East Greenland: the coastal flexure in the Kangerdlugssuaq area and the NE-trending line of plutonic centres (Fig. 2). This mega-lineament is approximately 1000 km long and has been named the 'Initial Magmatic Lineament' by Nielsen (in press). Geophysical investigations of the East Greenland shelf (Larsen 1984) indicate that the line of plutonic centres continues an additional 150 km towards the NE, and does not, as proposed by Haller (1971) and Noe-Nygaard (1976), shift towards the NNE to link up with the Kap Boer Ruys pluton.

The coastal flexure of the Kangerdlugssuaq area was first described by Wager and Deer (1938), who interpreted the flexure as a simple, coast-parallel monoclinal folding. Nielsen (1975) and Nielsen and Brooks (1981) interpreted the flexure as a half-graben structure dominated by antithetic block rotation. Their model implies considerable crustal extension over the width of the flexure which appears as a gentle warp approximately 50 km wide with a total vertical displacement of 8 km. The coastal flexure follows broadly the Blosseville coast and passes to the north out to sea, probably bounded to the north by the Scoresby Sund Fracture Zone (SCFZ) (Fig. 2). South of Kangerdlugssuaq only the Precambrian basement is present and possible southern continuation of the flexure is only indicated by the dip of the coastal dyke swarm. However, the flexure south of the Kangerdlugssuaq Fracture Zone (KAFZ) is much less developed than the Blosseville coast flexure (Larsen 1984).

These two prominent structural features meet in the Kangerdlugssuaq area where a roughly elliptical crustal dome, some 200 km across, occurs. This dome has been related to a mantle plume and a postulated triple junction of rifts (Brooks 1973; Burke and Dewey 1973). The two active arms of the junction follow the coastal flexure of the Blosseville coast and the line of plutonic centres south of Kangerdlugssuaq, respectively. According to Burke and Dewey (1973) these arms developed to the extent of the continental breakup about 60 Ma ago. This model does not, however, explain the whole length of the initial Magmatic Lineament which forms a natural continuation into the continental crust of the initial North Atlantic spreading ridge–ocean floor anomaly-24 (Fig. 2). This indicates that the continental crust represents a locked crustal zone (Courtillot 1982) where the propagating rifts of the North Atlantic did not succeed in crossing the disruption. Instead, the northward moving rift was reflected around the Blosseville coast by transform faults and joined with the southward propagating rift from north of Mesters Vig at the time of the ocean floor anomaly-6. The considerable crustal extension expressed by the coastal flexure along the Blosseville coast reflects the reaction of the locked crustal zones to the rifting forces. The dome centered on Kangerdlugssuaq therefore seems to be a result of the temporary standstill of the northward moving mantel plume as it entered the locked crustal area at Kangerdlugssuaq.

The Kangerdlugssuaq Fjord which cuts across the highest part of the dome is interpreted by Brooks (1973) as a fault-controlled feature. Salic rocks are particularly abundant within the area of the Kangerdlugssuaq dome and are largely generated by extensive melting of the continental crust (Brooks and Nielsen 1982). The distribution of salic rocks with their occurrence in distal regions to the initial spreading ridge, either along off-axial lineaments (the Kangerdlugssuaq Fjord) or as a continuation of the spreading ridge into the continental crust (the Mesters Vig line of plutonic centres), supports this hypothesis.

Table 3. Classification of mineralization in the porphyry-bearing metallogenetic provinces in the North Atlantic region

Type of mineralization	Dominant elements	Province[a]
Orthomagmatic	(1) Ni-Cu-sulphides	KA
	(2) Fe-Ti oxide$^{\pm}$P	OR
	(3) Nb + REE	OR, MV
Intramagmatic	(1) Mo	OR, KA, MV
	(2) Native Cu	OR, KA
Hydrothermal breccia/vein	(1) Mo	OR
Contact-metasomatic	(1) Fe oxide base metal sulphides	OR, MV
	(2) Base metal sulphides	OR, MV
Epigenetic		
Exocontact		
Vein	(1) Fe oxides	OR
	(2) Base metal sulphides	OR, MV, KA
	(3) Native Ag	OR
	(4) Nb + REE	MV
	(5) $^{\pm}$F$^{\pm}$Ba	OR, MV

[a] Classification of the Oslo Rift mineralization (after Ihlen 1986). OR = Oslo Rift; KA = Kangerdlugssuaq area; MV = Mesters Vig area.

3 Metallogeny of the Oslo Graben

A large number of mineral deposits are found in the Oslo Rift, including Fe oxide, molybdenum, base metal and native silver deposits as well as fluorite, barite and apatite deposits. These deposits have been classified by Ihlen and Vokes (1978) and Ihlen (1986), (Table 3, Fig. 1B).

The discovery of possible porphyry Mo mineralization in the Oslo Graben (Schønwandt 1975; Geyti and Schønwandt 1979) initiated renewed investigation by mining companies and research groups. Considerable resources of molybdenum were identified in the graben during this campaign, including the Nordli Mo deposit totalling approximately 200 million tons MoS_2 with an average grade of 0.14% (Pedersen 1986).

The above research has brought new knowledge of the different types of Mo mineralization and their associated alteration patterns as well as shedding light on the petrogenesis of the related rock suites. The molybdenite mineralizations are largely intramagmatic and can be grossly subdivided into caldera and batholith-related deposits (Schønwandt and Ihlen, in prep.) (Fig. 3).

3.1 Caldera-Related Deposits

Mo prospects have been located in several of the calderas in the Oslo Graben: Glitrevann, Ramnes and Sande. The Glitrevann caldera shows the most remarkable and varied Mo mineralization. This is an ash-flow caldera starting with ring fracture-

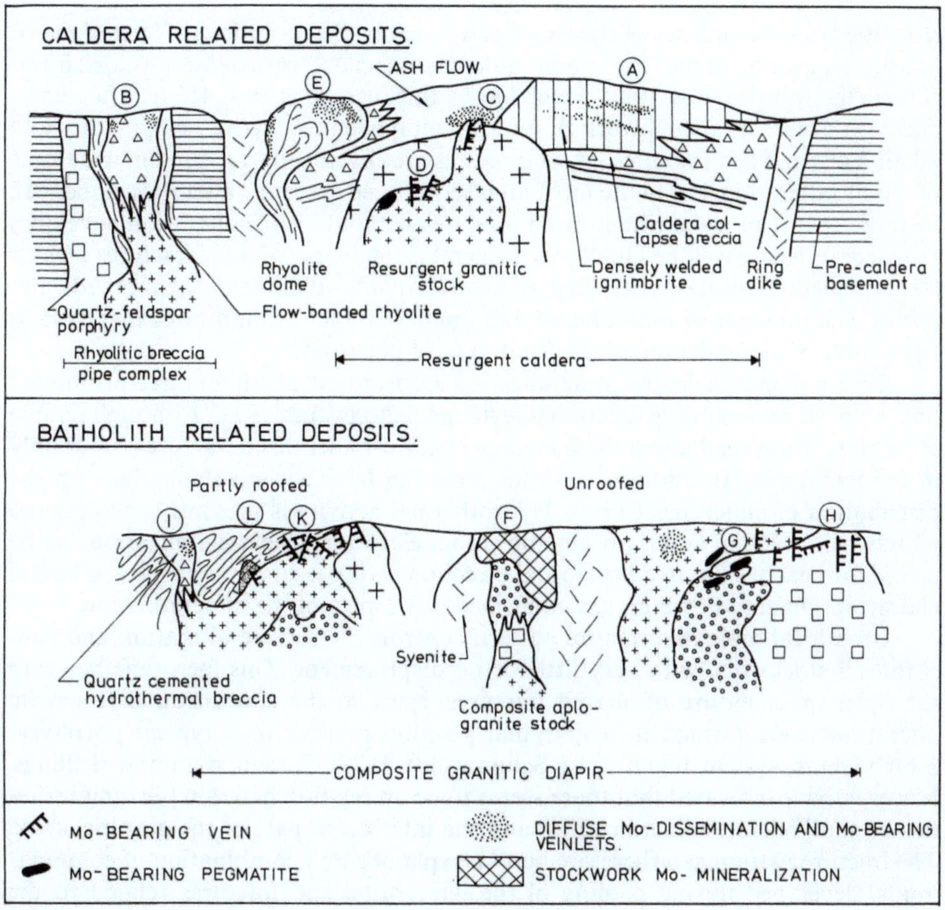

Fig. 3. Location of different types of Mo mineralization in the Oslo Rift

controlled eruptions of rhyolitic-trachytic dome complexes and ignimbrites. The eruptive history culminated with a central vent eruption of silicic ash flows forming a densely welded intracaldera ignimbrite. Subsidence of the caldera floor and deposition of caldera collapse breccias took place concurrently. Post-collapse activity comprises emplacement of a quartz-porphyry complex, including a rhyolitic breccia pipe along part of the ring fracture. This was followed by the intrusion of a syenitic ring-dyke complex, and the evolution terminated with resurgent emplacement of a central granite-aplite stock.

3.1.1 Glitrevann

The Mo mineralizations will be discussed according to their age of formation relative to the evolution of the caldera. A few Mo-mineralized veinlets occur within early rhyolitic dome complexes of the ring fracture-controlled eruptions. The first major Mo mineralization in the Glitrevann caldera is associated with densely welded intra-caldera ignimbrites (Fig. 3(A)), which have a thickness of at least 450 m. The mineralization occurs in a stockwork of quartz-, molybdenite-, pyrite-, magmetite- and alkali-feldspar-filled fractures. Most fractures are enveloped by a dm–cm wide zone of sericite alteration, but in the most intensely altered areas the alteration zones can be more than 1 m wide. Mineralized fractures without visible alteration envelopes cross-cut sericite-enveloped fractures. Mineral assemblages of these fractures include (1) molybdenite-quartz-pyrite, (2) pyrite and (3) quartz-alkali-feldspar-molybdenite-pyrite. The presence of both altered and unaltered fractures indicates that at least two events of molybdenite mineralization have occurred.

On a regional scale, the mineralization seems to occur where three prominent lineaments meet in a triple junction (Geyti and Schønwandt 1979). Although sericite alterations are spread along these lineaments, Mo-mineralized fractures occur only in the ignimbrite. In addition, altitude seems to have a controlling effect on the distribution of mineralized veins. Hydrothermal activity is presently encountered mainly below 450 m and is absent above that elevation. The mineralization can be traced for nearly 2 km in a horizontal direction giving it an apparent strata-bound character. Diamond drilling has confirmed this aspect of the mineralization.

On a local scale the fracture pattern controls the sericite alteration and constitutes a stockwork with very little or no displacement. This fact, together with the open space nature of altered fractures, leads to the conclusion that sericite alterations were formed in a marginal position relative to a typical porphyry-molybdenum system (Geyti and Schønwandt 1979). Recent diamond drillings, however, have indicated that there seems to be no relation between the ignimbrite-hosted molybdenite mineralization and the intersected part of the granitic stock. The fracture pattern can therefore best be explained as a combination of columnar joints, developed during cooling of the ignimbrite, and fractures related to the formation of the triple junction lineaments. The molybdenite occurrence is probably best understood as being related to fumarole activity during degassing and welding of the ignimbrite. This would classify the deposit as an auto-mineralization of a rhyolitic ignimbrite.

Minor Mo mineralization occurs in a rhyolitic breccia pipe (Fig. 3(B)) associated with the quartz-porphyry along the northwestern border of the Glitrevann caldera. Pyrite and sporadic molybdenite disseminations occur in the hydrothermally altered matrix of this breccia.

Widespread Mo mineralization in the Glitrevann caldera is also related to the annular aplogranite zone of the composite alkali-rich granite stock (Table 1). Molybdenite occurs here in open space fractures, together with quartz and pyrite. Only part of this fracture system is enveloped by sericite alteration. Molybdenite is also encountered in quartz-alkali-feldspar pegmatites and in miarolific cavities together with pyrite, quartz and alkali-feldspar. Finally, molybdenite sporadically occurs, disseminated in aplite granite (Fig. 3(D)).

Spatially related to the aplogranite are molybdenite occurrences hosted within lapilli-ash tuffs (Fig. 3(C)). The molybdenite occurs here disseminated in the ash fraction of a pyroclastic flow. High permeability of the host rock is probably important in controlling the distribution of this type of mineralization which is associated with pervasive sericite alteration.

3.1.2 Ramnes

Other caldera-hosted Mo mineralizations occur in the Ramnes and Sande calderas. Here, mineralization is associated with late to post-caldera domes of rhyolitic to trachytic composition (Fig. 3(E)). Most of these can be classified as cryptodomes. They occur both inside and outside the caldera and up to 10 km from the caldera complex. Generally, the domes follow the regional N–S trending structures of the graben and several of the domes have an elongated dykelike form, indicating a relationship to the regional fractures. Molybdenite occurs in veinlets together with quartz and pyrite and is associated with pervasive sericitic alteration of the domes, especially along their border zones.

3.2 Batholith-Related Deposits

The Nordli deposit is economically the most important Mo occurrence in the Oslo Graben (Fig. 3(F)) and shows striking similarities to the Climax-type porphyry-Mo deposits. The deposit was discovered in 1978 by Norsk Hydro and subsequent exploration included 10–200 m of diamond drilling. Descriptions of the Nordli deposit are given by Stougaard (1983), Schønwandt and Petersen (1983) and Pedersen (1986).

The Nordli deposit is related to a composite stock of highly fractionated, alkali-rich granites (Table 1). The stock, which has a diameter of about 400 m and a vertical extension of at least 1000 m, intruded an earlier intrusive breccia which was apparently formed by phreatic explosion of granite porphyries. It is nested within a composite biotite-syenite and granite batholith (Schønwandt and Petersen 1983). At the present level of exposure, the complexes have subcircular outlines and may represent the remnants of a deeply eroded, composite caldera. Regional fracture systems seem to have a controlling effect on the emplacement of the stock.

The Nordli complex consists of three phases. An uppermost granophyre (1) which is replaced by aplogranite (2) 150 m below the present surface. About 450 m below, this unit is substituted by microgranite (3). These granite units seem to be genetically associated with three mineralization events, each having a characteristic type of veinlet (Pedersen 1986). The granophyre is associated with quartz, molybdenite, pyrite and sericite veinlets, whereas the aplogranite has ribbon-type veinlets of quartz, molybdenite, pyrite, sericite, biotite and alkali-feldspar. The microgranite is related to cavernous veinlets with quartz, molybdenite, K-feldspar, calcite, magnetite and pyrite. A dominant feature of the mineralization is a zone of steeply dipping veinlets, indicating that regional fractures have a controlling effect on the distribution of the mineralization.

The Nordli deposit is surrounded by a complex alteration halo. An early formed pervasive K-feldspar + hematite alteration is genetically and spatially associated with the granophyre. The remaining part of the alteration pattern (quartz-sericite-pyrite zone and argillic zone) is suggested to have been generated by hydrothermal convection involving both meteoric and magmatic solutions (Pedersen 1986).

Epigenetic intraplutonic quartz-molybdenite veins have long been recognized in the Oslo Graben, as well as molybdenite-bearing pegmatites and sporadic disseminations of molybdenite in deep-seated granites (Fig. 3(G)) (Bugge 1963; Vokes and Gale 1976). Molybdenite veins are confined to the granite porphyries of the Drammen biotite-granite pluton (Fig. 3(H)). These occur as steeply dipping, sheeted veins and rarely exceed 5 cm in width. They are coarse grained and occasionally contain alkali-feldspar and/or beryl, indicative of a pegmatitic character (Ihlen and Martinsen 1986). Wall rock alteration is generally absent, but widespread topaz, sericite and albite alteration preceded the molybdenite mineralization.

Finally, an open stockwork of quartz-molybdenite-pyrite veinlets associated with sericite alteration may occur in partly roofed batholiths in both an endo- and exocontact position (Fig. 3(K)) (Olerud and Sandstad 1983; Ihlen pers. comm.). A halo of weak propylitization generally encloses the mineralization. Within this halo molybdenite-bearing, quartz-cemented hydrothermal breccias have been located (Fig. 3(I)). Contact metasomatic Fe oxide deposits and skarn-hosted Zn deposits sometimes contain accessory amounts of molybdenite (Fig. 3(L)) (Ihlen pers. comm.).

4 Metallogeny of the Tertiary Province of East Greenland

4.1 Kangerdlugssuaq

The Kangerdlugssuaq area has only been prospected superficially and the main activity has concentrated on the Tertiary igneous rocks. The only known pre-Tertiary mineralization in the area is minor disseminated molybdenite associated with amphibolite bands in Precambrian gneiss. The mineralization related to the Tertiary igneous activity in SE Greenland is presented in Table 3.

The Flammefjeld Complex (Fig. 2). Stockwork Mo mineralization occurs as fragments in an intrusive breccia of the Flammefjeld Complex. The complex is situated near the contact of the Kangerdlugssuaq syenite intrusion and covers an area of 500 × 800 m (Geyti and Thomassen 1984). The complex consists of two main units: (1) an older intrusive breccia and (2) a younger quartz-feldspar porphyry.

Fragments of basalt and syenite dominate in the older breccia, but fragments of acidic composition are common locally. Typical stockwork molybdenite mineralization occurs in some of these acid fragments which comprise granite, quartz-feldspar porphyry and granophyre. Molybdenite, quartz and pyrite constitute the mineral assemblage in the stockwork veinlets. Analysis of one of these fragments yielded 0.45% MoS_2. Some of the veinlets are enveloped by sericite alteration, whereas other fragments are completely altered by either sericitization or silicification. The intense Fe-oxide staining due to oxidation of pyrite and argillic alteration is characteristic within and around the breccia complex and provided a name for the occurrence: Flammefjeld (flame mountain).

The younger quartz-feldspar porphyry is alkali-rich and composed of quartz, orthoclase and minor plagioclase phenocrysts in a dense quartz-feldspar matrix. It seems to form a sheet like intrusion in the breccia. Different types of acid dykes comprising rhyolite, quartz-feldspar porphyry and aplite dykes are related to the complex which probably forms a nearly vertical pipe like body.

4.2 The Mesters Vig Area

The line of plutonic-subvolcanic centres in this area is characterized on Landsat MSS images by outstanding colour anomalies due to hydrothermal alteration and oxidation, primarily of pyrite. Landsat image-analysis techniques have been used in mapping over 50 significant anomalies in this province (Conradsen and Harpøth 1984). Mineralization related to the Tertiary igneous activity of the Mesters Vig area is presented in Table 3. Two types of intramagmatic molybdenite mineralization can be distinguished: (1) mineralization hosted in composite granite stocks and (2) volcanic-hosted deposits.

4.2.1 Granite Stocks

In the Werner Bjerge Alkaline Complex two molybdenite mineralized areas are known: the Malmbjerget porphyry-Mo deposit and the Mellempas occurrence (Fig. 4). The Werner Bjerge Complex is roughly circular with a diameter of about 17 km. Bearth (1959) subdivided the complex into three lithological units: (1) a basic unit in the southeastern part; (2) an alkali-syenite-granite which dominates the northern part; (3) a nepheline syenite in the southwest. The Malmbjerget Mo deposit and the Mellempas occurrences are temporally and spatially related to the syenite-granite unit.

The basic complex is the oldest of the three complexes, whereas the age relations between the two other units remain uncertain. Nepheline syenite inclusions in the lower part of the Malmbjerget stock indicate that the syenite-granite unit is younger than the nepheline syenites. Radiometric dating of syenite yields a whole rock Rb/Sr age of 30 ± 2 Ma (Rex et al. 1979). K–Ar ages of the Malmbjerget granite stock range from 26 ± 1.1 Ma to 21.1 ± 0.9 Ma) (Schassberger pers. comm.), indicating that the granite stock represents one of the youngest events in the Werner Bjerge Complex. Geochemistry points towards a comagmatic origin for the Werner Bjerge Complex (Bearth 1959). The mineralogy of the rocks supports this conclusion (Brooks et al. 1982).

Malmbjerget. Malmbjerget is a typical porphyry Mo deposit of Climax type. The deposit was discovered during systematic mapping of the Werner Bjerge Complex by Bearth (1959). From 1954 to 1979 a total of 22 877 m was drilled and 1329 m of adit has been excavated to investigate the deposit. An orebody of 150 million tons

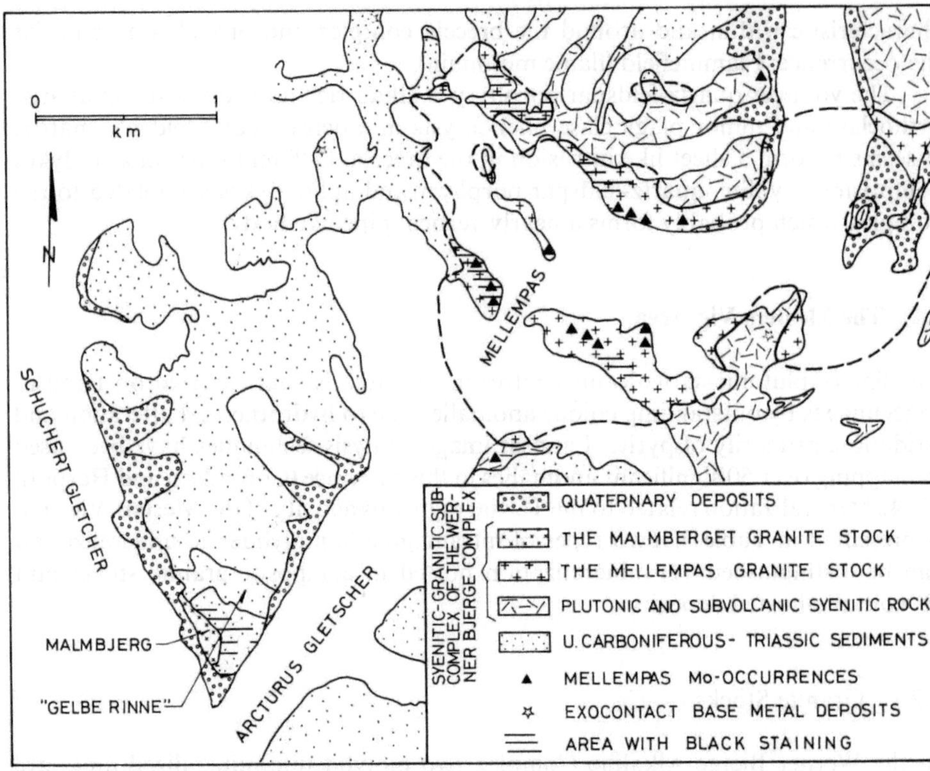

Fig. 4. Simplified geological map of the northeastern part of the Werner Bjerge Complex showing location of the Malmbjerget deposit and the molybdenum occurrences of Mellempas

with 0.23% MoS_2 at a cut-off of 0.16% MoS_2 has been proven. Details of the exploration history were published by Harpøth et al. (1986).

The Malmbjerget granite stock occurs isolated from other Werner Bjerge intrusions and is exposed over 1 km along the Schuchert glacier (Fig. 4). The granite intrusion is a composite stock consisting of three main lithological units, each of which can be further subdivided into texturally different units.

The uppermost part of the stock consists of an alkali-rich perthite granite with a quartz-feldspar porphyry roof phase (Fig. 5 and Table 1). Then follows a heterogeneous porphyritic aplite (Table 1). In the lowest part of the stock two texturally different porphyritic granites occur (not shown in Fig. 5). The uppermost part of the porphyritic granite has an equigranular to slightly porphyritic texture with rounded quartz, whereas the lower part of the porphyritic granite has plagioclase and orthoclase phenocrysts in a fine- to medium-grained groundmass. The porphyritic aplite cross-cuts the perthite granite forming the upper part of the stock, whereas the contact zone between the aplite unit and the porphyritic granite of the lower part of the stock is less obvious with regard to age relations.

The molybdenite mineralization appears as an inverted bowl-shaped body (Sudgen 1963) and is mainly located within the perthite granite and its equiva-

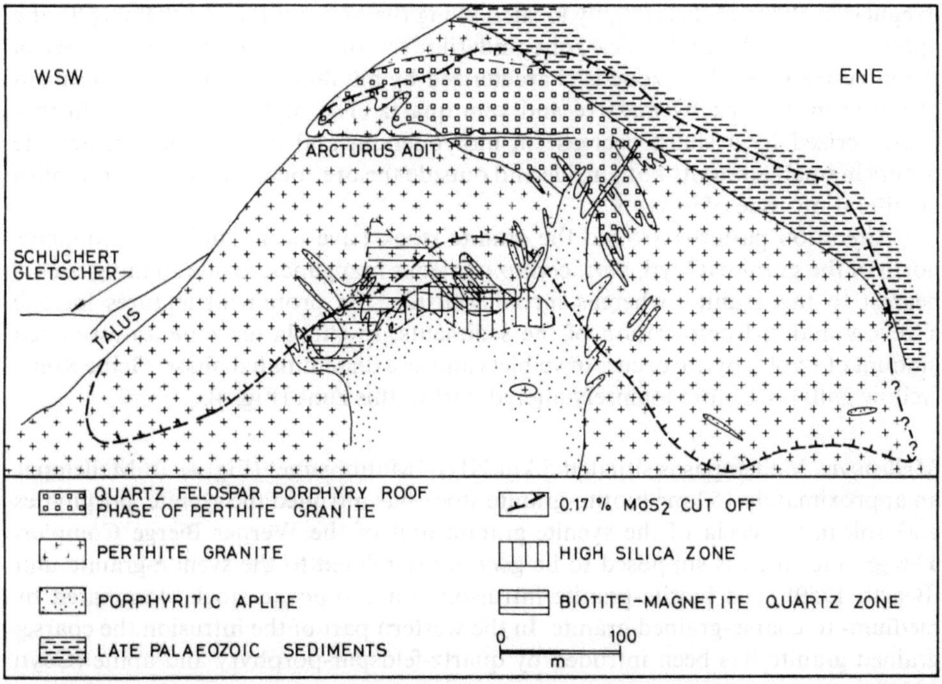

Fig. 5. WSW–ENE section through the central part of the Malmbjerget Mo deposit showing generalized geology, ore zone and alteration zones (After Schassberger and Galey 1975)

lent roof phase. Sediments immediately above the granite cupola are also mineralized. The rest of the mineralization is hosted in the heterogeneous porphyritic aplite unit. Molybdenite occurs in at least three generations of quartz veinlets and several assemblages with biotite and fluorite. The veinlets range in thickness from hairline size to about 5 cm. They form a stockwork of mutually offsetting veinlets. The pyrite content in the mineralized zone is generally less than 1%. A small, high-silica zone occurs below the main molybdenite mineralization (Fig. 5).

Flat-lying greisen veins cut the stockwork molybdenite mineralization. However, a few quartz-molybdenite veinlets cross-cut the greisen zone indicating that the greisen event took place at the termination of the stockwork formation. The greisen assemblage includes topaz, wolframite, fluorite, molybdenite, beryl, siderite, pyrite, sphalerite and chalcopyrite. Up to three generations of topaz and pyrite have locally been recognized in the veins (Kirchner 1964). Minor base metal veins overprint greisen and stockwork mineralization.

Two pronounced alterations have been established by Schassberger and Galey (1975): the biotite-magnetite-quartz zone and the high silica zone (Fig. 5). The biotite-magnetite-quartz zone is characterized by vein and disseminated biotite and magnetite, up to 40 and 20 vol% respectively. Quartz occurs as veinlets and as quartz flooding, and makes up to 40 vol% of the zone of alteration. The zone is irregular in shape and is roughly located along the upper surface of the porphyritic aplite subunit. The high silica zone underlies the ore zone in the central part of the intrusive mass. This zone is characterized by an almost complete replacement of the primary rock by massive and vein quartz. The high silica zone is further characterized by a sharp decrease in molybdenum and potash content. Sericite occurs in small amounts both inside and outside the ore zone, but does not establish an alteration zone of its own.

Alteration patterns around the granite stock have been studied in sediments north of the granite (Geyti, pers. comm.). Intense manganese-oxide staining occurs nearest to the granite, whereas further outside, the prominent features include a yellow and red colouration of the sediments. In this latter zone an estimated amount of 1 vol% pyrite occurs in veinlets and as a disseminated phase. Gelbe Rinne (yellow gully) is a more intensely altered part of this zone (Fig. 4).

Mellempas. Mellempas is situated 5 km NE of Malmbjerget (Fig. 4). In Mellempas an approximately 15 km^2 biotite-granite stock has intruded the syenite, porphyries and volcanic breccia of the syenite-granite unit of the Werner Bjerge Complex. The granite stock is supposed to be genetically related to the syenite-granite unit (Bearth 1959). The biotite-granite intrusion is a composite stock, dominated by medium- to coarse-grained granite. In the western part of the intrusion the coarse-grained granite has been intruded by quartz-feldspar-porphyry and aplite (Geyti 1981).

Molybdenite occurs as disseminated grains in coarse-grained granite and in vugs and pegmatites, especially in the quartz-feldspar porphyry. The mineral associations in the vugs and pegmatites are molybdenite, quartz, K-feldspar, fluorite and pyrite. Strictly speaking, these molybdenite occurrences are orthomagmatic. However, true epigenetic molybdenite mineralization occurs in the central aplite mass as moly paint and fine-grained molybdenite in veinlets which are occasionally enveloped by quartz-sericite-pyrite alteration. The most prominent alteration of the Mellempas area is reddish-brown staining of iron oxide due to decomposition of pyrite and a black manganese-oxide staining. The alterations are roughly centred around the central aplite mass with the black staining forming a ring like pattern within the red-brown stained area.

4.2.2 Volcanic-Hosted Deposits

The Kap Simpson Complex (Fig. 2) was mapped by Schaub (1938, 1942) and has since been investigated locally by various Nordmine and Amax prospecting groups (Schassberger and Newall 1980; Geyti 1980; Damtoft and Grahl-Madsen 1982). The Kap Simpson Complex has intruded Late Palaeozoic and Mesozoic sediments and can be subdivided into a northwestern alkali-syenite intrusion and the so-called Dreibuchten zone towards the southeast. The Dreibuchten zone is surrounded by a discontinuous syenite ring dyke. The area inside the ring dyke is composed of different sedimentary, volcanic and subvolcanic rocks. The area encircled by the ring dyke has apparently subsided, indicating that the Dreibuchten zone is a caldera structure.

The evolution of the presumed caldera is only known in broad outline. The caldera-related igneous activity can be divided into three episodes: (1) an early volcanic event including ash-flow tuffs, rhyolite flows, ignimbrites and minor stocks of feldspar porphyries, (2) intrusion of syenite and granite, (3) a late volcanic episode comprising rhyolite dykes and plugs and intrusive feldspar porphyries.

Altered areas up to several km^2 in size occur in the southern part of the Dreibuchten zone. Rocks within these areas are variously argillized and silicified. Fluorite is a characteristic mineral of these areas and occurs mostly in cm-thick veins, sometimes with quartz and pyrite. Scattered molybdenite grains have been found in a few quartz veinlets and disseminated in granite. Anomalous Mo contents (65–645 ppm) have been found in several fluorite-bearing veins where no Mo minerals have been identified.

The overall impression of the mineralization and alteration pattern in the Dreibuchten zone suggests that fumarolic activity may be responsible for the major part of the alteration. Mineralization and alteration are locally clearly related to granite intrusions.

5 Conclusions

1. Porphyry-Mo deposits in the North Atlantic region occur in intraplate magmatic provinces, appearing either as a line of plutonic centres or in a failed rift.
2. The deposits are associated with syenite-granite complexes, but they are spatially related to highly fractionated, alkali-rich granite intrusions (Table 1).
3. The occurrences show striking similarities with the Climax deposit in Colorado, both in style of mineralization and alteration as well as in the associated granite intrusion.
4. The wide variety of Mo mineralizations in these provinces indicate the great capacity of the mineralizing processes to precipitate molybdenite. However, only a subvolcanic environment seems to favour generation of porphyry-type deposits.
5. The porphyry-Mo deposits form part of a metallogenetic province characterized by orthomagmatic and epigenetic hydrothermal deposits. The absence of stratiform 'red-bed'—like mineralizations probably reflects the lack of a sedimentary depocentre, because the provinces remained areas of regional uplift for a considerable time.

Acknowledgements. The support of the European Economic Community under *contract No. MSM*-117-*DK* is acknowledged. I wish to thank the exploration staff of Norsk Hydro A/S and Nordisk Mineselskab A/S for help during field work and permission to use some of their data. I thank Drs. J.S. Petersen, J.R. Wilson and J. Korstgård for reviewing the manuscript. This work is published with permission of the Director of the Geological Survey of Greenland.

References

Arthaud F, Matte P (1977) Late Paleozoic strike-slip faulting in southern Europe and northern Africa: result of a right-lateral shear zone between the Appalachians and the Urals. Geol Soc Am Bull 88:1305–1320

Bearth P (1959) On the alkali massif of the Werner Bjerge in East Greenland. Medd Grønl 153:63

Brooks CK (1973) Rifting and doming in southern East Greenland. Nature Phys Sci 244:23–25

Brooks CK (1979) Geomorphological observations at Kangerdlugssuaq, East Greenland. Medd. Grønl Geo Sci 1:24

Brooks CK, Nielsen TFD (1982) The E Greenland continental margin: a transition between oceanic and continental magmatism. J Geol Soc (Lond) 139:265–275

Brooks CK, Pedersen AK, Larsen LM, Engell J (1982) The mineralogy of the Werner Bjerge Complex, East Greenland. Medd Grønl Geosci 7:

Bugge A (1963) Norges molybdenforekomster. Geol Surv Norway 217:134

Burke K, Dewey JF (1973) Plume-generated triple junctions: key indicators in applying plate tectonics to old rocks. J Geol 81:406–433

Bylund G, Patchett PJ (1977) Palaeomagnetic and Rb–Sr isotopic evidence for the age of the Særna alkaline complex, western central Sweden. Lithos 10:73–79

Conradsen K, Harpøth O (1984) Use of landsat multispectral scanner data for detection and reconnaissance maping of iron oxide staining in mineral exploration, central East Greenland. Econ Geol 79:1229–1244

Courtillot V (1982) Propagating rifts and continental breakup. Tectonics 1:239–250

Damtoft JS, Grahl-Madsen L (1982) Geological and VLF-geophysical mapping of the Bredehorn Ba–Pb prospect and preliminary mapping of SE Traill Ø for porphyry molybdenum mineralization. Internal report Nordisk Mineselskab A/S

Dons JA (1978) Terminology and history of investigation. In Dons JA, Larsen BT (eds) The Oslo paleorift. A review and guide to excursions. Geol Surv Norway 337:9–16

Geyti A (1980) Drilling and mapping in the EGMO concession area. Internal report Nordisk Mineselskab A/S

Geyti A (1981) Porphyry molybdenum in the Malmbjerget-Mellempas area. Internal report Nordisk Mineselskab A/S

Geyti A, Schønwandt HK (1979) Bordvika — a possible porphyry molybdenum occurrence within the Oslo rift, Norway. Econ Geol 74:1211–1220

Geyti A, Thomassen B (1984) Molybdenum and precious metal mineralization at Flammefjeld, southeast Greenland. Econ Geol 79:1921–1929

Haller J (1971) Geology of the East Greenland Caledonides. Wiley, New York, p 413

Harpøth O, Pedersen JL, Schønwandt HK, Thomassen B (1986) Mineral occurrences in central East Greenland. Medd Grønl Geosci 17:138

Ihlen PM (1986) The geological evolution and metallogeny of the Oslo paleorift. In Olerud S, Ihlen PM (eds) Metallogeny associated with the Oslo paleorift. Geol Surv Sweden Ser Ca 59:6–17

Ihlen PM, Martinsen M (1986) Ore deposits spatially related to the Drammen granite batholith. In: Olerud S, Ihlen PM (eds) Metallogeny associated with the Oslo paleorift. Geol Surv Sweden Ser Ca 59:6–17

Ihlen PM, Vokes FM (1978) Metallogeny. In Dons JA, Larsen BT (eds) The Oslo paleorift. A review and guide to excursions. Geol Surv Norway 337:75–90

Kirchner G (1964) Die Molybdänlagerstätte des Erzberges bei Mesters Vig, Östgrönland. Berg- Hüttenmänn Monatsh 109:162–173

Larsen HC (1984) Geology of the East Greenland shelf. In Spencer AM (ed) Petroleum geology of the North European margin Graham & Trotman London. pp 329–339

Mutschler FE, Wright EG, Ludington S, Abbott JT (1981) Granite molybdenite systems. Econ Geol 76:874–897

Nielsen TFD (1975) Possible mechanism of continental break-up in the North Atlantic. Natur 253:182–184

Nielsen TFD (1987) Tertiary alkaline magmatism in East Greenland: A review. In Fitton JG, Upton BGJ (eds) Alkaline igneous rocks. Geol Soc Spec Publ 30:489–515
Nielsen TFD, Brooks CK (1981) The E Greenland rifted continental margin: an examination of the coastal flexure. J Geol Soc (Lond) 138:559–568
Noe-Nygaard A (1976) Tertiary igneous rocks between Shannon and Scoresby Sund, East Greenland. In: Escher A, Watt WS (eds) Geology of Greenland. Geol Surv Greenl Copenhagen:386–402
Olerud S, Sandstad JS (1983) Geology of the Skrukkelia molybdenite deposit, northwestern margin of the Oslo Graben, Norway. Geol Surv Norway 387:1–20
Pedersen FD (1986) An outline of the geology of the Hurdal area and the Nordli granite molybdenite deposit. In Olerud S, Ihlen PM (eds) Metallogeny associated with the Oslo Paleorift. Geol Surv Sweden Ser Ca 59:18–23
Pegrum RM (1984) The extension of the Tornquist zone in the Norwegian North Sea. Nor Geol Tidsskr 64:39–68
Ramberg IB (1976) Gravity interpretation of the Oslo Graben and associated igeneous rocks. Geol Surv Norway 325:194
Ramberg IB, Larsen BT (1978) Tectonomagmatic evolution. In: Dons JA, Larsen BT (eds) The Oslo rift. A review and guide to excursions. Geol Surv Norway 337:55–73
Ramberg IB, Spjeldoes N (1978) The tectonic history of the Oslo region. In Ramberg IB, Neumann E-R (eds) Tectonics and geophysics of continental rifts. NATO ASI Ser C 37:167–194
Rex DC, Geldhill AR, Brooks CK, Stenfelt A (1979) Radiometric ages of Tertiary salic intrusions near Kong Oscars Fjord, East Greenland. Rapp Geol Surv Greenl 95:106–109
Schassberger HT, Galey JT (1975) Report on the 1974 core relogging Erzberg project, East Greenland. Amax Internal report Nordisk Mineselskab A/S
Schassberger HT, Newall GC (1980) The East Greenland molybdenum project. Amax Internal report Nordisk Mineselskab A/S
Schaub HP (1938) Zur Vulkanotektonik der Inseln Traill und Geographical Society (Nord-Ost Grønland). Medd Grønl 114:29–44
Schaub HP (1942) Zur Geologie der Traill Insel (Nord-Ost Grønland). Eclog Geol Helv 35:1–54
Schønwandt HK (1975) Oslofeltets Mo-mineraliseringer. Internal Report Norsk Hydro A.S., Oslo
Schønwandt HK, Petersen JS (1983) Continental rifting and porphyry-molybdenum occurrences in the Oslo region, Norway. Tectonophysics 94:609–631
Sillitoe RH (1980) Types of porphyry molybdenum deposits. Mining Mag 142:550–553
Stougaard S (1983) Geologiske karakteristika og konsekvenser omkring Nordli porfyr-Mo forekomsten. Thesis University of Aarhus p 143
Sudgen FJ (1963) Erzberg project East Greenland. Summary report of 1962 exploration program. Amax Internal report Nordisk Mineselskab A/S
Touret J (1970) Deep-seated volcanism along the major Precambrian breccia in South Norway I: Degernes at lake Vegår. Norsk Geol Tidsskr 50:249–252
Vokes FM, Gale GH (1976) Metallogeny related to global tectonic in southern Scandinavia. Geol Assoc Canada Spec Pap 14:413–441
Wager LR, Deer WA (1938) A dyke swarm and crustal flexure in East Greenland. Geol Mag 75:49–56
Watson J (1976) Eo-Europa: the evolution of a craton. In: Ager DV, Brooks M (eds) Europe from crust to core. Wiley, New York, pp 59–78
Westra G, Keith SB (1981) Classification and genesis of stockwork molybdenum deposits. Econ Geol 76:844–873
White WM, Bookstrom AA, Kamilli RJ, Ganster MW, Smith RP, Ranta DE, Steininger RC (1981) Character and origin of climax-type molybdenum deposits. Econ Geol 75:270–316
Ziegler PA (1978) North Sea rift and basin development. In: Ramberg IB, Neumann E-R (eds) Tectonics and geophysics of continental rifts. NATO ASI Ser C 37:249–277

Niobium-Tantalum Mineralisation in the Motzfeldt Centre of the Igaliko Nepheline Syenite Complex, South Greenland

T. TUKIAINEN[1]

Abstract

The Motzfeldt Centre (1310 ± 10 Ma) is one of the major, central complexes in the Gardar Province of alkaline igneous activity. The Motzfeldt Centre is made up of multiple intrusions of syenite with a wide range of textural and compositional characteristics. The syenites were emplaced as two main igneous phases into the Proterozoic Julianehåb Granite and the unconformably overlying Gardar supracrustal rocks.

The syenites of the early igneous phase occur as isolated bodies which are truncated by the syenite units of the main igneous phase, the Motzfeldt Ring Series. The bulk of the Motzfeldt Ring Series constitutes three major, steep-sided, outward dipping intrusions of predominantly peralkaline syenite and nepheline syenite which young inwards. The apparent intrusion mechanism was a combination of ring fracture and block subsidence.

Large quantities of the roofing sandstone and volcanics have been incorporated to the outermost unit of the Motzfeldt Ring Series, the Motzfeldt Sø Formation, where they are preserved as large rafts. The sandstone has, however, been largely assimilated and has given the outer zone of the Motzfeldt Sø Formation its unique (for the centre) quartz normative character. The Motzfeldt Sø Formation underwent an extreme in situ differentiation, probably due to effective crystal fractionation, which resulted in the formation of a peralkaline residuum rich in volatile and incompatible elements. The peralkaline residua gave rise to a complex of late peralkaline sheets of microsyenite and pegmatite and hydrothermal alteration with associated Th–U–Nb–Ta–Zr–REE mineralisation which increases in intensity towards the margins, and especially towards the roof of the intrusion.

The mineralisation is probably the result of the combination of an incompatible element/volatile-enriched magmatic residuum and an influx of silica and meteoric water, which resulted in a dramatic increase in oxygen fugacity, acidity and hydrothermal activity. A volatile-saturated outer shell developed which facilitated the migration, accumulation and precipitation of the incompatible elements.

The Motzfeldt Sø Formation hosts zones of economically interesting pyrochlore enrichment. The pyrochlore, characterized by the enrichment of LREE, Ta and U,

[1] Geological Survey of Greenland, Øster Voldgade 10, DK-1350 København K, Denmark

Mineral Deposits within the European Community
(ed. by J. Boissonnas and P. Omenetto)

shows a marked compositional variation depending on its relative depth in the Motzfeldt Sø Formation. The Ta content and the Nb/Ta ratio of the pyrochlore varies within wide limits, from 1.5 to 10.0% Ta_2O_5 and from 8 to more than 50 respectively. The pyrochlore at the deeper levels is enriched in Ta and Ca, whereas the pyrochlore at the higher levels of the igneous column is more enriched in Nb, U and LREE.

1 Introduction

The alkaline intrusive rocks of the Motzfeldt area (Fig. 1) make up one of the major manifestations of the Mid-Proterozoic (ca. 1320–1120 Ma) alkaline magmatism in South Greenland (Blaxland et al. 1978). The area affected by this event, the Gardar Province, displays a variety of alkaline rocks which show a compositional continuum from basaltic to salic alkaline compositions (Upton and Emeleus, in press).

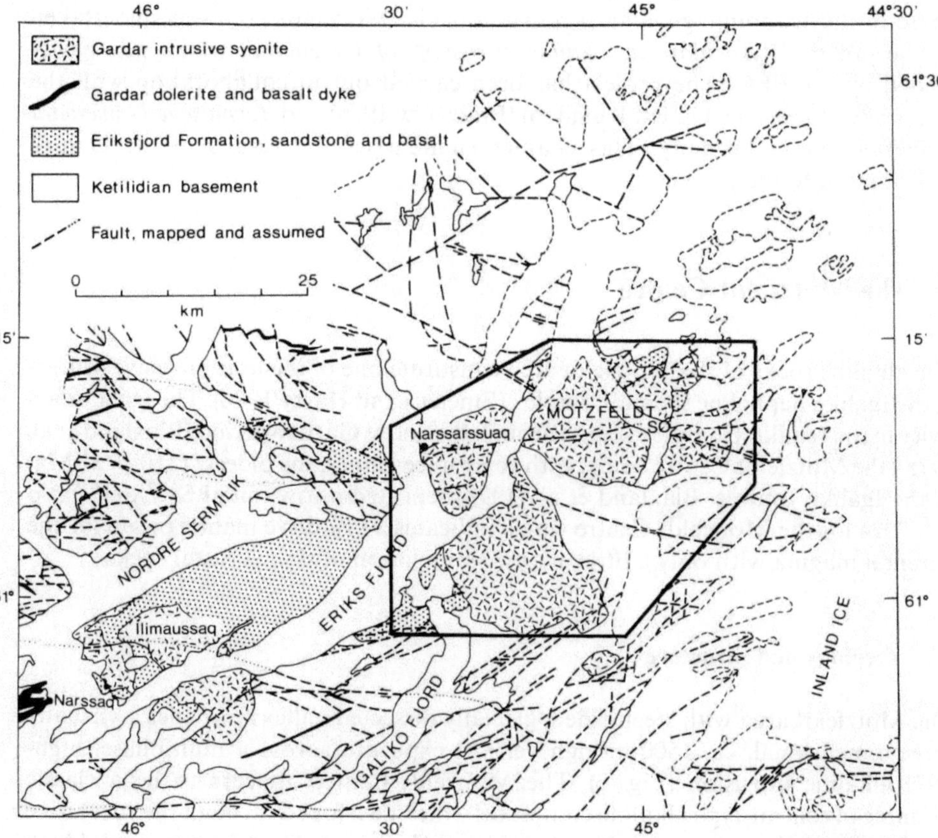

Fig. 1. Sketch map showing the locations of the Igaliko Complex (*outlined*) and the surrounding geology

The Gardar igneous activity has been related to repetitive rifting with the Mid-Proterozoic supercontinent of North America, Greenland and Europe (Piper 1982)

The Gardar activity commenced with a large-scale uprise of transitionally alkaline basalts and hawaiites which, together with sandstone, are preserved in the downfaulted units (Eriksfjord Formation). The basement and the unconformably overlying Eriksfjord Formation were intruded by a variety of alkaline dykes and about ten central complexes which vary in size from a few hundred metres across to ca. 45 km in diameter.

The mineralisation in the Motzfeldt area was discovered in 1979 by the drainage geochemical and airborne gamma-spectrometric reconnaissance surveys of the *Syduran project* (Armour-Brown et al. 1982, 1983), which was partly financed by the EEC during the period 1st Dec. 1979–1st Dec. 1980. The geological and radiometric mapping of the Motzfeldt area was commenced in 1982 as part of the extended *Syduran project* under the *Danish Ministry of Energy*'s Research Programmes of 1981 and 1982 (Armour-Brown et al. 1983). The purpose of this study was to make a detailed geological and radiometric map of the alkaline rocks of the Motzfeldt area with a view to provide a reliable reference framework for the evaluation of the economic mineral potential of these rocks (Tukiainen et al. 1983). The latest phase in the economic geological research of the Motzfeldt area was undertaken by the project '*Pyrochlore in alkaline intrusions of Greenland*' which was partly funded by the EEC. The project has been carried out in collaboration with the *University of Durham*, United Kingdom (Bradshaw 1985) and *Technische Universität München*, West Germany. This chapter summarises the results of the Danish contribution to the project.

2 The Motzfeldt Centre

The alkaline rocks of the Motzfeldt area constitute one of the four intrusive centres of the Igaliko nepheline syenite complex (Emeleus and Harry 1970). The igneous activity in the Igaliko Complex spans through the entire Gardar period (Blaxland et al. 1978); the Motzfeldt Centre being, with some uncertainty, the oldest (1310 $\pm$ 31 Ma) of the Igaliko centres. Blaxland et al. (1978) reported a low initial Sr^{87}/Sr^{86} ratio of 0.7024 for the Motzfeldt Centre which indicates a primitive mantle origin for the parental magma with only a little, if any, assimilation of sialic crustal rocks.

2.1 Geology and Structure

The Motzfeldt area with steep-sided, glacially dissected valleys provides a virtually three-dimensional, ca. 1500-m-high vertical exposure across a multiphase, high-level alkaline intrusion (Fig. 2). The Motzfeldt Centre appears to be a classic example of central-type alkaline complexes which developed through the successive emplacement of syenite magma guided by a combination of ring fractures and block subsidence.

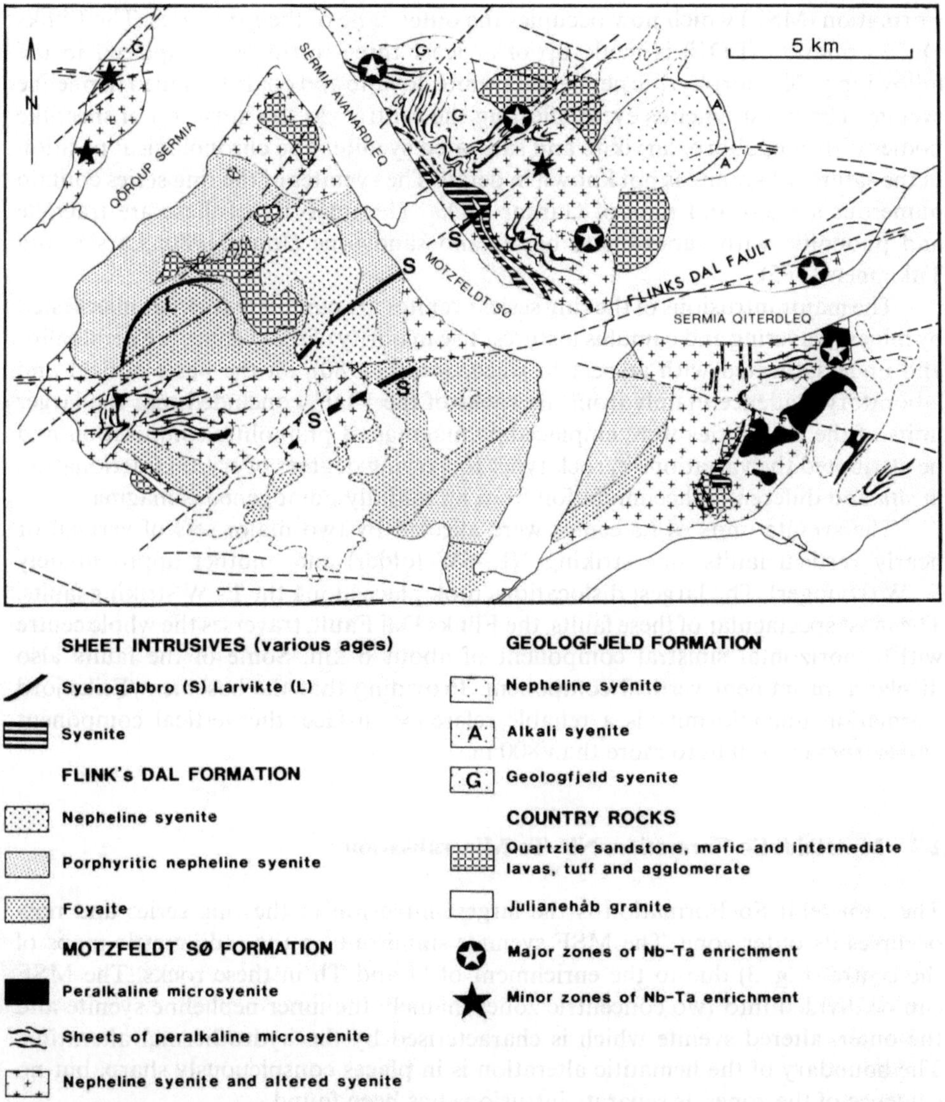

Fig. 2. Simplified geological map of the Motzfeldt Centre

The Motzfeldt Centre developed through two major phases of igneous activity (Tukiainen et al. 1983). The syenite units of the early igneous phase, the Geologfjeld Formation, originally comprised a number of separate syenite stocks which are now truncated by the syenite units of the main igneous phase, the Motzfeldt Ring Series (MRS).

The Motzfeldt Ring Series is made up of concentric, steep-sided intrusions of syenite whose marginal contacts dip outwards. The intrusions young inwards. The main igneous phase commenced with the emplacement of the Motzfeldt Sø

Formation (MSF) which now occupies the outer zone of the ring series. The Flinks Dal Formation (FDF) is made up of at least three intrusions, emplaced in the following order: porphyritic nepheline syenite, foyaite and coarse-grained nepheline syenite. The latest igneous event affecting the centre was the intrusion of sheetlike bodies of syenogabbro, larvikite and saturated syenite. The chronological position of the saturated syenite is not known in detail. The syenites of the ring series contain numerous screens and rafts of Gardar rocks. The majority of these are trachyte and phonolite with subordinate basalt and sandstone (Jones 1980; Larsen and Tukiainen 1985).

The major intrusions of the ring series syenites are cumulate rocks as manifested by mineral-layering and cumulus textures. The majority of the syenites are peralkaline and undersaturated with respect to silica. Jones (1980), on the basis of field and laboratory evidence mainly from the rocks of the FDF, concluded that the larger units of the ring series were emplaced as magmas of phonolitic composition and he attributed the variations in rock types to varying degrees of crystal fractionation *in situ* and differential accumulation from an initially homogeneous magma.

The syenite units of he centre were affected by two major sets of vertical or nearly vertical faults, one striking NE–SW (older) and another approximately E–W (younger). The largest dislocations took place along the E–W striking faults. The most spectacular of these faults, the Flinks Dal Fault, traverses the whole centre with a horizontal sinistral component of about 6 km. Some of the faults also display a prominent vertical component. Providing that the basement/Eriksfjord Formation unconformity is a reliable reference surface, the vertical component varies from ca. 100 m to more than 800 m.

2.2 Motzfeldt Sø Formation. Nb–Ta Mineralisation

The Motzfeldt Sø Formation is the largest intrusion in the ring series and now occupies its outer zone. The MSF syenites stand out on the radiometric maps of the centre (Fig. 3) due to the enrichment of U and Th in these rocks. The MSF can be divided into two concentric zones: namely the inner nepheline syenite and the outer, altered syenite which is characterised by the hydrothermal alteration. The boundary of the hematitic alteration is in places conspicuously sharp, but no evidence of the zones as separate intrusions has been found.

The MSF hosts a complex of late peralkaline sheets (Figs. 1 and 4). The geometry of the sheet complex is a kind of cupola which apparently conforms to the 'bell jar' — shape of the intrusion roof. The frequency and thickness of the sheets increase upwards in the MSF. The inner part of the MSF, the nepheline syenite, contains merely a series of more or less irregular pegmatite sheets with subordinate microsyenite. At higher levels, mainly in the altered syenite, the geometry of the sheet complex becomes more persistent, although the sheets are still reminiscent of zoned pegmatite bodies with a microsyenitic interior. At the highest levels of the MSF the microsyenite component of the sheets predominates and the sheets display pronounced intrusive relationships with the older rocks. The northeastern part of the centre displays the most complete section across the sheet complex where the

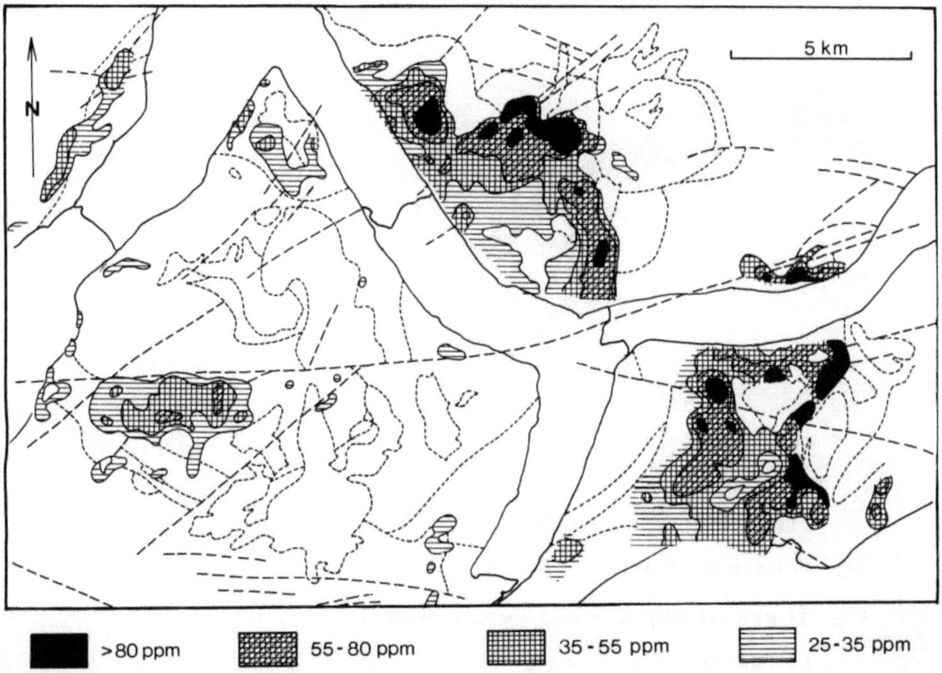

Fig. 3. Thorium airborne radiometric map of the Motzfeldt Centre. Contours $\geq$ 25 ppm Th are shown

aggregate maximum thickness of the sheet complex is ca. 700 m including screens of the older syenite.

The alteration is strickingly restricted to these rock types, and this indicates that the hydrothermal activity was intimately related to the igneous evolution of the MSF. It invades both the MSF-syenite ('altered syenite') and peralkaline microsyenite. The country rocks, the basement granite and the Eriksfjord Formation are virtually unaffected by the alteration/mineralisation.

2.3 Petrographical Characteristics of the Motzfeldt Sø Formation

The following petrographical description is in part based on the data which was kindly provided by C. Bradshaw, *University of Durham.*

The MSF *nepheline syenite* is a coarse-grained rock containing abundant pegmatitic patches which range from fist-sized closed vugs to major sheetlike bodies. The rock often displays a synneusis texture having amphibole, pyroxene, Fe oxides and apatite clustered together. Mesoperthitic alkali feldspar (60%), nepheline (25%) and Na–Ca pyriboles (15%) are the essential minerals.

The alkali feldspar grains, 1–10 mm in length, are randomly orientated and subhedral-granular in habit. Albite becomes more abundant in the more evolved rocks. Both amphibole, which is the major mafic phase, and pyroxene show distinct zonation and a wide range of compositions. The main amphibole phase in the less

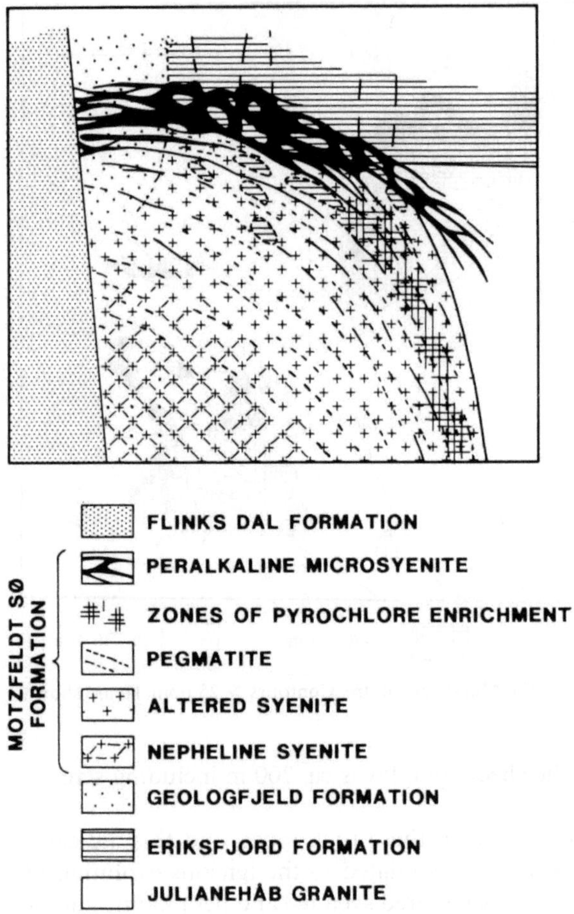

Fig. 4. Schematic cross-section of the Motzfeldt Sø Formation

evolved rock is ferroedenitic hornblende/ferroedenite with only a narrow rim of katophorite. In more evolved rocks the predominant amphibole is katophorite, although the larger katophorite crystals have discrete cores of ferroedenitic hornblende/ferroedenite. The zonation is discontinuous with the core being rimmed by small orange biotite flakes and Fe–Ti oxides, all enclosed by katophorite. The pyroxene centres range in composition from salite to ferrosalite, the grains being rimmed by aegirine-augite or aegirine-hedenbergite. Aegirine occurs as individual crystals in the more evolved and pegmatitic rocks.

The MSF *altered syenite* ranges from leucotypic to mesotypic and is characterised by a reddish-brown discolouration. The altered syenite is texturally heterogeneous; the textural complexity increases towards the margin of the MSF. The heterogeneities are due to the extreme variations in grain size which varies from aplitic to coarsely pegmatitic, and irregularities in distribution of mafic minerals. Miarolitic cavities, variably filled with quartz and other minerals, are a standard feature. The primary Na–Ca pyriboles characteristic of the nepheline syenite are virtually absent in the MSF-altered syenite. The predominating mafic silicate is arfvedsonite which is both optically and compositionally homogeneous. A variety of the MSF-altered syenite rich in arfvedsonite (up to 60% of the mode) occurs sporadically in the margin of the MSF. In the alkali feldspar, the first manifestation of the alteration is the appearance of finely disseminated Fe-oxide dust and micaceous alteration products. The increasing degree of alteration is accompanied by the appearance of specular hematite and fresh albite replacing the alkali feldspar. In the extreme cases of the albitisation there are merely braids or patches of the original perthite left in the feldspar grains.

Fluorite is a rock-forming mineral in the MSF-altered syenite and can constitute up to 10% of the mode. In the more intensively altered rocks zircon, thorite and pyrochlore become important constitutents. Zircon and thorite are ubiquitous and pyrochlore is enriched in patchy, discontinuous zones adjacent to the intrusion margin.

The *peralkaline microsyenite* ranges from leuco- to mesotypic and is texturally and compositionally very variable. The mesotypic rocks are of microsyenite grain size (less than 1 mm), the leucotypic varieties are coarser grained and often pegmatitic. The peralkaline microsyenite displays a well-developed banding with alternating bands, ranging from a few centimetres up to several metres, of leucocratic and mesocratic microsyenite. The banding is irregular, displaying small folds and pinch and swell structures coupled with rapid fluctuations in composition. The texture of the mesocratic rocks varies from massive to laminated and the rocks are characteristically porphyritic with phenocrysts of orthoclase up to 5 mm in length in a matrix of orthoclase, albite, arfvedsonite, aenigmatite, eudialyte, analcime and fluorite. Varieties rich in nepheline and aegirine occur sporadically. Nepheline is in most cases, however, heavily altered to analcime and micas. Sphalerite, låvenite, astrophyllite, zircon, pyrochlore and micaceous alteration products are characteristic accessories. Apart from the coarser grain size and higher feldspar content, the mineralogy of the leucotypic rocks is similar.

The general scheme of the alteration process in the peralkaline microsyenite is very similar to that of the MSF syenite. Eudialyte is replaced by zircon, unidentified Zr–REE silicates, bastnaesite, catapleite, albite and fluorite already in the early stages of alteration. In the most intensively altered rocks thorite, hydrous varieties of zircon, pyrochlore and REE-carbonate become essential constituents.

2.4 Geochemical Characteristics of the Motzfeldt Sø Formation

Table 1 shows some geochemical values typical of the syenite units in the Motzfeldt Centre. The unaltered nepheline syenites of the MRS have very similar major element compositions and agpaitic ratios close to unity. The fractionation index also shows a narrow range of values from 88 to 89. The data for the unaltered peralkaline microsyenite in Table 1 summarise the characteristics of the mesotypic, fine-grained varieties of this suite. Their silica saturation varies from saturated to undersaturated, whereby the rocks become agpaitic.

As to the trace element concentration levels, the values of the high field strength (HFS) elements (Zr, Nb, Ta, Th and U) are fairly similar in all the unaltered nepheline syenites of the MRS. The low field strength (LFS) elements (Ba, Sr and Rb) show more pronounced variation. The unaltered peralkaline microsyenite is strongly enriched in the HFS elements which is first of all due to the variably high content of eudialyte. The pegmatitic varieties of this rock suite are in places very rich (up to 40% of the mode) in eudialyte.

The changes in the major element geochemistry along with the increasing alteration of the MSF are essentially due to an increase in Si which is counter-balanced by a decrease in Na, K and Al. Such characteristics as agpaitic ratio,

Table 1. Typical geochemical values from the syenite units of the Motzfeldt Centre[a]

Formation Unit[b]	FDF PNeSy	FDF Foy	FDF NeSy	MSF NeSy	MSF Nesy[c]	MSF Pms	MSF Pms[c]
Major elements (wt%)							
SiO_2	55.8	55.5	55.2	56.0	56–66	55–58	55–70
Al_2O_3	19.9	20.3	20.5	19.4	8–16	15.1	8–15
MgO + FeO	6.1	6.0	5.3	6.6	4–9	9.0	5–11
$Na_2O + K_2O$	14.2	14.2	15.2	14.1	10–12	12.2	10–12
(Na + K)/Al	1.03	1.04	1.04	1.02	0.8–1.2	1.21	0.8–1.4
Norm. neph.%	21	20	23	20	0–2	0–22	0–1
FI[d]	89	88	88	88	60–88	88–95	63–86
Trace elements (ppm)							
Rb	197	269	189	214	200–600	300–400	300–1000
Sr	384	95	371	284	40–280	40–100	40–100
Ba	652	137	565	453	0–700	10–120	10–90
Nb	198	238	193	203	200–200000	400–800	400–15000
Ta	12	12	10	9	10–2000	10–50	10–300
Zr	693	910	693	800	800–13000	1500–3000	1500–18000
Y	65	76	48	62	60–700	100–250	100–1300
Th	10	12	8	13	10–5000	10–50	10–10000
U	4	4	4	5	5–1500	5–30	5–2000
La	129	123	105	124	100–4000	190–400	190–3000
Ce	253	236	197	232	200–8000	400–700	400–6500
Pb	12	12	12	12	10–600	10–80	10–1700
Zn	188	183	151	182	100–2000	250–500	250–2500
Ga	34	43	34	35	30–60	30–50	30–90
Cl	941	1294	427	452	100–400	200–300	90–300
S	161	104	71	112	50–100	50–100	10–2000

[a] Single values are arithmetic means of representative samples.
[b] NeSy = nepheline syenite; Foy = foyaite; PNeSy = porphyritic nepheline syenite; Pms = peralkaline microsyenite.
[c] Hydrothermally altered/mineralised.
[d] FI, fractionation index.
Analyst: University of Copenhagen, Department of Mineralogy.

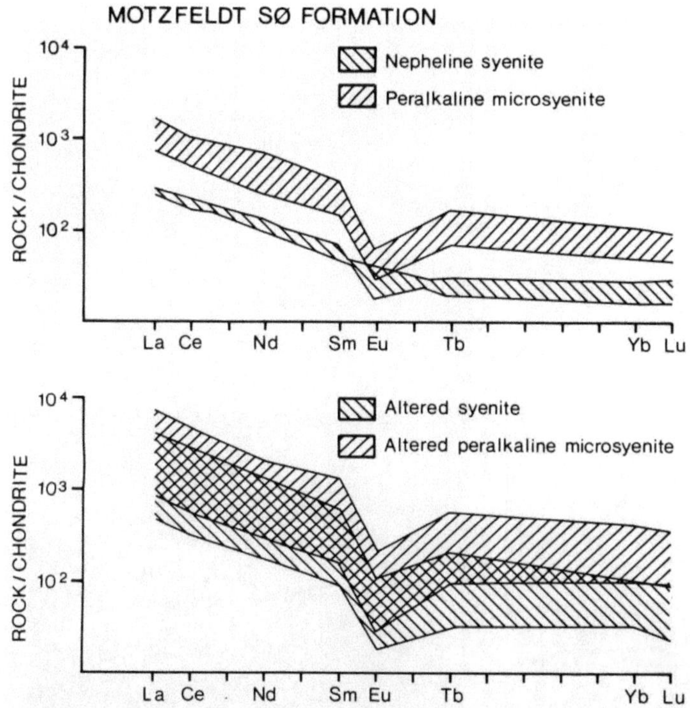

Fig. 5. Chondrite normalised REE patterns from the intrusive units of the Motzfeldt Sø Formation. The patterns show the variation within the units

differentiation index and normative composition are of questionable value for the altered rocks because the element concentrations are to a varying extent related to hydrothermal processes. For the sake of comparison they are given in Table 1. The trace element content of the altered syenite is characterised by a dramatic increase in the content of HFS elements and a decrease in the LFS elements, whereby some of the HFS elements are locally enriched in economically interesting amounts.

The REE distribution patterns (Fig. 5) demonstrate the overall enrichment of the LREE. There is a relative enrichment of HREE from the altered syenite through peralkaline microsyenite to altered microsyenite. The REE spectra are characterised by the increasing magnitude of the negative Eu anomaly from the MSF nepheline syenite, which shows only a slight negative anomaly or no anomaly at all, to the MSF-altered syenite and microsyenite.

2.5 The Pyrochlore Mineralisation

Apart from being a common accessory mineral in the altered syenites, pyrochlore occurs in zones of economically interesting enrichment in the altered MSF sye-

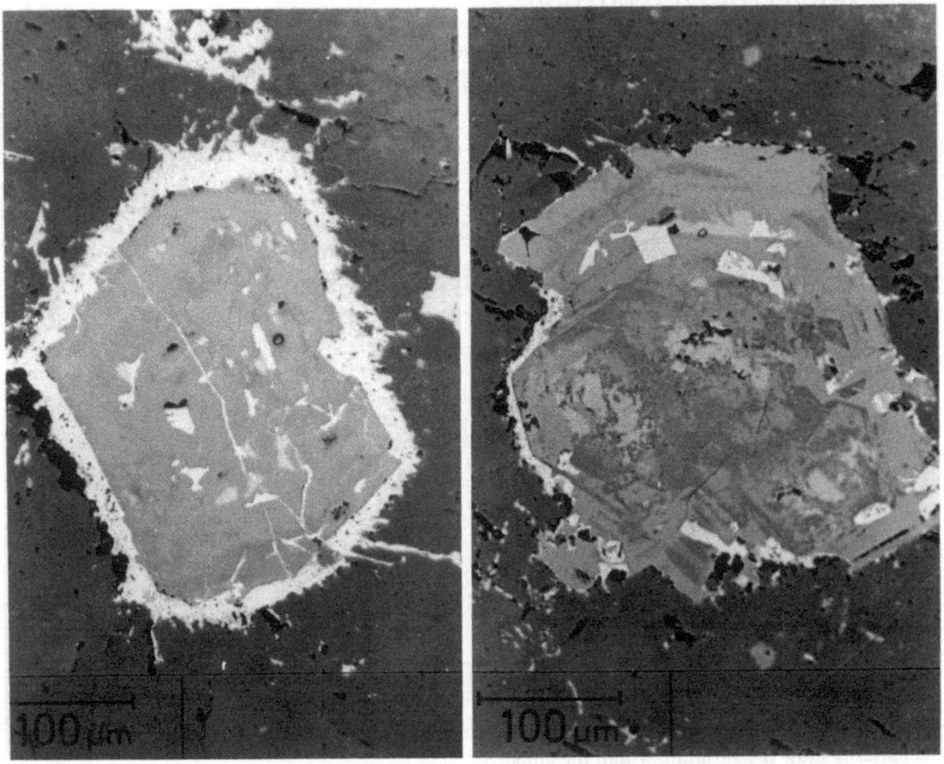

Fig. 6. Microphotographs of pyrochlore. *Left*: Characteristic hematite coating on the pyrochlore. *Right*: Altered pyrochlore surrounded by a later pyrochlore phase (areas of higher reflectance). Reflected light

nite and altered peralkaline microsyenite. The mineralisation in the altered MSF syenite is considered more attractive from an exploration point of view because of its simpler mineralogy and higher Ta content. The locations of the known major, occurrences are indicated on the geological map (Fig. 1). The preliminary estimates of the size and grade of the best-known zones, based on the interpretation of the airborne radiometric data and the available rock analyses, indicate that the mineralisation in the altered MSF syenite totals some 50 million tonnes of rock whose Ta_2O_5 and Nb_2O_5 content vary from 0.03 to 0.1 and from 0.4 to 1.0% respectively. These rock volumes may contain minor bodies of higher grade with 0.08 to 0.15% and 1.0–1.5% Ta_2O_5 and Nb_2O_5 respectively.

The pyrochlore occurs as euhedral grains which often have a thin coating of hematite (Fig. 6). The grains may contain inclusions of hematite, Fe–Ti oxide, feldspar and fluorite. The fresh grains are characteristically straw yellow in transmitted light. The grains are, however, often variably altered and thereby dark, reddish-brown in colour. Almost opaque or opaque grains are not uncommon. Vague zoning is often seen both in transmitted and reflected light (Fig. 6), but the microprobe assays of the zoned grains do not reveal any distinct compositional zoning. The Ta content tends to be slightly higher in the cores of the grains. The grains are metamict due to the high uranium content. Representative analyses of pyrochlore and other Nb–Ta-bearing minerals from the Motzfeldt Centre are given in Table 2.

The pyrochlore from the Motzfeldt Centre shows a marked compositional variation both in A- and B-cation roups of the pyrochlore formula $A_2B_2X_7$. The most important cations are: A = Ca, Na, LREE, U; B = Nb, Ti, Ta; X = OH, F, Cl. The assayed grains do not contain detectable amounts of Sr and Ba. Zirconium is invariably present, but its content is often much below 1.5% ZrO_2. The pyrochlore contains sporadically some lead (up to 4% PbO) in the heavily altered grains. The iron content is very variable and the bulk of it is probably related to the Fe–Ti oxide 'dust' in the grains.

In the A-cation group $LREE^{3+}$ and U^{4+} substitute for Ca^{2+} and, to a lesser extent, Na^{1+}. The variations in the B-cation group concern mainly the substitution of Nb by Ta. Relatively high Nb contents are also accompanied by higher Ti values.

The Motzfeldt pyrochlores have considerable contents of LREE and U and Ta. The pyrochlores can be roughly divided into two groups, those in the altered MSF syenite and those in the altered microsyenite respectively.

The pyrochlore in the altered MSF syenite is characterised by higher Ta contents than in altered peralkaline microsyenite. The Ta contents vary from 4.5 to 10.0% and 1.5 to 2.3% Ta_2O_5 respectively. The A-position of the pyrochlore in the microsyenite is dominated by LREE and U (up to 15.0% RE_2O_3 and 9.0% UO_2 respectively), whereas Ca is the main cation in the A-position of the pyrochlore in the altered MSF syenite.

The low analysis totals are probably due to both primary and secondary deficiencies in the A-cation group. The primary deficiencies are due to the substitution of Ca and Na by ions with higher valencies (REE and U). The electrochemical stability is compensated by vacant places in the A-cation position and/or simultaneous substitution of Nb and Ta by ions with lower valencies, such as Ti, Zr, Fe^{3+} etc. (Borodin and Nazarenko 1957a). The loss of A-group cations due to leaching and/or hydration ('secondary deficiencies'; Borodin and Nazarenko 1957b) has affected the pyrochlore in the altered MSF syenite in particular as indicated by the back-scattered electron assays of the pyrochlore grains from this unit.

The Nb/Ta ratio in the pyrochlore of the altered MSF syenite varies from 8 to 14 which generally coincides with the Nb/Ta ratios of the whole rock analyses indicating that the bulk of Nb and Ta in the altered syenite is incorporated by the pyrochlore. Altered varieties of the pyrochlore in places show an anomalously low Nb/Ta ratio (analysis 10, Table 2).

The Nb/Ta ratio in the pyrochlore of the altered microsyenite is highly variable and ranges from ca. 18 up to 50 or even more. The distribution of Nb and Ta in the altered microsyenite, especially in the silicified varieties of this rock suite, is complicated. The main carrier, again, is pyrochlore but substantial amounts of Nb and Ta are incorporated in zircon, thorite, Fe–Ti oxides and columbite (Table 2, analyses 1–6). These are compositionally very variable and are often extremely fine-grained mixtures of several mineral phases.

Table 2. Representative microprobe analyses of the minerals from the mineralised microsyenite (1–6) and altered MSF syenite (7–12) (values as wt%; WDS, wavelength dispersive system; EDS, energy dispersive system)

Method	1 WDS	2 EDS	3 EDS	4 WDS	5 WDS	6 EDS
Nb_2O_5	53.23	71.74	4.00	2.01	2.02	1.02
Ta_2O_5	2.22	1.00	0.00	0.25	0.08	—
TiO_2	3.40	4.25	0.50	0.27	0.21	—
Fe_2O_3	2.04	12.76	80.33	5.19	1.45	—
MnO	—	7.96	0.00	—	—	—
La_2O_3	4.00	—	—	0.00	0.06	29.48
Ce_2O_3	4.77	—	—	0.61	0.30	33.12
Pr_2O_3	0.36	—	—	0.12	0.27	0.25
Nd_2O_3	0.84	—	—	0.48	0.81	4.71
Sm_2O_3	0.00	—	—	0.17	0.36	—
Y_2O_3	0.23	—	—	0.00	1.84	—
Na_2O	0.00	—	—	0.34	0.00	—
CaO	2.19	—	—	1.16	1.20	—
ThO_2	0.22	0.20	0.00	42.86	2.89	—
UO_2	8.11	0.30	0.00	4.18	0.69	—
ZrO_2	0.60	—	—	12.80	49.23	—
SiO_2	5.22	1.00	2.00	15.36	26.54	—
P_2O_5	—	—	—	—	—	30.04
Total	87.43	99.21	86.83	85.80	87.95	98.62

1 Pyrochlore 2 Columbite 3 Fe oxide 4 Thorite
5 Zircon 6 Monazite

Method	7 WDS	8 WDS	9 WDS	10 EDS	11 WDS	12 WDS
Nb_2O_5	53.97	56.78	52.68	28.21	0.20	0.00
Ta_2O_5	6.53	4.20	3.36	14.85	0.00	0.00
TiO_2	5.98	4.52	4.42	3.79	0.00	0.02
Fe_2O_3	0.35	2.83	1.63	—	0.00	0.00
La_2O_3	0.84	0.71	1.47	2.69	0.20	17.62
Ce_2O_3	3.16	3.84	3.70	2.30	0.40	33.77
Pr_2O_3	0.32	0.04	0.47	—	0.10	2.85
Nd_2O_3	1.02	1.98	1.01	0.41	0.20	10.68
Sm_2O_3	0.26	0.53	0.13	—	0.00	1.66
Y_2O_3	0.00	0.37	0.00	—	1.90	0.34
Na_2O_3	3.78	0.88	3.19	0.00	0.00	0.00
CaO	13.81	9.82	12.78	1.23	0.21	2.33
ThO_2	0.12	0.18	0.22	0.38	0.20	0.64
UO_2	4.75	4.13	5.47	1.33	0.00	0.33
ZrO_2	0.15	0.71	0.60	1.49	62.50	0.00
SiO_2	0.53	1.88	0.37	—	33.00	0.10
Total	95.39	93.40	91.50	56.68	98.91	70.34

7 Pyrochlore 8 Pyrochlore 9 Pyrochlore 10 Altered pyrochlore
11 Zircon 12 Bastnaesite

3 Discussion and Conclusions

The geochemical characteristics of the MRS can be accounted for by two processes. The primary element concentrations are related to the igneous processes such as fractional crystallisation and gravitative settling. Furthermore, secondary element distribution took place during late-stage hydrothermal activity whoe role as a concentrator of for instance Th is well exemplified by Fig. 3.

3.1 Evolution of the Motzfeldt Sø Formation

Chilled margins comparable to those in the intrusions of the FDF have not been found in the MSF. Consequently, there is no direct evidence on the liquid composition of the MSF magma. Whether the more mafic marginal facies of the MSF is indicative of a more basic primary liquid is not known. Based on the composition of the unaltered inner parts of the MSF (nepheline syenite), whose geochemical characteristics are very similar to those of FDF where more direct evidence on the liquid composition is available, it is assumed that the bulk composition of the primary liquid was phonolitic.

The rock types of the MSF, their field relationships, mineralogical and chemical characteristics, all indicate an extreme internal differentiation of the phonolitic magma. The role of crystal fractionation as an effective mechanism to produce rocks of considerable variation in the Gardar central complexes has been emphasised (e.g. Emeleus and Upton 1976; Upton and Emeleus, in press). The REE distribution patterns of the MSF rocks indicate that crystal fractionation has contributed to the evolution of the MSF. The evolution of the MSF thus commenced with the crystallisation of Ca-rich mineral phases (plagioclase, clinopyroxene) and the liquid became impoverished in Ba, Sr and Eu. The MSF nepheline syenite, now representing the lowest part of the exposed igneous column, shows only slightly negative Eu anomalies or no Eu anomaly at all. The syenite can be interpreted as the topmost part of a predominantly cumulate floor sequence which was formed by crystallisation along the sides and top of the chamber and settling of the crystals to the floor. Along with the progressive crystal fractionation the incompatible elements and volatile components were increasingly concentrated at the top of the chamber. The peralkaline magma was able to dissolve large amounts of volatiles (Kogarko 1974) and there may have been a gradual transition into a hydrothermal fluid, as first seen in the experiments of Tuttle and Bowen (1958) and Tuttle (1961). The formation of the sheet complex could be related to the formation of such a fluid phase. The upward migration of the fluid phase was probably further intensified by the formation of subhorizontal zones of weakness in the more or less completely solidified MSF. These zones could be related to the reduction of volume due to the crystallisation.

3.2 Significance of Volatile Components

The most important volatile components in the initial magma were fluorine and water. The enrichment of water could be at least in part due to the assimilation of country rocks, especially from the Eriksfjord Formation. The bulk of the Eriksfjord Formation adjacent to the centre is made up of sandstone, but inclusions of country rocks other than volcanics are virtually absent in the MSF. This indicates that large-scale selective assimilation of the sandstones may have taken place. Whatever the case, the buildup of volatiles must have affected the physical and chemical properties of the residual magma/fluid. Enrichment of volatiles is known to increase the general acidity of the system (Kogarko 1974). The general decrease in the content of more basic cations, as exemplified by Ca and Na, in the Nb–Ta, Zr and REE mineral phases at the higher levels of the igneous column, could be attributed to a decreased chemical activity of the more basic cations.

The intense and pervasive formation of hematite and other Fe–Ti oxides implies that the fugacity of oxygen (fO_2) must have increased dramatically towards the end of the hydrothermal stage. An influx of meteoric water could have resulted in an increase of fO_2 because the water would dissociate and H_2 would diffuse out of the system much more rapidly than O_2. Sato and Wright (1966) have shown that in silica glass H_2 diffuses approximately a million times faster than O_2.

3.3 Outline of the Mineralising Process

The migration and precipitation of Th, U, Nb, Ta, Zr and REE have probably been complicated processes which were controlled by the acidity and fO_2 of the system. It can be expected that the migration involved a continuous readjustment (precipitation/leaching) of the mineral phases with the proceeding hydrothermal activity, which was accompanied by a relative and absolute enrichment of the high field strength elements due to the higher stability and mobility of their complexes. The lower field strength of Ta probably explains the lower mobility and earlier precipitation of Ta at the lower levels of the igneous column.

The formation of such a volume of peralkaline residuum with the extreme enrichment of volatiles and incompatible elements as seen in the outer part of the MSF implies that the fractionation must have been particularly effective. The reason why this happened only in the MSF whose primary magmatic characteristics were very similar to those in the FDF is a reflection of the different physico-chemical conditions of the two magma suites. In comparison to the MSF, the FDF magmas had a lower fO_2 and activity of silica. The development of chilled margins and the paucity of pegmatitic segregations in the FDF attest to the relatively 'dry' and mobile nature of the FDF magmas (C. Bradshaw, pers. comm.). They were highly charged with dissolved depolymerising volatiles such as fluorine and chlorine, which enhanced fluidity and facilitated the development of cumulate structures. The lack of the late-stage enrichment of volatiles/mineralisation in the FDF may imply that the volatile phase had left the system as a consequence of, for example, volcanic activity.

The abundance and increase of the content of hydrous mineral phases along with the increasing degree of alteration/mineralisation in the MSF attest to the importance of water in the mineralisation process. It is proposed that the high content of volatiles, especially water, as a result of the influx of meteoric water, which lowered the viscosity of the melt and thus facilitated convection and liquid-stage differentiation as well as extended the crystallisation range of the rocks, was the decisive factor for the effective fractionation and the development of the extensive hydrothermal regime in the MSF. The extreme enrichment of the volatile components in the altered MSF syenite indicates that the system was effectively closed until the very last stages of the hydrothermal activity. The overall geometry of the hydrothermally altered MSF displays many similarities to the H_2O-saturated outer shell or 'carapace' in the granite and granodiorite intrusions where porphyry copper and molybdenum deposits formed. According to Burnham (1979), such a "carapace" acted "as a barrier to the migration of volatiles, either outward to the wallrocks or inward from the wallrocks".

Acknowledgements. The project was supported by the *European Economic Communities* through *Contract No. MSM-118-DK.* The author is grateful to C. Bradshaw, who kindly put his data at my disposal and to L. van Wambeke from the EEC, for constructive discussions in the course of the project. The manuscript benefitted from comments and criticism by L.M. Larsen. This work has been published with the permission of the Director of the *Geological Survey of Greenland.*

References

Armour-Brown A, Tukiainen T, Wallin B, Bradshaw C, Emeleus CH (1983) Uranium exploration in South Greenland. Rapp Grønlands Geol Unders 115:68–75

Blaxland AB, Breemen, O van, Emeleus CH, Anderson JG (1978) Age and origin of the major syenite centres in the Gardar province of south Greenland: Rb–Sr studies. Bull Geol Soc Am 89:231–244

Borodin LS, Nazarenko II (1957a) Chemical composition of pyrochlore and isomorphic substitution in the molecule $A_2B_2X_7$. Geochem no. 4, Moscow, pp 278–295

Borodin LS, Nazarenko II (1957b) Deviations of minerals in the pyrochlore group from the type formula $A_2B_2X_7$ and the role of water of constitution in the crystal structure of the pyrochlore. Dokl Akad Nauk SSSR 115:783–786

Bradshaw C (1985) The alkaline rocks of the Motzfeldt Centre; progress report on the 1984 field season. Rapp Grønlands Geol Unders 125:62–64

Burnham CW (1979) Magmas and hydrothermal fluids. In: Barnes LH (Ed) Geochemistry of hydrothermal ore deposits. Wiley, New York, pp 71–136

Emeleus CH, Harry WT (1970) The Igaliko nepheline syenite complex. General description. Bull Grønlands Geol Unders 85:116 (also Meddr Grønland 186:3)

Emeleus CH, Upton BGJ (1976) The Gardar period in southern Greenland. In: Escher A, Watt WS (eds) Geology of Greenland, 154–181. Copenhagen: Geol Surv Greenland

Kogarko LN (1974) Role of volatiles. In: Sørensen H (ed) The alkaline rocks. Wiley-Interscience, London, pp 474–487

Larsen LM, Tukiainen T (1985) New observations on the easternmost extension of the Gardar supracrustals (Eriksfjord Formation), south Greenland. Rapp Grønlands Geol Unders 125:64–66

Piper JDA (1982) The Precambrian palaeomagnetic record: the case for the Proterozoic supercontinent. Earth Planet Sci Lett 59:61–89

Sato M, Wright TL (1966) Oxygen fugacities directly measured in volcanic glasses. Science 153:1103–1105

Tukiainen T, Bradshaw C, Emeleus CH (1983) Geological and radiometric mapping of the Motzfeldt Centre of the Igaliko Complex, south Greenland. Rapp Grønlands Geol Unders 120:.78–83

Tuttle OF (1961) 'Residul solution formed by crystallization', in Physico-chemical problems of the genesis of ores and rocks (in Russian, English summary). Izd AN SSSR, Moscow, pp 647–653

Tuttle OF, Bowen NL (1958) Origin of granite in the light of experimental studies in the system $NaAlSi_3O_8$–$KAlSi_3O_8$–SiO_2–H_2O. Geol Soc Am Mem 74:153

Upton BGJ, Thomas JE, MacDonald R (1971) Chemical variation within three alkaline complexes in south Greenland. Lithos 4:163–184

Upton BGJ, Emeleus CHE (in press) Mid-Proterozoic alkaline magmatism in southern Greenland: the Gardar province. Geol Soc Lond Spc Publ

Part II
Chromite and Platinum-Group Elements

Structural Controls on the Location and Form of the Vourinos Chromite Deposits

S. Roberts[1], A. Rassios[2], L. Wright[1], I. Vacondios[3], G. Vrachatis[3], E. Grivas[2], R.W. Nesbitt[1], C.R. Neary[4], T. Moat[1], and L. Konstantopolou[5]

Abstract

The development of new methods of exploration for chromitite requires an understanding of the most crucial elements involved in the concentration and preservation of the ore deposits. Following their formation at constructive plate margins, chromite deposits are subject to deformation, within the upper mantle, as the enclosing lithosphere spreads away from the ridge axis, and during decoupling from the oceanic plate under a steadily failling P–T regime. The construction of a comprehensive emplacement model for the Vourinos complex shows that emplacement fabrics are prevalent throughout the chromite-bearing mantle sequence. Detailed studies of two key localities on Vourinos indicate that both mantle and emplacement structures are preserved within the chromite deposits, although their respective role in the present form and distribution of the chromite deposits differs between the two localities. This study suggests a new exploration strategy for the Vourinos complex which relies on more expansive structural models than previously applied in ophiolitic terrain. These models evaluate each phase of deformation and address its effect commencing with post-emplacement structures and working backwards to the earliest ridge tectonics. They are then modified to suite the style of deformation which is predominant in a particular mineralized area.

1 Introduction

It is now broadly accepted that ophiolites represent the land-bound fragments of the former oceanic lithosphere (Coleman 1977; Gass and Smewing 1981). Studies of chromite deposits commonly contained within ophiolitic mantle sequences must therefore involve an understanding of the complex interplay between magmatic and

[1] Department of Geology, Southampton University, Highfield, Southampton, UK
[2] I.G.M.E., Kozanis Branch, Kamvounion 13, Kozani, Greece
[3] I.G.M.E., Messoghion 70, Athens, Greece
[4] B.G.S, Keyworth, Nottingham, UK
[5] Department of Geology, University of Athens, Panepistimopolis, Athens, Greece

Mineral Deposits within the European Community
(ed. by J. Boissonnas and P. Omenetto)

tectonic processes responsible for the formation of the oceanic lithosphere at constructive plate margins and its subsequent obduction. Ophiolitic chromite deposits are considered to represent the early fractionates of rising basaltic/picritic melts in small transient magma chambers, within the upper mantle, beneath ridge axes (Neary and Brown 1979; Harkins et al. 1980; Lago et al. 1982). High temperature deformation of the deposits is usually attributed to mantle flow as the host lithosphere spreads away from the ridge axis (Leblanc et al. 1980). Studies of this type have produced classifications of ophiolite chromite bodies based on the orientation of the deposits in relation to the prevailing mantle structures (Cassard et al. 1981; Christiansen 1986).

Many ophiolite mantle sequences appear to be emplaced as crystalline blocks, little affected by emplacement tectonics, except at the base of the sequence where significant emplacement overprints have been identified (Spray 1984). In contrast, our work on Vourinos indicates that emplacement structures are widely distributed. This chapter considers some of the relationships between mantle flow processes and emplacement tectonics and the relative importance of each process in determining the distribution and form of chromite deposits in the Vourinos ophiolite.

2 Setting

The Vourinos complex, situated in northern Greece, represents an almost complete, but tectonically disrupted, ophiolite which comprises part of the Tethyan ophiolite belt. Remnants of an extrusive unit, a sheeted dyke complex, a cumulate sequence and an ultramafic tectonite unit are all present but only the lowermost parts of the ophiolite are well exposed (Moores 1969).

Ultramafic rocks form about 85% of the Vourinos complex (Fig. 1), and include massive harzburgite, harzburgite with interlayered dunite, harzburgite containing discrete dunite bodies, with or without chromite segregations and minor websterite (Moores 1969). The mantle sequence rocks exhibit a pervasive planar linear fabric defined by flattened and elongate spinel and/or blocky elongate orthopyroxene grains. Ross et al. (1980) showed that the harzburgites and dunites recorded various degrees of deformation and mylonitization. The least deformed rocks are coarse-grained peridotites which develop shear textures and eventually mylonitic textures with increasing deformation. The fabric of the coarse-grained rock is considered by Ross et al. (1980) to have been generated in a rising mantle diapir between 75–40 km depth at temperatures between 1200°–900 °C. The mylonitic fabric was superimposed upon this earlier fabric at a shallower depth (30 km) and lower temperatures (770°–823 °C), at some distance from the ridge axis/spreading centre.

Previous studies on the dunite and chromitite bodies of the Vourinos mantle sequence have located more than 700 separate chromite occurrences throughout the complex (Vrachatis and Grivas, 1980). These show a variety of different ore textures including disseminated, schlieren and nodular ore (Zachos 1964). Complex folding of many of the deposits was documented by Ayrton (1968) and interpreted by Moores (1969) as flow layering after Thayer (1963). Christiansen (1986) noted

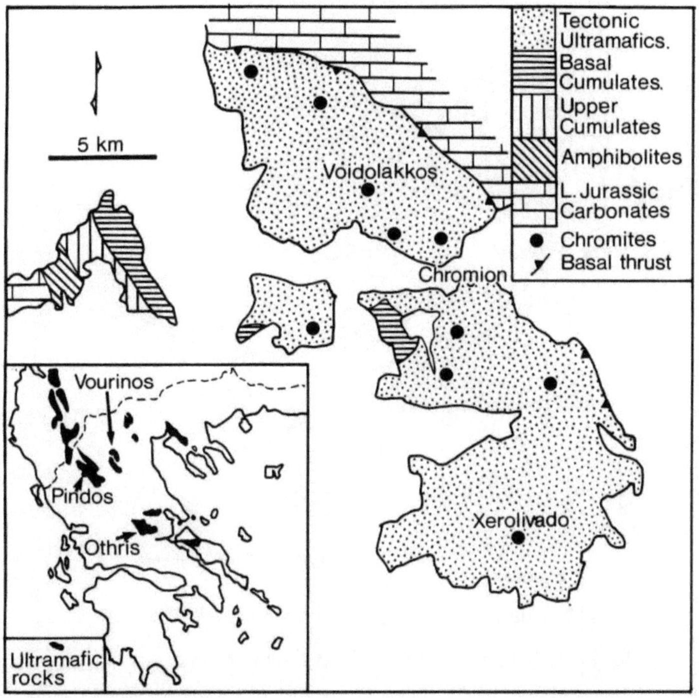

Fig. 1. Simplified geological map of the Vourinos complex showing the location of the major chromite deposits

large and small scale isoclinal fold structures within the chromite deposits, with axial planes parallel to the regional and local foliation and fold axes parallel to the regional and local lineation. Burgath and Weiser (1980) considered the chromite deposits of the Vourinos complex to represent the early fractionates of rising basaltic melts which are subsequently affected by "mantle flow" tectonics.

This study of the Vourinos complex indicates that both mantle and emplacement tectonics play a part in the present location and form of the chromite deposits. Following a brief outline of the emplacement structures, detailed descriptions of two localities are presented. Xerolivado, the first locality described, is a deposit where the present shape and form of the ore is controlled by high temperature mantle fabrics. In contrast, the second area described, Voidolakkos, shows a deposit where the present form of the ore has been significantly affected by emplacement tectonics.

3 Emplacement

The Vourinos ophiolite sequence tectonically overlies a thick SW dipping, greenschist facies, metasedimentary sequence of well-jointed marbles, and locally, a

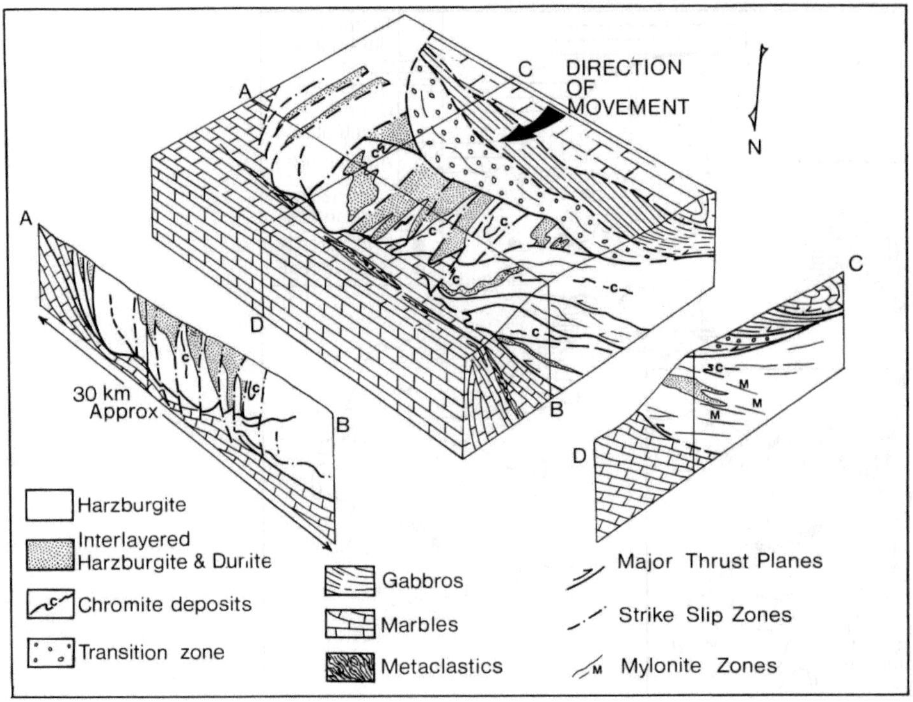

Fig. 2. Schematic block diagram of the Vourinos complex showing the geometry of the emplacement structures

thinner metaclastic sequence. Dating of the metamorphic sole (Spray and Roddick 1980) indicates that ophiolite obduction occurred during the early to mid-Jurassic (170–180 m.y.). The direction of ophiolite emplacement is controversial. Naylor and Harle (1976) and Smith (1977) proposed that emplacement of the ophiolite was from the SW, whereas Vergely (1976) suggested that the ophiolite was emplaced from the Vadar region to the NE.

A combination of structural observations from the basal thrust and the interior of the ophiolite supports a movement direction of the ophiolite towards 050 from the SW. Figure 2 shows this NE movement direction with the basal thrust overriding the marbles and shales. In this model the shales absorb much of the deformation, although the more competent lithologies within the shales show stretching lineations plunging towards 230. Alternatively, they form boudins, around which the geometry of the shale structures define the movement direction (Fig. 3A). Massive harzburgite near the basal contact often shows complementary 140 trending, 50 SW dipping, imbricate brittle faults.

The ophiolite itself comprises a complex series of imbricate sheets, bounded by thrust shear planes of both a brittle and ductile nature with similar orientations and movement directions to the basal thrust (Fig. 3C). The curvature of the basal thrust, formed during the later stages of emplacement, is accommodated by a series of strike-slip structures within the ophiolite; these are evident in the central part of the

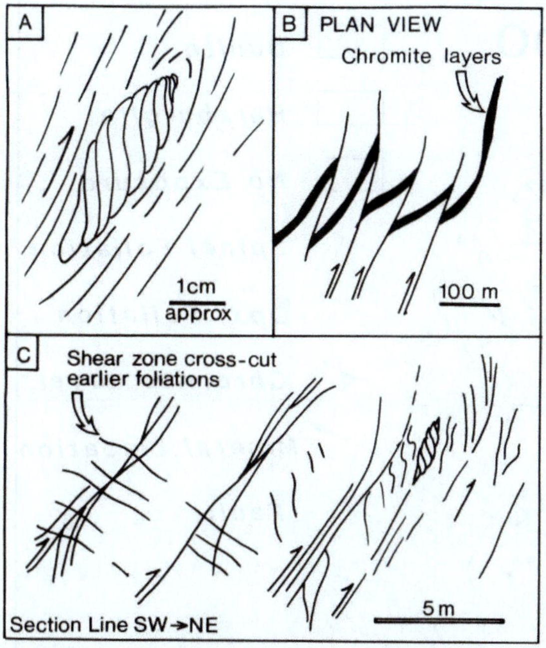

Fig. 3A–C. Sketches of field relations. **A** Tectonic fish or boudins defining movement direction. Drawing shows a large pebble in a thin conglomeratic band. **B** Plan view of the chromite layer of the Aetoraches mine showing the importance of dextral strike-slip zones in redistributing the ore (source Aetoraches mine plans). **C** East and northeast-directed shear zones south of Voidolakkos. The shear zones post-date the earlier foliation and preferentially shear along dunite layers

complex and in particular the Aetoraches area (Fig. 3B). Parts of the southermost ultramafic sequence and the transitional and cumulate sequences appear to represent a series of late stage out-of-sequence imbricate thrusts.

Throughout the ophiolite, the entire spectrum of structures from plastic to brittle are preserved and these show a more or less constant movement direction towards 050. This encourages the view that deformation is incremental, with ductile deformation zones, faults and thrusts developing over a period of time, although mostly related to a single deformation event. Given the emplacement model outlined above, it is anticipated that the chromite deposits may preserve a variety of structures including high temperature fabrics produced at the ridge axis and later ductile to brittle structures developed during emplacement. Two contrasting chromite localities within the ophiolite are now considered in an attempt to discern the key structural parameters involved in the control of the shape and form of the chromite bodies within individual localities.

4 Structural Studies

Field studies of the Vourinos complex conducted for the present study included detailed mapping of two areas of chromite mineralization at a variety of scales (1:10 000, 1:2000 and 1:500). The Xerolivado area, in the southern part of the complex, contains the largest chromite mine of Vourinos with some 3 million tonnes of ore mined to date and 3 million tonnes of proven reserves being exploited at

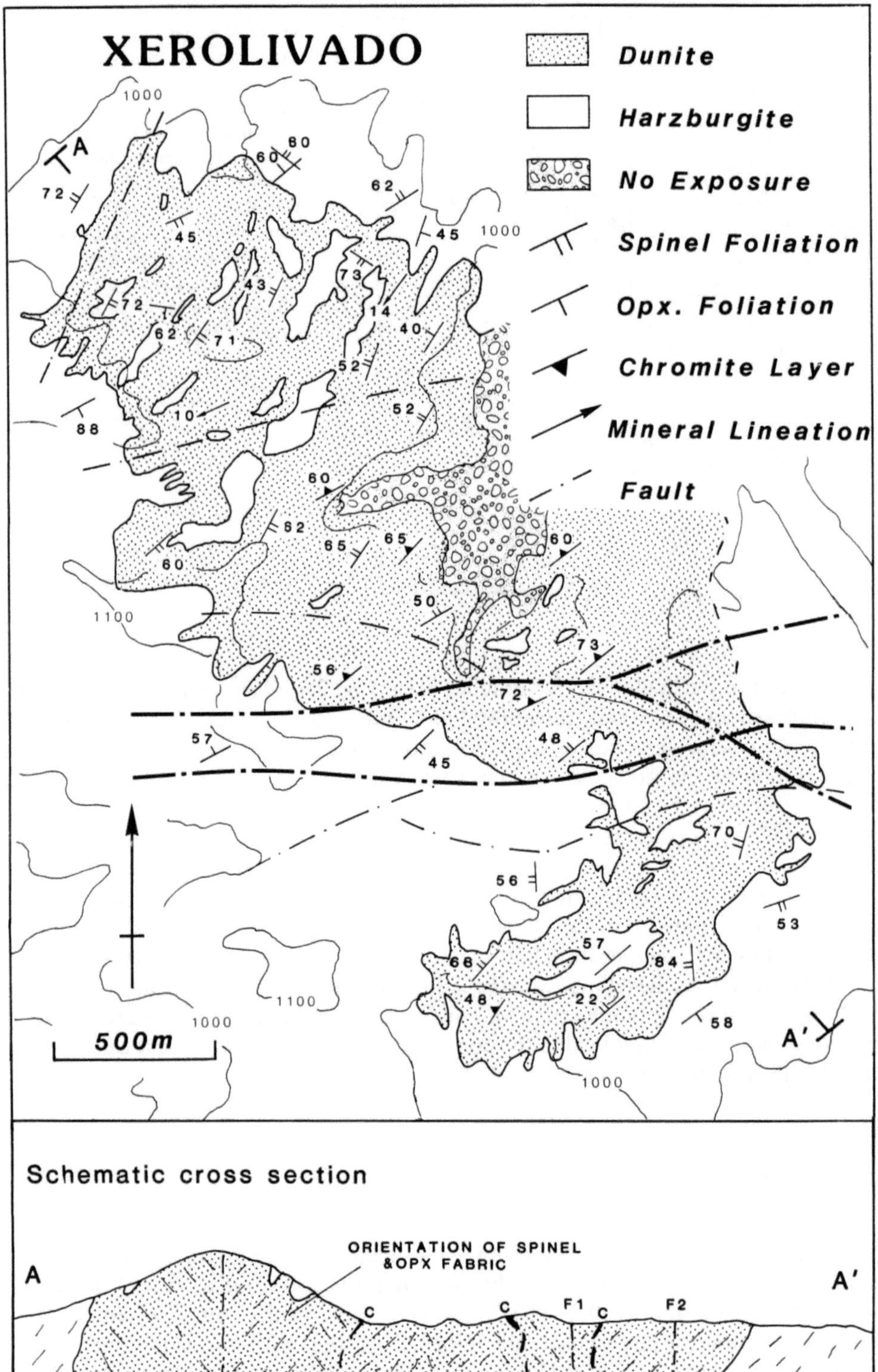
XEROLIVADO
Dunite
Harzburgite
No Exposure
Spinel Foliation
Opx. Foliation
Chromite Layer
Mineral Lineation
Fault
500m
A
A'
Schematic cross section
ORIENTATION OF SPINEL &OPX FABRIC
C
F1
F2

Fig. 4

present. The second area of interest is in the northern part of the complex. The Voidolakkos locality is a site of former chromite exploitation (2000 tonnes of massive ore extracted between 1938–52, Apostilides 1977) and an area well known for its contrasting style of deformation in comparison with Xerolivado. The basic geology and ore development in both areas were established prior to the present study by IGME and Hellenic Ferro-chrome Company geologists (Vrachatis and Grivas 1980).

4.1 Xerolivado

The Xerolivado area comprises a large dunite body with an approximate surface exposure of 3 km by 0.5 to 1 km, enclosed by harzburgite tectonite (Fig. 4). Seven centrally located chromite layers are contained within the dunite body striking NE–SW and steeply dipping towards the NW, which reach a maximum thickness of 5 m and extend downdip for up to 150 m (Apostolides et al. 1980; Stamoulis 1984). The layers are predominantly of schlieren ore, which is an association of disseminated chromite and thin discontinuous massive chromite layers set in an olivine matrix (Plate 1). Deformation of the chromite layers is characterized by the

Plate 1. Small flexures of schlieren ore from the open pit of the Xerolivado mine

←

Fig. 4. Geological map of the Xerolivado area with a schematic cross-section

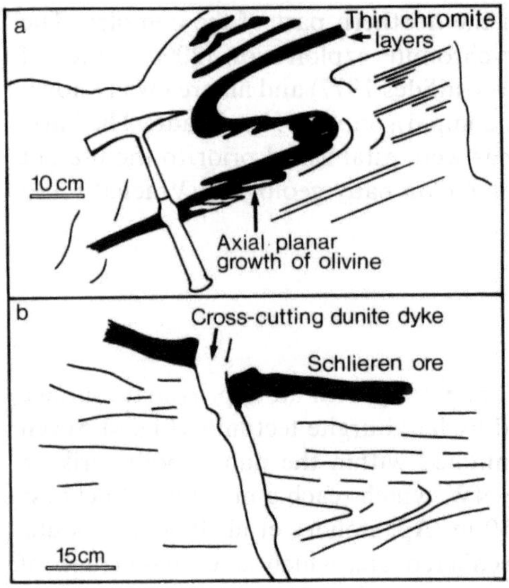

Fig. 5. a Small-scale fold with a chromite layer within the Xerolivado mine with axial planar growth of olivine grains. **b** Schlieren ore within the mine cross-cut by a dunite dyke

elongation of the chromite grains and small-scale flexures and folding of the layering. Small-scale folds are often developed within individual layers with an axial planar fabric defined by large elongate or rod-shaped olivine grains (Fig. 5a). Within the mine a variety of deformed and folded ore textures are preserved within the ore. No relict igneous textures are apparent and the chromite layers are often cross-cut by dunite dykes (Fig. 5b).

Chromite layers define a girdle around a shallow SW plunging fold axis (Fig. 6), which coincides with the direction of the mineral lineations within the chromite layer. However, despite the presence of mesoscopic open folds and flexures of the chromite layering, which implies earlier folding, no large-scale closures have been recorded either on the surface or within the mine. The dunite host to the chromite layers shows a fabric defined by spinel grains which coincides with the foliation within the chromite layers (Fig. 6). Harzburgite tectonite hosts the dunite body but also outcrops within the dunite body as elongate masses, up to 300 m in long dimension, striking NE–SW. Unlike the chromite layers, the contacts between the harzburgite masses and the dunite host are locally flat lying. All harzburgitic lithologies are characterized by a fabric of orthopyroxene and accessory chromite grains which for the most part reflect the foliation and lineation measurements recorded within the chromite layers and dunite body (Fig. 6). All the lithologies in the area are cross-cut by NW to SE trending 3–15 cm thick pyroxenite dykes, which appear in the field to be undeformed.

A variety of brittle structures are observed dominated by two E–W striking normal faults which downthrow the dunite body and chromite layers to the south (Fig. 4) (Stamoulis 1984). Two sets of brittle faults are commonly observed, one parallel, the other normal to the chromite layering, often trapping massive ore at points of intersection. Rare thrust ramps are encountered trending 140 with a

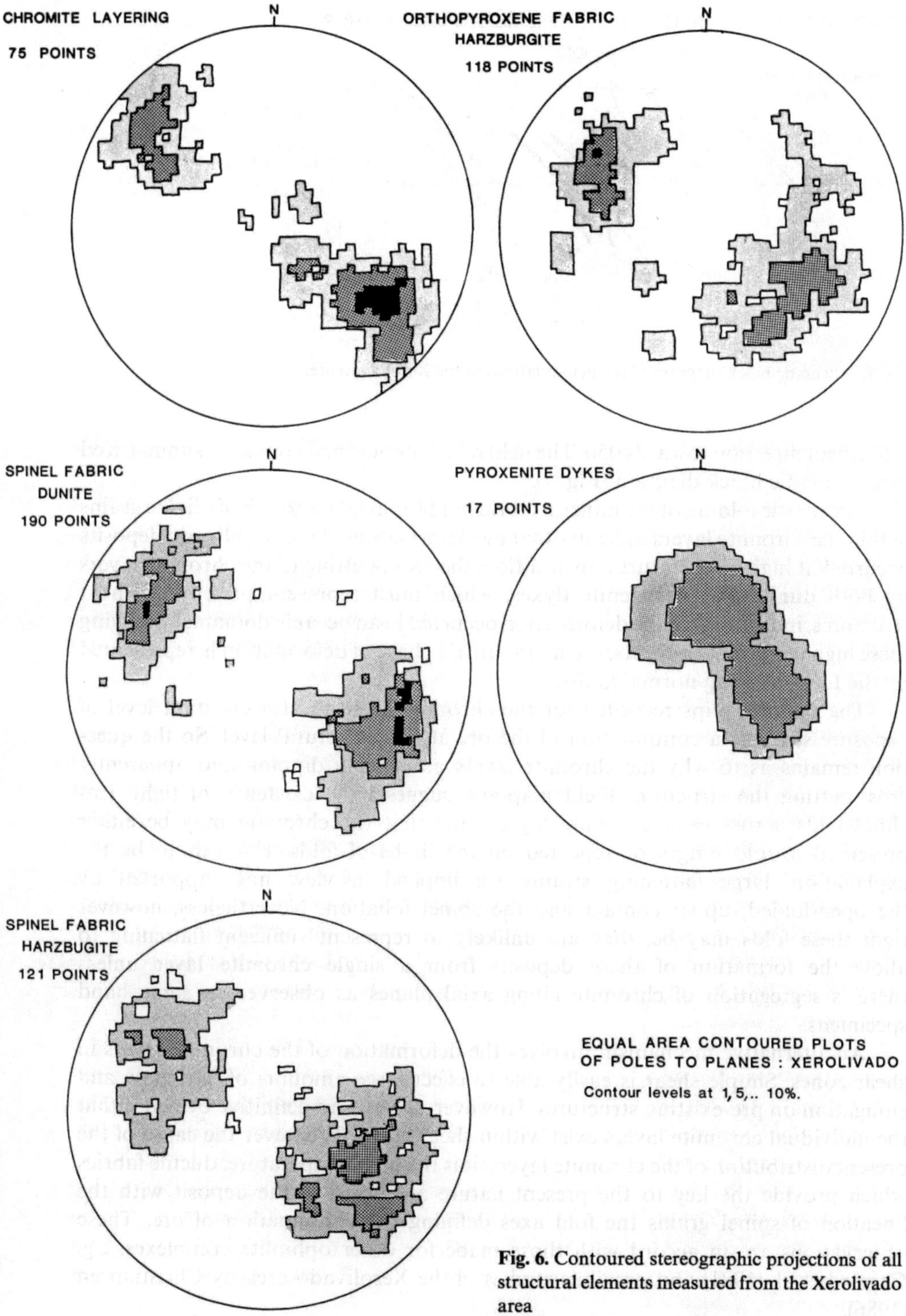

Fig. 6. Contoured stereographic projections of all structural elements measured from the Xerolivado area

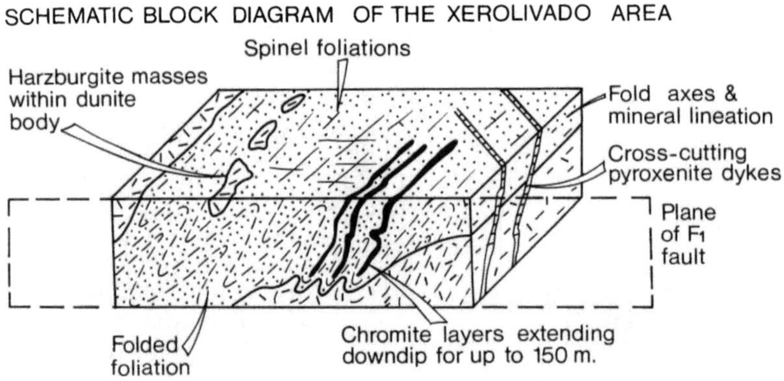

Fig. 7. Schematic block diagram of the field relations in the Xerolivado area

movement direction towards 050. The field relations outlined above are summarized in a schematic block diagram (Fig. 7).

The plastic folding of the mineral fabrics and the axial regrowth of olivine grains within the chromite layers indicates that the deformation of the Xerolivado deposits occurred at high temperatures. In addition, the cross-cutting of the chromite layers by both dunite and pyroxenite dykes, which must represent pre-emplacement structures, indicates that the deformation occurred in an oceanic domain. Following these high temperature events, the major brittle phase of deformation is represented by the E–W striking normal faults.

The opposed dips recorded for the chromite layering at a constant level of exposure suggests a continuation of the ore at this structural level. So the question remains as to why the chromite layers are steeply dipping and apparently cross-cutting the structure. Field mapping suggests the existence of tight, isoclinal folds across each chromite layer, such that the chromite may be either preserved in fold hinges or repeated on the limbs of folds. For this to be the explanation, large flattening strains are implied, a view not supported by the open-folded, upper contact and the spinel foliation. Nevertheless, however tight these folds may be, they are unlikely to represent sufficient flattening to allow the formation of these deposits from a single chromite layer unless there is segregation of chromite along axial planes as observed in some hand specimens.

An alternative mechanism involves the deformation of the chromite layers in shear zones. Simple shear is easily able to effect large amounts of flattening and elongation on pre-existing structures. However, there is no definitive evidence that the individual chromite layers exist within shear zones. Whatever the cause of the present distribution of the chromite layers, it is the high temperature, ductile fabrics which provide the key to the present nature and form of the deposit with the lineation of spinel grains the fold axes defining the continuation of ore. These observations are in accord with those made for other ophiolite complexes, e.g. Cassard et al. (1981) and previous studies of the Xerolivado area by Christiansen (1986).

4.2 Voidolakkos

The Voidolakkos valley is situated within the northern part of the complex (Fig. 1). Large dunite masses containing chromite concentrations are exposed on both sides of the valley, although the detailed description presented here relates only to the northern half. A detailed map of the area shows an irregularly shaped dunite body enclosed by harzburgite tectonite with a number of dunite fingers extending out of the dunite body trending NW–SE (Fig. 8). Chromite occurrences, of both schlieren and massive ore types, are exposed within the dunite body. The largest chromite exposure is a centrally located, 5 m wide layer of schlieren ore which strikes NW–SE and dips approximately 40° to the SW, referred to as number 37. The schlieren ore is deformed, with a spinel fabric generally subparallel to the layer, but in parts folded around a SW-plunging fold axis (Fig. 9). Towards the SE corner of the area mapped, in particular within the dunite fingers, massive ore is exposed. Commonly, the massive ore, often up to 50-cm-thick, has a very narrow dunite envelope, often <5 cm, and contains fold structures with axes plunging towards the SW (Plates 2 and 3).

Two spinel fabrics are often recognized within the host dunite, the first is subparallel to the chromite layers, the second strikes NE–SW and dips to the NW. In addition, an intersection lineation within the dunite plunges towards the SW. Both the spinel and orthopyroxene fabrics of the harzburgite are more strongly defined approaching the dunite body, locally becoming mylonitic. This allows the recognition of two types of harzburgite, referred to as "blocky" and "mylonitic harzburgite". The mylonites were first described by Ross et al. (1980). They occur throughout the area as distinct zones from 5 cm to 5 m wide (Plate 4) which often develop into brittle thrusts displacing the chromite deposits in a NE direction. In the area mapped mylonites are concentrated into a 150 m wide belt in the SE, striking NW–SE and dipping around 50 SW, where there is a distinct spatial relationship between the presence of mylonitic host rocks and subparallel layers of massive ore.

In contrast to the Xerolivado area, the pyroxenite dykes are visibly deformed by the mylonite zones, although away from the mylonitic areas they appear to show a more normal appearance and distribution. Superimposed on all the earlier structures across the whole area is a brittle cleavage which becomes particularly intense towards the mylonite zones. Flat-lying thrust planes often terminate the chromite exposures, e.g. at the base of the No. 37 exposure and at the top of an exposure of massive ore. To the SE, flat-lying thrusts represent a late, out-of-sequence thrusting event. The southermost part of the area mapped terminates against a major fault which occupies the Voidolakkos valley. A schematic block diagram of the field relations encountered in the Voidolakkos area is presented in Fig. 10.

The best preserved layer of schlieren ore, No. 37, appears to be situated in a less deformed lense between two mylonite zones, preserving the high temperature structures which include isoclinal folds within the chromite layers. In contrast, the common association of massive ore with the mylonitic rocks suggests that recrystallization associated with mylonitization plays an important part

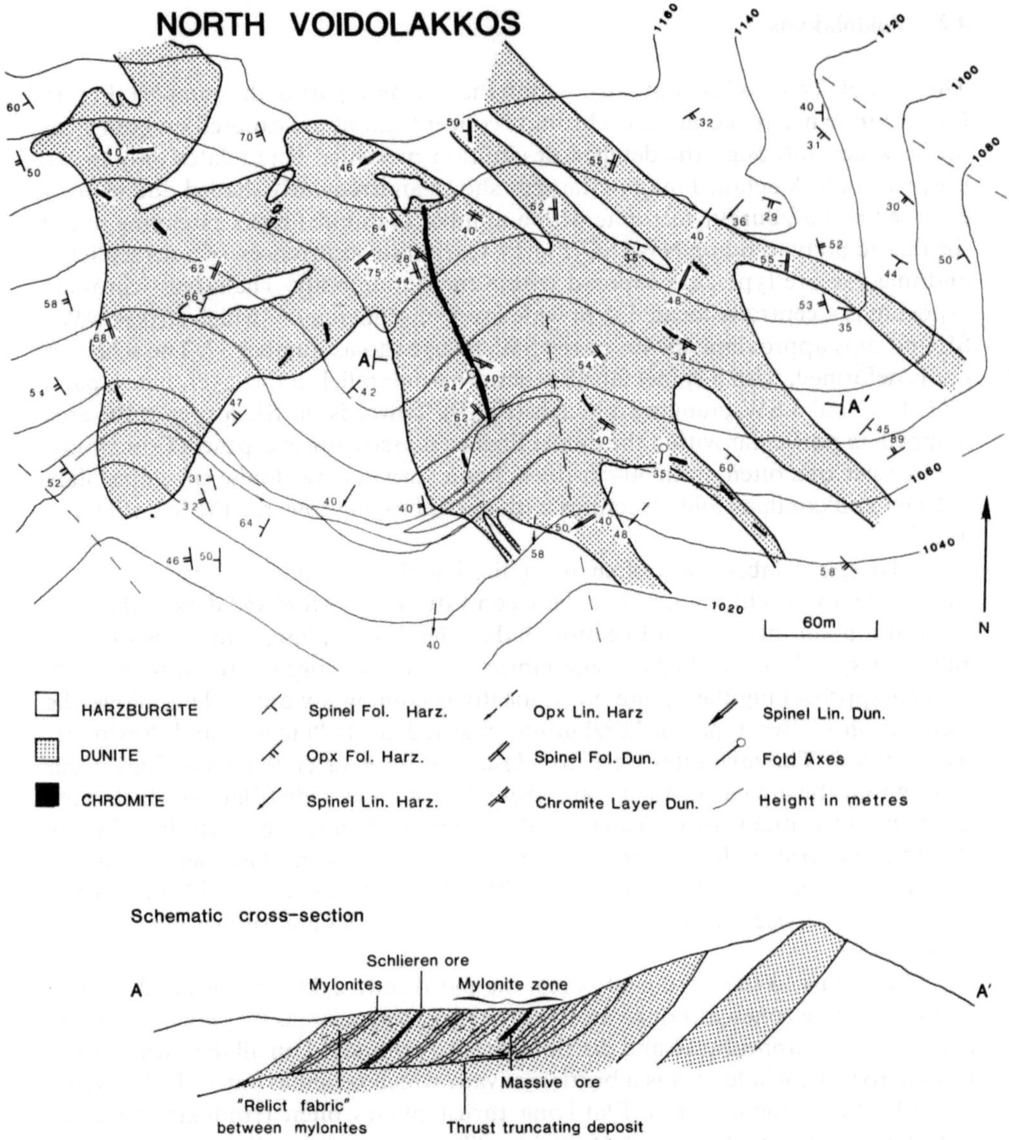

Fig. 8. Map of the Voidolakkos deposit with schematic cross-section

in the formation of massive ore. The demise of the ductile/brittle structures into the harzburgite host suggests that the dunite body provided an original zone of weakness on which the later ductile-brittle structures nucleated. The complete range from plastic through brittle structures are observed within the Voidolakkos area and, in contrast to the Xerolivado area, these ductile/brittle structures play a major part in the present distribution and form of the chromite deposits.

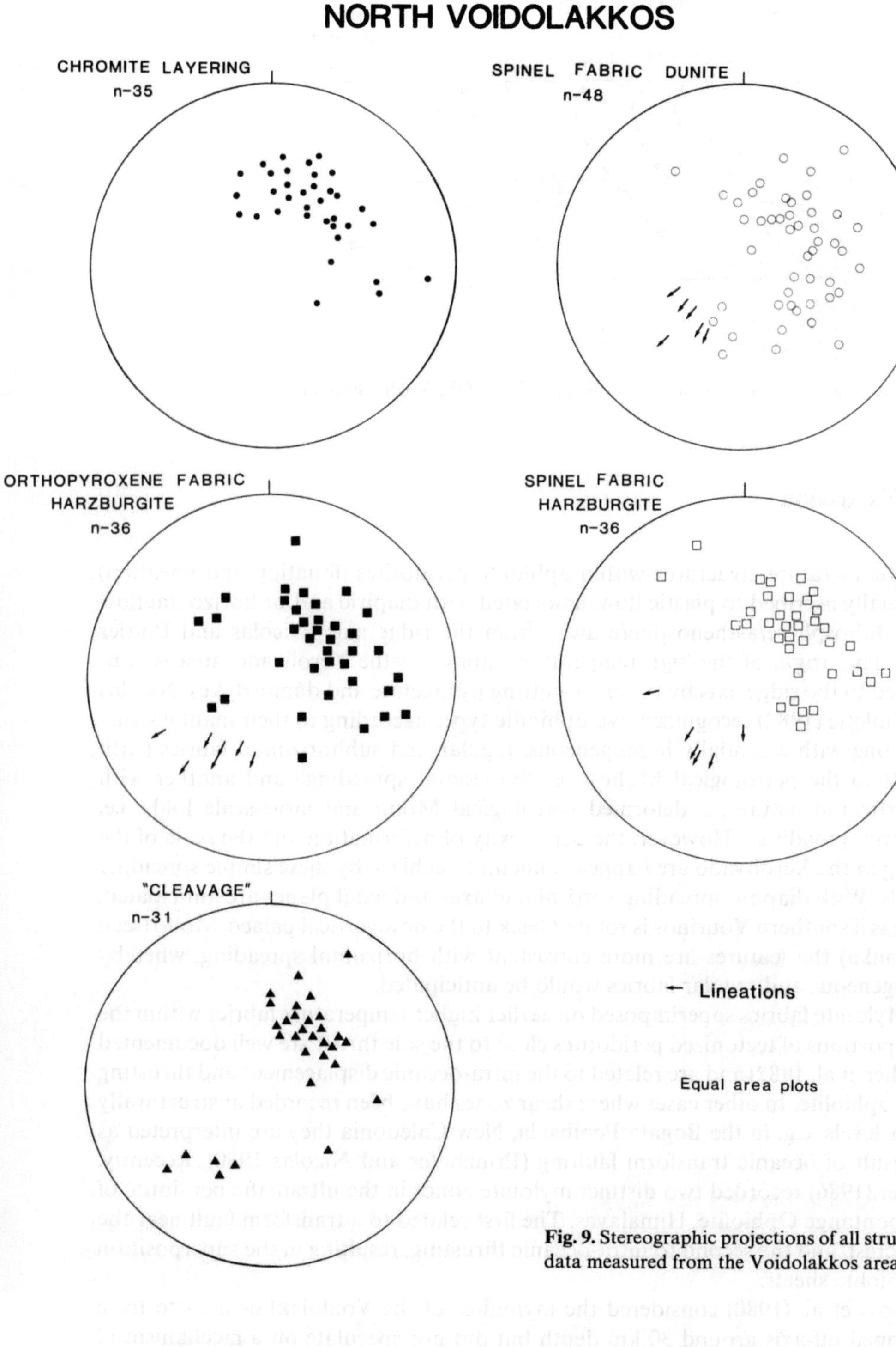

Fig. 9. Stereographic projections of all structural data measured from the Voidolakkos area

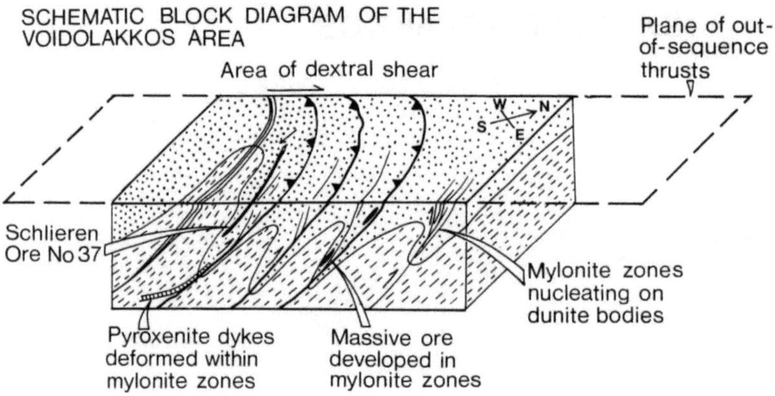

Fig. 10. A schematic block diagram of the field relations of the Voidolakkos area

5 Discussion

High temperature structures within ophiolite peridotites (foliation and lineation) are usually ascribed to plastic flow, associated with diapiric and/or horizontal flow of the lithosphere/asthenosphere away from the ridge axis (Nicolas and Poirier 1976). The origin of the high temperature fabrics of the Xerolivado area is constrained to the ridge axis by the cross-cutting pyroxenite and dunite dykes. Nicolas and Violette (1982) recognized two ophiolite types according to their mantle structures: one with essentially homogeneous, regular and subhorizontal fabrics (with respect to the petrological Moho), i.e. "horizontal spreading" and another with subvertical structures, a deformed petrological Moho and large-scale folds, i.e. "diapiric spreading". However, the complexity of deformation and the scale of the folding in the Xerolivado area appear difficult to achieve by these simple spreading models. With diapiric spreading vertical fold axes and axial planes are anticipated, whereas if southern Vourinos is rotated back to the now vertical palaeo-Moho (seen at Tsouka) the features are more consistent with horizontal spreading, whereby homogeneous and regular fabrics would be anticipated.

Mylonite fabrics superimposed on earlier higher temperature fabrics within the basal portions of tectonized peridotites close to the sole thrust are well documented (Boudier et al. 1982) and are related to the intra-oceanic displacement and thrusting of the ophiolite. In other cases where shear zones have been recorded at structurally higher levels, e.g. in the Bogata Peninsula, New Caledonia they are interpreted as the result of oceanic transform faulting (Prinzhofer and Nicolas 1980). Recently, Reuber (1986) recorded two distinct mylonite zones in the ultramafic peridotite of the Spontange Ophiolite, Himalayas. The first related to a transform fault near the ridge crest, and the second to intra-oceanic thrusting, resulting in the superposition of ophiolite sheets.

Ross et al. (1980) considered the mylonites of the Voidolakkos area to have developed off-axis around 30 km depth but did not speculate on a mechanism of

Plate 2. Thin band of massive ore in Voidolakkos subparallel to the mylonites and spaced cleavage

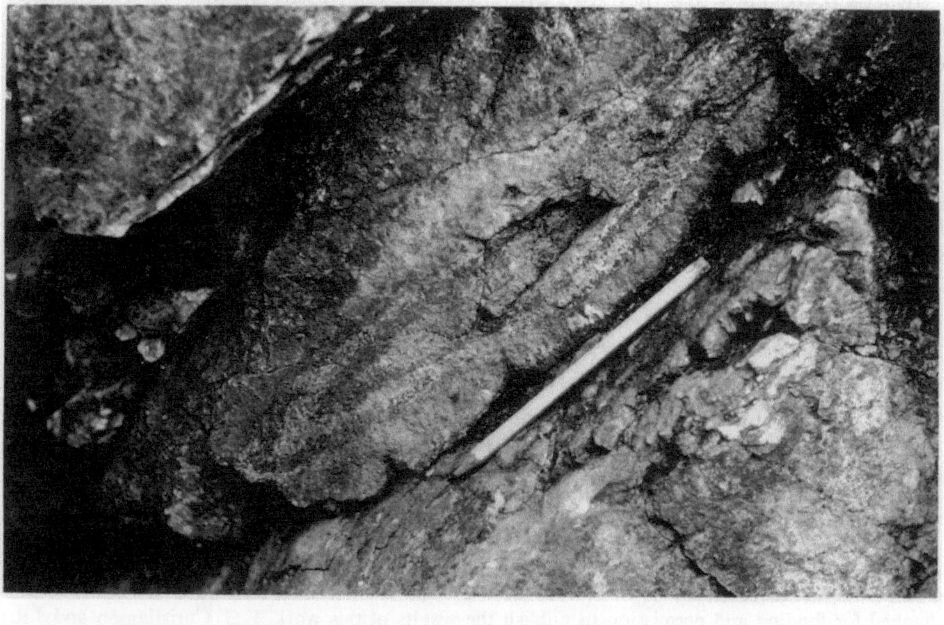

Plate 3. Fold structures within massive ore of the Voidolakkos area

Plate 4. Mylonites of the Voidolakkos area

formation. This study indicates that the mylonite zones of the Voidolakkos area deform both the higher temperature fabrics and the pyroxenite dykes and are in turn associated with brittle thrusts all showing a similar movement direction (towards 050). They are therefore probably related to the earliest (intra-oceanic) stages of ophiolite emplacement. It is interesting to speculate that the displacement of an ophiolite near the ridge axis may develop fabrics within the mantle sequence, at the base of an incipient thrust, indistinguishable from mantle flow structures. In theory, ophiolite displacement close to the ridge axis could account for the high temperature folding of the mantle fabrics in the Xerolivado area.

Irrespective of the actual origin of the tectonism there can be little doubt that the present nature and form of the Vourinos chromitites are structurally controlled. For example, the importance of plastic structures (of mantle origin) is clear in the Xerolivado area, whereas the ductile to brittle fabrics (of emplacement origin) control the present nature and form of the Voidolakkos orebodies. Furthermore, these studies suggest that structural classifications based on high temperature fabrics alone, e.g. Cassard et al. (1981), do not fully appreciate all the structural elements concerned. Christiansen (1986) produced a structural classification which included the onset of emplacement tectonics; this study suggests that more expansive models of this nature are required.

Acknowledgements. This chapter presents data recently acquired within the framework of a collaborative project between the Institute of Geology and Mineral Exploration (IGME), Athens and the University of Southampton (*EEC Contracts Nos. MSM. 135. GR* and *MSM. 136. UK*). IGME and the EEC are thanked for funding and permission to publish the results of this work. F.G. Christiansen and J.R.

Andrews are thanked for fruitful discussion and critical reviews of the manuscipt. Anthea Dunkley, John Taylor and Barry Marsh are thanked for cartographic and photographic services.

References

Apostolidis G (1977) Study of the potential for chromite exploration in Greece. IGME Int Rep 98 pp

Apostolides G, Mastoris K, Vgenopoulos AS (1980) Exploration of the Xerolivado chromite deposits and their chemical mineralogical and physical properties. In: An international symposium on metallogenesis of mafic and ultramafic complexes. UNESCO, Athens 1:1–20

Ayrton S (1968) Structures isoclinales dans les péridotites du mont Vourinos, un exemple de déformation de roches ultrabasiques. Bull Suisse Miner Petrog 48:734–750

Boudier F, Nicolas A, Bouchez L (1982) Kinematics of oceanic thrusting and subduction from basal sections of ophiolites. Nature 296:825–828

Burgath K, Weiser T (1980) Primary features and genesis of Greek podiform chromite deposits. In: Panayiotou (ed) Ophiolites. Proc Int Oph Symp Cyprus 1979, pp 675–690

Cassard D, Moutte J, Rabinovitch M, Nicolas A, Leblanc M, Prinzhofer A (1981) Structural classification of podiform chromite deposits from New Caledonia. Econ Geol 76:805–831

Christiansen FG (1986) A structural classification of ophiolitic chromite deposits. In: Gallagher MT, Ixer RA, Neary CR, Prichard (eds) Metallogeny of basic and ultrabasic rocks. Institute of Mining and Metallurgy, London, 279–290

Coleman RG (1977) Ophiolites: ancient oceanic lithosphere? Springer Berlin Heidelberg New York, 229 pp

Gass IG, Smewing JD (1981) Ophiolites: obducted oceanic lithosphere. In: Emiliani C (ed) The Seas, vol 7. Wiley, New York, pp 339–361

Harkins ME, Green HW, Moores EM (1980) Multiple intrusive events documented for the Vourinos ophiolite complex, Northern Greece. Am J Sci 280A: 284–295

Lago BL, Rabinowicz M, Nicolas A (1982) Podiform chromite orebodies: a genetic model. J Pet 23:103–126

Leblanc M, Dupuy C, Cassard D, Moutte J, Nicolas A, Prinzhoffer A, Rabinovitch M, Routhier P (1980) Essai sur la genèse des corps podiformes de chromitite dans les péridotites ophiolitiques. In: Panayiotou A (ed) Etude des chromites de Nouvelle-Calédonie et comparaison avec celles de Méditerranée orientale. Proc Int Oph Symp Cyprus, 1979, pp 691–702

Moores EM (1969) Petrology and structure of the Vourinos ophiolite complex, northern Greece. Geol Soc Am Spec Pap (Reg Stud) 118:74

Naylor MA, Harle TJ (1976) Palaeogeographic significance of rocks and structures beneath the Vourinos ophiolite — northern Greece. J Geol Soc (Lond) 132:667–675

Neary CR, Brown MA (1979) Chromites from the Al Ays complex, Saudia Arabia, and the Semail complex, Oman. In Al Shanti (ed) Evolution and mineralisation of the Arabian shield. IAS Bull 2:193–205

Nicolas A, Poirier JP (1976) Crystalline plasticity and solid state flow in metamorphic rocks. Wiley, New York

Nicolas A, Vialette J-F (1982) Mantle flow at oceanic spreading centres: models derived from ophiolites. Tectonophysics 81:319–339

Prinzhofer A, Nicolas A (1980) The Bogota Peninsula, New Caledonia: a possible oceanic transform fault. J Geol 88:387–398

Reuber I (1986) Geometry of accretion and oceanic thrusting of the Spontang ophiolite, Ladakh-Himalayas. Nature 321:592–596

Ross JV Mercier J-CC, Ave Lallement HG, Carter NL, Zimmerman J (1980) The Vourinos ophiolite complex Greece: the tectonite suite. Tectonophysics 70:63–83

Smith AG (1977) Othris, Pindos and Vourinos ophiolites and the Pelagonian zone. Proc Colloq Geol Aegean Region 6:1369–1374

Spray JG (1984) Possible causes of upper mantle decoupling and ophiolite displacement. In Ophiolites and oceanic lithosphere. Spec Publ Geol Soc Lond 13:255–268

Spray JG, Roddick JC (1980) Petrology and Ar40/Ar39 geochronology of some Hellenic sub-ophiolite metamorphic rocks. Contrib Miner Pet 72:43–55

Stamoulis K (1984) Reserve estimates of the southern sector of the Skoumtsa mine area. Hellenic Ferro-chrome company. Int Report, 15 pp

Thayer TP (1963) Flow layering in Alpine peridotite gabbro complexes. Miner Soc Am Spec Pap 1:55–61

Vergely P (1976) Chevauchement vers l'Ouest et rétrocharriage vers l'Est des ophiolites; deux phases tectoniques au cours du Jurassique supérieur–Eocrétacé dans les Hellénides internes. Bull Soc Geol Fr 18:231–244

Vrachatis G, Grivas I (1980) The geology and ore deposits of the Vourinos-Flambouros region (1:10000 scale). IGME Int Rep, 80 pp

Zachos K (1964) Chromite exploration in Greece. In: Woodtli R (ed) Methods of prospection for chromite. OECD, Paris, pp 55–59

Comparative Study of Chromite Deposits from Troodos, Vourinos, North Oman and New Caledonia Ophiolites

T. Augé and Z. Johan[1]

Abstract

The comparative study of chromitites from Vourinos (Greece), Troodos (Cyprus), Semail (Oman), Tiébaghi and Massif du Sud (New Caledonia) ophiolitic complexes reveals that all studied chromite deposits are surrounded by dunite whose extent bears no relation to the size of the orebody. The dunite-chromitite contact may be gradational or sharp; in Oman, its sharpness is emphasized by the presence of interstitial silicates (plagioclase, pyroxenes) in chromitite which are missing in the surrounding dunite. The occurrence of PGM inclusions in disseminated chromite from dunite in the Vourinos and Tiébaghi ophiolites infers a genetic relationship between chromitites and their dunitic envelopes. In chromite deposits from Vourinos and Massif du Sud, olivine is the only interstitial phase. The Oman chromitites exhibit a complex interstitial assemblage with olivine, clinopyroxene, plagioclase and pargasitic amphibole. Silicate inclusions (olivine, orthopyroxene, clinopyroxene, plagioclase, pargasite, phlogopite, sodium equivalent of phlogopite, nepheline) are commonly observed in chromitites. Their distribution within the host crystal indicates that they were trapped during chromite crystallization. Tabular crystals of both pyroxenes found in some chromite bodies might result from epitaxial intergrowth or from exsolution.

No difference in composition of olivine from wall dunite and barren dunite bodies was recorded, except for Vourinos, where some dunites consist of olivine Fo 94.4. Spinels from dunite have a lower XMg and are slightly enriched in Fe^{3+} than those from chromitites. Their Ti content approaches that of chromitites, varying from 0.20 (Oman) to 0.05 wt% (Massif du Sud).

Analyses of chromitite show that a given deposit is characterized by a very constant chromite composition. Chromitites from Vourinos, Tiébaghi and Massif du Sud exhibit very narrow compositional ranges. In Oman, Cr-rich chromites come from deposits in the deep part of the mantle sequence, those rich in Al occurring in the cumulate hosted deposits.

Olivine associated with chromite ore is systematically Mg-richer than that of dunites: Fo 95.7–97.6. Interstitial and included orthopyroxene shows En 92–94 and is characterized by low Cr and Al contents. Included clinopyroxene is a chromium

[1] GIS BRGM–CNRS CRSCM, 1A, rue de la Férollerie, 45071 Orléans, Cedex 02, France

Mineral Deposits within the European Community
(ed. by J. Boissonnas and P. Omenetto)

diopside slightly enriched in Na_2O (0.4 wt%) and impoverished in Al_2O_3 (1.4 wt%). Amphibole is a potassium-free pargasitic hornblende with XMg 0.95–0.97. Its TiO_2 content varies from 1.1 (Semail) to 0.27 wt% (Vourinos). Sodium-phlogopite inclusions were observed in chromitites from Tiébaghi, Massif du Sud and Semail.

Platinum-group minerals (PGM) occur in chromitites from all the studied ophiolites, except in the northern part of Oman. Rutheniridosmine, iridosmine, ruthenosmiridium, osmiridium, ruthenian osmium, laurite, erlichmanite, xingzhongite, osarsite and irarsite were identified. Significant differences in parageneses appear from one ophiolite to another: in the Massif du Sud, Os–Ir–Ru alloys predominate, while in chromitites from the southern part of the Oman ophiolite and in Troodos, laurite is the only PGM found. In the Tiébaghi massif, a complex association of PGM including laurite, erlichmanite, xingzhongite and an unknown Ir, Cu-sulphide, was observed. There is strong evidence of a very early crystallization of PGM, preceding the precipitation of chromite and even the crystallization of some silicates included in chromite. Variations in the nature of PGM from one deposit to another indicate the temperature and/or the S_2 activity variation in the ore-forming system.

1 Introduction

Ophiolitic complexes represent an important but difficult target for chromite prospecting. The size of individual chromite orebodies is highly variable and their distribution within a given complex seems extremely irregular. Structural studies have indicated that most deposits are located in the top kilometre of the mantle sequence, but this level cannot always be recognized. Furthermore, the model of chromitite formation is still not entirely understood.

In order to contribute to the understanding of the genesis of ophiolitic chromitites, the following complexes have been studied: Vourinos (Greece), Troodos (Cyprus), Semail (Oman), Tiébaghi and Massif du Sud (New Caledonia).

In a recent paper, Johan and Augé (1986) discussed the main characteristics of ophiolitic mantle sequences and showed their diversity. The results presented here concern mineral chemistry of chromitites and wall dunites, as well as the associated phases, including platinum-group minerals (PGM).

The Vourinos ophiolitic complex, emplaced in the Upper Jurassic, is located in northwestern Greece and covers some 450 km^2. The mantle sequence, formed by harzburgite tectonite, interlayered with dunite, constitutes about 80% of the complex (Moores 1969; Harkins et al. 1980). This major unit is locally overlain by dunite forming the base of the cumulate sequence and grading up progressively into wehrlite and gabbro. Quartz dolerite, granophyre and plagiogranite complete the succession (Brunn 1956). Pillow lavas of island arc affinities have also been recognized (Noiret et al. 1981; Beccaluva et al. 1984).

Most of the chromite orebodies in the Vourinos complex occur in the mantle sequence; only a few of minor importance are located within the dunitic cumulates.

The Cenomanian Oman ophiolite occurs in a 700 km-long belt extending along the southwestern coast of the Gulf of Oman. In its northern part, a complete sequence, about 14-km-thick, is exposed, the lowermost part of which (approximately 7-km-thick) consists of harzburgite with minor dunite and represents the mantle sequence. The latter is overlain by a well-developed oceanic crustal sequence comprising from the base to the top: layered ultramafic and mafic cumulates, isotropic gabbros, diorites, trondjemites, a sheeted complex, and finally, a pillow lava succession (Smewing 1980).

Three different environments were distinguished for the chromite orebodies, according to their stratigraphic position within the ophiolite: (1) thin chromite orebodies of very limited extent, located in the basal ultramafic cumulates; (2) deposits occurring in the uppermost part of the mantle sequence; these orebodies are the most common and the largest known; (3) deposits located in deeper parts of the mantle sequence, generally small and highly deformed (Brown 1982; Augé 1986; Christiansen 1986).

The Troodos complex, of Upper Cretaceous age (Vine et al. 1973), consists of a complete ophiolitic sequence which covers about 3000 km^2 of the island of Cyprus. The mantle sequence, composed of harzburgite with associated dunite, crops out in the centre of the Troodos mountains and is overlain by a 350-m-thick dunite unit (main dunite), interpreted as the base of the cumulate sequence (Greenbaum 1977; Johan et al. 1982). The upper part of the cumulate series is formed by wehrlite, clinopyroxenite and gabbro, overlain by a sheeted dyke complex and pillow lavas.

The most important chromite deposits (Kokkinorotsos, Kannoures and Hadjipavlou) are located in the uppermost part of the mantle sequence, very close to the contact with the main dunite. Small chromite occurrences are known near the base of the main dunite. Greenbaum (1977) reported minor chromite occurrences from clinopyroxene-bearing dunite, stratigraphically situated between the main dunite and wehrlite cumulates.

The New Caledonian ophiolite nappe, obducted during the Upper Eocene (Paris 1981), covers more than 8000 km^2. Its lithological sequence is very incomplete and most of the ophiolitic massifs consist only of the harzburgitic upper mantle series (Guillon 1975; Rodgers 1976; Prinzhofer and Nicolas 1980; Prinzhofer et al. 1980). In the Tiébaghi massif, however, plagioclase and spinel lherzolites are associated with harzburgites (Moutte 1982; Nicolas and Dupuy 1984; Johan and Augé 1986). Cumulates crop out only in several discontinuous patches preserved in the Massif du Sud, the largest being Montagne des Sources (Nicolas and Prinzhofer 1983), consisting of a sequence of dunite, wehrlite and gabbro.

The distribution of chromite deposits within the nappe is very irregular. The Tiébaghi massif, which represents less than 2% of the total peridotite cover, has supplied more than 80% of New Caledonian chromite production. The Massif du Sud is characterized by the presence of numerous small chromite deposits, while the so-called intermediate massifs are barren. Except for some small showings in cumulate dunites of the Montagne des Sources, all chromite orebodies are located within the mantle sequence.

2 Chromite Deposits and Associated Dunites

2.1 Chromite Orebodies

Some of the main characteristics of chromite orebodies, investigated during this study of ophiolites, include their diversity in size, shape, structure (internal and relation to host rocks), ore type, nature and proportion of gangue silicates and included phases, as well as the importance of the dunitic envelope. Despite this diversity, some distinctive features can be observed for a given ophiolitic complex, as summarized in Table 1.

Table 1. Characteristics of chromite ore from different ophiolitic complexes[a]

	SEMAIL	VOURINOS	TIEBAGHI	TROODOS	MASSIF DU SUD
ore type					
massive	●		●	●	●
schlieren		●		●	
nodular		●	●	●	
orbicular				●	
disseminated	●	●			●
layered		●	●	●	●
interstitial sili					
olivine	●	●	●	●	●
clinopyroxene	●	•	•	•	
orthopyroxene	●	•	•		
plagioclase	●				
amphibole	•				
included silicates					
olivine	•	•	●	●	●
clinopyroxene	●	●	●	●	●
orthopyroxene	•	•	•	●	•
plagioclase	•				
amphibole	●	●	•		●
nepheline					•
mica	•			•	•

[a] The size of the symbol is proportional to the abundance of the various silicates. Note that the distribution of included and interstitial silicates may vary from one deposit to another. The mica mentioned for Troodos has been found in spinel from wall dunite.

Cassard et al. (1981) proposed a structural classification of chromite deposits based on the study of the Massif du Sud in New Caledonia; they distinguished concordant, subconcordant and discordant orebodies whose attitude with respect to the harzburgite foliation depends on the degree of mantle deformation. This classification has been discussed by Johan (1986) and by Christiansen (1986). Christiansen proposed a new structural classification based on the study of ophiolitic complexes reported in this chapter. He suggested six orebody types, according to their shape, orientation and stratigraphic position: (1) layers within the cumulate sequence, (2) tabular, subconcordant paleo-horizontal bodies, (3) tabular, concordant bodies, (4) folded concordant bodies, (5) tabular, discordant paleo-vertical bodies, (6) irregular, podlike bodies in serpentinite. Types 2 to 6 occur within the mantle sequence.

Primary magmatic textures, common in types 1 and 2, include occluded silicates and silicate clots, but the ore texture ranges from submassive to disseminated ore (Roberts 1986). These types have only been found in the Oman and Troodos ophiolites. Generally, the deposits belonging to type 1 are rather small, while those of type 2 show large size variations. Most of the investigated deposits belong to type 3. The ore texture is variable, including nodular, disseminated (with massive patches), submassive and coarse-grained ore types. Folded bodies (type 4) are common in the Vourinos and Tiébaghi complexes, but uncommon in Oman, Troodos and the Massif du Sud. The largest orebody in Vourinos, the Xerolivado deposit, belongs to this category. Again, nodular, submassive, banded, coarse-grained, and "schlieren type" ore textures can be observed.

Type 5, the discordant orebodies of Cassard et al. (1981), has only been recognized in New Caledonia, and is located in shear zones (Christiansen 1986). Type 6, highly deformed orebodies, are characterized by massive, often altered ore and are located close to syn- and post-emplacement faults and thrusts.

2.2 Wall Dunites

All studied chromite deposits are surrounded by dunite whose extent bears no relation to the size of the orebody, varying strongly within a given deposit from a few centimetres to several metres (e.g. the Tiébaghi mine, New Caledonia). The only exceptions are deposits showing tectonic contacts with harzburgite, where dunite could be missing. This systematic presence of dunite envelopes indicates a genetic relationship between chromite deposition and the formation of dunite. The understanding of this relationship is thus fundamental for the comprehension of the process of chromite concentration. However, in ophiolitic complexes, most of the dunite occurrences within the upper mantle sequence are barren. Hence, the important question is, do barren dunites derive from a process similar to that for those associated with chromite orebodies? Johan and Augé (1986) concluded that barren dunites in ophiolite mantle sequences result from a "metasomatic-like" process, leading to the destabilization of orthopyroxene under a relatively high water pressure (Bowen and Tuttle 1949; Dungan and Avé Lallemant 1977). Could a similar process be envisaged for the origin of dunitic envelopes?

The dunite-harzburgite contact may be gradational or sharp. Johan (1986), in New Caledonian deposits, described pervasive replacement of orthopyroxene by olivine, occurring along the harzburgite foliation plane, gradational at Marais Kiki, sharp at Anna Madeleine. Augé (1985b) reported mostly sharp contacts for Semail and Vourinos orebodies. Similarly, the transition between dunite and ore may be either sharp or gradational and marked by a progressive increase in chromite crystals. From the structural relations between dunite envelope and harzburgite, on the one hand, and the textural relationship of olivine and chromite, on the other hand, Johan (1986) suggested that the dunite results from in situ metasomatic replacement of harzburgite, due to circulation of fluids favoured by the existence of shear zones, and concluded that chromite precipitation is slightly later than the formation of wall dunites.

In the Oman ophiolite, the sharpness of the chromitite-dunite contact is emphasized by the presence of interstitial silicates in chromitite (plagioclase, pyroxenes) which have never been found in the surrounding dunite. This could argue against a direct relationship between these two units (Augé 1987). On the other hand, the discovery of inclusions of platinum-group minerals (PGM) in disseminated chromite from dunite in the Vourinos and Tiébaghi ophiolites, whose paragenesis is similar to that found in chromitites, and which is characteristic of a given complex (see below), has led Augé (1988) to infer a genetic relationship between chromitites and their dunitic envelopes. Furthermore, the same silicate inclusions were observed in chromitites and disseminated chromium spinels from dunites. The fact that PGM and silicate inclusions were found in all types of dunite, including barren dunite, seems to indicate that all dunite bodies occurring within an ophiolitic mantle sequence result from a single process.

2.3 Associated Silicates

Depending on the ore textures, the proportion of gangue silicates varies from 1–5 vol% (massive ore) to 95 vol% (disseminated ore). In all the chromite deposits from Vourinos and southern New Caledonia, olivine is the only interstitial phase. There are, however, small chromite bodies in Vourinos and in the southernmost part of the Tiébaghi complex with olivine + orthopyroxene + clinopyroxene as interstitial phases.

The Oman chromitites, in contrast, exhibit a complex silicate assemblage. In decreasing order of abundance, olivine, clinopyroxene, orthopyroxene, plagioclase and pargasitic amphibole were observed, together or as single grains. Rarely, large variations in the nature of interstitial silicates may occur at the scale of a thin section (Augé 1985b).

Besides interstitial silicates, silicate inclusions are commonly observed in chromitites, generally as euhedral to subhedral crystals, but sometimes also as irregular grains. They are either randomly distributed within the host crystal or, more commonly, concentrated in the core of chromite crystals or distributed along growth zones. This indicates that they were trapped during chromite crystallization. Olivine, orthopyroxene, clinopyroxene, plagioclase, pargasite, phlogopite, a sodium equivalent of phlogopite and nepheline have been observed. As shown in Table 1, their distribution varies from one ophiolitic complex to another. Note that in all cases, their nature, composition and relative proportion are different from those of interstitial silicates. Thin tabular crystals of both pyroxenes have also been observed in some chromite bodies (especially in the Anna Madeleine deposit, Massif du Sud). Their orientation with respect to the chromite host crystal suggests that they result from epitaxial intergrowth or from exsolution (Johan 1986).

3 Mineralogy and Mineral Chemistry of the Ore System

3.1 Wall Dunites

There is no difference in composition between olivine from wall dunites and that of barren dunite bodies. Similar Fo contents were recorded in all the complexes investigated, except for Vourinos, where some dunites consist of Fo-rich olivine (up to 94.4 mol%). Compared to olivine from harzburgite, the compositions of olivine from dunite are more scattered but, except for Vourinos, with similar average Fo contents.

Spinels from dunite always have a lower XMg ratio than those from chromitites. In the Vourinos and Troodos complexes, the Cr/Al ratio of spinel from dunites is relatively constant and close to that of chromitite spinels; it is only slightly higher in the Massif du Sud. In the Tiébaghi and Semail complexes, dunite spinels have a rather large Cr/Al variation which, in the case of Semail, plots in the field of chromitites (Fig. 1). Tiébaghi is the only example where most of the spinels from wall dunites exhibits lower Cr/Al than those of chromitites. However, even in this case, Cr-rich terms cover the chromitite field (Fig. 1). This is also the only case studied where two populations of dunitic spinels can be distinguished on the basis of their Cr/Al ratio: those occurring in barren dunite bodies and in wall dunites are characterized by higher Cr/Al, and those from dunites forming "layers" 10- to 100-cm-thick are concordant with the clinopyroxene-bearing harzburgite foliation. It is interesting to note that the Cr/Al ratio of spinel from these dunite layers fits exactly with the field of spinel from harzburgites and lherzolites occurring in the Tiébaghi massif (Fig. 1; Johan and Augé 1986).

In most cases, spinels from dunite are slightly enriched in Fe^{3+} compared to those from chromitites. In contrast, their TiO_2 content approaches that of chromitites and, in spite of a large standard deviation, the same average as for chromitites is observed, Oman: TiO_2 0.20 (σ 0.09), Vourinos: 0.11 (0.06), Troodos: 0.09 (0.05), Tiébaghi: 0.08 (0.04), Massif du Sud: 0.05 (0.04).

Silicate inclusions similar to those in chromitites were also found in disseminated spinels from wall dunites. However, due to the rarity of spinel, they are exceptional. In their detailed study of Troodos wall dunites, Johan et al. (1982) reported clinopyroxene (common), orthopyroxene and nepheline.

Compositions of clino- and orthopyroxenes included in disseminated spinel from dunite are close to those of pyroxene inclusions in chromitite; in particular,

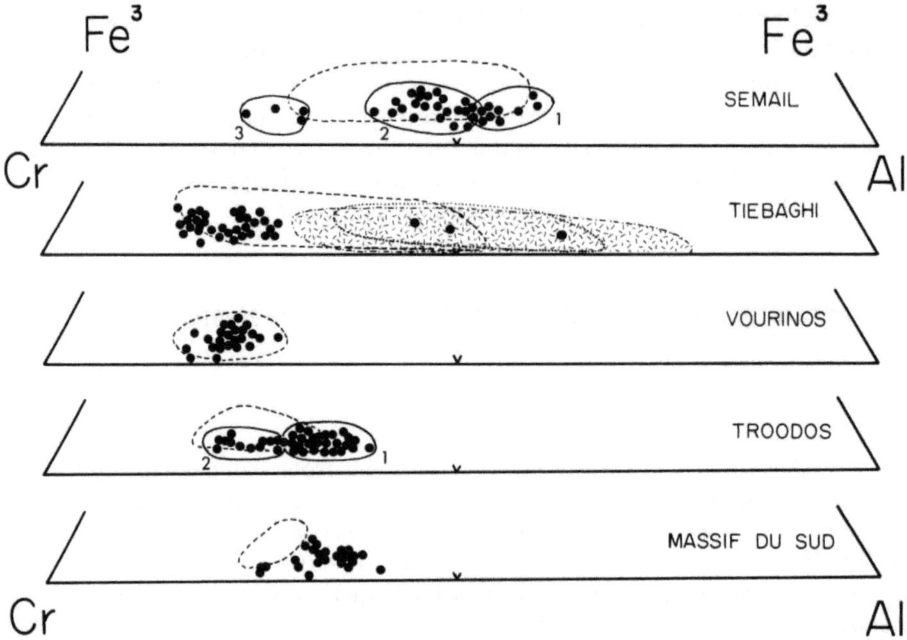

Fig. 1. Chemical composition of spinel (atomic proportions of Cr, Al, Fe^{3+}) from chromite ore (*dot*) and dunite (*broken line*). *Semail*: Fields *1*, *2* and *3* correspond respectively, to deposits from the cumulate series, from the top, and from the deeper part of the mantle sequence; *Tiébaghi*: *broken line* corresponds to wall dunite and barren dunite bodies, while the *dotted line* indicates dunite interlayered with pyroxene peridotite. *Hatched area* represents spinel from harzburgite and lherzolite; *Troodos*: fields *1* and *2* correspond respectively, to deposits from the northern and southern part of the massif

the clinopyroxene is enriched in MgO and Na_2O and impoverished in Al_2O_3. Orthopyroxene is enriched in MgO and impoverished in CaO. Like nepheline from chromitites, that from the Troodos dunite is purely sodic (Johan et al. 1982).

3.2 Chromitites

3.2.1 Chromite Composition

Compositions of chromite from chromitites in the ophiolites studied are reported on the Cr–Al–Fe^{3+} diagram (Fig. 1). From Fig. 1, the following observations can be made: (1) a given chromite deposit is characterized by a very consistent chromite composition; (2) chromitites from the Vourinos, Tiébaghi and Massif du Sud complexes show very narrow compositional ranges; (3) Tiébaghi and Vourinos chromitites are very similar in composition, corresponding to a metallurgical ore (Table 2). The few points of Al-rich chromite from the Tiébaghi massif correspond to the small orebody with pyroxenes as interstitial silicates mentioned above; (4)

Table 2. Selected microprobe analyses of spinel from chromite deposits[a]

	1	2	3	4	5	6	7	8	9	10	11	12	13	14	15
SiO_2	0.02	0.06	0.03	0.00	0.00	0.00	0.06	0.00	0.00	0.00	0.00	0.04	0.07	0.09	0.13
TiO_2	0.04	0.10	0.12	0.02	0.06	0.08	0.13	0.10	0.00	0.02	0.00	0.05	0.00	0.06	0.05
Al_2O_3	11.04	10.75	14.16	10.00	32.07	13.99	17.15	10.14	20.28	15.05	16.67	16.99	17.18	19.53	18.02
Cr_2O_3	58.49	58.00	55.52	60.71	31.96	54.44	50.98	59.54	51.11	57.12	54.53	52.71	53.04	51.71	53.74
Fe_2O_3	3.98	4.37	3.56	2.67	5.85	3.86	4.39	3.64	1.82	2.25	3.04	3.61	3.13	1.49	2.36
FeO	12.87	12.14	11.40	12.24	17.78	13.42	12.11	11.96	9.89	11.01	9.71	11.10	9.80	12.82	10.46
MgO	13.68	13.94	14.89	13.74	12.64	13.47	14.74	13.98	16.49	15.43	16.34	15.28	16.03	14.56	16.10
MnO	0.04	0.15	0.09	0.25	0.14	0.04	0.03	0.23	0.00	0.14	0.06	0.25	0.22	0.22	0.13
NiO	0.13	0.00	0.13	0.00	0.14	0.00	0.00	0.00	0.17	0.00	0.04	0.00	0.16	0.00	0.00
Total	100.29	99.51	99.90	99.63	100.64	99.30	99.59	99.59	99.76	101.02	100.39	100.03	99.63	100.48	100.99
Si	0.005	0.014	0.009	—	—	—	0.015	—	—	—	—	0.011	0.018	0.022	0.033
Ti	0.007	0.019	0.022	0.004	0.011	0.015	0.024	0.019	—	0.004	—	0.008	—	0.011	0.011
Al	3.337	3.267	4.203	3.050	8.934	4.217	5.046	3.088	5.837	4.383	4.830	4.965	5.009	5.647	5.168
Cr	11.860	11.829	11.055	12.421	5.972	11.009	10.062	12.164	9.816	11.162	10.601	10.333	10.374	10.029	10.342
Fe^3	0.768	0.848	0.675	0.520	1.041	0.743	0.828	0.708	0.332	0.419	0.563	0.673	0.583	0.276	0.433
Fe^2	2.760	2.618	2.401	2.649	3.515	2.870	2.526	2.585	2.008	2.276	1.997	2.302	2.026	2.630	2.130
Mg	5.228	5.360	5.591	5.301	4.453	5.135	5.485	5.384	5.969	5.685	5.988	5.647	5.910	5.323	5.836
Mn	0.009	0.032	0.019	0.054	0.029	0.010	0.006	0.050	—	0.029	0.012	0.053	0.047	0.047	0.027
Ni	0.026	—	0.027	—	0.026	—	—	—	0.018	—	0.007	—	0.031	—	—
Σ	24.000	23.987	24.002	23.999	23.981	23.999	23.992	23.998	23.980	23.958	23.998	23.992	23.998	23.985	23.980
XMg	0.655	0.672	0.700	0.667	0.559	0.642	0.684	0.675	0.748	0.714	0.750	0.710	0.745	0.669	0.733
YCr	0.743	0.742	0.694	0.777	0.375	0.689	0.631	0.762	0.614	0.699	0.663	0.647	0.650	0.629	0.649
YAl	0.209	0.205	0.264	0.191	0.560	0.264	0.317	0.194	0.365	0.275	0.302	0.311	0.314	0.354	0.324
YFe^{3+}	0.048	0.053	0.042	0.032	0.065	0.047	0.052	0.044	0.021	0.026	0.035	0.042	0.036	0.017	0.027

[a] Vourinos: *1* Tsouka deposit, *2* Xerolivado; Tiébaghi complex, New Caledonia: *3* Tiébaghi mine, *4* Alpha; North Oman: *5* Aleya (cumulative series), *6* Rayy (mantle sequence); Troodos: *7* Kokkinorotsos, *8* East Hadjipavlou; Massif du Sud, New Caledonia: *9* Anna Madeleine, *10* and *11* Marais Kiki, *12* and *13* GR2H deposit, *14* Bon Secours, *15* Dyne.

chromitites from the Massif du Sud are relatively poor in Cr_2O_3; (5) a large compositional range is observed for the Oman ophiolite chromitites. However, three fields, corresponding to three different environments of the orebodies may be distinguished: the Cr-rich chromites come from deposits located in the deep part of the mantle sequence, while those rich in Al occur in cumulate hosted deposits; intermediate compositions are those of chromitites located near the top of the mantle sequence.

The plot of Troodos chromitites shows also a relatively large spread. Two groups with no overlap can be distinguished (Fig. 1). The first, characterized by high Cr/Al ratios, corresponds to the occurrences from the southernmost part of the harzburgite unit, and the second, with Cr/Al similar to that of the Massif du Sud deposits, includes the largest orebodies known in this area (Kokkinorotsos, Kannoures). Johan et al. (1982) noted this feature but gave no explanation. They noted that the Cr/Al ratio in the overlaying dunite cumulates follows an inverse trend, decreasing from north to south in the complex.

Generally, chromitites are characterized by a large variation in XMg [$Mg/(Mg + Fe^{2+})$]. Johan and Augé (1986) showed that in dunite and harzburgite, the YCr [$Cr/(Cr + Al + Fe^{3+})$] of disseminated chromite is negatively correlated with XMg. For chromite ores, however, large variations recorded in the YCr (especially for Oman chromitites) are not correlated with XMg (Table 2).

In most cases, the chromite ore is characterized by higher XMg than co-existing dunite spinel. Generally, the XMg increases when the proportion of interstitial silicates decreases. This feature might be partly explained by a re-equilibration process between olivine and spinel (Lehmann 1983). In this case, higher XMg values observed for massive ores would correspond to the assemblage which has undergone a low degree of re-equilibration and could then reflect the primary compositions unlike disseminated spinels.

The TiO_2 content (wt%) in chromite from chromitites varies from one complex to another: the highest values, TiO_2 0.22 (0.13) were observed for Oman chromitites with common exsolved rutile, followed by Vourinos: 0.12 (0.04), Troodos: 0.10 (0.04), Tiébaghi: 0.09 (0.05) and the Massif du Sud: 0.04 (0.03). Similar values were recorded for spinel from associated dunite, whilst the TiO_2 content of spinel from any harzburgite is generally below the microprobe detection limit.

3.2.2 Interstitial and Included Silicates

Olivine associated with chromite ore is systematically Mg-richer than that of associated dunite. Moreover, its Fo content increases with the grade of chromite ore, massive chromitites including olivines with the highest Fo content. Thus, the average olivine composition (in Fo mol%) for chromite-poor ores is 94.84 (0.53) and 93.48 (1.09) for Vourinos and Semail respectively, and 95.56 (0.54) and 94.84 (0.74) for massive ores from the same ophiolitic complexes. Due to a higher degree of serpentinization, only four values were obtained for the Tiébaghi chromitite: average 94.56 (1.03), and none for Troodos. In the Massif du Sud, the interstitial olivine from disseminated ores has the composition 94.79 (1.25).

The nickel content of interstitial olivine is slightly higher than that from other ophiolitic units: 0.48 (0.12), 0.54 (0.05) and 0.40 (0.14) wt% NiO for Vourinos, Tiébaghi and Semail respectively (Table 3). Olivine included in chromite ore is even Mg- and Ni-richer as summarized below:

	Fo (mol%)		NiO (wt%)	
	$\bar{x}$	σ	$\bar{x}$	σ
Oman	95.96	0.83	0.60	0.20
Vourinos	97.65	0.21	0.64	0.11
Tiébaghi	96.78	0.78	0.63	0.18
Massif du Sud	95.83	0.79	0.63	0.15
Troodos	95.75	0.17	0.42	0.17

Interstitial clinopyroxene is a chromiferous diopside, $En_{49.9}Fs_{2.6}Wo_{47.4}$, pyroxene in harzburgite. The average composition for Oman is $En_{92.2}Fs_{6.4}Wo_{1.4}$; furthermore, low Cr_2O_3: 0.49 (0.12) and Al_2O_3: 1.76 (0.21) wt% contents are noteworthy. The included orthopyroxene is richer in MgO, Cr_2O_3 and impoverished in CaO and Al_2O_3 with a low Al(IV) site occupancy. The average composition is $En_{94.2}Fs_{5.5}Wo_{0.3}$ (Table 3).

Interstitial clinopyroxene is a chromiferous diopside, $En_{49.9}Fs_{2.6}Wo_{47.4}$, slightly enriched in Na_2O: 0.25 (0.08) wt% for Semail and 0.53 (0.08) for Tiébaghi, and containing from 0.0 to 0.3 wt% TiO_2. The clinopyroxene from Tiébaghi is richer in Cr_2O_3: 1.52 (0.08) and Al_2O_3: 2.79 (0.12) compared to Semail: 0.87 (0.13) and 2.38 (0.02) wt% respectively. Included clinopyroxene is slightly enriched in MgO, Na_2O (0.4 wt%), but impoverished in Al_2O_3 (1.4 wt%) (Table 3).

Amphibole is the most common included phase, in contrast with its exceptional occurrence as an interstitial mineral in chromitites. In all cases, it is a pargasitic hornblende (Table 3) characterized by high XMg ranging from 0.95 to 0.97. It is always K_2O-free; its Na_2O content (wt%) seems to vary from one complex to another: 2.3 (0.4) for Semail, 3.2 (0.3) for Vourinos 2.2 (0.8) for Tiébaghi and 3.0 (0.6) for the Massif du Sud. Pargasite from Semail, like the host chromite, shows the highest TiO_2 content: 1.1 (0.6) which decreases to 0.49 (0.27) for the Massif du Sud, 0.39 (0.12) for Tiébaghi and 0.27 (0.03) for Vourinos. The NiO concentration in pargasite from all the complexes studied averages 0.1 wt%. No Cl or F have been detected.

Sodium-phlogopite inclusions were observed in chromitites from the Massif du Sud, Tiébaghi and Semail (Table 3). In the latter complex, K-phlogopite was also found. Like all the ferromagnesian silicates included in chromite, these micas are characterized by high XMg (0.96–0.97). Their TiO_2 content is variable, ranging from 0.1 to 4.1 wt%. The Cr_2O_3 concentration is high, varying from 1.6 to 2.7 wt%. A relatively high NiO content, up to 0.9 wt% has also been observed in some grains. Except in the K end member, the Na/(Na + K) ratio of sodium phlogopite varies from 0.56 to 1. Microprobe analyses totals ($\sim$9.6%) confirm the absence of higher hydrates described by Carman (1974). Neither Cl nor F were detected.

Plagioclase was identified only in chromitites from Semail, where it is a common interstitial phase but has been found only once as an inclusion ($An_{91.4}$). Unlike the other silicates associated with chromitite, the plagioclase composition varies within

Table 3. Selected microprobe analyses of various silicate inclusions in chromitites[a]

	1	2	3	4	5	6	7	8	9	10	11	12	13	14	15	16	17
SiO_2	41.88	58.69	53.95	45.41	44.52	41.32	41.00	46.60	42.23	42.37	54.56	40.38	56.22	55.13	42.29	55.99	60.49
TiO_2	0.00	0.00	0.17	0.08	1.72	0.46	0.23	0.74	1.58	0.02	0.02	0.00	0.00	0.00	0.00	0.00	0.00
Al_2O_3	0.00	0.55	1.93	35.01	11.41	19.02	15.81	11.05	16.50	35.32	1.14	0.00	1.00	0.64	0.00	0.41	0.92
Cr_2O_3	0.73	0.91	1.30	0.81	2.54	1.77	2.25	2.93	2.29	0.75	1.59	0.38	2.05	1.62	0.58	0.53	0.95
FeO	2.50	4.65	1.63	0.29	2.93	2.10	1.19	1.64	1.23	0.22	1.10	2.98	2.51	0.90	2.31	0.64	3.44
MnO	0.04	0.15	0.03	0.00	0.06	0.12	0.00	0.04	0.00	0.04	0.00	0.03	0.13	0.00	0.12	0.00	0.00
MgO	54.70	36.34	17.44	0.05	19.07	25.05	26.16	20.34	25.96	0.03	17.20	54.70	37.40	18.97	54.29	18.41	34.69
CaO	0.00	0.13	24.51	17.83	11.75	0.10	0.39	12.07	0.05	0.75	24.63	0.06	0.19	23.36	0.04	23.99	0.43
Na_2O	0.00	0.02	0.00	0.93	3.60	6.97	1.19	2.29	6.20	20.21	0.42	0.00	0.00	0.18	0.00	0.03	0.02
K_2O	0.00	0.00	0.00	0.00	0.14	0.25	8.86	0.08	0.36	0.03	0.00	0.00	0.02	0.00	0.00	0.00	0.00
NiO	1.07	0.07	0.00	0.00	0.06	0.28	0.32	0.13	0.64	0.00	0.07	0.72	0.13	0.10	1.63	0.05	0.16
H_2O					2.11	4.51	4.37	2.17	4.50								
Total	100.92	101.51	100.96	100.41	99.91	101.95	101.77	100.09	101.54	99.74	100.73	99.25	99.65	100.90	101.26	100.05	101.10
Si	0.990	1.980	1.941	2.086	6.322	5.492	5.622	6.521	5.621	2.008	1.967	0.973	1.928	1.973	0.997	2.010	2.028
Al	—	0.022	0.082	1.896	1.910	2.980	2.555	1.822	2.589	1.973	0.048	—	0.041	0.027	—	0.017	0.036
Ti	—	—	0.005	0.003	0.184	0.046	0.024	0.077	0.158	0.001	—	—	—	—	—	—	—
Cr	0.014	0.024	0.037	0.030	0.286	0.186	0.244	0.324	0.240	0.028	0.045	0.007	0.055	0.046	0.011	0.015	0.025
Fe	0.049	0.131	0.049	0.011	0.348	0.234	0.136	0.192	0.136	0.009	0.033	0.060	0.072	0.027	0.046	0.019	0.096
Mn	0.001	0.004	0.001	—	0.008	0.013	—	0.005	—	0.002	—	0.001	0.004	—	0.002	—	—
Mg	1.928	1.827	0.935	0.004	4.036	4.963	5.347	4.242	5.512	0.002	0.924	1.965	1.912	1.012	1.908	0.985	1.734
Ca	—	0.005	0.945	0.878	1.788	0.015	0.058	1.809	0.007	0.038	0.951	0.001	0.007	0.896	0.001	0.923	0.015
Na	—	0.002	—	0.083	0.991	1.796	0.317	0.620	1.602	1.857	0.029	—	—	0.012	—	0.002	0.001
K	—	—	—	—	0.026	0.043	1.549	0.014	0.062	0.001	—	—	0.001	—	—	—	
Ni	0.020	0.002	—	—	0.007	0.030	0.035	0.016	0.068	—	0.002	0.014	0.003	0.003	0.031	0.001	0.004
OH					2.000	4.000	4.000	2.000	4.000								
Σ	3.002	3.997	3.995	4.991	17.906	19.798	19.887	17.642	19.995	5.919	3.999	3.021	4.023	3.996	2.996	3.972	3.939
XMg	0.975	0.933	0.950	—	0.921	0.955	0.975	0.957	0.974	—	0.965	0.971	0.964	0.974	0.976	0.981	0.947

[a] *1* Tiébaghi, *2–7* North Oman, *8–15* Massif du Sud, *16–17* Troodos, olivine: *1, 12, 15*; clinopyroxene: *3, 11, 14, 16*; orthopyroxene: *2, 13, 17*; pargasite: *5, 8*; sodium phlogopite: *6, 9*; phlogopite: *7*; plagioclase: *4*; nepheline: *10*.

a single deposit from $An_{81.0}$–$An_{97.2}$. Variations up to 4 mol% An were recorded at the scale of a thin section for unzoned crystals.

Nepheline inclusions were found in the GR2H deposit (Massif du Sud) where it occurs associated with pargasite and clinopyroxene (Johan 1986). It is K_2O-free and has a CaO content lower than 0.5 wt%.

3.2.3 Platinum-Group Minerals

The recent discovery of platinum-group minerals (PGM) as discrete inclusions in chromitite from the Troodos ophiolite (Constantinides et al. 1980) led us to undertake a systematic study of PGM in ophiolitic chromitites. Most of the results on individual ophiolite complexes have already been published (Johan and Legendre 1980; Legendre 1982; Johan et al. 1982; Augé 1985a, b, 1986; Legendre and Augé 1986; Johan 1986; Prichard et al. 1986; Prichard and Lord, this volume). The purpose of this chapter is to summarize and compare existing and new data (Table 4).

PGM were found in chromitites from all the studied ophiolites, except in the northern part of the Oman ophiolitic nappe. They occur as minute grains, from 1 to 10 μ in the largest dimension and are generally euhedral, but also form anhedral, dropletlike grains. PGM appear either as single crystals within the host chromite, remote from fractures, or as composite grains. The latter comprise the PGM + silicate association (rarely, PGM can be completely included in silicate), the PGM + base metal sulphide association, or an assemblage of two PGM which may be formed by sulphide + alloy, two different sulphides, two different PGE-sulpharsenides, etc. Contrary to occurrences in the Shetland ophiolite (Prichard et al. 1986), no PGM has been found interstitially to chromite.

Significant differences appear in the nature and composition of PGM from one ophiolite to another, as summarized in Table 5. In the Massif du Sud, Os–Ir–Ru alloys (mainly rutheniridosmine, rarely ruthenian osmium and osmiridium) predominate over PGE sulphides (laurite). However, a laurite crystal commonly occurs adjacent to a large PGE alloy. In chromite deposits from the southern part of the Oman ophiolite, as in the Troodos complex, laurite is the only PGM found. In the Tiébaghi massif, PGE sulphides have been found, including laurite (the most common), erlichmanite and two Ir–Cu sulphides. PGE alloys occurring in the Tiébaghi chromitite comprise ruthenosmiridium and osmiridium but they represent only 3% of the total PGM found. The PGM assemblage determined in the Vourinos chromitite consists of laurite (55%), alloys (30%; iridosmine, rutheniridosmine and osmiridium) and sulpharsenides (osarsite and irarsite).

The average composition of rutheniridosmine in the Massif du Sud is $(Os_{0.48}Ir_{0.32}Ru_{0.20})$. In the Vourinos and Tiébaghi chromitites these alloys are poorer in Ru, averaging $(Os_{0.57}Ir_{0.37}Ru_{0.06})$ for Vourinos (Fig. 2). Note that alloys impoverished in Ru co-exist with Ru sulphides (Fig. 3). Alloys in the Tiébaghi deposit contain a minor amount of Cu which was not detected in alloys from other complexes.

No major differences in laurite composition were recorded among the four ophiolites studied (Fig. 3). Except for Oman, where it tends to contain less Ru, laurite

Table 4. PGM assemblages in spinel from chromite deposits. Figures correspond to the number of grains seen

	TIEBAGHI	SEMAIL	VOURINOS	MASSIF DU SUD	TROODOS
RUTHENIRIDOSMINE (Os,Ir,Ru)			● 1	● 21	
IRIDOSMINE (Os,Ir)			● 3		
RUTHENOSMIRIDIUM (Ir,Os,Ru)	● 1				
OSMIRIDIUM (Ir,Os)	● 1		● 1	● 1	
RUTHENIAN OSMIUM (Os,Ru)				● 3	
LAURITE (RuS_2)	● 61	● 14	● 9	● 2	● 22
ERLICHMANITE (OsS_2)	● 10				
XINGZHONGITE (Ir,Cu)S	● 2				
SULPHIDE Ir, Cu $(Ir,Cu)_2S_3$	● 4				
OSARSITE (OsAsS)			● 1		
IRARSITE (IrAsS)			● 1		

from all other complexes shows the same compositional range, extending from rather pure RuS_2 towards a composition approaching $(Ru_{0.6}Os_{0.3}Ir_{0.1})S_2$. The maximum solubility of Ir in laurite corresponds approximately to 1.5 to 3.3 at% Ir. Note that like alloys, laurites from the Tiébaghi deposit contain Cu (up to 6.3 wt%, average 0.8 wt%). Relatively high concentrations of As, up to 5.5 wt%, were found in laurites from Oman.

Copper becomes a major element in two Ir-rich sulphides identified in the Tiébaghi chromitite, one corresponding to xingzhongite (Ir, Cu)S, and the other to an unnamed mineral with the ideal formula $(Ir, Cu)_2S_3$. This latter phase can also contain Rh (up to 12 wt%). PGM of similar composition was described by Stockman and Hlava (1984).

Table 5. Electron microprobe analyses of platinum-group mineral inclusions in chromitites[a]

	1	2	3	4	5	6	7	8	9	10	11	12
S[b]	34.34	31.82	28.54	24.18	16.77	20.89	0.03	0.00	0.06	0.00	35.05	35.55
As	0.00	0.18	0.35	0.00	0.00	0.00	0.00	0.00	0.00	0.10	0.00	0.00
Os	2.17	17.37	33.92	51.39	0.44	0.00	49.58	68.16	48.25	47.12	1.73	12.56
Ru	49.00	31.22	18.65	3.42	0.02	0.00	10.17	13.27	6.60	10.78	54.09	40.23
Ir	3.15	12.48	9.26	9.52	49.34	57.65	32.91	10.43	31.89	37.37	1.34	6.81
Rh	2.30	0.00	0.00	0.11	0.21	0.20	0.30	0.14	0.15	0.53	0.18	0.12
Pt	0.00	0.00	0.00	0.00	0.44	1.73	0.00	0.00	0.00	0.44	0.00	0.00
Pd	0.00	0.11	0.00	0.07	0.08	0.04	0.00	0.00	0.00	0.00	0.00	0.00
Ni	0.17	0.09	0.79	0.06	0.09	0.41	0.00	0.42	0.05	0.10	0.05	0.02
Cu	0.00	0.10	1.48	0.00	12.48	9.58	0.00	0.00	0.00	0.00	0.00	0.00
Fe	1.42	0.65	1.32	1.48	4.60	1.20	0.84	0.93	1.63	0.75	0.70	0.48
Cr	5.07	2.59	3.57	7.32	25.49	6.06	2.68	0.98	7.73	1.17	2.26	1.25
Total	97.62	96.60	97.89	97.54	109.96	97.75	96.51	94.34	96.36	98.36	95.40	97.02
S[c]	37.69	34.08	30.63	27.24	21.00	23.08	0.08	—	0.04	—	37.92	37.31
As	—	0.19	0.38	—	—	—	—	—	—	0.10	—	—
Os	2.38	18.60	36.40	57.90	0.55	—	50.87	72.40	55.45	48.69	1.87	13.18
Ru	53.77	33.44	20.02	3.86	0.03	—	12.45	15.15	7.58	11.14	58.52	42.22
Ir	3.45	13.37	9.94	10.72	61.78	63.70	36.27	11.14	36.64	38.61	1.45	7.15
Rh	2.52	—	—	0.13	0.26	0.23	0.32	0.05	0.17	0.55	0.19	0.12
Pt	—	—	—	—	0.55	1.91	—	—	—	0.46	—	—
Pd	—	0.11	—	0.08	0.10	0.04	—	—	—	—	—	—
Ni	0.19	0.10	0.85	0.07	0.11	0.45	—	0.44	0.06	0.10	0.05	0.02
Cu	—	0.10	1.58	—	15.63	10.58	—	—	—	—	—	—
Fe	—	—	0.19	—	—	—	—	0.82	—	0.35	—	—
S[d]	66.57	67.79	66.14	67.91	53.09	58.16	0.41	—	0.37	—	66.38	68.88
As	—	0.16	0.35	—	—	—	—	—	—	0.23	—	—
Os	0.71	6.24	13.25	24.33	0.23	—	45.73	62.28	51.89	43.83	0.55	4.10
Ru	30.13	21.10	13.71	3.05	0.02	—	21.06	24.53	13.53	18.87	32.49	24.72
Ir	1.02	4.43	3.58	4.46	26.05	26.77	32.26	9.48	33.92	34.39	0.42	2.20

Table 5. (*continued*)

	1	2	3	4	5	6	7	8	9	10	11	12
Rh	1.39	—	—	0.10	0.21	0.18	0.53	0.08	0.29	0.91	0.10	0.07
Pt	—	—	—	—	0.23	0.79	—	—	—	0.40	—	—
Pd	—	0.07	—	0.06	0.08	0.03	—	—	—	—	—	—
Ni	0.18	0.11	1.00	0.09	0.16	0.62	—	1.23	0.18	0.29	0.05	0.02
Cu	—	0.11	1.73	—	19.93	13.46	—	—	—	—	—	—
Fe	—	—	0.24	—	—	—	—	2.39	—	1.08	—	—

[a] Laurite: *1*, *2*, *11*, *12*; erlichmanite: *3*, *4*; xingzhongite: *5*; Ir–Cu sulphide *6*; Ru–Ir–Os alloys: *7* to *10*. Analyses *1* to *6* are from Tiébaghi; *7* to *10*: Massif du Sud; *11* and *12* are from Troodos.

[b] Microprobe analyses in wt%.

[c] Analyses recalculated to 100% after subtraction of chromite.

[d] Composition in at% obtained from recalculated analyses.

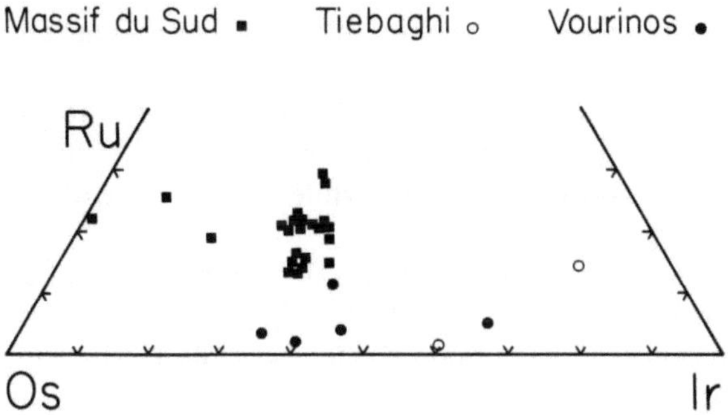

Fig. 2. Chemical composition of PGE alloys (atomic proportion of Ru, Os, Ir) from Massif du Sud, Tiébaghi and Vourinos chromitites

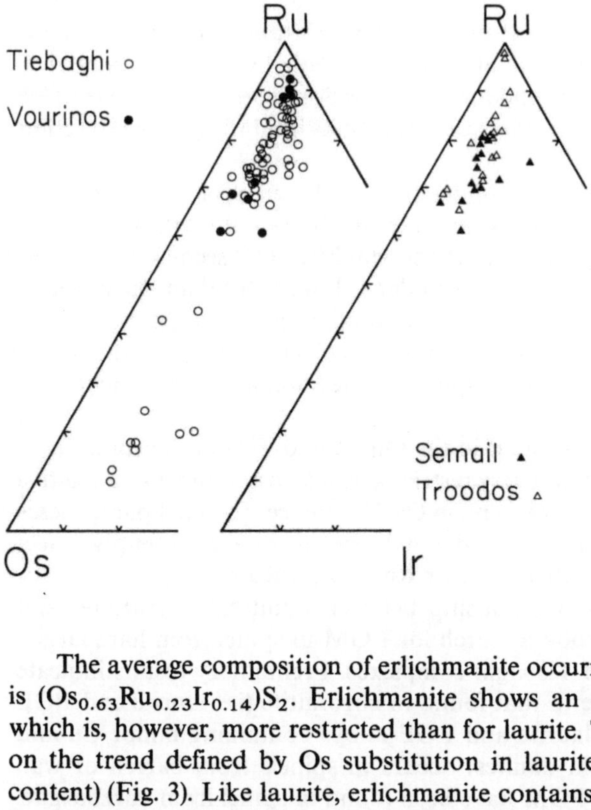

Fig. 3. Chemical composition of PGE sulphide (laurite and erlichmanite; atomic proportion of Ru, Os, Ir) from Tiébaghi, Vourinos, Semail and Troodos chromitites

The average composition of erlichmanite occurring in the Tiébaghi chromitite is $(Os_{0.63}Ru_{0.23}Ir_{0.14})S_2$. Erlichmanite shows an extensive Os–Ru substitution which is, however, more restricted than for laurite. The erlichmanite analyses plot on the trend defined by Os substitution in laurite (with a low and constant Ir content) (Fig. 3). Like laurite, erlichmanite contains minor Cu (up to 3.1 wt%), As (average 0.44 wt%), Rh (0.23), Pd (0.15) and Ni (0.20 wt%).

The composition of single PGM grains shows no differences from that of PGM in composite inclusions. Note that xingzhongite has been found everywhere associated with erlichmanite.

In conclusion, we can stress the fact that only minor differences in composition of a given PGM were observed from one complex to another. On the other hand, major differences appear in PGM parageneses, characterized by the predominance of alloys in the Massif du Sud, the association of alloys with laurite in Vourinos, and the predominance of sulphides in Oman, Troodos and Tiébaghi, with a large variety for the latter (laurite, erlichmanite and rare Ir-bearing sulphides).

The difference in PGM assemblages in Vourinos and Tiébaghi chromitites contrasts with the similarity of chromite compositions. On the other hand, similar PGM associations were observed in Troodos and Oman, while the chromite composition in these two ophiolites is very different.

Thus, no relationship appears to exist between chemical characteristics of a PGM paragenesis and the host chromite. Furthermore, PGM may exhibit large compositional variations within a single chromite orebody (for example, the laurite composition in the Tiébaghi deposit).

It is now widely accepted that PGM were trapped during the growth of the chromite crystals. Consequently, their composition and paragenesis are very useful to estimate depositional conditions. However, an important parameter remains unknown, that of the initial PGE content of the magmatic liquid. The abundance of Ru, Os and Ir-bearing minerals could indicate a predominance of these elements over Rh, Pd and Pt but it could also reflect the conditions inhibiting their extraction from the magma.

Considering sulphide-alloy equilibrium curves (Toma and Murphy 1977; Stockman and Hlava 1984), it becomes possible to obtain comparative data on temperature and sulphur activity during PGM crystallization. It seems obvious that chromitites containing only alloys evolved under a lower aS_2 than those where PGM sulphides and alloys co-exist or even those containing only sulphides. Hence, in terms of increasing aS_2, the Massif du Sud must be placed before Vourinos. The presence of erlichmanite and Ir-bearing sulphide in Tiébaghi indicates a higher aS_2 for which PtS should be stable.

The fact that only laurite was found in the Oman and Troodos chromitites is difficult to interpret. It may indicate a very restricted range of aS_2 and T or a relative impoverishment in Ir and, to a lesser extent, in Os. Differences in PGM parageneses between Vourinos, Troodos, Massif du Sud and Tiébaghi could also simply reflect the variation in aS_2 for similar initial Os/Ir, Ir/Ru and Os/Ru ratios.

In order to understand the relationship between dunites, harzburgites and chromitites, Augé (1988) undertook a search for PGM in spinel from harzburgite and dunite in the Vourinos and Tiébaghi complexes. Preliminary results indicate that a similar PGM assemblage to that found in chromitite (i.e. laurite and PGE alloys in the Vourinos dunite, laurite and PGE alloys containing minor Cu and Ir–Cu sulphides in the Tiébaghi dunites) occurs in spinels from barren or wall dunites. On the other hand, no PGM have been found in spinel from harzburgite. These observations led Augé (1988) to conclude that dunite and chromitite originate from similar liquids and that massive chromitites could be formed by accumulation of spinel disseminated in dunite.

4 Discussion

The comparative study of chromitites, the results of which are reported here, reveals similar compositions of chromite ores, except two populations from the Semail ophiolite, exhibiting much lower Cr/Al ratios (Fig. 1). The study of mantle sequences (Johan and Augé 1986) showed that Oman, Vourinos, Troodos and Massif du Sud ophiolites are characterized by strongly depleted harzburgites, unlike Tiébaghi which has a high proportion of spinel- and plagioclase lherzolites. Assuming the magmatic origin of chromite orebodies, or at least their genetic relationship with a high-temperature ore-forming system directly related to the evolution of a magma chamber, different types of liquids must have existed, generating chromite deposits in the uppermost part of the mantle sequence and near the base of the cumulate series within the Semail ophiolite. According to Ohnenstetter (1985), Al-rich chromitites are related to magmatic liquids rich in Al_2O_3 and TiO_2, giving rise to cumulates with early plagioclase. The presence of interstitial plagioclase in chromitites from Semail only, and the presence of exsolved rutile in massive chromite from Semail deposits, confirm Ohnenstetter's conclusions. It is noteworthy that the Semail chromitites are impoverished in platinum-group element concentration with respect to the other deposits studied here.

The origin of dunites within the harzburgitic upper mantle is still poorly understood. For the Tiébaghi complex, the extent of Cr/Al spinel variation in some dunites shows a remarkable correlation with that in associated harzburgites (Johan and Augé 1986). On the other hand, most wall dunites and some barren dunitic bodies contain spinels whose composition is very close to that of chromitites.

It seems then that the ophiolitic mantle sequences contain several types of dunite of different origin, though they are difficult to distinguish. A metasomatic origin, in the presence of a fluid phase, could be evoked for some of them, others could be genetically related to penetrating magmatic liquids, still others could have a high-temperature hydrothermal origin (Bowen and Tuttle 1949; Dungan and Avé Lallemant 1977; Johan et al. 1983).

The paragenesis of silicate inclusions in chromitites is remarkable and enables a better understanding of the unusual chemical and thermodynamic characteristics of the ore-forming system. First, the presence of sodium-rich hydrous phases indicates a high Na activity in the system and implies a temperature limit corresponding to the maximum thermal stability of pargasite and sodium phlogopite. Furthermore, the abundance of clinopyroxene in inclusions is evidence of high calcium activity. Note that the association of clinopyroxene with pargasite and sodium phlogopite indicates a lower temperature of crystallization than that which would be expected for wall dunites and interstitial olivine. Generally, the assemblage of included silicates is entirely different from that of interstitial phases, which again reflects the unusual chemical character of the ore-forming system. Consequently, the succession of phenomena leading to the formation of a chromite orebody within the mantle sequence is of extreme importance. In some deposits of the Massif du Sud, Johan (1986) observed chromite deposition slightly later than wall dunites and reported the existence of late dunite dikes cross-cutting and metasomatically replacing the mineralization. There is no difference in the chemical composition of mineral phases between these two dunite generations.

As mentioned above, there is strong evidence of a very early crystallization of PGM before the precipitation of chromite and probably even earlier than the crystallization of some silicates included in chromite. This is demonstrated by laurite inclusions in orthopyroxene included in chromite (Legendre 1982). Considering the relatively high temperature of crystallization of orthopyroxene En_{94} with respect to other included silicates, especially hydrous phases (pargasite, sodium phlogopite), the PGM regardless of their nature (alloys or laurite) appear to be formed before the bulk of silicates included in chromite.

The commonly observed association PGM + silicate could result from mechanical trapping of PGM crystals during the crystallization of silicates. Note that the euhedral habit of PGM contrasts with dropletlike sulphide assemblages in places associated with PGM. This variation of shape could reflect much lower crystallization temperatures of base metal sulphides which might still be liquid when incorporated by growing chromite crystals. In this case, these droplets of sulphide liquid armoured by chromite may represent closed microsystems undergoing extensive subsolidus evolution.

Legendre's (1982) observation that laurite is late with respect to Os–Ir–Ru alloys, and important variations in the nature of PGM from one deposit to another reported above (Augé 1988), indicate the temperature and/or the S_2-activity variation in the ore-forming system. It is interesting to note that in the Tiébaghi chromitites, which originated from a system characterized by relatively high sulphur fugacity compared to other deposits studied, Os-rich laurite and erlichmanite occur, which are compatible with lower temperatures of crystallization (Stockman and Hlava 1984). Taking into account the values of free energy of formation for PGE sulphides (Toma and Murphy 1977), namely for RuS_2, it appears that laurite becomes stable slightly below 1100 °C. This temperature is compatible with the maximum thermal stability of pargasite and sodium phlogopite (1050 °C) (Holloway 1973; Carman 1974) but incompatible with the crystallization temperature of olivine Fo_{95} or orthopyroxene En_{94}, later than PGM, from a magmatic liquid undersaturated in water.

The discovery of PGM in disseminated spinel from dunite in harzburgite (Augé 1988) undeniably indicates a convergence between processes generating some of the dunites, on the one hand, and the ore-forming system, on the other hand. Besides the presence of PGM, the abundance of exsolved clinopyroxene in olivine of wall dunites and of clinopyroxene inclusions in associated chromitites also constitutes an important common denominator (Johan 1986).

Acknowledgements. This work has benefitted from discussion with A.J. Naldrett, University of Toronto and D.H. Watkinson, Carleton University, Ottawa. J. Kemp, BRGM, improved the English of the manuscript. We express our gratitude to the E.E.C. Commission which provided the financial support for this research project (*Contract No. MSM-023-F*).

References

Augé T (1985a) Platinum-group-mineral inclusions in ophiolitic chromitite from the Vourinos complex, Greece. Can Miner 23: 163–171

Augé T (1985b) Chromitites et minéraux du groupe du platine dans les complexes ophiolitiques. Caractérisation des séries hôtes. Thesis Univ Orléans, p 270

Augé T (1986) Platinum-group-mineral inclusions in chromitites from the Oman ophiolite. Bull Minér 109:301–304

Augé T (1987) Chromite deposits in the northern Oman ophiolite: mineralogical constraints. Miner Dep 21:1–10

Augé T (1988) Platinum-group minerals in the Tiébaghi and Vourinos ophiolitic complexes: genetic implications. Can Miner 26: in press

Beccaluva L, Ohnenstetter D, Ohnenstetter M, Paupy A (1984) Two magmatic series with island arc affinities within the Vourinos ophiolite. Contrib Miner Pet 85:253–271

Bowen NL, Tuttle OF (1949) The system $MgO–SiO_2–H_2O$. Geol Soc Am Bull 60:439–460

Brunn JH (1956) Etude géologique du Pinde septentrional et de la Macédoine occidentale. Ann Géol Pays Hellén 7:358

Carman JH (1974) Synthetic sodium phlogopite and its two hydrates: Stabilities, properties and mineralogic implications. Am Miner 59:261–273

Cassard D, Nicolas A, Rabinovitch M, Moutte J, Leblanc M, Prinzhofer A (1981) Structural classification of chromite pods in southern New Caledonia. Econ Geol 76:805–831

Christiansen FG (1986) Structural classification of ophiolitic chromite deposits. In: Gallagher MJ et al. (eds) Metallogeny of basic and ultrabasic rocks. The Institution of Mining and Metallurgy, London, pp 279–290

Constantinides CC, Kingston GA, Fisher PC (1980) The occurrence of platinum group minerals in the chromitites of the Kokkinorotsos chrome mine, Cyprus. In: Panayiotou A (ed) Ophiolites. Proc Int Ophiolite Symp Cyprus, pp 93–101

Dungan MA, Avé Lallemant HG (1977) Formation of small dunite bodies by metasomatic transformation of harzburgite in the Canyon Mountain ophiolite, northeast Oregon. Oreg Dept Geol Miner Ind Bull 96:109–128

Greenbaum D (1977) The chromitiferous rocks of the Troodos ophiolite complex, Cyprus. Econ Geol 72:1175–1194

Guillon JH (1975) Les massifs péridotitiques de Nouvelle-Calédonie. Type d'appareil ultrabasique stratiforme de chaîne récente. Mém ORSTOM 76:120

Harkins ME, Green HW II, Moores EH (1980) Multiple intrusive events documented from the Vourinos ophiolite complex, northern Greece. Am J Sci 280A:284–295

Holloway JR (1973) The system pargasite-$H_2O–CO_2$: a model for melting of a hydrous mineral with a mixed-volatile fluid. I. Experimental results to 8 kbar. Geochim Cosmochim Acta 37:651–666

Johan Z (1986) Chromite deposits in the Massif du Sud ophiolite, New Caledonia: genetic considerations. In: Chromites. Theophrastus, Athens, pp 311–339

Johan Z, Augé T (1986) Ophiolitic mantle sequences and their evolution: mineral chemistry constraints. In: Gallagher MJ et al. (eds) Metallogeny of basic and ultrabasic rocks, The Institution of Mining and Metallurgy, London, pp 305–317

Johan Z, Legendre O (1980) Minéralogie des platinoïdes dans les chromites massives du feuillet ophiolitique de la Nouvelle-Calédonie. Rés Sci Tech SGN, BRGM, Orléans, p 80

Johan Z, Dunlop H, Le Bel L, Robert JL, Volfinger M (1983) Origin of chromite deposits in ophiolitic complexes: evidence for a volatile- and sodium-rich reducing fluid phase. Fortschr Miner 61:105–107

Legendre O, Augé T (1986) Mineralogy of platinum-group-mineral inclusions in chromitites from different ophiolitic complexes. In: Gallagher MJ et al. (eds) Metallogeny of basic and ultrabasic rocks, The Institution of Mining and Metallurgy, London, pp 361–372

Lehmann J (1983) Diffusion between olivine and spinel: application to geothermometry. Earth Planet Sci Lett 64:123–138

Moores EM (1969) Petrology and structure of the Vourinos ophiolitic complex of northern Greece. Geol Soc Am Spec Pap 118:74

Moutte J (1982) Chromite deposits of the Tiébaghi ultramafic massif, New Caledonia. Econ Geol 77:576–591

Nicolas A, Dupuy C (1984) Origin of ophiolitic and oceanic lherzolites. Tectonophysics 110:177–187

Nicolas A, Prinzhofer A (1983) Cumulative or residual origin for the transition zone in ophiolites: structural evidence. J Pet 24:188–206

Noiret G, Montigny R, Allegre CJ (1981) Is the Vourinos complex an island arc ophiolite? Earth Planet Sci Lett 56:375–386

Ohnenstetter M (1985) Classification pétrographique et structurale des ophiolites, écho de la dynamique des zones de transition croûte-manteau. Incidence sur la nature et la disposition des corps de chromite associés. CR Acad Sci Paris 301:1413–1418

Paris JP (1981) Géologie de la Nouvelle-Calédonie: un essai de synthèse, vol 113. Mém BRGM, p 278

Prichard HM, Neary CR, Potts PJ (1986) Platinum-group minerals in the Shetland ophiolite. In: Gallagher MJ et al. (eds) Metallogeny of basic and ultrabasic rocks. The Institution of Mining and Metallurgy, London, pp 395–414

Prinzhofer A, Nicolas A (1980) The Bogota peninsula, New Caledonia: a possible oceanic transform fault. J Geol 88:387–398

Prinzhofer A, Nicolas A, Cassard D, Moutte J, Leblanc M, Paris JP, Rabinovitch M (1980) Structures in the New Caledonia peridotite-gabbros; implications for oceanic mantle and crust. Tectonophysics 69:83–112

Roberts S (1986) The role of igneous processes in the formation of ophiolitic chromitite. Thesis, The Open University, Milton Keynes, p 261

Rodgers KA (1976) Ultramafic and related rocks from southern New Caledonia. Bull BRGM sect IV 1:33–55

Smewing JD (1980) Regional setting and petrological characteristics of the Oman ophiolite in North Oman. Ofioliti Spec Issue 2:335–377

Stockman HW, Hlava PF (1984) Platinum-group minerals in alpine chromitites from southwestern Oregon. Econ Geol 79:491–508

Toma SA, Murphy S (1977) The composition and properties of some native platinum concentrates from different localities. Can Miner 15:59–69

Vine FJ, Paster CK, Gass IG (1973) Aeromagnetic survey of the Troodos igneous massif, Cyprus. Nature Phys Sci 244:34–38

The Shetland Ophiolite: Evidence for a Supra-Subduction Zone Origin and Implications for Platinum-Group Element Mineralization

H.M. PRICHARD and R.A. LORD[1]

Abstract

Chromite-rich samples from the ultrabasic and basic igneous complex in the Shetland Islands contain exceptionally rich PGE concentrations previously unrecorded in other ophiolite complexes and geochemically similar to those in stratiform complexes. This chapter re-assesses the Shetland complex, with particular emphasis on the chrome spinels and dykes, confirming that it forms the lower part of a Penrose ophiolite assembly. The podiform chromite is shown to have ophiolitic compositions and new electron microprobe analyses from chromite-rich samples indicate that there is nothing unusual about the compositions of the chrome spinels in PGM-rich samples, which have similar compositions to barren samples. The uppermost part of the ophiolite complex consists of a swarm of dykes in gabbro and their geochemistry has boninitic affinities possibly indicating an origin for the complex from a hydrous magma source in a supra-subduction zone (SSZ). The relative roles of primary and secondary fluids in the concentration of the PGE in Shetland are difficult to determine because the chromite-rich lithologies characteristically lie in altered, highly serpentinized zones. Dunites associated with some of the PGE-rich chromite in lenses in the mantle harzburgite have been found to be PGE-rich and significantly these are some of the least altered of the ultramafic rocks. This suggests that the PGE concentrations are not solely related to alteration zones or secondary processes.

1 Introduction

The presence of platinum-group elements (PGE) in the basic and ultrabasic igneous complex on Unst and Fetlar (Shetland Isles, NE of Scotland) was first recorded in assays of heavy mineral concentrates taken from the chromite ore crushing mill on Unst (Hitchen 1929). More recently, during a re-evaluation of chromite deposits on Unst (Gass et al. 1982), Ru-, Ir- and Os-bearing platinum group minerals (PGM)

[1] Department of Earth Sciences, The Open University, Walton Hall, Milton Keynes, Buckinghamshire MK7 6AA, UK

Mineral Deposits within the European Community
(ed. by J. Boissonnas and P. Omenetto)

were discovered in chromite-rich samples (Prichard et al. 1981). This led to renewed interest in the PGE mineralization and a systematic study of PGM in chromite-rich samples from the complex was undertaken for the EC Raw Materials Programme (1982–1985). This research benefited from the use of beta-autoradiography, a new technique capable of systematically locating grains of selected PGM in polished thin sections (Potts 1984, 1986). Two localities were discovered, at Cliff and Harold's Grave, that are exceptionally rich in the number and diversity of PGM (Prichard et al. 1984; Neary et al. 1984). These discoveries were subsequently confirmed in a survey by the British Geological Survey (Gunn et al. 1985).

Whole rock PGE analyses of selected samples have established that the concentration of each of the PGE (Os, Ir, Ru, Rh, Pt, Pd) exceeds 1 ppm, and a detailed mineralogical investigation has led to the identification of a large number of PGM in which all six PGE singly form the dominant component of individual PGM (Prichard et al. 1986). These results are quite unexpected in relation to previous studies of ophiolite complexes which are traditionally thought to be solely Ru-, Ir- and Os-bearing with whole rock PGE concentrations two to three orders of magnitude lower than those found in Shetland (Constantinides et al. 1980; Legendre 1982; Page et al. 1982a and b, 1984; Agiorgitis and Wolf 1978; Augé 1985). Indeed, the high values of Pt, Pd and Rh found in Shetland are much more typical of layered stratiform complexes such as the Bushveld (Prichard et al. 1986).

The geology of the basic and ultrabasic complex on Unst and Fetlar was described as early as 1821 by Hibbert, with further work by Heddle (1879) and Phillips (1927). The complex was mapped by Read (1934) and further studies on the igneous petrology of the complex were published by Amin (1954) and Flinn (1970). The suggestion that the complex is ophiolitic was first made by Garson and Plant (1973) in a review of ultramafic rocks in Scotland. More recently the area has been studied in detail to evaluate the ophiolitic nature of the complex (Gass et al. 1982; Prichard 1985; Flinn 1985). In view of the PGE discoveries, which are thought to be unusual for an ophiolite complex, it is appropriate to review the evidence for an ophiolitic origin for this basic and ultrabasic complex. Geochemical evidence published here for the first time is presented to support this proposition and further to suggest that the complex formed in a supra-subduction zone (SSZ) environment.

2 The Ophiolite Complex

Based on its igneous stratigraphy it is possible to identify the complex as the lower part of an ophiolite as defined by the GSA Penrose Conference on ophiolites in 1972. Tectonized harzburgite of the mantle sequence is overlain by a crustal sequence. Dunite grades up through wehrlites and clinopyroxenites and above these are gabbros intruded at their highest exposed level by a dyke swarm (Figs. 1 and 2). The ophiolitic nature of the complex is substantiated by the geology and geochemistry of (1) the dykes and (2) the chromite.

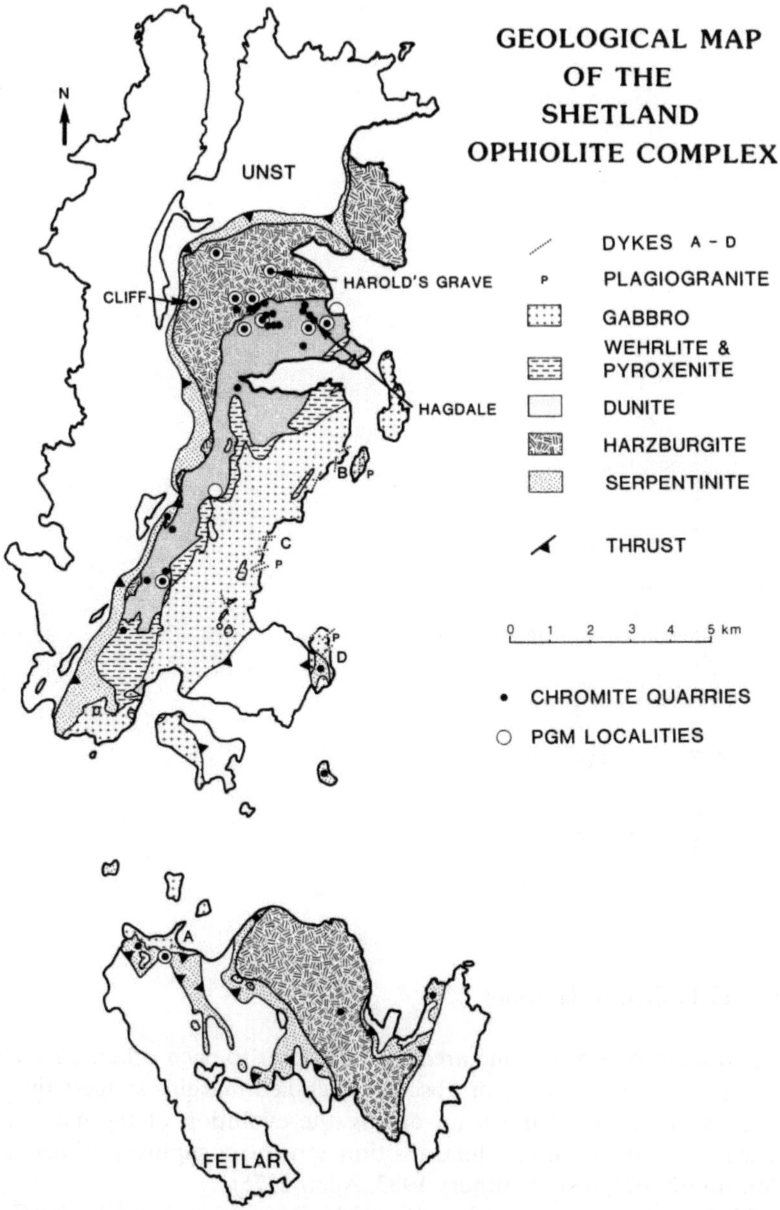

Fig. 1. Geological map of the Shetland ophiolite complex showing the location of the dykes (A–D) and disused chromite quarries. Localities where PGM have been discovered are also shown

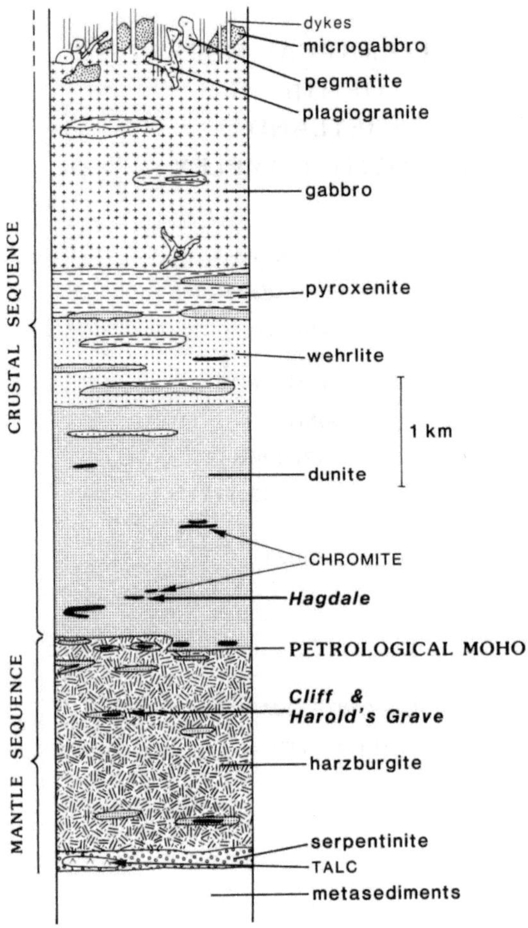

Fig. 2. Geological column showing the relationship of the lithologies within the ophiolite complex and the stratigraphical position of the rich PGE concentrations

2.1 The Dykes and Their Geochemistry

The dykes occur in gabbro and in some areas are adjacent to each other. Cross-cutting relationships and the presence or absence of chilled margins suggest that there were several generations of intrusion during the evolution of the gabbro complex. These features correspond to the transition zone from gabbro to sheeted dykes in other ophiolite complexes (Rothery 1983; Allen 1975).

Samples of 23 dykes from four localities (Fig. 1A–D) were analyzed by XRF. The method used is described by Potts et al. (1984) and the results are listed in Table 1. Selected data for the elements Ti, Zr, Y and Cr are plotted on the 'classical' discrimination diagrams of Pearce et al. (1981) (Fig. 3). These diagrams delineate the fields of island arc, within-plate and mid-ocean ridge type basalts. The Shetland dyke data plot mainly in the island arc field depleted in Ti, Zr and Y. The most convincing discrimination is in the diagram of Cr versus Y where samples are characterized by a wide range of Cr values and relatively constant Y. If decreasing

Table 1. Major and trace element analyses of dykes from Shetland[a]

Dykes from locality A				Dykes from locality B						
Sample No.	df 2	df 4	df 5	du10	du11	du12	du13	du14	du15	du16
SiO_2	48.36	52.51	55.84	51.49	50.48	50.62	53.91	51.76	52.17	55.02
TiO_2	0.83	0.37	0.50	0.69	0.67	0.51	0.71	0.81	0.66	0.39
Al_2O_3	14.83	19.89	17.88	16.79	17.69	18.29	19.41	15.48	17.14	18.93
Fe_2O_3	10.61	7.76	8.78	7.89	7.99	7.03	6.68	8.93	7.77	6.58
MnO	0.19	0.10	0.13	0.11	0.11	0.12	0.13	0.16	0.12	0.09
MgO	8.72	4.96	3.56	5.84	5.49	6.07	3.61	5.86	5.52	4.34
CaO	10.64	8.99	7.24	9.39	8.93	9.62	7.32	8.95	9.10	6.82
Na_2O	3.11	5.12	4.92	4.92	4.90	4.80	6.37	4.53	5.22	5.94
K_2O	0.03	0.01	<	<	<	0.01	<	0.02	0.05	<
P_2O_5	0.08	0.09	0.08	0.06	0.10	0.06	0.16	0.04	0.07	0.03
S	<	0.03	0.04	1.65	1.37	0.99	0.70	0.58	0.91	<
L.o.i.	1.85	1.97	2.11	2.39	2.38	2.53	2.28	1.60	2.27	1.78
Total	99.25	101.80	101.08	101.22	100.11	100.65	101.28	98.72	101.00	99.92
Rb	2	2	3	<	2	2	2	2	2	2
Sr	120	170	202	194	216	191	216	161	188	117
Y	19	9	12	16	12	7	24	19	12	7
Zr	50	54	61	45	58	42	115	64	53	59
Nb	3	2	3	2	3	3	3	3	3	3
Cu	31	36	21	206	155	68	109	67	59	46
Zn	73	50	67	45	54	46	67	63	53	42
Ni	130	38	1	32	26	49	31	61	49	21
Cr	378	30	24	59	71	90	54	140	83	37

Table 1. (*continued*)

Dykes from locality C						Dykes from locality D							
Sample No.	du2	du3	du4	du5	du6	du17	du18	du19	du20	du21	du22	du23	du24
SiO_2	49.10	48.16	49.09	50.43	53.68	48.72	47.71	51.99	50.23	52.81	54.50	49.99	50.96
TiO_2	0.79	0.63	0.61	0.37	0.27	0.15	0.12	0.29	0.22	0.32	0.42	0.39	0.20
Al_2O_3	15.96	15.34	15.02	16.12	11.88	8.66	9.41	12.67	11.86	15.04	14.67	16.22	11.55
Fe_2O_3	8.72	9.60	9.20	9.30	9.49	10.56	10.72	9.57	10.10	10.14	10.00	9.73	9.97
MnO	0.14	0.16	0.15	0.15	0.18	0.19	0.21	0.16	0.18	0.11	0.12	0.16	0.20
MgO	6.18	9.17	8.93	8.02	9.76	16.96	17.65	9.93	12.37	7.98	5.15	6.52	14.89
CaO	10.43	10.28	10.03	9.52	9.74	9.00	8.37	7.45	8.00	8.54	9.13	9.70	7.13
Na_2O	3.69	3.12	3.19	3.78	3.68	1.58	1.39	3.88	3.14	1.98	0.89	3.28	2.78
K_2O	0.10	0.14	0.05	<	0.03	0.10	0.02	0.30	0.01	<	0.04	<	<
P_2O_5	0.09	0.06	0.04	0.08	0.06	0.04	0.03	<	0.03	<	0.04	<	<
S	1.60	0.72	0.97	0.03	0.02	<	<	<	0.02	0.02	0.30	0.02	<
L.o.i.	2.47	2.88	2.63	2.39	1.18	3.16	3.71	2.20	2.71	3.33	2.88	2.41	3.53
Total	99.27	100.26	99.91	100.19	99.97	99.12	99.34	98.17	98.87	100.27	98.14	98.42	101.21
Rb	4	4	<	2	3	4	2	2	3	2	4	2	2
Sr	179	147	154	137	84	6	10	48	42	154	188	103	16
Y	21	15	13	8	7	4	3	7	6	7	10	8	5
Zr	55	46	56	29	24	24	21	28	28	27	36	29	23
Nb	2	2	1	3	3	3	2	3	3	3	2	3	3
Cu	104	74	36	68	33	3	5	30	10	58	69	170	33
Zn	57	69	52	65	65	76	97	61	71	51	52	69	83
Ni	54	156	27	72	73	451	457	186	285	42	38	37	311
Cr	135	444	542	330	680	1827	1896	676	1167	173	155	130	1400

[a] Localities of dykes (A–D) are shown in Fig. 1. Analytical results were obtained using energy-dispersive XRF at the Open University following the procedures of Potts et al. (1984). Major elements are given as wt.% oxides and trace elements are in ppm; < below detection limit. L.o.i = Loss on ignition.

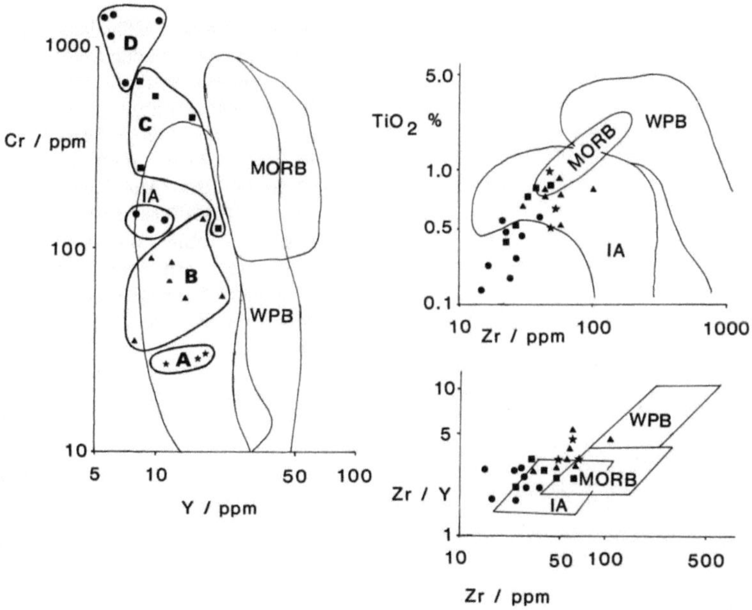

Fig. 3. The geochemistry of the dykes is illustrated on TiO_2, Zr, Y and Cr discrimination diagrams (After Pearce et al. 1981). Dyke localities are shown in Fig. 1 (*A* = ★, *B* = ▲, *C* = ■, *D* = ●). *IA* Island arc; *MORB* mid-ocean ridge basalt; *WPB* within plate basalt

Cr content is taken to represent increasing fractionation of the magma forming the dyke, then the results in Fig. 3 indicate a wide range of magma differentiation. It is also apparent that dykes from the same locality give analyses which cluster in distinct subfields. Some dykes appear to be particularly basic and Y-depleted and plot outside the island arc field. These dykes are very altered, but still contain chrome spinels with high Cr/(Cr + Al) ratios (ca. 0.8). They have high Mg, Cr, Ni and Si contents but low Ti and Zr and therefore have a boninitic composition. Such dykes have been described from other ophiolite complexes. Their compositions are similar to those from the Betts Cove ophiolite complex (Coish et al. 1982), which is considered to have SSZ characteristics (Pearce et al. 1984). Boninitic dykes have also been described from the Troodos ophiolite complex, Cyprus (Murton 1986). The geochemistry of these Shetland dykes suggests an island arc-type ophiolite (Pearce et al. 1981). Based on their geochemistry this type of ophiolite was redefined as a SSZ type forming by hydrous melting above a subduction zone (Pearce et al. 1984).

2.2 The Chromite

The ultrabasic part of the complex contains chromite-rich lenses. Mining records suggest that at least 52,000 tons of chromite ore have been quarried from chromite-rich lenses within the lower ultramafic parts of the complex, principally on the island

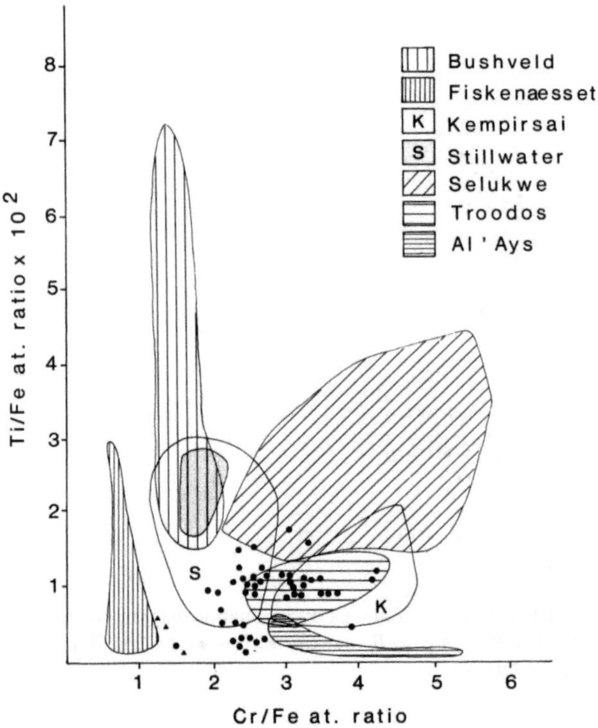

Fig. 4. A comparison of Shetland electron microprobe analyses of chrome spinels with those from ophiolite and stratiform complexes is shown on a graph of Ti/Fe × 10^2 atomic ratio vs. Cr/Fe atomic ratio (After Neary and Brown 1979). The Shetland analyses of chrome spinels in chromite-rich samples from disused quarries (●) and chrome spinels from the dykes (▵) are shown

of Unst. Exploration and quarrying for chromite have been documented by Hitchen (1929), Rivington (1953) and Johnson et al. (1980). This chromite has been described as ophiolitic (Prichard and Neary 1981, 1982) on two main lines of evidence. Firstly, the chromite-rich layers form discontinuous lenses or pods either surrounded by envelopes of dunite in the harzburgite or in dunite overlying the harzburgite. Such lenses, grading from massive to disseminated chromite, are described as podiform (Thayer 1960; Dickey 1975) and are typical of ophiolite complexes. Secondly, a number of geochemical parameters have been used to distinguish ophiolitic chromite from other types, including stratiform complexes (Neary 1974; Dickey 1975). The Shetland chromites have similar compositions to other ophiolitic chromites (Prichard and Neary, 1981, 1982). These conclusions are supported by Data plotted in Fig. 4. Ratios of the elements Ti/Fe and Cr/Fe are plotted on the fields of data shown by Neary and Brown (1979) to discriminate between chromite from different geological settings. Data from Shetland plotted on this diagram substantially overlap the results of Neary and Brown from the Troodos ophiolite complex (Cyprus). Furthermore, the Shetland values are quite distinct from the field of data for the Bushveld, a stratiform-type deposit. Chromite-rich samples from Shetland were divided into PGM-rich, PGM-bearing and barren using beta-autoradiography to systematically locate PGM. Chromite grains from these specimens were analyzed by wavelength-dispersive electron microprobe in

Table 2. Typical wavelength-dispersive electron microprobe analyses of chrome spinel for PGM-rich samples from Cliff (1), Harolds Grave (2), PGM-bearing samples (3) and barren samples (4) illustrating the similarity between their compositions[a]

	(1)	(2)	(3)	(4)
Wt% Oxide				
SiO_2	0.11	0.08	0.08	0.08
TiO_2	0.15	0.15	0.18	0.15
Al_2O_3	20.10	19.44	18.11	18.83
FeO^T	17.63	18.71	16.44	15.55
MnO	0.13	0.15	0.15	0.15
MgO	13.50	12.41	14.07	14.06
CaO	0.01	0.01	0.01	0.01
Cr_2O_3	48.38	48.91	50.77	50.93
NiO	0.04	0.07	0.10	0.13
TOTAL	100.05	99.93	99.91	99.89
Formula units (32 oxygens)				
Si	0.0274	0.0201	0.0201	0.0200
Ti	0.0281	0.0284	0.0340	0.0281
Al	5.9068	5.7701	5.3546	5.5378
Fe	3.6753	3.9395	3.4482	3.2442
Mn	0.0267	0.0329	0.0312	0.0309
Mg	5.0152	4.6564	5.2589	5.2272
Ca	0.0027	0.0027	0.0027	0.0027
Cr	9.5338	9.7348	10.0660	10.0440
Ni	0.0080	0.0142	0.0202	0.0261
TOTAL	24.2241	24.1991	24.2357	24.1610
Recalculated analysis				
Octahedral sites				
Si	0.0271	0.0199	0.0199	0.0199
Al	5.8522	5.7226	5.3025	5.5009
Cr	9.4456	9.6547	9.9681	9.9771
Fe^{III}	0.5866	0.5223	0.6224	0.4236
Ti	0.0278	0.0282	0.0337	0.0279
Tetrahedral sites				
Mg	4.9688	4.6181	5.2078	5.1924
Fe^{II}	3.0547	3.3848	2.7922	2.7990
Ni	0.0079	0.0141	0.0200	0.0259
Mn	0.0265	0.0326	0.0309	0.0307
Ca	0.0027	0.0027	0.0027	0.0027

[a] Analyses are recalculated to include ferric iron on the basis of apparent cation surplus.

order to determine differences in major and minor element composition. Typical analyses for each variety, including samples from Cliff and Harold's Grave, are given in Table 2. Figure 5 is an enlarged portion of the Ti/Fe vs. Cr/Fe discrimination diagram shown in Fig. 4 with these new analyses plotted. There is an obvious overlap between analyses of each variety within the typical Shetland field indicating that there are no unusual features in the composition of PGM-rich chromite

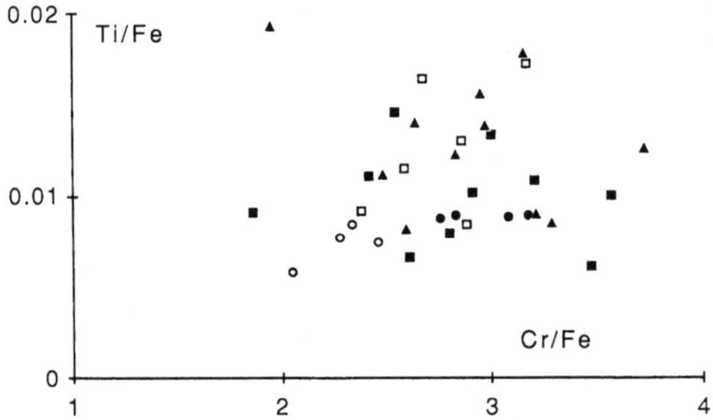

Fig. 5. Shetland chrome-spinel analyses are plotted on an enlarged part of the Ti/Fe vs. Cr/Fe graph shown in Fig. 4. The data represent the mean of three analyses of chrome spinel from one chromite-rich sample. The PGE-rich chrome spinels from Cliff and Harold's Grave are shown as ● and ○ respectively, PGM-bearing chrome-spinels as ■, barren chrome spinels from PGM bearing localities as □ and barren chrome-spinels from localities where no PGM have been found as ▲

samples and suggesting that the processes controlling the chromite composition are not important in influencing the degree of PGE concentration.

3 The Platinum Group Mineralization

The details of the PGM mineralogy, their distribution in the ophiolite complex and whole rock PGE analyses are given elsewhere (Prichard et al. 1986; Prichard and Tarkian 1986). The following paragraphs describe the relationship of the PGE concentrations to the igneous stratigraphy of the complex and to altered, serpentinized zones.

PGM are located in chromite-rich samples at all levels within the ultramafic rocks: in the mantle sequence in dunite lenses within the harzburgite; and in the overlying crustal sequence in discontinuous chromite-rich layers in the dunite and one chromite-rich lens in the wehrlite (Fig. 1). The two small chromite quarries at Cliff and Harold's Grave with unusually large PGE values are situated in the mantle sequence 1 km from the dunite-harzburgite boundary in lenses of dunite within the harzburgite. In addition, mining records suggest that the PGE-rich concentrations recorded by Hitchen (1929) were obtained from Hagdale. This is the largest chromite quarry in the complex and is situated in dunite in the crustal sequence. It thus appears that the PGE-rich values are not restricted to the mantle sequence but were also present in the crustal sequence.

All the ultramafic rocks in the complex have been serpentinized to some extent but fresh primary olivine, clinopyroxene, orthopyroxene and chrome spinels are present. Locally the silicate minerals in the ultramafic rocks are totally altered, especially along the basal contact and along shear zones. In addition, chromite-rich lenses are always situated in highly altered zones and the primary silicate minerals interstitial to the chromite grains have been completely altered to chlorite and serpentine. This is the case for the silicates in the chromite-rich samples from Cliff but the adjacent chromite-poor dunites are some of the freshest and least deformed ultramafic rocks in the complex, with abundant relict olivine preserved. The initial study which led to the discovery of the unusually rich PGE concentrations in the Shetland ophiolite complex was restricted to chromite-rich samples which always have totally altered interstitial silicates. A very limited study of the host silicates has subsequently shown that some of these relatively fresh dunites forming an envelope around the chromite-rich lenses at Cliff are also rich in PGE. The PGM in these dunites include laurite (RuS_2), hollingworthite (RhAsS) and sperrylite ($PtAs_2$) and whole rock analyses for Ir give some of the richest values recorded (up to 15 ppm).

4 The Role of Fluids in the PGE-Rich Lithologies

SSZ-type ophiolite complexes are thought to form by seafloor spreading and melting of hydrated mantle above a descending subduction zone. They are characterized by depleted mantle sequences, podiform chromite deposits, the crystallization of clinopyroxene before plagioclase in the crustal cumulate sequences and Ti- and Y-depleted extrusives (Pearce et al. 1984). The dyke geochemistry, the presence of harzburgite rather than Lherzolite, podiform chromite and wehrlite rather than troctolite, all suggest that the Shetland complex is a SSZ-type ophiolite. Some of the dykes from Shetland show boninitic affinities and Cameron et al. (1979) suggest that boninites form from a depleted mantle, under hydrous conditions, above the sinking slab of oceanic crust in a subduction zone. Pearce et al. (1984) suggest that boninites are formed during the initial stage of subduction by partial melting of hydrated oceanic lithosphere. If the magmas which formed the Shetland ophiolite complex have boninitic affinities and melted under hydrous conditions in a SSZ-type situation then there would have been a volatile component to the magma. There is much debate concerning primary processes that control PGE concentration in stratiform complexes. The role of volatiles is emphasized by Ballhaus and Stumpfl (1985), the role of sulphides as PGE collectors is favoured by Naldrett and Duke (1980) and magmatic processes are thought to be most important by Irvine et al. (1983). Many of the controversies concerning PGE distribution are also applicable to the Shetland ophiolite mineralization.

In Shetland it is difficult to assess the role of magmatic and hydrothermal processes on the distribution of the PGM because secondary alteration is superimposed on primary lithologies. Much lower-temperature hydration caused serpentinization of the ultramafic rocks, and there is evidence of a further hydrothermal event, possibly linked with acid igneous activity and talc formation (Neary and Prichard, 1985). The silicates in the chromite-rich samples are totally altered. This may be a consequence of the greater competence of the chromite-rich lenses than the surrounding silicates during deformation of the complex, which would have led to the creation of pathways for fluids, facilitating preferential alteration. Several phases of deformation have affected the complex including obduction, emplacement of the ophiolite and regional tectonics during the Caledonian orogeny.

It is significant therefore that there are PGM concentrations in the dunite at Cliff because until the discovery of abundant PGM in the dunite it was not clear whether the PGE concentrations in the chromite-rich zones were associated with the chromite or with the alteration zone. The presence of PGM in the less altered and less deformed silicates suggests that the PGE are not necessarily concentrated in the alteration zone but are associated with the primary chromite-rich pod and enclosing dunite envelope. This implies that the PGE are concentrated by primary processes, and further that their discovery in the dunite indicates that they are not only found chromite-rich rocks but may be present in other primary igneous lithologies of the ophiolite.

5 Conclusions

The geochemistry of the dykes and podiform chromite in the Shetland basic and ultrabasic complex confirms its ophiolitic origin and indicates formation in a SSZ. For the first time, therefore, a very varied PGM assemblage similar to stratiform complexes is recorded in a SSZ-type ophiolite complex. Enriched PGE concentrations occur in chromite-rich samples at all levels in the ultramafic part of the complex, including chromite-rich lenses in dunite pods in mantle harzburgite and in overlying crustal sequence dunite and wehrlite. PGM have also been located in relatively fresh dunite adjacent to a PGE-enriched chromite in a lens in the mantle sequence. The PGE concentrations are apparently independent of chromite composition but the primary lithological association suggests the influence of igneous rather than secondary alteration processes. Late magmatic fluids may therefore influence PGE mobility and hydrous melting conditions proposed for the formation of SSZ ophiolite complexes may be important for PGE concentration. The investigation of PGE in ophiolite complexes is still at a preliminary stage. Although the grade of individual samples from Shetland is very high the tonnage of Shetland ore has not been established. Further work is in progress to determine the extent of the PGM distribution and to elucidate the processes that concentrated the PGE. This Shetland example suggests that other ophiolite complexes with SSZ-type geochemistry may have PGE-rich concentrations and may represent exploration targets.

Acknowledgements. This work has been funded from the EEC Raw Materials Programmes which have supported two projects entitled "The chromite of the Shetland ophiolite. A re-appraisal in the light of new theory and techniques" (*contract No. MPP-043-UK*), and "The development of techniques for the determination of platinum-group elements in ultramafic rock complexes of potential economic significance: mineralogical studies" (*Contract No. MSM-089-UK*). Sincere thanks are also given to Mr. S. Owers who gave access to old chromite mining records, to Prof. I.G. Gass and Dr. P.J. Potts who critically and constructively read the manuscript, to Dr. C.R. Neary for valuable discussion, to J.S. Watson for the XRF analyses, to Dr. A. Tindle for help with microprobe analyses, to G.R. Ward for cartographic work and for checking the manuscript and to C. Whale for clerical services.

References

Agiorgitis G, Wolf R (1978) Aspects of osmium, ruthenium and iridium contents in some Greek chromites. Chem Geol 23:267–272

Amin MS (1954) Notes on the ultrabasic body of Unst, Shetland Island. Geol Mag 91:399–406

Augé T (1985) Platinum-group-mineral inclusions in ophiolitic chromite from the Vourinos complex, Greece. Can Miner 23:163–171

Ballhaus CG, Stumpfl EF (1985) Occurrence and petrological significance of graphite in the Upper Critical Zone, western Bushveld complex, S Africa. Earth Planet Sci Lett 74:58–68

Cameron WE, Nisbet EG, Dietrich VJ (1979) Boninites, komatiites and ophiolitic basalts. Nature 280:550–553

Coish RA, Hickey R, Frey FA (1982) Rare earth element geochemistry of the Betts Cove ophiolite, Newfoundland: complexities in ophiolite formation. Geochim Cosmochim Acta 46:2117–2134

Conference Participants (1972) Ophiolites Penrose field conference. Geotimes 7:24–25

Constantinides CC, Kingston GA, Fisher PC (1980) The occurrence of platinum group minerals in the chromitites of the Kokkinorotsos chrome mine, Cyprus. In: Panayiotou A (ed) Ophiolites Proc Int Ophiolite Symp Cyprus 1979, Cyprus Ministry of Agriculture and Natural Resources, pp 93–701

Dickey JS Jr (1975) A hypothesis of origin for podiform chromite deposits. Geochim Cosmochim Acta 39:1061–1074

Flinn D (1970) Some aspects of the geochemistry of the metamorphic rocks of Unst and Fetlar, Scotland. Proc Geol Assoc 81:509–527

Flinn D (1985) The Caledonides of Shetland. In Gee DG, Sturt BA (eds) The Caledonide Orogen-Scandinavia and related areas John Wiley pp 1159–1172

Garson MS, Plant JA (1973) Alpine type ultramafic rocks and episodic mountain building in the Scottish Highlands. Nature 242:34–38

Gass IG, Neary CR, Prichard HM, Bartholomew ID (1982) The chromite of the Shetland ophiolite. A re-appraisal in the light of new theory and techniques. A Report for the Commission of European Communities, Milton Keynes, The Open University

Gunn AG, Leake RC, Styles MT, Bateson JH (1985) Platinum-group element mineralisation in the Unst ophiolite, Shetland. Rep Br Geol Surv 73:115

Heddle MF (1879) The geognosy and mineralogy of Scotland, Orkney and Shetland Island of Unst Miner Mag 2:12–35

Hibbert S (1821) Discovery of chromate of iron in Shetland. Trans Soc Lond Encour Arts Manuf Commer 38:23–26

Hitchen CS (1929) Unst and its chromite deposits. Mining Mag 40:18–24

Irvine TN, Keith DW, Todd SG (1983) The J.M. platinum-palladium reef of the Stillwater complex, Montana. II. Origin by double-diffusive convective magma mixing and implications for the Bushveld complex. Econ Geol 78:1287–1334

Johnson CE, Smith CG, Fortey NJ (1980) Geophysical investigation of chromite-bearing ultrabasic rocks in Baltasound-Hagdale area Unst Shetland Islands Rep Inst Geol Sci 35:70

Naldrett AJ, Duke JM (1980) Platinum metals in magmatic sulphide ores. Science 208:1417–1424

Neary CR, Brown MA (1979) Chromites from the Al'Ays complex, Saudi Arabia, and the Semail complex, Oman. In: Al-Shanti AMS (ed) Evolution and mineralization of the Arabian-Nubian Shield. AG Bull. 2:193–205

Neary CR, Prichard HM (1985) Molybdenum mineralisation on Unst, Shetland Isles. Scott J Geol 21:197–204

Neary CR, Prichard HM, Potts PJ (1984) Chromite, platinoids, gold and moly in the Shetlands. Mining Mag 151:559–560

Page NJ, Pallister JS, Brown MA, Smewing JD, Haffty J (1982a) Palladium, platinum, rhodium, iridium and ruthenium in chromite-rich rocks from the Semail ophiolite, Oman. Can Miner 20: 537–548

Page NJ, Cassard D, Haffty J (1982b) Palladium, platinum, rhodium, ruthenium and iridium in chromitities from the Massif du Sud and Tiebaghi Massive, New Caledonia. Econ Geol 77:1571–1577

Page NJ, Engin T, Singer DA, Haffty J (1984) Distribution of platinum group elements in the Bati Kef chromite deposit, Guleman-Etazig area, Eastern Turkey. Econ Geol 79:177–184

Pearce JA, Alabaster T, Shelton AW, Searle MP (1981) The Oman ophiolite as a Cretaceous arc-basin complex: evidence and implications. Philos Trans R Soc Lond A Math Phys Sci 300:299–317

Pearce JA, Lippard SJ, Roberts S (1984) Characteristics and tectonic significance of supra-subduction zone ophiolites. In: Kokelaal BP, Howells MF (eds) Marginal basin geology. Geol Soc Spec Publ 16:77–94

Phillips FC (1927) The serpentines and associated rocks and minerals of the Shetland Islands. QJ Geol Soc Lond 83:622–652

Potts PJ (1984) Neutron activation-induced beta-autoradiography as a technique for locating minor phases in thin section: application to rare earth element and platinum-group element mineral analysis. Econ Geol 79:738–747

Potts PJ (1986) Neutron activation induced beta-autoradiography as a technique for locating minor phases in thin section emulsion response characteristics. Mater Sci Forum 7:35–44

Potts PJ, Webb PC, Watson JS (1984) Energy dispersive X-ray fluorescence analysis of silicate rocks for major and trace elements, X-ray spectrum 13:1–15

Prichard HM (1985) The Shetland ophiolite. In: Gee DG, Sturt BA (eds) The Caledonide Orogen-Scandinavia and related areas Wiley, New York, pp 1173–1184

Prichard HM, Neary CR (1981) Chromite in the Shetland Islands ophiolite complex. In: An international symposium on metallogeny of mafic and ultramafic complexes UNESCO Project 169(3):343–360, Athens

Prichard HM, Neary CR (1982) Some observations on the chromite in the Shetland ophiolite complex. Ofioliti 2/3:455–466

Prichard HM, Tarkian M (1986) Platinum group minerals from two PGE-rich localities in the Shetland ophiolite complex. Abstract 14th General Meeting of the International Mineralogical Association California, p 204

Prichard HM, Potts PJ, Neary CR (1981) Platinum group element minerals in the Unst chromite Shetland Isles. Trans Inst Mining Metall (Sect B: Appt Earth Sci) 90:186–188

Prichard HM, Neary CR, Potts PJ (1984) Platinum and gold in the Shetland ophiolite Mining J 303:77

Prichard HM, Neary CR, Potts PJ (1986) Platinum group minerals in the Shetland ophiolite. In: Metallogeny of basic and ultrabasic rocks. (eds) Gallagher MJ, Ixer RA, Neary CR, Prichard HM, The Institution of Mining and Metallurgy, London, pp 395–414

Read HH (1934) The metamorphic geology of Unst in the Shetland Islands. Q J Geol Soc Lond 90:637–688

Rivington JB (1953) Recent chromite exploration in Shetland. Mining Mag 89:329–337

Rothery DA (1983) The base of a sheeted dyke complex, Oman ophiolite: implications for magma chambers at oceanic spreading axes. J Geol Soc 140:287–296

Thayer TP (1960) Some critical differences between Alpine-type and stratiform peridotite-gabbro complexes. Rep Int Geol Cong pt 13, Copenhagen, pp 247–259

Experimental Evidence on the Formation and Mineralogy of Platinum and Palladium Ore Deposits

M. MAKOVICKY, E. MAKOVICKY, and J. ROSE-HANSEN[1]

Abstract

The dry, condensed phase systems Pt–Fe–S and Pd–Fe–S were studied by means of laboratory syntheses at 900°, 500° and by long annealing at 300 °C. Reconnaissance runs were made in these systems with either Cu or Ni added. High solid solubility of Pt and Pd in pyrrhotite was established at 900 °C; it drops considerably towards 500° and 300 °C. Pentlandite acts as a concentrator of Pd but not Pt at 500 °C. Sulphide melts in the systems Pd–Fe–S and Cu–Fe–Pt (or Pd)–S act as concentrators of these PGE. Numerous applications of these observations to natural phenomena in PGE deposits are given.

1 Introduction

The relationship of platinum-groups elements (PGE) to base metal sulphides is central to most theories on the origin of major PGE deposits. However, until recently, field or experimental data on the distribution of PGE among these sulphides, PGE solubility in them and the mineralogy of PGE phases in such associations were not available.

The mineralogical study of PGE underwent an explosive development during the past 2 decades (Cabri 1981). It is only recently that the solubility of PGE in natural base metal sulphides has been studied by modern methods (e.g. Cabri and Laflamme 1976; Cabri et al. 1984); the previous work was primarily based on statistical treatment of analytical and recovery data (e.g. Vermaak and Henriks 1976). Some experimental data on the phase systems that involve PGE and base metal sulphides were obtained by Skinner et al. (1976), Distler et al. (1977) and Distler (1980). In the course of the current research we tried to collect more extensive quantitative data on the relationship of base metal sulphides and PGE at temperatures important for the formation and further development of PGE deposits. Such data are of basic importance not only for the theories of ore formation but also for the exploration, ore beneficiation and metallurgy of PGE.

[1] Geological Central Institute, University of Copenhagen, Østervoldgade 10, DK-1350 Copenhagen K, Denmark

Mineral Deposits within the European Community
(ed. by J. Boissonnas and P. Omenetto)

This chapter reports on the results obtained for the systems Fe–Pt–S and Fe–Pd–S as well as for those with additional Cu and Ni. These systems contain the most important base metal sulphides present in PGE deposits: pyrrhotite/troilite, pyrite, chalcopyrite/intermediate Cu–Fe–S solid solution and pentlandite.

These studies were performed at two distinct temperatures, 900° (related to the magmatic/submagmatic range) and 500 °C (related to post-magmatic/hydrothermal processes). Similar studies involving Ru and Rh were reported elsewhere (Makovicky et al. 1986).

The experimental results are compared with the PGE concentration data in natural sulphides and with the mineral associations observed in the deposits of PGE in mafic and ultramafic rocks in order to assess the relative importance of the observed phenomena in the formation of different types of PGE deposits.

2 Experimental

In all experiments, dry, condensed sulphide systems were studied. These were prepared by reacting weighed mixtures of pure elements in evacuated silica glass tubes. The 250-mg charges were pre-reacted at 300 °C and subsequently annealed, with one homogenization, for a total of about 2 months.

After quenching and polishing, the charges were examined in reflected light and by microprobe analysis in order to establish the phases present and the distribution of PGE in them. Further experimental details are described in Makovicky et al. (1986).

Appreciable smear of Pt over adjacent grains of sulphides and other extraneous substances was found when grains of Pt with variable Fe contents were present in a polished section. The false values can reach up to 0.5 wt% Pt in the sulphide grains immediately adjacent to the Pt grain and they fall off in all directions. Their presence in cases where the resulting charge was powderlike (from 500 °C), and the powder grains came in contact with the Pt grain accidentally during the embedding process, proves that they represent experimental artefacts. Such a phenomenon was not observed with other PGE.

Selected charges, in which high solubility of PGE in sulphide phases was found by means of microprobe analyses, have been resealed in evacuated silica glass tubes and re-annealed continuously for 9 to 11 months at 300 °C in order to study low-temperature exsolution and re-equilibration phenomena. These samples have been analyzed by microprobe with extensive use of back-scattered electron images at high magnifications.

3 System Pt–Fe–S

This system is of fundamental importance for the PGE deposits of Merensky Reef in the Bushveld Complex and the J–M Reef in the Stillwater Complex.

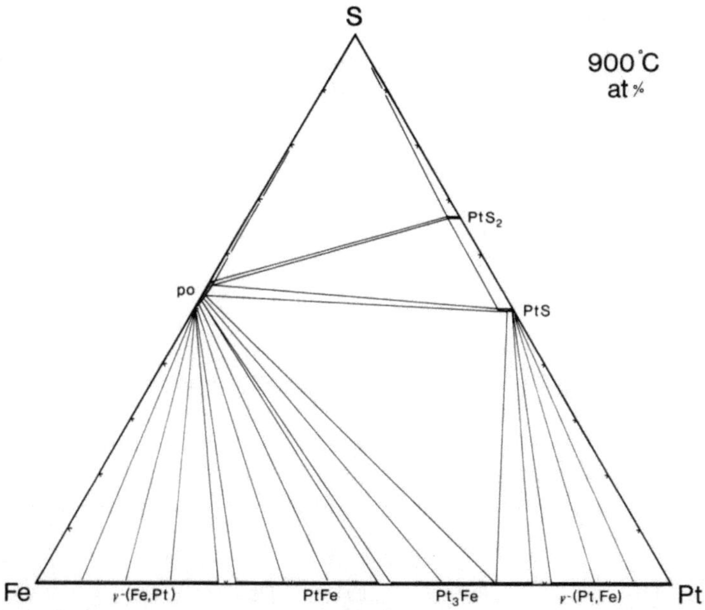

Fig. 1. The system Pt–Fe–S at 900 °C. Here, and in the following figures, the thickness of the composition fields of the sulphides of PGE and of the alloys is exaggerated, those of pyrrhotite (*po*) and pyrite (*py*) are realistic. For the alloys in these figures the data collected by Hansen and Anderko (1958), Elliot (1965), Shunk (1969), Cabri and Feather (1975) and Skinner et al. (1976) were cross-checked against our results and critically interpreted

At 900 °C pyrrhotite can dissolve up to 0.6 at% (2.7 wt%) Pt at its sulphur-richest compositions, associated with Fe-bearing PtS_2 and PtS (Fig. 1). This association has not yet been found in Pt deposits although its existence cannot be excluded.

In the above mentioned Pt deposits, the sulphur-poorer association pyrrhotite-isoferroplatinum (Pt_3Fe)-cooperite (PtS) is dominant over their large portions that originated under a quiescent regime outside the disturbed areas of potholes and away from dunite pipes. Our research indicates that at 900 °C in this association pyrrhotite dissolves 0.13 at% (0.57 wt%) Pt, PtS dissolves up to 1.1 at% Fe and the Pt–Fe alloy (isoferroplatinum) contains 27.5 at% (9.8 wt%) Fe. The values for isoferroplatinum are conspicuously identical to those from the natural associations (centered on 10 wt% Fe), even when cooperite was not detected (e.g. Stumpfl and Rucklidge 1982). As calculated by Elliot et al. (1982), without the sulphur buffer such Pt–Fe alloys would exist only at oxygen fugacities far above those measured by them for the rocks of the Bushveld suite. Therefore, sulphur fugacity appears to buffer the ore mineral associations in large portions of these layered intrusions, limiting them to the central portions of the Pt–Fe–S triangle. Cases of Pt–Fe alloys with Fe contents substantially higher or lower than 10 wt% must be connected with special conditions (low S fugacities, variable oxygen fugacities).

At 500 °C (Fig. 2), the solubility of Pt in pyrrhotite is below microprobe detection limits for all pyrrhotite compositions. Pyrrhotite coexists with Pt–Fe alloys,

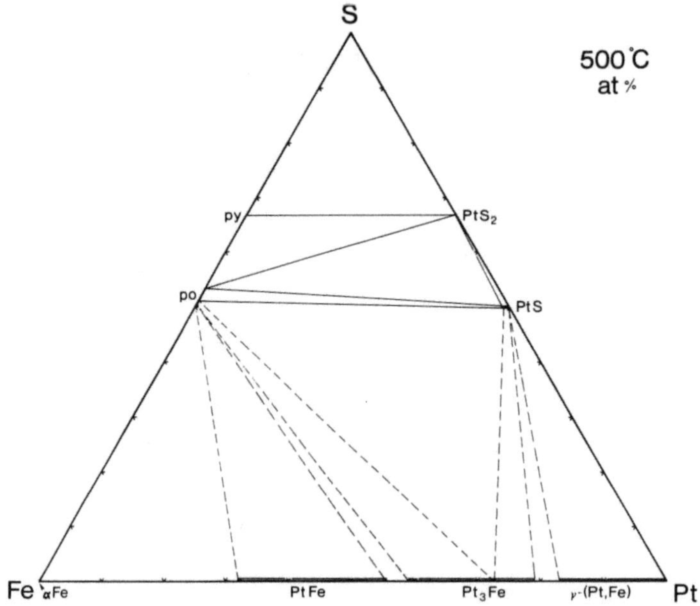

Fig. 2. The system Pt–Fe–S at 500 °C. Here, and in Fig. 4 the *dashed lines* indicate appreciable discrepancies in the published solid solution boundaries for the Fe–Pt and Fe–Pd alloys (see legend to Fig. 1)

PtS and PtS_2. PtS dissolves up to 1.9 at% (0.6 wt%) Fe, whereas PtS_2 is probably Fe-free at this temperature. It is difficult to obtain fully equilibrated samples at this temperature, especially in the case of Pt–Fe alloys. Pyrite dissolves no detectable amounts of Pt. However, marcasite associated with pyrite, PtS_2 and PtS at 500 °C, appears to dissolve more than 0.5 wt% Pt. It is not clear whether this is the result of platinum smear over the marcasite crusts on pyrite, these being more porous than the pyrite cores, or whether it is true solubility of Pt in marcasite.

4 System Pd–Fe–S

This system is of equally great importance for the PGE deposits as the previous one. The universal presence of braggite, (Pd, Pt, Ni) S, with widely different Pd/Pt ratios in these deposits, is significant. It contrasts, however, with the low Pd contents in the natural Pt–Fe alloys, suggesting that only the complete system Pd–Pt–Fe–S (one of our future projects) can give complete answers on the variability of Pd–Pt–Fe minerals in sulphide deposits.

The solubility of palladium in pyrrhotite is quite high at 900 °C: up to 4.7 at% (11.0 wt%) Pd dissolves in the sulphur-richest compositions of pyrrhotite, in association with sulphur vapour and with a sulphide melt that contains at least 24 at% Pd and up to 42 at% S (Fig. 3). However, the solubility of Pd in pyrrhotite drops

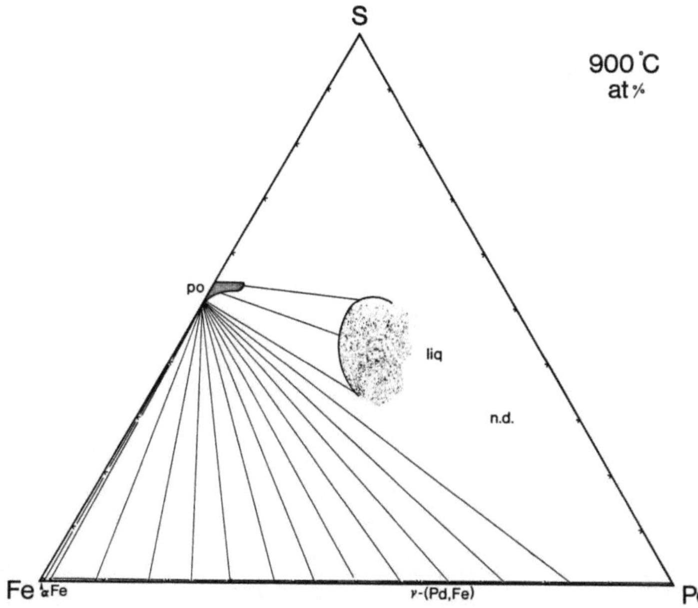

Fig. 3. The system Pd–Fe–S at 900 °C. n.d. denotes areas not investigated in detail

rapidly towards sulphur-poorer compositions until about 1 at% Pd is reached; it decreases more slowly afterwards and reaches zero at the troilite composition. The melt coexists with these pyrrhotite compositions down to a solubility of 0.17 at% (0.4 wt%) Pd.

Although the Pd–S join was not studied in detail, Fig. 3 shows that the liquid field prevents direct coexistence of Pd-enriched pyrrhotite with braggite at these temperatures. According to Skinner et al. (1976), this liquid field recedes with decreasing temperature from the Fe–FeS eutectic (at 988 °C) towards the ternary eutectic (570 °C) at 60 at% Pd, close to the Pd–S join. This melt should then act as a collector for PGE and as a host for a number of other rare elements (Skinner et al. 1976). We obtained a eutectic intergrowth of Fe-enriched Pd_4S with Fe-richer phases by the solidification of a sulphide melt (7.3 at% Fe and 67.7 at% Pd, i.e. 82.8 wt% Pd) quenched from 500 °C (Fig. 4). Thus, the melt extends to lower temperatures and Pd-richer compositions than previously assumed.

These results offer an explanation for the observations of Kingston and El-Dosuky (1982) in the Bushveld Complex who found widespread replacement of earlier cooperite by braggite and of the minerals with lower Pd contents by those with higher Pd contents. Such a replacement can be ascribed to the interaction of Pd-rich residual liquid with these phases, without influencing the original Pt–Fe alloys formed at higher temperatures.

Many important changes take place in the Pd–Fe–S system between 900° and 500 °C. As just mentioned, only a small residual liquid field exists in the Pd-richest portions of the phase system at 500 °C, situated between $Pd_{16}S_7$ and Pd_4S (Fig. 4). No solid ternary phases are present. The solubility of Pd in pyrrhotite drops

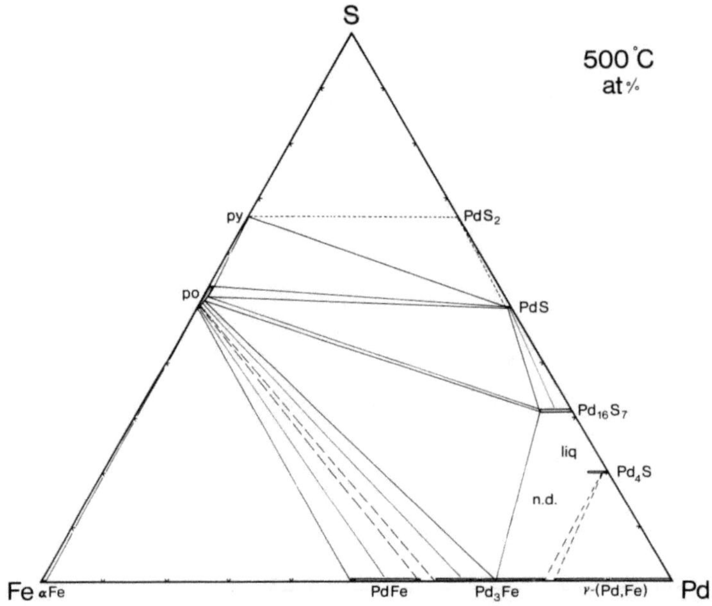

Fig. 4. The system Pd–Fe–S at 500 °C. n.d. denotes areas not investigated in detail. See also Figs. 1 and 2

dramatically. Its maximum for the S-richest portions of the pyrrhotite compositional field is only 0.2 at% (0.4–0.5 wt%) Pd, established from several runs. PdS dissolves up to 0.8 at% (0.6 wt%) Fe at this temperature. Fe-bearing $Pd_{16}S_7$ is in association with pyrrhotite (0.12 at% Pd) and Pd–Fe alloy (71.9 at% Pd = 82.3 wt% Pd). Pd_4S dissolves Fe as well (Fig. 4). The limited solubility of Fe in braggite (vysotskite, PdS) is in sharp contrast to the high solubility of Ni in this mineral (Cabri et al. 1978; Karup-Møller and Makovicky, 1987).

The solubility of Pd in pyrite, in association with pyrrhotite and PdS, is at the resolution threshold of microprobe measurements (0.03 at%, i.e. 0.08 wt% Pd). Finally, in an iron-free run at 500 °C we prepared by direct synthesis the controversial compound PdS_2 in association with PdS. Hitherto this compound has been prepared only indirectly (Hansen and Anderko 1958; Elliot 1965; Shunk 1969). A large exsolution gap appears in the Pd–Fe alloys at this temperature, between practically pure iron and the Pd–Fe alloys with more than 58 at% Pd (Fig. 4). Both are associated with troilite and Pd-poor, Fe-rich pyrrhotite.

The ternary assemblage pyrrhotite-PtS-alloy appears to dominate the relationships in the natural Fe–Pt–S systems down to low temperatures. The situation is different for the Fe–Pd–S system. PdS and the (Pd, Fe) alloys are first separated from each other by the intervening sulphide melt; at lower temperatures they become separated by two to three lower sulphides of Pd. These were not observed as minerals. Their possible low-temperature instability, their later alteration to PdS by changes in sulphur fugacity at lower temperatures, as well as their role being taken over by bismuthides or tellurides of Pd, or by the Pd-rich replacement products observed at Bushveld — all these represent potential reasons for their absence as minerals.

5 Observations at 300 °C

Extended annealing at 300 °C of the Pt- or Pd-richest pyrrhotite samples which were synthesized at 900° or 500 °C yielded interesting data on the processes which take place during the cooling of PGE deposits.

Already on quenching from high temperatures, the Pt- and Pd-enriched pyrrhotites exsolve surplus PGE in the form of fine particles dispersed throughout the pyrrhotite matrix; these are clearly discernible in back-scattered electron images (Fig. 5). In the case of Pd, they were partly oriented in pyrrhotite crystals, partly situated along the grain boundaries. These, and the unoriented Pt-rich particles dispersed in Pt-enriched quenched pyrrhotites, can vary in size to below the microprobe resolution at 700 × magnification, reflecting the local quenching rates. Thus, the high-temperature compositions had to be investigated analyzing representatively large areas of pyrrhotite crystals.

The Pt-enriched pyrrhotite does not appear to alter its total Pt content on annealing at 300 °C, and gives the same average Pt concentration as before annealing. The situation appears different for palladium. The average Pd contents left in pyrrhotite after annealing are 0.25–0.3 wt%, after an apparent drop from the original values of 0.5–1.5 wt%.

High-magnification, back-scattered electron images (Figs. 6, 7 and 8) taken from the samples annealed to 300 °C indicate extensive submicroscopic exsolution

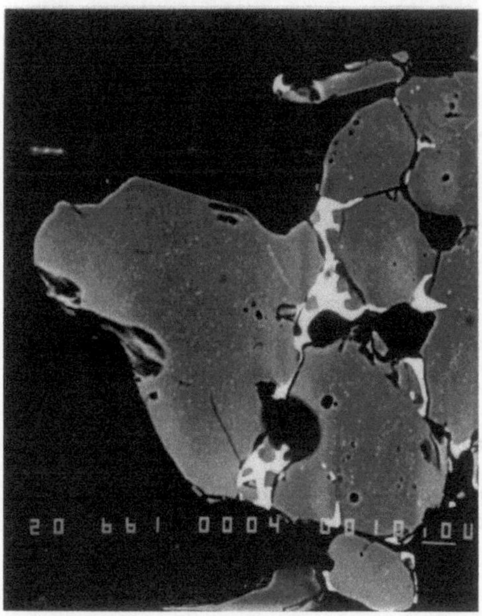

Fig. 5. Exsolution of Pd-rich particles (*bright dots*) on quenching from Pd-rich pyrrhotite (*dark grey*) prepared at 900 °C. *White areas*: Pd-rich sulphide melt. *Bar* indicates 10 μm. Back-scattered electron image

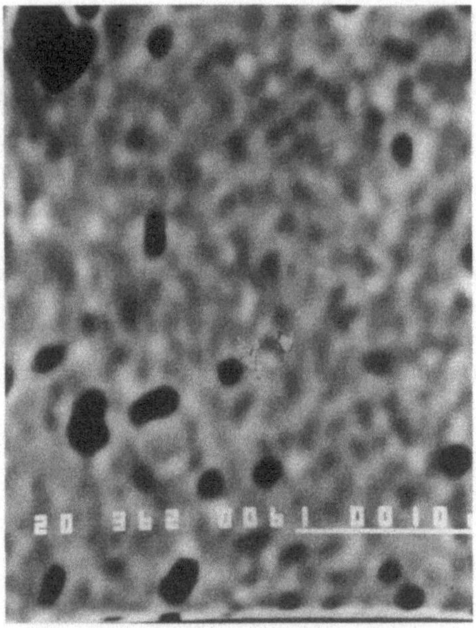

Fig. 6. Exsolution of palladium sulphide from Pd-containing pyrrhotite after annealing at 300 °C. The original phase contained ~0.26 wt% Pd at 500 °C. 0.8 wt% Pd was measured over the light portions, whereas only 0.06 wt% Pd over the dark portions of the exsolution intergrowth. Back-scattered electron image; *bar* indicates 10 μm

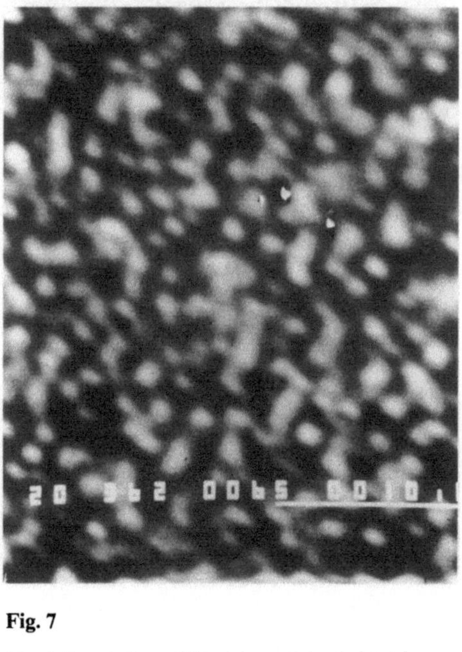

Fig. 7

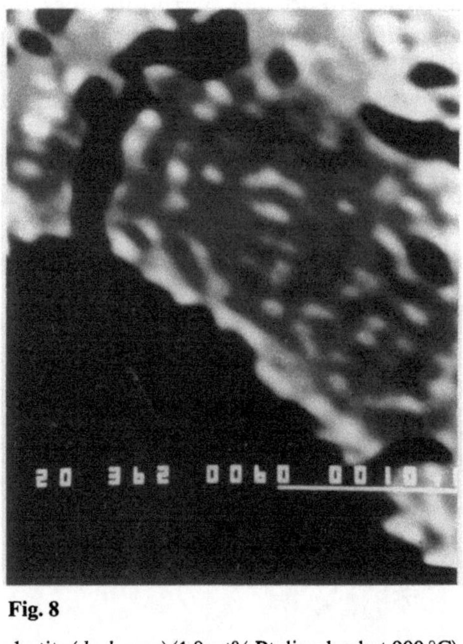

Fig. 8

Fig. 7. Exsolution of Pt-rich particles (*white*) from pyrrhotite (*dark grey*) (1.8 wt% Pt dissolved at 900 °C) on extended annealing at 300 °C. Back-scattered electron image; *bar* indicates 10 μm

Fig. 8. Pt-rich particles (*white*) in pyrite (*dark*, *centre*) formed by replacement of Pt-bearing pyrrhotite (*grey*, along the upper margins) from Fig. 7 on extended annealing at 300 °C

of Pt or Pd-rich particles. They are too small for exact analyses. PGE contents which represent several times the bulk PGE concentration were measured over them, whereas the PGE contents in the pyrrhotite areas between them represent only a fraction of the average PGE value (Fig. 6). Remarkably, pyrite, formed on annealing at 300 °C as a result of reduction of the width of the pyrrhotite compositional field, inherits the dispersed Pt-rich particles from the pyrrhotite it replaces (Figs. 7 and 8) and can display up to 0.8 at% (3.8 wt%) Pt in microprobe analyses.

Thus, PGE exsolve as small, submicroscopic particles inside pyrrhotite with a possible outward migration observed only for Pd. One of the possible reasons for such behaviour can be low diffusion rates for PGE in pyrrhotite at 300 °C. However, a much more serious reason for this behaviour might be the presence of numerous precipitation nuclei in the form of exsolution droplets formed when the samples studied were quenched to room temperature, before the annealing at 300 °C took place. Therefore, our samples may not correspond exactly to natural situations which presumably are distinguished by slow cooling rates and long diffusion times.

The solubility of Pt and Pd in pyrrhotite at high temperatures and the possible ways of their recycling when they are rejected from pyrrhotite on cooling (by residual melts, transporting fluids or PGE-precipitating fluids with As, Te, Sb, etc.) are very important for the formation of PGE deposits. Under conditions of rapid cooling or at temperatures too low for substantial outward diffusion, these PGE might stay as sulphide emulsions in pyrrhotite. This would imply the presence of an important "invisible", "colloidal" fraction of Pt in pyrrhotite found in Bushveld and elsewhere (Vermaak and Hendriks 1976; Kinloch 1982).

6 PGE-Enriched Pentlandite

Knop et al. (1976) prepared synthetic pentlandites with Pd, Ru and Rh, without investigating their stoichiometry. In the present study, Fe:Ni = 1:1 pentlandite compositions were weighed out with 1/3, 2/3, 3/3 and 4/3 PGE atom in the formula unit M_9S_8; these values also represent the expected occupancy of the octahedral metal position by PGE. All experiments were carried out at 500 °C.

For all four concentrations, *platinum* does not enter pentlandite within the analytical sensitivity of $\pm$0.05 wt%. The associated nickel-bearing pyrrhotite also does not dissolve detectable amounts of Pt. The Pt–Fe alloy present contains maximally 25.3 at% Fe (Fig. 9).

Palladium can enter pentlandite, filling practically the octahedral position (up to 5.3–5.5 at% Pd, i.e. 90–93% occupation of the octahedral position by Pd was observed, Fig. 9). The concentrations found suggest a solid solution towards Pd-free pentlandite; it has been ascertained down to about 3.2 at% (7.4 wt%) Pd (i.e. 55% occupancy of the octahedral position by Pd) and then again only below 0.15 at% (0.35 wt%) Pd, i.e. a low-Pd pentlandite. It is difficult to assess the importance of the compositional gap found in this study. The analyses of Pd in pentlandite by Genkin et al. (1973), Cabri and Laflamme (1979) as well as Laflamme (in Cabri 1981) indicate up to 0.65 wt% Pd (exceptionally 1.5 wt% Pd) in pentlandite from the

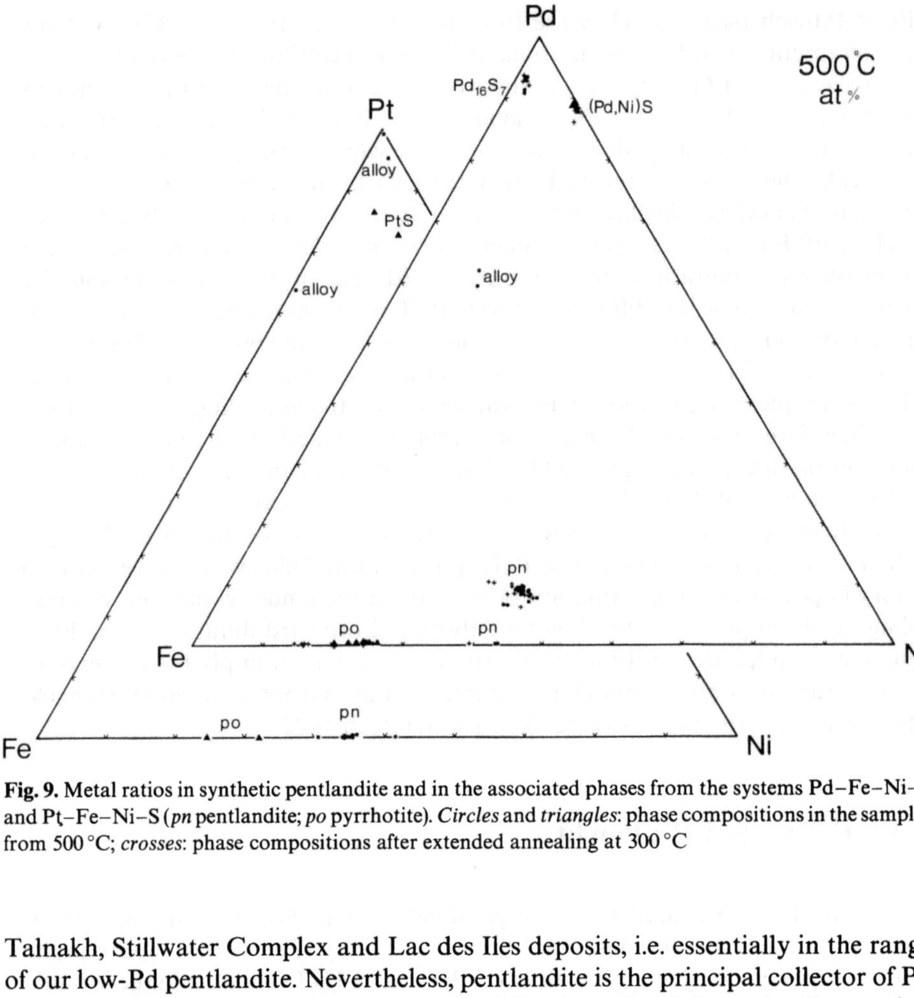

Fig. 9. Metal ratios in synthetic pentlandite and in the associated phases from the systems Pd–Fe–Ni–S and Pt–Fe–Ni–S (*pn* pentlandite; *po* pyrrhotite). *Circles* and *triangles*: phase compositions in the samples from 500 °C; *crosses*: phase compositions after extended annealing at 300 °C

Talnakh, Stillwater Complex and Lac des Iles deposits, i.e. essentially in the range of our low-Pd pentlandite. Nevertheless, pentlandite is the principal collector of Pd in many deposits, as stressed in numerous publications. Our Pd-pentlandite coexists with pyrrhotite with 0.1–0.35 at% Pd and other phases (Fig. 9).

No apparent exsolution of Pd takes place in Pd-rich pentlandite on annealing at 300 °C; its composition appears to enrich slightly in Fe. The same process takes place, but with much greater intensity, in the associated Ni-pyrrhotite (Fig. 9).

7 The Systems Cu–Fe–Pt–S and Cu–Fe–Pd–S

The positive correlation between copper and platinum, or palladium, is typical for various parts (especially offsets) of the Sudbury ore deposit. It represents the association of PGE minerals with chalcopyrite (Cabri and Laflamme 1976) but not the solid solution of PGE in chalcopyrite (Cabri et al. 1984). Such correlations exist also for the Noril'sk deposit (Genkin et al. 1981), the Kambalda deposits (Ross and

Keayes 1975) and, on a local to microscopic scale, also for Merensky Reef and the Stillwater Complex (Kingston and El-Dosuky 1982; Mostert et al. 1982; Bow et al. 1982). Several hypotheses were put forward about this correlation and are summarized by Naldrett et al. (1982). They are based on fractional crystallization of a sulphide melt or on successive portions of sulphide melts released from magma, on progressive partial melting of the monosulphide solid solution or on diffusion of PGE under thermal gradients.

In order to investigate experimentally the reasons for this correlation, exploratory runs with 1 and 5 at% PGE were positioned in the systems Cu–Fe–Pt–S and Cu–Fe–Pd–S so as to examine the potential solubility of these platinum metals in chalcopyrite and in the intermediate solid solution (iss) with variable M:S ratios.

No solubility of Pt or Pd was found in chalcopyrite or iss at 500 °C. In these runs, iss dissociated on quenching into intergrowths of iss products (± bornite). In the case of platinum, a quaternary S-rich compound $Cu_{2.0}Fe_{1.2}Pt_{2.6}S_8$ was found associated with chalcopyrite, pyrite and $Pt_{96.6}Cu_{3.4}Fe_{0.02}$.

At 900 °C, the behaviour of Pt and Pd is dominated by the presence of sulphide liquids along the central portions of the Cu–Fe–S subsystem. Although a complete delimitation of phase relationships would require many more experimental runs than were possible in this investigation, our studies adequately determined the fundamental PGE concentration processes in this system. These results are shown in Figs. 10 and 11. They are in wt% in order to express better the PGE concentrations. Circles denote the compositions of solid phases, squares those of liquids, whereas stars indicate selected ideal mineral compositions. Tie lines connect coexisting phases. Pt or Pd content is indicated as height over base. Full symbols represent phases that exist at 900 °C, whereas the open symbols are either the low-temperature exsolution products of high-temperature phases or they represent

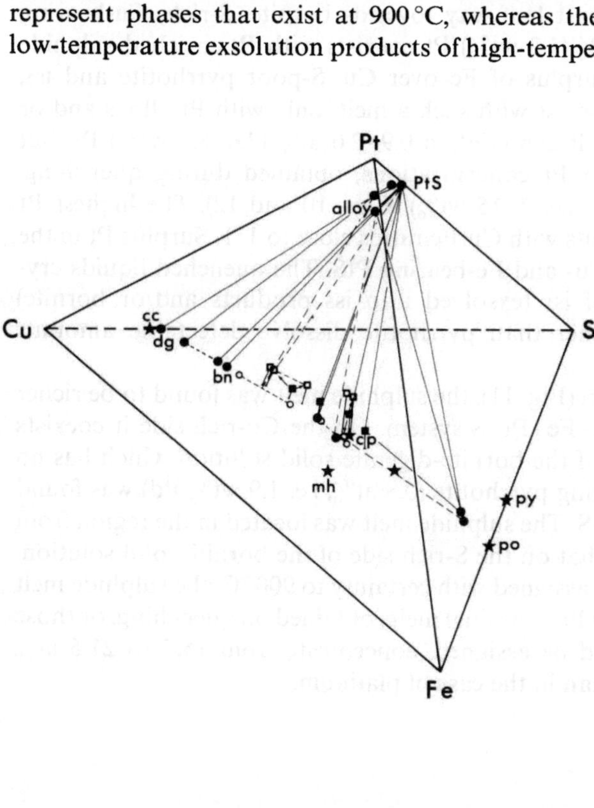

Fig. 10. Compositions of solid phases and liquids in the central portions of the phase system Pt–Cu–Fe–S at (and below) 900 °C. Explanations are given in the text

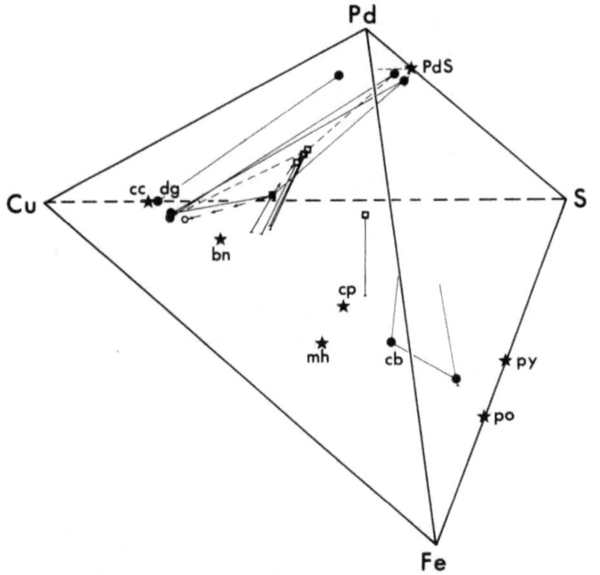

Fig. 11. Compositions of solid phases and liquids in the central portions of the phase system Pd–Cu–Fe–S at (and below) 900 °C. Explanations in the text

the compositions of final liquids at their solidification at some temperature during quenching (evolution indicated by arrows).

The Pt-containing sulphide liquids at 900 °C have Cu/Fe/S ratios corresponding to the compositions of chalcopyrite and of the S-rich to S-moderate intermediate solid solution and stretch at least half way towards bornite. S-rich, Cu-bearing pyrrhotite that contains 0.3 at% (1.3 wt%) Pt coexists with Pt-enriched sulphide melt which displays a small surplus of Fe over Cu. S-poor pyrrhotite and iss, digenite and bornite do not co-exist with such a melt, only with Pt alloys and/or PtS. At 900 °C the sulphide melt can contain 0.9–2.0 at% (3.6–8.0 wt%) Pt, but the final melts, with maximum Pt concentrations, obtained during quenching, contain from 1.8 to 4.0 at% Pt (i.e. 7–15 wt%) (Figs. 10 and 12). The highest Pt values were obtained for the melts with Cu/Fe ratios close to 1:1. Surplus Pt in the charge crystalizes as primary Cu- and Fe-bearing PtS. The quenched liquids crystallize as eutectic mixtures of iss (exsolved into iss products and/or bornite) with PtS. No solid phases other than pyrrhotite dissolve detectable amounts of Pt.

In the Cu–Fe–Pd–S system (Fig. 11), the sulphide melt was found to be richer in sulphur than that in the Cu–Fe–Pt–S system. On the Cu-rich side it coexists with an intermediate member of the bornite-digenite solid solution which has no detectable Pd content. Pd-bearing pyrrhotite (0.8 at%, i.e. 1.9 wt% Pd) was found to coexist with cubanite and PdS. The sulphide melt was located in the region from a Cu:Fe ratio close to 1:1 to that on the S-rich side of the bornite solid solution. In the only case which could be assigned with certainty to 900 °C, the sulphide melt concentrated 9.2 at% (14.6 wt%) Pd. The final melts obtained on quenching, or those to which no temperature could be assigned, concentrate from 15.5 to 21.5 at% (31–40 wt%) Pd, much more than in the case of platinum.

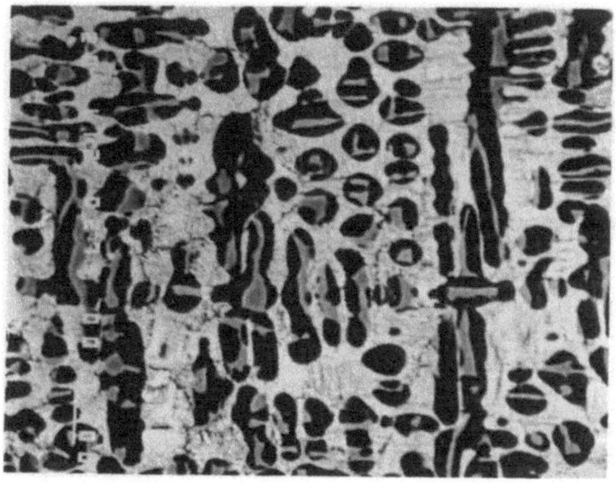

Fig. 12. Quenching products of a Pt-containing Cu–Fe sulphide liquid from 900 °C. Both the Pt-free dendrite and the Pt-enriched interstitial liquid decomposed into a bornite-iss integrowth (+PtS). Back-scattered electron image; *bar* indicates 10 μm

Thus, it is the concentration of Pt and Pd in the Cu, Fe-rich residual sulphide melts during the fractional crystallization of the initial sulphide melts that apparently accounts for the majority of observed cases. Pt, Pd and Cu will also be resupplied into this melt as their solubility in previously crystallized pyrrhotite decreases.

The solidification temperatures of these melts remain to be investigated. Their compositions close to chalcopyrite and iss are of obvious importance for many PGE deposits; their extension towards bornite is of interest for the Noril'sk deposit (Genkin et al. 1981), Skærgaard intrusion in Greenland (Wager et al. 1957) as well as for some areas of the Bushveld Complex (McLaren and De Villiers 1982). The concentration of Pt and Pd in these melts takes place in our experiments with only Cu, Fe, Pt or Pd and S present; there is no need for additional anions, cations or volatile components. In addition to PGE, Cu-rich sulphide liquids in nature concentrate As, Te, T1, rare chalcophile elements, K, Cl and other components (Hoffman et al. 1979; Genkin et al. 1981) and yield in their final stages very colourful mineral associations.

8 Summary

Our investigations of Pt- and Pd-containing sulphide systems at 900° and 500 °C allowed us to establish the following concentration mechanisms for these PGE:

1. Important solid solubility of Pt and Pd in pyrrhotite at 900 °C which exceeds by far their initial concentrations in nature. This solubility drops considerably with temperature, although it still remains important for Pd at 500 °C. However, in the case of stoichiometric troilite, Pt and Pd are concentrated in their alloys with Fe.
2. Progressive concentration of Pd in a sulphide melt in the system Fe–Pd–S between 900° and 500 °C.

3. Important collector role of pentlandite for Pd and the existence of palladian pentlandite close to Pd $(Fe, Ni)_8S_8$ at 500 °C. Pt did not concentrate in pentlandite at this temperature.
4. Progressive concentration of Pd and Pt in Cu, Fe-rich sulphide melts at and below 900 °C.

The application of these observations to natural conditions was discussed.

Acknowledgements. The project "Solubility and distribution of platinum group elements in base metal sulphides in platinum group deposits" was sponsored by the EEC under *Contract No. MSM-116-DK.* We would like to express our gratitude to Dr. M. Ghisler, Dr. S. Karup-Møller, Dr. J. Bailey, lecturers J.G. Rønsbo and E.S. Leonardsen, as well as Mr. J. Fløng, Mrs. B. Møller and Mrs. U. Koester for their support, interest and assistance in the project. The microprobe apparatus used was financed by the National Research Council of Denmark (Natural Sciences).

References

Bow C, Wolfgram D, Turner A, Barnes S, Evans I, Zdepski M, Boudreau A (1982) Investigations of the Howland Reef of the Stillwater Complex, Minneapolis Adit Area: stratigraphy, structure and mineralization. Econ Geol 77:1481–1492

Cabri LJ (ed) (1981) Platinum group elements: mineralogy, geology, recovery, vol 23. The Canadian Institute of Mining and Metallurgy

Cabri LJ, Feather CE (1975) Platinum-iron alloys: a nomenclature based on a study of natural and synthetic alloys. Can Miner 13:117–126

Cabri LJ, Laflamme JHG (1976) The mineralogy of the platinum group elements from some copper-nickel deposits of the Sudbury area, Ontario. Econ Geol 71:1159–1195

Cabri LJ, Laflamme JHG (1979) Mineralogy of samples from the Lac des Iles area, Ontario. Canmet Rep 79-27, Dept Energy, Mines, Resources, Ottawa

Cabri LJ, Laflamme JHG, Stewart JM, Turner K, Skinner BJ (1978) On cooperite, braggite and vysotskite. Am Miner 63:832–839

Cabri LJ, Blank H, El Goresy A, Laflamme JHG, Nobiling R, Sizgoric MB, Traxel K (1984) Quantitative trace-element analyses of sulphides from Sudbury and Stillwater by proton microprobe. Can Miner 22:521–542

Distler VV (1980) Solid solutions of platinoids in sulphides (in Russian). In: Sulphosalts, platinum minerals and ore microscopy. Proceedings of the General Meeting of the International Mineralogical Association, Novosibirsk 1978. IGEM AN SSSR, Nauka, pp 191–200

Distler VV, Malevskyi AY, Laputina IP (1977) Distribution of platinoids between pyrrhotite and pentlandite during the crystallization of sulphide melt (in Russian). Geokhimiya 11:1646–1657

Elliott RP (1965) Constitution of binary alloys, 1st suppl. McGraw-Hill Book, New York

Elliott WC, Grandstaff DE, Ulmer GC, Buntin T, Gold DP (1982) An intrinsic oxygen fugacity study of platinum-carbon associations in layered intrusions. Econ Geol 77:1493–1510

Genkin AD, Distler VV, Laputina IP, Filimonova AA (1973) Geochemistry of palladium in copper-nickel ores. Geokhimiya 9:1336–1343

Genkin AD, Distler VV, Gladyshev GC et al. (1981) Copper-nickel sulphide ores from the Noril'sk deposits (in Russian). IGEM AN SSSR, Nauka

Hansen M, Anderko K (1958) Constitution of binary alloys. McGraw-Hill Book, New York

Hoffman EL, Naldrett AJ, Alcock RA, Hancock RGV (1979) The noble metal content of ore in the Levack West and Little Stobie Mines, Ontario. Can Miner 17:437–451

Karup-Möller S, Makovicky E (1987) The Ni–Pd–S phase system at 900, 725,550 and 400 °C. Abstr 5th Magm Sulphides Field Conf, Harare, Zimbabwe, Aug 1987 (in press)

Kingston GA, El-Dosuky B (1982) A contribution on the platinum-group mineralogy of the Merensky Reef at the Rustenburg Platinum Mine. Econ Geol 77:1367–1384

Kinloch ED (1982) Regional trends in the platinum-group mineralogy of the critical zone of the Bushveld Complex, South Africa. Econ Geol 77:1328–1347

Knop O, Huang C, Reid KIC, Carlow JS (1976) Chalcogenides of transition elements. X-ray, neutron, Mössbauer and magnetic studies of pentlandite and the π phases, π (Fe, Co, Ni, S), Co_8 MS_8 and Fe_4 Ni_4 MS_8 (M = Ru, Rh, Pd). J Solid State Chem 16:97–116

Makovicky M, Makovicky E, Rose-Hansen J (1986) Experimental studies on the solubility and distribution of platinum group elements in base metal sulphides in platinum deposits. In: Gallagher MJ, Ixer RA, Neary CR, Prichard HM (eds) Metallogeny of basic and ultrabasic rocks. Proc Conf Edinburgh 9–12 April 1985. The Institution of Mining and Metallurgy, London, pp 415–425

McLaren CH, Villiers JPR de (1982). The platinum group chemistry and mineralogy of the UG-2 chromitite layer of the Bushveld Complex. Econ Geol 77:1348–1366

Mostert AB, Hofmeyr PK, Potgieter GA (1982) The platinum-group mineralogy of the Merensky Reef at the Impala Platinum Mines, Bophuthatswana, Econ Geol 77:1385–1394

Naldrett AJ, Innes DG, Sowa I, Gorton MP (1982) Compositional variations within and between five Sudbury ore deposits. Econ Geol 77:1519–1534

Ross, JR, Keayes RP (1975) Precious metals in volcanic-type nickel sulphide deposits in Western Australia. I. Relationship with the composition of the ores and their host rock. Can Miner 17:417–435

Shunk FA (1969) Constitution of binary alloys, 2nd suppl. McGraw-Hill Book, New York

Skinner BJ, Luce FD, Dill JA et al. (1976) Phase relations in ternary portions of the system Pt–Pd–Fe–As–S. Econ 71:1469–1475

Stumpfl EF, Rucklidge JC (1982) The platiniferous dunite pipes of the eastern Bushveld. Econ Geol 77:1419–1431

Vermaak CF, Henriks LP (1976) A review of the mineralogy of the Merensky Reef, with specific reference to new data on the precious metal mineralogy. Econ Geol 71:1244–1269

Wager LR, Vincent EA, Smaley AA (1957) Sulphides in the Skærgaard intrusion, East Greenland. Econ Geol 52:855–903

Part III
Base Metals, Phosphates, Placer Minerals

Part III
Base Metals, Phosphates, Placer Minerals

Metallogenic Models and Exploration Criteria for Buried Carbonate-Hosted Ore Deposits: Results of a Multidisciplinary Study in Eastern England

J.A. PLANT[1]*, D.G. JONES[2], G.C. BROWN[3], T.B. COLMAN[2], J.D. CORNWELL[2], K. SMITH[2]+, N.J.P. SMITH[2]+, A.S.D. WALKER[2], and P.C. WEBB[3]

Abstract

Exploration criteria for carbonate-hosted (Pennine and Irish-style) ore deposits buried beneath later cover have been developed based on new metallogenic models and methods of basin analysis usually applied to hydrocarbon exploration. New and existing data sets for eastern England were analysed and areas prospective for the two deposit types identified using an image analysis system.

Irish SEDEX (Zn–Ba–Pb) and Pennine (F–Pb–Ba–Zn) deposits can be related to different phases in the tectonic evolution of the Northern Foreland of the Hercynian orogen. Irish-style mineralisation formed over zones of high heat flow and tectonism in the waxing phase of a regime of crustal extension and the rise of hot asthenophere beneath the crust. The deposits are syngenetic/diagenetic in Tournaisian shelf carbonates. They were formed by the expulsion of hot, moderately saline fluids from half-graben basins, through basal sandstones, which reached the seafloor via listric faults.

In contrast, the epigenetic Pennine-style deposits were a product of tectonism following a period of declining geothermal gradients and crustal subsidence. The proposed model involves dewatering of Visean-Namurian shale basins, which were overpressured beneath later Carboniferous sediments, by seismic pumping related to Lower Permian (Hercynian) tectonism. Moderately acid, highly saline NaCl–$CaCl_2$ brines carrying hydrocarbons, Pb, Zn, Ba and F were driven into fracture systems in Asbian-Brigantian platform limestones with mineral deposition due to acid neutralisation and sulphate reduction reactions during fluid mixing. In the Northern Pennines buried, high heat production granites locally focussed ore fluids into hydrothermal convection cells so that mineral zones are spatially related to the subcrop of the granites.

[1] British Geological Survey, 154 Clerkenwell Road, London. EC1R 5DU, England
[2] British Geological Survey, Keyworth, Nottingham. NG12 5GG, England
[3] Department of Earth Sciences, The Open University, Walton Hall, Milton Keynes, MK7 6AA, England
* Present address: British Geological Survey, 64 Gray's Inn Road, London, WC1X 8NG
+ Present address: British Geological Survey, 19 Grange Terrace, Edinburgh, EH9 2LF

Mineral Deposits within the European Community
(ed. by J. Boissonnas and P. Omenetto)

1 Introduction

Metalliferous mineral exploration is generally based on the application of relatively simple combinations of geochemical and geophysical methods. Geochemical exploration is aimed mainly at the detection of anomalously high levels of metals, such as Cu, Pb, Zn, U or Mo, in surface soils and stream sediments associated with exposed or near-surface ore deposits. Most geophysical methods, such as induced polarisation, which is used for the location of sulphides, are also direct, and until recently, the overall assessment of ore deposits has been based on descriptive classifications, such as those of Smirnov (1976). This approach is useful in detecting metalliferous minerals undergoing erosion and leaching but, as pointed out by Woodall (1985), exploration is increasingly based on the scientific understanding of the ways in which ore deposits form. This is particularly true where concealed deposits are sought. For example, the Olympic Dam deposit in South Australia was located beneath 300 m of cover using predictive metallogeny (Roberts and Hudson 1983).

The detection of buried ore deposits requires an approach comparable to that used for hydrocarbon exploration, whereby exploration criteria, based on robust genetic models, are applied. The development of such methods is particularly important in northern and central Europe where exposed basement and 'older' cover formations often contain mineralisation, but where much of the land area is covered by barren Mesozoic-Recent sedimentary rocks. Because the cover is frequently thin, buried ore deposits are likely to occur at economically viable depths.

Since the late 1960s there has been great progress in understanding metallogenesis (e.g. Skinner 1983) and more recently important advances have been made in the application of image analysis computer systems (IAS) to mineral exploration. Such systems have been shown to provide a powerful method of processing a wide range of spatially related data including geological, geochemical and geophysical information (Guinness et al. 1983; Green 1984). The present study was designed to take advantage of these developments to devise methods of exploration for buried mineralisation based on criteria derived from metallogenic models using the Natural Environment Research Council I^2S image analysis system.

Carbonate-hosted deposits were chosen for the study since they are of considerable potential economic significance and they are most amenable to exploration based on methods of basin analysis developed for hydrocarbon exploration. The East Midlands of England was selected for the investigation because it contains the South Pennine Orefield and there is evidence of buried mineralisation in boreholes to the east, for example at Eakring (Lees and Taitt 1946; Ineson and Ford 1982). The orefield has been studied intensively over a long period and a large amount of mineralogical, isotope and geochemical data are available. Moreover, extensive subsurface information is available for the East Midlands; in addition to regional gravity and aeromagnetic surveys there is a wealth of seismic reflection and borehole data as a result of intense hydrocarbon exploration in recent years. The principal study region is indicated in Fig. 1. The North Pennine Orefields are

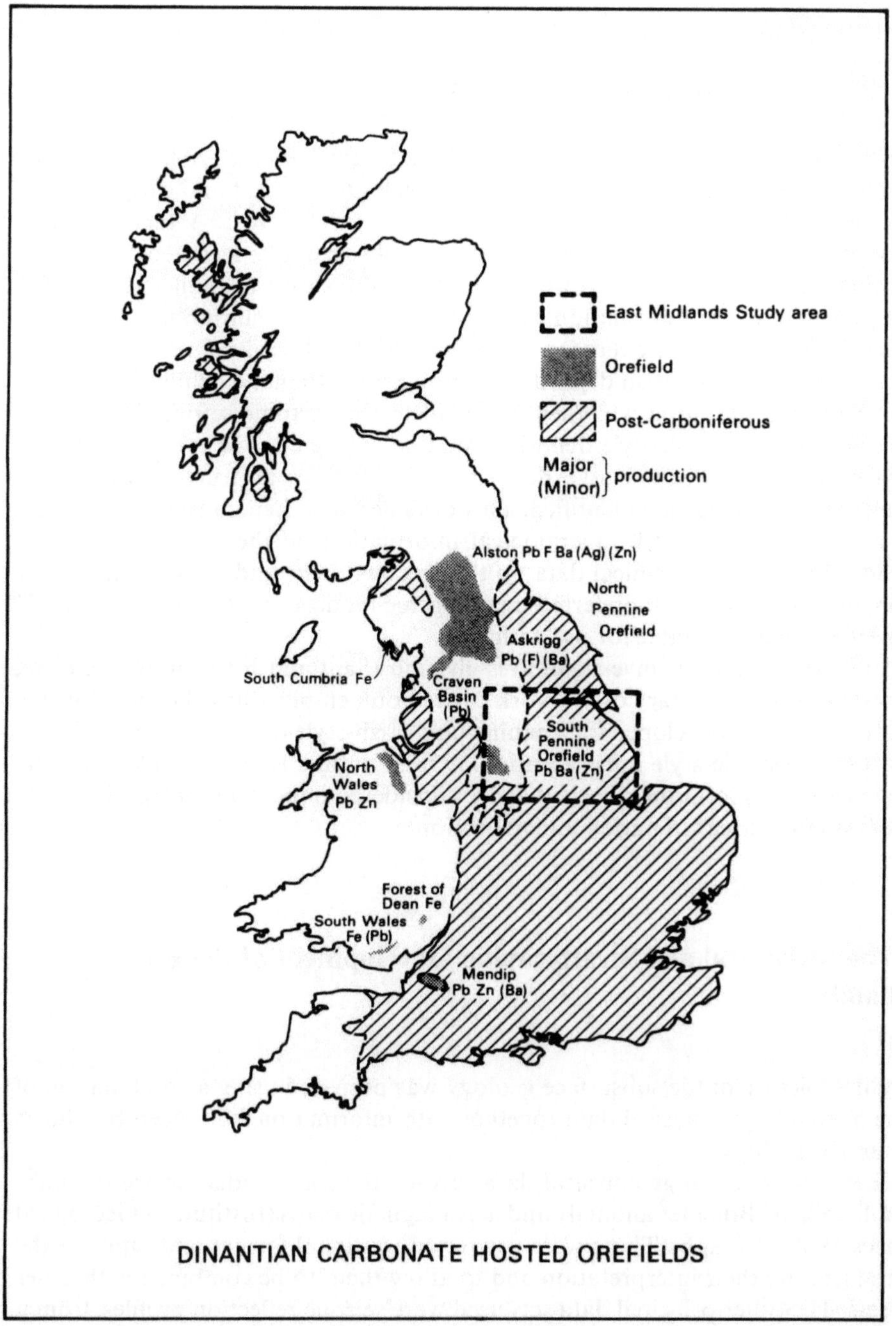

Fig. 1. British Dinantian carbonate-hosted orefields. The main East Midlands study area is *outlined*

also shown, since their metallogeny was considered in the formulation of a general model type for Pennine-style deposits.

Two types of carbonate-hosted deposits likely to occur in the Carboniferous 'older' cover rocks of eastern England were considered, those of Pennine and those of Irish-style. The mineralised limestones of the South Pennine Orefield rise more than 600 m above sea level in the Derbyshire Peak District, but in the east they are concealed beneath a mainly barren Silesian to Cretaceous sedimentary sequence. The approach adopted was therefore to use geophysical and deep geological data to elucidate the structural and stratigraphic evolution of the Carboniferous Pennine basin and to provide information at various levels in the subsurface. Irish-style deposits have not been discovered in England but rocks of similar Tournaisian facies to those which host the Irish deposits occur in central England (Phillips et al., this Vol.). Particular attention was therefore paid to the tectonic setting of the early Dinantian in which Irish-style deposits could have formed. The events leading to the formation of the Pennine-style deposits and the principal structural and stratigraphic controls were also identified. This evidence was then considered together with available isotope and mineralogical information on the Irish and Pennine orefields, and new geochemical data on the igneous rocks and shales of the study region, in order to develop preferred metallogenic models, exploration criteria and prospectivity maps suitable for use with IAS.

A full account of the investigation is given in Plant and Jones (in press). Here, we present only a summary of the work carried out, emphasising the metallogenic models which were developed for Pennine- and Irish-style deposits, with particular reference to Pennine-style mineralisation. A brief outline is also included of some of the regional exploration criteria that are considered important for the identification of potential areas of buried mineralisation.

2 The Structural and Stratigraphic Development of the East Midlands

A detailed picture of the subsurface geology was prepared using a combination of all the available geophysical data together with information from deep boreholes and surface geology.

The main regional geophysical data sets for the East Midlands are the published 1:250,000 Bouguer anomaly and aeromagnetic maps (Institute of Geological Sciences 1965, 1977a, b). These were converted to digital format and input to the IAS to facilitate their interpretation and to allow them to be combined with other data sets. The other principal data sets used were seismic reflection profiles, from a large BGS data base, mainly comprising confidential data acquired by hydrocarbon exploration companies, and stratigraphic and other logs from numerous deep boreholes in the area. A considerable amount of borehole information is available on Carboniferous and younger formations, but relatively few boreholes have penetrated Pre-Carboniferous basement. A recent compilation of heat flow measurements (Downing and Gray 1986) together with several seismic refraction profiles (e.g. Bamford et al. 1977; Whitcombe and Maguire 1981) were also used

Isopachyte maps, constructed from the seismic reflection and borehole data, were used to strip the gravity map of the effects of Post-Dinantian strata. Primary and derived geophysical data sets (including reduced to the pole magnetics and second vertical derivative and stripped gravity maps) were also prepared and these were analysed and integrated using the IAS. Detailed modelling of individual Bouguer gravity anomalies was performed using standard computer techniques. Further information on these procedures is given in Cornwell and Walker (in: Plant and Jones, in press) and Smith and Smith (in: Plant and Jones, in press). The principal features derived from interpretation of the geophysical data are shown in Fig. 2.

The information was used to provide a geological interpretation of the area to a depth of a few kilometres and to build up the structural and stratigraphic history of the region which is outlined below and summarised in Fig. 3.

2.1 Late Proterozoic-Silurian

Heterogeneities developed in the basement as a result of Precambrian and Caledonian orogenesis. To the S of the study area, Lower Palaeozoic rocks on the Midlands Microcraton remained essentially undeformed until Hercynian times. In Ireland, Wales, the Lake District and western England the structural grain of the basement has a NE–SW Caledonian trend, whereas in much of eastern England NW–SE trending basement structures have been inferred from gravity and aeromagnetic features (Fig. 2; Cornwell and Walker in: Plant and Jones, in press). The eastern England Caledonian structural province appears to be juxtaposed against that of the main Caledonian belt along a N–S zone approximately coinciding with the main Pennine axis. Early Caledonian intrusions, such as the Mountsorrel granodiorite and the South Leicestershire diorites, were probably emplaced concordantly with major structures. These bodies generally have a pronounced magnetic response but show little gravity expression and are not associated with significant mineralisation.

2.2 Lower Devonian

Low density, high heat production (HHP) granites such as Weardale and Wensleydale were emplaced, probably, as in the case of the Scottish Caledonides, discordantly and following the ending of orogenesis and uplift (Plant 1986). A suite of inferred low density granites, e.g. at Market Weighton, Newark and around The Wash (Fig. 2) also appear to cross-cut Caledonian structures. These plutons have similar geophysical characteristics to the Weardale and Wensleydale granites, which have a spatial relationship with ore deposits in the northern Pennines.

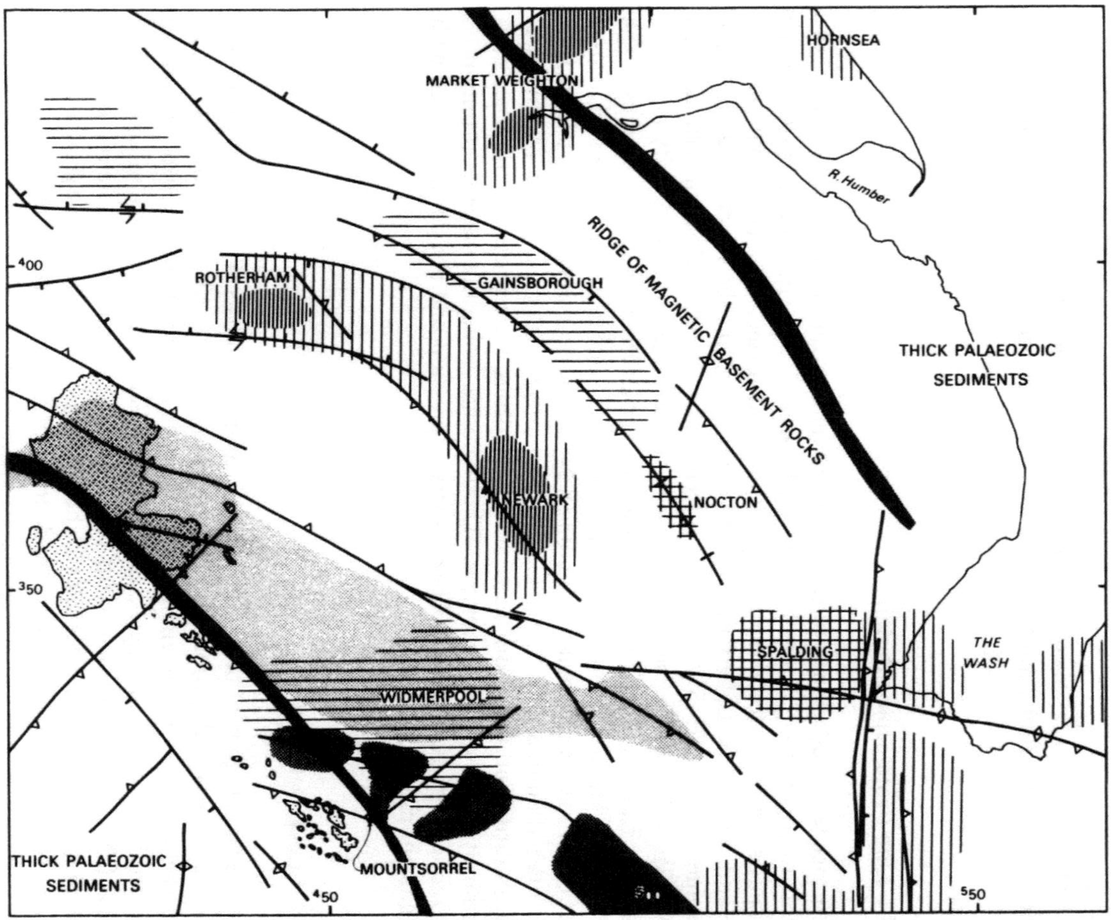

Fig. 2. Summary of the principal geophysical features of the East Midlands and their interpretation

2.3 Late Devonian-Tournaisian

Basement tilt blocks and half-graben were formed as a result of listric growth faulting during a period of crustal extension and thinning (Fig. 3a, b). The listric faults frequently follow the line of the earlier Caledonian lineaments and in places dextral displacement has been demonstrated (Arthurton 1984; Coller 1984). Low density granites appear to have played a role locally, throughout the Lower–Middle Carboniferous, by buoying up the crust to form basement highs on which little sedimentation took place. In Ireland the early onset of basin formation was related to the proximity of the Variscides foredeep, and there was widespread development of Waulsortian facies sedimentation. In the Pennine basin there is evidence of deep (>3 km) Early Dinantian basins as well as the thinner Tournaisian sequences of the East Midlands shelf.

The extension and thinning of the crust in Tournaisian times was followed in Ireland by the emplacement, especially in Limerick, of minor amounts of alkaline basic volcanics (Upton 1982). In Britain the most extensive volcanism occurred along Caledonian (NE–SW) trending structures such as the Northumberland Trough and Midland Valley of Scotland. In Derbyshire and eastern England volcanism may have been restricted because of more limited movement on the NW–SE trending structures in the predominantly dextral stress system (MacDonald et al. 1984; Leeder 1982). In Ireland unconformities developed at basin margins above the Tournaisian sequence as a result of intra-basinal submarine scarps which are possibly linked to the development of later Dinantian basins. Rapid subsidence of the basins followed with differential uplift of adjacent blocks.

2.4 Early Visean (Chadian-Arundian)

Carbonate sedimentation over a wide area of the Pennine basin was associated with the continuation of active growth faulting (such as that inferred in Derbyshire) controlling sedimentation with progressive onlap onto the culminations of tilt blocks (Fig. 3c). Waulsortian-type facies developed widely at this time, including reef knolls currently exposed in SW Derbyshire (Aitkenhead and Chisholm 1982). Argillaceous carbonates were deposited in the deeper water at the downslope end of the tilted blocks.

2.5 Mid-Visean (Holkerian)-Early Namurian

Some of the early Dinantian growth faults ceased to move at this time, perhaps because they were sealed by contemporaneous volcanism; the volcanic centres are generally aligned along these structures. A late Dinantian platform developed as a drape structure over the sealed early faults (Fig. 3d) and the apron reefs which mark the boundary between shelf and basin facies, therefore, have no direct relationship with the major basement faults. Growth on the remaining Dinantian faults is inferred to have produced erosional scarps on the newly stabilised blocks, analogous to those formed in the Arundian in Ireland. Upper Dinantian basinal

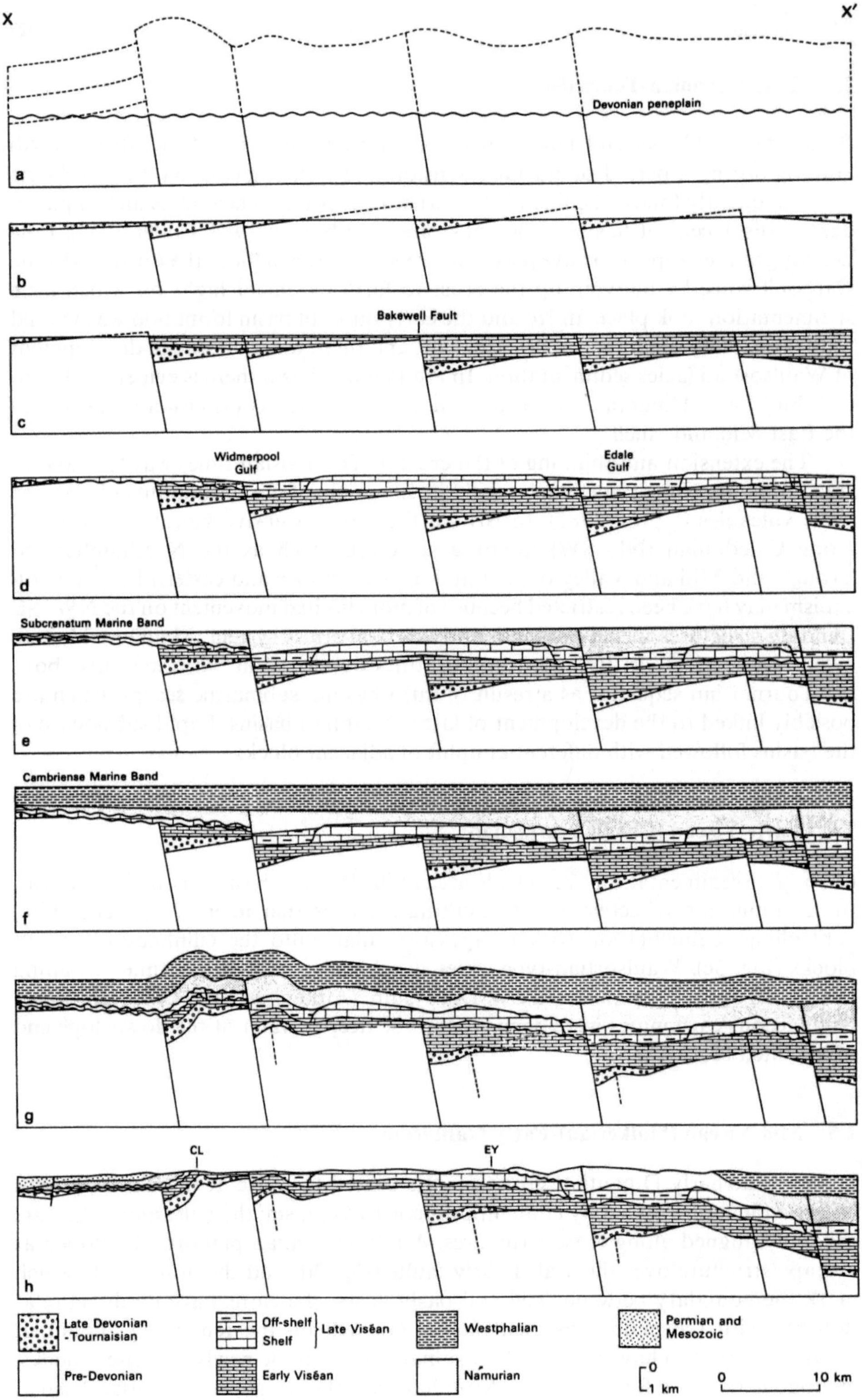
X
X'
Devonian peneplain
a
b
Bakewell Fault
c
Widmerpool Gulf
Edale Gulf
d
Subcrenatum Marine Band
e
Cambriense Marine Band
f
g
CL
EY
h
Late Devonian -Tournaisian
Off-shelf
Shelf
Late Visëan
Westphalian
Permian and Mesozoic
Pre-Devonian
Early Visëan
Namurian
0
1 km
0
10 km

Fig. 3a–h

facies sediments accumulated in the active half-graben and overlapped onto the eroded scarps at the block margins. In the early Namurian, sedimentation in the basins, such as the Gainsborough Trough, Edale Gulf and Widmerpool Gulf (Fig. 4), continued to be dominated by shales. Similar conditions are believed to have existed in Ireland where only minor late Dinantian and early Namurian areas have been preserved from the effects of later erosion.

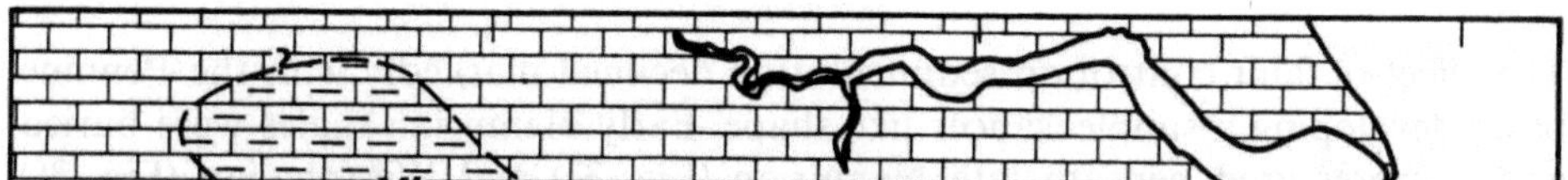

Fig. 4. Late Dinantian palaeogeography showing the position of deep basins (e.g. Edale Gulf) which persisted into the Lower Namurian. Outcropping apron reef belts are indicated; Castleton (*CA*), Earl Sterndale (*ES*) and Brassington (*BR*)

Fig. 3a–h. A model for the structural and stratigraphic evolution of the southern Pennine basin. The location of section XX^1 is shown in Fig. 4. Thicknesses of formations are schematic. **a** Late Silurian–early Devonian, deformed Caledonian basement is reduced to a peneplain. **b** Late Devonian–Tournaisian, tilted block structures, possibly controlled by pre-existing Caledonian fractures, are initiated during crustal extension. **c** Chadian–Arundian, tilted blocks continue to control thickness and facies variation. **d** Holkerian–Brigantian, some of the early Dinantian growth faults cease to move as the effects of crustal extension begin to diminish. Carbonate platforms form as drape structures across inactive faults. Intra-basin unconformities develop by scarp retreat at the newly established block margins. **e** End Namurian, Namurian deltas fill the fault-controlled basins and overlap the late Dinantian carbonate platforms. **f** Late Westphalian, the simple pattern of Westphalian thickness variation shows that major faulting was not important in controlling sedimentation. **g** Late Westphalian-early Permian, inversion of the Pennine basin and folding of the East Midlands carbonate shelf during Variscan deformation. Erosion preceded the deposition of the Basal Permian Sand in the Upper Permian. **h** Recent, continued uplift of the Pennine axis and the broad subsidence of the eastern England shelf produced the present-day outcrop pattern. The positions of deep boreholes at Caldon Low (*CL*) and Eyam (*EY*) are shown

The intensity of mineralisation in the South Pennine Orefield also increases towards the late Dinantian basin margin, which is inferred to exist beneath the Edale Gulf. Late Dinantian–early Namurian growth faulting effectively juxtaposed the shale basins against the carbonate rocks.

2.6 Late Namurian–Westphalian C

The effect of fault control on sedimentation declined markedly with the Pennine basin developing a simple, saucer-like shape. Early Namurian shales were buried and overpressured beneath late Namurian (Fig. 3e) and Westphalian (Fig. 3f) fluviodeltaic sediments. The centre of the Pennine basin is inferred to have reached its maximum depth in late Westphalian C times.

2.7 End Westphalian–Early Permian

Major transcurrent displacement occurred reactivating basement structures and generating regional ENE and WNW trending fracture systems in the cover as a result of an E–W directed stress system associated with the ending of the Variscan orogeny over Europe. The deformation reactivated basement faults with small-scale transcurrent movement taking place, in a dextral sense in NE–SW trending Caledonide basement, and in a sinistral sense in NW–SE trending basement. Inversion of the Pennine basin took place with the maximum uplift coinciding with areas of thickest Silesian deposition (Fig. 3g).

2.8 Upper Permian–Recent

Intermittent uplift and erosion along the Pennine axis continued from Permian to Recent times exposing the North and South Pennine Orefields (Fig. 3h). In Ireland uplift and erosion removed much of the post-Dinantian cover and exposed the Tournaisian rocks containing base metal mineralisation.

3 Geochemistry

Most models for the genesis of the Pennine mineralisation involve derivation of ore-forming fluids from shales or their interaction with igneous rocks (particularly buried granites or Lower Carboniferous basalts). In the case of the Irish deposits debate centres on the extent to which ore fluids have interacted with Caledonian basement. For the present study new systematic analyses were made on basement and Carboniferous shales and igneous rocks, from surface exposures and boreholes, in or near the study area (the term shale is used here in a broad sense to cover all argillaceous rocks). The results were interpreted with particular reference to sources of metals, ligands such as Cl (and F), the content of heat-producing elements U, Th and K and to provide information on the structural and stratigraphic evolution of the area. In the case of igneous rocks, mantle-normalised trace element plots and discriminant diagrams, based on the relatively immobile high field strength (HFS) elements, were used to deduce tectonic settings (Webb and Brown: in Plant and Jones, in press) while the more mobile large ion lithophile (LIL) elements were used, together with petrographic evidence, to identify the extent of water-rock interaction (Plant 1986). In the case of shales elements with low seawater-upper crust partition coefficients (K_y^{sw}) and long residence times in seawater (τ) which are more likely to be in the organic or carbonate fraction, are given most emphasis in considering ore fluid sources, while the elements with high K_y^{sw} and τ values, which are most likely to occur in refractory detrital phases, are used to provide information on the sedimentary provenance.

3.1 Geochemistry of the Basement

3.1.1 Igneous Rocks

A wide variety of intrusive and extrusive rocks, of late Precambrian to Caledonian age, occurs in the East Midlands as isolated outcrops at the surface, which geophysical and borehole data suggest are more extensive at depth.

Discriminant diagrams for granites such as Y–Nb (Fig. 5) indicate a trend towards increasing magmatic evolution from the Late Precambrian to the end of the Caledonian orogeny. Hence, the late Precambrian volcanics and diorites of the Charnwood-Leicester area probably formed in primitive volcanic arc settings (Fig. 5), but by the time the Caledonian diorites of South Leicestershire (Croft and

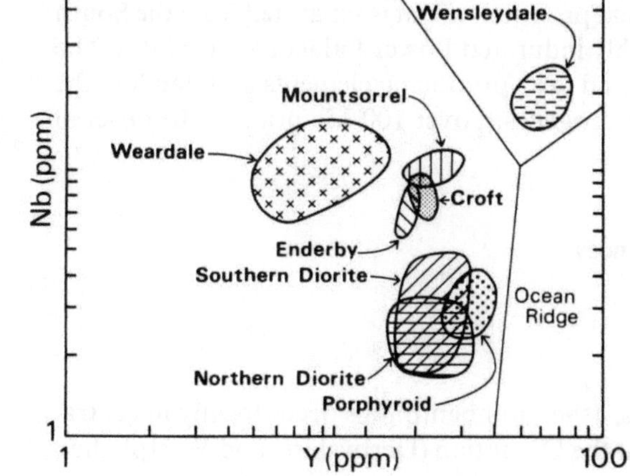

Fig. 5. Tectonic setting discrimination diagrams for granite compositions (Fields from Pearce et al. 1984) showing the affinities of late Precambrian rocks from Charnwood Forest (porphyroid, Northern and Southern Diorite) and Caledonian intrusions

Enderby) and the Mountsorrel granodiorite (Figs. 2, 5) were emplaced, the arc appears to have thickened and matured (cf. Le Bas 1982; Webb and Brown, in: Plant and Jones, in press). The Lower Devonian granites of the North Pennines (Weardale and Wensleydale) are evolved calc-alkaline suites.

The contents of the ore-forming metals (Pb, Ba and Zn) in these igneous suites are generally too low for them to be the source of the large tonnages in the orefields, although there is evidence of significant hydrothermal alteration, particularly in the more evolved Caledonian granites of Leicestershire and the North Pennines. Webb and Brown (in: Plant and Jones, in press) concluded that the granites were not the main source of ore metals, despite evidence of base metal mobilisation, particularly in the Weardale granite.

The Weardale and Wensleydale granites have similar high heat production values of 4.3 and 3.5 μWm^{-3} respectively, for their sampled upper levels (Webb and Brown, in: Plant and Jones, in press). However, since the Wensleydale intrusion is more highly fractionated, heat production may decline with depth, whereas the Weardale figure is more likely to be representative of the granite as a whole.

3.1.2 Shales

In the case of the basement shales the content of metals and heat-producing elements shows a marked difference between the Lower Cambrian–Tremadoc shales (Stockingford Shales) on the Midlands Microcraton, to the south and west of the study region (Fig. 6a) and those of the Caledonian tract of Lower Palaeozoic shales in the area itself (Fig. 6b). The former contain marked enrichments of As, Ba, Cu, Mo, Pb, S, U and V relative to average shales and average crustal abundance which, with the exception of Ba, appear to be associated with the organic fraction. The mean heat production value for the shales of 3.2 μWm^{-3} is 50% higher than that of average shale. To the north and east, however, the Lower Palaeozoic shales have only average, or below average, contents of ore-forming and heat-producing elements. The difference is attributable partly to the age and conditions of sedimentation of the two sequences and partly to the degree of induration and deformation.

The Cambrian sequence on the Microcraton represents a potential source of ore-forming elements, and has high heat production, but is separated from the South Pennine Orefield by 30–40 km of highly indurated Lower Palaeozoic argillites. The latter have low levels of ore-forming and heat-producing elements and had lost the bulk of their fluid during Caledonian orogenesis, over 100 Ma prior to the onset of Pennine mineralisation.

3.2 The Carboniferous Cover Sequences

3.2.1 Igneous Rocks

Carboniferous basaltic lavas and sills, ashes and bentonites occur locally in central and northern England, particularly in the Dinantian (Derbyshire) and Westphalian

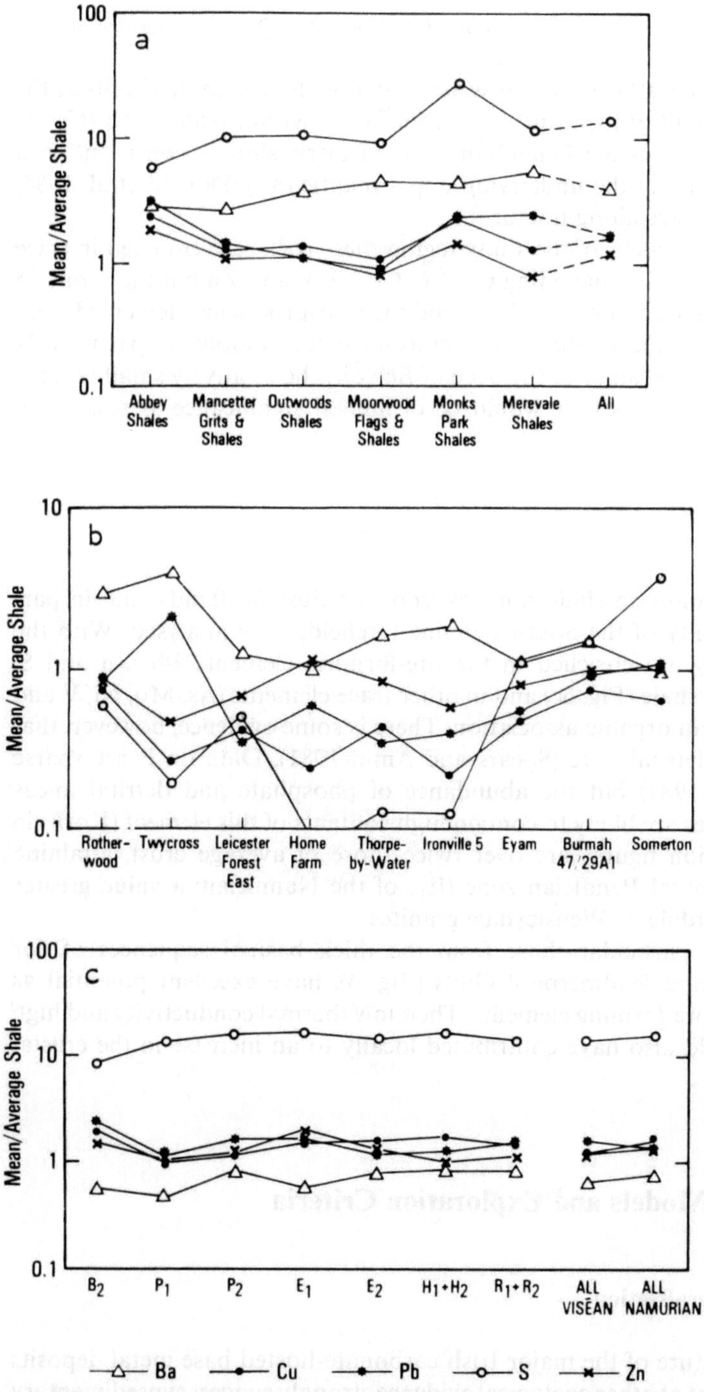

Fig. 6a–c. Variations in Ba, Cu, Pb, S and Zn contents in shales from eastern England normalised to average shale values of Turekian and Wedepohl (1961). **a** Stockingford Shales from Nuneaton, ranging in age from the Middle Cambrian (Abbey Shales) to the Tremadoc (Merevale Shales). **b** Other Lower Palaeozoic shales from eastern England of Upper Cambrian to Lower Silurian age. **c** Visean–Namurian shales from the East Midlands from the Asbian (B_2 zone) to the Marsdenian (R_2 zone)

(Vale of Belvoir). The basaltic activity was associated with deep-seated faults at the hinge lines between half-graben and is essentially of 'within-plate' type (Upton 1982). The geochemistry of the Dinantian lavas of Derbyshire is consistent with localised partial melting of the underlying upper mantle (MacDonald et al. 1984) and passage to the surface along fractures.

Compared to other rocks of the study region these rocks are enriched in trace elements of basic association including Cu, Co, Cr, Ni, V and Zn but they contain relatively low to average contents of Pb, Ba and the heat-producing elements U and Th. Although there is evidence that base metals have been mobilised, particularly from the Westphalian volcanics of the Vale of Belvoir, the relatively small volume (thichness <200 m) suggests they would be of limited significance as sources of metals.

3.2.2 Shales

Over 400 Visean–Namurian shale samples from the East Midlands, and in particular from the vicinity of the South Pennine Orefield, were analysed. With the exception of Ba they are enriched in the ore-forming elements Pb, Zn and S, compared to average shale (Fig. 6c) and in other trace elements (As, Mo, Ni, V and U) also indicative of an organic association. There is some evidence, however, that Zn may be held in detrital illite (Spears and Amin 1981). Data on F are sparse (Ramsbottom et al. 1981) but the abundance of phosphate and detrital micas indicate that the shales are likely to contain high contents of this element (Koritnig 1978). Heat production figures are over twice those of average crust, attaining 4.5 μWm^{-3} for the basal Pendleian zone (E_1) of the Namurian; a value greater than that of the Weardale or Wensleydale granites.

These shales, in particular those from the thick basinal sequences of, for example, the Edale and Widmerpool Gulfs (Fig. 4), have excellent potential as sources of fluids and ore-forming elements. Their low thermal conductivity and high heat production could also have contributed locally to an increase in the crustal geotherm.

4 Metallogenic Models and Exploration Criteria

4.1 Irish-Style Mineralisation

The strata-bound nature of the major Irish carbonate-hosted base metal deposits and a growing amount of other geological evidence strongly suggest synsedimentary or syndiagenetic ore emplacement during the Tournaisian. Model ages from Pb isotopes (Boast et al. 1981) support this, although interpretation of the Pb isotope age dates for Navan and Silvermines is difficult. The age and syngenetic style of the mineralisation differ from the younger, epigenetic orebodies of the Pennines. The

Table 1. Comparison of Pennine and Irish-style ore deposits

	Pennine	Irish
Style	Epigenetic veins	Syndiagenetic stratiform
Host rock	Late Viséan carbonates	Tournaisian carbonates
Mineralisation	End Westphalian	Tournaisian
Ore content	F > Pb > Ba > Fe > Zn	Fe > Zn > Pb and Ba
Salinity wt% NaCl	18–25	10–15
Fluids	Homogenous	Variable Neutral hypersaline to acid less saline
Fluid Temperatures	70–150 °C (higher in NPO)	100–250 °C
Lead isotopes	J-type	Normal
Igneous rocks	Not contemporaneous	Penecontemporaneous
Geothermal gradient	High (decreasing)	High (increasing)

Irish deposits also have lower salinities, generally higher temperatures (Samson and Russell 1983) and the ore assemblage has Zn > Pb > Ba with relatively little F (Table 1).

Unlike the Pennines, which are characterised by relative homogeneity of fluids, temporal changes from neutral hypersaline to acid, less saline solutions are indicated in the case of Irish deposits. Also, in contrast to the Pennine orefields, Pb isotope values vary from mine to mine reflecting changes in the Caledonian basement (H. Mills, pers. comm.) and are accompanied by systematic changes in fluid chemistry (e.g. in the Navan Mine). The Irish deposits are associated with tensional tectonics over reactivated deep faults of inherited Caledonian trend. Navan and Silvermines have been suggested to be located directly over the line of the Caledonian Iapetus suture (Phillips et al. 1976) and there is considerable structural and stratigraphic evidence for a direct association between faulting and mineralisation.

Irish-style deposits have been suggested by Russell (1978) and Boyce et al. (1983) to result from venting of warm, moderately saline brines onto the seafloor; the ore fluids being derived from the Caledonian basement by deeply excavating, convective hydrothermal systems. Deep circulation of fluids is required by this model because a normal geothermal gradient of 30°C km^{-1} was assumed. Lead isotope systematics tend to support interaction of fluids with Caledonian rocks while post-Caledonian cover formations have been considered to be relatively thin. Negative gravity anomalies have been interpreted mainly as buried granites (Phillips et al. this Vol.).

In contrast, Sangster (1986) has suggested that the Irish deposits are modified Mississippi Valley-type (MVT) deposits. Williams and Brown (1986) also prefer a basin dewatering model and present geophysical evidence for deep Upper Palaeozoic basins in some parts of the Irish Midlands. In the case of the Navan deposit, for example, a negative gravity anomaly to the southeast of the mine was previously interpreted as a granite (Murphy 1952). Williams and Brown (1986) reinterpret this as the infill of a deep sedimentary basin, partly intersected by the Trim No. 1 borehole which proved at least 1700 m of Carboniferous sediments (Sheridan 1972). If this basin were sufficiently deep, then it is possible that the temperatures of the Navan mineralising fluids may have been attained within the basin infill, particularly if the geothermal gradient were steepened at the time.

The Irish deposits were formed following a Tournasian marine transgression onto the margin of the Old Red Sandstone continent in a regime of regional thinning and extension of the crust (Anderton et al. 1979; Leeder 1982). According to the McKenzie (1978) model for sedimentary basin formation by lithospheric stretching, this early rifting stage of basin evolution is characterised by isothermal bunching, high heat flow and fracturing at the rifted shoulders of basins. This is followed by alkaline or tholeiitic basaltic magmatism, depending on the degree of stretching (Dewey 1982). The presence of a relict zone of high heat flow during the early Dinantian, coincident with the Iapetus suture across central Ireland, has been inferred by Leeder (1982). Evidence of alkali basaltic volcanism, for example in Limerick (Upton 1982), is consistent with a regime of high heat flow associated with crustal thinning, although at the late rifting stage of basin evolution.

The presence of zones of steep geothermal gradients in the Irish Tournaisian would have increased fluid circulation, particularly in the coarse basal clastic sequences in sedimentary basins and near extensional fault zones; permeability can increase from 0.05 mD to 300 mD where fault zones intersect sediment piles at 1 km depth (Barnes 1979). The lower salinity and hence relatively high buoyancy of fluids involved in Irish-style ore deposits may also have assisted fluid circulation. The ore assemblage with Zn > Pb > Ba and minor F is consistent with ore fluid compositions controlled mainly by dissolution-precipitation reactions with rocks of approximately intermediate calc-alkaline bulk composition. The ore fluid and Pb isotope chemistry of the deposits could thus be explained by interaction of the fluids with the Caledonian basement. Alternatively, they could reflect interaction with clastic sediments derived from the basement. The chemistry of the Old Red Sandstone basins of Scotland, for example, closely follows that of the Caledonian mountains from which they were derived (Plant et al. 1986). Interaction of ore fluids with Devonian or basal Carboniferous clastic sediments might thus account for the chemistry and Pb isotope systematics of the Irish-style deposits. Marked hydrocarbon anomalies associated with the mineralisation (Carter and Cazalet 1984) also suggest that the ore-forming fluids contained an important component of sedimentary basinal brines.

Overall, a model involving intra-basinal fluid flow, as a result of the heating, rapid subsidence and compaction of immature clastic sedimentary infills in tectonically controlled basins accounts for many of the geochemical, stratigraphic and tectonic features of the Irish deposits. Fluids from basins were channelled to the surface via growth faults during the Tournaisian, mineralising the thin Tournaisian successions at the basin margins (Fig. 7a). Continued movement on the main growth faults caused the submarine erosion of some of the Irish orebodies by scarp retreat in mid-Dinantian times (Boyce et al. 1983). Rapid subsidence of the basins followed ore deposition, with differential uplift of adjacent blocks. This is indicated by the dolomitic and chert breccias which overlie the orebodies at Silvermines and by boulder conglomerates, deposited by debris flows, over the ore zone at Navan, a feature which was important in preservation of the deposits (Andrew and Ashton 1985).

Such a model is consistent with the extensional tectonic regime and high geothermal gradient characteristic of the Tournaisian, and of zones of crustal thinning and extension in general. It differs from that of MVT deposits, and from geothermal fields such as Cheleken, which are characterised by normal geothermal gradients of 20–30 °C km^{-1} (Barnes 1979).

4.1.1 Exploration Criteria

In the case of Irish-style deposits the regional disposition of Carboniferous tilt blocks is one of the major factors controlling the location of deposits and listric faults, and areas immediately adjacent to them, have direct exploration significance. In Ireland mineralisation tends to be associated directly with the fault plane, e.g. Silvermines and Tynagh, or to occur in the small fault terraces at basin margins. Evidence of reactivation of Caledonian structures suggests that the identification of basement lineaments is also relevant to exploration. Listric faults and basement lineaments can be mapped using seismic and other geophysical data sets and prospectivity maps of the study area using these and other criteria are given in Plate 1.

The results of the present study suggest that the structural environments which were developed in the Pennine area during the early Dinantian were very similar to those in which Irish deposits formed (Fig. 7a). In England, uplift and erosion to bring the deposits to a level where they could be discovered by conventional geochemical or geophysical exploration has not occurred, however. In the Pennines, as in Ireland, there is evidence of deep early Dinantian sedimentary basins as well as the thinner Dinantian sequences of the East Midlands shelf. The deep basin infill sequences have yet to be proved by boreholes but considerable thicknesses (of the order of 3 km) of early Dinantian strata have been inferred in the Edale and Widmerpool Gulfs from seismic reflection profiles. The disposition of early Dinantian rocks in the East Midlands is shown in Plate la. As in Ireland, the generation of fluids of appropriate temperatures appears to have depended on the ambient geothermal gradient which was probably elevated in the inferred regime of crustal extension. The geothermal gradient of about 50 °C km^{-1} is thought to have persisted until Silesian times; in the Lower Dinantian it may well have been increased by thermal anomalies ahead of the eruption of Upper Dinantian volcanics. Both the Edale and Widmerpool Gulfs are controlled by major bounding faults at their southern boundaries, whereas the main fault controlling the Gainsborough Trough is at its northern boundary. The areas flanking these faults are considered to merit further investigation for Irish-style deposits.

4.2 Pennine-Style Mineralisation

The starting points for developing the metallogenic model presented here are the suggestion of Dunham (1983) that the Pennian mineralisation belongs to the

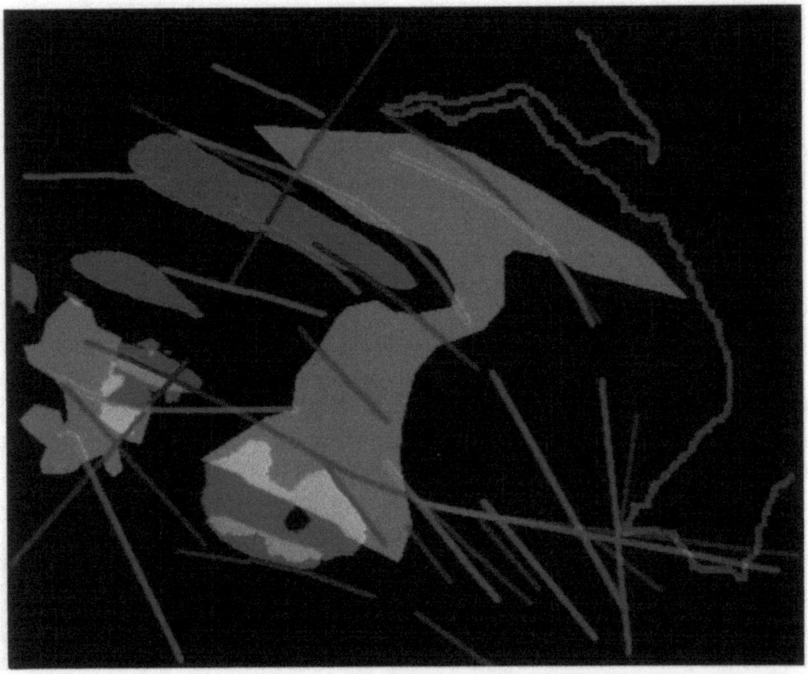

a

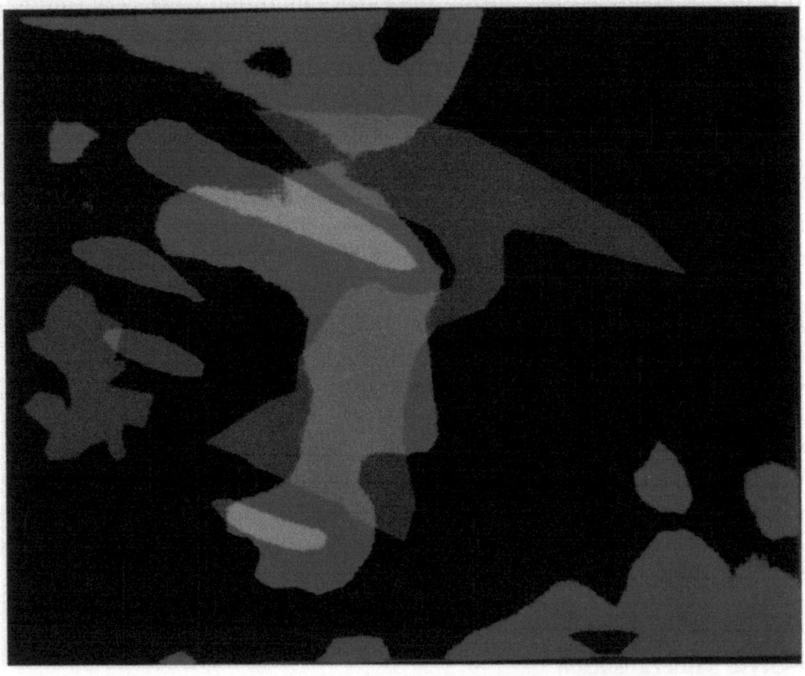

b

Plate 1. Prospectivity maps. *Irish-style mineralisation*: **a** Inferred Carboniferous igneous centres (*red*); speculative Dinantian basins (*green*); outcrop and subcrop of Brigantian limestones (*blue*) and geophysical lineaments (*red*, based on gravity evidence; *blue*, on aeromagnetic evidence). *Pennine-style mineralisation*: **b** Main Bouguer gravity lows (*red*); Dinantian basins (*green*); Brigantian limestones (*blue*). **c** As **b** with addition of faults determined from deep geological evidence. **d** As **b** with addition of geophysical lineaments (as in **a**)

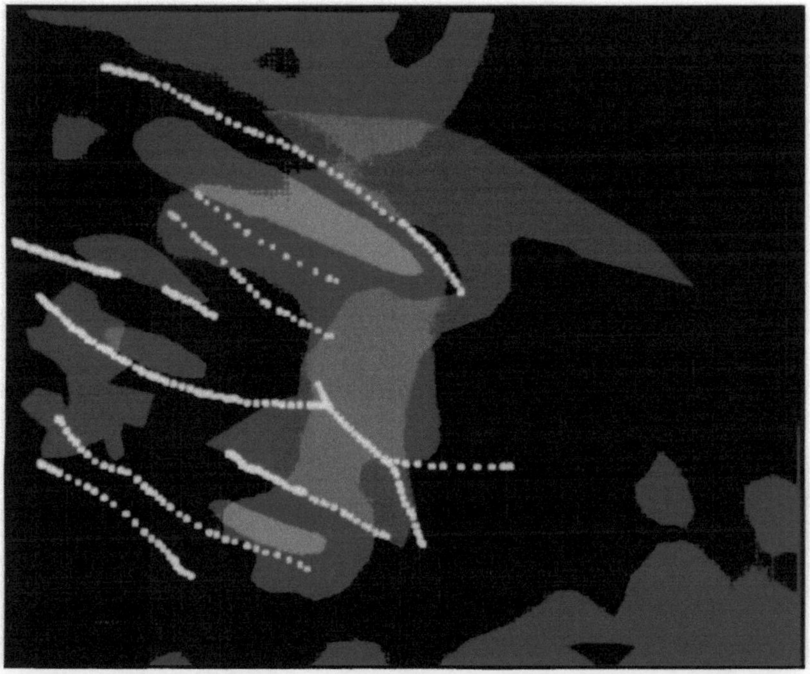

c

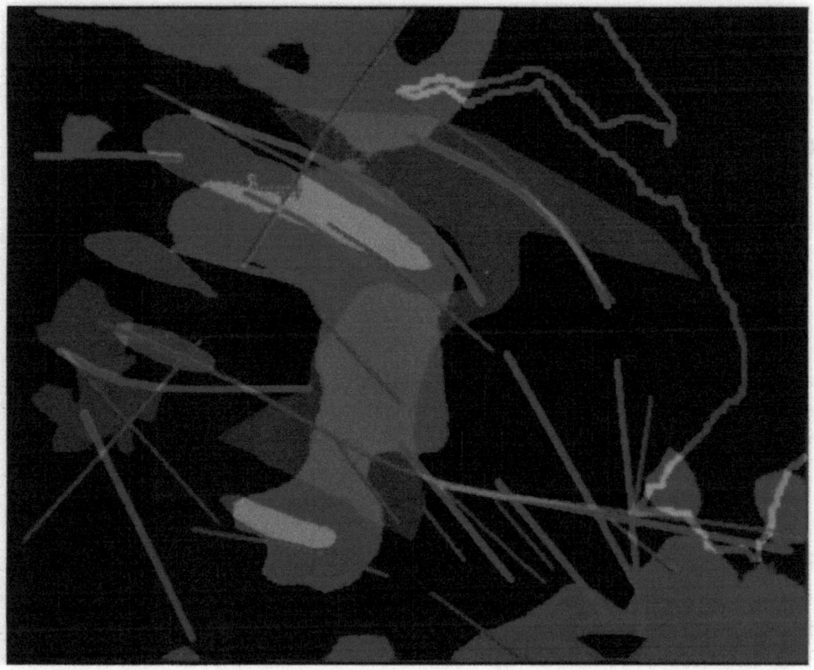

d

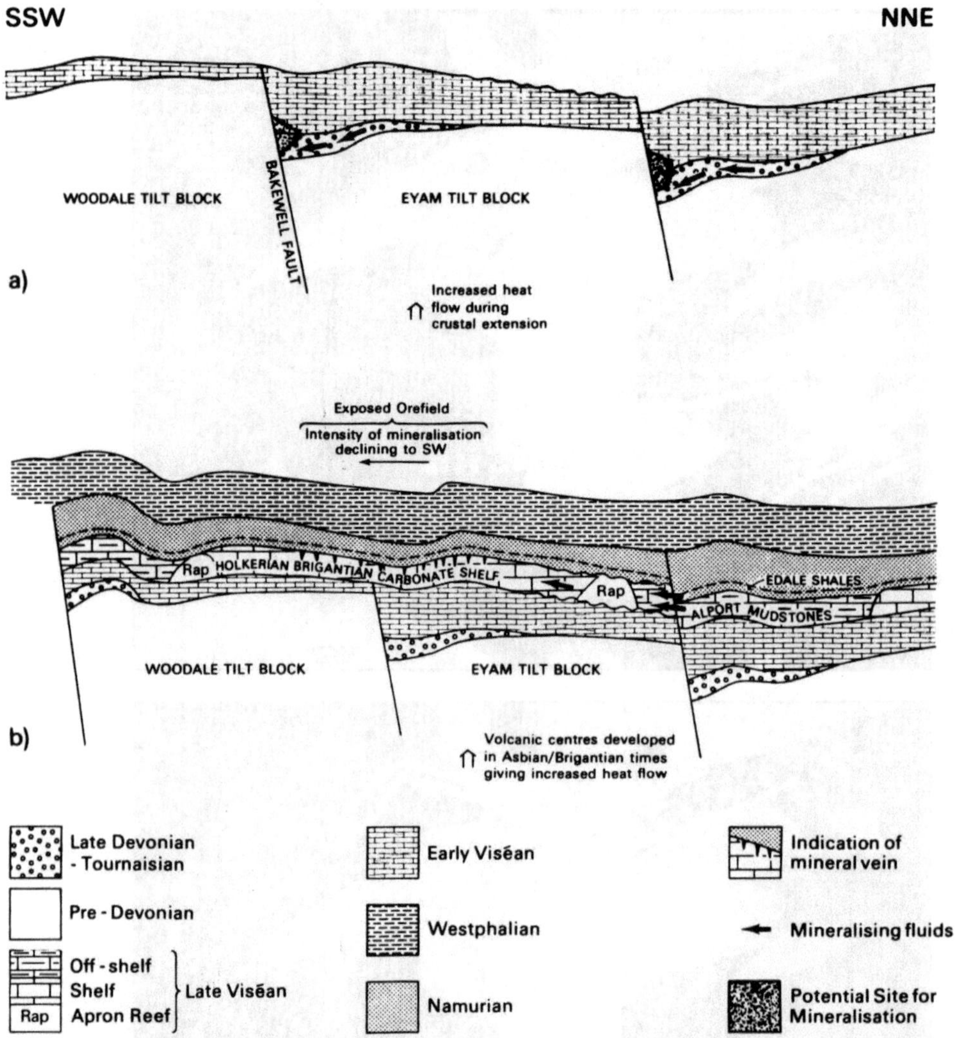

Fig. 7a. Proposed model for Irish-style mineralisation applied to the southern Pennines. **b** Proposed model for Pennine-style mineralisation applied to the southern Pennines. Ore fluids are expelled from Visean–Namurian shales by seismic pumping related to movement on reactivated listric faults and migrate into shelf carbonates

Mississippi Valley-type (MVT) genus of ore deposits (Table 2) and the basinal brine models for such deposits first proposed by White (1958), whereby they are attributed to metal deposition from warm brines, migrating from depth in sedimentary basins to their margins. The processes involved are controversial, with some authors favouring dewatering as a result of sediment compaction (e.g. Jackson and Beales 1967) whilst others envisage episodic dewatering with overpressuring driving sudden bursts of deep brines towards the basin margins (e.g. Cathles and Smith 1983).

Table 2. Characteristics of Mississippi Valley-type ore deposits

Characteristics of MVT deposits:
1. Mainly Phanerozoic age with maxima in Cambro-Ordovician and Carboniferous
2. Epigenetic deposition in non-orogenic settings in platform carbonate (sandstone) host rocks with normal geothermal gradients (20–30 °C km^{-1})
3. Close association with shale-filled basins.
4. Lack of association with contemporaneous igneous activity
5. Common association of Pb, Zn, Ba and F in ore deposits of simple mineralogy
6. Formation at low temperatures (mainly in the 90–100 °C range) from high salinity fluids
7. An association with hydrocarbons, especially petroleum
8. Low Ag content of galena and very low Au content

Table 3. Comparison of Pennine and Mississippi Valley-type ore deposits

	Pennine	Mississippi valley type
Style	Fracture-controlled veins	Strata-bound replacement
Ore content	F > Pb > Ba > Fe > Zn	Zn > Pb and Fe > F (F and Ba locally important in fracture controlled veins)
Fluid temperature	70–150 °C (up to 210 °C in NPO)	90–100 °C
Igneous rocks	Present	Generally absent
Geothermal gradient	High	Normal

An alternative model (e.g. Bethke 1986) holds that gravity-driven groundwater flow, due to topographic differences across the basin, carries warm fluids from deeply buried strata into shallow sediments. The application of each of these models to the Pennines is considered, together with more detailed evidence on the source of fluids, metals, heat and the processes involved in ore deposition. The characteristics of the Pennine orefields are compared with those of MVT deposits in Table 3.

Sangster (1983) and Anderson and MacQueen (1982) stress the diversity of MVT deposits which vary in their mineralogy, deposit type, host rock lithology and chronology within the overall scheme. The galena and/or fluorite-dominant deposits of the North American MVT orefields are the most similar to those of the Pennines (Dunham 1983). In the Pennines Pb:Zn ratios are as high as 100:1 and fluorite is a major component of the deposits.

4.2.1 Source of Fluids

Several lines of evidence suggest that the orefields were generated from shales in the late Dinantian–Namurian basins which are enriched in trace metals. The trace element suite, with enrichment in P_2O_5, and high U/Th ratios, suggests that the ore-forming metals are largely associated with the organic and/or carbonate fraction of the shale from which they could have been readily mobilised. High contents of Rb and U account for the enrichment of the ores in radiogenic Sr and Pb isotopes

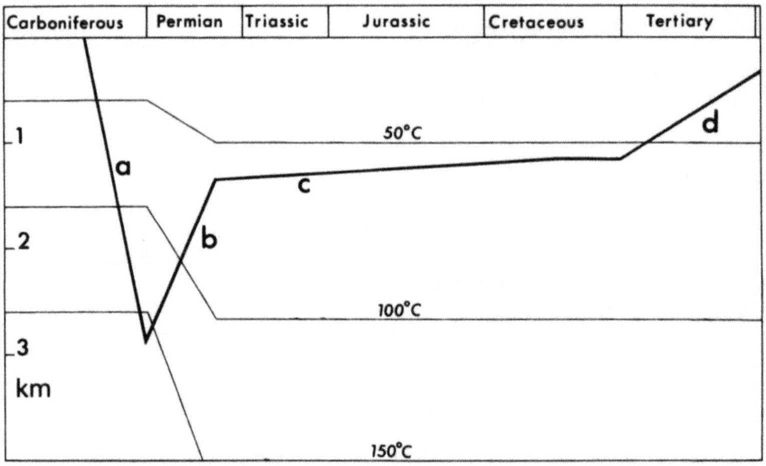

Fig. 8. Burial history of basal Namurian shales in the Edale Gulf. Isotherms were calculated assuming an elevated Carboniferous geothermal gradient which declined in Lower Permian times. *a* Silesian thickness decompacted using standard curves. *b* Enhanced compaction during Variscan deformation, but reduced depth due to uplift and erosion. *c* It is assumed that the area remained elevated through this interval. *d* Further Tertiary uplift

and the J-type model Pb ages of the ore assemblage. Moreover, calculated heat production values for the Visean–Namurian shales of 3.6 μWm^{-3} approximate those of the buried high heat production granites of Weardale and Wensleydale. Together with the thermal blanket effect of the flat-lying, low conductivity shales their high heat production may have assisted the maturation of hydrocarbons and ore fluids. The late Dinantian–Namurian shales also occur adjacent to the orefields and structural and stratigraphic analysis suggests that the evolution of the basins was such that ore fluids of appropriate temperatures could have been generated in late Westphalian—Lower Permian times.

This model for the derivation of the fluids was examined in detail for the Edale Gulf using a thermal history derived from decompaction procedures constrained by vitrinite reflectance measurements (Fig. 8).

The basin reached its maximum depth in Westphalian times, prior to Variscan deformation, when the base of the Namurian is calculated to have been at a depth of about 2.9 km. In order to generate ore fluids at this time at temperatures equivalent to the highest fluid inclusion temperatures observed in the South Pennine orefield (157 °C, Atkinson 1983), a geothermal gradient of about 50 °C km^{-1} would have been required. This figure is close to the Westphalian (pre-Whin Sill) geothermal gradient of 54 °C km^{-1} which has been proposed by Suggate (1981) for the Alston Block. The development of a similarly elevated gradient in the East Midlands may be partly related to the presence of active volcanism in the area throughout the Carboniferous and to high heat production, low conductivity shales in the basin. Calculations of the time and temperature index (TTI) for basal Namurian shales show that most of the Pennine area lay within the oil generative window in late Westphalian and early Permian times. In the deepest parts of the basin Namurian shales had already begun to generate gas during this interval (Kirby et al., in press).

The inversion of the Pennine area during Variscan deformation and the subsequent removal of most of the Westphalian cover ensured that high temperatures could not be attained subsequently within the Namurian shale succession without assuming unrealistically high geothermal gradients (exceeding 100 °C km^{-1}). A Post-Carboniferous rise in the geothermal gradient much above the present-day value of 25–30 °C km^{-1} can be eliminated because of the effect it would have had on the Permian and Mesozoic rocks which are preserved elsewhere in the East Midlands (Kirby et al., in press). Hence, high temperature fluids (160 °C) were likely to exist in the deeper parts of the East Midlands basins for only a short period of time after the end of the Westphalian.

Thermal modelling does not, therefore, support isotope and other evidence which indicate episodic mineralisation until at least Jurassic times (e.g. Ineson and Mitchell 1973; Ineson and Ford 1982). These events probably represent minor, lower temperature mineralisation or redistribution/re-equilibration of the ore/gangue mineral assemblage.

4.2.2 Composition of Ore-Forming Fluids

Overall there is strong evidence that one component of the ore-forming system was formational water derived from shales. Oxygen isotope data (enrichment in $\delta^{18}O$), the sulphide sulphur isotope data, δD values obtained on fluorites and the composition of fluid inclusions (which are in the Na–Ca–Cl system) are in the range of those from the main stage of Mississippi Valley mineralisation which are of undisputed oil field brine affinity (Taylor 1974). Sheppard and Langley (1984), on the basis of $\delta^{18}O$ and δD data and the concentration of major elements in modern Na–Ca–Cl brines (which contain 8360–197 000 mgl^{-1} of total dissolved solids) from collieries in northeast England, have suggested that ore fluids could be generated rapidly by reaction between meteoric water and evaporites and clays. Fluid inclusion studies (e.g. Rogers 1977; Atkinson 1983) and REE data on fluorites (e.g. Atkinson 1983; Shepherd et al. 1982) indicate fluid homogenisation throughout the source area of each orefield, however, with infinite reservoir conditions of fluid generation. Hence, a model involving equilibration between fluids and rocks is preferred to a simple meteoric recharge-discharge system to generate the evolved, highly saline brines involved in the Pennine mineralisation.

There remains the problem of Pb dominance, the quantities of fluorite in the orefields and the association with tectonism. Oil field brines encountered at 2.4–4 km depth (comparable to that proposed for the generation of the Pennine ore fluids) in central Mississippi carry up to 367 mg^{-1} Zn and 182 mg^{-1} Pb (Carpenter et al. 1974) and the modern brines described by Sheppard and Langley (1984) are also enriched in Zn. The enrichment of Zn relative to Pb in these fluids is consistent with the ratio of Zn:Pb of about 5:1 in the crust generally and in the Lower Palaeozoic and Carboniferous shales of the East Midlands. It is also consistent with the higher solubility of Zn than Pb in chloride brines generally (Barrett and

Anderson 1982). The high ratio of Pb:Zn throughout the Pennine orefields has been attributed to a fine control of pH during ore deposition, whereby Zn was retained in solution following deposition of Pb (Atkinson 1983). It is considered unlikely that Zn would remain in a system in which excess H^+ ions would have been readily consumed by reaction with limestone, however, as indicated by the presence of an outer calcite zone in the South Pennine Orefield (e.g. Firman and Bagshaw 1974). Moreover, mineral textures suggest an approach to equilibrium conditions (Atkinson 1983) while only relatively small amounts of sphalerite are found throughout the range of structural settings of mineralisation in the Pennines.

The evidence thus suggests a predominance of Pb:Zn in the ore fluid at or near the source which is difficult to account for by simply dissolution models. Following Hanor (1979) and Colman et al. (in: Plant and Jones, in press) we therefore suggest that the high content of Pb (and F) reflects the control of ore fluid composition by diagenetic phase transformations. In the case of the South Pennines, REE data on fluorites also suggest the probable involvement of clay minerals in these reactions. Geochemical data from Visean–Namurian shales, discussed earlier, indicate that Pb may have been associated with the organic fraction of the sediment, while Zn correlates with illite content and thus Pb may have been more readily available for leaching into ore fluids.

4.2.3 Fluid Transport

The stratigraphic and structural evolution of the Pennine region indicates that the Pennine axis has been a N–S zone of uplift from the Lower Permian (Variscan) tectonic event until the present. Hence, it is difficult to apply a gravity model of fluid movement since the ore fluids would tend to be driven away from the orefields towards the eastern seaboard of Britain. Moreover, the scale of the Illinois Basin, for which a numerical model of gravity flow was developed by Bethke (1986), is markedly different from that of the Pennines where fluids probably travelled distances of the order of only a few tens of kilometres. Geochemical evidence, discussed above, is also inconsistent with a simply basin recharge-discharge model of ore deposition. This model is therefore not considered further. The early diagenetic basin dewatering model generally applied to MVT deposits cannot be applied directly to the Pennine orefields either, however, because of the time interval between the development of the shale-filled basins (Dinantian–Namurian) and the main episode of mineralisation (Lower Permian). In the South Pennines the reconstruction of the structural–stratigraphic evolution of the area also indicates that mineralisation occurred when the host carbonate rocks were at depths of the order of 2 km.

In the Pennine orefields and those of central Kentucky and Illinois–Kentucky there appears to be an association between Pb dominance, a high fluorite content in the orefields generally and the fracture control of orebodies. Control of ore fluid compositions by diagenetic phase transformations may, in turn, reflect overpressuring and/or the heat production and thermal blanketing of shales in the sedimentary basins which are thought to be the source of ore fluids. In the Pennines release of these evolved basinal fluids appears to have depended on regional tectonism and may be one of the most important factors in the genesis of Pb–F dominant MVT deposits generally. Tectonism during the Lower Permian would not only explain the disposition of the mineralisation in regional vein systems, it is also consistent with time constraints, whereby fluids of appropriate temperature were available in the sedimentary basins for only a short time after the end of Westphalian. The amount of metals in the orefields can be accounted for; if the volume of the Visean–Namurian sediments in the Edale Gulf alone is calculated (assuming a conservative thickness of 2 km), it is necessary to extract only very small percentages of the Pb, Zn, Ba and F present to provide all the recorded output from the South Pennine Orefield (2, 0.1, 0.03 and 0.1% respectively). It does not, therefore, seem necessary to derive fluids from more distant sources such as the North Sea Basin, as has been proposed (e.g. Ford 1976; Ineson and Ford 1982).

4.2.4 Deposition of Mineralisation

Formational fluids from shale basins could have transported Pb, F, Ba, Zn and Cl, but the transport of base metals and sulphide species in the same solution requires temperatures in excess of 200 °C, or exceptionally low pH values, which are outside the ranges indicated for the Pennine orefields by fluid inclusion studies (e.g. Roedder 1967; Rogers 1977). The most likely potential sources of sulphur involve reduction of sulphate either (1) carried in the brines or (2) held in formational fluids in limestone or evaporates. The supply of SO_4^{2-} in the brines is difficult to reconcile with the quantity of baryte which has very limited solubility in SO_4^{2-} bearing solutions (Barnes 1979). Moreover, in the Pennine orefields, sulphide-rich zones are generally surrounded by oxidised SO_4^{2-} rich zones. Such a pattern is consistent with a low SO_4^{2-}, but Ba-enriched brine, encountering a region rich in H_2S.

The evidence thus favours a mixing model of ore deposition with the deposits formed by:

1. The interaction of metalliferous brine with a zone enriched in H_2S in the limestones in which the formational fluid contained SO_4^{2-}. Thermal degradation of organic sulphur compounds in the more deeply buried parts of the carbonate sequences could release H_2S (Hunt 1979). Or:
2. The interaction of metalliferous brines containing hydrocarbons with a sulphate formational fluid in the limestone to produce a sulphide zone surrounded by a sulphate zone, with extensive re-equilibration of the sulphide sulphur isotope systems with the basinal fluids.

A mixing model of ore deposition is also favoured by the distribution and composition of fluorite (and baryte). Although the ore solutions may be in equilibrium with carbonate during transport, deposition of sulphides from chloride complexes releases acid according to the reaction;

$$\underset{\text{metal chloride}}{MeCl_2} + H_2S = \underset{\text{metal sulphide}}{MeS} + 2HCl$$

The acid could not escape from the system without dissolving carbonate: the consumption of H^+ ions produced by sulphide precipitation involving neutralisation reactions of metal chloride brines of the type:

$$MeCl_2 + CaCO_3 + H_2S = MeS + CaCl_2 + H_2CO_3.$$

Hence, since neutralisation reactions are important in promoting sulphide precipitation, carbonate is the most likely host rock for the deposition of MVT and Pennine deposits.

The release of calcium would precipitate fluorite as in the case of geothermal waters in the western USA (Nordstrom and Jenne 1977). The chemistry of the fluid inclusions, the REE patterns in fluorites and $^{87}Sr/^{86}Sr$ ratios from the Pennines are most readily explained by the mixing of fluids with similar NaCl but different $CaCl_2$ contents consistent with limestone formational fluids. Deposition of baryte and fluorite by mixing is also consistent with the mineralogical, geochemical and isotope evidence (Colman et al. in: Plant and Jones, in press).

4.2.5 The Role of Granites and Carboniferous Basic Volcanics

In the case of the Alston Block a buried radiothermal granite appears to have played an important role in localising mineralisation (e.g. Solomon et al. 1971), although the ore-forming fluids are nevertheless thought to represent brines derived initially from adjacent Carboniferous basins (Dunham and Wilson 1985).

The Weardale and Wensleydale granites are high heat production granites, although Wensleydale is smaller and there is evidence that radioactive contents decline with depth. The Weardale granite appears to have been hydrothermally altered contemporaneously with its emplacement in Lower Devonian times. Regeneration of the Weardale hydrothermal system, which occurred as a result of tectonism and the higher than normal (but declining) geothermal gradient in Lower Permian times focused the ore fluids giving rise to the close spatial relationship between the mineralisation and the subcrop of the granite. In the case of the Wensleydale granite, which is relatively unaltered, it seems likely that it formed only a generalised hot spot and its main role in mineralisation may have been to buoy up the Askrigg Block.

Differences in such features as the REE patterns in fluorites, the presence of REE-bearing mineral phases in veins overlying the buried granite cupolas of the Weardale intrusion, the K/Na ratios of the fluids and the different temperatures of ore deposition between the orefields can be related to the extent to which the reaction occurred between buried HHP granites and the migrating ore-forming brines.

In the case of Carboniferous igneous rocks it has been suggested that these may have provided a source of F (Dunham 1983). This is unlikely, however, since the ore assemblage (low Cu, Co, Ni, etc.) is generally incompatible with extensive leaching of basic rocks. The main role of the volcanics appears to have been as permeability barriers to the ore fluids; they may also have provided a supply of bases for acid neutralisation reactions promoting ore deposition.

Hence, in common with MVT deposits generally, there appears to be no association between ore deposition and contemporaneous igneous activity, although the presence of igneous rocks in the zone of fluid flow and fluid/rock interaction has modified the MVT signature of the Alston Block (Brown et al. 1987).

The overall model developed here for Pennine-style mineralisation is illustrated in Fig. 7b and summarised in the final section.

4.2.6 Exploration Criteria

Exploration criteria for locating buried Pennine-style deposits are described in detail in Colman et al. (in: Plant and Jones, in press). In the East Midlands the most important criterion is the regional disposition of late Dinantian platform carbonates (the main host rock) and late Dinantian–early Namurian shale basins (the main source of mineralising fluids). At depth, the identification of basins can be made using Bouguer gravity anomalies, some of which relate to the significant density contrast between shales and Dinantian limestones, while the detailed internal structure of the basins can be interpreted from seismic reflection surveys.

Geological and geophysical data indicate that the late Dinantian–early Namurian shale basins in the Pennines and East Midlands are asymmetric graben or half-graben. In the modified dewatering model proposed the most intense mineralisation is predicted to occur in the carbonate platforms adjacent to the thickest part of the shale sequence, alongside the main bounding fault of the basin. In contrast, a basin dewatering model would predict mineralisation to occur updip of the deepest part of the basin. Gravity and seismic data can be used to map the disposition and form of shale basins and the second vertical derivative of the gravity data, and seismic interpretation, can be used to identify listric faults.

Buried HHP granites can be mapped using gravity data, although they can be difficult to distinguish from sedimentary basins. Platform carbonates are invariably developed in those parts of the Pennines which are situated above Caledonian granites. This relationship can be used to indicate the presence of a suitable carbonate host rock, but cannot be used to infer the presence of mineralisation. The position of the Wensleydale granite beneath the Askrigg Block, for example, does not indicate directly which parts of the overlying carbonate platform are mineralised. Similarly, the absence of a granite beneath a structural block, as in Derbyshire, does not preclude the development of mineralisation. The major contribution of the Caledonian granites is probably their structural role in controlling which parts of the pre-Carboniferous basement nucleated as stable blocks. Some large volume, low density granites, characterised by distinct Bouguer gravity lows, which are likely to have high heat production, can also focus hydrothermal processes.

Tectonic models based on stratigraphic and structural evidence are of particular value in reconstructing pathways of mineralising fluids. The measurement of fault, vein and joint trends, together with vein widths and lengths in the exposed orefields, can be used to predict the likely shape, size and trend of orebodies in the subsurface. ENE and EW vein systems, which are the most prospective in the

Pennines, originated as a result of pervasive east-west directed stress during Variscan deformation. At the time of mineralisation, the Derbyshire carbonate platform was a buried anticlinal drape, which culminated structurally above the concealed sites of the early Dinantian Bakewell and Bonsall growth faults. Mineralising fluids migrated towards these structural culminations beneath an impermeable cover of Namurian shales.

An approach based on modern basin analysis methods which are used in hydrocarbon exploration can thus be developed for Pennine-style mineralisation and MVT ore deposits generally. Structural and stratigraphic reconstructions can be used together with thermal maturity indicators, such as vitrinite reflectance data, to reconstruct the thermal history of the shale basins and carbonate host rocks, while the application of sedimentary decompaction models provides an estimate of basin depth variation with time. The geothermal histories of various parts of the basin can be compared with fluid inclusion data from the exposed orefields to provide an indication of the expected ranges of temperatures in the mineralising fluids. Calculations of the amount of fluid produced by the compaction of the source rocks can be linked with observations on the known metal contents of formational fluids to estimate the volume of potential ore deposits.

Other regional exploration criteria for Pennine-style mineralisation include the presence and type of basement lineaments, the composition of the sedimentary basins and the distribution of additional heat sources such as igneous rocks and radiothermal granites or shales. Carbonates promoting ore deposition as a result of acid neutralisation reactions are probably of more direct significance for Pennine and MVT deposits than for Irish-style mineralisation. These and the other exploration criteria are discussed in detail in Plant and Jones (in press).

The areas considered prospective for Pennine-style ore deposits are shown in Plate 1 b–d. They are based on a combination of gravity lows, considered indicative of buried high heat production granites, basement lineaments identified by gravity and aeromagnetic data (for which there is additional seismic/borehole evidence of listric movement in Carboniferous times) and Asbian/Brigantian shelf limestones juxtaposed by such faults against deep Visean–Namurian shale basins. The prospectivity map correctly identifies the eastern margin of the South Pennine Orefield. The value of combining a number of criteria is shown by the identification of buried granites around The Wash which lack appropriate Carboniferous cover and hence are not considered prospective for Irish-style or Pennine-style carbonate-hosted deposits.

5 Conclusions

It is clear that the formation of Irish-style deposits was associated with crustal rifting and conditions of high heat flow in the Lower Dinantian which transformed the margins of the newly consolidated Caledonian continent into deep marine basins. New evidence suggests that comparable rocks occur at depth in the study region. In contrast, MVT deposits generally form during periods of regional basinal subsidence under declining or normal geothermal gradients. In terms of the McKenzie (1978) model of basin formation Irish-style deposits are thus associated with the waxing phase of the cycle, as hot asthenosphere rises beneath thinned continental crust. MVT deposits are formed during the waning phase, associated with subsidence and isostatic adjustment of the crust following the decline of the thermal anomaly in the mantle. In the Pennines and East Midlands of England such deposits could have formed during Namurian-Westphalian times, when a major crustal sag is inferred to have developed in the central Pennines (Leeder 1982), but the formation of large replacement ore deposits, comparable to those of the Mississippi Valley, would depend on such factors as the availability of open structures for mineralisation and hence the depth of burial of likely host rocks.

The Pennine orefields are believed to be of a modified, tectonically-driven, Mississippi Valley type, in which Pb and F are dominant. Ore fluids were pumped from shale basins, such as the Edale Gulf, into regional ENE fracture systems. These formed in response to sinistral transcurrent movement along reactivated, NNW trending, early Carboniferous growth faults. The ore fluid chemistry was probably controlled by phase transformations during burial and overpressuring.

The main episode of mineralisation was probably in end Westphalian–early Permian (Variscan) times and there is a strong link between the formation of the Pennine orefields and the oil fields of the region. Thermal models of Pennine basin evolution preclude major episodes of mineralisation, related to basin dewatering, after the Variscan deformation. Ore deposition occurred following the incursion of Pb, Ba, F, Cl and Zn enriched fluids into limestones, where they mixed and reacted with SO_4 and $CaCO_3$. In the South Pennine Orefield basic volcanics acted as physical barriers to the ore fluids and supplied bases promoting ore deposition. In the North Pennines HHP granites locally reacted with and focused metalliferous brines.

The identification of exploration criteria based on metallogenic models provides a valuable new method of exploration and resource evaluation for Irish-, MVT- and Pennine-style mineralisation at the regional to district scale. The approach is particularly suitable for use with image analysis systems, where sets and subsets of exploration criteria can be used in interactive studies for exploration or resource evaluation in 'real time'. Further research and development of the methods are required, however, to develop more precise methods for detailed exploration and they should be tested and refined in other areas, particularly areas of greater structural complexity. Other types of data bases, such as high resolution remotely sensed data (e.g. Seasat, SPOT), thermal imagery and regional geochemistry, should be incorporated for modelling, exploration and resource evaluation for metalliferous minerals.

Acknowledgements. We would like to thank Drs. N. Aitkenhead, D. Holliday, D. Slater and A. Whittaker of the BGS for advice and information and Dr. P.A. Sabine, Dr. R. Howarth, Mr. P.J. Moore and Mr. D. Ostle for support and encouragement. We are particularly grateful to Dr. J. Boissonnas and Dr. L. Van Wambeke of the CEC for their advice and encouragement throughout the research programme. XRF analyses were carried out under the direction of Dr. B.P. Atkin and Dr. P.K. Harvey of Nottingham University and neutron activation analyses were performed at the HERALD Reactor Centre, AWRE,

Aldermaston, the London University Reactor Centre and ICI Billingham. The rare earth elements were determined within the BGS, under the supervision of Mr. B. Tait. Funding for the programme, by the EEC (*Contract No. MSM-90-UK*) and the British Department of Trade and Industry, is gratefully acknowledged. This work is published by permission of the Director of the British Geological Survey (NERC).

References

Aitkenhead N, Chisholm JI (1982) A standard nomenclature for the Dinantian formations of the Peak District of Derbyshire and Staffordshire. Inst Geol Sci Rep 82:8

Anderson GM, MacQueen RW (1982) Ore deposit models-6. Mississippi Valley-type lead-zinc deposits. Geosci Can 9(2):18–117

Anderton R, Bridges PH, Leeder MR, Sellwood BW (1979) A dynamic stratigraphy of the British Isles. Allen & Unwin, London

Andrew CJ, Ashton JH (1985) Regional setting, geology and metal distribution patterns of Navan orebody, Ireland. Trans Inst. Miner, Metall (Section B) 94:B66–93

Arthurton RS (1984) The Ribblesdale fold belt, NW England — a Dinantian-early Namurian dextral shear zone. In: Hutton DHW, Sanderson DJ (eds) Variscan tectonics of the North Atlantic region. Spec Publ Geol Soc London 14:131–138

Bamford D, Nunn K, Prodehl C, Jacobs B (1977) LISPB-III Upper crustal structure of northern Britain. J Geol Soc (Lond) 133:481–488

Barnes HL (1979) Geochemistry of hydrothermal ore deposits. Wiley, New York

Barrett TJ, Anderson GM (1982) The solubility of sphalerite and galena in NaCl brines. Econ Geol 77:1923–1933

Bas MJ Le (1982) Geological evidence from Leicestershire on the crust of southern Britain. Trans Leicester Lit Philos Soc 76:54–67

Bethke CM (1986) Hydrologic constraints on the genesis of Upper Mississippi Valley mineral district from Illinois Basin brines. Econ Geol 81:233–249

Boast AM, Swainbank IG, Coleman ML, Halls C (1981) Lead isotope variation in the Tynagh, Silvermines and Navan base-metal deposits, Ireland. Trans Inst Miner Metall (Section B) 90:B115–119

Boyce AJ, Anderton A, Russell MJ (1983) Rapid subsidence and early Carboniferous base-metal mineralisation in Ireland. Trans Inst Miner Metall (Section B) 92:B55–66

Brown GC, Ixer RA, Plant JA, Webb PC (1987) The geochemistry of granites beneath the North Pennines and their role in orefield mineralisation. Trans Inst Miner Metall (Section B) 96, B65–76

Carpenter AB, Trout ML, Pickett EE (1974) Preliminary report on the origin and evolution of lead and zinc-rich oil field brines in central Mississippi. Econ Geol 69:1191–1206

Carter JS, Cazelet PCD (1984) Hydrocarbon gases in rocks as pathfinders for mineral exploration. In: Prospecting in areas of glaciated terrain (Conference volume). Inst Miner Metall, London, pp 11–20

Cathles, LM, Smith AT (1983) Thermal constraints on the formation of Mississippi Valley-type lead-zinc deposits and their implications for episodic basin dewatering and deposit genesis. Econ Geol 78:983–1002

Coller DW (1984) Variscan structures in the Upper Palaeozoic rocks of west central Ireland. In: Hutton DWH, Sanderson DJ (eds) Variscan tectonics of the Northern Atlantic region. Spec Publ Geol Soc London 14:185–194

Dewey JF (1982) Plate tectonics and the evolution of the British Isles. J Geol Soc (Lond) 139:371–412

Downing RA, Gray DA (1986) Geothermal energy — the potential in the United Kingdom. Br Geol Surv London HMSO

Dunham KC (1983) Ore Genesis in the English Pennines: a fluoritic subtype. In Kisvarsanyi G, Grant SK, Pratt WP, Koenig JW (eds) International Conference on Mississippi Valley-type lead-zinc deposits. Proceedings volume: University of Missouri-Rolla Press, Rolla, Missouri, pp 86–112

Dunham KC, Wilson AA (1985) Geology of the northern Pennines Orefield: volume 2, Stainmore to Craven. Econ Mem Br Geol Surv

Firman RJ, Bagshaw C (1974) A re-appraisal of the controls of non-metallic gangue mineralisation in Derbyshire. Mercian Geol 5:143–161

Ford TD (1976) The ores of the South Pennines and Mendip Hills, England — a comparative study. In: Wolf KH (ed) Handbook of strata-bound and stratiform ore deposits. II. Regional studies and specific deposits, vol 5. Regional Studies. Elsevier, Amsterdam, pp 161–195

Green PM (1984) Digital Image Processing of integrated geochemical-geological information. J Geol Soc (Lond) 141:941–949

Guinness EA, Arvidson RE, Leff CE, Edward MH, Bindschadler DL (1983) Digital image processing application to analysis of geophysical and geochemical data for southern Missouri. Econ Geol 78:654–663

Hanor JS (1979) The sedimentary genesis of hydrothermal fluids. In: Barnes HL (ed) Geochemistry of hydrothermal ore deposits. Wiley New York, pp 137–172

Hunt, JM (1979) Petroleum geochemistry and geology. Freeman, San Francisco, pp 169–172

Ineson PR, Ford TD (1982) The South Pennine orefield: its genetic theories and eastward extension. Mercian Geol 8:285–304

Ineson PR, Mitchell JG (1973) Isotopic age determinations on clay minerals from lavas and tuffs of the Derbyshire orefield. Geol Mag 109:501–512

Institute of Geological Sciences (1965) Aeromagnetic map of Great Britain Sheet 2, 1:625000 Southampton, Ordnance Survey

Institute of Geological Sciences (1977a) Humber-Trent (provisional edition). Sheet 53° N–02° W. 1:250000 Bouguer gravity anomaly map series (London: Institute of Geological Sciences)

Institute of Geological Sciences (1977b) Humber-Trent (provisional edition). Sheet 53° N–02° W. 1:250000 aeromagnetic anomaly map series (London: Institute of Geological Sciences)

Jackson SA, Beales FW (1967) An aspect of sedimentary basin evolution: the concentration of Mississippi Valley-type ores during the lates stages of diagenesis. Bull Can Pet Geol 15:393–433

Kirby GA, Smith K, Smith NJP, Swallow PW Oil and gas generation in eastern England. In: Proceedings of the 3rd conference on the petroleum geology of North West Europe (In press)

Koritnig S (1978) Fluorine. In: Wedepohl KH (ed) Handbook of geochemistry, II-I, Elements H (1) to Al (13). Springer, Berlin Heidelberg New York

Leeder MR (1982) Upper Palaeozoic basins in the British Isles — Caledonide inheritance versus Hercynian plate margin processes. J Geol Soc (Lond) 139:479–491

Lees GM,Taitt AH (1946) The geological results of the search for oilfields in Great Britain. QJ Geol Soc London 101:255–317

MacDonald R, Gass KN, Thorpe RS, Gass IG (1984) Geochemistry and petrogenesis of the Derbyshire Carboniferous basalts. J Geol Soc (Lond) 141:147–159

McKenzie D (1978) Some remarks on the development of sedimentary basins. Earth Planet Sci Lett 40:25–32

Murphy T (1952) Measurements of gravity in Ireland: gravity survey of central Ireland. Geophysics Memoir 2, (part 3). Dublin: Institute of Advanced Studies

Nordstrom DK, Jenne EA (1974) Fluorite solubility equilibria in selected geothermal waters. Geochim Cosmochim Acta 41:175–188

Pearce JA, Harris NBW, Tindle AG (1984) Trace element discrimination diagrams for the tectonic interpretation of granitic rocks. J Pet 25:956–983

Phillips WEA, Stillman CJ Murphy T (1976) A Caledonian plate tectonic model. J Geol Soc (Lond) 132:579–609

Plant JA (1986) Models for granites and their mineralising systems in the British and Irish Caledonides. In: Andrew CJ, Crowe RWA, Finlay S, Pennell WM, Pyne JF (eds) Geology and genesis of mineral deposits in Ireland. Irish Assoc Econ Geol, Dublin pp 121–156

Plant JA, Jones DG (eds) Metallogenic models and exploration criteria for buried carbonate-hosted ore deposits: a multidisciplinary study in eastern England. Inst Miner Metall, London (In press)

Plant JA, Forrest MD, Hodgson JF, Smith RTS, Stevenson AG (1986) Regional geochemistry in the detection and modelling of mineral deposits. In: Thornton I, Howarth RJ (eds) Applied geochemistry in the 1980's. Graham & Trotman, London, pp 103–139

Ramsbottom WHC, Sabine PA, Dangerfield J, Sabine PW (1981) Mudrocks in the Carboniferous of Britain. Q J Eng Geol 14:257–262

Roberts DE, Hudson GRT (1983) The Olympic Dam copper-uranium-gold deposit, Roxby Downs, South Australia. Econ Geol 78:799–822
Roedder E (1967) Environment of deposition of stratiform (Mississippi Valley-type) ore deposits from studies of fluid inclusions. Econ Geol Monogr 3:349–361
Rogers PJ (1977) Fluid inclusion studies in fluorite from the Derbyshire orefield. Trans Inst Miner Metall (Section B) 86:B128–132
Russell MJ (1978) Downward excavating hydrothermal cells and Irish-type ore deposits: importance of an underlying thick Caledonian prism. Trans Inst Miner Metall (Section B) 87:B168–171
Samson IM, Russell MJ (1983) Fluid inclusion data from Silvermines base-metal-baryte deposits, Ireland. Trans Inst Miner Metall (Section B) 92:B67–71
Sangster DF (1983) Mississippi Valley-type deposits: a geological mélange. In: Kisvarsanyi G, Grant SK, Pratt WP, Koenig, JW (eds) International conference on Mississippi Valley-type lead-zinc deposits. Proceedings volume: University of Missouri-Rolla Press, Rolla, Missouri, pp 7–19
Sangster DF (1986) Age of mineralisation in Mississippi Valley-type (MVT) deposits: a critical requirement for genetic modelling. In: Andrew CJ, Crowe, RWA, Finlay S, Pennell WM, Pyne JF (eds) Geology and genesis of mineral deposits in Ireland. Irish Assoc Econ Geol, Dublin, pp 625–634
Shepherd TJ, Darbyshire DPF, Moore GR, Greenwood DA (1982) Rare earth element and isotopic geochemistry of the North Pennine ore deposits. In: Gites Filoniens Pb Zn F Ba de basse température du domaine varisque d'Europe et d'Afrique du Nord. Symposium Orleans. Bull BRGM II (2-3-4):371–377
Sheppard SMF, Langley KM (1984) Origin of saline formation waters in northeast England: application of stable isotopes. Trans Inst Miner Metall (Section B) 93:B195–201
Sheridan DJR (1972) The stratigraphy of the Trim No. 1 well, Co. Meath and its relationship to Lower Carboniferous outcrop in east-central Ireland. Bull Geol Surv Ireland 1:311–334
Skinner BJ (ed) (1983) Economic geology: seventy fifth anniversary volume 1905–1980. Econ Geol, El Paso, Texas
Smirnov VI (1976) Geology of mineral deposits. MIR, Moscow
Solomon M, Rafter TA, Dunham KC (1971) Sulphur and oxygen isotope studies in the northern Pennines in relation to ore genesis. Trans Inst Miner Metall (Section B), 81:B259–275
Spears DA, Amin MA (1981) Geochemistry and mineralogy of marine and non-marine Namurian black shales from the Tansley borehole, Derbyshire. Sedimentology 28:407–417
Suggate RR (1981) Coal ranks on the Alston Block, North-east England: a discussion. Proc York Geol Soc 43:451–453
Taylor HP (1974) The application of oxygen and hydrogen isotope studies to problems of hydrothermal alteration and deposition. Econ Geol 69:843–883
Turekian KK, Wedepohl KH (1961) Distribution of elements in some major units of the Earth's crust. Bull Geol Soc Am 72:175–192
Upton BGJ (1982) Carboniferous to Permian volcanism in the stable foreland. In: Sutherland DS (ed) Igneous rocks of the British Isles. Wiley, New York, pp 255–275
Whitcombe DN, Maguire PKH (1981) Seismic refraction evidence for a basement ridge between the Derbyshire Dome and the west of Charnwood Forest. J Geol Soc (Lond) 138:653–659
White DE (1958) Liquid of inclusions in sulfides from Tri-State (Missouri–Kansas–Oklahoma) is probably connate in origin. (Abstract.) Bull Geol Soc Am 69:1660
Williams B, Brown C (1986) A model for the genesis of Zn–Pb deposits in Ireland In: Andrew CJ, Crowe RWA, Finlay S, Pennells WM, Pyne JF (eds) Geology and genesis of mineral deposits in Ireland. Irish Assoc Econ Geol Dublin pp 579–590
Woodall R (1985) Limited vision: a personal experience of mining geology and scientific mineral exploration. Aust J Earth Sci 32:231–237

Structural Studies and Multidata Correlation of Mineralization in Central Ireland

W.E.A. PHILLIPS[1], A. ROWLANDS[2], D.W. COLLER[3], J. CARTER[4], and A. VAUGHAN[5]

Abstract

The zinc-lead deposits of Silvermines and Navan are stratiform ore bodies hosted in carbonate sediments of lower Dinantian (Courceyan) age. These deposits overlie a Caledonian basement close to the Caledonian plate collision boundary. Geophysical data plus lithofacies structural, and remote sensing data from the Dinantian rocks of both regions indicate that granitic plutons of Caledonian age are situated close to both deposits. Detailed structural analyses in both mines and in their surrounding regions have shown that mineralization occurred in dilation zones which developed during minor dextral strike slip reactivation of Caledonian fractures. At Silvermines, E–W-trending dextral shear terminated to the east against the rigid obstruction of a buried Caledonian granite, thus generating a complex dilation zone of late Courceyan age. At Navan, NE–SW-trending dextral shear culminated in late Courceyan times with an eastward termination against Caledonian granitic and syenite intrusions. In both areas, the Caledonian granites may well have provided a source of heat which generated hydrothermal circulation of saline fluids. These models for the structural control of mineralization at Silvermines and Navan indicate that further exploration should be guided by methods which can identify Caledonian shear zones which have been reactivated during the Courceyan and which can identify where such shear zones terminate against buried Caledonian granitic plutons or volcanic complexes. Combinations of analyses of geophysical, structural, lithological and remote sensing data can be used for this purpose.

1 Introduction

This paper summarizes a part of a major study of the structural controls of mineralization in Central Ireland which was carried out at Trinity College with

[1] Department of Geology, Trinity College, Dublin 2
[2] Maptec Limited, 5 South Leinster Street, Dublin 2
[3] E.R.A. 5 South Leinster Street, Dublin 2
[4] Mercury Hydrocarbons Ltd., Limerick, Ireland
[5] Department of Geology, Trinity College, Dublin 2

Mineral Deposits within the European Community
(ed. by J. Boissonnas and P. Omenetto)

support from two contracts with the European Commission, and contracts with the Irish National Board for Science and Technology and with mineral exploration companies.

Central Ireland contains some of the largest Pb–Zn deposits in Europe. The area is poorly exposed, with thick glacial and post-glacial deposits and extensive agriculture. The known base metal deposits are hosted in Lower Carboniferous carbonates and are stratabound in nature. All the major deposits are adjacent to faults and there is preferential development near to the Courceyan/Chadian boundary (Table 1). Further deposits are likely to exist beneath a cover of younger Carboniferous rocks and/or beneath thick Quaternary sediments. Such deposits are likely to be blind to conventional exploration methods.

The geology of Central Ireland (Fig. 1) is dominated by a widespread cover of late Devonian and Carboniferous sediments which rest with angular unconformity upon Cambrian-Silurian rocks which were deformed during the Caledonian orogeny (ca. 400 Ma.). Seismic refraction and wide angle reflection data (Jacob et al. 1985) indicate that crystalline basement, probably Precambrian gneisses, lies at depths of 3–4 kms beneath a line extending SW from Dundalk Bay to Loop Head (Fig. 1).

The Caledonian basement rocks crop out in a number of inliers which have been assigned into northern and southern units separated by the Iapetus Suture Zone (Phillips et al 1976). This major Caledonian structural element is therefore a collision suture marking the closure of the Iapetus Ocean during the Cambrian-Silurian. This plate collision, together with subduction, generated accretion at the southern margin of the northern (American) Plate, and produced Caledonian folding and cleavage with variable trends about the overall NE–SW to ENE–WSW trend of the suture. The Silvermines-Navan Fault Systems seen in the Carboniferous cover rocks are broadly parallel to and lie within the zone of the basement suture zone.

During the Upper Palaeozoic, Central Ireland represented a foreland shelf, lying to the north of the main Hercynian belt. Within Ireland the Hercynian orogenic front is generally placed between the Munster Basin, with its thick Upper Palaeozoic sedimentary fill and strong cleavage, folding and faulting, and Central Ireland, with its thinner cover and generally weaker, but more heterogeneous deformation (Fig. 1).

Most of Central Ireland is underlain by Devonian and Carboniferous rocks (Table 1). At the base lie fluviatile Old Red Sandstone sediments of late Devonian to early Dinantian (Courceyan) age with a thickness varying from 300 m in the south to 0 m in the north. The overlying Courceyan rocks consist of a complex series of shallow marine shelf sediments dominated by carbonate rocks. They record several transgressive cycles over a moderately stable shelf (Phillips Sevastopulo 1986), with a general upward trend towards deeper water environments. The Waulsortian reef limestones, of late Courceyan-Chadian age, probably represent the deepest water conditions; they usually reach a thickness of about 150 m in the Midlands, but thin northerwards passing laterally into shallower water carbonates. During the Courceyan, variable thicknesses indicate that there were uneven rates of subsidence with notably thicker successions developed in the Shannon and Dublin basins and in the

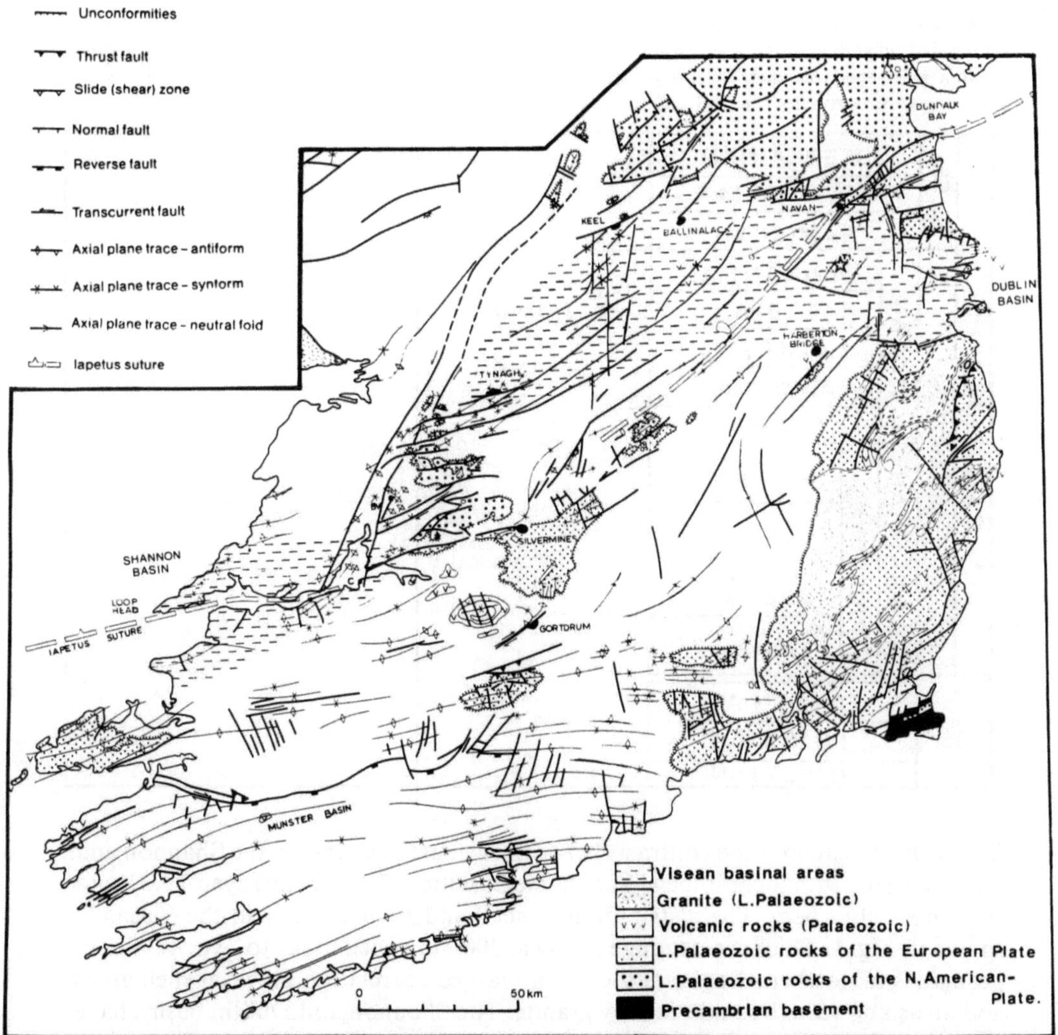

Fig. 1. Structural map of central Ireland. *BV* Ballyvergin; *C* Courtbrown (in the Shannon Basin)

Strokestown and Keel regions. In the Shannon basin there is an estimated maximum 800 m of Waulsortian reef limestones, approximately 300 m of subWaulsortian marine beds and an estimated 400 m of Old Red Sandstone, giving a total of 1.5 km in comparison to the ca. 550 m in much of the Central Midlands (Philcox, 1984). The thickest Courceyan succession recorded in the Irish Midlands is in the Trim No 1 well within the Dublin basin (Fig. 1) where the total estimated thickness of Courceyan rocks is 2.1 km (Sheridan 1972). More local sags occur at Keel and Strokestown (Philcox 1984.), while the Lower Palaeozoic inlier of SE Leinster, with its major Leinster Granite was partly emergent at this time (Sevastopulo, 1981). Differential rates of subsidence began to outstrip rates of sedimentation during the

Table 1. Summary of late Devonian and Carboniferous stratigraphy of Ireland and the distribution of mineralisation at Silvermines, Navan and Tynagh in terms of the age of their host rocks. V in the middle column represents volcanics. BVS represents the horizon of the Ballyvergin Shale Formation (After Phillips and Sevastopulo, Fig. 15 1986).

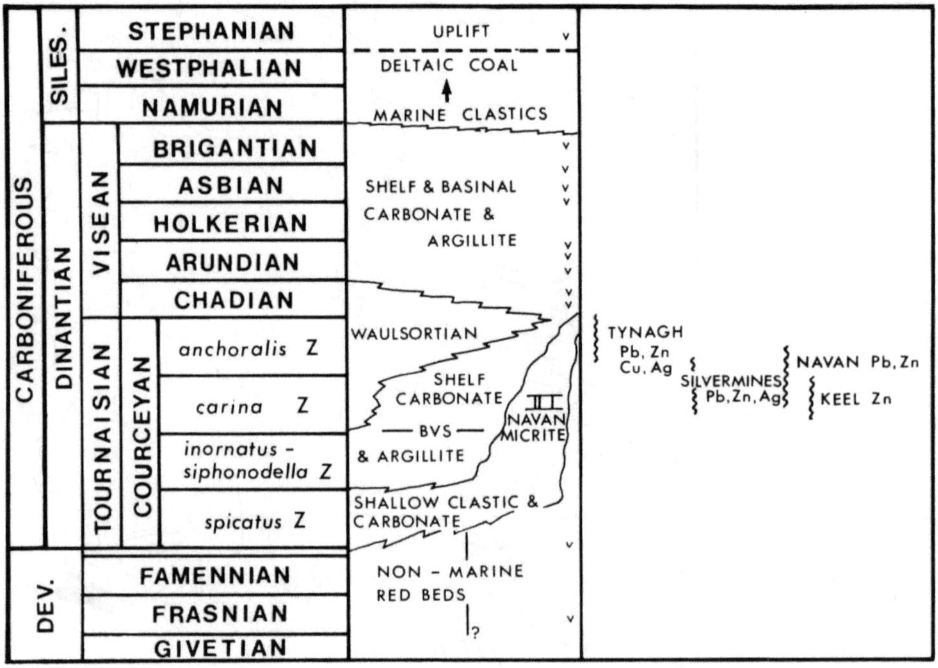

Visean, leading to marked contrasts between basinal limestones in the Shannon and Dublin basins with local somewhat alkaline basaltic volcanism (Strogen 1977), and shelf areas elsewhere. The distribution of shelf and basinal facies in the Visean is shown in Fig. 1. Thicknesses range from ca. 300 m in shelf areas to over 1.5 km in basins. As in northern England, there is a marked correlation between shelf areas and areas known to be underlain by granite. The Shannon and Dublin basins have a dominantly E–W trend which must reflect a N–S extension during the Dinantian. Small outliers of Namurian rocks in the Midlands record the continuation of these basins and shelf environments. Local turbidites developed within the Namurian of the Dublin and Shannon basins, with thinner shallower water clastic facies in areas such as Leinster, where gravity data suggests a continuation of the Caledonian Leinster Granite in the basement.

The Upper Palaeozoic rocks show a very heterogeneous pattern of deformation. During late Courceyan-Chadian time there was widespread minor faulting which played a most important part in controlling mineralization at Silvermines, Navan, Tynagh and Keel. More substantial deformation took place in the Hercynian orogeny of later Westphalian-early Permian age (Plant et al., this Vol.). The overall trend of folds is E–W to ENE–WSW, but these trends become modified adjacent to major faults and shear zones. Cleavage is heterogeneously developed throughout the area and it intensifies and swings in strike into shear zones. In the less deformed blocks between shear zones, cleavage is generally absent, bedding dips are low and joints predominate over veins. Minor epigenetic mineralization occurred at this time at Courbrown and Ballyvergin (Fig. 1).

2 Structural Controls of Known Mineralization

2.1 Silvermines

The Silvermines deposit of about 18 M tons grading 2.53% Pb and 6.43% Zn is hosted in Courceyan siliciclastic and carbonate sediments. Mining by Mogul Ireland ceased in 1982.

The relation of the deposit to deep structural features, defined by analyses of gravity and aeromagnetic data, is shown in Fig. 2. The deposit lies within a NNE-trending negative Bouguer gravity anomaly at the eastern end of an E–W-trending positive gravity ridge associated with inliers of Lower Palaeozoic rocks. Magnetic basement beneath the deposit is depressed to about 4.5 km (Brown and Williams 1985). The setting of the deposit in relation to Caledonian geology is summarized in Fig. 3. The Caledonian suture zone (Phillips et al. 1976, 1979) shows a swing in strike to a more NNE trend in the Silvermines area. Caledonian cleavage

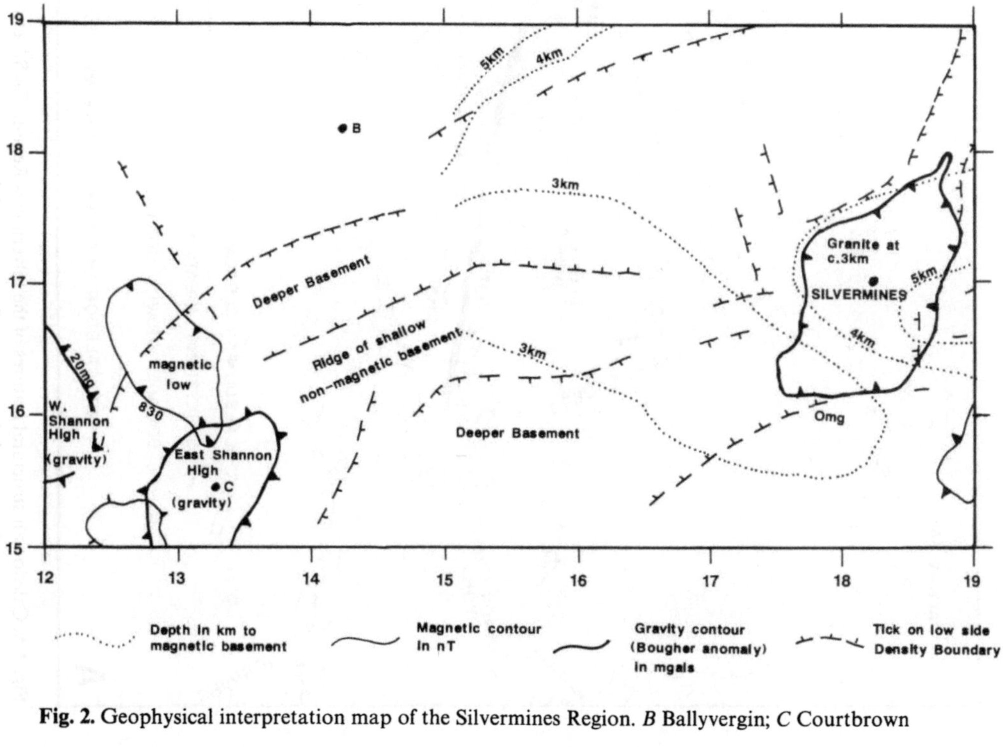

Fig. 2. Geophysical interpretation map of the Silvermines Region. *B* Ballyvergin; *C* Courtbrown

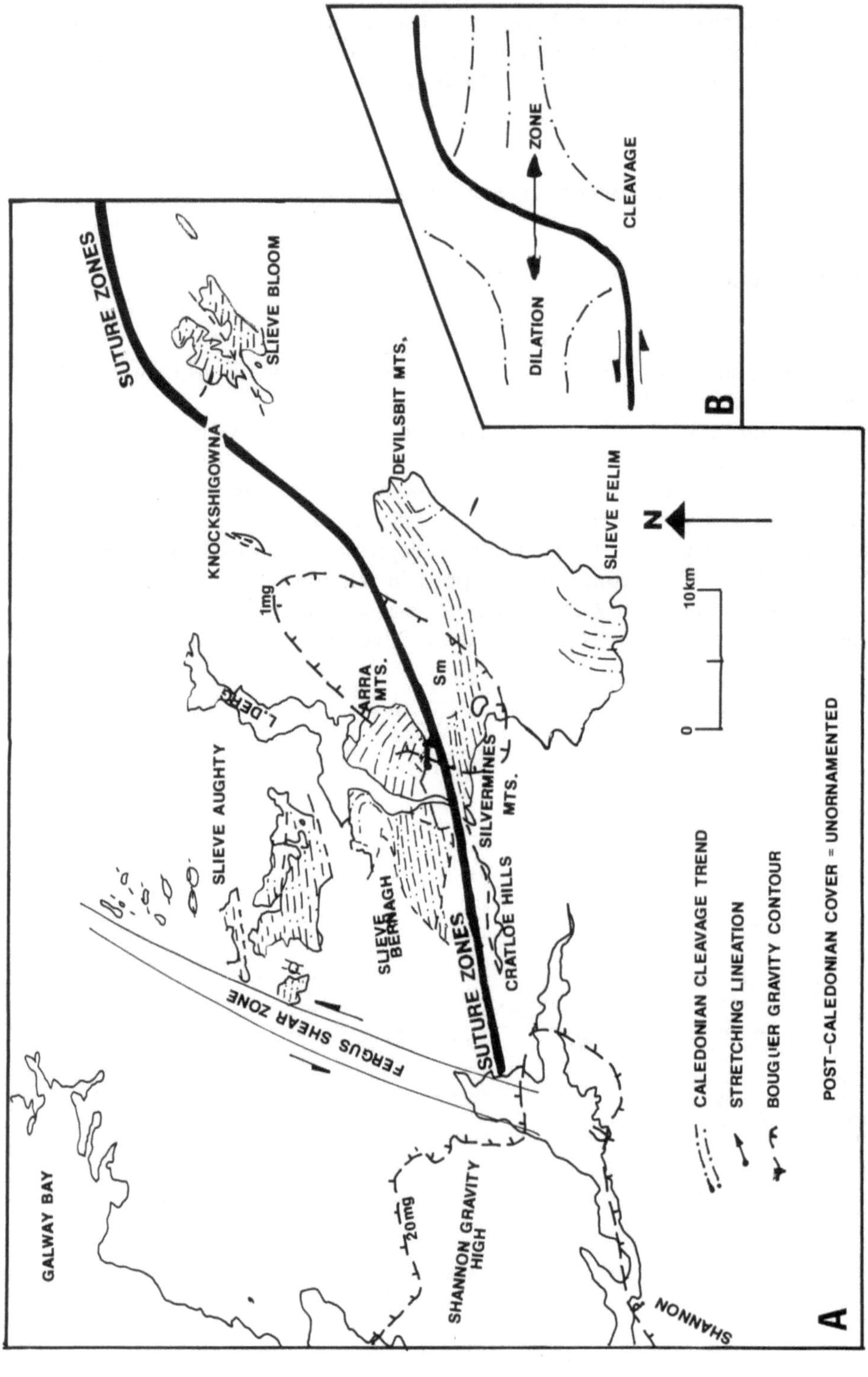

Fig. 3 A. Caledonian structural features of the Silvermines Region. *Sm* Silvermines deposit. **B** Structural model explaining the generation of a dilation zone on a releasing bend of the sinistral shear zone following the Caledonian suture zone

in the Lower Palaeozoic inliers adjacent to this strike swing shows a pattern of strike swings which can be explained by sinistral shear on the suture generating a dilation zone (releasing bend) in the NNE trending sector. (Fig. 3b). The sigmoidal shape of the gravity low here (Fig. 2) is appropriate for such a sinistrally generated dilation zone. Two-dimensional modelling of the anomaly in terms of a simple sphere suggests that it could be explained by a granitic body at a depth of about 3 km in the Lower Palaeozoic basement. Indirect evidence to support this interpretation comes from the presence of a shelf facies in the Visean limestones in the area of the gravity low and of more basinal facies away from it (Bruck 1982). There is also an exceptionally low level of strain in the Visean rocks above the inferred granite ("Nenagh Pluton") with low dips of bedding, an absence of cleavage and a dominance of joints over veins (Phillips 1982). The "Nenagh Pluton" appears therefore to have acted as a positive and rigid block during the Carboniferous in the same way as the North Pennine Block, underlain by Caledonian granite, acted at this time in Northern England (Robson 1980).

The setting of the Silvermines deposit in relation to Hercynian structures is shown in Fig. 4. It lies within a clockwise bend at the eastern termination of the Silvermines Shear Zone (SF on Fig. 4). Stratigraphic data show that there was downthrow to the north of this structure, however the intensification and clockwise rotation of folds and cleavage into this zone (Phillips 1982) show that there was an important component of dextral shear. At Silvermines, the shear zone bends clockwise and terminates in a broad zone of deformation. Strain is largely accommodated by movement on a series of fractures which strike clockwise of the main E–W shear zone and hence lie in the extensional field of the dextral incremental shear (Fig. 5). These fractures form a major dilational zone, the overall structure being a stepped half-graben with the Silvermines Fault forming the southern boundary fault (Figs. 3, 4, 5).

There are three important mineralized horizons. The Upper Devonian sandstones are hosts to breccia and vein style mineralization, with a high initial porosity and fluid pressure possibly facilitating hydraulic fracture. Secondly, vein and replacement mineralization (Lower zone) is present in the Lower Dolomite (bioclastic Middle Ballysteen Limestone), in which dolomitization generated much secondary porosity. The most complex and important stratigraphic control of mineralization is within dolomitized, in situ and transported breccias which overlie or interdigitate with Waulsortian mudbank reef limestones. The two largest stratiform orebodies (the Upper G and B zones) are hosted in this lithology.

All the known mineralization is within the limits of the dilation zone and is at least partially controlled by fault activity. A simplified model outlining the growth of the faulting and dilation zone at Silvermines in shown in Fig. 6. The Oblique Extension Faults (Fig. 5) were active during the Lower Carboniferous; metal zonation and textural studies suggest that they acted as feeders for mineral-bearing fluids, with mineralization decreasing away from the faults. This is the situation in the upper G and B zones, where faults control the development of debris flows producing the mineralized breccias. The maximum stratigraphic separation on the B fault zone is near its centre, where a large slump fold was generated. The breccias

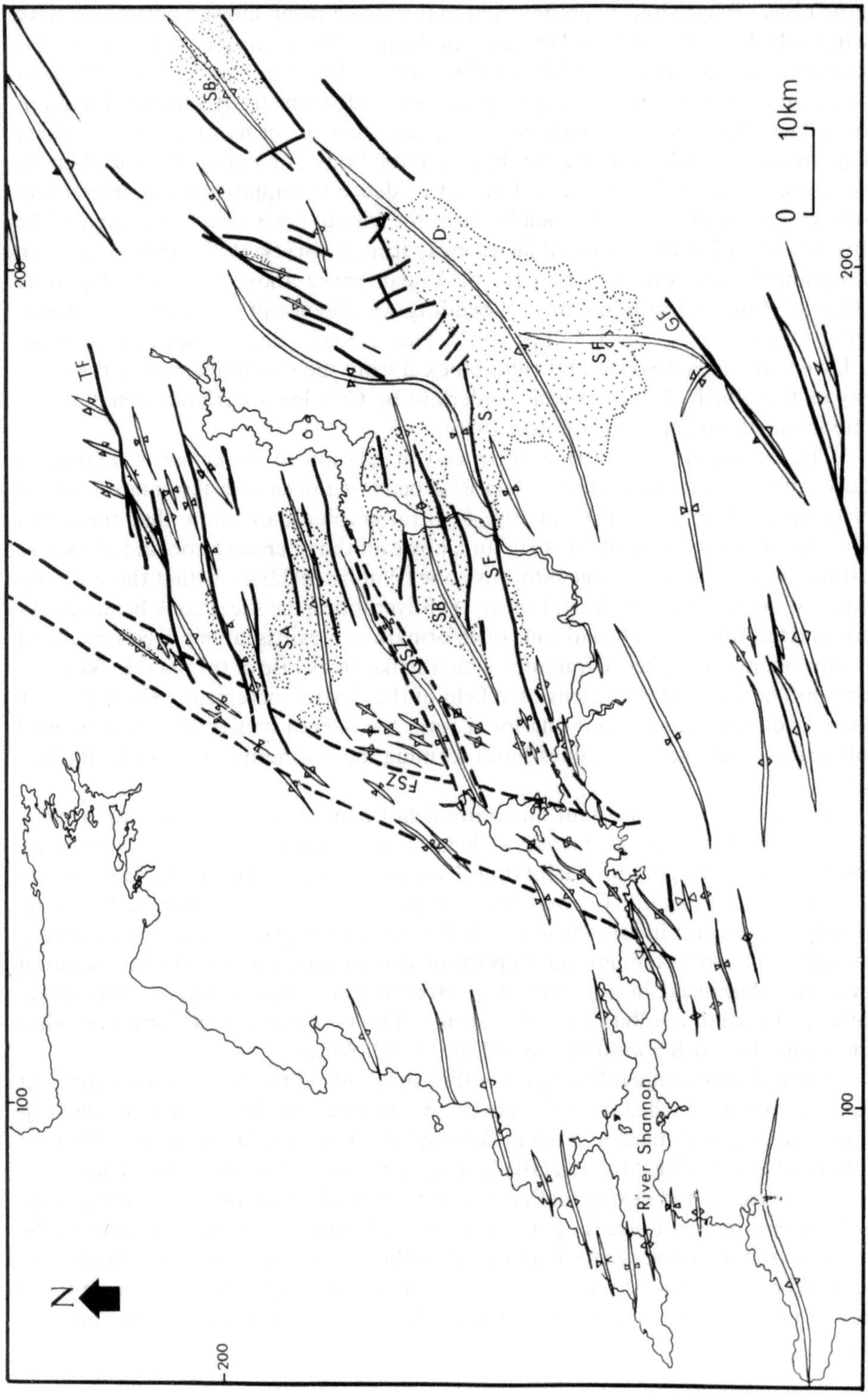

Fig. 4. Structural map of west central Ireland showing Hercynian folds, faults and shear zones. The *stippled areas* are Lower Palaeozoic inliers. *FSZ* Fergus Shear Zone; *TF* Tynagh Fault; *SF* Silvermines Fault; *GF* Gortdrum Fault; *QSZ* Quinn Shear Zone; *SA* Slieve Aughty; *D* Devilsbit Mountains

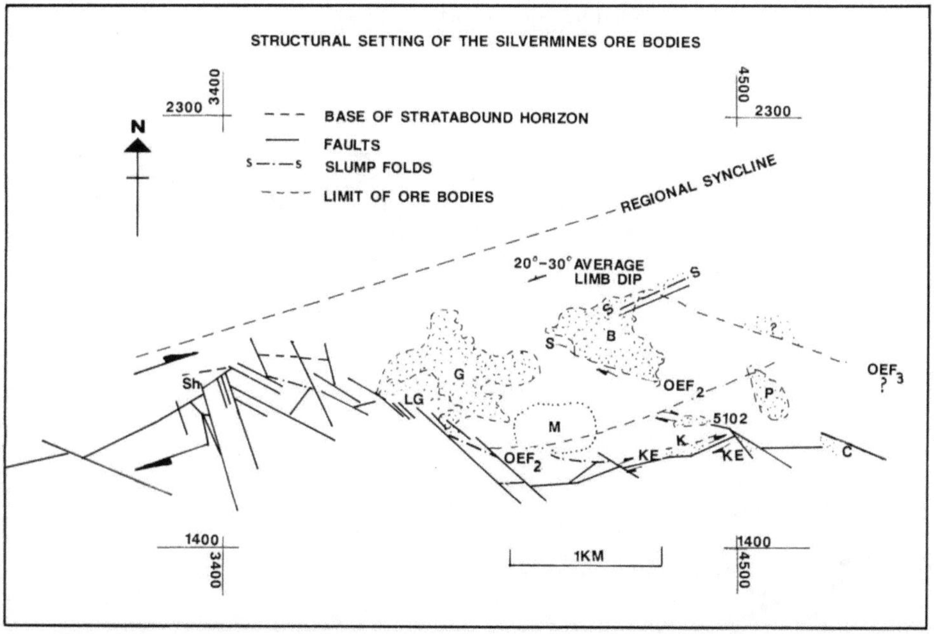

Fig. 5. Structural setting of the Silvermines orebodies. The three Pb–Zn stratabound orebodies (*LG* Lower G Zone; *G* G Zone; *B* B Zone) were controlled by the development of oblique dextral extension faults (*OEF*) at the termination zone of the Silvermines Fault. The feeder areas for *G* and *B* zones stratiform orebodies are located at the foci of the oblique extension faults at their points of maximum downthrow

and slump fold suggest that the fault acted as a "growth fault" during the early Visean.

Transcurrent movement on the oblique extension (OEF), principal shear (P), extension (E) and antithetic Riedel (R′) faults (see Fig. 7) generated local dilation zones at overlaps bends and terminations and these areas are often the sites of mineral concentration. The K Zone orebody is located on the Silvermines Fault where dextral movement at a clockwise bend in the fault would have produced a local dilation zone (Fig. 8). Extensional faults generated the veins and brecciation in the Shallee area and contribute to mineralization in many of the other orebodies.

Hydraulic fracturing associated with all the faulting is commonly localized along faults and resulted in fluid flow which contributed to the remobilization and concentration of ores. In addition to these direct controls of mineralization, open NE–SW-trending folds may have controlled the alignment and position of mud-banks which in turn strongly influence the distribution of stratabound ore. These folds are compatible with compression in a ENE dextral shear regime.

Thus, the Silvermines deposit is, as a whole, structurally controlled at a dilation zone at the termination of the Silvermines Shear Zone. Individual faults and resulting dilation zones control the individual orebodies within the area. The extension faults produced acted as feeders for the syngenetic stratiform mineralization. Synsedimentary faulting produced a thick development of dolomitized lime-

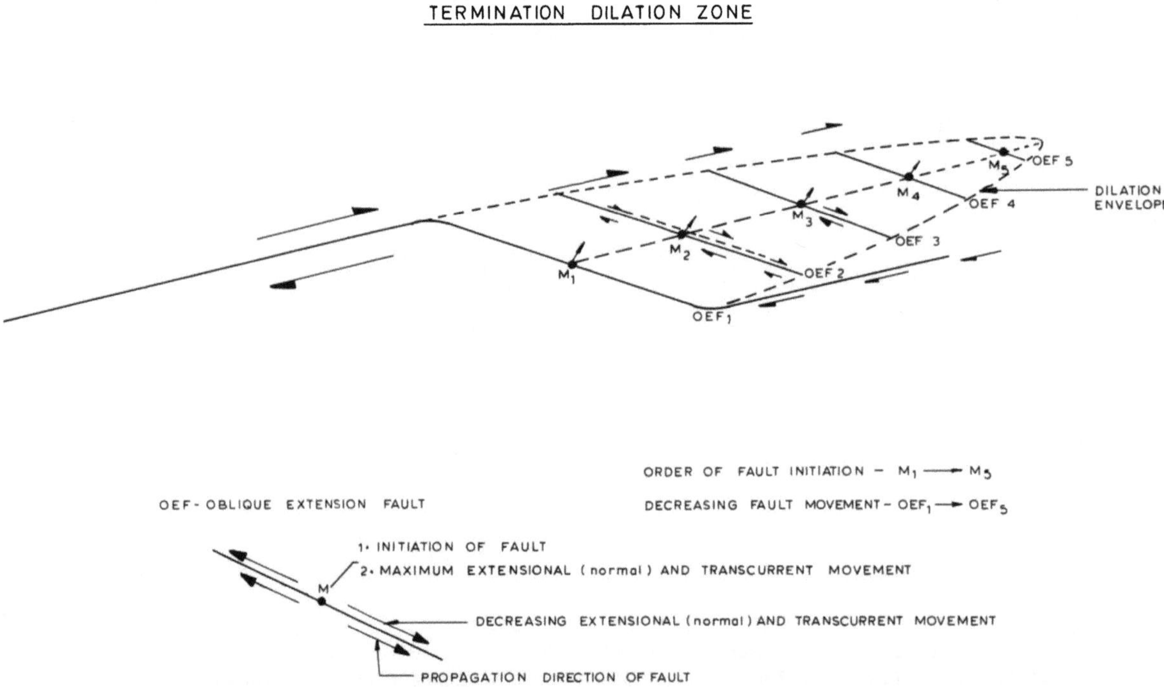

Fig. 6. A schematic plan summarizing the main fault orientations and movement histories in the Silvermines area. The initiation and early movement of the faults is compatible with dextral shear on the ENE trending principal shear — the Silvermines Fault

stones and breccias which act as hosts for the mineralization. Later movements and fracturing produced epigenetic mineralization in veins and breccias.

Figure 7 summarizes the stress histories inferred from mapping minor structures in the Silvermines mine. The early faulting associated with mineralization requires a NW–SE compression. The movement was probably terminated by the rigid behaviour of the buried granitic body which we have inferred. This Caledonian granite may, in addition to generating a dilation zone in which mineralization took place, have provided a source of heat to drive the hydrothermal system which was released into the Courceyan rocks at about 220 °C (Samson and Russell 1983; See also the model of Plant et al., this Vol.). Prolonged (ca. 60 Ma) burial of the granite during the Devonian and Courceyan would have allowed radiogenic heat to be retained in the pluton. The late Courceyan aged faulting, with local dilation, could then have allowed sea water to penetrate downwards scavenging base metals in chloride complexes (Russell 1986). A reversal to upward flow could have been caused by heating as the fluids penetrated towards the buried hot dry pluton. Studies of lead isotope ratios from lead-zinc deposits within the Carboniferous cover and in the Lower Palaeozoic basement (O'Keeffe 1986, Caulfield et al. 1986) indicate a deep crustal origin for this lead with no evidence for large-scale horizontal transport.

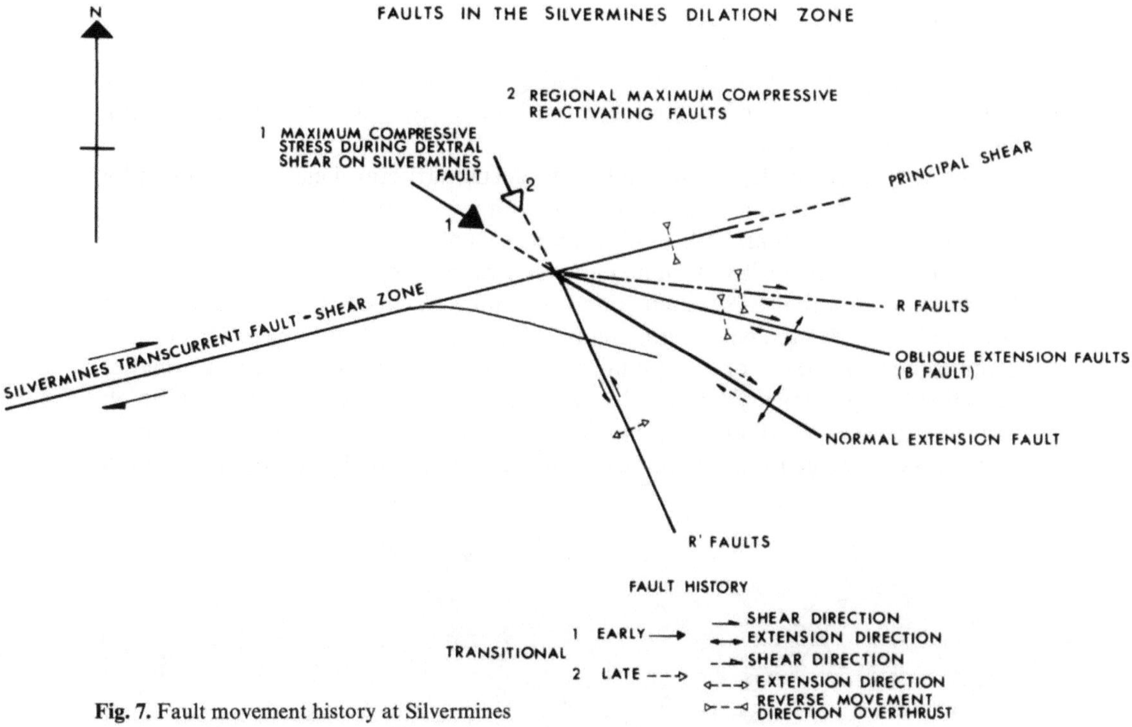

Fig. 7. Fault movement history at Silvermines

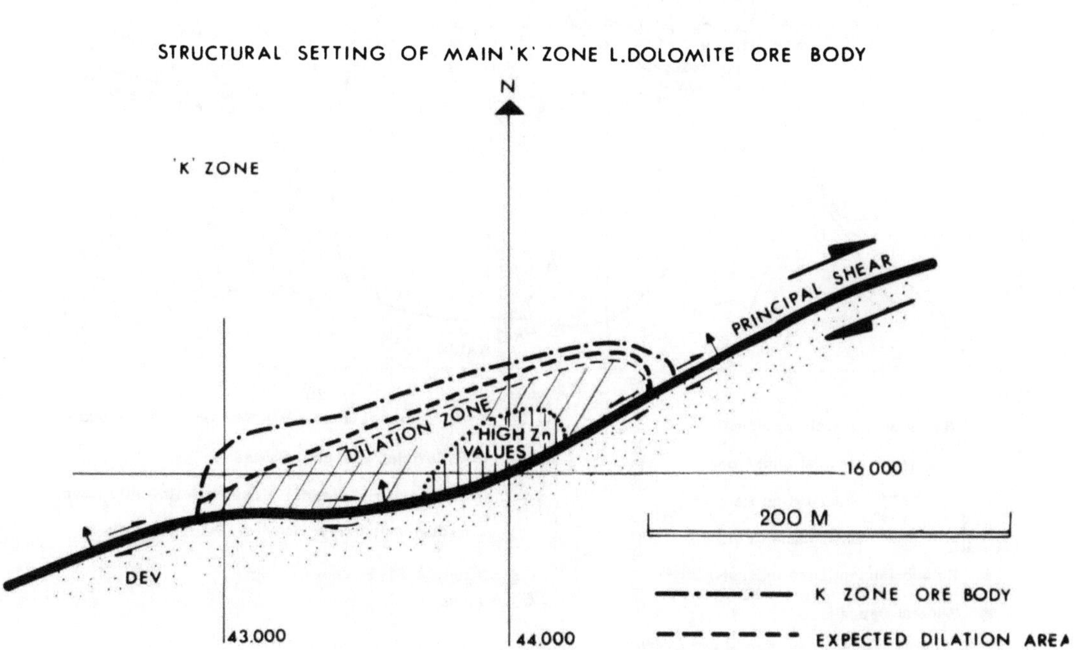

Fig. 8. Structural control of epigenetic K zone mineralization at a bend along a dextral wrench fault Silvermines Fault at Silvermines

2.2 Navan

The Navan deposit of at least 69.9 mt grading 10.1% Zn and 2.6% Pb is a complex stratabound deposit hosted in Courceyan carbonate sediments (Ashton et al. 1986). The geological setting is shown in Fig. 9.

The setting of the deposit in relation to deeper structural elements defined by analyses of gravity and aeromagnetic data, is shown in Fig. 10. The deposit lies at the confluence of E–W- and NE–SW-trending magnetic anomalies which terminate in the Navan area. A positive Bouguer anomaly ridge extends southwestwards from Navan, indicating that a basement ridge continues in this direction but lacks the associated NE–SW magnetic high seen NE of Navan over outcrops of Ordovician volcanic rocks. Analysis of horizontal derivative maps of the Bouguer anomalies has defined major density boundaries which are dominantly NE–SW, but which swing to a more E–W trend east of Navan. A conspicuous negative Bouguer gravity anomaly is centred at about 15 km east of Navan. This anomaly has been interpreted

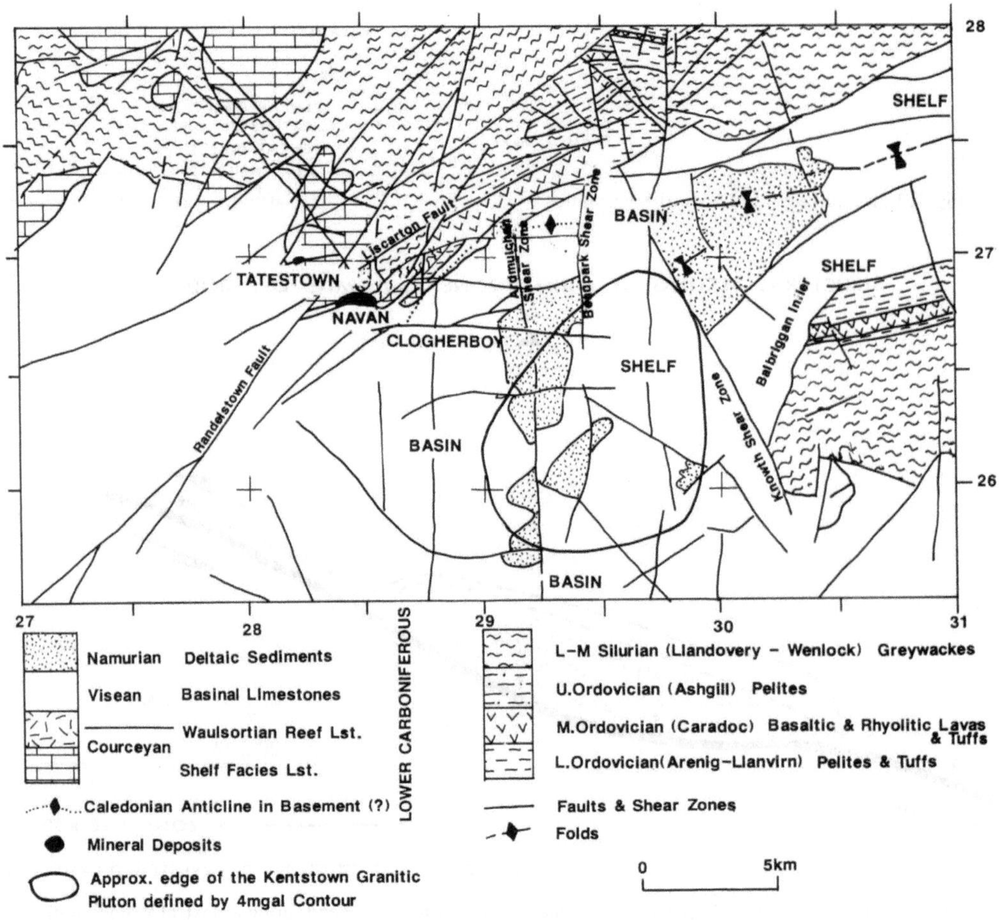

Fig. 9. Geological map of the Navan Region, showing areas of shelf and basinal facies during the Visean

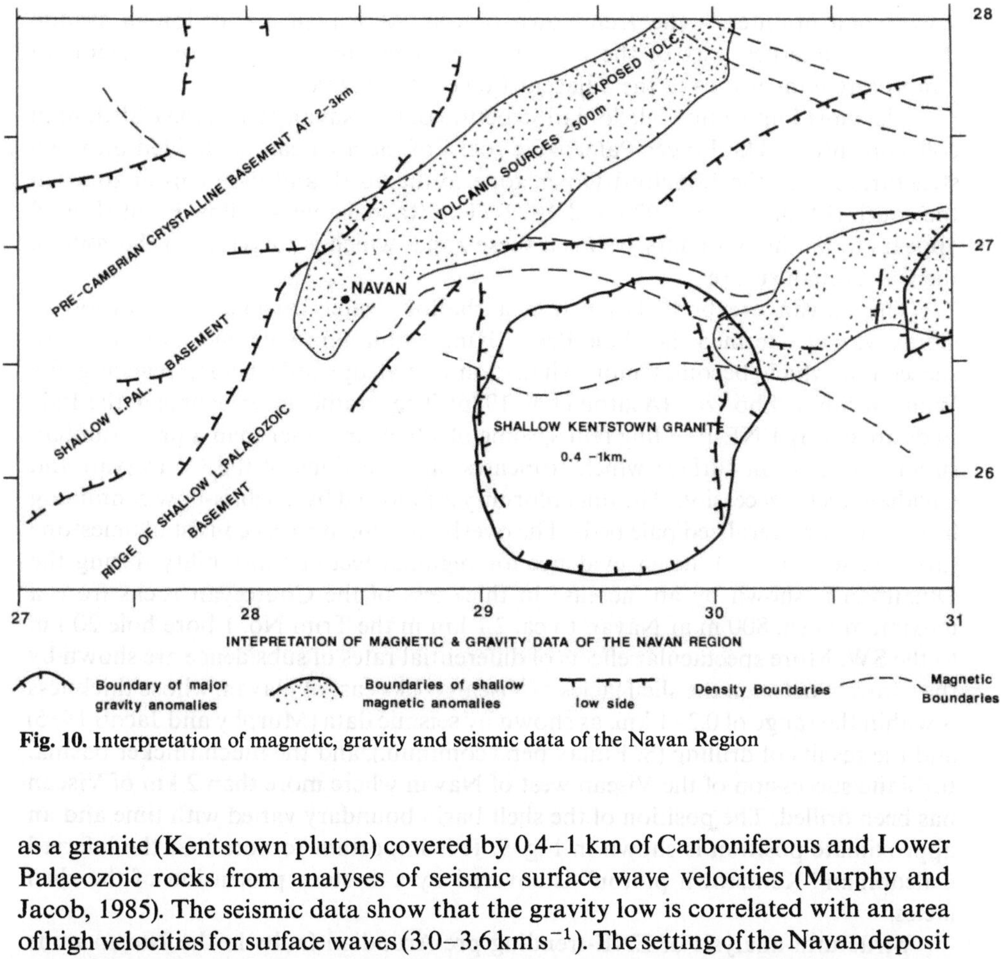

Fig. 10. Interpretation of magnetic, gravity and seismic data of the Navan Region

as a granite (Kentstown pluton) covered by 0.4–1 km of Carboniferous and Lower Palaeozoic rocks from analyses of seismic surface wave velocities (Murphy and Jacob, 1985). The seismic data show that the gravity low is correlated with an area of high velocities for surface waves (3.0–3.6 km s^{-1}). The setting of the Navan deposit is therefore comparable to that of Silvermines, lying in an area of strike swing of the Caledonian basement and adjacent to a largely buried granite. In contrast to Silvermines, the Navan ore body also lies above the SW end of two converging near-surface magnetic anomalies.

Both the magnetic and gravity anomalies can be directly related to Caledonian geology. The magnetic anomalies overlie outcrops of lower-middle Ordovician basaltic and silicic lavas and pyroclastic rocks which were derived from a series of explosive vents east of Navan. These rocks lie on the limbs of a Caledonian anticline, now covered by Carboniferous rocks, whose trend must swing from E–W to NE–SW in the Navan area (Figs. 9, 10). The positive gravity, non-magnetic, ridge extending SW from Navan probably reflects uplifted denser basement (Cambrian or Precambrian) in the core of this anticline. The negative gravity anomaly SE of Navan extends over the western part of the Lower Palaeozoic inlier of Balbriggan (Fig. 9) where there are a number of sheets of granite intruding Wenlock greywacke and deformed by the main Caledonian cleavage. There is also a local hornfelsing of shales. There is thus little doubt that the negative gravity anomaly represents a

Caledonian granite — the Kentstown pluton. At Navan a Caledonian syenite intrusion underlies part of the deposit and there are a number of Caledonian lamprophyre intrusions in the adjacent Lower Palaeozoic rocks.

The most important Caledonian structure of the Navan area is the Caledonian collision suture. The Lower Palaeozoic rocks of the area can be divided into two structural zones, the Longford-Down zone to the north and the Leinster zones in the south (Phillips et al. 1976 and 1979). The Randelstown-Collon Fault (Fig. 9) appears to be the boundary between these zones which is therefore in the zone of the Caledonian suture.

The Navan ore body lies within a shallow water carbonate succession of Courceyan age, termed the "Pale Beds", lying within up to 800 m of a Courceyan succession which becomes more shale-dominated upwards before reaching the Waulsortian reef horizon (Ashton et al. 1986). The stratiform ore lenses in the Pale Beds are cut by ENE-trending faults, some of which are older than a pre-Arundian submarine erosion surface which truncates large sections of the Courceyan and Chadian (reef) succession. The unconformity is followed by a debris-flow containing fragments of mineralized pale beds. The overlying Visean rocks consist of limestone turbidite sequences. Further evidence for regional tectonic instability during the Dinantian is shown by an increase in thickness of the Courceyan rocks from a maximum of ca. 800 m at Navan to ca. 2.1 km in the Trim No. 1 bore hole 20 km to the SW. More spectacular effects of differential rates of subsidence are shown by the contrast between the shelf facies of Visean rocks east of Navan, whose thickness is within the range of 0.2–1 km, as shown by seismic data (Murphy and Jacob 1985) and the results of drilling (S. Finlay pers. commun.), and the much thicker basinal turbidite succession of the Visean west of Navan where more than 2 km of Visean has been drilled. The position of the shelf-basin boundary varied with time and an approximate position is shown in Fig. 9. It is important to note that the inferred Caledonian "Kentstown pluton" is covered by a western protrusion of the shelf facies.

Structural analysis of ENE-trending pre-Arundian faults in the Navan mine (Phillips and Haughey 1982) has shown that movement involved components of dextral and normal displacement with the normal displacement becoming dominant as the faults swing clockwise in strike to the east. The overall structural pattern of these early faults is very similar to that seen at Silvermines, with normal faulting on E–W-trending sectors representing dilation at the termination of a NE–SW-trending dextral shear zone. The termination can be readily explained at Navan in view of the incoming of more rigid Caledonian basement in the form of an intrusive syenite and of silicic and basaltic volcanic Ordovician rocks. More substantial Hercynian aged shear zones trend NE through the Navan deposit with a dominantly dextral sense of shear. An important insight into the role of dextral shear in controlling mineralization comes from a detailed study of Hercynian extension veins in the Navan mine, of facies changes in the Courceyan rocks of the area, and of the distribution of satellite ore bodies.

Figure 11 summarizes the pattern of NW–SE-trending extension veins of Hercynian age, which are widespread in the Navan mine. Detailed structural study by Coller and Pepper (1984) showed that the vein system can be explained as a

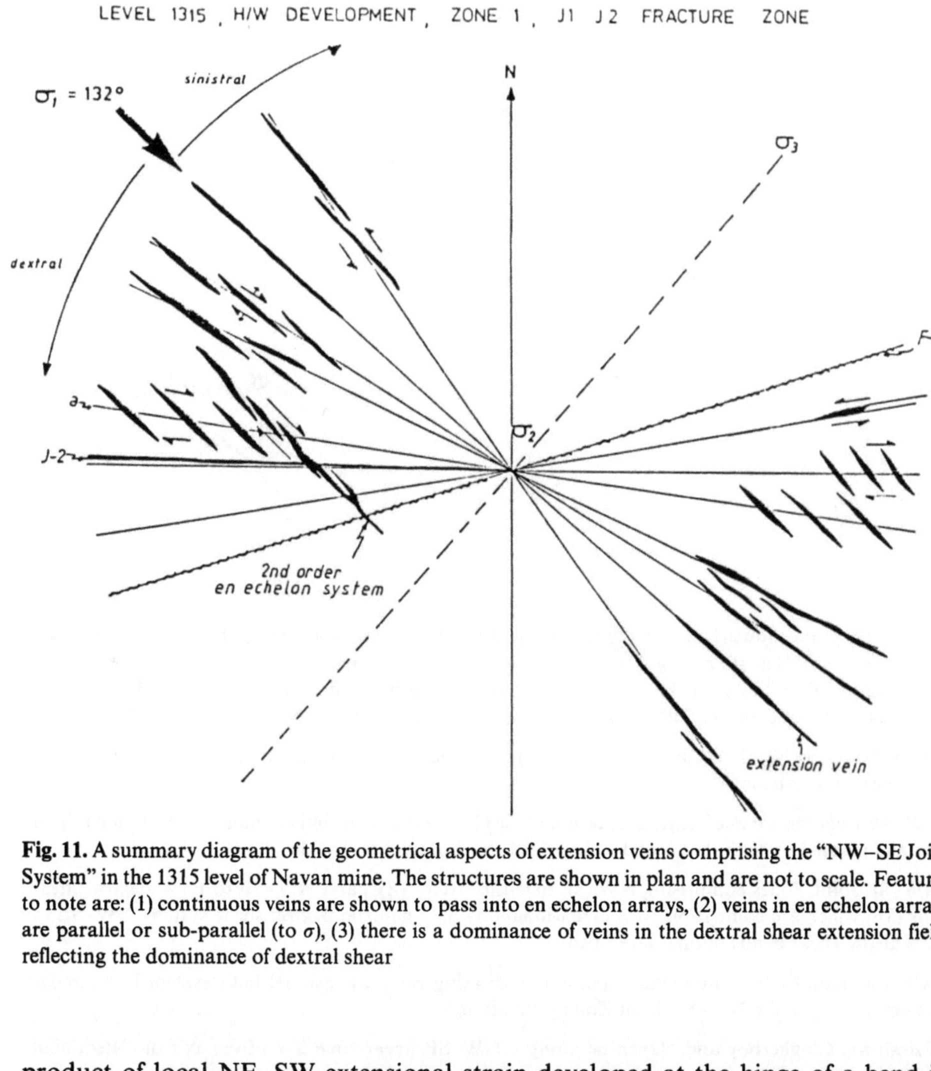

Fig. 11. A summary diagram of the geometrical aspects of extension veins comprising the "NW–SE Joint System" in the 1315 level of Navan mine. The structures are shown in plan and are not to scale. Features to note are: (1) continuous veins are shown to pass into en echelon arrays, (2) veins in en echelon arrays are parallel or sub-parallel (to σ), (3) there is a dominance of veins in the dextral shear extension field, reflecting the dominance of dextral shear

product of local NE–SW extensional strain developed at the hinge of a bend in strike of dextral shear zones. These shear zones swing from NE–SW to E–W as they are traced through the Navan area. Figure 12 summarizes a structural model which relates prolonged dextral shear on these zones to the formation of an early Courceyan extensional trough trending NW from Navan to the satellite deposit of Tatestown. The model also explains the siting of satellite deposits along this NW–SE axis in late Courceyan time. Finally, the extensional strain at the bend generated the Hercynian NW–SE veins and a series of N–S sinistral shear zones identified by statistical analyses of lineaments seen on aerial photographs. The strike swing of early faults and shear zones in the Navan deposit is paralleled by mineralized feeder veins (Andrew and Ashton, 1985). Figure 13 shows how the mapped pattern of dextral faults in the Navan region could have generated parallel extension veins as they moved because of the asymmetric orientation of the axis of their strike swing.

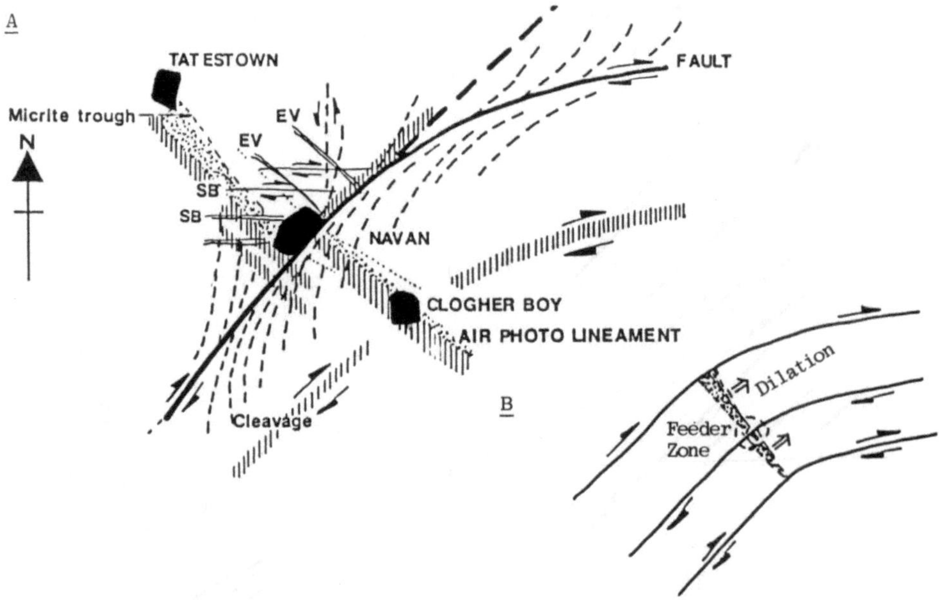

Fig. 12 A Diagram showing the structural setting of the Navan and satellite ore bodies. Dextral shear on Caledonian faults could have generated a local tensile stress regime at the SW "trailing edge" of the axis of regional Caledonian strike swing which may have been an important factor in controlling the location of the three deposits. Indicated on the diagram are features which support this model.

1. Dextral shear on NE–SW Caledonian-trending faults recognized from underground structure and air photo lineament statistics.

2. NW–SE-trending zones of sinistral rotation of air photo lineaments indicating a swing in Caledonian strike and parallelism with the second order structures and micrite troughs.

3. NW–SE Joint Zones comprising NW–SE extension veins (*EV*) and NW–SE to E–W dextral shear bands (*SB*) which are compatible with extensional flow in a co-axial dextral shear system. NW–SE to N–S sinistral shear bands poorly developed.

4. NW–SE trending "micrite trough" possibly indicating early extensional fault system in a similar structural setting to the NW–SE Joint Zone generation.

5. Tatestown, Clogherboy and Navan lie along a NW–SE linear zone SW of the axis of Caledonian strike swing.

B Summary diagram showing the position and size of dilation zones and of feeder zones for mineralization in the system of curved dextral shear zones of the Navan Region.

The position of this strike swing axis appears to be closely related to the N–S western margin of the Kentstown granitic pluton.

In summary the structural control of mineralization at Navan has been dominated by the presence of a Caledonian granitic pluton. This feature and an earlier Ordovician silicic and basaltic volcanic centre has caused deflection of small components of NE–SW trending dextral shear, particularly during late Courceyan times, when the resulting local dilation afforded an outlet for mineralizing fluids. As in the case of Silvermines, the presence nearly of major Caledonian shear zones

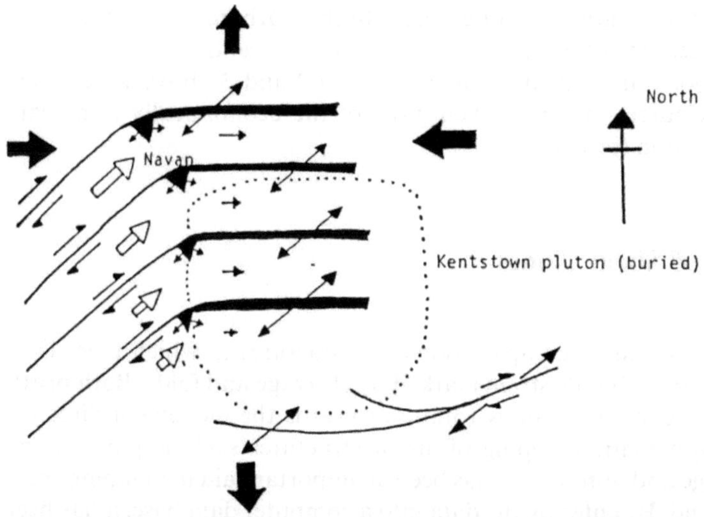

Fig. 13. Simplified structural model for the Navan Region showing how early Dinantian N–S extension and E–W compression generated local dilation at bends of dextral shear zones and in the area underlain by the Kentstown granite pluton

— the Caledonian suture zone — probably resulted in the subsequent tectonic activity during the Carboniferous. In addition to its important mechanical control, the Kentstown granite may also have provided the necessary thermal energy for driving the hydrothermal system which resulted in mineralization at Navan.

3 The Use of Knowledge of Structural Controls of Mineralization as a Guide for Further Exploration

Our studies of the structural controls of mineralization at Navan and Silvermines have lead to the conclusion that the following factors are important:

1. Presence of Caledonian shear zones in the basement.
2. Presence of more rigid rocks in the Caledonian basement such as granites or volcanic centres which inhibited reactivation of basement shearing during the Carboniferous.
3. Presence of Caledonian granites in the basement which influenced the siting of basin margins and which may also have provided a thermal source which drove hydrothermal systems.

In order to use these conclusions as guides for further exploration, it is necessary to devise methods for:

1. Identifying Caledonian shear zones which have been reactivated during the Carboniferous.

2. Identifying Caledonian granites buried beneath the Carboniferous cover.
3. Identifying the presence of buried Ordovician volcanic centres.
4. Identifying the position of dilation areas close to 2 and 3 above which were formed in late Courceyan times when most of the economically significant known mineralization formed.

4 Identification of Shear Zones

Shear zones generally occur as complex systems of fractures in which there is an increase in the intensity of ductile strain marked by cleavage and folds. Both brittle and ductile structures commonly show rotation towards the movement direction of the shear zone. Systematic mapping of minor structures such as joints, veins, minor shears, cleavage and minor folds has been an important aid in mapping shear zones in central Ireland. By entering the data into a computer data base, it has been possible to display on images and print-out maps of variable scales both individual data sets such as minor fold orientations or correlations between structural and geophysical data (Fig. 14). Figure 15 shows how simple statistical plots such as the mean orientation of a particular fracture system in 2 × 2 km cells for example can be used to define shear zones. A very wide range of techniques such as plots of entropy, atypicality or directional dominance have also been useful (Coller et al. 1986, Critchley et al. 1986).

Statistical analyses of the patterns of lineaments mapped from aerial photographs or suitably enhanced Landsat images have also been shown by Coller et al. (1986) to be a powerful method of mapping shear zones in areas of thick overburden.

The position of possible shear zones in the Caledonian basement has been studied by interpolating Bouguer anomaly data (provided by Professor T. Murphy of the Dublin Institute for Advanced Studies) onto a 500-m grid using a kriging method. First and second horizontal derivatives were then computed and the results displayed as vector maps (Fig. 16). Density boundaries were then identified from these maps (Fig. 10). By computing curves for the Hilbert transform (Nabighian 1972) along aeromagnetic flight lines, it was also possible to map the position and depth of edges to magnetic bodies (Fig. 10). By correlating such deep geophysical boundaries with plots of structural features such as folds in the Carboniferous cover (Fig. 14), Caledonian shear zones have been indicated which were reactivated during the Carboniferous. Evidence for the late Courceyan faulting, of such importance for mineralization, is, however, also dependant upon stratigraphic data.

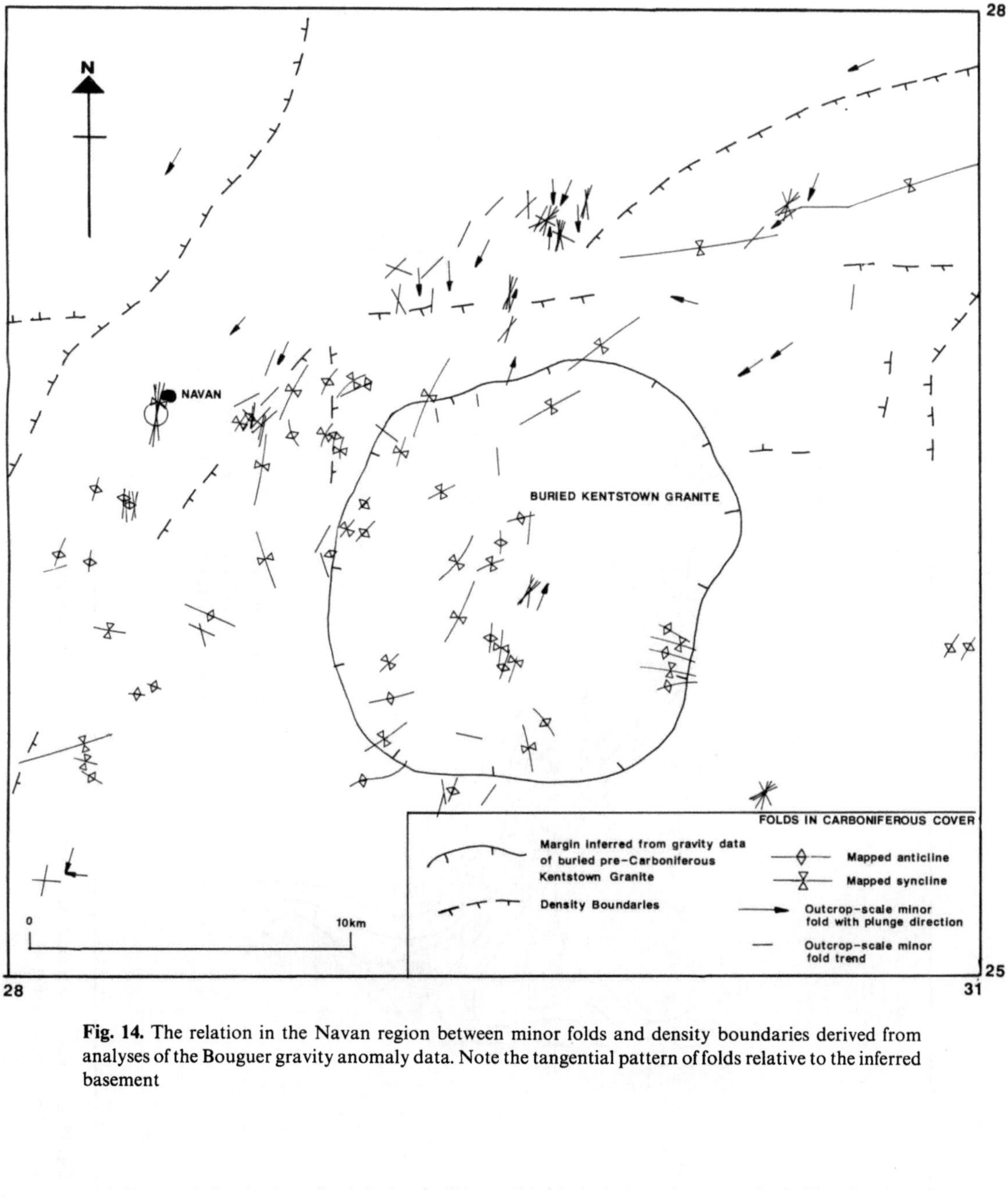

Fig. 14. The relation in the Navan region between minor folds and density boundaries derived from analyses of the Bouguer gravity anomaly data. Note the tangential pattern of folds relative to the inferred basement

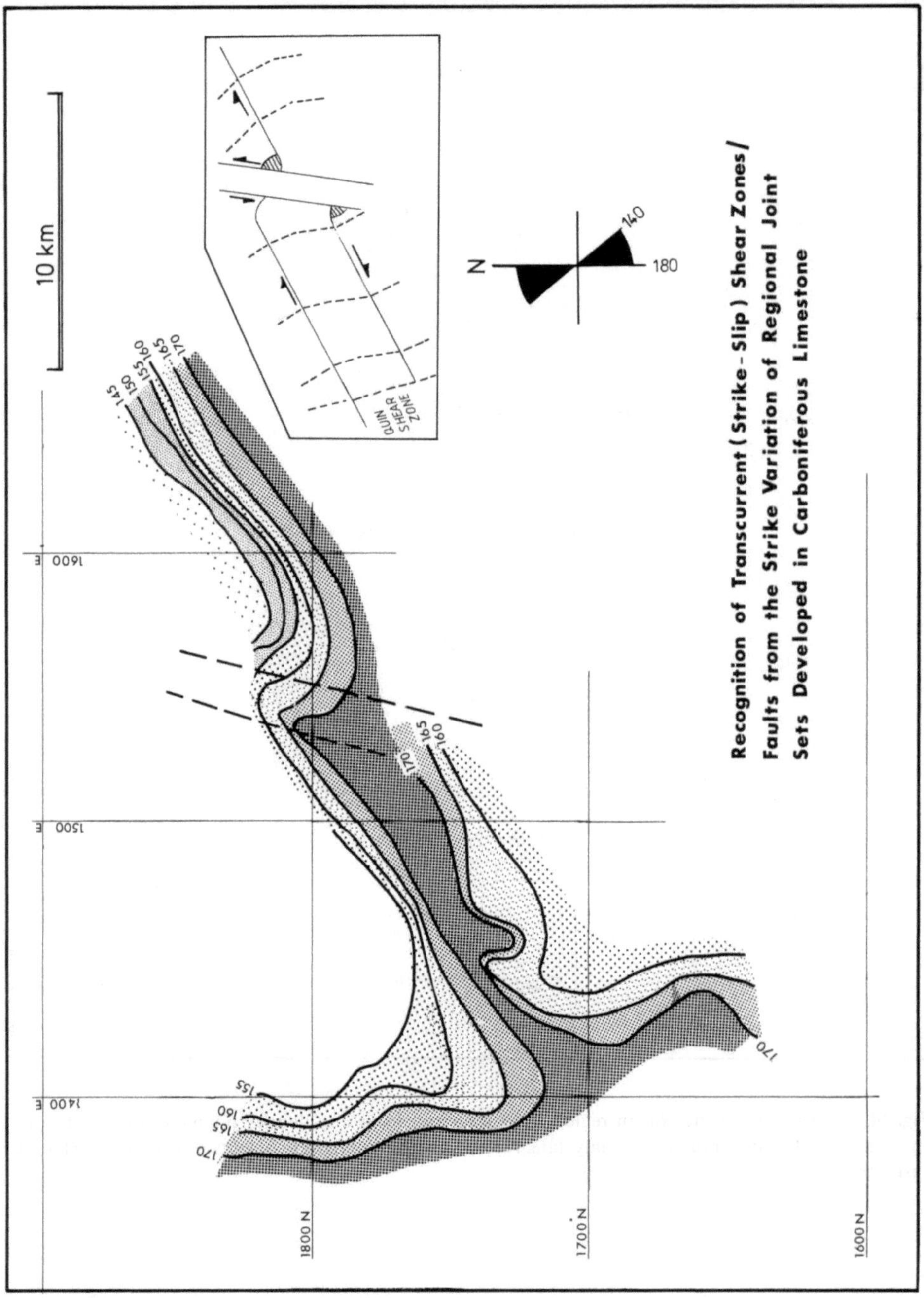

Fig. 15. Contoured group mean map showing the clockwise rotation of the regional NNW–SSE fracture system within the Quinn Shear Zone of the West Midlands (Fig. 4)

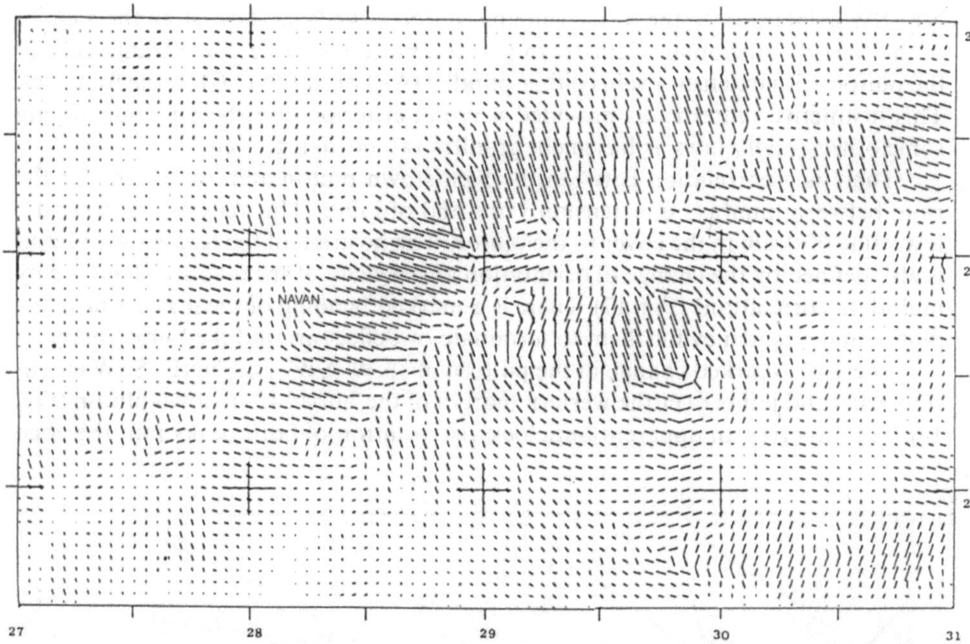

Fig. 16. Vector plot showing the direction and intensity of the second horizontal derivative of the Bouguer gravity field interpolated by kriging to a 500 m^2 grid in the Navan region. Density boundaries in the area are shown by low intensity vectors (*short lines*). Many of these are NE–SW in trend

5 Detection of Buried Granite in the Basement

The Bouguer gravity anomaly data of central Ireland (unpublished data kindly provided by Professor T. Murphy) shows a large number of negative anomalies. Brown and Williams (1985) have interpreted many of these in terms of NE–SW-trending sedimentary troughs bounded by a series of linear volcanic blocks in the Ordovician basement. An alternative interpretation, which we favour, is that many of the negative anomalies are caused by Caledonian granites in the pre-Carboniferous basement. Both structural and stratigraphic criteria can be used to decide in each case which is the more likely interpretation, for as we have shown, buried granites would have acted as buoyant crustal elements during deposition of the Carboniferous cover and they should therefore be correlated with positive blocks with a cover of thin shelf facies Visean rocks. The rigidity of buried granites should have protected the cover rocks from deformation, therefore they should show up as areas of low strain and of low density of fractures and of remotely sensed lineaments. Alternatively, if a gravity low represents a sedimentary trough as suggested by Brown and Williams, then there should be higher strain in the fill and its cover than on the margins, where more rigid Ordovician volcanic rocks are inferred by them. Figure 14 shows the complex pattern of Hercynian folds within the Carboniferous rocks of the Navan area. The regional pattern of folds is one of

NE–SW axial traces in the west, swinging to ENE–WSW in the east. In the vicinity of the buried Caledonian Kentstown granitic pluton, folds become trangential to the pluton with extension veins either parallel (AB) or at 90° (AC) to the fold axes. This distinctive pattern has provided a useful criterion elsewhere for identifying some other gravity lows as basement granites rather than grabens.

There is a correlation between the Kentstown pluton and low values for the ratio between numbers of extension veins and of joints. This reflects the generally low value of strain in the cover to the pluton. The position of the Kentstown pluton is also defined by the change of orientation and density of shear zones identified by multidata correlations (Fig. 17). There is a marked bending of shear zones near the pluton and a tendency for them to terminate or deflect around the buried body.

There is a marked change in the pattern of "NW–SW" and NNE–SSW extension veins and joints over the western margin of the buried Kentstown pluton (Fig. 18). This involves an apparent anticlockwise rotation of both peaks into the zone

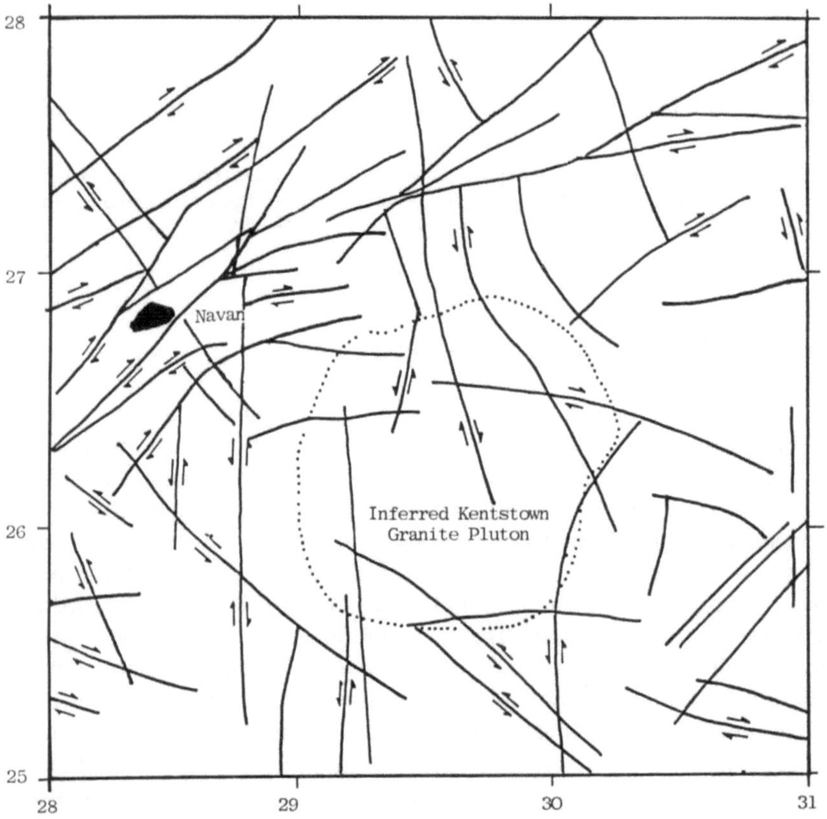

Fig. 17. Pattern of shear zones identified by correlating ground structural, geophysical and remote sensing data in the Navan Region. Note the deflections and low density of zones in the area underlain by the Kentstown granite pluton

of regional strike swing along the western side of the pluton. Such a pattern of change may again be useful elsewhere in identifying strike swings in the basement which may be linked to the presence of Caledonian plutons.

Buried Ordovician volcanic centres are clearly defined by aeromagnetic data. They may be distinguished from Carboniferous volcanic centres if the depth of anomalies is calculated using, for instance, the Hilbert transform method.

In conclusion, our work has shown that shear zones can be mapped using a combination of ground structural, geophysical and remote sensing analyses. The presence of granites buried beneath the Carboniferous cover can be detected by combining gravity data with structural, remote sensing and lithofacies data.

Potentially mineralized dilation zones can be identified on shear zones where they terminate close to basement granite plutons or basement volcanic complexes. These methods have been applied to a 25,000 km^2 area of central Ireland and the results are now being used to assist exploration programmes in this area.

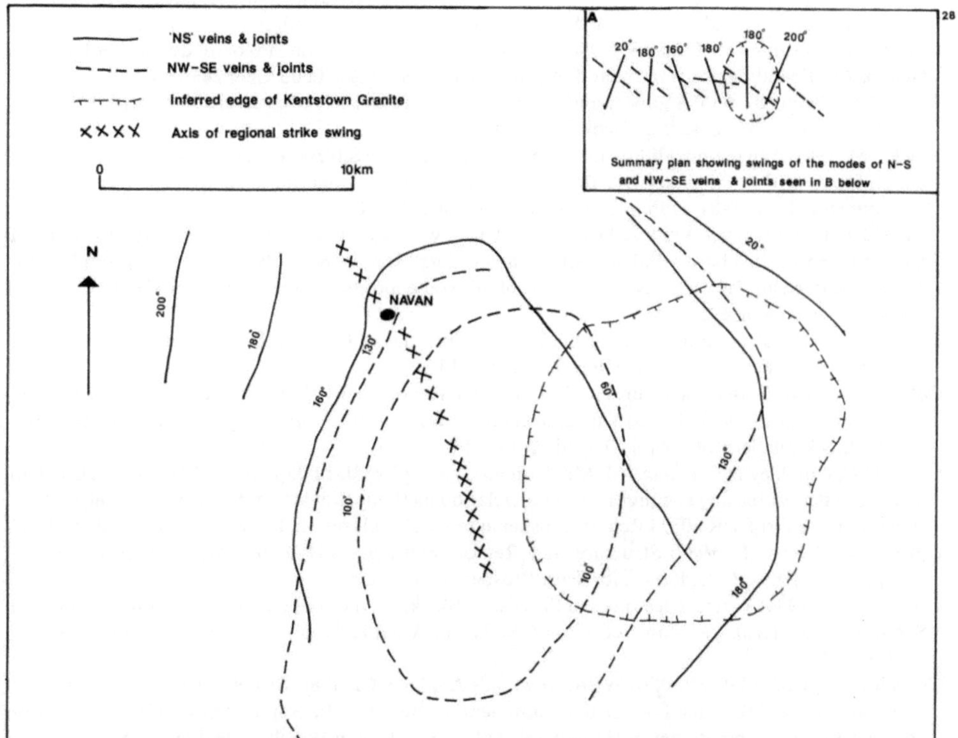

Fig. 18 A A summary of the changes in strike of extension veins and joints in Carboniferous rocks of the Navan region. The change occurs west of the buried Kentstown granite over the area of strike swing of the Caledonian basement. **B** Contour map showing details of the modes of extension veins and joints in Carboniferous rocks of the Navan region

Acknowledgements. The work summarized here was part of a more extensive study of mineralization in Central Ireland supported by the European Commission (*Contracts MPP-159-EIR and MSM-111-EIR*). It is a pleasure to acknowledge the valuable support and encouragement given by Dr. L Van Wambeke of the Commission during our work. The relevance of the project to the aim of improving methods of mineral exploration was dependent upon a close working relationship being established with all the active mineral exploration companies in Ireland. We gratefully acknowledge the financial and intellectual support which we received from: Tara Exploration and Development Plc. Tara Mines Plc. Aquitaine Minerals (Ireland) Limited. Billiton Minerals (Ireland) Limited. Cominco Ireland Limited. Irish Base Metals.

Professor T. Murphy of the Geophysical section of the Dublin Institute for Advanced Studies provided access to all his Bouguer anomaly data for the study area. His generosity help and advice were greatly appreciated. We would also like to thank Dr C.E. Williams, Director of the Geological Survey of Ireland, for allowing us to use the aeromagnetic data belonging to G.S.I., which was acquired with EEC Funding (1978–1981 research programme).

References

Andrew CJ and Ashton, JH (1985) The regional setting, geology and metal distribution patterns of the Navan Orebody, Ireland. Trans. Inst. Min. Metall. (Sect. B: Appl. Earth sci) 94:B66–93

Ashton JH, Downing DT, Finlay S. (1986) The geology of the Navan Zn–Pb orebody. In: Geology and Genesis of Mineral Deposits in Ireland. Andrew et al. (eds). Ir Ass Econ Geol 243–280

Brown C, Williams B (1985) A gravity and magnetic interpretation of the structure of the Irish Midlands and its relation to ore genesis. J. Geol. Soc. London, 142:1059–1075

Bruck PM (1982) The regional lithostratigraphical setting of the Silvermines zinc-lead and the Ballynoe barite deposits, Co. Tipperary. In: Brown, AG (ed) Mineral exploration in Ireland: Progress and developments 1971–1981. Dublin Ir Assoc Econ Geol 162–170

Caulfield JBD, Le Hurray AP, Rye DM (1986) A review of lead and sulphur isotope investigations of Irish sediment-hosted base metal deposits with new data from the Keel, Ballinalack, Moyvoughly and Tatestown deposits. In: Geology and genesis of mineral deposits in Ireland. Andrew CJ et al. (eds). Ir Assoc for Econ Geol. 591–615

Coller DW (1984) Structural studies of the Navan mine, Navan Co. Meath — Structure of the Basalt Sheet in 1 Zone. Report for Tara Mines Limited 1–14

Coller DW (1984) Variscan structures in the Upper Palaeozoic rocks of west central Ireland. In: Hutton, D.H.W. and Sanderson, D.J. (eds) Variscan tectonics of the North Atlantic region. Geol Soc Spec Publ No. 14. Blackwell Scientific Publ. Oxford pp 185–194

Coller DW, Critchley MF, Dolan JM, MacDonaill C, Murphy CJ, Phillips WEA, Sanderson DJ (1986) Structural remote sensing and multivariate correlation methods as aids to mineral exploration, Central Ireland. In: Report EUR 10334 Remote sensing in mineral exploration. L. Van Wambeke (ed) pp 1–41

Coller DW, Pepper S (1984) Structure and Remote Sensing study of the Athboy-Navan area — Prospecting License 1439. Report for Tara Prospecting Limited

Critchley MF (1984) Variscan tectonics of the Alston Block, northern England. In: Hutton, D.H.W. and Sanderson, D.J. (eds). Variscan tectonics of the North Atlantic Region: Geol Soc London Spec Pub 14:139–146

Critchley MF, Coller DW, Phillips WEA, Rowlands AS, Daly CJ (1986) Structural, remote sensing and multivariate correlation methods as aids to mineral exploration. In: Report EUR 10511. First European workshop on remote sensing in mineral exploration. L. Van Wambeke (ed) pp 33–84

Jacob AWB, Kaminski W, Murphy T, Phillips WEA, Prodehl C (1985) A crustal model or a northeast-southwest profile through Ireland. Tectonophysics 113:75–103

Keil M, Sengpiel KP, Mollat H (1985) Test of modern methods of helicopter geophysics for prospecting for Pb–Zn ores in Ireland. Trans Inst Min Met (B), 94:14–19

Murphy NP, Jacob AWB (1985) Detailed S-wave structure in the Dublin Basin and its northern margins. Geophys Roy Astron Soc 83:803–7

Nabighian MN (1972) The analytic signal of two-dimensional magnetic bodies with polygonal cross-section: its properties and use for automated anomaly interpretation. Geophysics 37:507–517

O'Keeffe WG, (1986) Age and postulated source rocks for mineralization in Central Ireland, indicated by lead isotopes. In: Geology and genesis of mineral deposits in Ireland. Andrew et al. (eds). Ir Assoc Econ Geol 617–624

Philcox ME (1984) Lower Carboniferous lithostratigraphy of the Irish Midlands. In: Ir Assoc Econ Geol 89

Phillips WEA (1982) Correlation of geological, geochemical and geophysical data with satellite imagery, west-central Ireland. Final report EEC Contract MPP-159-81-EIR(H). 236 p

Phillips WEA, Stillman CJ, Murphy T (1976) A Caledonian plate tectonic model. J. Geol Soc Lond 132:579–609

Phillips WEA, Haughey N (1982) Structural studies of the Tara mine, Navan, Co. Meath. Report for Tata Mines. 1–94

Phillips WEA, Coller DW, Critchley MF, Skelly F (1983) Progress report no. 1 on analysis of the structural and remote sensing data of the Navan-Kells area. Report for Tara Exploration and Development Limited. pp 1–60

Phillips WEA, Sevastopulo GD (1986) The Stratigraphic and Structural setting of Irish Mineral deposits. In: Geology and genesis of mineral deposits in Ireland. Andrew CJ et al. (eds). Ir Assoc Econ Geol 1–30

Robson DA (1980) The geology of north east England. Special Publication Nat Hist Soc Northumbria Hancock Museum, Newcastle. 1–96

Russell MJ (1986) Extension and convection: a genetic model for the Irish Carboniferous base metal and barite deposits. In: Geology and genesis of mineral deposits in Ireland. Andrew CJ et al. (eds). Ir Assoc Econ Geol 545–554

Sevastopulo GD (1981) Lower Carboniferous. In: Holland, C.H. (ed) A geology of Ireland. Scottish Academic Press, Edinburgh pp 147–172

Sheridan DJR (1972a) The stratigraphy of the Trim No. 1 well Co. Meath and its relationship to Lower Carboniferous outcrop in east central Ireland. Geol. Surv. Ireland Bull. 1:311–334

Strogen P (1977) The evolution of the Carboniferous volcanic complex of southeastern Limerick, Ireland. J. Geol Soc London 133:409–10

Lithogeochemical Investigations in the Navan Area, Ireland

P. Van Oyen and W. Viaene[1]

Abstract

The Lower Carboniferous strata, which host the Navan Pb–Zn deposit, have been analyzed in an area NW of the Navan mine. They are shallow water carbonates characterized by complex carbonate diagenesis. Lithogeochemical data for Fe, Mn, Mg, Pb, Hg, Zn, As, Ba, Cu and Sr have been evaluated in relation to changes in lithology, carbonate diagenesis and occurrence of sulphides.

Several anomalous lithological and geochemical trends occur towards a subeconomic mineralization at Tatestown (eastern part of the study area) and at Clonabreany (western part of the study area). Both the thickness and the nature of the Lower Pale Beds, which contain the mineralization, represent a lithological anomaly: the thickness strongly increases towards Tatestown (and Navan), while the strata hosting the mineralizations are heterogeneous, high energy sediments. A Mn enrichment in some carbonate cements is present; it may be related to the mineralizations but cannot be detected in the bulk analyses. An anomalous population of Zn, Pb, Ba and Hg has been statistically separated from the background population. This anomalous population is related to the presence of sulphides and shows clear trends towards the mineralizations.

1 Introduction

The search for economic ore deposits in the Carboniferous of central Ireland has become increasingly difficult for several reasons, one of which is the depth of burial. The Navan deposit (County Meath, Ireland) was discovered in 1970 by means of soil geochemistry (Mining Magazine, July 1975), following a decade of successful exploration. Recent discoveries, albeit of only subeconomic mineralizations, confirm the potential for finding larger carbonate-hosted base metal deposits in the Lower Carboniferous of Ireland.

[1] Katholieke Universiteit Leuven, Fysico-chemische Geologie, Celestijnenlaan 200 C, B-3030 Heverlee, Belgium

Mineral Deposits within the European Community
(ed. by J. Boissonnas and P. Omenetto)

Rock geochemistry has been considered in the search for deep-seated ore deposits. From the literature and from the results of several projects, funded by the EC in the first R&D program on primary raw materials (1978–1981), it is clear that considerable variations exist in the geochemistry of carbonate sediments. Some lithogeochemical patterns are known to be associated with base metal deposits (e.g. Russell 1974; Gwosdz and Krebs 1977; Chabot 1982; Viaene 1982; Erickson et al. 1983). However, little is known about the nature of the patterns and interference may occur from geochemical variations due to stratigraphy, facies and lithology.

Sites of ore deposits in carbonate environments are often characterized by features related to facies and palaeogeography. Depending on the type of deposits and on the metallogenetic concept, several diagnostic criteria have been suggested. For the 'Irish style' deposits several features have led to metallogenetic models involving active synsedimentary faulting and rapid subsidence (e.g. Russell 1978; Deeny 1981; Boyce et al. 1983; Large 1983; Plant et al., this Vol.). Sediment petrography can be an important tool assisting in locating mineralizations in a direct or indirect way by a better evaluation of lithogeochemical patterns. For this purpose, the Tournaisian carbonate succession of the Navan area, host of several deposits, was investigated.

The Navan Pb–Zn deposit is generally believed to have a sedimentary-exhalative origin (Boast et al. 1981; Russell 1983). It is hosted in Lower Carboniferous shallow water carbonates. The geological setting has been described by Andrew and Ashton (1985) while textures of the ore have been discussed by the same authors (1982) and by Kucha and Wieczorek (1984). Lithogeochemical studies in a small area around the Navan mine have shown vertical and lateral trace element zonations (Finlay et al. 1984). The presence of subeconomical mineralizations in the area supports the potential of the Lower Carboniferous.

Complete cores from 14 boreholes from an area NW of Navan (Fig. 1A) have been studied. The area contains both the Tatestown-Scallanstown prospect, which is considered to be a satellite deposit of the Navan orebody (Andrew and Poustie 1984), and the subeconomic mineralization of Clonabreany in the west.

The aim of the study was to evaluate new lithogeochemical data in relation to such factors as sediment petrography, diagenesis and mineralogy. Furthermore, an attempt has been made to find large-scale geochemical trends which indicate the mineralization.

2 Materials and Methods

The location of the different drill holes is shown in Fig. 1A. The cores were split vertically and 6-m intervals of half-cores were prepared for chemical analyses. The insoluble residue (IR) was obtained after dissolution with aqua regia. The solution was analyzed for Zn, Pb, Fe, Mn, Mg, Sr, Cu, Ba, As and Hg by atomic absorption spectrometry. The detailed analyses can be obtained from the authors.

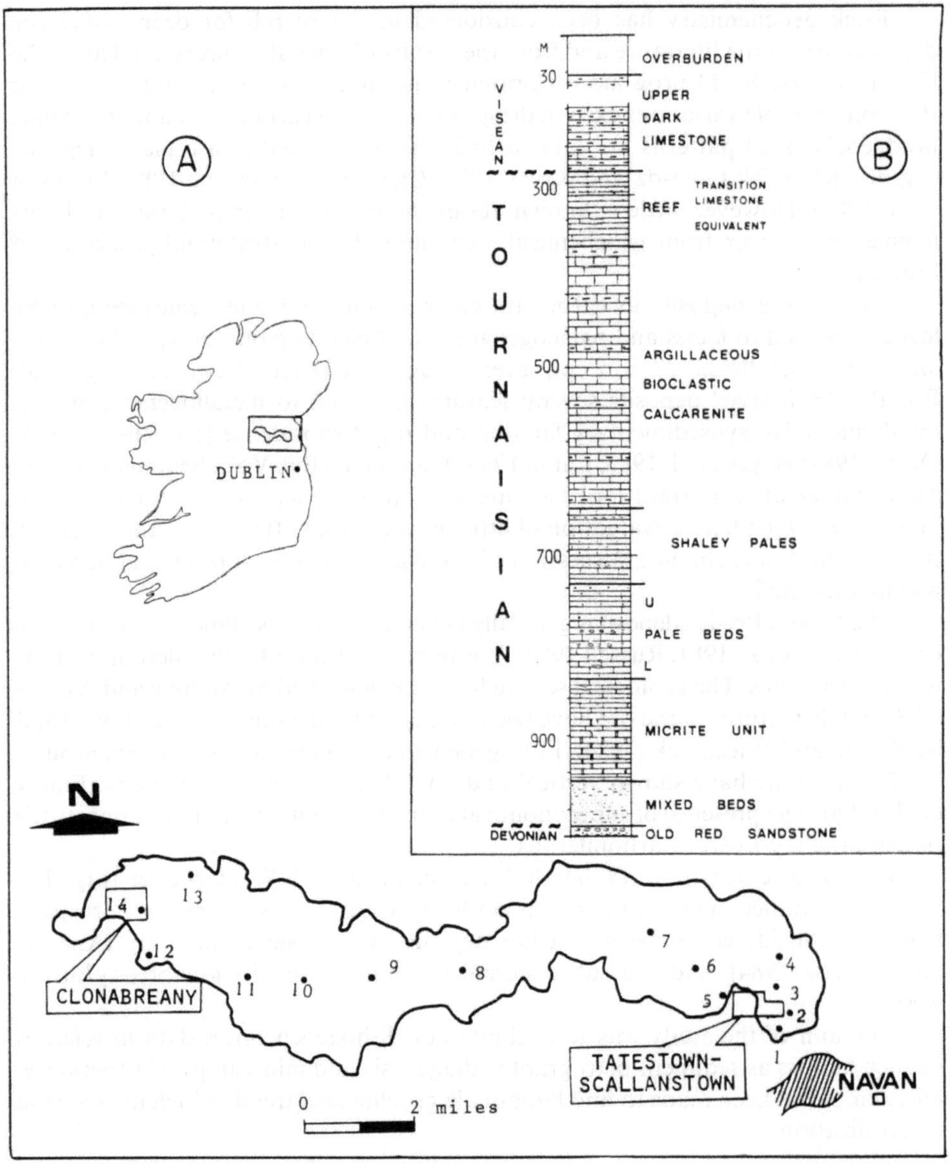

Fig. 1A, B. Location of the study area and drill holes **(A)** and general lithostratigraphy based on data of hole 7 **(B)**

The second half of the core was used for petrographical analysis by means of stained acetate peels and thin sections. Unstained thin sections were studied with a cold-cathodoluminescense microscope. Polished sections were prepared for the sulphide study. Microprobe analyses on carbonates were performed using carbonate standards for Zn, Pb, Mn, Mg, Sr, Ca and Fe.

3 Lithostratigraphy

In contrast to the host rock of the Tynagh deposit (Clifford et al., this Vol.) the lithological sequence of the Lower Carboniferous in the Navan area is characterized by a rapid change of facies and lithology, which is due to the transgression of the Carboniferous shelf sea onto the Lower Palaeozoic Old Red Sandstone continent. A general stratigraphical sequence for the area is shown in Fig. 1B, based on the data of hole 7 (see also Phillips et al., this Vol.). The continental Red Beds (Old Red Sandstone) are separated from the marine Shaley Pales and overlying units of the main Carboniferous transgression by a sequence of lagoonal sediments and shallow water carbonates with varying amounts of sand, clay and bioclasts.

The Mixed Beds, which contain an important detrital component comprise about 40 m of sandstones, siltstones and mudstones developed in an open intertidal environment. The Muddy Limestone unit, representing a fossiliferous calcsiltite with a lagoonal facies and reaching a thickness of ~ 15 m at Navan (Andrew and Ashton 1982) is absent from the study area.

In this study special attention has been paid to the Micrite Unit (MU) and to the Lower Pale Beds (LPB), which host the massive sulphide lenses at Tatestown and at Navan. The Micrite Unit is mainly composed of algal micrite, with algal clasts and bird's eye textures. A restricted intertidal to supratidal sedimentation environment with limited detrital influx is indicated. The Lower Pale Beds are composed of an alternation of oolitic and bioclastic limestones, with a decreasing oolite content towards the younger lithologies and with some sand or clay interlayers. No in situ fauna has been found in these layers.

The Upper Pale Beds (UPB) consist of an alternation of sandy biosparite (crinoids and brachiopods) and calcareous sandstones. The facies points to a subtidal environment in which the bioclastic levels developed between sandbars.

The Shaley Pales and the Argillaceous Bioclastic Calcarenite reflect a gradual deepening of the environment. The overlying Waulsortian Limestone and Upper Dark Limestone have not been investigated in detail.

4 Influences on the Lithogeochemistry

4.1 Lithology

The distribution of the lithostratigraphical units and their petrographical features demonstrate an anomaly in the thickness of the Micrite Unit and the Lower Pale Beds. In the western part of the study area the thickness of the Micrite Unit is constant (~ 80 m) and the unit is immediately overlain by sandstones and crinoidal limestones from the Upper Pale Beds. Eastwards from hole 7 the Micrite Unit becomes thinner (~ 25 m at Tatestown and less than 15 m at Navan) but the Lower Pale Beds thicken very strongly (from 0 to ~ 120 m over a distance of about 6 km).

The nature of the Lower Pale Beds has also been shown to be anomalous. The oolitic layers are heterogeneous and badly sorted and they contain lithoclasts from the underlying Micrite Unit. Micrite sedimentation or cementation is completely absent. These layers, the components of which have probably formed in shallow water, were deposited after considerable transport. The first appearance of these high energy sediments coincides with the lowermost important mineralization at Navan and Tatestown. A detailed description is given by Tijskens et al. (in press).

The Lower Pale Beds of hole 7 are almost exclusively composed of bioclastic limestone. Broken bioclasts, coated grains and skeletal detritus indicate at least some transport and deposition in an energetic environment. These layers alternate with the oolitic layers in all other boreholes. The oolite content increases towards Tatestown.

Lithogeochemical confirmation of these lithological variations is difficult because of the varying amount and nature of the insoluble residue (clays, sand, etc.) from one lithology to another and because of the 6-m sampling lengths of the analyzed cores. Microprobe analyses have shown, however, that the trace element content of the carbonate components, i.e. oolites and lithoclasts of the Lower Pale Beds, is similar to the content in the Micrite Unit and Upper Pale Beds. Lithostratigraphical limits have been shown to correspond to significant changes in the amount of insoluble residue. Hence, the mean value of the geochemical data should be corrected for the different proportions of insoluble residue.

4.2 Diagenesis: Carbonate Phases

The diagenesis of the Micrite Unit and Lower Pale Beds sediments is represented in Table 1. The normal diagenetic sequence shows early lithification of the shallow water carbonates with a rim cement and a dogtooth calcite. A blocky calcite characterizes further cementation in the meteoric phreatic zone. The different phases have been analyzed by microprobe (Table 2A). In the blocky calcite one or more Mn-rich zones are sometimes present; their emplacement, however, varies from one sample to another. The zones sometimes etch the earlier cements and their Mn content can be as high as 1%. The influence on the Mn content of bulk samples is small and depends on the initial porosity since the Mn enrichment required an open diagenetic system.

Several types of dolomite are developed. Microprobe results are given in Table 2B. Early dolomitization is irregular and never pervasive. As can be seen from Table 2B, it has only a minor influence on the lithogeochemical data of Zn, Mn or Fe. A late-diagenetic 'baroque' dolomite is mainly associated with iron calcite veining. It is zoned in Fe and Mn; its content can be as high as 7.3% Fe and 1% Mn.

Iron calcite veining is irregularly developed and composes only a minor proportion of a bulk sample; the contents of Fe and Mn vary from 1750 to 8000 ppm and from 1000 to 1300 ppm respectively.

Table 1. Diagenetic sequence in non-mineralized sections of the eastern part of the study area

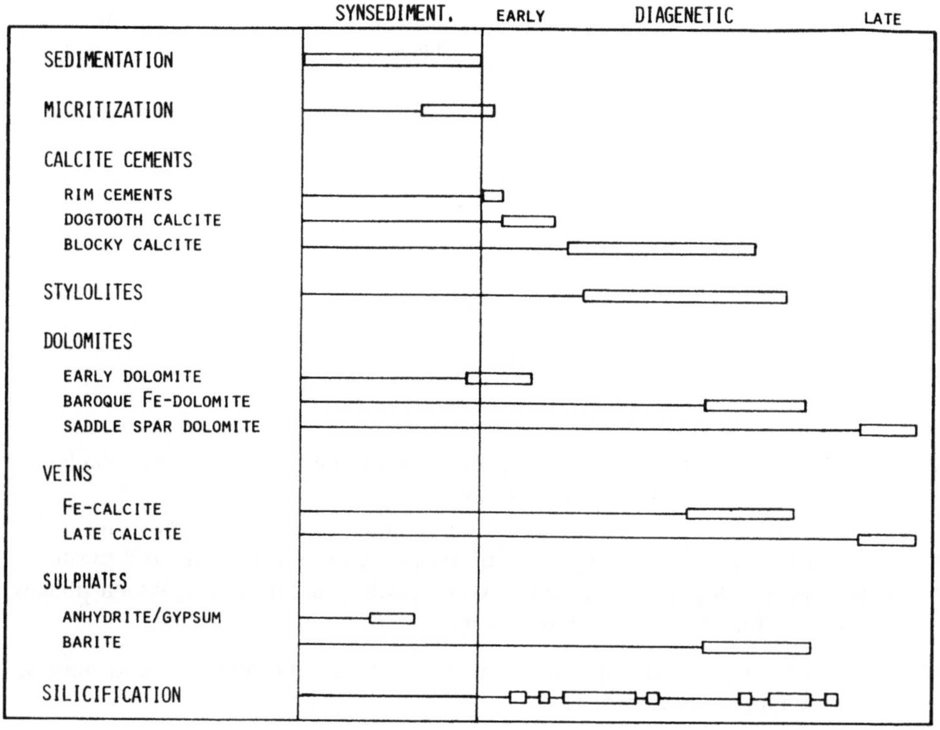

4.3 Sulphides

Pyrite framboids are widely dispersed throughout the area. They are associated with strata, rich in clay and in organic matter; they are homogeneously distributed over the area with the exception of the mineralized sections of Tatestown (and Navan) where their concentration is much higher.

The geochemical data from the different boreholes show an association of anomalous Pb, Zn, Ba and Hg values. These anomalies correspond with the presence of macroscopic sulphides of Pb and Zn, which can be detected by careful examination of the half-cores. Observations on these sulphides (Viaene et al. 1986), reveal that their occurrence and their textures form halos around the mineralization of Tatestown (and Navan). The distribution of the sulphides in the boreholes near the mineralized sections is thought to be due to remobilization of pre-existing mineralizations because their occurrence is always associated with newly created porosity and permeability (i.e. in veins or stylolites, in or near faults or in dolomitized sections).

Variations in the content of the analyzed elements are thus due to original sedimentary features and to diagenetic changes. They are:

1. The irregular distribution of lithological units of different types and with different thickness;

Table 2. Microprobe results[a]

	Fe (ppm)	Mn (ppm)	Zn (ppm)	Mg (ppm)
A. Ranges for calcite phases				
Dogtooth calcite	<d–1640	<d–780	<d–1300	1800–5670
Blocky calcite	500–3600	600–3000	<d–780	<d–3020
Mn-rich zones	3500–5500	7000–10500	<d	<d
Fe calcite (veins)	1750–8000	1040–1280	<d–930	1750–4340
B. Ranges for the dolomites				
Early dolomite	2890–4370	<d–2630	<d	n.a.
Baroque dolomite	2630–7.3 wt%	2630–10000	<d–3820	n.a.

[a] <d = Below detection limit; n.a. = not analyzed.

2. The large variations in the carbonate and insoluble residue content, the latter being partly composed of Fe-containing illite;
3. The occurrence of a baroque dolomite influencing the Fe, Mg and Mn data;
4. The distribution of macroscopic sulphides, i.e. galena, sphalerite and pyrite;
5. The complex diagenetic sequence of carbonate cementation in which phases occur with different Fe and Mn contents.

Several of these features are spatially associated with mineralization and may be related to the mineralizing process.

5 Treatment of the Geochemical Data

The treatment of the geochemical data requires the filtering of the influences governing the geochemical distribution. It can be partly achieved by distinguishing different populations. The geochemical data have been treated by the SAS program (Statistical Analysis System, SAS Institute Inc.). The Micrite Unit and Lower Pale Beds have been treated separately from the other lithologies because they host the mineralization and because of their similar carbonate content.

5.1 Univariate Statistics

Groups of elements which have a similar behaviour are recognized:

a) Fe, Mn and Mg have a complex distribution (e.g. Fe and Mn in Fig. 2A), reflecting lithological variations. Separation of the different populations producing these histograms was impossible, even after log-transformation of the data. Only in the case of Mn was it possible to distinguish a small anomalous population; it is related to dolomitized samples (baroque dolomite).

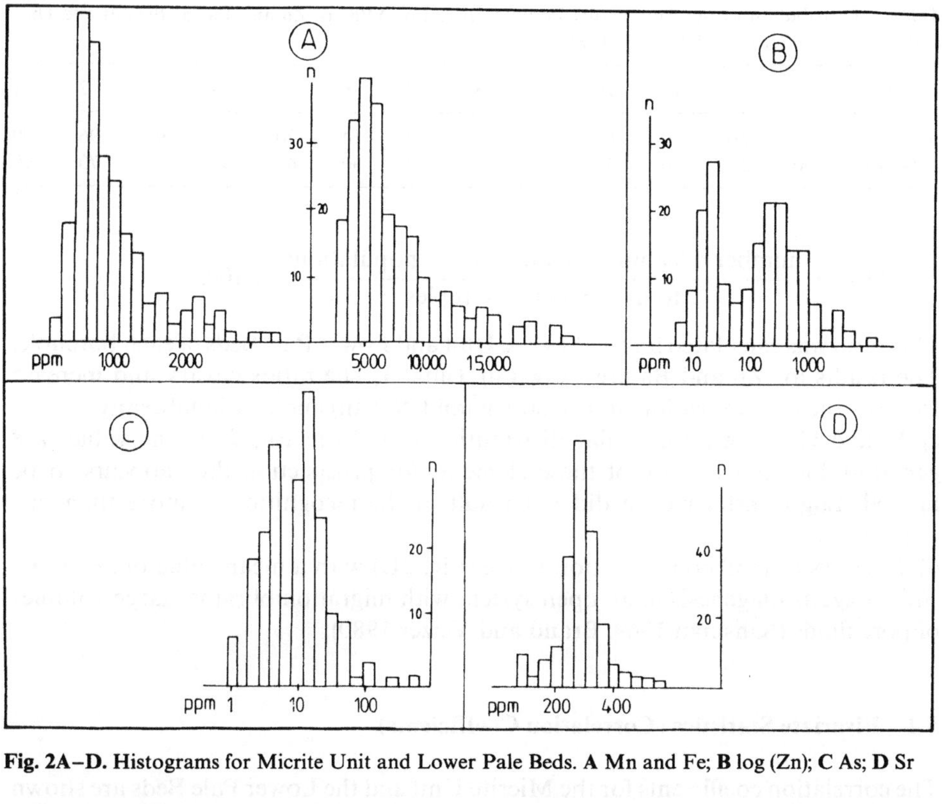

Fig. 2A–D. Histograms for Micrite Unit and Lower Pale Beds. **A** Mn and Fe; **B** log (Zn); **C** As; **D** Sr

b) The histograms of Zn, Pb, Ba and Hg are irregular. The logarithmically transformed data, however, show a clear bimodal distribution (e.g. Zn in Fig. 2B). The log-transformation of the data is justified for further treatment:

1. Two populations are separated which correspond to the different mode of occurrence of the elements;
2. Normal populations are obtained, which is a condition for further statistical analysis;
3. The anomalously high values are incorporated and do not have to be omitted for further treatment.

The first population of these elements is characterized by the following mean values and upper limits: Zn (20 ppm/80 ppm), Pb (30 ppm/150 ppm), Hg (60 ppb/200 ppb) and Ba (200 ppm/600 ppm). These populations are interpreted as background populations due to the adsorption on clays or incorporation in carbonates.

The second anomalous population of these elements shows values as high as 8000 ppm Zn, 3400 ppm Pb, 1750 ppb Hg and 4000 ppm Ba. These values are related to the presence of galena and/or sphalerite and of barite.

To check the use of this anomalous population, an empirical 'Anomalous Population Ratio' (APR) was defined as

Table 3. Distribution of the Anomalous Population Ratio (APR) of Zn and Ba for the Micrite Unit and Lower Pale Beds (no data for hole 8)

Drill hole	13	12	11	10	9	8	7	6	5	4	3	2	1
APR (Zn)	25	50	0	63	33		10	14	70	84	92	96	70
APR (Ba)	50	50	50	66	30		10	30	40	94	45	70	90

$$APR = \frac{\text{number of samples of anomalous populations}}{\text{total number of samples}} \times 100.$$

This ratio was calculated for the Micrite Unit and Lower Pale Beds of each borehole. The results for Zn and Ba are shown in Table 3. The ratios covary and increase towards the mineralization at Tatestown (and Navan) and at Clonabreany.
c) As and Cu have an irregular distribution (e.g. As in Fig. 2C). The values are generally low and the use of these elements for prospection thus appears to be limited. Log-transformation did not result in the recognition of more than one population.
d) Sr shows a small concentration range (Fig. 2D) with a mean value of 280 ppm. This suggests diagenesis in an open system with migration of rather large volumes of pore fluids (Kinsman 1969; Brand and Veizer 1980).

5.2 Bivariate Statistics (Correlation Coefficients)

The correlation coefficients for the Micrite Unit and the Lower Pale Beds are shown in Table 4. There are significant correlations between Zn, Pb, Ba and Hg, which are the elements occurring in sulphides and barite, and also between Fe, Mn and Mg, the metals held in carbonates. This group correlates negatively with Sr, indicating leaching of Sr by diagenetic processes. The illites, which are a major component of the insoluble residue, are iron-rich, as indicated by the correlation between Fe and the insoluble residue (IR). This correlation is accentuated by the presence of pyrite framboids in shaley layers.

Table 4. Correlation matrix for the Micrite Unit and Lower Pale Beds ($L = Log_{10}$, $n = 219$ and $P < 0.0001$)

Fe	Fe										
Mn	0.74	Mn									
LPb			LPb								
LZn			0.69	LZn							
LBa			0.37	0.33	LBa						
LHg	0.24	0.24	0.49	0.49	0.31	LHg					
LAs							LAs				
LCu								LCu			
Mg	0.84	0.78				0.27			Mg		
Sr	−0.56	−0.37			0.38				−0.60	Sr	
Ir	0.29		−0.35								Ir

Table 5. Rotated factor pattern of the factor analysis for the Micrite Unit and Lower Pale Beds ($L = Log_{10}, n = 219$)

	Rotated Factor Pattern			
	Factor 1	Factor 2	Factor 3	Factor 4
LZN	−0.083	0.881	0.293	−0.123
LPB	0.089	0.593	0.501	−0.328
LBA	−0.140	0.120	0.872	0.045
LHG	0.378	0.294	0.597	−0.091
LCU	0.565	−0.523	0.390	0.173
LAS	0.668	0.485	0.076	0.219
MG	0.928	−0.066	−0.007	−0.023
MN	0.899	−0.184	0.110	−0.132
FE	0.906	0.090	−0.051	0.232
IR	0.095	−0.163	−0.020	0.950
% Variance Explained	34.4	18.0	16.3	11.9

5.3 Multivariate Analysis

5.3.1 Factor Analysis

Factor analysis of the data from the Micrite Unit and Lower Pale Beds has been carried out with varimax rotation after principal component analysis. Sr has been omitted because it has a very narrow range. The results of the rotated factor pattern are shown in Table 5.

The first factor is composed mainly of Fe, Mn and Mg but it also explains part of the variance of Cu, As and Hg. The value of these elements in prospection is limited since they are included in this carbonate factor. Moreover, the variance of Fe is almost totally explained by this factor, suggesting that the pyrite framboids cannot be filtered out statistically.

Pb and Zn are distributed between two factors which may indicate different mineralizing fluids in time or space. Most of the variance of Pb and Zn is explained by factor 2. In factor 3 Pb and Zn are associated with Ba, Hg and Cu. The contributions of factors 2 and 3 to the total variance are almost equal.

The observations from the uni- and bivariate analyses are confirmed by factor analysis: the elements are grouped according to their distribution and their behaviour with respect to the other elements.

5.3.2 Calculation of the Scores

The scores for factors 2 and 3 are calculated and plotted vs. the depth of some boreholes (Micrite Unit and Lower Pale Beds) in Fig. 3. It is clear that the scores of factor 2 are almost exclusively positive in the eastern part of the area (east of

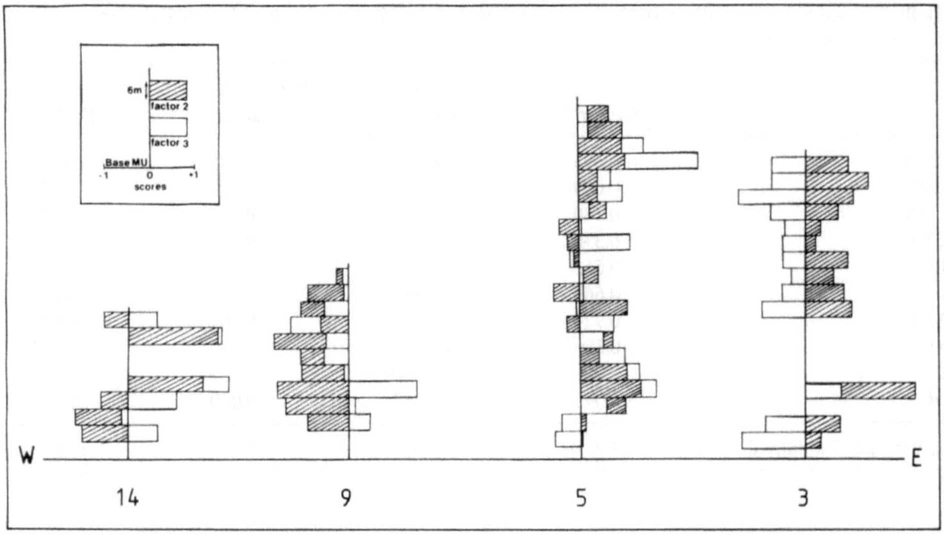

Fig. 3. Distribution of the scores of factors 2 and 3 of some boreholes in the Micrite Unit and Lower Pale Beds

hole 8). For factor 3, the positive scores are concentrated in the western part (west of hole 11). Both parts are separated by a 'barren' area.

The highest scores for factors 2 and 3 correspond to the presence of sulphides. Moreover, chalcopyrite has been found in holes 12 and 13 (Cu in factor 3) and early-diagenetic barite is present in some levels of the Clonabreany area (Ba in factor 3).

6 Implications for Prospection and Conclusions

Significant trends of value in prospection to 'Irish style' carbonate-hosted mineralizations have been found using a few elements only (Pb, Zn, Ba, Mg, Mn, Fe, IR). Elemental values or calculated ratios of Cu, Hg, As or Sr showed no detectable trends towards the mineralization.

Mn enrichment in calcite cements was detected, although this is not reflected in the Mn data of whole rock analyses. This diagenetic anomaly is possibly related to the mineralization. Mn halos around similar deposits have been described at Meggen (Gwosdz and Krebs 1977) and Tynagh (Russell 1974).

There is a clear relationship between the Anomalous Population Ratio (APR) of Zn and Ba and the distance to the mineralizations at Tatestown and Clonabreany; this can be explained by an increasing sulphide content towards the mineralization.

Both the nature and the thickness of the host rock of the mineralization significantly vary towards Tatestown (and Navan). This lithological anomaly corresponds to an allochthonous lithology which Large (1983) described in his model for sediment-hosted massive sulphide deposits.

The study area can be divided into three zones each having its own sediment petrographical and geochemical features.

Acknowledgements. This study has been supported by the Ministerie van Wetenschapsbeleid of Belgium (*contract No. MP/CE/14*) and the European Economic Community (*contract No. MSM-076-B*). The authors would like to thank the management of Ennex International plc. for permission to publish this work and for research facilities. Special thanks are due to J. Clifford, A. Poustie, Dr. R. Steiger and C. Andrew for information and constructive discussions. We were fortunate to have stimulating discussions with Dr. J. Boissonnas, Dr. R. Swennen, Dr. H. Kucha, E. Tijskens, Ph. Muchez and Dr. J. Decleer. Microprobe facilites were provided by the Department of Metallurgy and by C.A.M.S.T. Technical assistance was given by G. Vanden Eynde, C. Moldenaers and M. Bressinck.

References

Andrew CJ, Ashton JH (1982) Mineral textures, metal zoning and ore environment of the Navan orebody, Co. Meath, Ireland. In: Brown AG, Pyne J (eds) Mineral exploration in Ireland: progress and developments 1971–1981 (Wexford Conference 1981). Irish Association for Economic Geology, pp 35–46

Andrew CJ, Ashton JH (1985) The regional setting, geology, and metal distribution patterns of the Navan orebody, Ireland. Trans Inst Miner Metall (Sect B, Appl Earth Sci) 94:66–93

Andrew CJ, Poustie A (1984) Syndiagenetic or epigenetic? The evidence from the Tatestown prospect, Co. Meath. In: Geology and genesis of mineral deposits in Ireland (abstracts volume) I.A.E.G. conference 14–16 sept. 1984 (Dublin). Irish Association for Economic Geology, pp 29–30

Boast AM, Swainbank MA, Coleman ML, Halls C (1981) Lead isotope variation in Tynagh, Silvermines and Navan base metal deposits, Ireland. Trans Inst Miner Metall (Sect B, Appl Earth Sci) 90:115–119

Boyce AJ, Anderton R, Russell MJ (1983) Rapid subsidence and early Carboniferous base-metal mineralizations in Ireland. Trans Inst Miner Metall (Sect B, Appl Earth Sci) 92:55–66

Brand U, Veizer J (1980) Chemical diagenesis of multicomponent carbonate systems. I. Trace elements. J Sediment Pet 50:1219–1236

Chabot A (1982) Search for new geochemical metallotects in carbonate environments. Final report EEC-project 030-79-7 MPPB, part 2. Geochemistry of lithofacies (carbonate microfacies).

Deeny DE (1981) An Irish Carboniferous metallogenetic model. Trans Inst Miner Metall (Sect B, Appl Earth Sci) 90:183–185

Erickson RL, Mosier EL, Viets JG, Odland SK, Erickson MS (1983) Subsurface geochemical exploration in carbonate Terrane-midcontinent, USA. In: Kisvarsanyi G et al. (eds) Proceedings volume of the International Conference on Mississippi Valley Type lead-zinc deposits, Rolla, Missouri, 1983, pp 575–583

Finlay S, Romer DM, Cazalet PCD (1984) Lithogeochemical studies around the Navan Zn–Pb orebody, Ireland. In: Prospecting in areas of glaciated terrain. Spec Vol Inst Miner Metall London, pp 35–55

Gwosdz W, Krebs W (1977) Manganese halo surrounding Meggen ore deposit, Germany. Trans Inst Miner Metall (Sect B, Appl Earth Sci) 86:73–77

Kinsman DJ (1969) Interpretation of Sr^{2+} concentrations in carbonate minerals and rocks. J Sediment Pet 39(2):486–508

Kucha H, Wieczorek A (1984) Sulphide-carbonate relationship in the Navan (Tara) Zn–Pb deposit, Ireland. Miner Dep 19:208–216

Large DE (1983) Sediment-hosted massive sulphide lead-zinc deposits: an empirical model. In: Sangster DF (ed) Short course in sediment-hosted stratiform lead-zinc deposits. MAC, pp 1–24

Russell MJ (1974) Manganese halo surrounding the Tynagh ore deposit, Ireland: a preliminary note. Trans Inst Miner Metall (Sect B, Appl Earth Sci) 83:65–66

Russell MJ (1978) Downward-excavating hydrothermal convection cells and Irish type ore deposits: importance of an underlying thick Caledonian prism. Trans Inst Miner Metall (Sect B, Appl Earth Sci) 87: 168–171

Russell MJ (1983) Major sediment-hosted exhalative zinc + lead deposits: formation from hydrothermal convection cells that deepen during crustal extension. In: Sangster DF (ed) Short course in sediment-hosted stratiform lead-zinc deposits. MAC, pp 251–282

Viaene W (1982) Prospection for Pb–Zn mineralizations in the Dinantian carbonate rocks of Belgium based on lithogeochemical dispersion pattern. Final report EEC-project 033-79-MPPB

Viaene W, Van Oyen P, Clifford J (1986) Sediment petrography of Lower Carboniferous sediments in the Navan area (Ireland) and its relationship to lithogeochemistry and base metal mineralization. Final report EEC-project MSM-076-B and Wetenschapsbeleid MP/CE/14

Lithogeochemistry, Its Applicability to Base Metal Exploration in a Carbonate Environment

J.A. Clifford[1], H. Kucha[2], and A.H. Meldrum[1]

Abstract

Lithogeochemical studies of the elemental distributions in the carbonate limestones around the Tynagh Zn–Pb orebody confirm the presence of a geochemical halo. From an exploration standpoint the effective size of the target is thereby increased. The methodology of the study is described and the results summarised.

1 Introduction

A series of Zn–Pb discoveries beginning with the Tynagh deposit in 1961 and culminating with the major Navan orebody in 1970 led to the recognition of the Irish Carboniferous as a world-ranking metallogenic province. These discoveries were due primarily to geochemical methods with supporting geophysics and diamond drilling (Schultz 1971). By the end of the 1970s, however, it had become evident that the shallow soil geochemical methods had reached the practical limit of their effectiveness. This is due to the fact that the entire area of the Carboniferous has been systematically sampled to an extent that any major, near-surface geochemical expression relating to an orebody is unlikely to have been missed. In fact, a detailed analysis of the discovery record during this period clearly shows that many of the discoveries did not outcrop, had no clearly defined surface expression and were primarily the result of geologically directed diamond drilling. Research on the mines and prospects established that the deposits are not evenly distributed through the Lower Carboniferous stratigraphic sequence. Particularly favourable units came to be recognised, notably the Waulsortian Bank or Reef limestone and the Navan Group sequences (Morrissey et al. 1971; Sevastopulo 1979; Hitzman and Large 1986). Notwithstanding this increased geological understanding the prospective areas remain extremely large and beyond the budgetary capacity of any individual company, or consortium, to test on a grid-drilling basis. Hence, it became clear that methods would have to be developed which increased the effective

[1] Ennex International plc, 162 Clontarf Rd., Dublin 3, Ireland
[2] Institute of Geology & Mineral Deposits, 30-059 Krakow, Av. Mickiewicza 30, Poland

Mineral Deposits within the European Community
(ed. by J. Boissonnas and P. Omenetto)

size of the target being sought. As a result, Ennex International plc., an Irish-based natural resource company, and the Economic Community, jointly funded a study of the Lower Carboniferous lithogeochemistry in Ireland and its applicability to the search for concealed mineral deposits (Clifford and Hughes 1987).

2 Study Methodology

The primary objective of mineral exploration is to find mineral deposits of economic significance. Any method used in exploration should be judged on how effectively it contributes to achieving that objective. Hence, in designing the rock geochemical studies the development of an efficient exploration methodology and of a tool to detect blind orebodies are paramount. This means that the method should be simple to apply in the field and that the results can be interpreted without a necessity for sophisticated computer manipulation.

In selecting the elements to be analysed the dominant view is that the method would have to be independent of any genetic constraint and that the best elements to analyse are those being explored for. This judgement is confirmed in this study.

Three methods of sampling drill cores were evaluated: intermittent, geologically controlled, select sampling; continuous groove sampling; and continuous half-core sampling. In the intermittent, geologically controlled, select sampling pieces of core 0.5–1 ft in length were taken at intervals of 5–7 ft. Analytical values derived from these proved to have an extremely irregular distribution (Fig. 1a, c). This was due to individual samplers making judgements on what was representative in any particular section. Hence, a standardised procedure did not prove possible to implement. Groove samples also gave unreliable results. What is most apparent with this method is the consistently higher values obtained when compared to other methods (Fig. 1b). Following a detailed review of procedures the problem was traced to the fact that the drill core being sampled had been stored in a mine environment for a number of years. As a result it was concluded that continuous half-core sampling was the most reliable method. When fully operational a productivity in the order of 1000 ft/man week was achieved using this method.

In the study 60% of the direct costs were associated with sample preparation and analysis, and 40% with the actual taking of the sample. Hence, the sampling interval was extremely important. Initially, samples were taken in 5-ft lengths. However, as the study developed comparisons were made with other sampling lengths. It was found that similar patterns could be recognised in both 5-ft and 20-ft sampling lengths (Fig. 2). As a result sampling of the longer length became standard practice, giving a 75% decrease in the number of samples for analysis.

3 Tynagh Case History

One of the first areas studied for its lithogeochemical signature was the Tynagh area.

The Tynagh orebody was hosted in Waulsortian Bank (Reef) limestone in the hanging wall of an east-west normal fault (Fig. 3a) (Derry et al. 1965; Clifford et al. 1986). The main minerals were sphalerite, galena, pyrite and barite. The Waulsortian facies consists of well-developed massive pale to medium-grey reefoid biomicrites which are enveloped and overlaid by crinoidal biomicrites and by off-bank slump breccias containing grey micrite clasts in a dark, limey, shaley matrix. The reef flank biomicrites and biomicrudites thin to the north and grade into reef equivalent beds (Fig. 3b). In the mine area the reef can be subdivided into lower and upper tongues separated by reddened limestones and breccias. The Iron Formation interdigitates with the Reef horizon at this stratigraphic level. The Grey Calp is also considered as a facies of the Waulsortian. It is an intraclastic biomicrite with argillaceous laminae. Nodular cherty beds typically develop at its upper and lower contacts.

The spatial relationship between faulting and mineralisation has long been recognised (Russell 1968; Moore 1975). Assuming that the reef breccias were triggered by fault movement, then two main episodes of fault activity may be suggested. The first at the time line of the Grey Calp–Reef Equivalent contact, which is the same as that between the upper and lower reef tongues. The second phase of movement is considered to have occurred between the deposition of the Grey and Black Calp, where massive and large-scale development of debris flow breccias and turbidites formed on the slopes of the Upper Waulsortian Reef mound. If these fault movements are temporarily related to mineralisation, then it could be inferred that the first movement is related to the syndiagenetic Zn–Pb mineralisation, with the second related to the epigenetic Cu–Pb–Ba mineralisation which have been recognised by Boast et al. (1981).

During the 1970s, a number of rock-geochemical studies were undertaken by various academic workers (Russell 1974, 1975) and in-house Ennex personnel. Notwithstanding the fact that many of these studies were based on limited sampling and had little lithostratigraphic control, it was clear that the method might offer a way forward. Given the large data bank of drill cores available at Tynagh, it provided a unique opportunity to evaluate the relevance of rock geochemistry to mineral exploration.

A number of these studies, and in particular those by Russell, had indicated the presence of a manganese anomaly peripheral to the Tynagh orebody. This study confirms that in the vicinity of the orebody manganese concentrations are typically 500 ± 200 ppm in the Black Calp, rise to over 10,000 ppm in the upper Grey Calp, maintain a level of $5{,}000 \pm 3{,}000$ ppm through the lower Grey Calp and Reef intervals and fall off gradually to 400 ± 200 ppm in the subreef limestones (Fig. 4). It is noted that the peak of the Mn anomaly is at the Grey Calp–Black Calp contact, coincident with an episode of fault movement which has an inferred relationship with epigenetic Cu–Pb–Ba mineralisation.

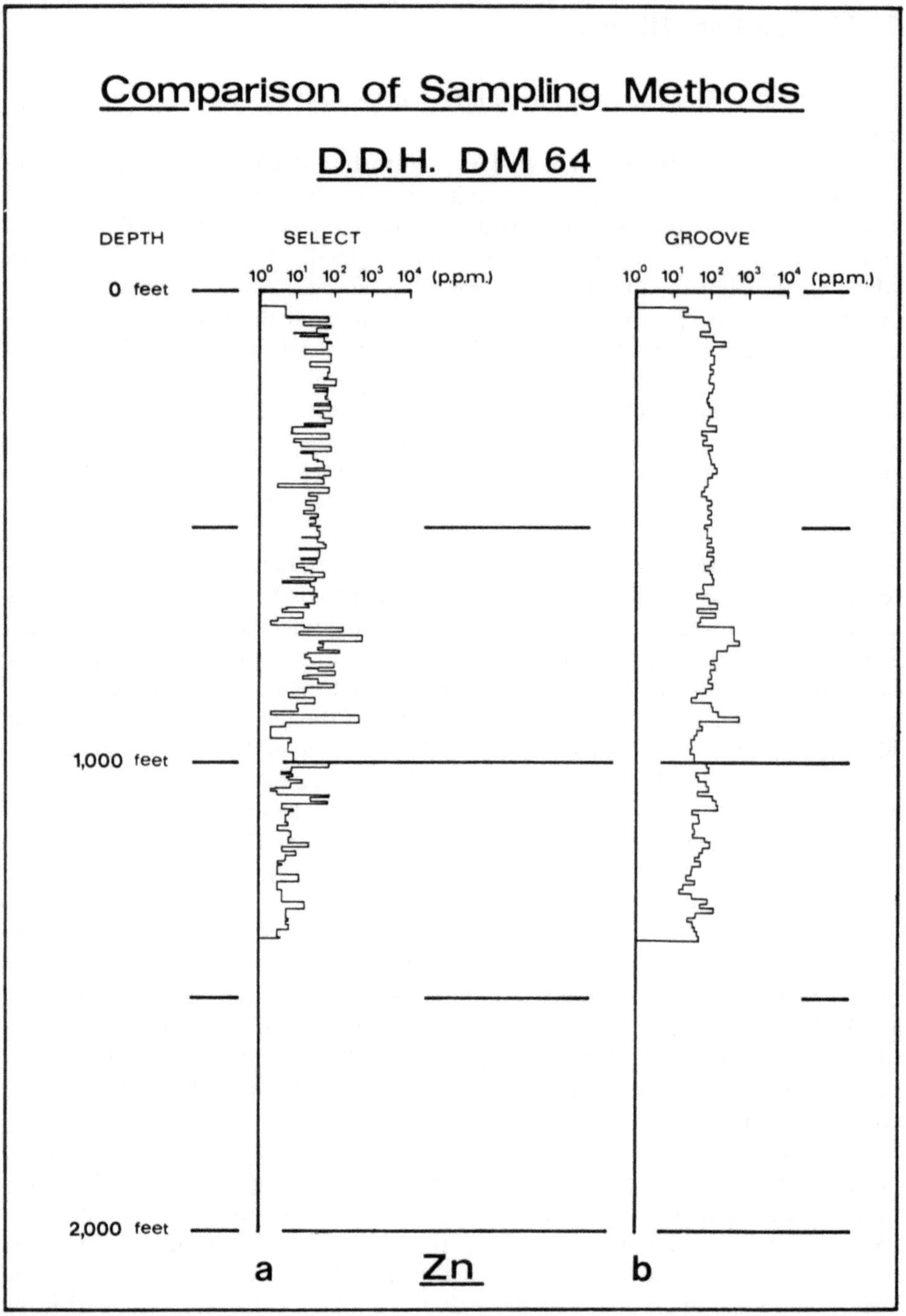

Fig. 1a,b

Fig. 1a,b. Comparison of sampling methods D.D.H. DM64; **c, d** comparison of sampling methods D.D.H. 435

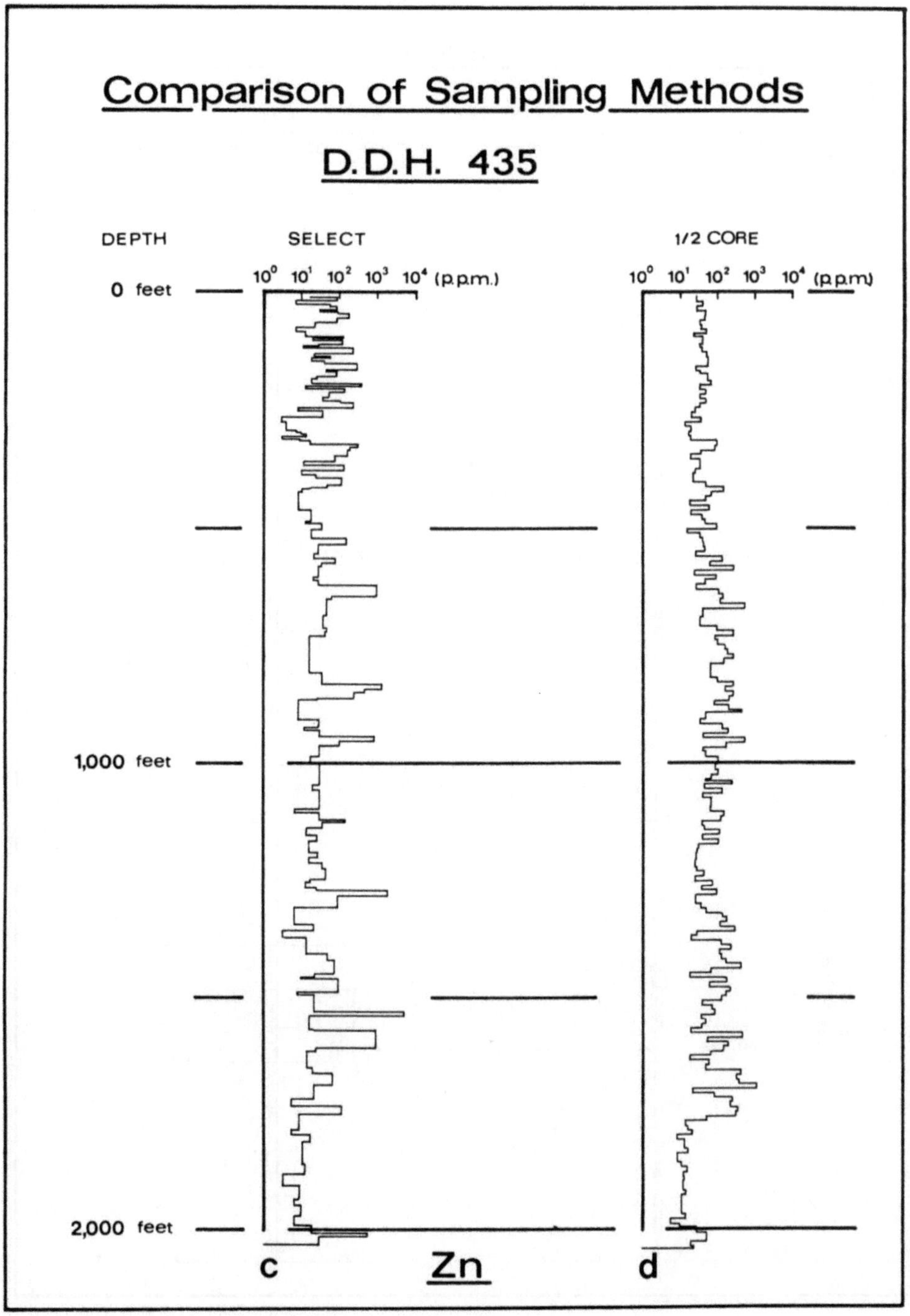
Comparison of Sampling Methods
D.D.H. 435
DEPTH
SELECT
1/2 CORE
10⁰ 10¹ 10² 10³ 10⁴ (p.p.m.)
10⁰ 10¹ 10² 10³ 10⁴ (p.p.m.)
0 feet
1,000 feet
2,000 feet
c
Zn
d

Fig. 1c,d

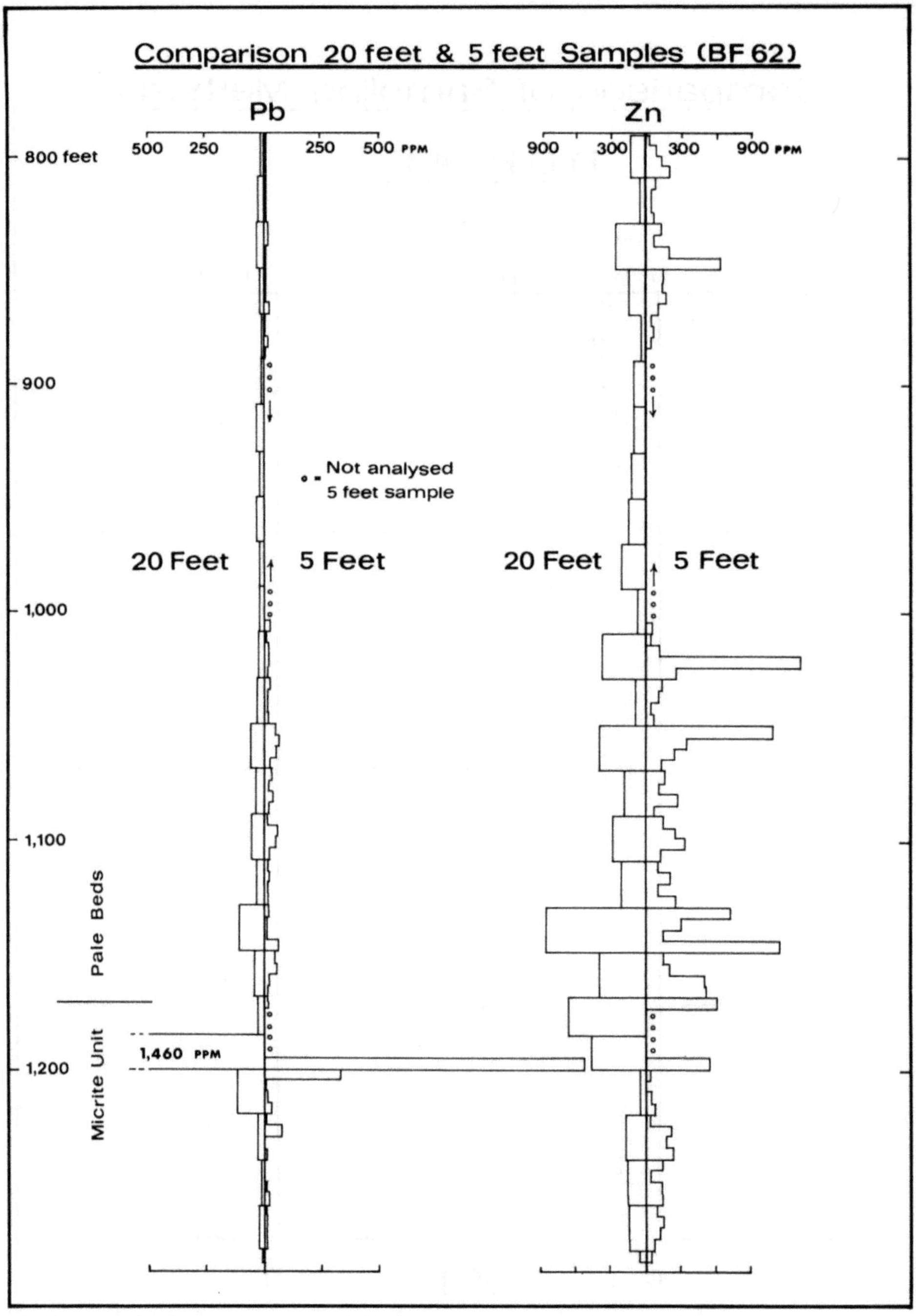

Fig. 2. Comparison of 20-ft and 5-ft samples (BF 62)

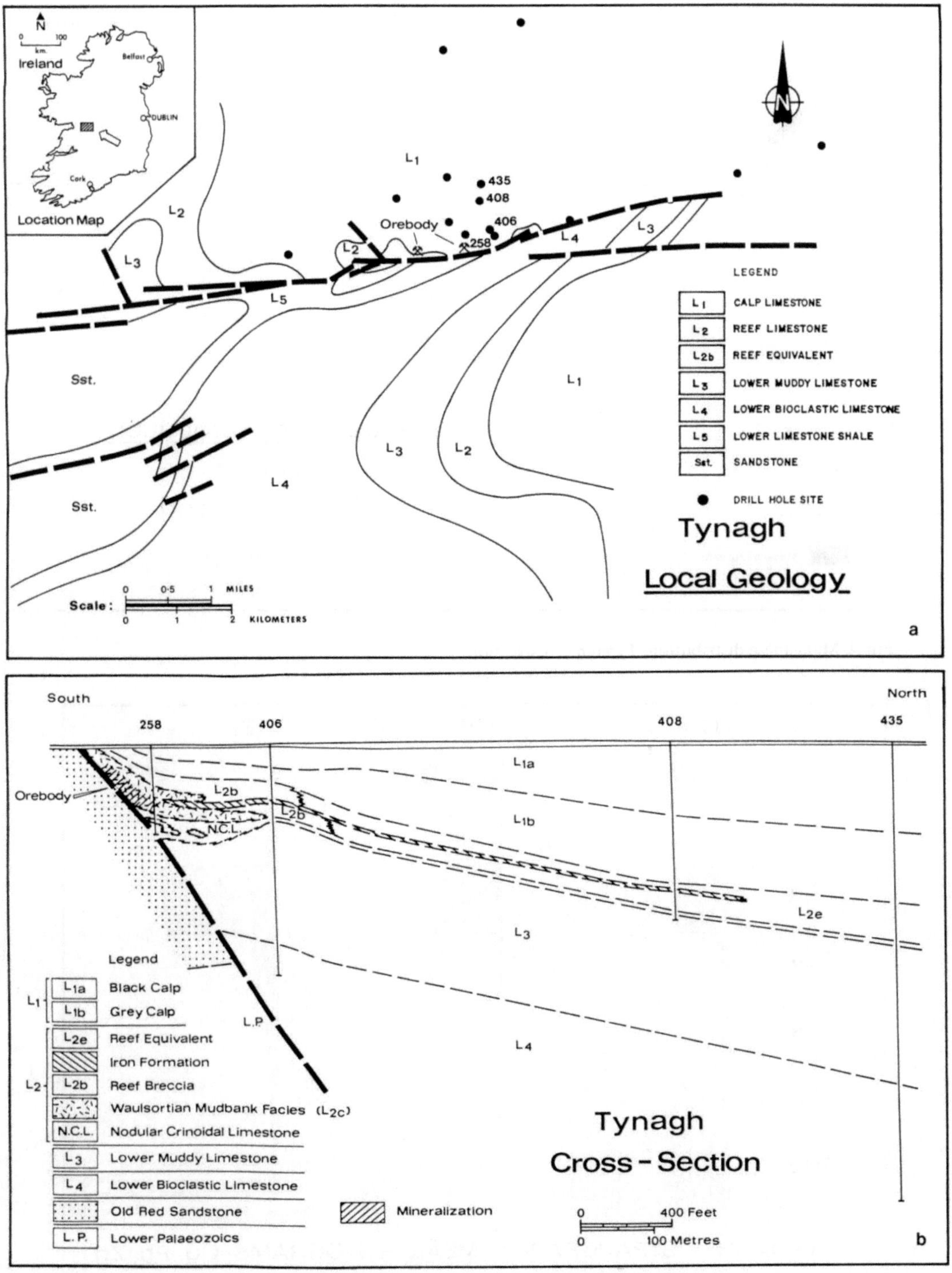

Fig. 3a. Tynagh, local geology; **b** Tynagh, cross-section

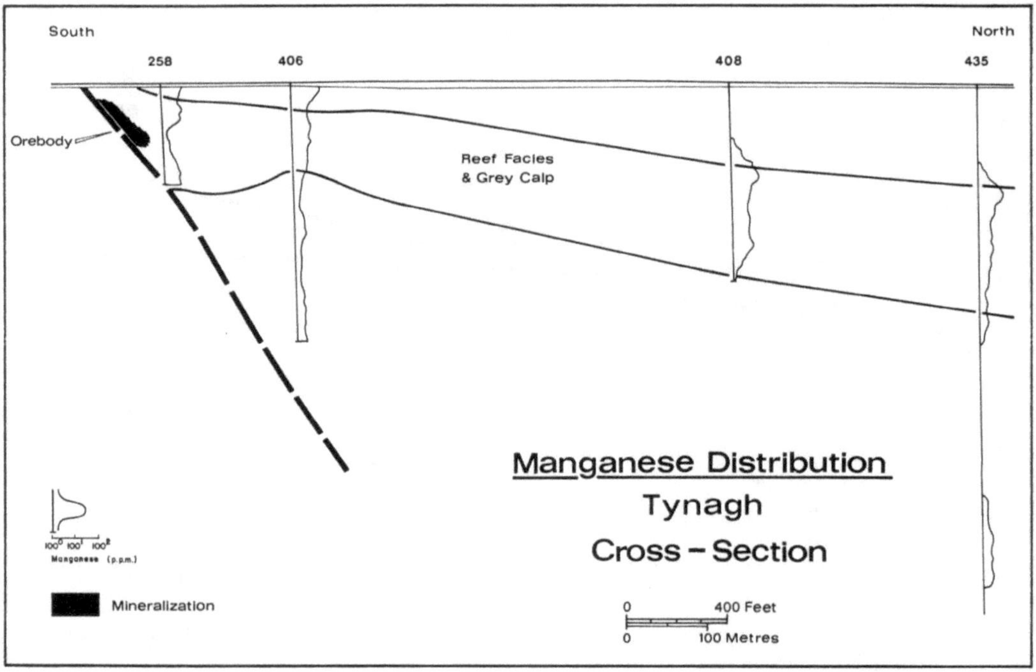

Fig. 4. Manganese distribution, Tynagh cross-section

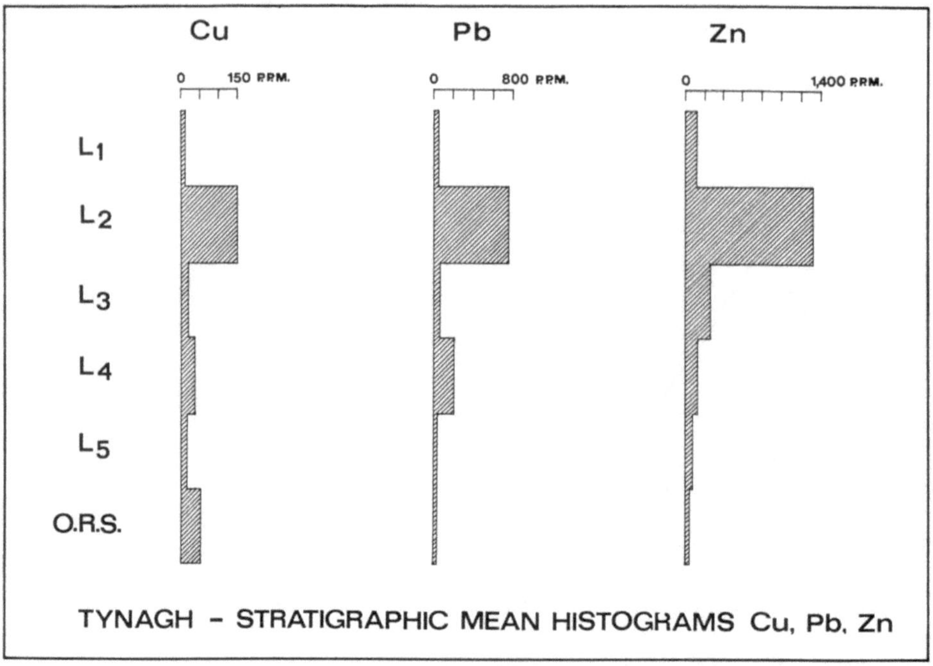

Fig. 5. Tynagh-stratigraphic mean histograms for Cu, Pb and Zn

Table 1. Geochemical statistics of Waulsortian Reef at Tynagh with increasing distance from the orebody (n: number of samples per drill hole)

Hole No.	Distance from orebody (m)	n	Concentrations (ppm) Cu	Pb	Zn	Fe	Mn	Mg	Ba	As	Hg[a]	Sr	Co	Ni
258	135	26	130	5268	5528	7012	2638	4973	14550	78	2313	859	32	106
277	170	13	25	255	820	13285	3008	3331	1111	224	909	270	9	41
406	290	6	15	25	181	17583	3900	5967	673	57	27	260	9	31
408	850	13	21	56	278	22631	3092	5900	146	11	35	321	14	76
435	1160	19	16	32	148	14679	2316	5674	98	7	12	293	13	66
Waulsortian Reef Background			16	23	192	13340	2446	6284	136	7	31	NC[b]	NC[b]	NC[b]

[a] Hg in ppb.
[b] NC, Not calculated.

However, given the basic philosophy regarding genetic interpretation, it was concluded that while the presence of a manganese anomaly was of explorations significance, its absence need not necessarily be a negative factor. Hence, the distribution of other elements, and in particular the ore elements, was examined.

A review of the elemental means for Cu, Pb and Zn by stratigraphic unit clearly shows that these elements reach their peak values in the stratigraphic unit which hosts the nearby ore deposit (Fig. 5). Simple plots of actual values show a clear increase towards the mineralised body with the anomalous values enveloping the prospective horizon (Table 1 and Fig. 6a,b). It should be noted that on the sections there are two peaks, one coincident with the Iron Formation that is at the horizon of the main body of syndiagenetic mineralisation. The second peak occurs at the base of Reef horizon at which colloform and granular pyrite of inferred early diagenetic origin has been located. It should also be noted that the variation of the elemental values are much more extreme proximal to the known mineralisation, for example comparison of holes 406 and 435. It should be pointed out that the anomalous Pb values at the base of hole 435 reflect mineralisation of a different style to that in the Tynagh orebody. Maps showing the average distribution within the prospective horizon show a similar spatial relationship of values to the known mineralised body (Fig. 7a,b).

Since haloes are often subtle, methods of enhancing the anomaly contrast were investigated. As a first attempt additive and multiplicative formulae using the ore elements were tested (Fig. 8). These were ratioed in some instances by elements known to be antipathetic to the mineralisation (Fig. 9). For example, the Zn–Pb mineralisation at Tynagh is peripheral to dolomitisation. These ratios improve the contrast. That the ratios have some geological validity is shown by a comparison of those with and without Cu and As (Fig. 10a,b). Both show the same pattern within the Reef horizon. However, the ratio which includes Cu and As is much more anomalous adjacent to the fault, possibly reflecting the influence of the late-stage Cu–Pb–Ba mineralisation.

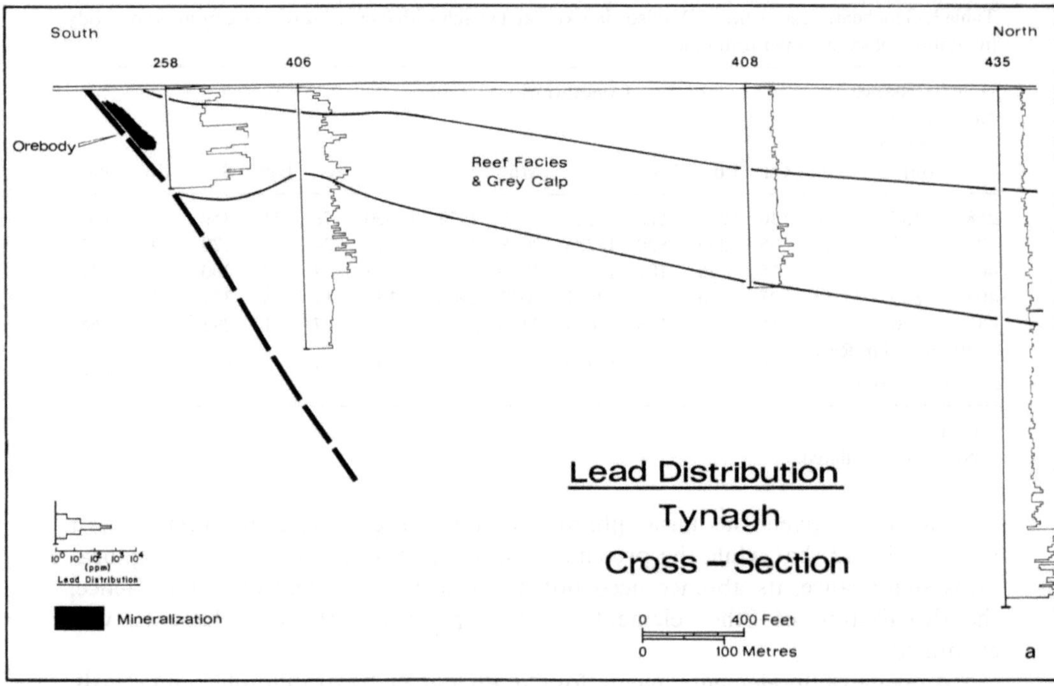

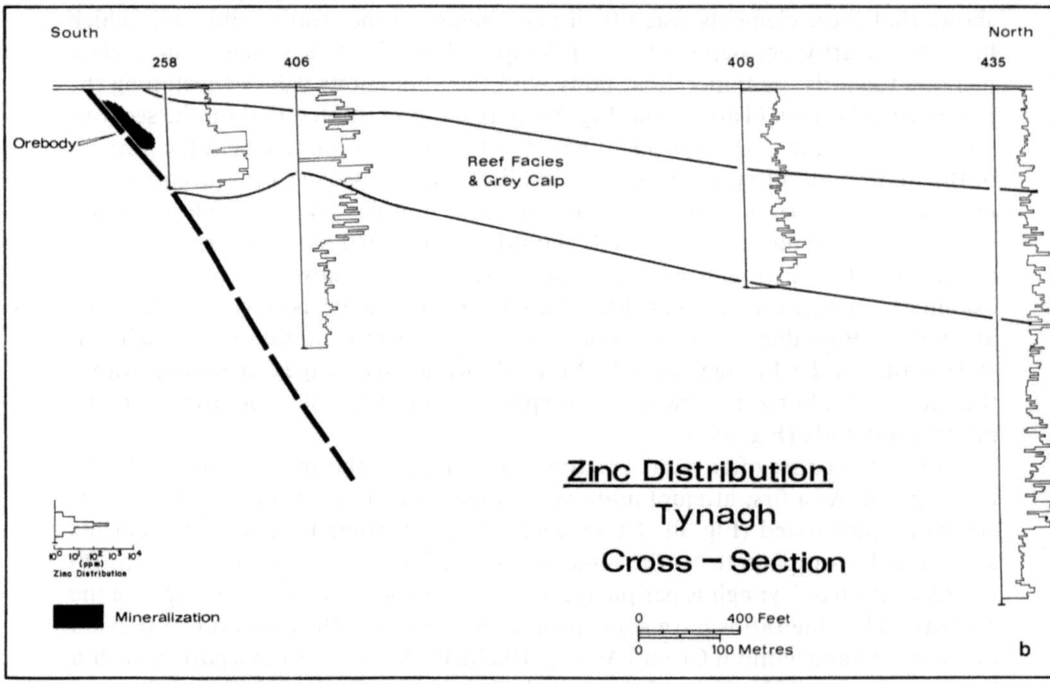

Fig. 6a. Lead distribution, Tynagh cross-section; **b** zinc distribution, Tynagh cross-section

Fig. 7a. Tynagh lead distribution; **b** Tynagh zinc distribution

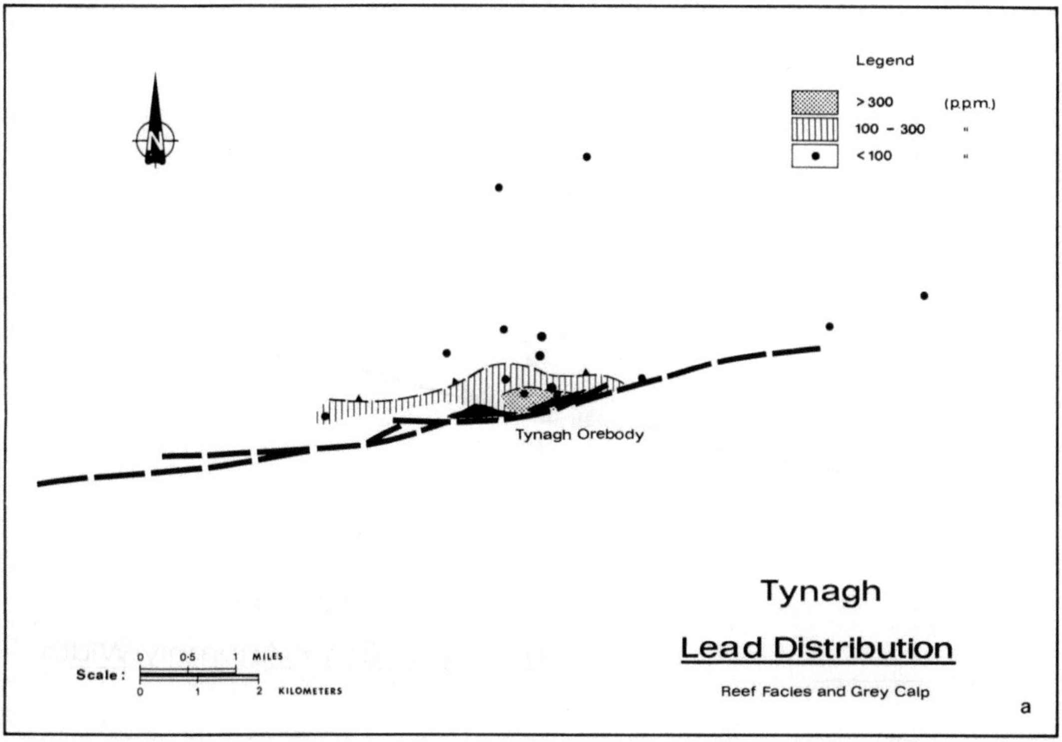
Legend
> 300 (p.p.m.)
100 - 300 "
< 100 "
N
Tynagh Orebody
Tynagh
Lead Distribution
Reef Facies and Grey Calp
Scale :
0 0·5 1 MILES
0 1 2 KILOMETERS
a

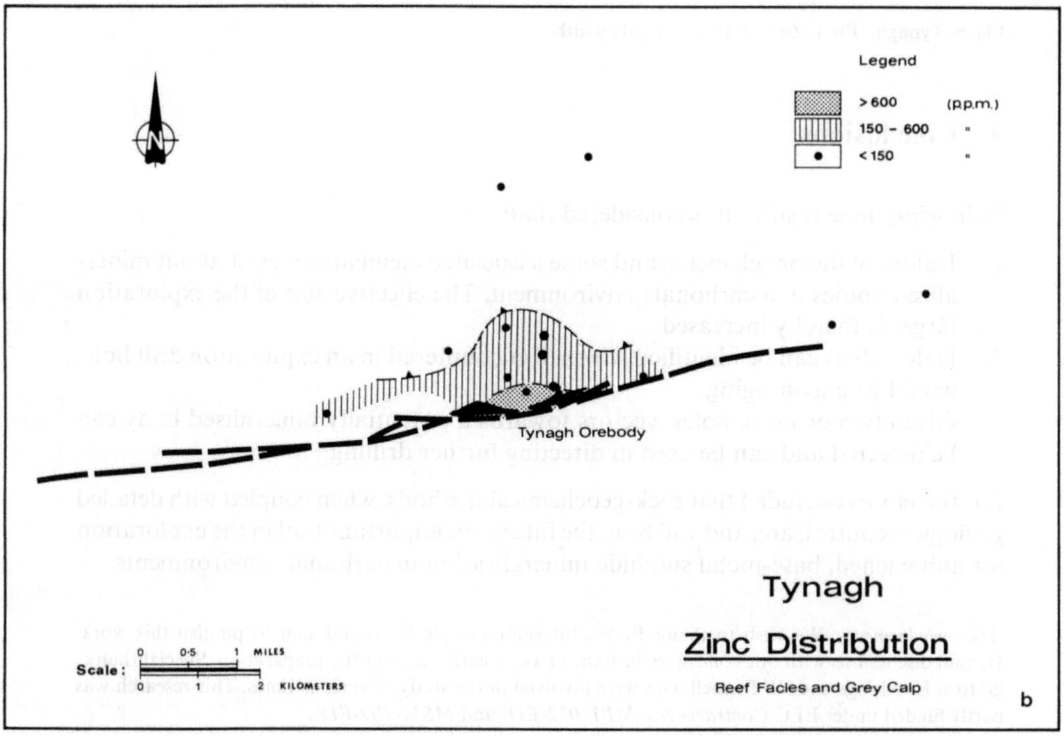
Legend
> 600 (p.p.m.)
150 - 600 "
< 150 "
N
Tynagh Orebody
Tynagh
Zinc Distribution
Reef Facies and Grey Calp
Scale :
0 0·5 1 MILES
0 1 2 KILOMETERS
b

Fig. 7a,b

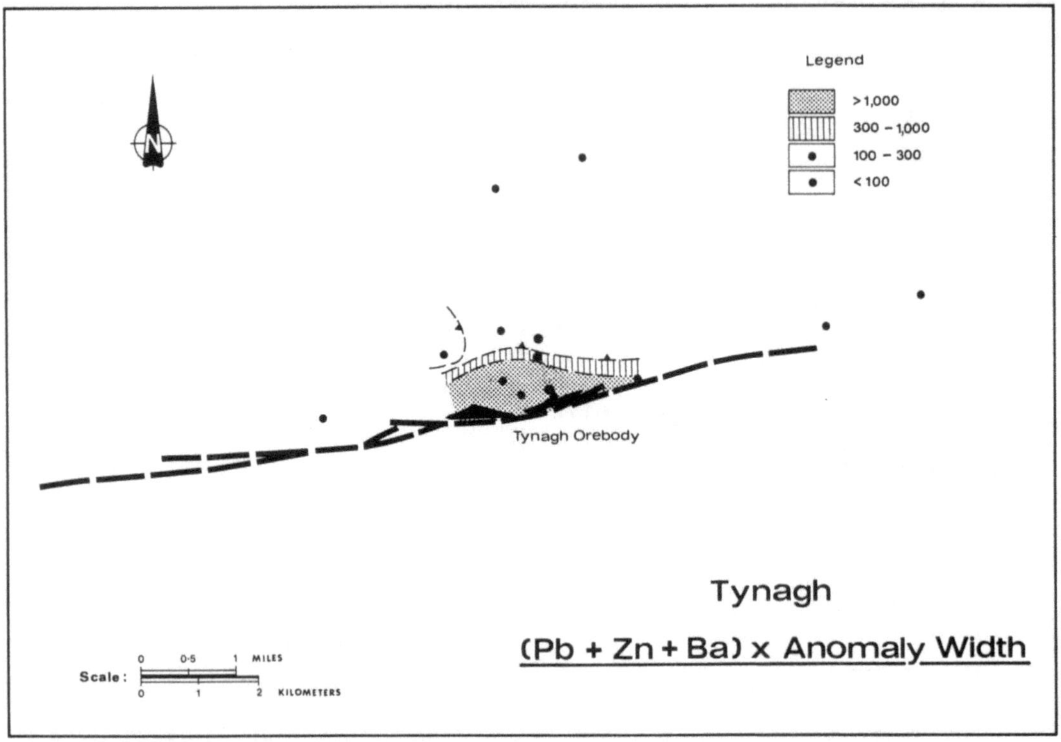

Fig. 8. Tynagh, (Pb + Zn + Ba) × anomaly width

4 Conclusions

Following these results, it is considered that:

1. Haloes of the ore elements and some associated elements do exist about mineralized bodies in a carbonate environment. The effective size of the exploration target is thereby increased.
2. Halo values can be identified which, if encountered in an exploration drill hole, would be encouraging.
3. Given two or more holes, vectors towards a potentially mineralised body can be detected and can be used in directing further drilling.

It is therefore concluded that rock-geochemical methods, when coupled with detailed geological control, are, and will be in the future, an important tool in the exploration for unbreached, base-metal sulphide mineralisation in carbonate environments.

Acknowledgements. We wish to thank Ennex International plc for permission to publish this work. Helpful discussions with our colleagues in Ennex have greatly aided in the preparation, Special thanks go to J. Hutchings and W. Pennell who were involved in the study at various times. This research was partly funded under EEC *Contracts No. MPP-072-EIR* and *MSM-109-EIR*.

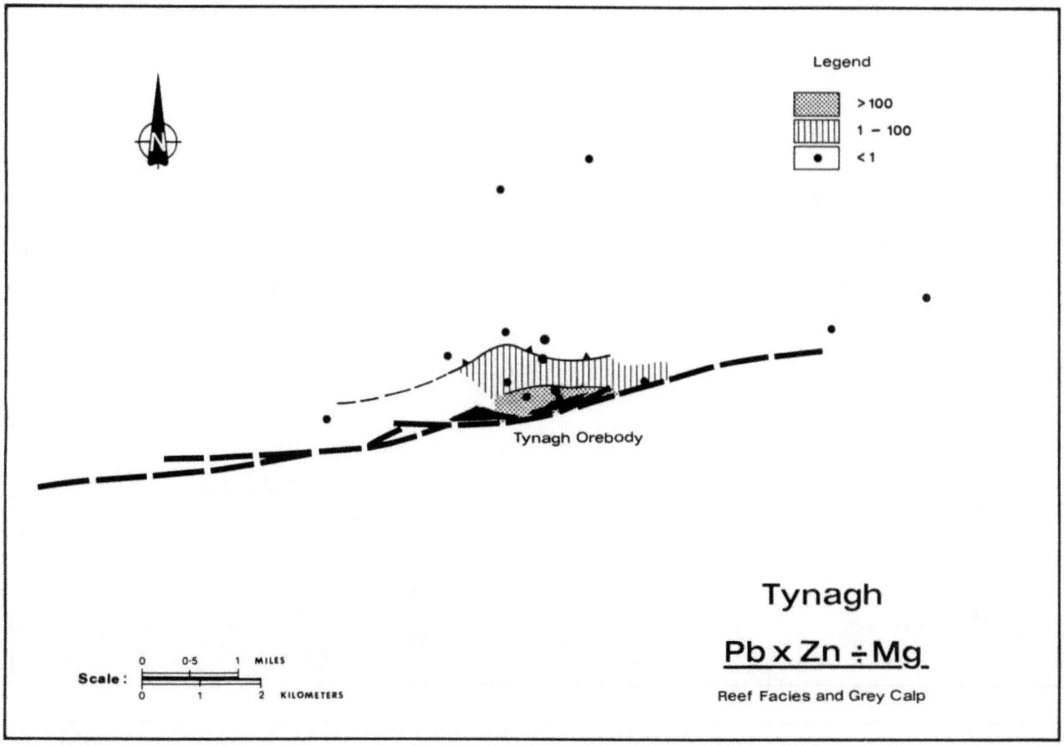

Fig. 9. Tynagh, Pb × Zn ÷ Mg

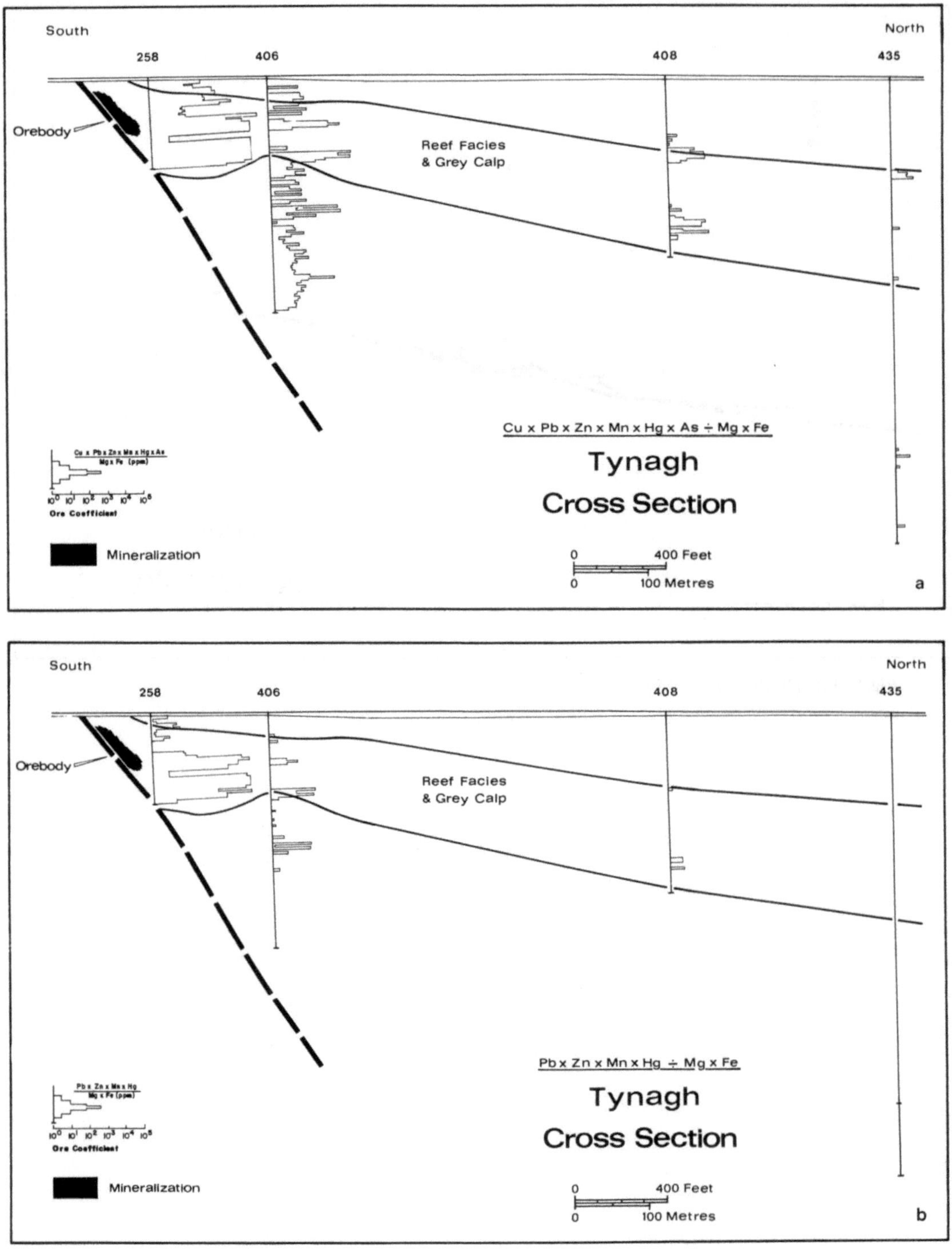

Fig. 10a. Tynagh cross-section, Cu × Pb × Zn × Mn × Hg × As ÷ Mg × Fe; **b** Tynagh cross-section, Pb × Zn × Mn × Hg ÷ Mg × Fe

References

Boast AM, Coleman ML, Halls C (1981) Textural and stable isotopic evidence for the genesis of the Tynagh base metal deposit, Ireland. Econ Geol 76:27–55

Clifford JA, Kucha H, Ryan P (1986) Geology of the Tynagh orebody. In: Geology and genesis mineral deposits in Ireland. Dublin Ir Assoc Econ Geol

Clifford JA, Hughes DH (1987) The geochemistry of Lower Carboniferous sediments in Ireland, its relationship to base metal deposits and applicability th the search for concealed deposits. Unpubl. Final report of EEC contract MSM-109-MIR, Westland Exploration Ltd, Dublin.

Derry DR, Clarke GR, Gillatt N (1965) The Northgate base metal deposit at Tynagh, Co. Galway, Ireland. Econ Geol 60:1218–1237

Hitzman M, Large D (1986) A review and classification of the Irish carbonate hosted base metal deposits. In: Geology and genesis mineral deposits in Ireland. Dublin Ir Assoc Econ Geol

Moore J McM (1975) Fault tectonics at Tynagh Mine, Ireland. Trans Inst Miner Metall (Sect B, Appl Earth Sci) 84:B141–B145

Morrissey CJ, Davis GR, Steed GM (1971) Mineralization in the Lower Carboniferous of central Ireland. Trans Inst Miner Metall (Sect B, Appl Earth Sci) 80:B174–185

Russell MJ (1968) Structural controls of base metal mineralization in Ireland in relation to continental drift. Trans Inst Miner Metall (Sect B, Appl Earth Sci) 77:B117–128

Russell MJ (1974) Manganese halo surrounding the Tynagh ore deposit, Ireland. Preliminary Note. Trans Inst Miner Metall (Sect B, Appl Earth Sci) 83:B865–866

Russell MJ (1975) Lithogeochemical environment of the Tynagh base metal deposit, Ireland and its bearing to ore deposition. Trans Inst Miner Metall (Sect B, Appl Earth Sci) 84:B128–133

Schultz RW (1971) Mineral exploration practice in Ireland. Trans Inst Miner Metall (Sect B, Appl Earth Sci) 80:B238–258

Sevastopulo GD (1979) The stratigraphical setting of base metal deposits in Ireland. In: Prospecting in areas of glaciated terrain 1979. IMM, London

Light Hydrocarbon Gases and Mineralization

J.S. CARTER, P.C.D. CAZALET[1], and J. FERGUSON[2]

Abstract

This chapter describes the results of an investigation into the possible use of light (C1–C5) hydrocarbons as geochemical pathfinders for mineral exploration, carried out by groups at Imperial College, London, and at Mercury Hydrocarbons Ltd., Limerick, Ireland.

The work involved two main areas of study:

1. A series of orientation surveys around known mineral deposits and over background areas, to demonstrate the existence of hydrocarbon gas anomalies associated with mineralization.
2. Experimental and theoretical studies to develop a better understanding of light hydrocarbon geochemistry and its relationships to mineralization.

The results of the orientation surveys have shown that there are major variations in the hydrocarbon gas contents of rocks which are closely related to the presence of mineralization. The origin of these changes is still not fully understood, but it seems likely that they reflect large-scale migration of hydrocarbons within the hydrothermal systems which produced the sulphide deposits. These hydrocarbons could have been carried passively in the ore-forming solutions. Alternatively, they could have been generated at the site of mineralization by chemical reactions or by thermal decomposition of kerogens in the surrounding rocks.

Further research is needed, particularly to develop the analytical techniques and methods of processing the data, but the studies reported here have established an association between light hydrocarbons and mineralization, and suggest that there is considerable potential for using hydrocarbon gas geochemistry as an aid to mineral exploration.

1 Introduction

The association between hydrocarbons and certain types of mineralization is well established and has been widely studied, but has been based mainly on the observa-

[1] Mercury Hydrocarbons Ltd, Raheen Industrial Estate, Limerick, Ireland
[2] Geology Department, Imperial College, London SW7 2AZ, England

Mineral Deposits within the European Community
(ed. by J. Boissonnas and P. Omenetto)

tion of heavier petroleum hydrocarbons and bitumens within mineral deposits. To judge from published information, the lighter gaseous hydrocarbons have received very little attention.

The study described here developed from two pieces of research carried out in the mid-1970s. First, some theoretical work by Dunsmore and Shearman (Dunsmore 1975; Dunsmore and Shearman 1977) suggested that methane might be generated as an end-product of sulphate reduction (a chemical reaction which can account for the precipitation of metal sulphides). Second, a geochemical survey of northern England, carried out by the staff of the British Gas Corporation in the early 1970s (Ferguson 1985) discovered anomalous amounts of methane in rocks close to the lead-zinc veins of the Northern Pennine Orefield, consistent with the predictions of the sulphate reduction theory.

These ideas were investigated further as a PhD topic by Carter (1981). This research showed that there was a distinct enrichment in methane in the rocks around the Northern Pennine Orefield and also produced the initial evidence for similar changes around major lead-zinc deposits in Ireland.

The results suggested that hydrocarbon gas geochemistry could have potential both as a mineral exploration technique and as a more general research tool for investigating the origins of lead-zinc sulphide mineralization.

The research described here investigated these possibilities in more depth. It was carried out as a joint project involving teams at Imperial College and Mercury Hydrocarbons Ltd., with two objectives:

1. To investigate the feasibility of using light hydrocarbons as geochemical indicators for mineral exploration.
2. To develop a better understanding of the mechanisms of emplacement of lead-zinc and associated mineralization.

2 Methodology

The EC project involved three main areas of research:

1. The development of techniques to extract and measure light hydrocarbons in rocks and to process the resulting data.
2. Orientation surveys involving the analysis of samples collected from surface exposures and boreholes around known mineral deposits and from background areas.
3. Experimental studies of carbonate diagenesis to investigate the geochemistry of light hydrocarbons in rocks and ore-forming systems.

3 Analysis of Light Hydrocarbons in Rocks

3.1 Heat Extraction Method

Most of the analytical work was carried out using the simple heating technique developed by Carter (1981). This involved heating a coarsely crushed rock sample inside a sealed bottle at 210°C for 2.5 h and then analyzing the head-space gases by gas chromatography.

A typical chromatogram (Fig. 1) shows that under the standard operating conditions used at Mercury Hydrocarbons, a total of 12 component gases can be detected. These are listed in Fig. 1, together with an ID code, used for convenience

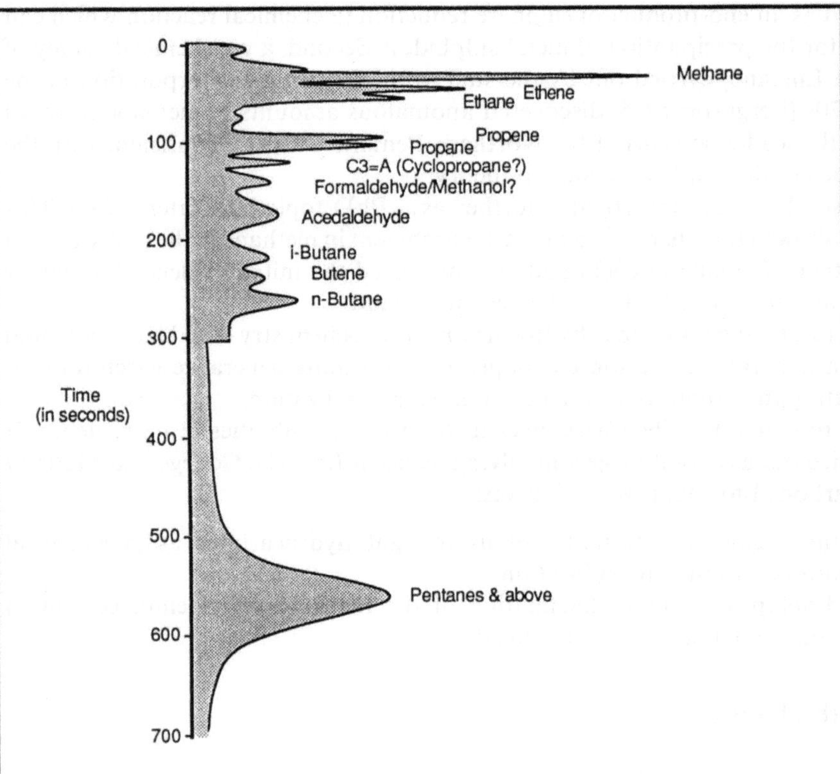

Retention Time	ID	Name of Compound
33	C1	Methane
49	C2=	Ethene
54	C2	Ethane
102	C3=	Propane
108	C3	Propane
120	C3=A	Cyclopropane
141	C3=B	Formaldehyde/methanol?
175	170	Acedaldehyde
216	IC4	i-Butane
234	C4=	Butene
260	NC4	n-Butane
560	C5+	Pentanes and above

Fig. 1. Example of chromatogram produced by the equipment used at Mercury Hydrocarbons Ltd., together with a listing of the 12 main components detected

within the laboratory, and their retention time (the time taken to pass through the gas chromatograph). They include a full range of saturated and unsaturated paraffins from methane (C1) to pentanes and above (C5+). Also present are other compounds which have been tentatively identified as cyclopropane (C3 = A), formaldehyde or methanol (C3 = B) and acetaldehyde (170).

The amounts of each component are measured by an electronic integrator which calculates the area of each peak in arbitary integrator units, proportional to the weight of the hydrocarbon present and converts them to approximate concentrations (ppb or 10^{-9} g of gas/g of rock). These values are then fed into a microcomputer which is used to process the data, for example to calculate the relative proportions of hydrocarbons as percentages or ratios or to plot maps and scatter diagrams.

The heat extraction method was chosen principally because it was simple and fast and therefore well suited to the analysis of large numbers of samples. However, it is far from an ideal analytical technique. Heating does not release all the hydrocarbons held by a sample and hence the results are not directly comparable with measurements obtained using other techniques. Furthermore, like many partial extraction techniques, the precision of the measurements is relatively poor (between 20% and 50% for most components).

During the course of the research, experiments with heat extraction led to changes to the procedures which considerably improved the quality of the data obtained, but the method would still not be satisfactory for applications where measurement of the absolute amounts of hydrocarbon was required.

For mineral exploration, the limitations of the method are outweighed by the advantages of speed and low cost which it offers. Here, the primary interest is in detecting differences between samples and, although the poor precision may prevent some of the more subtle changes from being identified, the variation found close to mineralization are large enough to be clearly detectable.

3.2 Dry Grinding and Solution Procedures

Recent work at Imperial College has studied dry grinding and limestone solution as alternatives to the heat extraction procedure, with encouraging results.

The grinding method involves pulverizing a known weight of sample chips (15–30 g) in a stainless steel swing-mill pot with a port sealed with a silicon rubber disc. After grinding for about 30 min the head-space gas is sampled by syringe (Ferguson 1985 p. 70).

The solution procedure also under development at Imperial College is a major advance on previous chemical extraction procedures using dilute acids which generated CO_2 gas and required a cold trap to collect released hydrocarbons. In this new procedure, Na–EDTA is used to dissolve the limestone and a small admixture of NaOH solution absorbs any CO_2 released. The procedure requires about 18 h to dissolve 2 g limestone (Ferguson 1986, unpub. research).

Both techniques are slower than the heat extraction method, and at present it has not been established to what extent hydrocarbons may be retained by rock

Table 1. Comparison of results obtained by various methods of gas extraction, showing reproducibility obtained from a single sample (values in 10^{-9} cm^3/g sample)

	Dry grinding		EDTA		Heat
	1	2	1	2	
Methane	92.97	75.91	119.55	112.58	18.61
Ethene	6.50	4.92	0.00	0.00	11.59
Ethane	19.79	17.65	31.93	30.30	1.68
Propylene	4.21	3.16	0.00	0.00	10.76
Propane	9.74	8.32	14.54	11.35	10.00
I-butane	1.82	4.55	7.93	13.56	0.00
N-butane	5.06	4.87	6.65	9.23	13.68
1-butene	5.85	5.73	0.00	0.00	57.30

powder (grinding) or in solution (EDTA). However, initial data comparing the results for grinding, EDTA and heat extraction suggest that the new procedures may give better reproducibility than the heat extraction technique (Table 1). Another advantage may be that the results are less influenced by vegetation contamination than the heat extraction procedure (Ferguson 1986, unpub. research).

4 Results of Orientation Surveys

A large number of orientation surveys have been carried out, involving the analysis of several thousand rocks samples. The majority of these were around known sulphide deposits and across background areas within the Lower Carboniferous of central Ireland, but some surveys were also completed over different styles of mineralization in other geological settings in France, Britain, Sardinia, Spain and Scandinavia.

A variety of different sampling strategies were investigated in the course of these surveys. In most cases effort was concentrated on collecting a regional suite of samples, with rock samples being collected from surface exposures at as near an even density as possible over an area extending 5 to 10 km from the recorded mineralization. In addition, more detailed sampling was carried out wherever possible, using cores collected from boreholes. This included some very detailed work involving the analysis of 80 and 90 samples respectively, from two boreholes in Ireland (Carter and Cazalet 1985). The results from the borehole studies and from the analysis of multiple samples collected at the same surface sites show that there can be large fluctuations in the gas content of rocks over distances of a few metres or less. These small-scale variations introduce significant noise into the results but are not large enough to mask the major regional variations associated with mineralization.

The results from Ireland clearly demonstrate the existence of gas anomalies (Carter and Cazalet 1984, 1985). Orientation surveys were carried out around most of the major lead-zinc deposits and across several background areas (Fig. 2). The

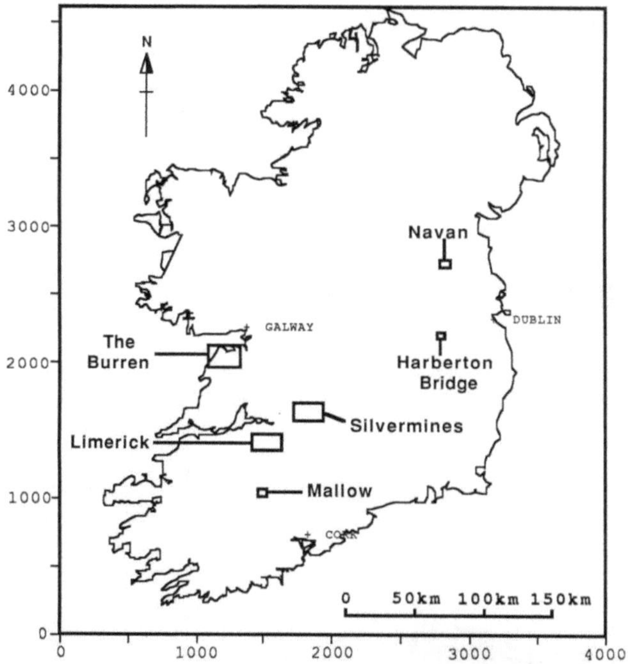

Fig. 2. Map of Ireland showing location of the study areas described in the text

results show distinct changes in the gas content of the limestones around all the mineral deposits studied, contrasting strongly with the relatively small variations found within similar rocks from background areas.

These differences between mineralized and background areas can best be demonstrated by scatter diagrams like those in Figs. 3 or 6, which plot the amounts of methane (C1) vs ethene + ethane + propene + propane (C2 = +C2 + C3 = +C3); C3 = A (cyclopropane?) vs C3 = B + 170 (formaldehyde? + acedaldehyde?); and butanes and butenes (Tot C4) vs all heavier hydrocarbons (C5+).

The scatter plot in Fig. 3a shows the variations in the amounts of gas released by a set of samples from a background area in the county Limerick. These were collected at a density of approximately 1 sample/km^2 across an area of about 200 km^2 and represent a full variety of carbonate lithologies, ranging from basal Carboniferous micrites through Waulsortian 'reef' to Visean shelf limestones. Despite the considerable differences in rock type, all samples released similar, relatively small amounts of hydrocarbon.

In contrast, samples collected close to mineralization show much larger variations in hydrocarbon gas composition. The rocks around the Silvermines Pb–Zn deposit in the county Tipperary are very similar to those in the county Limerick, but many of the samples released much larger amounts of gas (particularly of methane and the C3 = A component) clearly forming a very different pattern (Fig. 3b).

When plotted in map form, the variations in the amounts of hydrocarbon can be seen to correlate with mineralization. At Silvermines, the most conspicuous

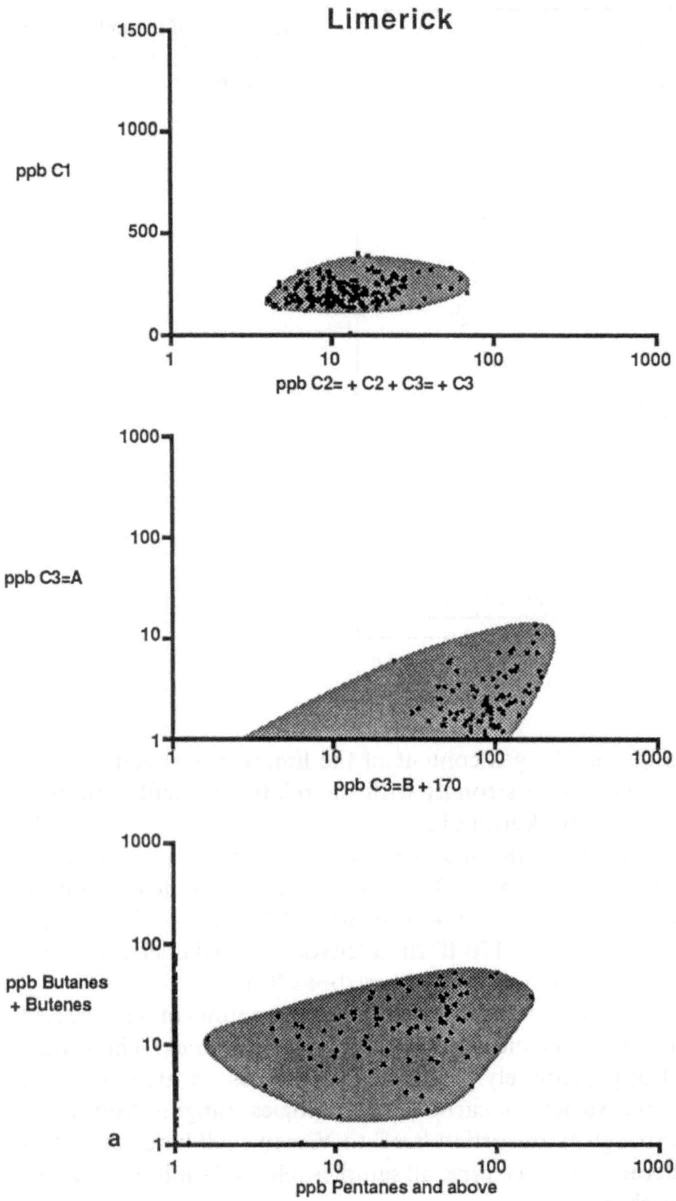

Fig. 3a

Fig. 3a,b. Scatter diagrams comparing the amounts of hydrocarbon [methane (C1) vs ethene + ethane + propene + propane (C2 = +C2 + C3 = +C3); C3 = A (cyclopropane?) vs C3 = B + 170 (formaldehyde? + acetaldehyde?); and butanes and butenes (Tot C4) vs all heavier hydrocarbons (C5+)] released by samples from a background area in the county Limerick (**a**) and around the Pb–Zn deposit at Silvermines, county Tipperary (**b**). Note the much higher levels of gas, particularly methane and C3 = A released by the samples obtained close to mineralization

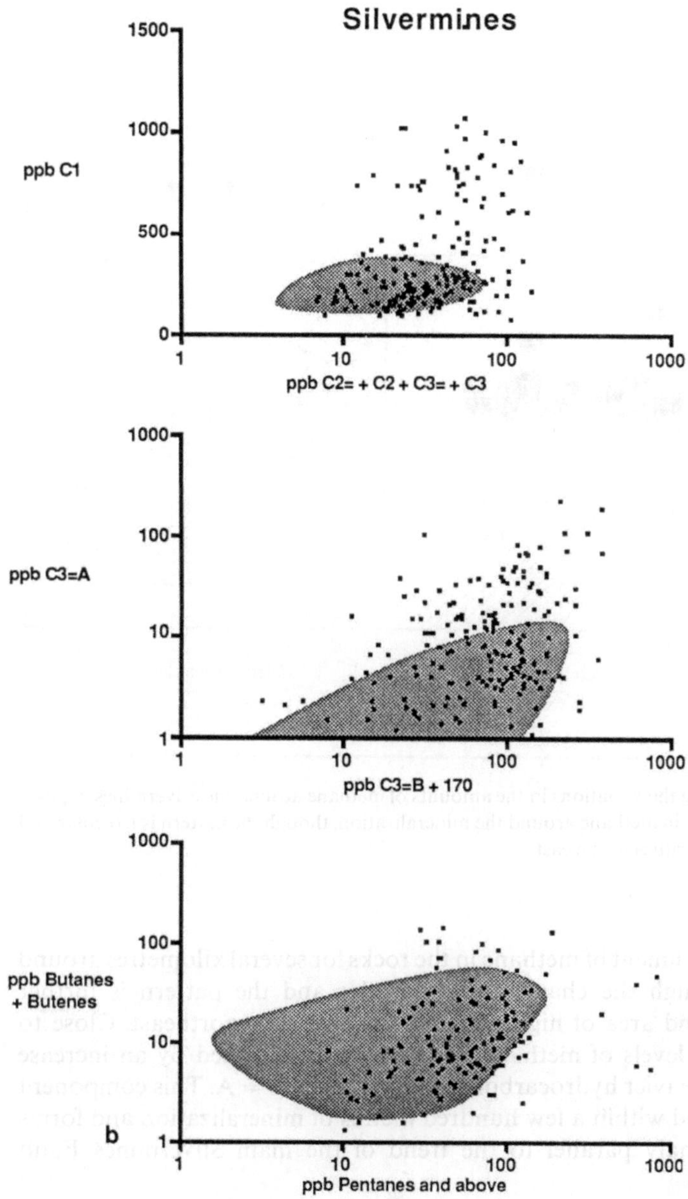
Silvermines
ppb C1
1500
1000
500
0
1
10
100
1000
ppb C2= + C2 + C3= + C3
ppb C3=A
1000
100
10
1
1
10
100
1000
ppb C3=B + 170
ppb Butanes
+ Butenes
1000
100
10
1
b
1
10
100
1000
ppb Pentanes and above

Fig. 3b

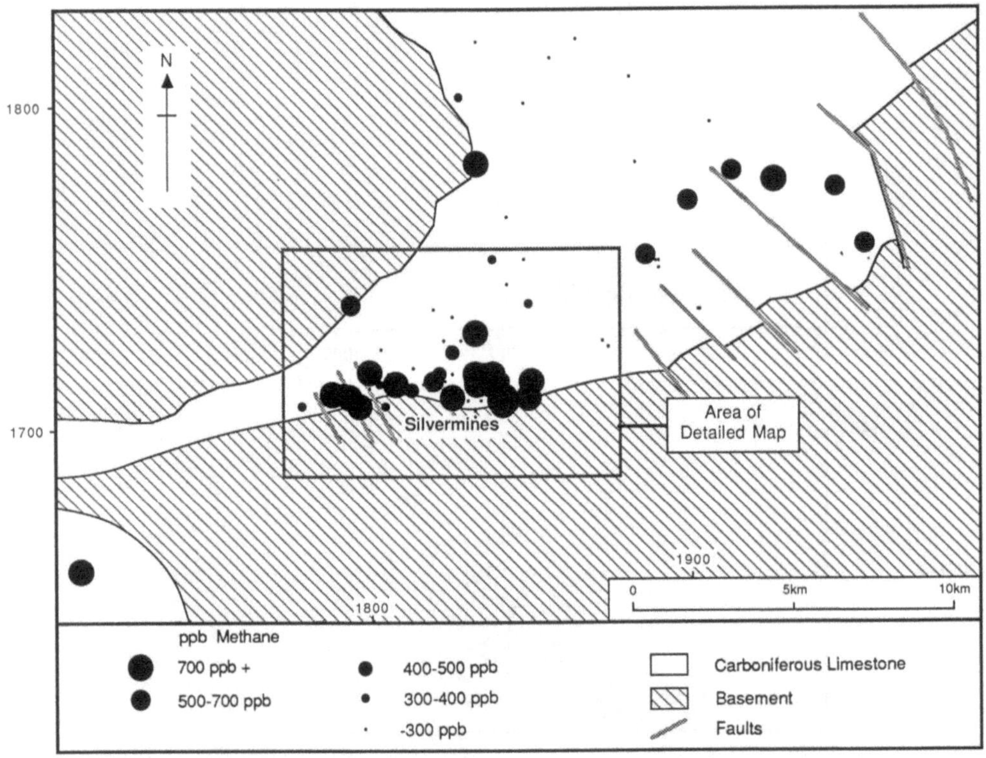

Fig. 4. Anomaly map showing the variations in the amounts of methane around the Silvermines deposit. There is a general enrichment in methane around the mineralization, though the pattern is irregular and complicated by further high values to the east

feature is a general enrichment of methane in the rocks for several kilometres around the mine (Fig. 4), though the changes are irregular and the pattern is further complicated by a second area of high methane values to the northeast. Close to the mineralization the levels of methane fall, but this is balanced by an increase in the amounts of the heavier hydrocarbons, in particular C3 = A. This component is very strongly enriched within a few hundred metres of mineralization and forms a distinct, linear anomaly parallel to the trend of the main Silvermines Fault (Fig. 5).

The results from Silvermines are typical of those found around the Irish deposits. In all cases the general form of the anomalies seems to be similar to extensive areas of methane enrichment (which may be several kilometres across the case of major deposits like Silvermines or Navan), surrounding a central concentration of heavier hydrocarbons.

Although the overall patterns are the same, the detailed composition of the anomalies do appear to change across the country. In the southwest, around Silvermines and the Cu–Ag deposit at Mallow (Carter and Cazalet 1985), the anomalies are very simple, consisting of an enrichment in methane and C3 = A only, but

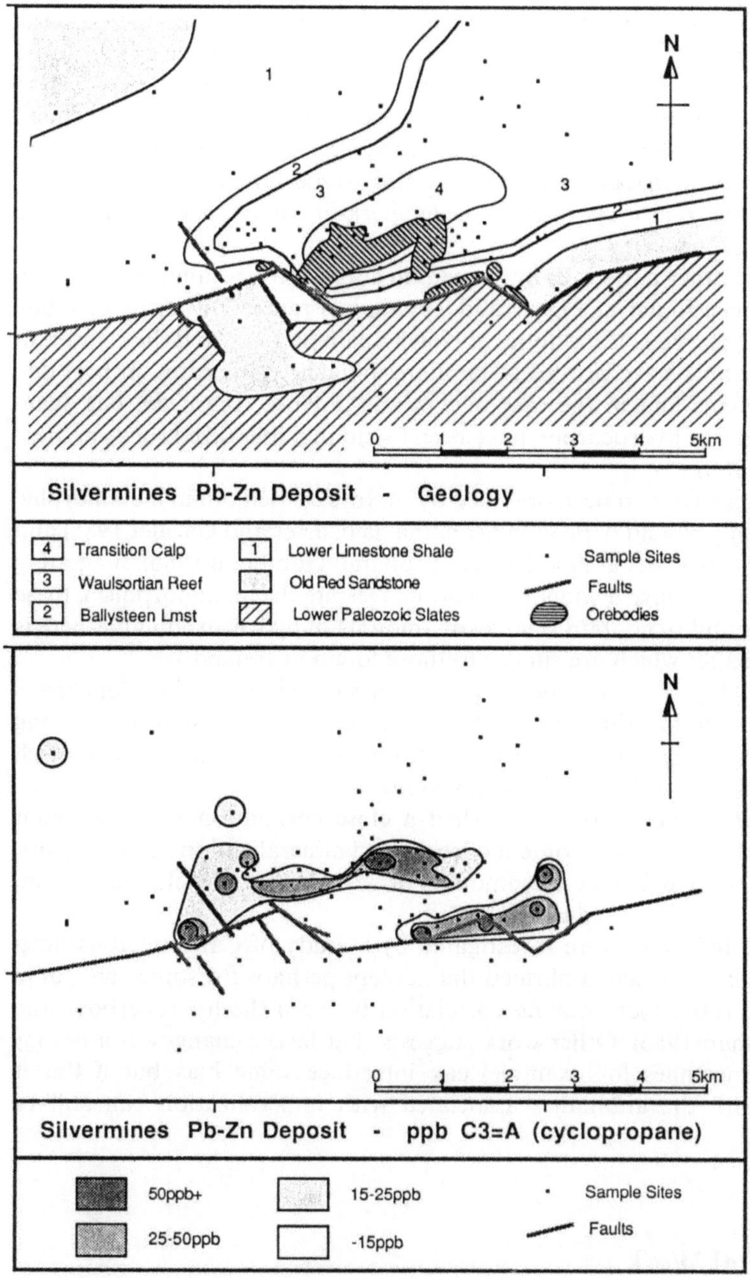

Fig. 5. Map showing the variations in the amounts of C3 = A (cyclopropane?) close to the Silvermines deposit. There is a strong linear zone anomaly which correlates closely to the distribution of mineralization

further east, around deposits such as Navan or Harberton Bridge, the patterns are more complex, with much higher concentrations of the heavier saturated hydrocarbons (ethane, propane, butane and pentane) as well as methane.

This difference can be clearly seen in the scatter plots in Fig. 6a and b which shows the variations in the amounts of hydrocarbon released by sample sets from Navan and Harberton Bridge. It seems to reflect some fundamental change in conditions across the country, perhaps in the temperature of the rocks or the source of the mineralizing fluids.

The surveys in other parts of Europe seem to confirm the results from Ireland. Anomalies have been found at Figeac and Trèves, in France (Disnar et al. 1986). Results from the Cretaceous Limestones of North Spain (Réocin) and the Cambrian Limestones of Sardinia have proved to be more difficult to interpret, in part due to sampling problems and in the case of Sardinia, intense tectonism, but again anomalous amounts of hydrocarbon have been found in mineralized rocks in these areas (Ferguson 1985).

Positive results have also been obtained by surveys carried out in metamorphic rocks, around a copper-gold deposit in Scandinavia (Carter and Cazalet 1985) and the barite-lead-zinc prospect at Aberfeldy, in Scotland (Goodman 1986), suggesting that the technique may have a wider application. Despite the metamorphism, these rocks have been found to contain significant amounts of hydrocarbon and there is evidence for anomalies which are similar to those found in Ireland.

For example, Fig. 7 shows the variations in the amounts of cyclopropane (C3 = A) across a section through the Scandinavian deposit. There is a strong enrichment of cyclopropane close to the mineralization forming an anomaly which is comparable with that at Silvermines or Mallow.

Together, these results have established a close correlation between major changes in the hydrocarbon gas content of rocks and mineralization. Other factors, such as lithology, may influence the amounts of gas which a sample releases but cannot account for the observed anomalies.

The effects of lithology were investigated by a study of Carboniferous limestones in SW England, which confirmed that, except perhaps for some changes in the alkane/alkene ratio, there was no correlation between the hydrocarbons and rock type (Goodman 1986). Other work suggests that larger changes in lithology (sandstones vs limestones for example) can introduce some bias but if this is taken into account, the anomalies associated with mineralization can still be resolved.

5 Experimental Work

In parallel with the field studies, a series of laboratory experiments were carried out. These have been of two types:

1. Studies of the extraction of hydrocarbons from rocks.
2. Experiments to simulate the diagenesis of carbonate rocks.

5.1 Experiments with Extraction Techniques

A range of experiments were carried out to study the extraction of hydrocarbons from rocks. These were mostly concerned with heat extraction and aimed primarily at discovering the optimum conditions for routine analysis (Carter and Cazalet 1985), though work on other extraction procedures (for example crushing in sealed containers) was carried out at Imperial College (Ferguson 1985). Studies were also made of the release of hydrocarbons from kerogen and the possible interferences due to surface organic matter and grease (Carter and Cazalet 1985).

Together, the results of these experiments give some insight into the ways in which the light hydrocarbons are held in the rock and the processes involved in their release.

For example, crushing experiments, in which the amounts of gas released by the different size fractions of the same rock were measured, showed different responses from different components (Fig. 8). The results suggest that the gases are held in different forms within the rock, some fractions being held lightly enough so that they are released during crushing, while other fractions are bound so tightly that they are only able to escape from the smallest rock particles.

In the example shown in Fig. 8, methane seems to be mostly present in the lightly bound fraction, while the butane is much more strongly bound to the rock. However, this pattern is not consistent and very different responses were obtained from other samples, indicating that particular hydrocarbons may be held in different forms in different rocks.

A second example of the experimental work, in this case an experiment aimed at identifying the gases held in the different fractions of the rock, is shown in Fig. 9. This histogram compares the amounts of hydrocarbon released by 5 g of a coarsely crushed (1–4 mm) limestone with those extracted from 5 g of the finely crushed (−0.05 mm) fraction; from the residue of clays and silicate left after dissolving 5 g of the same powder in HCl acid; and from the kerogen alone after the HCl residue was dissolved in HF acid.

The results reveal a number of interesting patterns which provide further support for the concept that the light hydrocarbons are held in a variety of different forms within the rock. In particular, note the very large amounts of the heaviest hydrocarbons released by kerogen on its own compared to the amounts released by the coarsely crushed rock, suggesting that hydrocarbons are evolved by kerogen during heating (either by decomposition of heavier molecules or the release of previously generated hydrocarbons), but that they are not normally able to escape from the rock and so do not contribute significantly to the gases measured by the standard technique. Note also the small amounts of methane released by the kerogen suggesting that most of this gas is held in the carbonate and clay fractions.

The release of hydrocarbons is therefore a very complex process and the yield of each species will depend on the physical and chemical composition of the sample, and the extraction procedure used.

Heated extraction apparently desorbs some of the hydrocarbon which is tightly held within the rock (probably mostly on the surfaces of mineral grains), but appears to be quite inefficient, only releasing a small proportion of the total gas present.

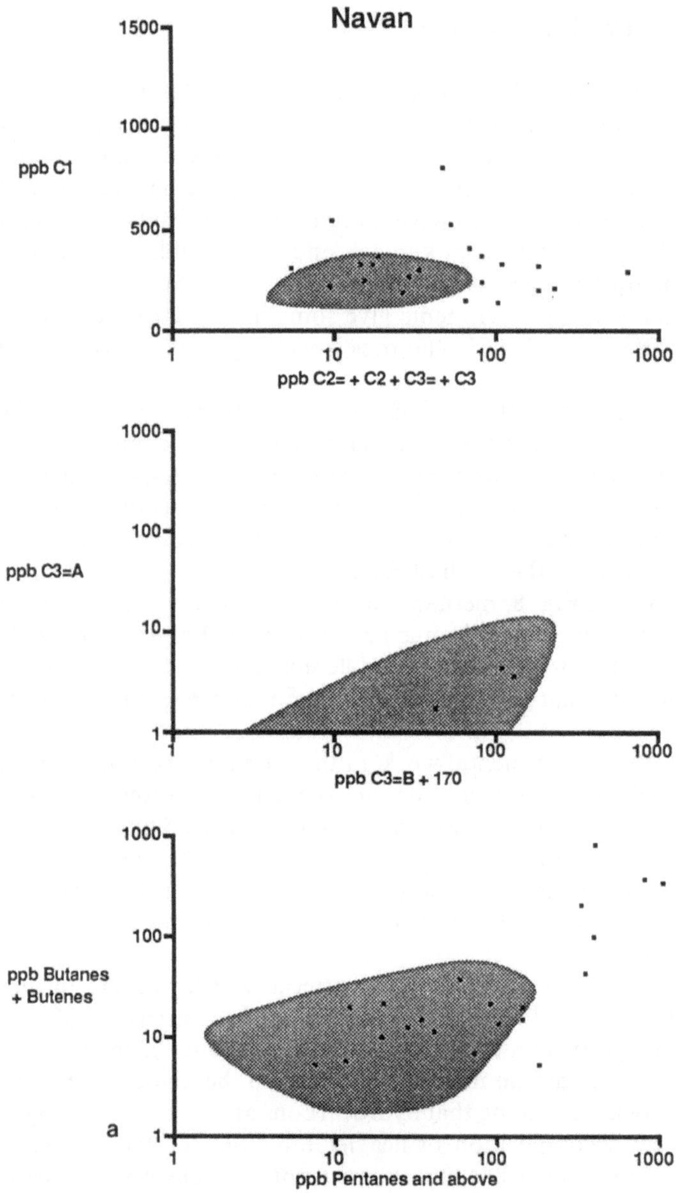

Fig. 6a

Fig. 6a,b. Scatter diagrams plotting the amounts of hydrocarbon released by limestone samples from around the Pb–Zn deposits at Navan (**a**) and Harberton Bridge (**b**). In both cases the lithologies were similar to those from Silvermines and county Limerick, and those samples collected away from mineralization released similar levels of gas as the background rocks from county Limerick (represented by *shaded areas* on the diagrams). Compared to this background, the samples obtained close to mineralization are clearly anomalous, but the pattern is different from that at Silvermines with higher levels of the heavier, saturated hydrocarbons (ethane, propane, butane and pentane) and less C3 = A

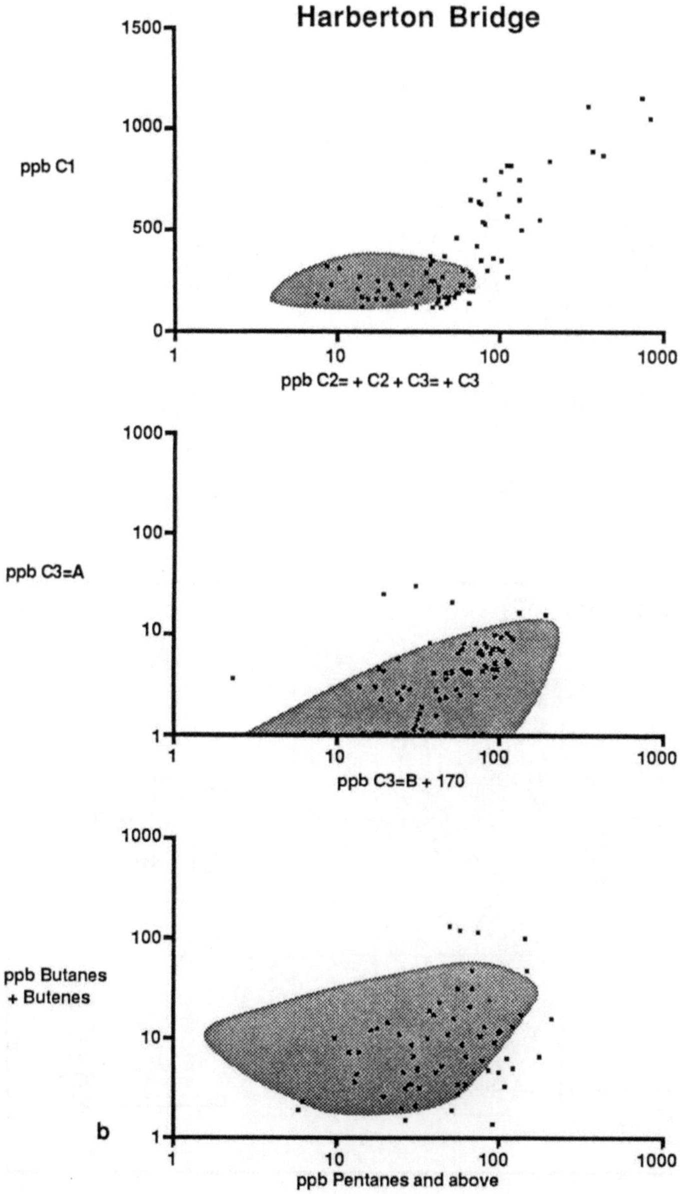
Harberton Bridge
1500
1000
500
0
ppb C1
1
10
100
1000
ppb C2= + C2 + C3= + C3
1000
100
10
1
ppb C3=A
1
10
100
1000
ppb C3=B + 170
1000
100
10
1
ppb Butanes
+ Butenes
b
1
10
100
1000
ppb Pentanes and above

Fig. 6b

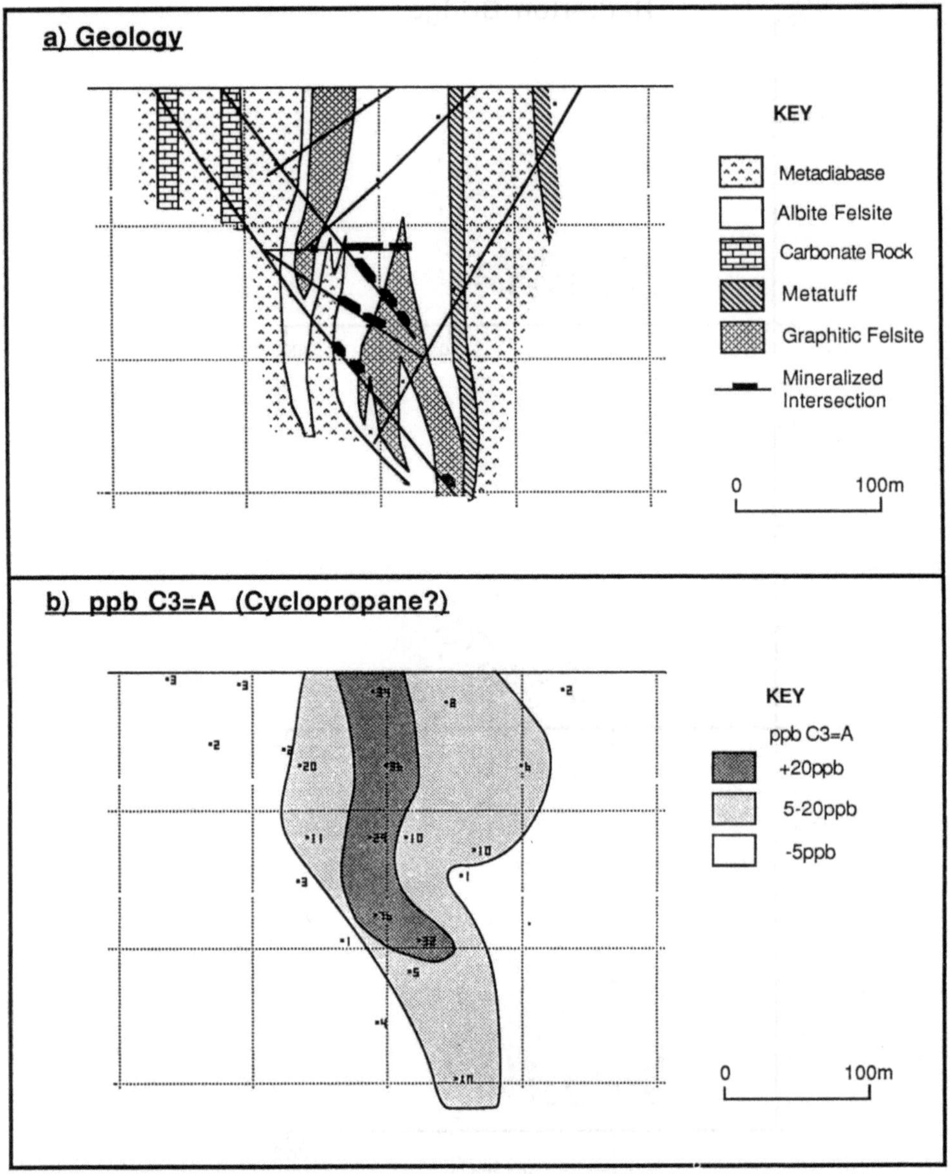

Fig. 7a,b. Cross-section through a Cu–Au deposit in Scandinavia, showing the variations in the amounts of C3 = A (cyclopropane?). There is a strong anomaly, similar to that found at Silvermines (Fig. 5), coinciding with the mineralized zone

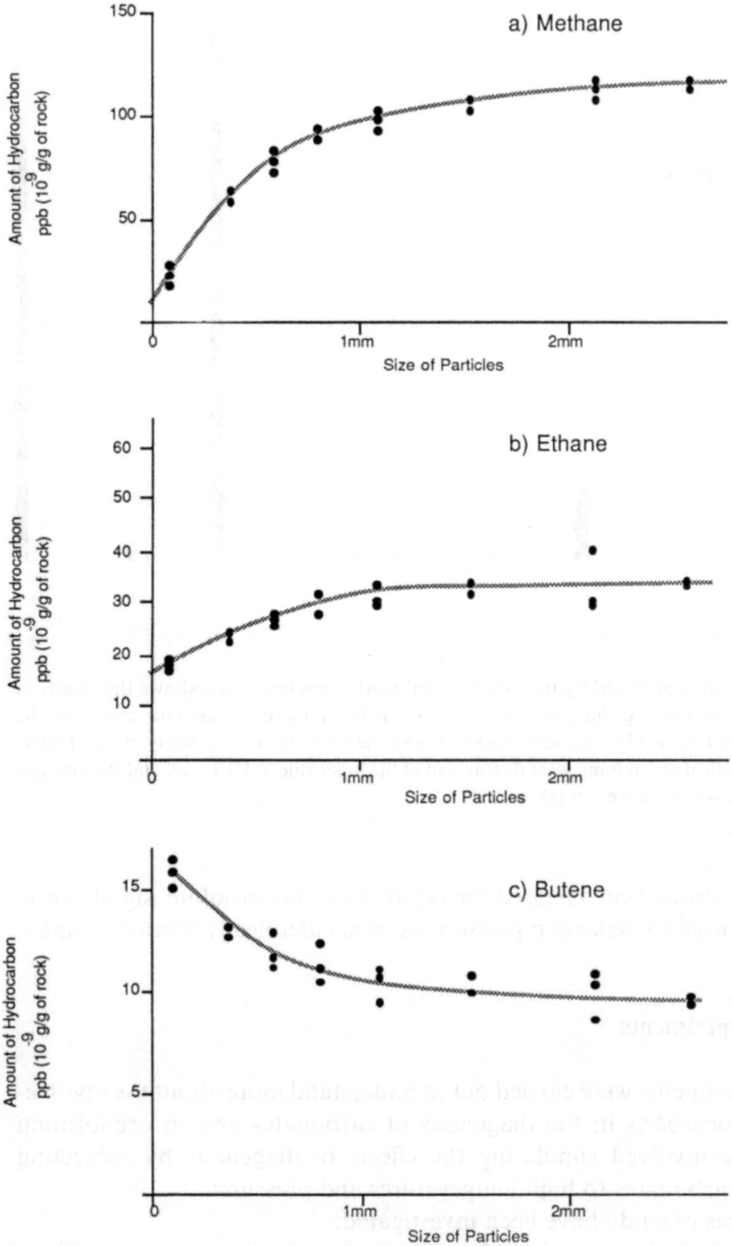

Fig. 8a–c. Example of results obtained by experimental work. This graph shows the amounts of **a** methane; **b** ethane; and **c** butene; released by heating different size fractions of the same limestone sample

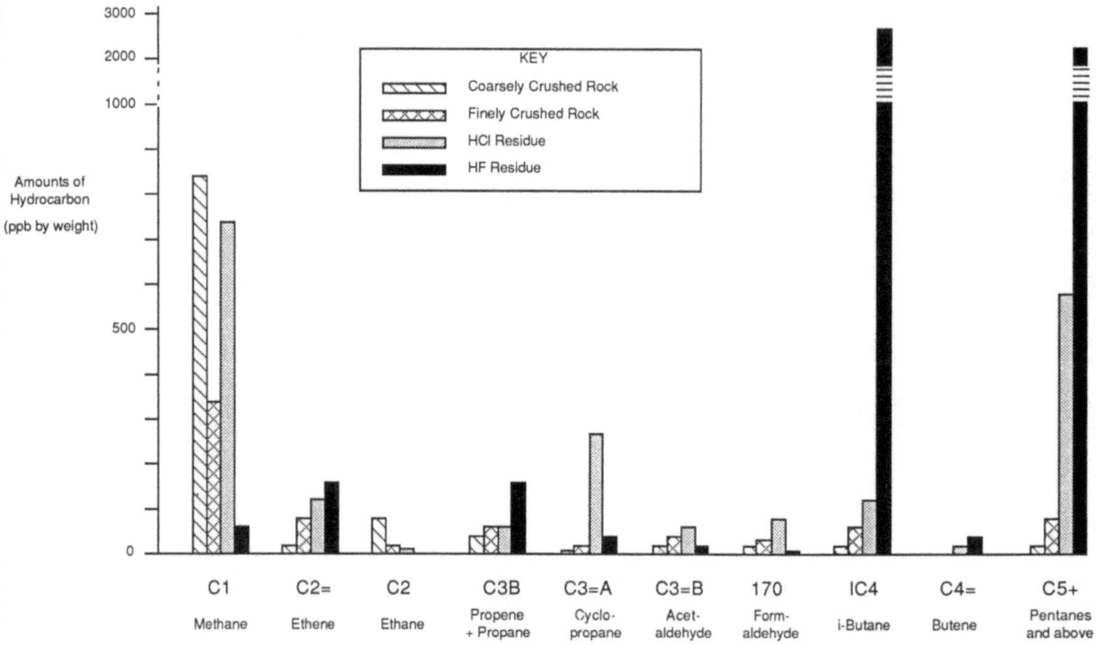

Fig. 9. Example of the results obtained by the experimental work. This histogram shows the results of an experiment aimed at identifying the gases held in different fractions of a rock, and compares the amounts of hydrocarbon released by coarsely crushed limestone with those from finely crushed rock; the residue of clays and silicate remaining after dissolution of the limestone in HCl acid; and the kerogen left after the HCl residue was dissolved in HF acid

The amounts of hydrocarbon released therefore have no absolute significance, though they can be used to make comparisons between identically treated samples.

5.2 Simulation Experiments

The simulation experiments were carried out to understand more about the involvement of light hydrocarbons in the diagenesis of carbonates and in ore-forming systems. They have involved simulating the effects of diagenesis by subjecting samples of Recent carbonates to high temperatures and pressures.

Three main areas of study have been investigated:

1. The examination of the gases released during simulated diagenesis.
2. The possibility of precipitating lead or zinc sulphides in the pore spaces of a cemented, Recent carbonate sample.
3. The study of the role of bacteria and/or algae in the topmost layers of sediment under laboratory conditions.

For the simulation experiments, samples of Recent carbonates were placed in the chamber of a pressure vessel which was then flooded with standard seawater and placed in an oven. Conditions within the vessel were then raised to 200 °C and 1000 p.s.i. and maintained for periods of several weeks. The apparatus has a reservoir for fluids, allowing the removal of water samples without drastically altering the chemistry of the system.

After some time (normally about 3 weeks), cementation of the sediment begins, marked by a sharp decrease in porosity and rise in pressure, and at this point the experiment was normally closed down to allow samples of gas to be taken for analysis by gas chromatography and the cemented carbonate material to be removed for detailed examination. Samples of the original sediment and the cemented rock generated by the experiment were also analyzed for light hydrocarbons using the standard extraction procedure.

The study of the gases released by the experiments has provided some interesting results, in particular showing a fall in the 170 component (acetaldehyde) balanced by an increase in methane, propene and possibly butene. This is in line with what might be expected if the 170 component was the degradation product of thermally immature organic material which was gradually transformed into hydrocarbon species in the course of the experiments.

Since the original carbonate material used contained some organic matter, largely algal in origin, analysis of the gases generated during the simulation showed a close correspondence with those found in carbonate rocks and considered to be the background gases. It was also possible to show that the gases were trapped in the newly generated carbonate cement.

Simulating lead precipitation was less satisfactory since although analysis of the brines showed that lead was being removed from the system, it was not possible to locate the lead compounds in the carbonate matrix (Goodman 1986).

5.3 Analysis of Calcites and Ore/Gangue Minerals

In addition to the simulation experiments above, Imperial College has also investigated the distribution of hydrocarbons in specific mineral phases, as against measurement of the hydrocarbon content of the whole rock used in the regional studies, the gases being released by grinding (Ferguson 1986, unpub. research).

The results for the calcite samples (Table 2) appear to reflect different stages in the diagenesis and tectonic evolution of the rocks from which they were collected; for instance, calcite sample 6 probably represents the hydrocarbon signature of material deposited during the tectonic evolution of the area, whereas the result for dripstone sample 4 is that of calcite dissolution and redeposition associated with the present groundwater regime. The data for calcite samples 1–3, however, show the effect of mineralization.

It is argued that studies of hydrocarbon contents of calcites could well throw light on hydrocarbon regimes operating at different geological times, and lead to a better understanding of the possible sources of the hydrocarbons associated with mineral deposits. Similarly, studies of variations in hydrocarbon content of ore and gangue minerals (Table 2, samples 10–15) may lead to further ideas on the mechanisms by which minerals retain gases. For instance, if it can be shown that gases

Table 2. Comparison of results obtained from samples of calcite and limestone, from mineralized and non-mineralized areas and some ore and gangue minerals. Analysis of gases released by grinding (values in 10^{-9} cm^3/g sample)

	Mineralized[a]			Non-mineralized[b]						Ore and gangue[c]					
	1	2	3	4	5	6	7	8	9	10	11	12	13	14	15
Methane	2850.8	1020.6	3426.7	12.8	407.0	793.5	247.0	81.3	16.5	91.0	113.2	173.1	15.4	3318	7781
Ethene	0.6	0.9	2.9	1.4	1.7	1.1	3.6	2.5	1.3	0.4	0.3	1.0	0.0	0	40
Ethane	153.2	29.3	55.8	1.3	60.4	100.5	29.8	2.4	0.8	1.4	1.1	2.1	0.2	453	1020
Propylene	0.4	6.8	2.4	0.3	0.9	0.0	1.6	1.6	0.8	0.2	0.0	0.4	0.0	0	16
Propane	71.6	0.0	6.0	1.0	55.8	86.9	18.2	1.1	0.4	0.5	0.0	0.8	0.0	376	690
I-butane	12.0	1.3	0.0	0.0	14.2	23.7	9.0	0.0	0.0	0.0	0.0	0.3	0.9	100	100
N-butane	20.6	1.1	0.9	0.2	20.7	30.3	11.0	0.4	0.5	0.0	0.3	0.8	0.2	252	217
1-Butene	0.0	0.0	1.5	0.0	0.0	0.0	0.0	0.0	0.3						

[a] *1* Calcite vein; Carboniferous, Great Limestone; Fourstones, Northumberland. *2* Calcite vein; Carboniferous, Great Limestone; Killhope, county Durham. *3* Calcite vein with mineralization; Lower Carboniferous shales; just south of Cullernose Point, Northumberland.

[b] *4* Drip stone; Carboniferous, Great Limestone; High Whittle, Northumberland. *5* Limestone; Carboniferous, Great Limestone; High Whittle, Northumberland. *6* Calcite vein; Carboniferous, Great Limestone; Lowick, Northumberland. *7* Limestone; Carboniferous, Great Limestone; Lowick, Northumberland. *8* Calcite vein; Lower Carboniferous, Sandbanks Limestone; Dunstanburgh, Northumberland. *9* Calcite; Jurassic, Bathonian/Bajocian limestone; near Figeac, France.

[c] *10* and *11* Fluorite; Greenhough, North Yorkshire. *12* Fluorite; Cambokeils Mine, Weardale, county Durham. *13* Galena; Combokeils Mine, Weardale, county Durham. *14* Fluorite; Mexico. *15* Fluorite; Windy Knoll, Derbyshire.

are trapped during mineral deposition, then the data should lead to information on the nature of gases accompanying mineralization and hence to the chemical reactions which may be responsible.

6 Origin of Hydrocarbon Gas Anomalies

The occurrence of hydrocarbon gas anomalies around mineral deposits suggests that light hydrocarbons were released into the rocks during the emplacement of the mineralization. There are three possible mechanisms which could account for this:

1. Release of Hydrocarbons from Solution. Light hydrocarbons are soluble in water (Conybeare 1970) and could have been transported in the same solution which carried the metals to the site of mineralization. Most hydrothermal solutions which originated in or passed through sediments probably contain hydrocarbons (Hanor 1980) and even those associated with igneous activity are likely to carry at least some of the lighter species, as is suggested for example by the measurement of high concentrations of methane in the modern hydrothermal systems of the East Pacific Rise (Welham and Craig 1979).

2. Generation of Hydrocarbon Gases at the Site of Mineralization. An alternative to the transport of light hydrocarbons in mineralizing fluids is that the gases were generated at the site of mineralization, by the breakdown of heavier, petroleum hydrocarbons carried in the brines. This could occur as a by-product of sulphate reduction, as suggested by Dunsmore and Shearman (Dunsmore 1975; Dunsmore and Shearman 1977), and would produce a closer association between anomalies and mineral deposits.

3. Generation of Hydrocarbons in the Surrounding Rocks. The mineralizing event could also have caused gases to be generated in the surrounding rocks through the breakdown of pre-existing organic matter. The most likely mechanism for achieving this would be an increase in temperature. If the metalliferous brines were hotter than the surrounding rocks (at a temperature of at least 100 °C) the breakdown of immature kerogen in sedimentary host rocks would take place, providing they had not already been heated to higher temperatures during burial, and providing the system operated for a considerable period of time. The outcome would be a thermal anomaly, perhaps like that described at Pine Point by Macqueen and Powell (1983), which would show up as a light hydrocarbon gas anomaly in the affected rocks.

6.1 Controls on the Form of Anomalies

The gases released during mineralization would tend to migrate outwards into the surrounding rocks, driven by a combination of:

1. Diffusion (movement from areas of high concentration to areas of low concentration).

2. Effusion (movement from areas of high pressure to areas of low pressure).
3. Circulation of the mineralizing brines and connate waters.

In the case of diffusion and effusion, the rate of migration depends on the molecular weight of the hydrocarbon, the lighter species, like methane and ethane being able to move through the rock more rapidly than heavier ones like butane or pentane (Leythaeuser et al. 1982; Starobinets et al. 1981). Simple computer simulation experiments of this process (Carter and Cazalet 1985) suggest that the hydrocarbon would segregate, with methane moving large distances away while the heavier components remained closer to the source, producing a halo pattern.

This general pattern is consistent with the hydrocarbon distribution observed around mineralization and on a smaller scale in individual boreholes. In practice, the anomalies are much more irregular, but this can be explained by local geological controls on the migration of the gases, the hydrocarbons migrating preferentially along permeable beds and up faults. Also, after its formation, an anomaly would be vulnerable to deformation, overprinting or partial erasure by later geological events.

6.2 Anomalies not Associated with Mineralization

The relationship between gas anomalies and mineralization depends on which of the several possible mechanisms were involved in the release of the hydrocarbons. If the gases were generated as a direct result of sulphide deposition by sulphate reduction, for example, a very close correlation between the two would be expected. On the other hand, processes such as the breakdown of in situ hydrocarbons in response to a local increase in temperature might produce anomalies unrelated to mineralization, as might a slow outward diffusion of accumulated hydrocarbons from a hydrocarbon reservoir.

Whatever the precise mechanisms involved, very large anomalies covering several square kilometres could only result from the large-scale migration of fluids and the centres of such anomalies would always seem to be logical targets for mineral exploration.

7 Conclusions

The study of light hydrocarbons associated with mineralization is still a very new field of research and much more work is needed to develop it fully. However, the studies completed so far show promising results. It has been possible to establish the existence of large variations in the gas content of rocks around a variety of different mineral deposit types, forming distinct anomalies which have close spatial relationships with mineralization.

Although it is not clear how these anomalies might have formed, it seems likely that they reflect the migration of large volumes of hydrocarbon gases released by some process associated with mineralization. The observations are consistent with Dunsmore and Shearman's prediction that methane would be generated at the site of mineralization as a by-product of sulphate reduction, but it is equally possible that hydrocarbons are carried passively in the mineralizing solutions or generated in the surrounding rocks by the thermal maturation of kerogens.

Whatever the precise mechanisms involved, it is clear that some major geological event (most probably the development of a major hydrothermal system) would be required to produce large gas anomalies on the scale of those which have been observed. Anomalies should therefore always mark favourable sites for mineralization.

From a practical viewpoint, the anomalies found around major mineral deposits such as Silvermines or Navan, in Ireland, are much larger than those expected with conventional trace element geochemistry (the outer zone of methane enrichment can extend many kilometres away from mineralization) and present large targets for exploration.

In particular, the technique would seem to be well suited to reconnaissance exporation, since it should be possible to use a relatively low density of samples (perhaps $1-2/km^2$) to to screen large areas of ground quickly and to identify general areas where the levels of gas are anomalous for more detailed investigation.

Acknowledgements. The authors would like to thank the many mining companies which supported the regional studies. Most of the research was carried out under *Contract No. MSM-106-UK* with the EEC. The current research at Imperial College is funded by a 3-year grant from Cominco (Europe) Ltd., and their assistance and that of research student Sean Mullshaw is gratefully acknowledged.

References

Carter JS, Cazalet PCD (1984) Hydrocarbon gases in rocks as pathfinders for mineral exploration. In: Prospecting in areas of glaciated terrain 1984 (Glasgow Symposium, 1984). London. Institution of Mining and Metallurgy, pp 11–20

Conybeare CEB (1970) Solubility and mobility of petroleum under hydrodynamic conditions, Surat basin, Queensland. Geol Soc Aust 16:667–681

Disnar J-R, Gauthier B, Carter JS (1986) Utilization des hydrocarbures gazeux en prospection régional des gîtes Pb–Zn de couverture: application au gîte de Trèves (Gard, France). Chronique de la recherche minière. BRGM, Orléans 482:67–85

Dunsmore HE, Shearman DJ (1977) Mississippi Valley-type lead-zinc orebodies: a sedimentary and diagenetic origin. In: Garrard P (ed) Proceedings of the forum on oil and ore in sediments. Imperial College, London, pp 189–205

Hanor JS (1980) Dissolved methane in sedimentary brines: potential effects on the PVT properties of fluid inclusions. Econ Geol 75:603–609

Leythaeser D, Schaefer RG, Yukler A (1982) Role of diffusion in primary migration of hydrocarbons. Bull Am Assoc Pet Geol 66(4):408–429

Macqueen RW, Powell TG (1983) Organic geochemistry of the Pine Point lead-zinc ore field and region, Northwest Territories, Canada. Econ Geol 78(1):1–25

Starobinets IS, Tikhomirova YS, Stativko GS, Starovoytov V, Litvinova VN, Tarasov IA (1981) influence of intrusive traps on the distribution of hydrocarbon gases. Int Geol Rev 23(12):1424–1430

Welham JA, Craig H (1979) Methane and hydrogen in East Pacific Rise hydrothermal fluids. Geophys Res Lett 6(11):829–831

Metallogenesis and Geodynamic Context in the Lower-Middle Cambrian of Montagne Noire (France) and Sardinia (Italy)

P. COURJAULT-RADÉ[1] and A. GANDIN[2]

Abstract

A sedimentological comparison of Lower-Middle Cambrian sequences indicates that southwestern Sardinia was closer to the Precambrian African shield and more stable than Montagne Noire. In both regions the main tensional event straddles the Lower-Middle Cambrian boundary, when shallow water carbonates were gradually replaced by deeper-water deposits. Sedimentological evidence, coupled with metallogenic data, suggests that repeated phases of extensional tectonics in the Cambrian induced recurrent circulation of metal-rich hydrothermal brines. These produced early epigenetic or syngenetic-to-syndiagenetic mineralizations. Cambrian metallogeny in the Asturian-Sardinian province appears to reflect the geodynamic evolution of the eastern Gondwana passive margin in the Early Paleozoic, and to be controlled by the initial paleogeographic setting: polymetallic mineralizations (Pb, Zn, Ba–As, Cu, Au, Bi) are scattered at various stratigraphic levels in the external mobile areas (Montagne Noire), whereas in internal and more stable areas such as southwestern Sardinia, lead-zinc sulphides (with associated barite) occur essentially in connection with the main tectonic phase (Lower-Middle Cambrian transition).

1 Introduction

Montagne Noire and Southwestern Sardinia are segments of the Southern European Variscan belt (Fig. 1). Paleomagnetic data (Scotese et al., 1979) suggest that both regions belonged to the same Lower Paleozoic paleogeographic province.

A number of dominantly lead-zinc deposits (with several mines still in operation) characterize the Cambrian sequences of the two areas. The economic potential of the Iglesiente-Sulcis (about 8 mt of combined Pb + Zn metal according to Brusca and Dessau, 1968) largely dominates that of Montagne Noire (+/−0.7 mt combined metal, Aubague et al. 1977). This fundamental discrepancy also reflects

[1] Laboratoire de Minéralogie, U.A. n° 67 du CNRS, 39 allées J. Guesde, 31400 Toulouse, France
[2] Dipartimento di Scienze della Terra Via delle Cerchia, 3-53100 Siena, Italy

Mineral Deposits within the European Community
(ed. by J. Boissonnas and P. Omenetto)

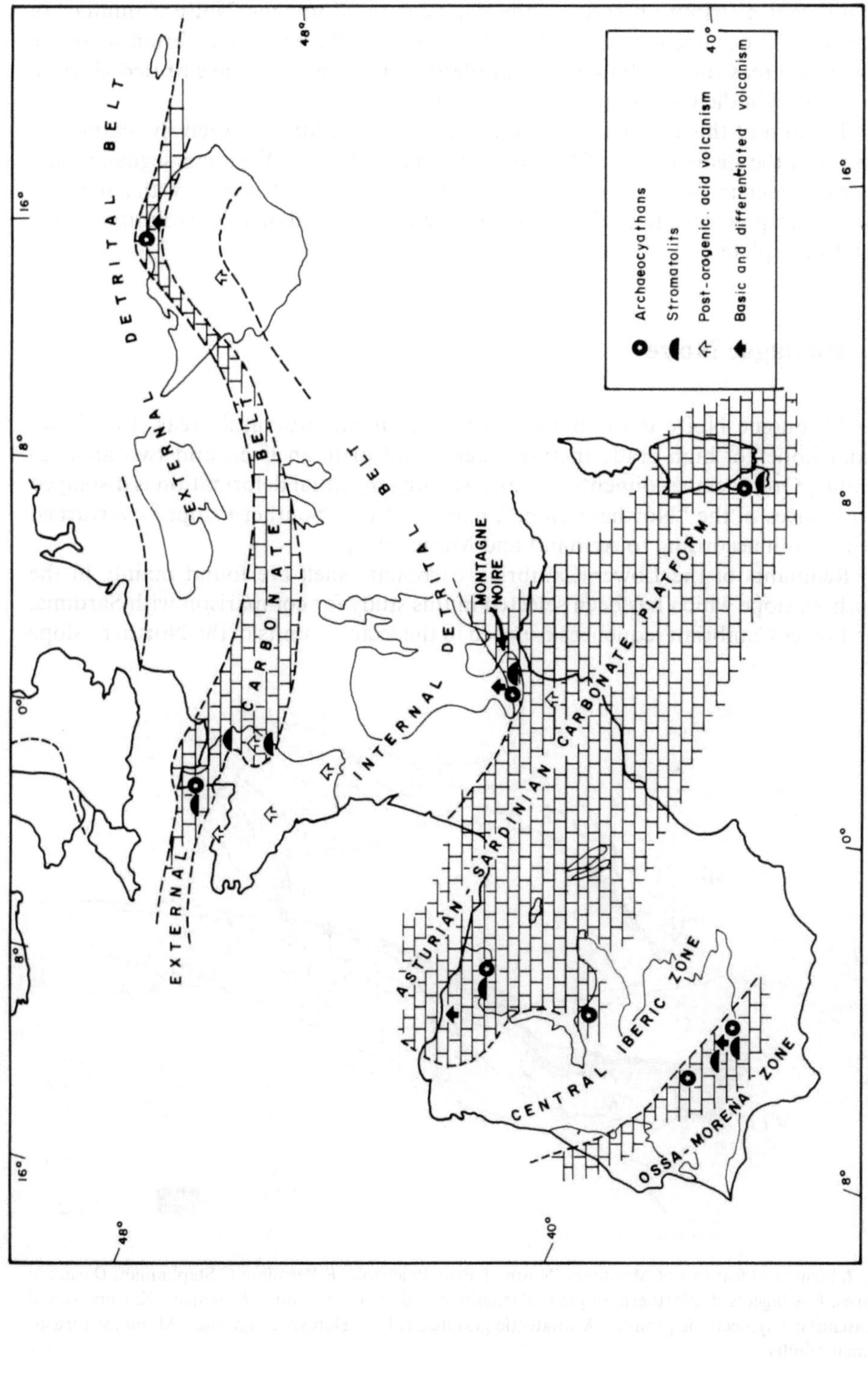

Fig. 1. Cambrian paleogeographic areas (post-Cambrian tectonics are not taken into account) (Doré, 1977)

the different distribution style of the deposits: in Montagne Noire a number of small Pb–Zn occurrences are scattered throughout the whole Cambrian sequence (Fig. 3) whereas the Sardinian high grade/tonnage ores are concentrated close to the Lower-Middle Cambrian boundary (Fig. 5).

The aim of this paper is to establish the relationships between the carbonate facies and the distribution of the ore deposits in the two Variscan segments and to give evidence that regional metallogeny reflects simultaneously the regional paleogeographic setting and the proto-Caledonian geodynamic evolution of the Gondwana plate.

2 Montagne Noire

The Montagne Noire is subdivided into three main structural areas (Fig. 2): an "Axial Zone" of high-grade metasediments and orthogneisses, and two areas of Cambro-Ordovician sediments in which strong tangential deformation is displayed (thrust slices of the "Northern slope", nappes of the "Southern slope" overturned to the South according to Arthaud and Matte 1977).

Remnants of the Lower Cambrian carbonate shelf are found mainly in the Southern slope which has been selected in this study for comparison with Sardinia. The Lower Cambrian sequences exposed in the tectonic units of the Northern slope

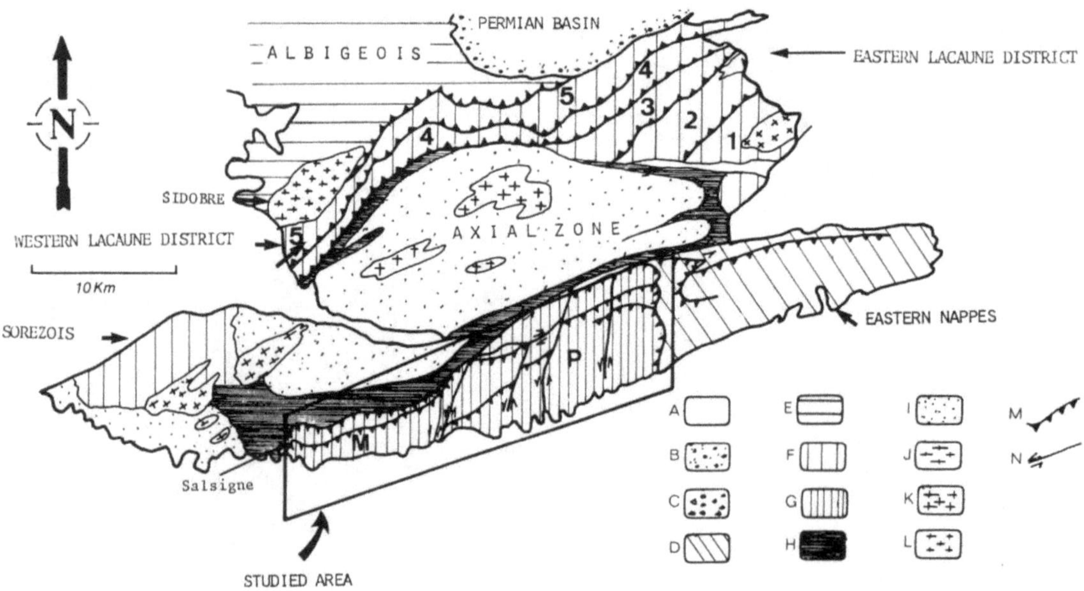

Fig. 2. Structural features of Montagne Noire. *A* Post-Paleozoic; *B* Permian; *C* Stephanian; *D* eastern nappes; *E* Albigeois; *F* Northern Slope; *G* Pardailhan and Minervois units; *H* Schistes X; *I* gneiss and micaschists; *J* syntectonic granites; *K* anatectic granites; *L* Late-Hercynian granites; M, major thrusts; N, major faults

display other sedimentologic and metallogenic features which are presented in this Volume by Lescuyer et al.

2.1 Dynamic Evolution of the Carbonate Shelf (Southern Slope Sequences)

The Lower-Middle Cambrian succession of the Southern Slope is a carbonate dominated complex ($\pm$700–800 m) underlain by a thick terrigenous suite ("Marcory": author) and overlain by a Middle-to-Upper Cambrian greenish fine-grained siliciclastic sequence. It comprises three formations[1]: Orbiel formation (Lower Cambrian), La Clamoux formation (Lower Cambrian) and Barroubio formation (Middle Cambrian) (Courjault-Radé, 1984), details of which are shown in Fig. 3.

This carbonate dominated sequence is divided in four main tectonosedimentary episodes, E1 to E4 (Courjault-Radé and Gandin, 1986).

Episode E1

1. E1-1. The Lower Orbiel Member begins with hematitic feldspathic sandstones (up to 100 m thickness) containing microconglomeratic (or locally conglomeratic) lenses (Boyer 1962) indicative of an updoming phase. Further up, the deposition of siltstones and shales appears to be related to a quick increase of subsidence.

Carbonates gradually developed as the subsidence decreased. At first, they occur as small, scattered limestone lenses embedded within shales, siltstones and/or very fine-grained sandstones. Their extension and thickness increase upwards, but they remain lenticular and interfingered with clastics (= "alternances inférieures").

2. E1-2. Tabular carbonate bodies were deposited in regular alternances with siliciclastics (= "alternances supérieures"). Volcanites and volcanoclastites are found within this mixed sequence exposed in the Salsigne mining district (Northern Minervois; Lépine et al. 1984). The episode ends with the uplift and emersion of the carbonate shelf and subsequent local karstification (Courjault-Radé 1984). Tectonic instability appears to be the key factor responsible for the alternances. Recurrent vertical movements triggered clastic progradation, which repeatedly interrupted carbonate deposition (Courjault-Radé 1984).

Episode E2

Episode E2 is represented in the two-thirds of the Lower La Clamoux Member (= "Dolomies massives": author) which consists mostly of dark homogeneous early

[1] According to the international stratigraphic rules (International stratigraphic Guide — IUGS Stratigraphic Commission, 1976. J. Wiley & Sons, Inc.) the term "Gneiss d'Orbiel" introduced in 1986 by Demange et al. cannot be retained, in order to avoid confusion with the previously established Orbiel Formation (Courjault-Radé 1984 and Courjault-Radé and Gandin 1984).

Similarly, the "Formation schisto-gréseuse de Barroubio" introduced by Gèze (1949) is not conformable with the international stratigraphic rules (ibidem). Barroubio Formation, as established by Courjault-Radé (1984), includes Gèze's unit as its younger member.

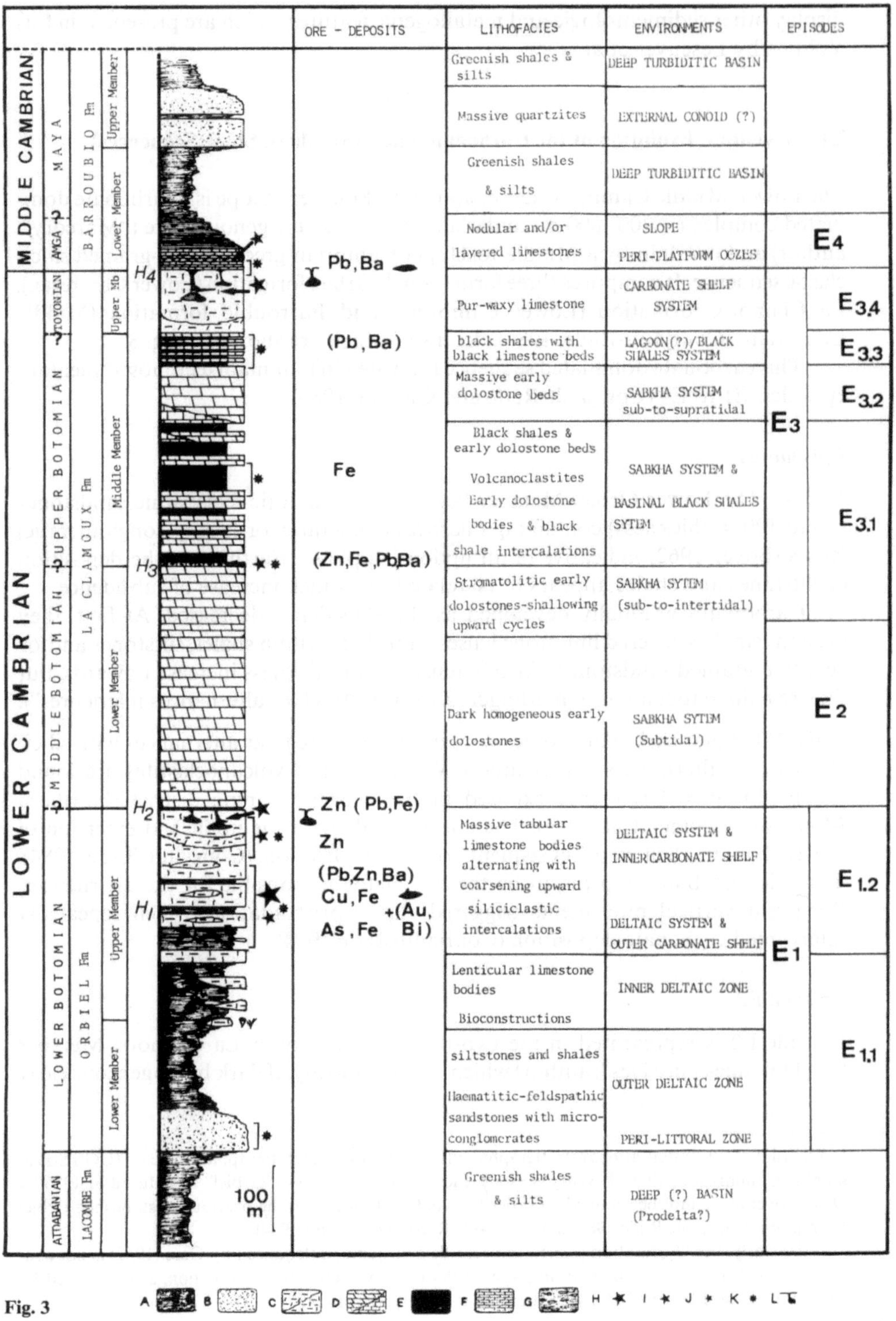
ORE - DEPOSITS
LITHOFACIES
ENVIRONMENTS
EPISODES
MIDDLE CAMBRIAN
LOWER CAMBRIAN
MAYA
AMGA ?
TOYONIAN
UPPER BOTOMIAN
MIDDLE BOTOMIAN
LOWER BOTOMIAN
ATDABANIAN
BARROUBIO Fm
LA CLAMOUX Fm
ORBIEL Fm
LACOMBE Fm
Upper Member
Lower Member
Upper Mb
Middle Member
Lower Member
Upper Member
Lower Member
H4
H3
H2
H1
100 m
Pb, Ba
(Pb, Ba)
Fe
(Zn, Fe, Pb, Ba)
Zn (Pb, Fe)
Zn
(Pb, Zn, Ba)
Cu, Fe
+(Au, Bi)
As, Fe
Greenish shales & silts
Massive quartzites
Greenish shales & silts
Nodular and/or layered limestones
Pur-waxy limestone
black shales with black limestone beds
Massive early dolostone beds
Black shales & early dolostone beds
Volcanoclastites
Early dolostone bodies & black shale intercalations
Stromatolitic early dolostones-shallowing upward cycles
Dark homogeneous early dolostones
Massive tabular limestone bodies alternating with coarsening upward siliciclastic intercalations
Lenticular limestone bodies
Bioconstructions
siltstones and shales
Haematitic-feldspathic sandstones with micro-conglomerates
Greenish shales & silts
DEEP TURBIDITIC BASIN
EXTERNAL CONOID (?)
DEEP TURBIDITIC BASIN
SLOPE
PERI-PLATFORM OOZES
CARBONATE SHELF SYSTEM
LAGOON(?)/BLACK SHALES SYSTEM
SABKHA SYSTEM sub-to-supratidal
SABKHA SYSTEM & BASINAL BLACK SHALES SYTEM
SABKHA SYTEM (sub-to-intertidal)
SABKHA SYTEM (Subtidal)
DELTAIC SYSTEM & INNER CARBONATE SHELF
DELTAIC SYSTEM & OUTER CARBONATE SHELF
INNER DELTAIC ZONE
OUTER DELTAIC ZONE
PERI-LITTORAL ZONE
DEEP (?) BASIN (Prodelta?)
E4
E3,4
E3,3
E3,2
E3
E3,1
E2
E1,2
E1
E1,1

Fig. 3

diagenetic dolomite, deposited in a quiet environment. The lack of intertidal-supratidal features indicates that subsidence was equal to sedimentation rate. This was a period of relative pause of tectonic instability.

Episode E3

This episode is dominated by a complex succession of tensional events and precedes the final collapse of the carbonate shelf (Courjault-Radé, 1985). It is characterized by intercalations of black shales, and comprises four stages (Fig. 3):

Stage 1 (E3-1, = "Schisto-dolomitique": author): alternances of early dolomites (arranged in regressive cycles) and black shales. In the middle part of the Lower La Clamoux Member, intercalations of black shales attain a thickness of 20–25 m and display evidence of volcano-clastic supply. Alternances are believed to reflect recurring phases of subsidence.

Stage 2 (E3-2, = "Silico-dolomitique": author): deposition of massive carbonates (essentially early dolomite), reflecting a relative and momentary pause of tectonics. The onset of carbonate deposition may be due to a phase of regional upwarping.

Stage 3 (E3-3, = "Calcaires à Ferralsia": author): renewed subsidence, leading to the deposition of black homogeneous limestone with intercalations of black shales.

Stage 4 (E3-4, = Calcaire blanc": author): massive pure limestones, again reflecting a relative pause of tectonic activity. This unit appears to correlate with the Ceroide Limestone of the Gonnesa Formation in SW Sardinia (Courjault-Radé and Gandin 1986).

Episode E4

Drowning of the carbonate shelf marks the Lower-Middle Cambrian boundary (Courjault-Radé, 1985). A succession of tensional pulses led to the gradual replacement of shallow-water carbonates by deeper-water terrigenous deposits (lower half of the Lower Barroubio Member). An initial phase of regular subsidence was followed abruptly by large-scale positive vertical movements. Differential subsidence of carbonate blocks resulted in isolated platform- and basin morphology. Karstic dissolution took place on the emerged uplifted blocks.

The co-existence of uplifted and collapsed areas (as observed in the gradual transition between the la Clamoux and Barroubio Formations) supports the hypothesis of block faulting. The deposition of shales with intercalations of red nodular limestone (Lower Barroubio Member) records the transition from shelf to slope.

Fig. 3. Lower to Middle Cambrian sequence of Southern Montagne Noire: depositional environments, tectono-sedimentary episodes (E1 to E4) and ore-deposits distribution. *A* Shales and silts; *B* fine-grained sandstones; *C* diagenetic dolostone (*a*) massive, pure limestone (*b*); *D* early dolostone (*a*), fine-grained limestone (*b*); *E* black shales; *F* layered and/or nodular limestone; *G* shales with calcareous nodules; *H* mines; *I* ore-deposits; *J* small ore-deposits; *K* disseminations; *L* karstic fillings

The presence of nodular facies, recalling the "Ammonitico rosso", is indicative of synsedimentary instability. The transition from the shales ("Schistes à nodules": author) to the overlying laminated greenish shales and siltstones is marked on a regional scale by deep-purple shales containing small calcareous nodules. This marker horizon appears to be related to the final tectonic pulse which resulted in drowning of the last surviving karstified blocks.

2.2 Stratigraphic Distribution of Ore Deposits (Southern Slope)

As shown in Fig. 3, there are several ore-bearing horizons (H1 to H4), each one characterized by its dominant metal association and morphologic features.

H1 (= upper Orbiel Mb; = E1-1 episode) consists of disseminated pyrite with associated lead-zinc sulphides occurring within the Upper Orbiel Member. Massive deformed lenses of pyrite/pyrrhotite and arsenopyrite/chalcopyrite (with native gold and bismuth) are embedded in the lower half of the mixed sequence of the Salsigne mining district (northern Minervois), and spatially associated with basic volcanites and volcanoclastics (Lépine and al. 1984, 1986; Courjault-Radé and Tollon 1985). Small scattered lenses of iron-bearing sphalerite and galena also occur in this mine district (Crouzet and Tollon 1980).

H2 (= Orbiel Fm./La Clamoux Fm transition, = E1/E2 boundary). The last meters of the Orbiel carbonate sequence are regionally characterized by high (up to 1000 ppm) Zn geochemical anomalies (Aubague et al. 1977) and are locally overlain by black shales slightly mineralized in Pb, Fe and associated Zn. These shales form the last detrital layer occurring at the Orbiel/La Clamoux transition (= "niveau d": author). Further diagenetic sphalerite-rich concentrations appear as karstic fillings at that same transition (Courjault-Radé, 1984).

H3 (= Lower/Middle La Clamoux Mb transition, = E2/E3 boundary) is characterized by disseminations and small Hercycnian epigenetic veins made up of sphalerite and pyrite with minor amounts of recrystallized galena located within silicified early diagenetic dolomite.

H4 (La Clamoux Fm/Barroubio FM transition, = E3/E4 boundary) consists of disseminations, karstic fillings, and small massive lenses of argentiferous galena and pyrite. In the southeastern district of Minervois, this horizon can be followed over 15 km ("Tête Rousse du Minervois", Boyer and Routhier 1958). Slightly mineralized internal fillings (mainly galena) and collapse breccias derived from karstic dissolutions are observed in the late Lower Cambrian massive limestone (Upper La Clamoux Mb.), whereas small stratiform lenses of silver-bearing galena and/or pyrite occur within the overlying shales (base of the Barroubio Fm.). Early-diagenetic microcrystalline silicification of the host rocks (carbonates and shales) is systematically associated with mineralization (Boyer and Routhier, 1958). It probably represents a signature of hydrothermal alteration, as it is usually observed in sediment-hosted lead-zinc dominated ore-bodies (Large 1980).

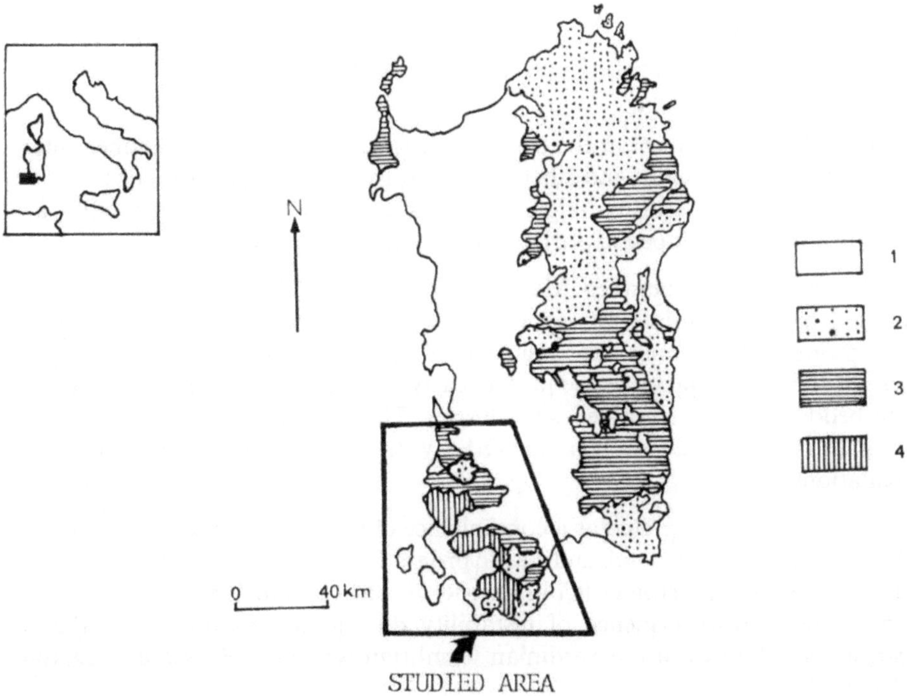

Fig. 4. Structural features of Southwestern Sardinia. *1* Post-paleozoic; *2* Late-Hercynian granites; *3* Ordovician; *4* Cambrian

3 Southwestern Sardinia

3.1 Dynamic Evolution of the Carbonate Shelf

The Lower Cambrian of southwestern Sardinia (Fig. 4) consists of a complex (1000–1700 m) underlain by a thick terrigenous suite (Bithia Fm) and overlain by a Middle Cambrian to Arenig fine-grained siliciclastic sequence (cf. references in Gandin 1987).

It has been subdivided into three formations, from bottom to top:

— Nebida Formation (Lower Cambrian)
— Gonnesa Formation (Lower Cambrian)
— Cabitza Formation (Middle Cambrian-Arenig)

The main geological features of southwestern Sardinia are shown on Fig. 4.

As in Montagne Noire, the dynamic evolution of the Lower-Middle Cambrian sequence developed during four tectono-sedimentary episodes (Courjault-Radé, and Gandin 1986), as follows (Fig. 5):

Episode E1

It corresponds to the Nebida Formation and is characterized by two tensional episodes:

1. E1-1 (Matoppa Member): at the base, it consists in feldspathic sandstones, alternating with greenish shales and siltstones. Upwards, the proportion of sandstone increases and Archaeocyathan-algae buildups appear. Traces of instability are lacking or poorly developed (Courjault-Radé and Gandin 1986).

2. E1-2: the boundary between Matoppa Member and Punta Manna Member is marked by the Oolitic Unit (up to 100 m). It is followed by fine-grained or coarse-grained oolitic limestones regularly alternating with sandstones and silty shales. At the top of the mixed sequence, early dolomite, replacing the limestone, is interbedded with fine-grained siliciclastics. The boundary with the overlying Gonnesa formation (E2–E3) is marked by the disappearance of siliciclastic intercalations.

Episode E1 can be interpreted as a consequence of recurrent carbonate onlaps, extending towards emerged land, and clastic progradations on a coastal environment, induced by tensional tectonics (Courjault-Radé and Gandin 1986). Syndiagenetic deformations provide evidence of instability during this episode. Nevertheless volcanism is unknown in the Sardinian Cambrian sequence (Brusca and Dessau 1968; Gandin 1987).

Episode E2

A relative pause of tectonic activity allows the deposition of pure carbonates ("Dolomia rigata": author). They mainly consist of early dolomite, with typical sabkha features (Gandin and al. 1974; Gandin 1987).

Episode E3

The transition between the early dolomite unit ("Dolomia rigata": author) and the Ceroide Limestone unit ("Calcare ceroide": author) is regionally marked by the occurrence of a diagenetically dolomitized horizon ("Dolomia blu": author). At the western margin of the carbonate shelf (Buggerru area), the transition is also marked by alternations of homogeneous limestone and algal-laminated early dolomite, corresponding to regressive cycles, the genesis of which can be related to recurrent subsidence oscillations connected with the resumption of tensional tectonics (Gandin 1987). According to stratigraphic correlations (Courjault-Radé, and Gandin, 1986), the transition between the "Dolomia rigata" and the Ceroide Limestone appears to be synchronous with the E3-1, E3-2 and E3-3 episodes recognized in Montagne Noire. The basic difference between Montagne Noire and SW Sardinia consists in the lack of black shale intercalations in the upper part of the "Dolomia rigata" (Figs. 3 and 5).

Episode E4

Collapse of the Lower Cambrian carbonate shelf has modified sedimentation in a similar way as in Montagne Noire.

The main tensional event straddling the Lower Cambrian (Gonnesa Fm)/Middle Cambrian (Cabitza Fm) boundary appears to be synchronous with the E4 episode of Montagne Noire. In Sardinia it consists of a succession of tensional pulses gradually leading to the replacement of shallow-water carbonates by deeper-water terrigenous sediments.

As in Montagne Noire a first main tensional pulse has resulted in block faulting, followed by exposure and karstification of the uplifted blocks (Gandin 1987).

Subsequent tectonics pulses resulted in sinking of the platform and resumption of terrigenous input on the shelf. This event is marked by the abrupt unconformity of the Cabitza formation over the Ceroide limestone. A gradual transition exists only in the distal parts of the carbonate shelf. Pollution by "terra rossa" is widely expressed in the siliciclastic deposits of Cabitza Fm (Gandin 1987).

The rather abrupt disappearance of carbonates in the Cabitza Fm and the deposition of laminated, greenish shales and siltstones (distal turbidites) suggest a last tectonic pulse, as in Montagne Noire (Figs. 2 and 5).

3.2 Stratigraphic Distribution of Ore Deposits

The main characteristic of Sardinian lead-zinc ores is that most of them are located at the Lower-Middle Cambrian transition (Brusca and Dessau 1968; Boni 1985). Nevertheless it is possible to define several mineralized stratigraphic horizons (Fig. 5), from bottom to top as follows:

H1 coincide with the Punta Manna Member where several massive pyrite lenses occur. The deposits of Campo Pisano, Sedda-Modizzis and Genna Luas, located at the very top of the Nebida Formation (Brusca and Dessau 1968) belong to this group. Synsedimentary and/or syndiagenetic deformations affect most of these concentrations (Boni 1985). Lead-zinc geochemical anomalies or galenasphalerite occurrences are frequently associated with the massive pyritic lenses (Brusca and Dessau 1968). The deposit of Canalgrande (NW Iglesiente) appears to be mostly composed of iron-bearing sphalerite with minor galena (Boni 1985). It is located in the middle part of Punta Manna Member where limestones are dominant. Finally small stratabound occurrences of barite are common in southern and central Sulcis at the top of the Nebida Formation (Gandin and al. 1974; Gandin 1987).

H2 consists in rare massive lenses of lead-zinc sulphides with associated barite and pyrite, which occur within the black dolomite located at the transition between the Laminated Dolomite unit and the Ceroide Limestone (Fig. 5). These are essentially observed in the Monte San Giovanni mining district (Fanni and al. 1982).

H3 occurs in the uppermost part of "Metalliferro" (upper part of Gonnesa Fm) (Fig. 5) and at the transition with the overlying Cabitza Formation. It represents the most continuous and most economically important ore-bearing horizon of SW Sardinia (Brusca and Dessau 1968).

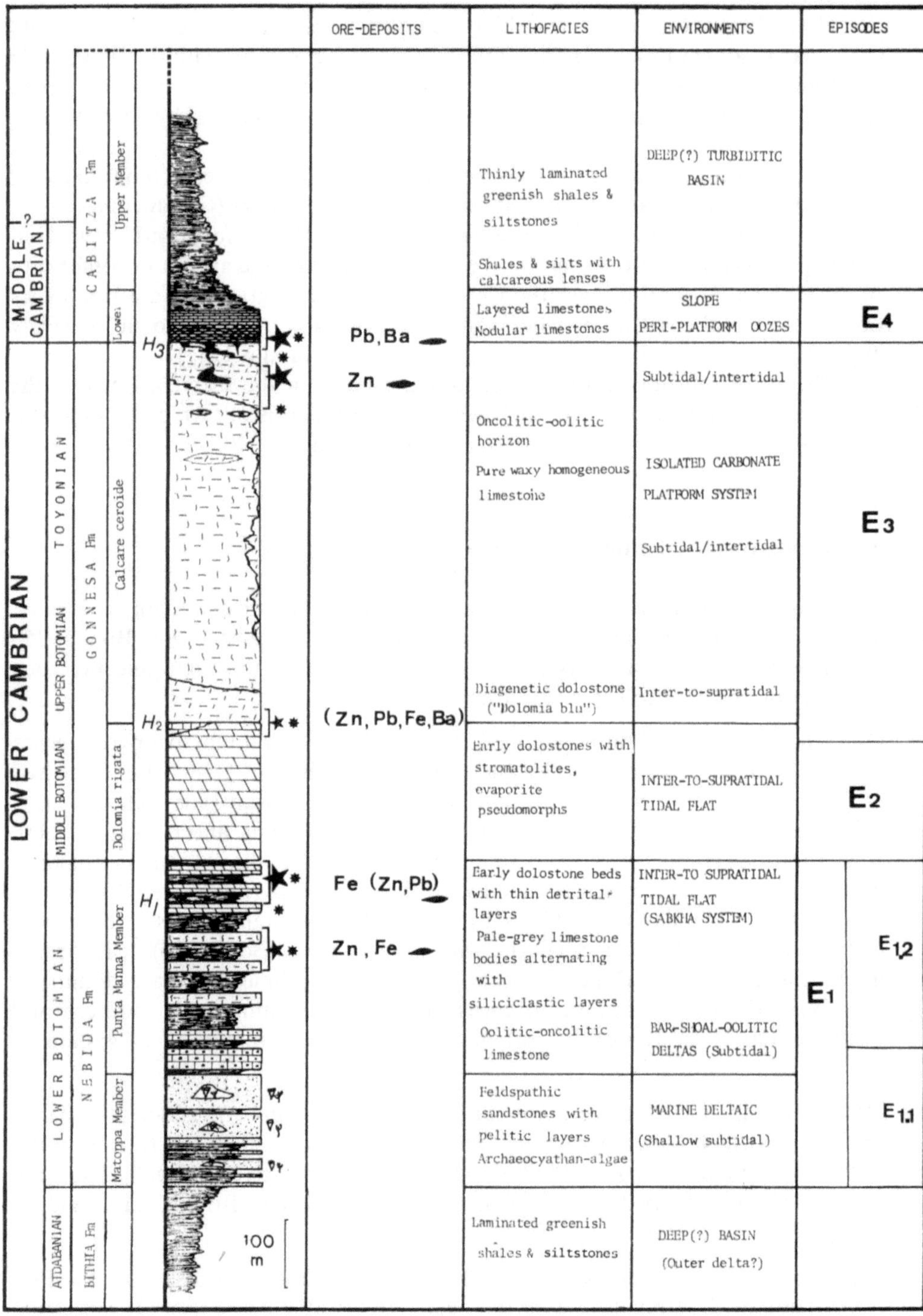

Fig. 5. Lower-to-Middle Cambrian sequence of Southern Sardinia (Iglesiente/Sulcis): depositional environments, tectono-sedimentary episodes (E1 to E4) and ore-deposits distribution. (For symbols see Fig. 3)

The deposits which are still mined (Monte San Giovanni, Monteponi and Acquaresi-Marx) are enclosed within this horizon. In the Monte San Giovanni Mine, regarded as the standard mineralized section, massive fine-grained sphalerite lenses first appear ("blendosi" ore: author) in the upper third of the Ceroide Limestone. Slightly argentiferous galena (up to 300 ppm; Brusca and Dessau 1968), associated with pure colloform barite in lenses and/or tabular beds ("Contatto" ore: author) is located at the top the Ceroide Limestone or locally at the base of the overlying mixed sequence (Cabitza Fm).

4 Geodynamic Setting and Metallogenic Relationships: Discussion

4.1 Comparison of Dynamic Evolutions in Montagne Noire and SW Sardinia

Biostratigraphic and sedimentological data indicate that the different tensional episodes were synchronous in the two regions (Courjault-Radé and Gandin 1986). Furthermore, a comparison of sedimentary evolutions throughout the Lower-Middle Cambrian shows that southwestern Sardinia was a more littoral depositional area than Montagne Noire (cf. Figs. 3, 5, and 6). Yet their precise relative position is not known, although Sardinia appears to have been closer to the Pan African Shield

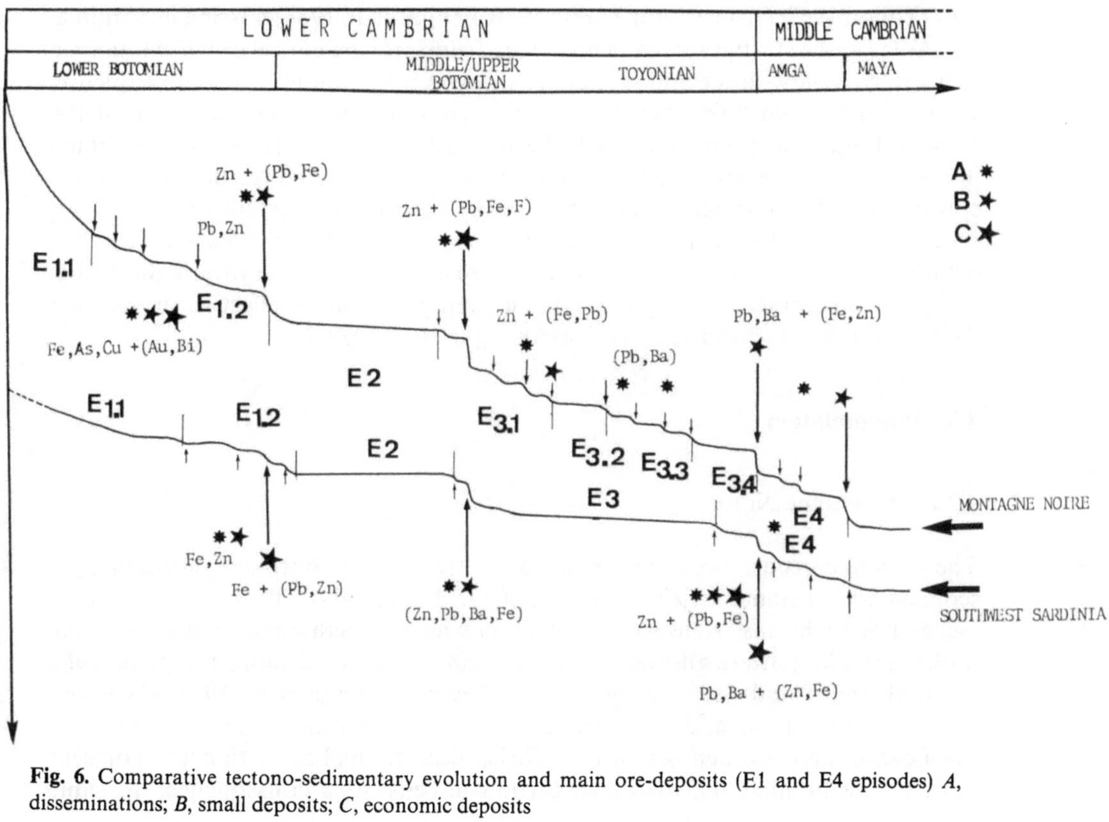

Fig. 6. Comparative tectono-sedimentary evolution and main ore-deposits (E1 and E4 episodes) *A*, disseminations; *B*, small deposits; *C*, economic deposits

(Courjault-Radé and Gandin 1986). The significant differences of depositional conditions noted during each episode, imply that Montagne Noire was also a more mobile area than southwestern Sardinia.

The tensional phases responsible for gradual sinking of the carbonate platforms (= E3 and E4 episodes) can be related to global tectonics affecting the Gondwana continental plate (Courjault-Radé and Gandin 1986). In fact, when considering other segments belonging to the Gondwana plate, such as Spain (Zamareño 1976), the Cevennes (Ortenzi 1986), Turkey (Dean et al. 1986), the Lower-Middle Cambrian transition appears to coincide with a major phase of instability, essentially resulting in the replacement of shallow-water carbonates by deeper-water deposits.

According to paleomagnetic data (Scotese et al. 1979), Sardinia and Montagne Noire were part of the eastern margin of the Cambro-Ordovician Protoatlantic ocean ("Iapetus"): early opening of this ocean, confirmed by the appearance of Lower Ordovician ophiolites (Dunning and Krogh 1985), could account for very active tensional tectonics during the Lower-Middle Cambrian.

4.2 Elements of Metallogenic Comparison

Metallogenic patterns in Montagne Noire and Sardinia reveal important differences. The volcanogenic, lead-zinc poor, gold and bismuth-bearing ores of the Salsigne area (Tollon 1969; Crouzet and Tollon 1980; Lépine et al. 1984) missing in Sardinia.

As far as the stratigraphically higher lead-zinc ores are concerned, in Montagne Noire they are low-grade and scattered throughout the whole carbonate sequence, whereas in SW Sardinia, they are *mainly* high-grade stocks concentrated at the Lower-Middle Cambrian boundary. The first-order parameter is broadly confirmed in both areas by some paragenetic persistences and by a spatial relation to the same episodes of instability along the stratigraphic column. The extent to which later processes have influenced these concentrations remains to be discussed [e.g., the deformation pattern: folds in Sardinia vs. thrusts and nappes in Montagne Noire; or karstic enrichment during the Triassic, apparently more effective in Sardinia (Boni 1985) than in Montagne Noire (Aubague et al. 1977)].

4.3 Interpretation

4.3.1 Montagne Noire

The lead-zinc occurrences are restricted to stratigraphic horizons reflecting syn-sedimentary/tensional tectonic episodes (E1-2, E3 and E-4). They are confined to "second-order basins" (sensu Large 1980) in which synsedimentary block-faulting took place. This pattern allowed metal-rich brines to circulate along marginal faults (evolved from Cambrian paleogeographic flexures: Lépine et al. 1984, Chen and Courjault-Radé 1986, and now seen as late-Hercynian strike-slip faults), inducing silicification and ore deposition (or geochemical anomalies) within the adjacent more or less lithified and disrupted carbonate beds. As a consequence, the same ore-bearing horizons, with a common Cambrian lead isotope signature (Brévart et al. 1982), are to be found in the different structural domains of the whole Hercynian belt.

4.3.2 Sardinia

The basic sequence of events looks very similar to that of Montagne Noire. The higher tonnages and grades of mineralization related to the E4-episode are probably explained by the greater availability of porous/brecciated rock volumes, involved in a convective hydrothermal system from which early (Boni 1985) to late diagenetic Pb–Zn ores were deposited. As in Montagne Noire, ore deposition was genetically related to the Lower-Middle Cambrian tectono-sedimentary evolution. A Cambrian age for mineralization is indicated by lead isotopic data (Boni and Köppel 1985).

Acknowledgements. This research was supported by EEC *Contract MSM-128-F*. We thank Dr. Boissonnas, Prof. Cocozza and Prof. Omenetto for critically reading the manuscript.

References

Arthaud F, Matte (1977) Synthèse provisoire sur l'évolution tectonique et les raccords entre les segments hercyniens situés autour du bassin Nord-Baléares (Sud de la France, Espagne, bloc Corsosarde). In: Chaîne varisque d' Europe moyenne et occidentale. Colloque interne CNRS, Rennes, 243:497–513

Aubague M, Orgeval JJ, Soulié M (1977) Les gîtes minéraux de la terminaison méridionale du Massif Central et sa bordure languedocienne. Bull BRGM (2) II, 3:333–363

Boni M (1985) Les gisements de type Mississippi valley du Sud-Ouest de la Sardaigne: une synthèse. Chron Rech Minière 479:7–34

Boni M, Köppel V (1985) Lead isotopic pattern from Iglesiente and Sulcis (SW Sardinia). The problem of remobilization of metals. Min Deposita 20:185–193

Boyer F (1962) Successions caractéristiques et niveaux repères dans le Paléozoïque de la région de Carcassonne à Saint Pons (Montagne Noire, Aude, Hérault). Bull Soc Géol Fr (7) IV:572–575

Boyer F, Routhier P (1958) Observations sur deux niveaux minéralisés dans le Paléozoïque inférieur des Monts du Minervois (Montagne Noire, Aude). Bull Soc Géol Fr (8):257–266

Brévart O, Dupré B & Allègre CJ (1982) Metallogenic provinces and the remobilization process studied by lead isotopes: lead-zinc ore deposits from the Southern Massif Central, France Econ Geology 77, 3:564–575

Brusca G, Dessau G (1968) I giacimenti piombo-zinciferi di S Giovanni (Iglesias) nel quadro della geologia del Cambrico sardo. Industria Mineraria 19:1–53

Chen SL, Courjault-Radé P (1986) Apports de la télédetection à compréhension du contexte métallogénique de la Montagne Noire (Massif Central). 11ème RST, Congr Clermont-Ferrand, Abstract, 35

Courjault-Radé P (1984) Modalités d' installation de la plateforme carbonatée Cambrien inférieur de la Montagne Noire (Massif Central, France). IAS IV Reg Congr, Marseille, Extended Abstract, pp 122–124

Courjault-Radé P (1985) Comparaison de l'évolution des séquences du Cambrien inférieur et moyen dans les versants sud et nord (unité de Brusque) de la Montagne Noire (Massif Central). CR Acad Sc, Paris, t 301, II, 1:43–48

Courjault-Radé P, Gandin A (1984) Sedimentary evolution of the Cambrian sequence in Montagne Noire (France) and Sardinia Italy: first tentative correlation. Intern Ass Sediment, IV Reg Congr Sedim Marseille, Extended Abstract, pp 124–125

Courjault-Radé P, Gandin A (1986) Comparative Early Paleozoic geodynamic evolution in Montagne Noire (France) and Sardinia (Italy). 12th IAS Congress, Camberra, Abstract, 63

Courjault-Radé P, Tollon F (1985) Economic ore concentration in a tenonal tectono-sedimentary context-An example from Lower Cambrian of Southern Montagne Noire (Salsigne mine district-Pardailhan/Minervois nappes). Fortschritte der Mineralogie, 63, 1, Abstract, 46

Crouzet J, Tollon F (1980) Le gîsement stratiform e et filonien de Salsigne, Aude-Au, As (Ag, Cu, Bí). 25 th International Geological Congress, Paris. Gîsements français, 54 p.

Dean WT, Monod G, Gunay Y (1986) Lower Paleozoic stratigraphy in the southern and central Amanos Mountains, South Central Turkey. Geol Mag 123(3):215–226

Demange M, Issard M, Perrin M (1986) Rapports entre la zone axiale de la Montagne Noire et les nappes du versant sud au sud-ouest du massif (Minervois et Cabardes-Aude, Hérault). Bull BRGM, Géologie de la France, 3:281–298

Doré F (1977) L'europe moyenne cambrienne, les modèler sédimentaires, leur zonalité, leurs contrôles. In: La chaîne varisque d'Europe moyenne et occidentale coll. intern. CNRS, Rennes; n° 243, (1977), pp 143–155

Dunning GR, Krogh TE (1985) Geochronology of ophiolites of the Newfoundland Appalachians. Can J Earth Sci 22:1659–1670

Fanni S, Gandin A, Grillo SM, Lippi F, Marras G, Salvadori G, Tocco S (1982) La piattaforma carbonatica cambrica nella Sardegna sud-occidentale: sedimentazione e deposizioni metallifere. Mem Soc Geol It 22(1981):123–138

Gandin A (1987) Depositional and paleogeographic evolution of the Cambrian in South-Western Sardinia. In: IGCP Project n° 5, Sassi FP (ed) Newsletter 7

Gandin A, Padalino G, Violo M (1974) Correlation between sedimentation environment and ore-prospecting. Sedimentological and ore genesis studies of "Cambrian arenarie" and "dolomia rigata" formations (Sardinia, Italy): deposition and concentration of barite in an evaporitic environment. Rend Soc It Min Petrol 30:251–303

Gèze B (1949) Etude géologique de la Montagne Noire et des Cévennes méridionales. Mem Soc Géol Fr 29(62):1–215

Large DL (1980) Geological parameters associated with sediment-hosted exhalative Pb–Zn deposits: an empirical model for mineral exploration. Geol Jb D 40:59–129

Lépine J, Courjault-Radé P, Crouzet, Tollon F (1984) Présence d'une zone haute au Cambrien inférieur dans le secteur de Salsigne (versant sud de la Montagne Noire). Manifestations volcaniques et hydrothermales associées et conséquences métallogéniques. CR Acad Sc, Paris 279 II 7:347–350

Lépine J, Crouzet J, Talayssat M, Tollon F (1986) Les minéralisations cambriennes sulpho-arseniées aurifères de la mine de Salsigne (Aude). Coll PIRSEM Facteurs de concentration de: matières premières minérales. Montpellier, Résumé, 63

Ortenzi A (1986) Proposition d'un modèle paléogéographique des Cévennes méridionales au Cambrien. Corrélations avec la Montagne Noire CR Acad Sc Paris, 303 II 11:1029–1034

Scotese CR, Bambach RK, Barton C, Van Der Voo R & Ziegler AM (1979) Paleozoic base maps Geol 87:217–272

Zamareno I (1976) Tipos y distribucion de facies en el nivel carbonatado del cambrico de Espana. In: Geologica de la parte norte del Macizo Iberico Vol homenaje I. Parga Pondal, Guadernos del seminario de estudio ceràmicos de Sargadelos, (1978), pp 289–311

Various Types of Cambrian Carbonate Hosted Zn–Pb Mineralization in the Northern Montagne Noire, Massif Central, France. Ages and Mechanisms of Concentration

J.L. Lescuyer[1], D. Giot[2], M. Donnot[2], and P. Beziat[3]

Abstract

Various types of lead-zinc concentrations from the Cambrian of the northern side of the Montagne Noire are discussed and situated in their paleogeographic environment.

Our study confirms the presence of early Zn–Pb mineralization which is related to the structuration of the Asturian-Sardinian platform and its northern unstable margin. The latter is characterized in the lower Cambrian by periodic volcanic activity and alternating open sea and confined sedimentation environments ("black series"). Stratiform zinc-dominated mineralization of probable hydrothermal sedimentary origin (as exemplified at Brusque and Lardenas) was emplaced at various levels of the black shaly-calcareous formations.

The southern part of the platform displays more classical types of lead-zinc concentrations in a shallow carbonate environment where breaks in the sedimentary cycles and diagenetic evolution played a major role, as demonstrated by the comparative study of the southern slope of the Montagne Noire and of SW Sardinia (see Courjault-Radé and Gandin, this Vol.).

Other lead-zinc concentrations of economic importance appear to be related to late Hercynian magmatic (La Rabasse) and tectonic (St. Salvy)activities. For instance, the La Rabasse mine may be classified as a carbonate replacement deposit showing a complex Fe-As-Zn–Pb (Ag–Au–Sn–Bi) paragenesis, and which was emplaced in Lower Cambrian barren dolomites at the periphery of a subvolcanic calc-alkaline intrusion.

Contrary to the nearly Cevennes region, the Mesozoic cover in the Montagne Noire does not contain notable lead-zinc concentrations. The only small occurrences observed with Si, Ba and silver-bearing tetrahedrite appear to be related to the Triassic-Liassic distension.

---

[1] Direction des activités minères, Département Gîtes minéraux B.R.G.M., BP 6009–45060 Orleans Cedex

[2] Service géologique national, Département Carte géologique et géologie générale B.R.G.M., BP 6009–45060 Orleans Cedex

[3] Direction des activités minières, Direction locale Sud-Ouest, B.R.G.M. Cité Lameilhé–Immeuble Dampierre–51100 Castres

Mineral Deposits within the European Community
(ed. by J. Boissonnas and P. Omenetto)

1 Introduction

1.1 Objectives and Methodology

This study describes the controls of lead-zinc and associated mineralization in part of the Cambrian-Ordovician province of Asturia-Sardinia (Fig. 1 in Courjault-Radé and Gandin, this Vol.), with particular emphasis on possible mechanisms of ore concentration. It is restricted to the northern side of Montagne Noire, whereas Courjault-Radé and Gandin focused their investigations on the southern side and on Sardinia.

Earlier authors (Aubague et al. 1977, Beziat 1977, Courjault-Radé 1982, Donnot and Milesi 1982) advocated early metallic concentrations in some horizons of the Cambrian. Based on these ideas, an attempt is made here to correlate the paleogeographic evolution of Montagne Noire during the Lower Cambrian with the history of Pb–Zn concentrations hosted in these formations. Two specific approaches are used:

— a study of the Cambrian sedimentation environment in the various structural units situated on the north side of Montagne Noire (e.g., the units of Avene-Mendic, Melagues, Brusque, Merdelou-Lacaune and Ouyre-Barre; Fig. 1);
— a metallogenic study of selected lead-zinc deposits of apparently very diverse types (La Rabasse, Brusque, Lardenas).

1.2 Geological Setting

The Montagne Noire forms the southwestern tip of the French Massif Central (Fig. 1). In this southern area of the Hercynian belt, thick weakly metamorphosed series of the Paleozoic are folded and overthrust on either side of an axial zone composed of high grade metasediments and orthogneisses.

On both sides of the axial zone, the series begin with ill-dated [Upper Proterozoic (?) to Lower Cambrian] fine-grained detrital sediments, near the top of which acid to intermediate volcanics occur locally. Thick Lower Cambrian calcareous sequences develop continuously on the south slope (see Courjault-Radé and Gandin, this Vol.) and in the southeasternmost thrust slices of the north side. To the north, farther away from the axial zone, shaly-calcareous facies called black series are distinguished in stratigraphic equivalence of the carbonates. Widespread clayey-sandy detrital sedimentation re-occurs in the Middle Cambrian and extends during Ordovician times. Sedimentation continues locally up to the Visean on the south side, but on the north side the last exposed sediments are Silurian in age.

The present structuration of the Montagne Noire was mainly acquired during the Hercynian orogenesis. Synkinematic granites of the axial zone are dated at 330 ma. The Paleozoic on either side is intensely deformed: on the southern side, large nappes are piled up over a sub-autochtonous unit termed schistes X, whereas in the sediments of the northern side, six distinct structural units were formed as a result of synschistose folding and subsequent thrust faulting towards the S.E. (Fig. 1).

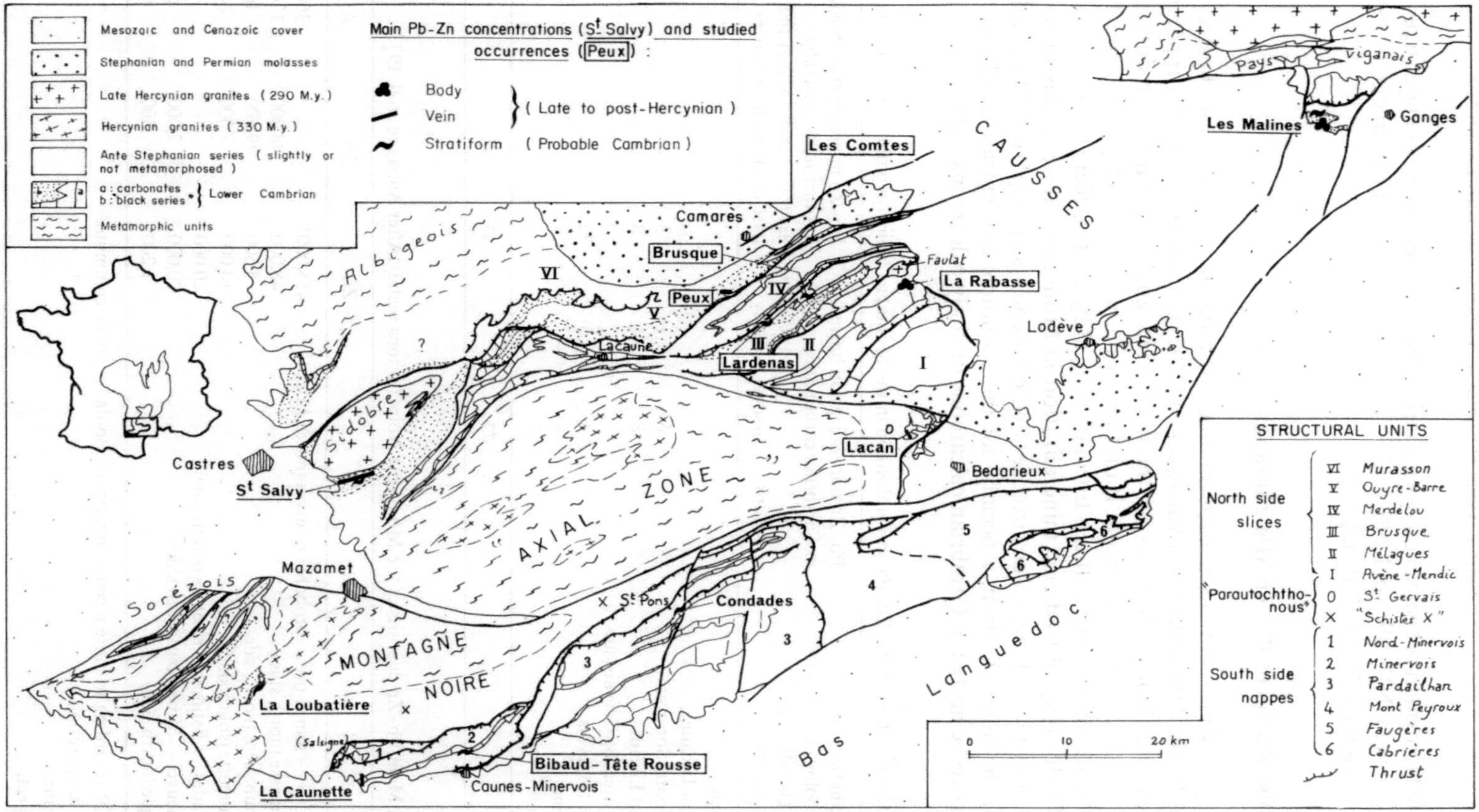

Fig. 1. Montagne Noire and Southern Cevennes: geological situation and location of the study areas. (After Aubague et al. 1977; Donnot 1982)

Late Hercynian orogenic activity is marked by the reactivation of major wrench faults oriented NE–SW and NS and is accompanied by emplacement of calc-alkaline granites (285 to 290 ma).

1.3 Distribution of Pb–Zn Mineralization

Montagne Noire contains a large number of mineral concentrations hosted in various levels of the Paleozoic (Aubague et al., 1977). Several domains are apparent: tungsten and fluorine domains on the north side, gold and manganese domains on the south side, and a zinc and lead domain on both sides. The latter is related to the "lead-zinc belt of southern Europe" defined by P. Routhier (1980).

The diversity of lead-zinc mineralization in Montagne Noire is illustrated in Table 1 in which all known occurrences and ore deposits are reported. The greater part of the mineralization, especially all the economic concentrations (Table 2), appears to be hosted in Cambrian calcareous or shaly-calcareous facies.

Table 1. Distribution of Pb–Zn (Ba) mineralization in the Montagne Noire (in number of occurrences)

Areas \ Schematic typological classification	Pb–Zn mineralization hosted in Cambrian carbonates			Pb–Zn mineralization hosted in other units: generally fissural or stratiform at the base of the mesozoic cover
	stratiform	fissural	massive body	
North side				
West of Lacaune, Sidobre, Sorezois	8	6	1	3
East of Lacaune	5	8	1	8
Eastern area: St-Gervais-sur-Mare, Lodévois	1	4	5	22
Axial zone	—	—	—	11
South side	29	7	2	3
Total		77		47

Table 2. Main Pb–Zn deposits of the Montagne Noire (in tons metal) (After Aubague et al. 1977)

	Zn	Pb	Ag
[b] St. Salvy — mined ore + probable reserves (north side)	455,000[a]	—	360
[c] La Rabasse (north side)	48,000	48,000	120
[c] La Loubatière (north side)	5,500	49,500	10
[c] La Caunette (south side)	14,000	7,000	84
[d] Condades — Estimated reserves (south side)	20,000	—	—
[c] Villeneuve — Minervois (south side)	20,000	2,000	2
[c] Brusque (north side)	1,200	300	0.3

[a] 575,000 according to more recent estimates (J. Bouladon, pers. commun).

[b] Mine in operation.

[c] Old mine.

[d] Prospect.

As last noted by Courjault-Radé and Gandin in this Volume, the economic potential (+/−0.7 Mt Pb–Zn) is appreciably lower than in similar series of southwest Sardinia (±8 Mt metal).

2 Paleogeographic Setting of Lower Cambrian Deposits and Early Pb–Zn Mineralization

The study area is located on the northern margin of a vast platform which, during the lower Cambrian, was part of the Asturian-Sardinian province extending over northern Spain, the Pyrenees, the Montagne Noire and Sardinia (see Fig. 1 in Courjault-Radé and Gandin, this Vol.).

This series which are exposed on the north side of the Montagne Noire were investigated by means of reference cross-sections (Fig. 2), mainly in the east area where the main sedimentary variations were observed.

The southern structural units of the north side (I and II in Fig. 2), which are composed of very thick and rigid calcareous deposits, represent undeformed relative autochthonous formations corresponding to the platform domain. Northeastwards, although more intensely deformed, the structural units which correspond to the platform margin show the development of a volcanic arc (Merdelou ridge, structural unit IV) which acted as a high zone behind a confined subsident trough (Donnot and Guérangé 1978; Gachet 1983). In the western area, although less intensively marked, a similar arrangement of troughs and ridges accounts for the development of alternating dark sulfide-bearing calcareous and clayey facies displaying external platform characteristics. The northern Albigeois zone, which is not included in this study, represents a distal marine domain.

2.1 Identification of Paleoenvironments and Paleogeographic Reconstruction

2.1.1 Sedimentological Observations

The Cambrian calcareous facies underwent intense polyphase diagenetic recrystallization in the eastern area and metamorphism in the western area. Among the relicts of sedimentary structures identified, gray, stratified, and very fine-grained dolomites (dolomicrite and dolomicrite-sparite) often display indications of shallow environments: algal mats, stromatoliths, oncolites, intraclastic breccias, birdseyes, and pseudomorphs after sulfates. Undestructive dolomitization suggests early transformation in a tidal to supratidal environment. The massive sparitic dolomites have practically never revealed preserved sedimentary structures. However, relicts of cross-bedding outlined by a system of lined-up dissolutions are indicative of former calcarenite deposits believed to correspond to a barrier-setting. Oolithic formations were preserved locally in particular silicified facies.

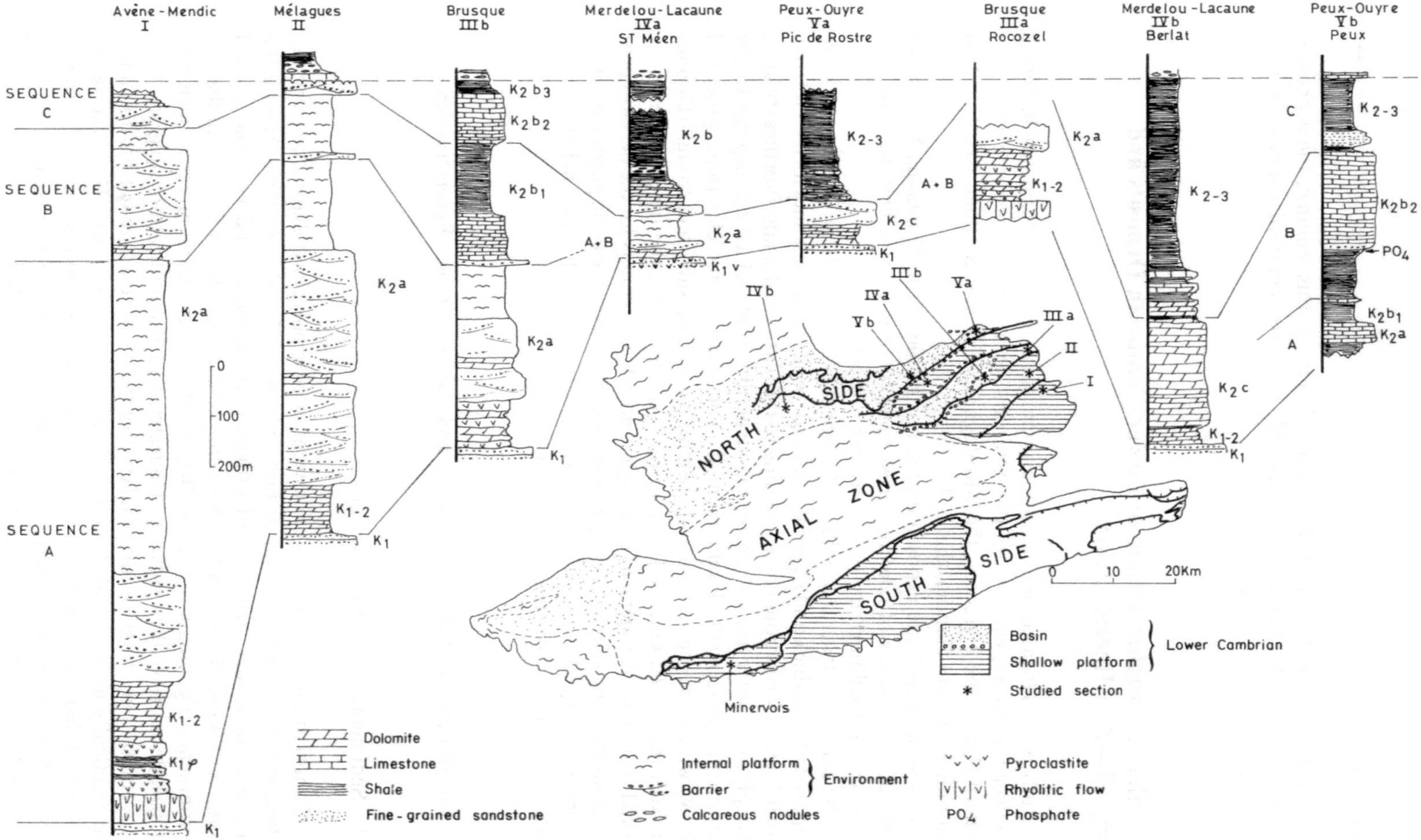

Fig. 2. Sequential evolution observed in the Lower Cambrian (K_2) calcareous succession on the north side of the Montagne Noire

Characterization of proximal to distal environments is generally inferred from lack of evidence of tidal environment. In the most favorable cases, bioclastic elements, particularly crinoids or phosphate grains, clearly indicate open marine conditions. Sediments generally display a regular rhythmic succession shown by decimeter to meter-thick bedding, common clayey-silty intercalations, and micro-bedding generated by sorting of fine detrital particles. Except for dedolomitization limestones, which are observed locally in particular in the Brusque unit and are of late diagenetic origin, the currently preserved dark color is always associated with the open sea domain.

Dolomitization also developed at the expense of marine limestones, generally in the form of dark bedded sparites, which can be easily distinguished from light-colored sparites recorded in the barrier domain and gray dolomicrites in the internal platform domain.

2.1.2 Sequential Interpretation and Geodynamic Consequences

After identification of deposits in terms of sedimentation setting, the vertical successions have been interpreted in terms of sequences (Fig. 2). The latter illustrate either a regressive evolution towards internal platform environments, or a transgressive evolution towards deep marine environments.

The fine-grained detrital and volcano-detrital facies (K_1) underlying the studied series were emplaced by rhythmic settling or by laminar currents and are locally affected by slumping. Carbonate sedimentation occurs progressively as intercalations in detrital or volcano-detrital material in an open sea environment. These alternations (K_{1-2}) are considered to be the base of sequence A, the lower member of which was deposited in a proximal infratidal environment, as indicated by the presence of intraclasts, oolites, and reworked volcanoclastic material. The overlying massive calcareous facies (K_2a), deposited in continuity, show at their base light coarse-grained dolomites assigned to the high energy platform zone. At their top, they grade to algal internal platform deposits which mark the end of the regressive sequence A. This succession, identified in the southeastern structural units I and II of the Montagne Noire north side, is more difficult to distinguish in the northern units where deeper marine conditions were prevailing.

A second regressive sequence, called sequence B, has been distinguished. Only sediments from barrier and internal platform environments developed in units I and II, thereby emphasizing the perenniality of the platform. In the open-sea domain, a chronostratigraphic equivalence is proposed with a sequence composed of black shales (K_2b_1) overlain by bedded limestones locally phosphatic at the base (K_2b_2).

On the Montagne Noire south side, in the Minervois region, the Lower Cambrian series, although reduced in thickness (400 m), may well be represented by the two superimposed sequences, A and B.

In the Brusque unit, an anomal sedimentary evolution, characterized by the reversed polarity of sequence B, is probably related to the development of a local subsident trough.

The third sequence, C, shows a radical change in sedimentary polarity. At its base, it displays intertidal deposits (calcarenite) on the platform and ridge and infratidal limestones or sandstones elsewhere, grading upwards to basin calcareous or clayey facies. Transgression is irregular on the platform: erosional unconformities (sedimentary gaps) occur in several places, as well as facies heterogeneity indicating structural instability linked to the subsidence of the platform.

The end of carbonate platform conditions is characterized by the deposit of trilobite-bearing nodular shales which marks the beginning of the extensive Middle-Cambrian detrital sedimentation.

If compaction rates and inferred bathymetry are taken into account to estimate subsidence, it is possible to reconstruct the geodynamic history of the Lower Cambrian platform in terms of sequential distribution (Fig. 3).

During the deposition of sequence A, the volcanic ridge of Merdelou was a major structural high. It served as a boundary for a probably deep marine northern domain in which sedimentation was reduced (about 400 m thick). In the southeast, the strong subsidence was compensated by high productivity of carbonates under shallow water and resulted in the building of a 1000-m-thick platform.

During the deposition of sequence B, the ridge was still relatively stable. In the open sea domain, subsidence accelerated simultaneously with sedimentation rate (500 to 600 m). The major phenomenon occurred south of the ridge: whereas the carbonate platform remained stable with a reduced subsidence of 250 m, a very subsident trough developed (Brusque unit) with sedimentation rate comparable to that of the basin environment. This trough appears to be asymmetrical: it passes into the platform northeastwards and is probably open southwestwards towards the open-sea domain.

Such troughs, developed at the margin of a platform in the immediate vicinity of a probably still active volcanic arc (Donnot and Guérangé 1978), appear to be controlled by synsedimentary tectonic activity. This structural environment is believed to be directly correlated to the emplacement of stratiform Pb–Zn mineralization on the north side of Montagne Noire.

During the deposition of sequence C, subsidence contrasts between the different paleogeographic domains become progressively indistinct due to the general sinking of the platform.

2.2 Examples of Early Zn–Pb Mineralization on the Northern Side: the Brusque Deposit and the Lardenas Occurrences

Stratiform Pb–Zn concentrations which are associated with black shales and limestones corresponding to external platform domains are known in the Brusque syncline (structural unit III) and in the St. Salvy mine environment in the northwest of the Montagne Noire (structural unit V).

→

Fig. 3. Interpreted paleogeographic evolution of the Lower Cambrian platform (northern margin)

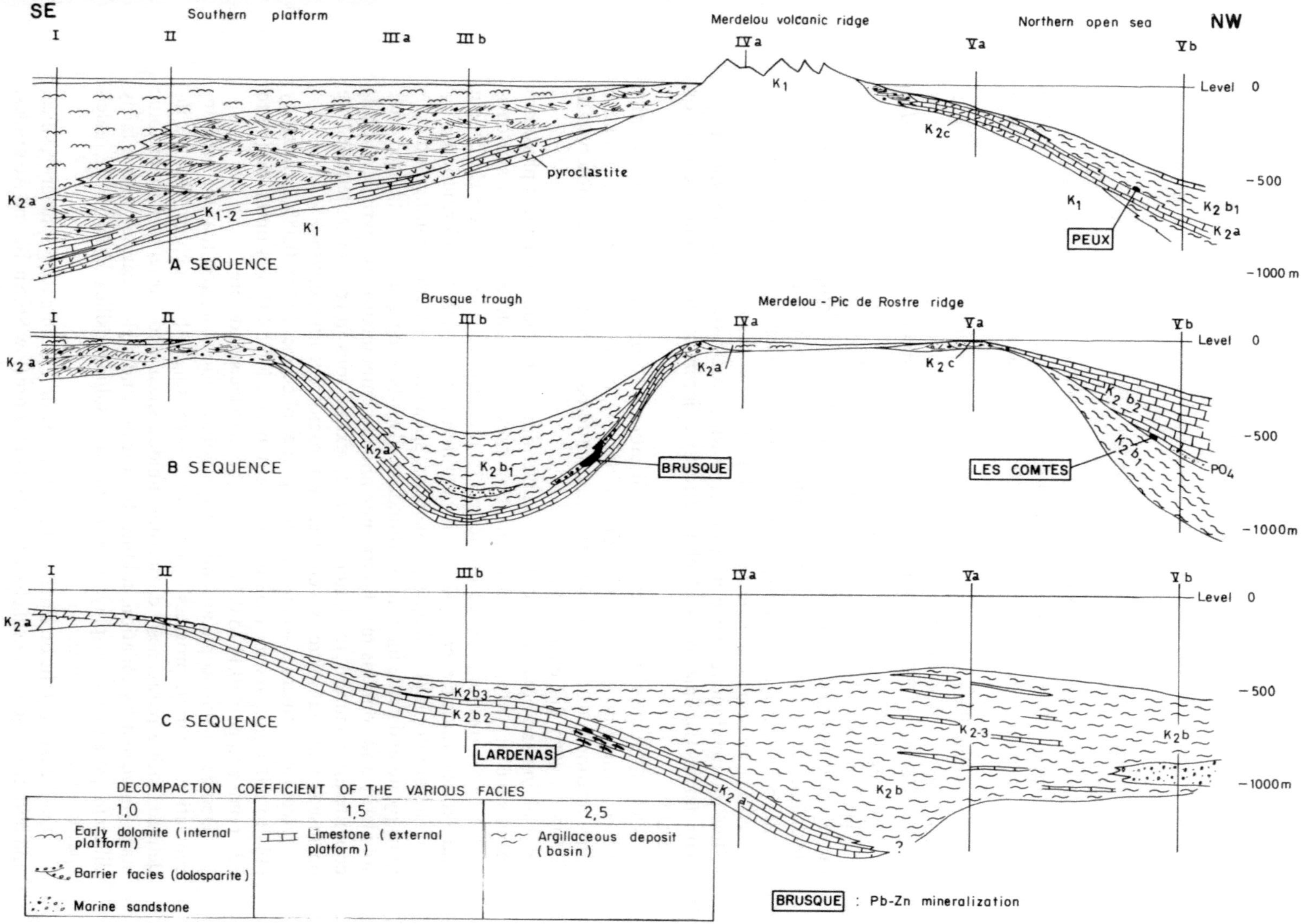

SE
Southern platform
Merdelou volcanic ridge
Northern open sea
NW
I
II
III a
III b
IV a
V a
V b
Level 0
-500
-1000 m
K1
K2a
K1-2
K2c
K2 b1
pyroclastite
PEUX
A SEQUENCE
Brusque trough
Merdelou - Pic de Rostre ridge
K2 b2
PO4
BRUSQUE
LES COMTES
B SEQUENCE
K2b3
K2b2
K2b
K2-3
LARDENAS
C SEQUENCE
DECOMPACTION COEFFICIENT OF THE VARIOUS FACIES
1,0
1,5
2,5
Early dolomite (internal platform)
Barrier facies (dolosparite)
Marine sandstone
Limestone (external platform)
Argillaceous deposit (basin)
BRUSQUE : Pb-Zn mineralization

Two examples have been selected in the Brusque syncline on the basis of their well-known paleogeographic position.

2.2.1 Brusque Deposit

The mine is located west of the village of Brusque (Fig. 4) on the left bank of the Sanctus brook. Mining works of limited extent are still easily accessible. Mining was interrupted around 1928; about 1200 t of zinc and 300 t of lead have been exploited.

Despite swelling and constrictions, the ore layer is remarkably continuous (Fig. 5). The siliceous ore steadily follows the contact where limestone (K_2a) is thrust over black shales (K_2b_1). The ore layer is visible over a distance of about 100 m and is at most 4 m thick.

The host rock of the ore layer underwent intense Hercynian tectonics. However, the original internal sedimentary organization has not completely disappeared. Limestones at the hanging wall of the mineralization are very finely bedded: microsparitic beds displaying crinoid ghosts alternate with dark beds containing organic matter. The additional presence of a detrital microquartzose phase, associated with yellow fractured microsphalerite and pyrite which are disseminated in small beds, characterizes the limestones which lie directly adjacent to the mineralized layer or laterally.

Despite intense subsequent recrystallization, the petrographic study permits to assign these peripheral sulfide disseminations, as well as an episode of barite concentration (subsequently silicified) and scattered microsilicification observed in the ore, to early diagenetic phases.

Folding and brecciation of the ore layer during the main Hercynian phase are responsible for two episodes of silicification, polymorphic dolomitization, synschistose calcsparitization, and complete recrystallization of sulfides. Late Hercynian and Triassic-Liassic events played only a minor part in these transformations.

The ore displays a simple paragenesis: sphalerite is predominating and contains pyrite and rutile inclusions; abundant silver-bearing galena crystallized after sphalerite and contains rare bournonite and electrum inclusions. Pyrite and chalcopyrite are common, the latter occurs as exsolutions in sphalerite; tetrahedrite enclosed in galena is rare as well as early automorphic arsenopyrite.

Accessory minerals account for low grades of copper (0.1 to 0.38%), antimony (484 to 661 ppm), arsenic (252 to 333 ppm), bismuth (158 to 194 ppm), tin (59 to 333 ppm) and gold (0.11 to 0.38 ppm) in the massive sulfide ore.

Detailed macroscopic and microscopic observations suggest that the ore concentration occurred progressively from an early stage, the epoch of which is ill-defined, with a paragenesis displaying sulfides, silica, and barite, to a syntectonic stage (indicated by foliation, folding, fracturing, and stylolitization) with a fairly different paragenesis displaying recrystallized sulfides, silica, dolosparite, calcisparite, and accessory barite.

On the basis of the originally simple composition (dominating sphalerite, silica, and barite) and geochemical characteristics of the ore (low Sb, Sn, Bi, and Au grades)

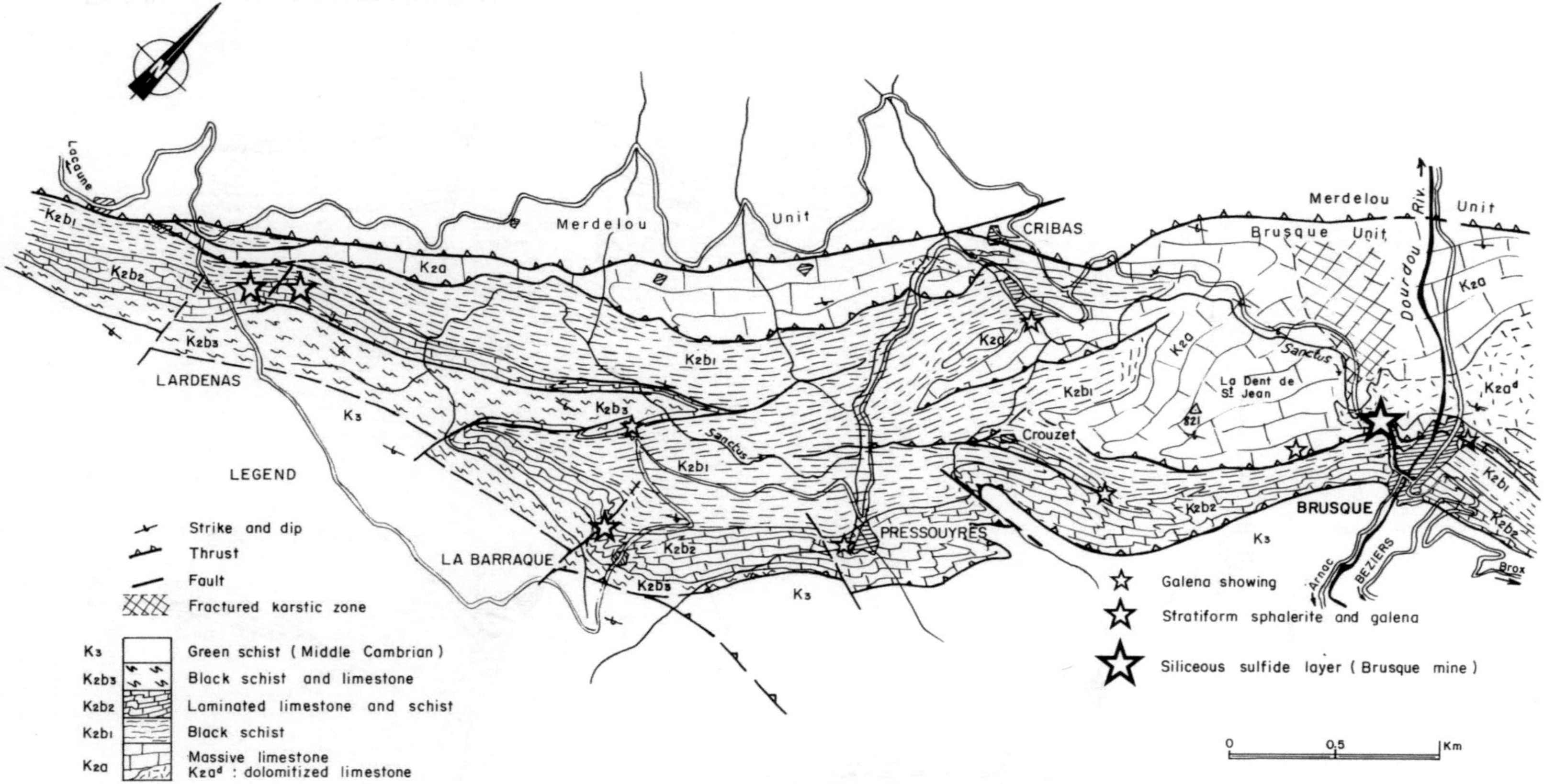

Fig. 4. Zn–Pb showings in the Lower Cambrian of the Brusque structural unit

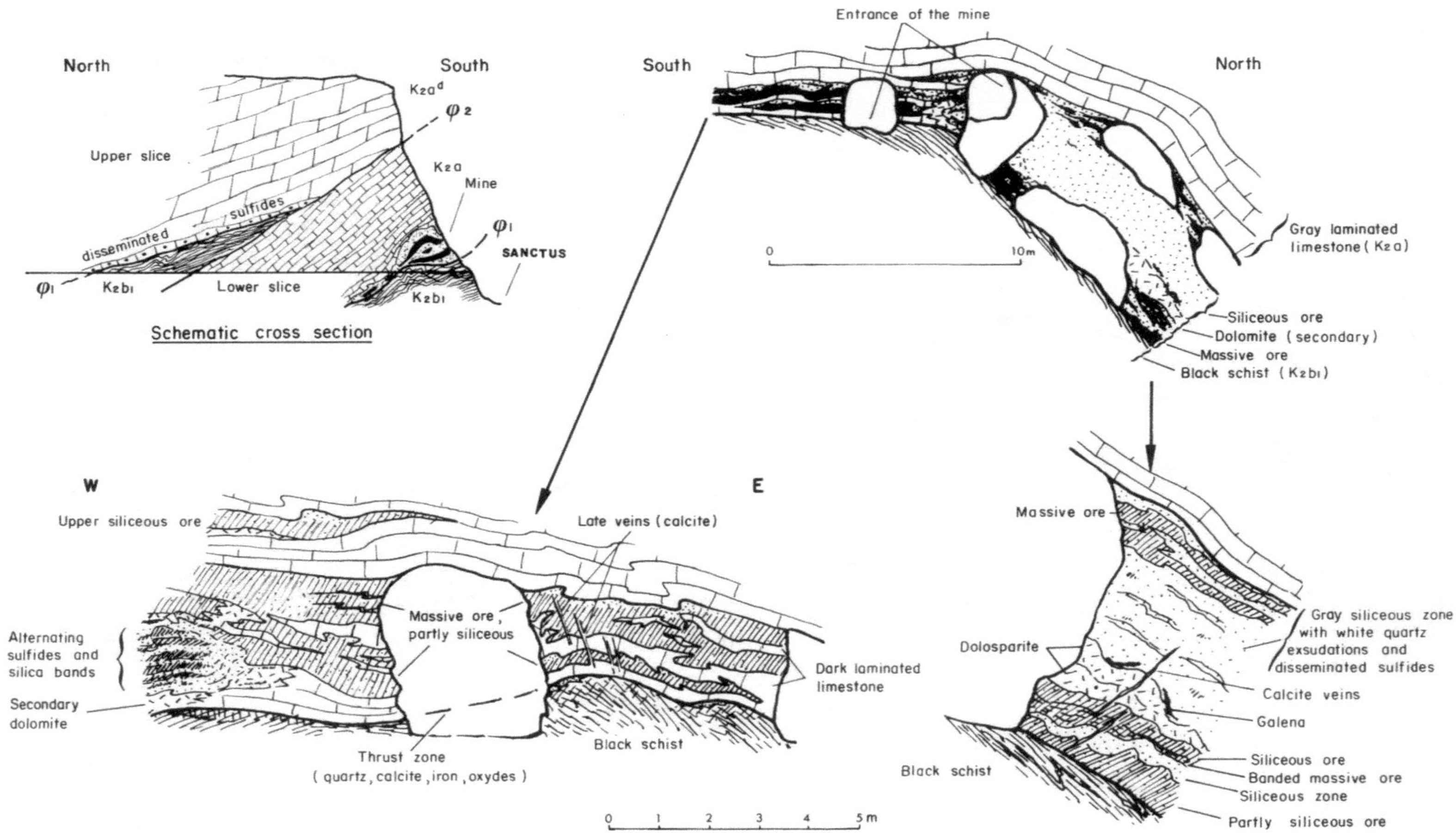

Fig. 5. Brusque mine: details of the mineralized layer

a hydrothermal sedimentary origin of possible Lower Cambrian age is proposed for the Brusque mineralized layer. Our model is based on that suggested by G. Pouit (1978) for some Paleozoic zinc-bearing mineralization in the Pyrenees. This hypothesis is supported by the paleogeographic and paleostructural position of the Brusque trough, in which the mineralization was early emplaced (Fig. 3 and Sect. 2.1.1).

At La Loubatiére, on the western side of the Montagne Noire and at Les Malines, in the southern Cevennes (Lacerda and Bernard 1984), some stratiform Pb–Zn concentrations exhibit the same characters and environment as those of the Brusque deposit and might derive from a similar mineralizing process.

2.2.2 Lardenas Occurrences

The Lardenas disseminated zinc-bearing mineralization located in the southwest of Brusque, in the same structural unit (Fig. 4) show important genetic affinities with the Brusque deposit.

The Lower Cambrian (Fig. 2) is chiefly represented here by the black series, composed of pyrite-bearing ampelite shales (K_2b_1), laminated limestones and calc-schist (K_2b_2) hosting the mineralization, and calcareous pelites (K_2b_3). The lower calcareous facies (K_2a) has been extensively reduced, partly by overthrust tectonics. The paleogeographic conditions are the same as those at Brusque, but the trough appears to be more marked.

Gray platy limestones and calc-schists which correspond to the base of K_2b_2 show over 30 m thickness about ten small stratiform mineralized bands, each corresponding to a few-millimeter-thick bed of microsphalerite associated or not with fine-grained galena and pyrite. The mineralized showings were observed in the field over 500 m, but the geochemical anomaly extends over 5 km. Considering the other anomalies and showings detected at the same stratigraphic level, the overall mineralized area before folding was 13 × 5 km.

The platy limestone host rocks show crystalline facies rich in crinoids and silty quartz as well as small phosphate nodules (Vallier, pers. commun.). Organic matter is commonly concentrated in laminae and stylolites. The sedimentary bedding is intensely affected by foliation which generated the common development of almond-shaped calcisparitic and quartzose crystallizations, locally with galena reconcentration.

The reddish microsphalerite mineralization is disseminated in sparite laminae, corroded by sparitization and microfissured. It occurs as an early-formed constituent of the sediment.

The mineralogical study has shown that sphalerite may exhibit chalcopyrite and pyrrhotite exsolutions and less common fine-grained disseminated galena. Pyrite and marcassite are rare. Neither early silicification or dolomitization processes have been observed, nor sulfates, thereby confirming the great poverty of the paragenesis.

The geochemical study has corroborated the above observations: Zn and Pb grades are low and only traces of Ag (1 ppm), As (120 ppm) and Mo and P were revealed.

The stratiform mineralization hosted in the black series are generally believed to be genetically associated with a *confined environment sedimentation.* This is well illustrated by the ampelite shales K_2b_1 which display a geochemical anomaly of some hundred ppm Zn + Pb.

However, the only noticeable concentrations are located at the hanging wall (as at Brusque) and at the footwall (as at Lardenas) of calcareous formations enclosing the black shales K_2b_1. Sedimentation conditions are the same in both sites: a distinctly infratidal (probable distal) marine environment. Moreover, striking similarities have been recorded between the Lardenas mineralization and the disseminated facies observed on the edges of the Brusque ore layer.

Consequently, common genetic hypotheses are put forward:

— either preconcentration in the black shales K_2b_1, followed by migration, then diagenetic fixation (which was not petrographically detected) in the calcareous layers at the footwall and hanging wall;
— or the same model as at Brusque with a probable distal, exhalative-sedimentary system occurring in an active tectonic setting.

The intensity of the subsequent tectonic activities makes it difficult to decide for either of the two early concentration models. The origin of the extensive stratiform Zn–Pb–Fe mineralization discovered at various levels of the black clayey, sandy calcareous unit (K_2b) in the host rocks at the St. Salvy mine is also not clearly established (Beziat 1977).

3 Late Pb–Zn Concentrations

The examples studied in the previous chapter have shown that the major Hercynian tectonic phase responsible for the commonly intense recrystallization of the early mineralization and their carbonate host rock played only a limited part in their *reconcentration.* However, the late Hercynian fracturing and associated magmatism on the north side generated the major Zn–Pb concentrations in the Montagne Noire, as exemplified at St. Salvy and at La Rabasse.

3.1 Zinc-Bearing Mineralization Related to late Hercynian Fracturing: St. Salvy Vein and Peux Occurrences

The St. Salvy vein-type deposit represents the most extensive zinc concentration in the Montagne Noire (Table 2). It has not been studied within the framework of this project, since the mechanisms of its formation are well known by the work of one of the authors (Beziat 1977), the conclusions of which are summed up in this paragraph.

The mined deposit is located within a major fracture zone oriented E–W to NE–SW. This structure, which extends over some kilometers, is related to a late Hercynian shear zone on the southern edge of the Sidobre granite dated 290 ma.

The ore is mainly composed of Ag–Ge–Cd sphalerite in a siderite and quartz matrix, with subordinate galena in the upper part of the vein. The indirect affiliation of the economic vein concentration with stratiform sphalerite, pyrite, galena and siderite mineralization developed in the host black series, Lower Cambrian in age (cf. Sect. 2.2.), has been demonstrated. Granite played a minor role, which appears to be mainly *mechanical*, in the reconcentration. Only an uneconomic paragenesis consisting of arsenopyrite, bismuthinite, scheelite, pyrrhotite, chalcopyrite, millerite, and indium-bearing sphalerite is directly related to the influence of granite in hornfels occurring at the hanging wall of the vein-type structure.

The zinc occurrences observed in the Ouyre-Barre structural unit in the Peux area show the same arrangement. Sphalerite is also located in brecciated, dolomitized, and silicified zones close to east-west trending senestrial wrench faults. The mineralogical association is composed of dominating sphalerite, iron sulfides, galena, chalcopyrite and traces of nickel and cobalt minerals. The brecciated structures appear to be mineralized only at the hanging wall of the Lower Cambrian limestones (K_2a), thereby suggesting the existence of a stratiform host rock containing disseminated Zn–Pb mineralization of the Brusque (distal) or Lardenas type (Fig. 3).

3.2 La Rabasse, an Example of Carbonate Replacement Deposit

The La Rabasse mine (Figs. 6 and 7), which was closed in 1954, was one of the most productive in Montagne Noire, with a total production of about 100,000 t Pb + Zn. Massive sulfide bodies are located at the hanging wall of the Lower Cambrian dolomites of the Melagues unit, close to the Faulat subvolcanic intrusion (290 ma) and in the vicinity of Triassic transgressive layers which are characterized by siliceous, baritic zones (Fig. 9).

The hypothesis of a pre-Hercynian metallic concentration suggested by some authors (Rouchy 1974) has not been confirmed by our investigations. Although an unconformity was revealed at the boundary between the Lower Cambrian dolomites (local erosion of sequence C, cf. Sect. 2.1.2.) and the clayey-silty formation of the Middle Cambrian, the base of which is locally calcareous (marmorean limestones), no early-formed mineralization of the Bibaud-Tête Rousse type, developed on the Southern side of Montagne Noire in a similar environment (cf. Fig. 1), has been evidenced.

The lithological contact between Lower Cambrian dolomites and Middle Cambrian shales played, however, an important part in the emplacement of thick rhyolite sills which represent apophyses of the Faulat late Hercynian intrusion. The silicified and pyrite, pyrrhotite, and arsenopyrite mineralized edges of the sills form the hanging wall of the La Rabasse massive sulfide deposits and of the Fontserenne Pb–Zn impregnated dolomites.

The mined ore-shoots were mainly formed of massive sulfides, but disseminated ore in a dolomitic or siliceous material was also sampled in old mining works. The main geochemical characteristics of these different kinds of ore are summarized in Table 3.

Table 3. Comparative geochemistry of some La Rabasse–Fontserenne ores

Type of ore	Metals mined				Associated elements					Matrix	
	Zn %	Pb %	Ag ppm	Au ppm	As %	Sb %	Cu %	Sn ppm	Bi ppm	SiO_2 %	CaO + MgO %
Massive ore (locally oxidized)	0.04 → 7.2	0.6 → 19	14 → 300	<0.1 → 6.3	0.6 → 21	0.02 → 0.3	0.03 → 0.17	4.8 → 670	19 → 256	<1 → 10	⩽1.2
Disseminated ore in dolomitic matrix	0.06 → 4.5	0.4 → 4.4	9 → 89	0.3 (1 analysis)	0.05 → 2	<0.09	0.03 → 0.2	34 → 209	16 → 228	<1 → 17	25 → 47
Siliceous ore (often oxidized)	0.005 → 0.4	0.03 → 1	2 → 244	0.2 → 1.4	0.4 → 2.2	0.02 → 0.1	0.008 → 0.3	11 → 289	<10 → 656	20 → 66	<1

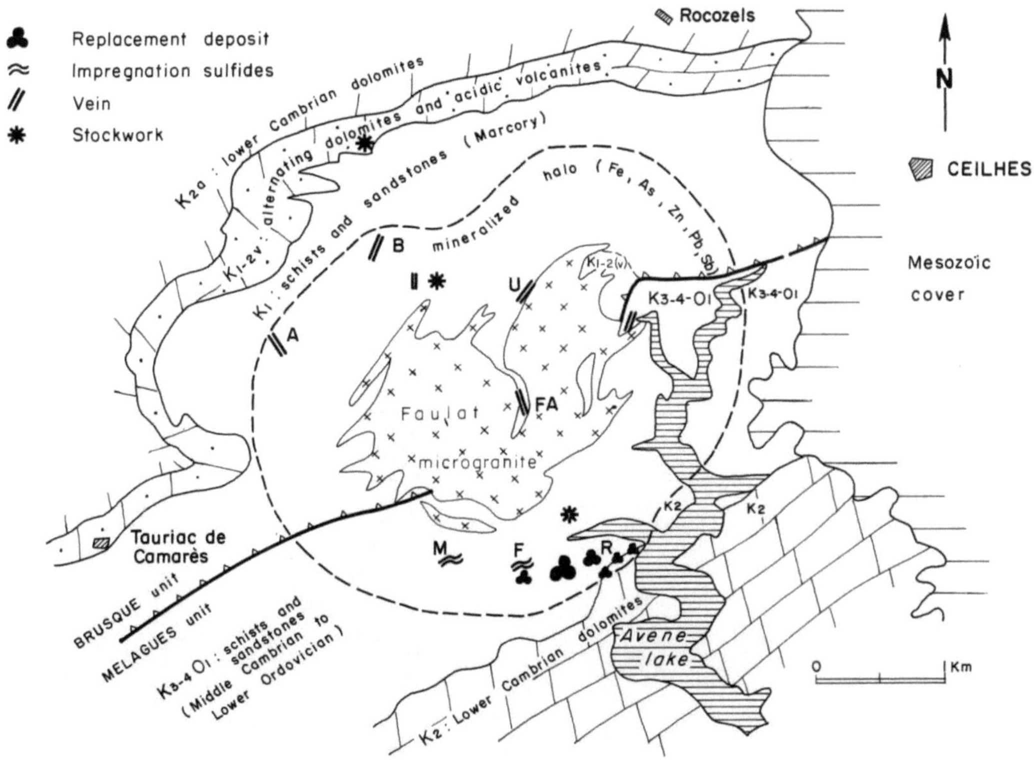

Fig. 6. Polymetallic mineralization around the Faulat subvolcanic intrusion — old mines: *A* Argenneuves; *B* Bournac; *F* Fontserenne; *Fa* Fontaine des Allemands; *M* Mathet; *R* La Rabasse; *U* Ravin des Usclades (Lascours)

The mineralogical study has shown that the high temperature mineralization was emplaced in three successive pulsations: (1) arsenopyrite-pyrrhotite–pyrite with traces of chalcopyrite, enargite, and cubanite; (2) ferriferous sphalerite with traces of stannite and cassiterite; (3) galena with traces of various sulfosalts. It generally occurred with silicification, and was followed by recrystallization of the dolomitic host rock. Periods of unstability (pyrrhotite → marcasite) and fracturation (between phases 2 and 3) emphasize the various emplacement phases.

Polymetallic mineralization displaying a similar paragenetic evolution was investigated on the margin of the Faulat intrusion. All these Fe, As, (Zn), (Pb), and (Sb) occurrences and deposits are distributed within a 800-m-wide halo containing extensive concentration of rhyolite veins around the microgranite body (Fig. 6).

The history of the mineralization appears thus to be intimately related to that of the late Hercynian calc-calkaline magmatism: at the end of the Carboniferous, the subvolcanic intrusion of Faulat was emplaced in a brittle zone at the boundary between the Melagues and the Brusque structural units, and was accompanied by the injection of rhyolite veins in the Cambrian-Ordovician host series. Accompanying tectonic movements induced local brecciation of the sills walls and the sub-

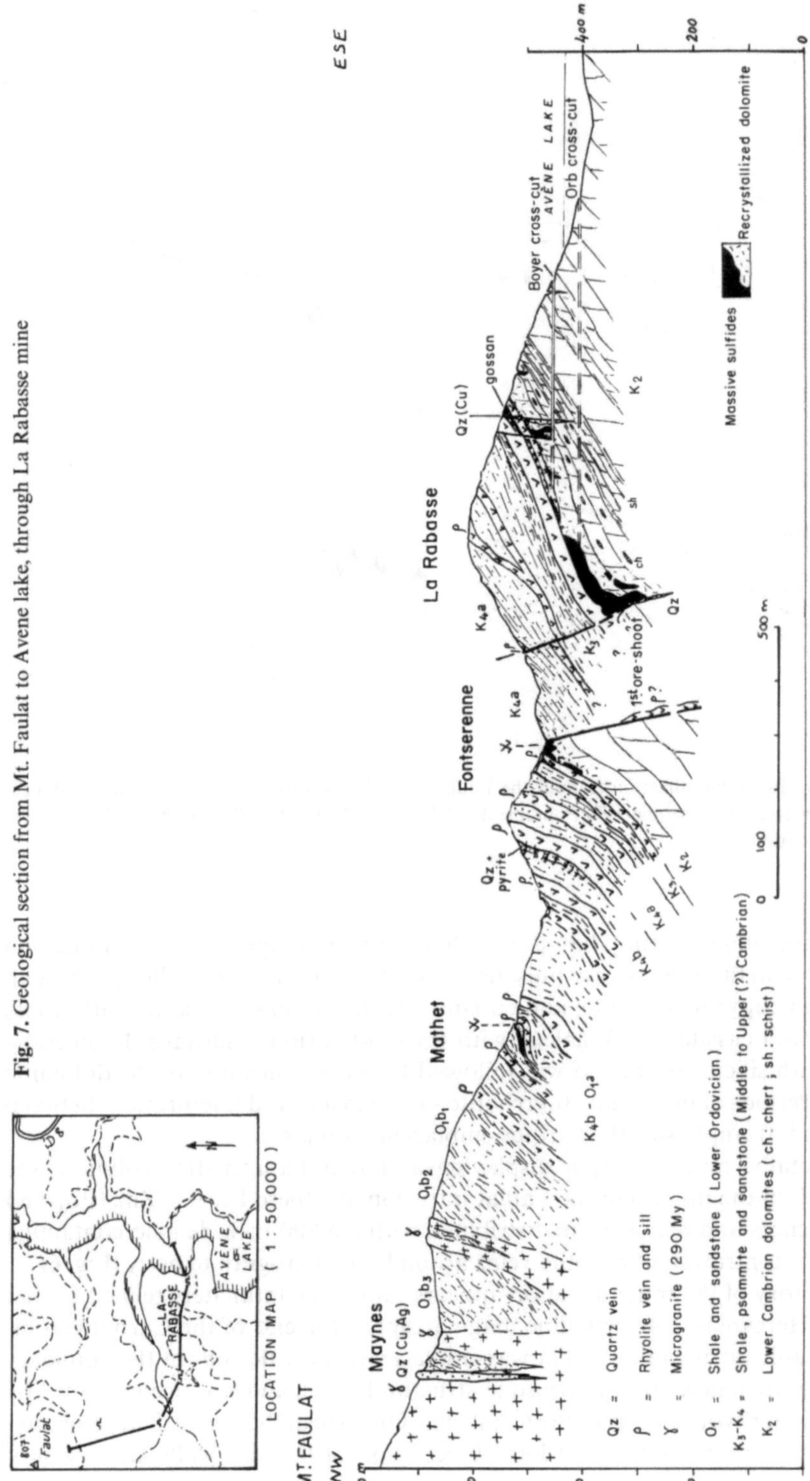

Fig. 7. Geological section from Mt. Faulat to Avene lake, through La Rabasse mine

meridian fracturing of the Paleozoic basement. These brecciated and fractured structures acted as *drains* for mineralized solutions of relatively high temperatures (probably in the 300°–400 °C range), which are detectable by localized hydrothermal alteration processes of the quartz-sericite to propylite types. The deposition of polymetallic mineralization occurred near hydrothermal circulation zones by various processes, depending on the nature of the host rock:

— precipitation in open fissures within competent beds such as in the Lower Cambrian sandy pelites, exemplified by the Bournac vein with a final stibnite phase (Munoz, 1981), and the Middle Cambrian and Ordovician sandstones.
— impregnation in the pelites containing calcareous nodules or layers at the base and top of the Middle Cambrian exemplified by the Mathet occurrence.
— carbonate replacement in hydraulic trap conditions near small submeridian wrench faults at the top of Lower Cambrian dolomites as at La Rabasse and Fontserenne. This type of mineralization generated the largest metallic concentrations.

3.3 Late Mineralization Related to the Mesozoic Transgression

Cross-cutting or conformable silicifications affect the base of the Mesozoic cover and its Paleozoic basement from the La Rabasse area to the West Bedarieux region (Figs. 8 and 9). These processes appear to be linked to the Triassic-Liassic distension, which is well marked on the margin of the Jurassic gulf of the Causses.

In the Avene area, the silicification processes are only associated with barite, commonly secondarily silicified, and with silver-bearing tetrahedrite. Although veins belonging to this late phase cut across the La Rabasse mineralization, zinc or lead sulfides from the remplacement deposit were not concentrated in secondary factures.

Further south, in the structural unit of St. Gervais (Fig. 1), relatively uneconomic Pb–Zn mineralization in veins or small bodies at the base of the transgressive Triassic, appear to be related to this Mesozoic structuration. For instance, the barite-sphalerite-galena vein at Lacan (about 200 t Zn) is hosted in Lower Cambrian dolomites which do not display evidence of early Pb–Zn mineralization.

4 Conclusions

This study has permitted the reconstruction of the genetic history of some Pb–Zn deposits and occurrences which are hosted in the Lower Cambrian series of the Montagne Noire. Various types of ore concentration have been distinguished, some of them occurring at an early stage (Cambrian), whereas the others are mainly related to Late Hercynian events.

The sedimentological study has resulted in a better definition of the development and sinking conditions of the Cambrian carbonate platform. The northern

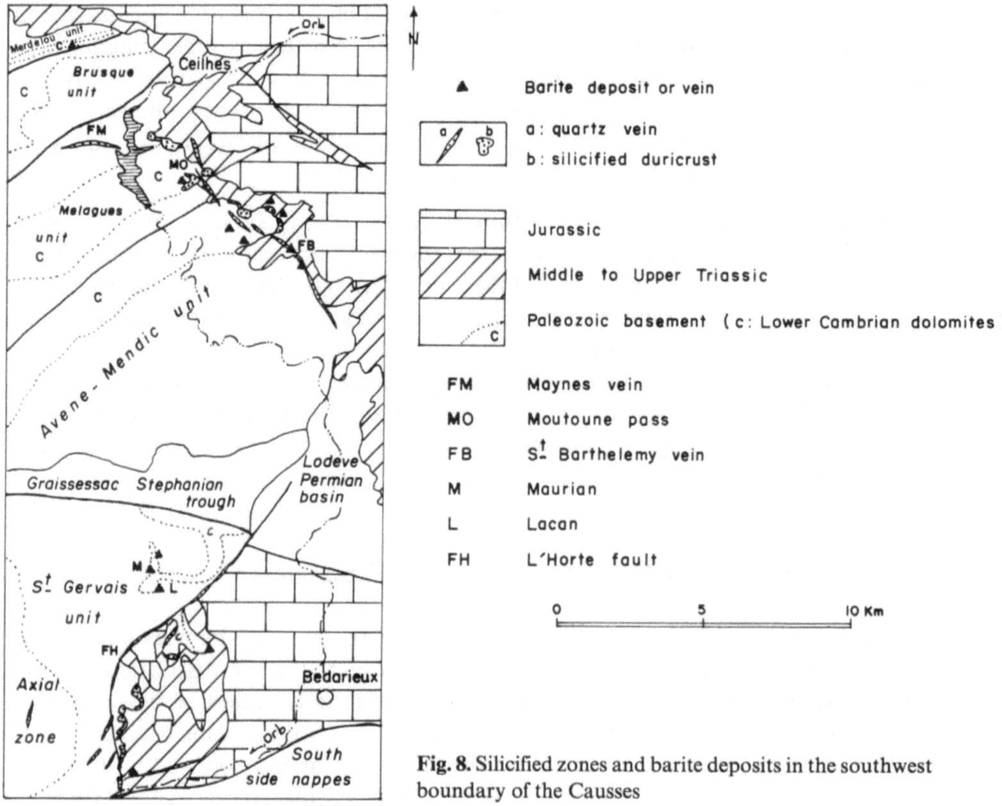

Fig. 8. Silicified zones and barite deposits in the southwest boundary of the Causses

margin of the platform is characterized by a volcanic ridge which forms a nearly perennial high. Subsidence parallel to the high axis led to the formation of confined troughs in which shaly-calcareous facies (black series) host disseminated Zn–Pb mineralization developed on a regional scale. The mineralizing process is thought to be related to the tectonic unstability of the platform margin, and probably indirectly to subcontemporaneous volcanism. A hydrothermal sedimentary origin is therefore suggested for the stratiform mineralization of the black series as exemplified by study of the Brusque and Lardenas occurrences.

The main Pb–Zn concentration processes occurred on the north side of the Montagne Noire during Late Hercynian tectonic phases. At La Rabasse, for instance, Pb–Zn carbonate replacement deposits are recorded in the Fe–As–Zn–Pb–Sb mineralized halo of a subvolcanic intrusion dated 290 ma, whereas no metallic preconcentration was found in the Lower Cambrian host dolomites.

On the other hand, vein-type Zn deposits (St. Salvy mine) are linked to the reconcentration, in Late Hercynian fractures, of an earlier mineralization hosted by Lower Cambrian black series.

Finally, it must be noted that the Triassic-Liassic distension did not induce in the Montagne Noire noticeable Pb–Zn metallogenesis as it did in the neighboring Cevennes.

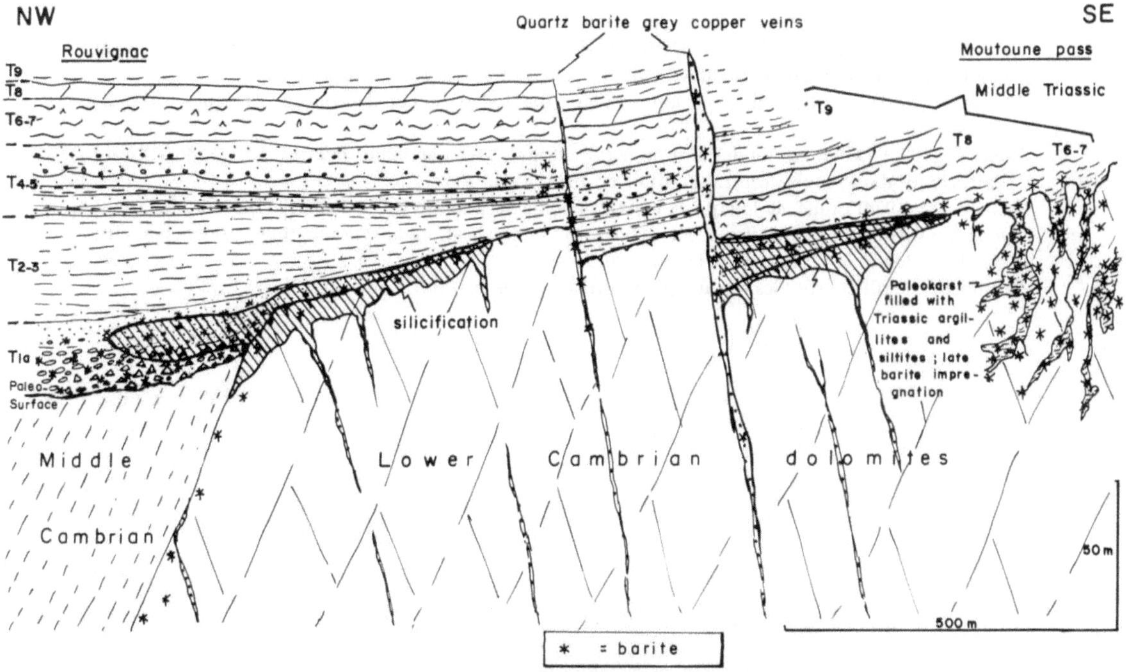

Fig. 9. Silicification and barite mineralization east of the Avene lake. (After Rouchy 1974)

This nonexhaustive study on the relationships between the Montagne Noire Zn–Pb mineralization and their Cambrian carbonate host-rock has shown that simple geometric features may hide complex concentration processes, of various types and ages.

Acknowledgement. This study was partly funded under EEC *Contract MSM-130-F*. We wish to thank J. Boissonnas, administrator of the EEC programme, F. Tollon, scientific coordinator of the project, J. Bouladon and M. Aubague for their suggestions on metallogenic aspects of the research.

References

Aubague M, Orgeval JJ, and Soulie M (1977) Les gîtes minéraux de la terminaison méridionale du Massif central et de sa bordure languedocienne. Bull BRGM (2), II, 3:139–181

Beziat P (1977) Joint action "Brassac". Recherche des facteurs lithostratigraphiques, pétrographiques et tectoniques en relation avec les concentrations plombo-zincifères, dans le cadre du district allant de Castres à Brassac (Tarn). Analyse du modèle Saint-Salvy de la Balme (Tarn) BRGM rep 77 RDM 029 PE

Courjault-RADE P (1982) Environnements sédimentaires des minéralizations (Pb, Zn) liées aux strates de la plate-forme carbonatée du Cambrien inférieur de la Montagne Noire. 3rd cycle thesis, Toulouse

Donnot M, Guerange B (1978) Le synclinorium cambrien de Brusque. Bull BRGM (2), I, 4:333–363

Donnot M, Milesi JP (1982) Cadre géologique, inventaire et typologie des Monts à l'Est de Lacaune. In: B.R.G.M. report 82 SGN 840 GMX, joint action Polymetallic Cu–Pb–Zn mineralization in the French Paleozoic

Gachet L (1983) Volcanisme cambrien des unités de Brusque et du Merdelou. 3rd cycle thesis, Lyons

Lacerda H, and Bernard AJ (1984) Existence de minéralisations plombozincifères syngénétiques du substratum cambrien du district des Malines (Gard, France). Min Dep 19:152–157

Munoz M (1981) Les minéralisations antimonifères du champ filonien de Bournac (Hérault). 3rd cycle thesis, Toulouse

Pouit G (1978) Différents modèles de minéralisations "hydrothermale sédimentaire" à Zn (Pb) du Paléozoïque des Pyrénées centrales. Min Dep 13:411–421

Rouchy JM (1974) Etude géologique et métallogénique de la haute vallée de l'Orb (Hérault). Relations socle-couverture. Problème des silicifications et des minéralisations barytiques. Bull Museum Nat Hist Nat Paris 3e Ser 214. Sciences de la Terre 34, 93 pp

Routhier P (1980) Où sont les métaux pour l'avenir? Mémoire du BRGM, no. 105

Isotope (Sr, C, O, and S) Tracing of Diagenetic Ore Formation in Carbonate-Hosted Ore Deposits Illustrated on the F–(Pb–Zn) Deposits in the Alpujarrides, Spain and the San Vicente Zn–Pb Mine, Peru

L. FONTBOTÉ[1] and H. GORZAWSKI[1,2]

Abstract

The diagenetic evolution of gangue and ore minerals of selected F–(Pb–Zn) deposits in the Alpine Triassic of the Alpujarrides, Spain and the San Vicente Zn–Pb Mine, Pucará basin (Upper Triassic-Liassic), central Peru has been isotopically characterized (Sr, C, O, and S).

These ore deposits occur mainly in dolomite horizons deposited within a carbonate platform at the margin of an emerged continent and are associated with a narrow facies span comprising (1) the inner part of the lagoon (with tidal flat facies and evaporite molds), (2) the lagoon sensu stricto, and (3) the barrier facies.

The isotope ratios display similar trends in both districts and can be summarized as follows. (1) No isotope differences between ore samples and host rock are recognized. (2) According to the position in the diagenetic crystallization sequence, systematic differences in the isotope composition are observed. (3) The strontium ratios in the first generations are less radiogenic than those of the last generations. Despite the clear trends observed, the absolute differences between $^{87}Sr/^{86}Sr$ ratios are always very small and plot close to those assumed for Triassic/Liassic ocean water. (4) The composition of the stable isotopes of C, O, and S changes in a similar way. With a few exceptions the later generations are slightly depleted in the heavier isotopes compared to the first generations.

A model of relatively early diagenetic ore formation is proposed for the San Vicente Mine. This model is thought to be also applicable, with variations, to other similar ore deposits, including the F–(Pb–Zn) deposits in the Alpujarrides.

1 Introduction

The present chapter reflects part of the results obtained within the framework of an EC project "Petrographic and geochemical indicators for the exploration of

[1] Mineralogisch-Petrographisches Institut, Universität Heidelberg, FRG
[2] Max-Planck-Institut für Chemie, Mainz, FRG

Mineral Deposits within the European Community
(ed. by J. Boissonnas and P. Omenetto)

hidden ore deposits in sedimentary rocks". Because of space limitation only a selection of the main results is presented.

More comprehensive descriptions and background information can be found in Fontboté and Gorzawski (1987) and in several additional publications presently in preparation. This project was part of a joint project with participation also of the Bureau de Recherches Géologiques et Minières (B.R.G.M.), Orléans (J.F. Sureau, D. Giot, Y.M. LeNindre, A. L'Homer), and the Service Géologique de Belgique (L. Dejonghe, M. Mardaga).

2 Scope. Methodical Approach

Numerous studies in carbonate-hosted ore provinces allow one to establish a quite general association of strata-bound Zn–Pb–(F–Ba) ore deposits with definite shallow-water facies. These investigations have shown that ore accumulation is essentially a diagenetic process. One of the central genetic aspects with immediate application in the exploration strategy, is to know why such a strict correlation of the ore occurrences with certain lithofacies (and paleogeographic positions) exists. Two extreme possibilities may be considered (Amstutz and Fontboté 1983):

1. The lithofacies of the host rock is important merely insofar as it is favorable or not, to receive "mineralizing solutions". The ore-forming process(es) would then take place during a *later stage* than deposition and early diagenesis of the host rock and would be *independent* of the depositional environment.
2. The lithofacies of the host rock is significant because it reveals a paleoenvironment in which ore-forming processes could take place during deposition and early diagenesis of the host rock.

It is evident that for each of these two extreme possibilities the exploration methodology should be completely different, as different as, for example, for exploration of hydrocarbons in (1) structural and sedimentary traps, or in (2) bituminous shales.

In fact, these two extreme hypotheses can be summarized in terms of time. Are the ore-forming processes mainly sedimentary, early diagenetic[1], late diagenetic, or postdiagenetic? In other words, we should be able to determine the time of formation of the ore minerals with respect to the general diagenetic evolution of the enclosing rock.

[1] The distinction between early diagenesis and late diagenesis represents, of course, a terminological problem. In the present work "early diagenesis" includes processes taking place through early stages of compaction, cementation, and burial (up to several tens of meters), exchange phenomena between the sediment and the sedimentary environment are still possible. (See terminologic discussion in Dunoyer de Segonzac 1968, and the terms early diagenesis, syndiagenesis, and syngenesis in Bates and Jackson 1980).

With this purpose in mind, detailed petrographic and geochemical investigations have been carried out on two Zn–Pb–(F–Ba) districts in which a clear relationship between deposition environment of the host rock and ore occurrence exists. The selected ore districts are the following (Fig. 1).

1. F–(Pb–Zn) district in the Alpine Triassic of the Alpujarrides (Betic Cordillera, southern Spain).
2. Zn–Pb Mine San Vicente in Liassic carbonate rocks of the Pucará Group, central Peru.

This approach is based on previous petrographic observations on a large amount of carbonate-hosted ore districts (see, among others, Amstutz et al., 1964; Amstutz and Bubenicek 1967; Amstutz and Park 1971) which have shown that (1) sulfides and other ore minerals can be assigned a definite paragenetic position within diagenetic crystallization sequences, (2) in virtually all nonmetamorphic ore provinces basically identical paragenetic sequences have been found, and (3) the same sequences are recognized in different types of textures and structures. These observations led to the working hypothesis that a fractional crystallization differentiation takes place during diagenesis.

In addition, the diagenetic crystallization process, by itself, is able to produce manifold megascopic textures and structures, for example, geodelike textures and *d*iagenetic *c*rystallization *r*hythmites (DCRs, Fontboté and Amstutz 1983). In DCRs the crystallization generations are megascopically visible (Fig. 2), and a clean separation of subsequent diagenetic generations is possible. They are therefore suitable objects for the geochemical (mainly isotope) investigations described below.

A total of 60 samples (mainly DCRs and related textures) from the districts mentioned above, as well as, for comparison, from other comparable ore districts have been selected. Systematic determinations of the isotope composition of Sr, C, O, and S with a total amount of over 200 isotope determinations have been carried out (Fontboté and Gorzawski 1987).

The material for isotope analyses was obtained with a slightly modified dental drill. An additional separation of the samples was carried out with the help of a binocular microscope to guarantee clean separation between different crystallization generations. The soluble fraction (with cold HCL 6N) of the same material was analyzed with atomic absorption spectrometry for several trace elements.

The strontium isotope ratios were determined at the B.R.G.M/Orléans (J.Y. Calvez). The carbon and oxygen analyses were carried out at the Max-Planck-Institut für Chemie/Mainz (H. Gorzawski). The sulfur isotope analyses were performed at the Oregon State University/USA (Dr. C.W. Field).

3 Main Geologic Paleogeographic, and Facies Features of the Selected Ore Districts

Both districts, despite the different ore parageneses, display many similarities in their tectonic, paleogeographic, and facies characteristics, as well as in the geometry

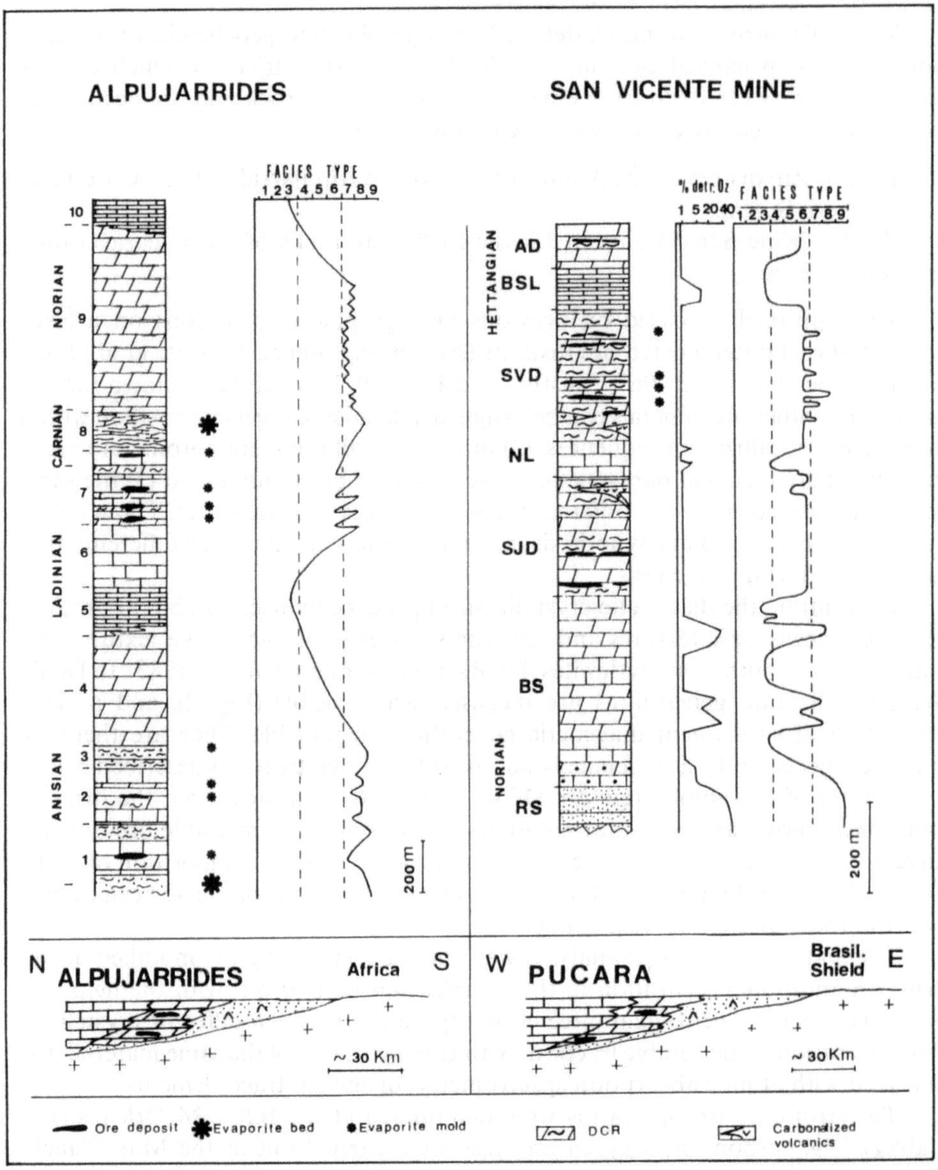

Fig. 1. Comparison of the main paleogeographic and lithostratigraphic features of the two ore districts on which this chapter focuses. Under *facies type* an evolution curve of the deposition environment has been represented. (*1* open basin; *4* slope; *5* platform edge; *6* barrier; *7* lagoon; *8* tidal flat; *9* continental). *Numbers* and *abbreviations* are referred to in the text. (Alpujarrides section after Martín et al. 1987, geologic ages in the San Vicente sequence according to Prinz 1985)

Fig. 2A, B. Typical examples of a dolomite (**A** Alpujarrides, Spain) and a sphalerite-dolomite DCR (**B** San Vicente Mine, Peru). Shown are the hand specimens as well as microphotographs of the corresponding thin section (**A**) or polished section (**B**). The subsequent diagenetic generations *I*, *II*, and *III* with dolomite (*do*) and sphalerite (*sl*) are indicated as well as the results of the isotope and trace element analyses

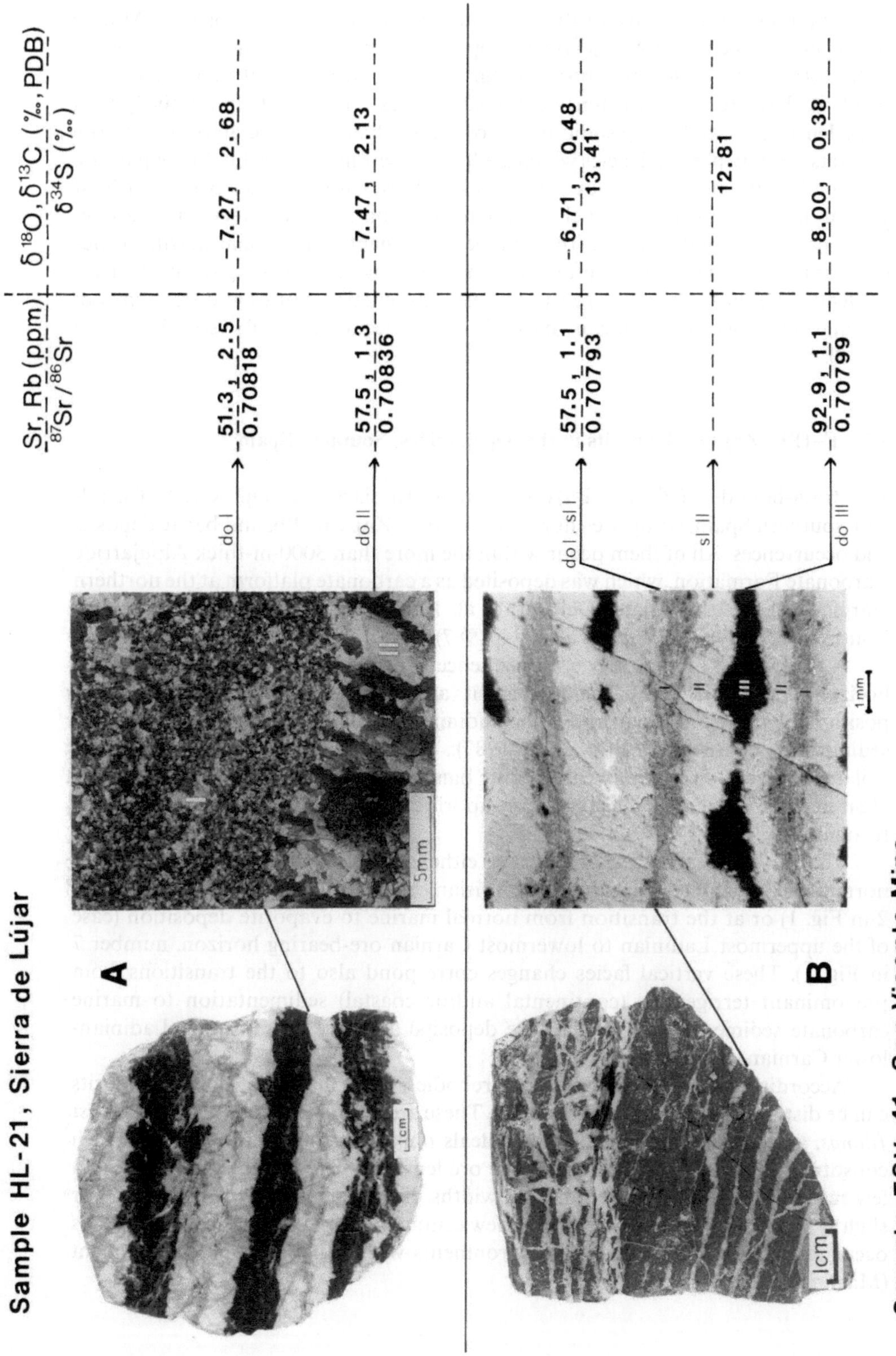
Sample HL-21, Sierra de Lújar
A
5mm
1cm
do I
do II
Sr, Rb (ppm)
$^{87}Sr/^{86}Sr$
51.3, 2.5
0.70818
57.5, 1.3
0.70836
δ¹⁸O, δ¹³C (‰, PDB)
δ³⁴S (‰)
-7.27, 2.68
-7.47, 2.13
Sample FSV-41, San Vicente Mine
B
1cm
1mm
do I sl I
sl II
do III
57.5, 1.1
0.70793
92.9, 1.1
0.70799
-6.71, 0.48
13.41
12.81
-8.00, 0.38

Fig. 2A, B

of the orebodies (essentially stratiform) and in their stratigraphic position (Middle to Upper Triassic in the Alpujarrides, Upper Triassic to Liassic in San Vicente). In both districts the geotectonic position during the deposition of the enclosing sediments is characterized by extensive block tectonics. This results in relatively high subsidence rates and in episodic basic to intermediate volcanic activity. The ore deposits occur in peritidal facies which are located within thick carbonate sequences. The ore-bearing peritidal facies constitutes, in both cases, a facies belt, of at least 100-km length, which is parallel to an emerged continent. Between the shallow-water carbonate platform and the emerged continent, transitional detrital facies with evaporite and carbonate intercalations are recognized. Among the most striking features in both districts are the complex diagenetic textures and structures of ore and host rock. A significant part of the ore occurs as DCRs (see above and Fig. 2).

3.1 F–(Pb–Zn) Ore Deposits in the Alpujarrides, Southern Spain

The strata-bound F–(Pb–Zn–Ba) district in the Alpujarride Complex, Betic Cordillera, southern Spain, comprises numerous F–(Pb–Zn), Zn–Pb, and barite deposits and occurrences. All of them occur within the more than 3000-m-thick Alpujarride Carbonate Formation, which was deposited as a carbonate platform at the northern margin of the African plate (Delgado et al. 1981; Martín and Torres Ruiz 1982; Fontboté et al. 1983; Martín et al. 1984, 1987).

A generalized lithostratigraphic sequence with the location of the ore-bearing horizons is shown in Fig. 1. The ore deposits are linked to two defined stratigraphic positions (within the Anisian and at the Ladinian-Carnian transition) and to definite sedimentary contexts (Martín et al. 1987). These are highly restricted lagoons isolated from the open sea by calcarenitic barriers, with noticeable development of algal mats in their inner margins. Preevaporitic deposition conditions are usually recognized.

The ore-bearing horizons are located either at the transition from evaporitic to normal marine deposition (case of the Anisian ore-bearing horizon, numbers 1 and 2 in Fig. 1) or at the transition from normal marine to evaporite deposition (case of the uppermost Ladinian to lowermost Carnian ore-bearing horizon, number 7 in Fig. 1). These vertical facies changes correspond also to the transitions from predominant terrigenous (continental and/or coastal) sedimentation to marine carbonate sedimentation (Anisian ore deposits) and vice versa (upper Ladinian-lower Carnian).

According to the geometry of the orebodies, three main types of ore deposits can be distinguished (Martín et al. 1987). These are *stratiform* orebodies, *paleokarst fillings*, and *vein fillings*. This chapter deals only with the stratiform ores which constitute the most important type. The ore lenses display lengths ranging from a few meters to several hundred meters, widths of the same order of magnitude or slightly smaller, and thicknesses from a few centimeters up to 10–15 m. They always occur in one of the following subenvironments within the lagoonal environment (Martín et al. 1987):

1. Inner margin of the lagoon (dolomitized tidal flat/beach facies 8 in Fig. 1).
2. Lagoon sensu stricto (lagoonal subenvironment, likewise in dolomitized as in nondolomitized facies 7 in Fig. 1), and
3. Outer margins of the lagoon (dolomitized barrier facies 8 in Fig. 1).

In the present investigation samples from stratiform fluorite deposits in the Sierra de Gádor, the Sierra de Baza, the Sierra de Lújar and the Benínar Mine have been selected. The latter corresponds to facies deposited in the lagoon sensu stricto and the samples from Sierra de Gádor and Sierra de Lújar are from horizons occurring in barrier facies. The ore paragenesis is very simple in the three selected localities. Fluorite is the main ore mineral (fluorite contents up to 95% in the studied samples). Galena and sphalerite occur also and have been formerly mined.

3.2. The San Vicente Zn–Pb Mine, Peru

The strata-bound Zn–Pb deposit of San Vicente lies about 328 km by road east of Lima, in the Chanchamayo area, 10 km south of San Ramon in the tropical rain forest of the Ceja de Selva. This ore deposit is the best-known example, and at present the only one in production of the strata-bound Zn–Pb–(Ba) belt in the eastern part of the Upper Triassic-Liassic Pucará basin (Levin 1975; Levin and Amstutz 1973; Lavado 1980; Fontboté 1981; González and Fontboté 1986).

Figure 1 shows a generalized lithostratigraphic profile of the San Vicente mining area. It presents two well-defined parts which correspond to different types of basin evolution. The *lower one* is a typical transgressive sequence, which ranges from continental sedimentation ("Red Sandstone", in Fig. 1 abbreviated as RS) to marine conditions with carbonate deposition and relatively abundant detrital material ("Basal Serie", BS). The *upper part* of the sequence comprises three units deposited in a peritidal carbonate platform virtually free of detrital components ("San Judas Dolomite, SJD", "San Vicente Dolomite, SVD" and "Alfonso Dolomite, AD"), which are separated by two episodes of deeper sedimentation ("Neptuno Limestone, NL" and "Bituminous Silty Limestone, BSL"). The transition between platform and basin sedimentation is clearly documented by slope breccias.

The shallow-water carbonate platform recognized in San Vicente can be followed more or less continuously, including also similar ore occurrences, between Norian and Sinemurian over a distance of at least 200 km extending from north to south in the eastern part of the Pucará basin. This facies belt most likely interfingers eastwards with the detritic sediments containing evaporite and carbonate intercalations of the Lower Sarayaquillo Formation, which occurs along the western margin of the Brazilian shield.

All three dolomitic units mentioned above are ore-bearing and display very similar petrographic and geochemical characteristics. The main ore horizons occur in the San Vicente Dolomite. The three dolomite ore-bearing units have been deposited in a peritidal platform comprising essentially the same subenvironments as described for the fluorite deposits in the Alpujarrides. The barrier subenvironment (facies 6 in Fig. 1) is predominant and is represented by completely dolomitized oolitic packstones and grainstones. Detailed profiles show that the ore horizons are mainly bound to layers of dolomitized mudstones and pellet-grainstones with abundant cryptalgal lamination deposited in lagoon and tidal flat subenvironments (facies 7 and 8 in Fig. 1). Molds of evaporites are recognized in tidal flat ore-bearing horizons ("manto 3t").

The ore occurs as lenticular bodies. Sphalerite and galena are the only ore minerals. The Zn content may range, in certain layers, up to 30%; however, in the mined horizons it lies generally between 8 and 16%. The Zn:Pb ratio is 10:1 on the average.

An extensive lithogeochemical investigation has been carried out on the whole sequence (Gonzalez and Fontboté 1986; Fontboté and Gorzawski 1987). Among other results, it has been found that the ore-bearing dolomite members are not anomalous in Zn (geometric mean of the "ore-free samples" in the San Vicente Dolomite is 42.5 ppm Zn). Another interesting result concerns the Mn analyses. As in other Zn–Pb strata-bound ore deposits, the Mn values are significantly higher in the ore-bearing units (mean = 1693 ppm Mn in the San Vicente dolomite, mean = 1544 ppm Mn in the San Judas dolomite), and also higher than in other

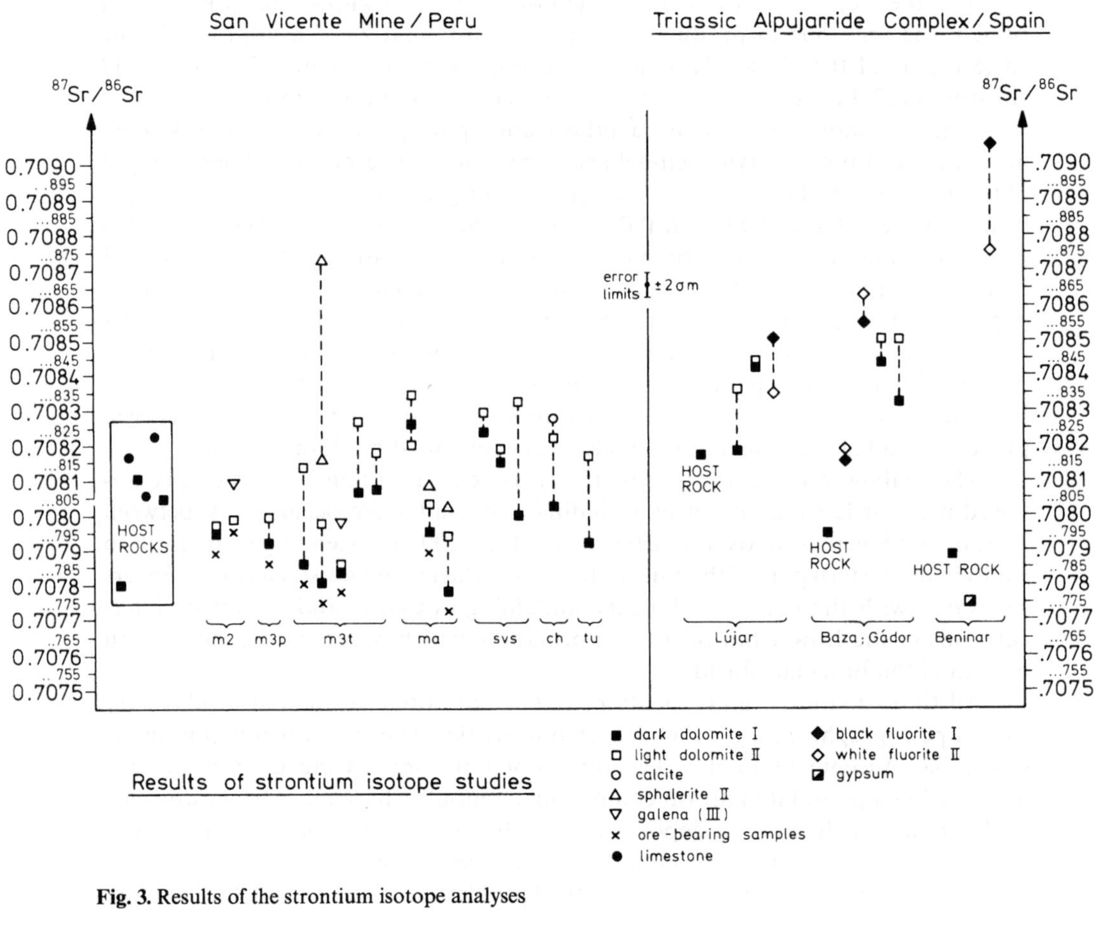

Fig. 3. Results of the strontium isotope analyses

shallow-water dolomitic rocks. These high Mn values bear no relation with volcanic activity and are clearly linked to the process of dolomitization.

4 Isotope Trends Reflecting the Evolution of the Diagenetic Crystallization Process

4.1 Alpujarrides

The samples analyzed are from the following localities: Sierra de Lújar, Sierra de Gádor, Sierra de Baza (all Ladinian–Carnian), and the Benínar Mine (Anisian). In all carbonate samples strontium, oxygen, and carbon isotope as well as trace element analyses have been carried out. $^{87}Sr/^{86}Sr$ ratios were also determined in fluorite DCRs and in a laminated gypsum sample from Benínar (Figs. 3, 4, 5).

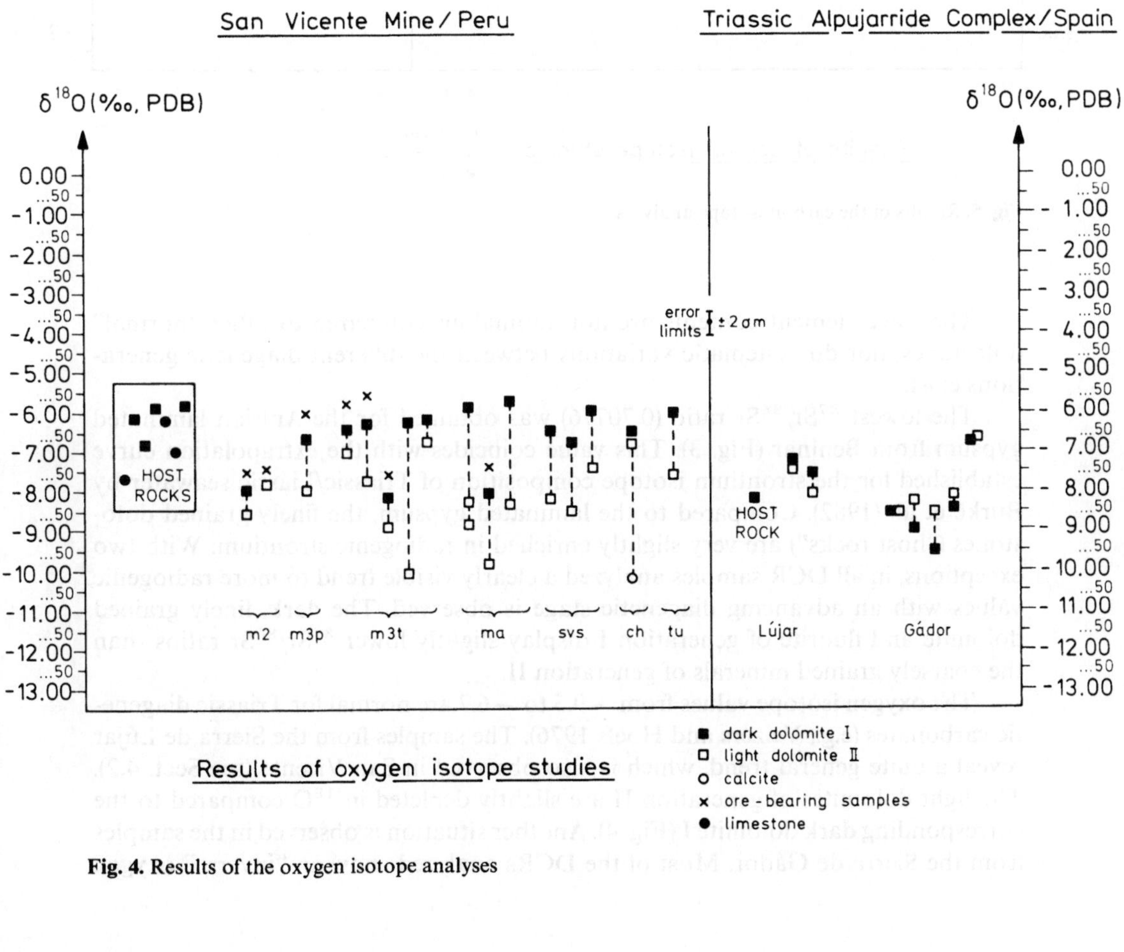

Fig. 4. Results of the oxygen isotope analyses

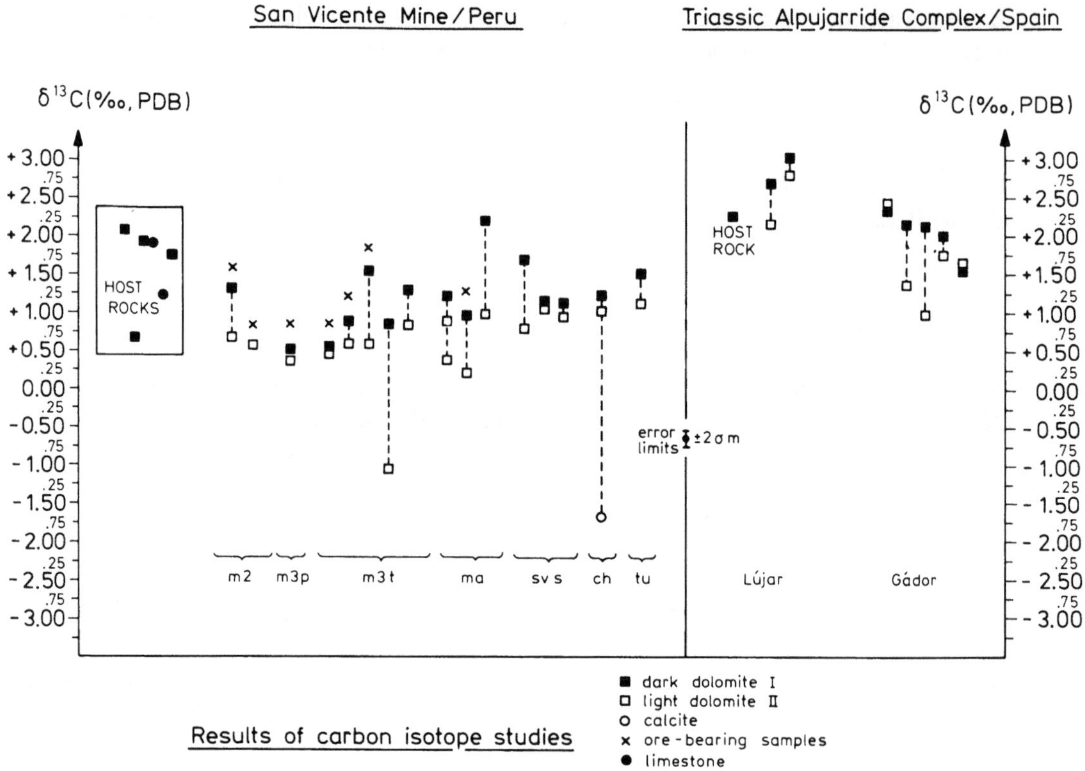

Fig. 5. Results of the carbon isotope analyses

The trace element contents are not anomalous compared to other "normal" dolostones, nor do systematic variations between the different diagenetic generations exist.

The lowest $^{87}Sr/^{86}Sr$ ratio (0.70776) was obtained for the Anisian laminated gypsum from Benínar (Fig. 3). This value coincides with the extrapolation curve established for the strontium isotope composition of Triassic/Liassic seawater by Burke et al. (1982). Compared to the laminated gypsum, the finely grained dolostones ("host rocks") are very slightly enriched in radiogenic strontium. With two exceptions, in all DCR samples analyzed a clearly visible trend to more radiogenic values with an advancing diagenetic stage is observed. The dark, finely grained dolomite and fluorite of generation I display slightly lower $^{87}Sr/^{86}Sr$ ratios than the coarsely grained minerals of generation II.

The oxygen isotope values from −9.5 to −6.7 are normal for Triassic diagenetic carbonates (e.g., Veizers and Hoefs 1976). The samples from the Sierra de Lújar reveal a quite general trend, which is also observed in San Vicente (see Sect. 4.2). The light dolomites of generation II are slightly depleted in ^{18}O compared to the corresponding dark dolomite I (Fig. 4). Another situation is observed in the samples from the Sierra de Gádor. Most of the DCRs analyzed show an "inverse" oxygen isotope trend, the dolomite of generation II being isotopically slightly heavier than the dolomite of generation I.

The carbon isotope compositions measured lie in a range between +0.97 and +2.99, which is again normal for Triassic dolostones, whereby the samples from the Sierra de Gádor show slightly lower values than the samples from the Sierra de Lújar (Fig. 5). Regarding the different dolomite generations, a diagenetic trend is clearly visible. In nearly all samples generation II displays a carbon isotope composition slightly depleted in ^{13}C compared to that of generation I.

4.2 San Vicente

The selected samples belong to the following groups: host rocks, diagenetic breccias and veinlets, DCRs without ore, ore-bearing DCRs, and sulfide minerals (Figs. 3, 4, 5). Strontium, oxygen, carbon, and sulfur isotope analyses have been carried out. In addition, most of the carbonate samples were analyzed with AAS for trace element contents.

The strontium and rubidium contents vary systematically from dolomite of generation I to dolomite of generation II. The first generation contains generally less strontium and rubidium than the corresponding later generation. Regarding the strontium isotope compositions, it should be underlined that there are no differences between host rocks and ore samples of the mined horizons (Fig. 3). All the $^{87}Sr/^{86}Sr$ ratios lie in a narrow field, which coincides fairly well with that of seawater during Upper Triassic and Liassic (Burke et al. 1982). Despite the narrow variation range of the $^{87}Sr/^{86}Sr$ ratios, two systematic trends, which do not depend on the presence or absence of ore minerals, are clearly recognized:

1. The light dolomite of generations II or III is in all samples slightly more radiogenic than the dark dolomite of generation I.

2. A facies dependence of the $^{87}Sr/^{86}Sr$ ratios is recognized. Detailed petrographic studies show that samples deposited in a lagoonal or tidal flat environment display generally lower strontium ratios than those deposited in the oolitic barrier (Fig. 6).

The oxygen isotope values (Fig. 4) of the host rocks and the dolomites of generation I plot very closely in the range of about −7.8 to −5.5. This coincides with reported values for Liassic carbonates. A clear trend is visible in the oxygen isotope evolution. The carbonates of generation II or III show values which are always slightly depleted in ^{18}O compared to those of generation I.

The carbon isotope composition of all samples varies from +0.5 to +2.2 which are again normal values for marine Liassic carbonates. However, a clear diagenetic trend is also visible. The light dolomite of generation II is always depleted in the heavier isotope compared to the corresponding dolomite I (Fig. 5).

A similar evolution is observed with regard to the sulfur isotope composition. The delta ^{34}S values obtained lie in a very narrow range between +10 and +13.5 in sphalerite and between +6 and +8 in galena. This confirms the results of Schulz (1971), who attributed this homogeneity of the sulfur isotope values to a later homogenization process. However, the present investigation allows the recognition

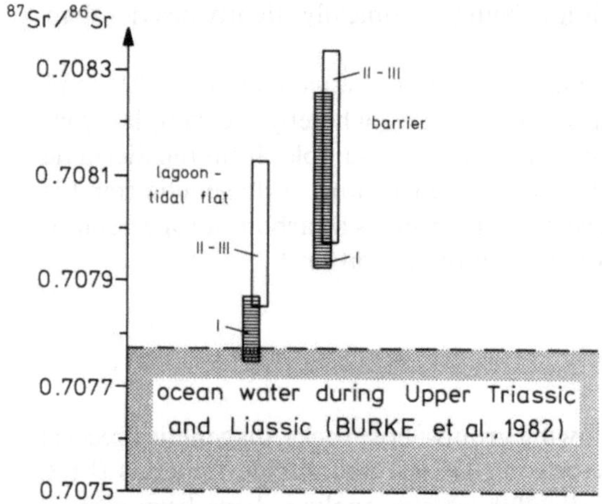

Fig. 6. The variation ranges of the strontium isotope ratios according to the facies position of the analyzed samples in the San Vicente Mine, Peru *I* generation I, *II* and *III* generations II and III (Fontboté and Gorzawski 1987)

of small isotope differences between subsequent sphalerite generations. The first sphalerite generation shows slightly higher delta ^{34}S values than generations II or III. The detection of this trend makes a later homogenization process as proposed by Schulz (1971) improbable.

These results are wholly consistent with the isotope trend observed for oxygen and carbon. A general trend toward lighter, stable isotope values in later diagenetic stages is recognized.

4.3 Comparison of the Isotope Results

The isotope ratios display similar trends for both areas. They can be summarized as follows.

1. No systematic differences between ore samples and host rock are recognized.
2. Systematic differences in the isotope composition according to the position in the diagenetic crystallization sequence are observed. In other words, a strict correlation between the isotope composition and the position in the crystallization sequence exists.
3. The strontium ratios in the first generations are less radiogenic than those of the last generations. It should be underlined that despite this clear trend, the absolute differences between $^{87}Sr/^{86}Sr$ ratios are always very small. This applies to differences between subsequent generations in each individual sample and also to the absolute differences in each district. The lowest $^{87}Sr/^{86}Sr$ ratios coincide fairly well with those being assumed for ancient ocean waters.
4. The isotope compositions of C, O, and S change in a similar way. With a few exceptions the later generations are slightly depleted in the heavier isotopes compared to the first generations.

In summary, it can be said that the minerals which form at the end of the crystallization sequence are slightly enriched in radiogenic Sr and depleted in the heavier isotopes of O, C, and S compared to those forming at the beginning. This is also in agreement with other isotope investigations carried out in non ore-bearing carbonate provinces (e.g., Dickson and Coleman 1980; Moore 1985; Meyers and Lohmann 1985; Woronick and Land 1985).

The investigations in the Alpujarrides and in San Vicente have demonstrated that the crystallization sequence observed by petrographic means is wholly matched by the isotope evolution. The isotope determinations represent therefore an excellent tool to trace the diagenetic evolution of ore and gangue minerals. In particular, the strontium isotope compositions, which deliver a sort of relative time scale of diagenetic events, have proved to be extremely useful to piece together the diagenetic evolution.

The trends observed in these areas can be interpretated as follows.

1. The fact that the strontium ratios become increasingly radiogenic during diagenesis implies the existence of an influx of fluid enriched in radiogenic strontium. An hypothesis which explains this concept is presented in Section 5.
2. The trend to lighter C isotopes can indicate fractionation processes produced by a reservoir effect in a system becoming progressively closed by an advancing diagenetic stage (Hannah and Stein 1984). An alternative explanation would be that minor amounts of organic carbon are incorporated in the carbonates. The decrease of ^{18}O is consistent with increasing T during diagenesis.
3. The sulfur isotope ratios in San Vicente are relatively heavy. Two possibilities could explain this. (a) Thermal degradation of sulfur-bearing hydrocarbons, (b) dissolution of preexisting evaporite minerals. The first possibility would require high temperatures during diagenesis. At the present stage of the investigation, the second possibility seems to apply better.

5 Presentation of a Model for "Early Diagenetic Ore Formation"

A conceptual genetic model for the San Vicente Mine which is consistent with the observational criteria is proposed. This model, which suggests an early diagenetic age for the main ore formation stage, is thought also to be applicable, with variations, to other areas. Specifically, the F–(Pb–Zn) deposits in the Alpujarrides display, as seen above, many similarities with respect to the paleogeographic position, facies associations, diagenetic features, and isotope evolution.

5.1 Constraints for the Proposed Model

This model (Fig. 7) bases on the following geologic, petrographic, and geochemical constraints observed in the San Vicente Mine (detailed descriptions of the observations listed below can be found in the references mentioned in Sect. 3.2, and particularly in Fontboté and Gorzawski (1987).

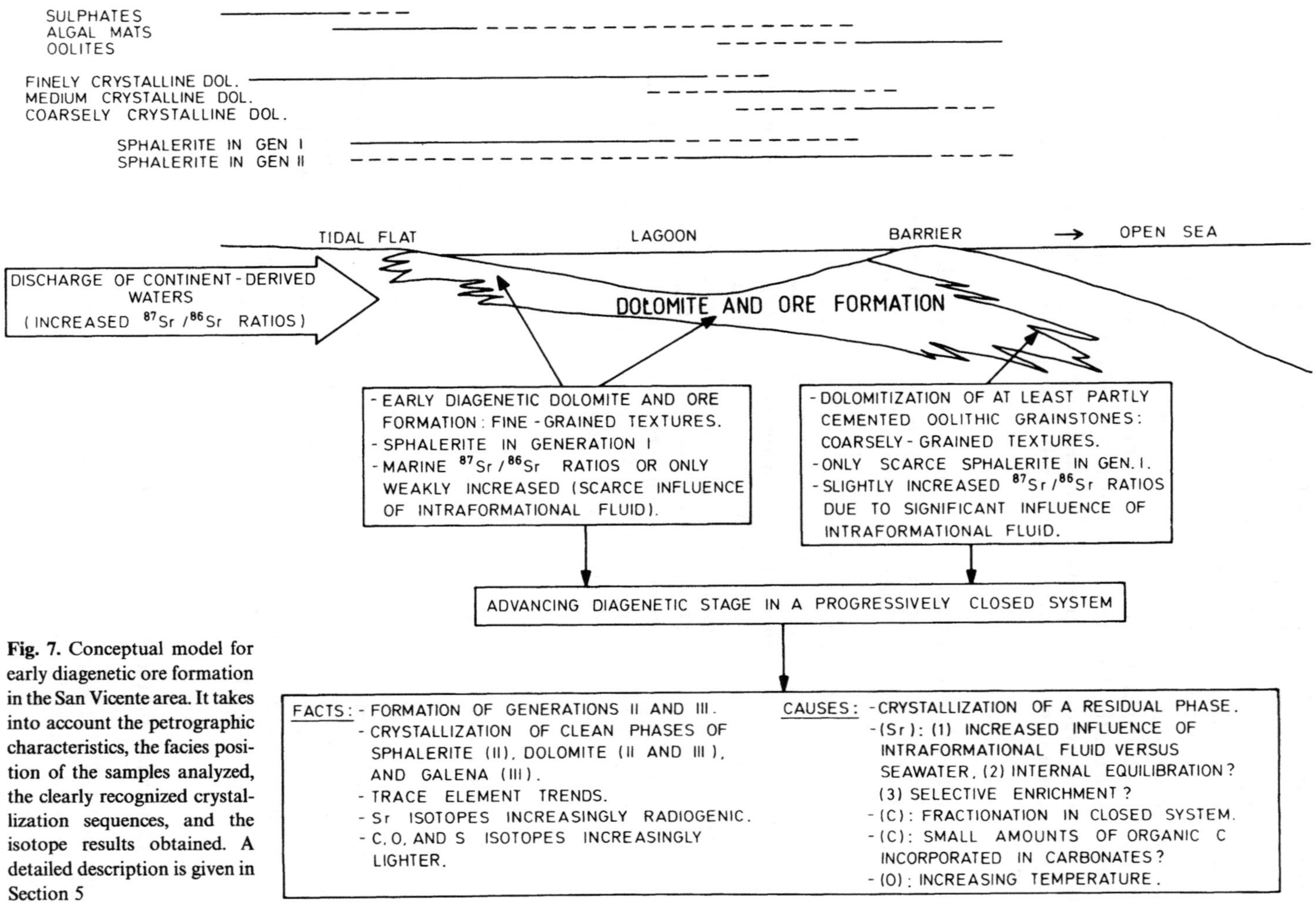

Fig. 7. Conceptual model for early diagenetic ore formation in the San Vicente area. It takes into account the petrographic characteristics, the facies position of the samples analyzed, the clearly recognized crystallization sequences, and the isotope results obtained. A detailed description is given in Section 5

1. Link to a definite paleogeographic position.
2. Link to definite facies within a peritidal environment.
3. The orebodies are essentially stratiform at the mine scale. In detail, cross-cutting structures are observed. The morphology of the orebodies can be compared to that of early dolomitic horizons. In the barrier facies the orebodies can present slightly oblique contacts with respect to the bedding, similarly to dolomitization fronts.
4. Sphalerite I (i.e., sphalerite in generation I of DCRs) is clearly intergrown with dolomite I. No corrosion features are observed in generation I. Thus, it can be concluded that the start of the ore formation process is contemporaneous with the dolomitization process.
5. The isotope data support also the conclusion that sphalerite I and dolomite I crystallized from the same fluid. The Sr isotope composition of dolomite of generation I in ore-bearing samples of the lagoon and tidal flat is the lowest measured in the San Vicente Mine. These compositions are only very slightly higher than that of ocean water in Liassic times. In two sphalerite samples where the measured strontium is considered to be contained as a trace element (not in carbonate inclusions), the $^{87}Sr/^{86}Sr$ ratios are identical to those of the intergrown dolomite I. Even if the measured strontium in these two samples could correspond partly to relicts of carbonates not leached in the sphalerite, the fact remains that *no difference in the isotope composition between ore-bearing and ore-free samples is observed.* A hypothetical ore-bearing fluid with different isotope composition should have dissolved and reprecipitated carbonate phases with other isotope compositions than those actually measured (it should also be kept in mind that dolomite of generation I does not show signs of corrosion). We must consider, therefore, that the same fluid producing the dolomitization of generation I must also be responsible for the contemporaneous precipitation of sphalerite I.
6. The lithogeochemical investigations provide additional constraints for this fluid. Striking are the high Mn values in the dolomite. It is known that Mn^{2+} moves readily only in an acid-reducing environment, for example, organic swamps and bogs. It can therefore be proposed that this fluid circulates through an acid and reducing environment. By increasing the pH value (necessary to precipitate dolomite), Mn is trapped in the dolomite lattice.
7. The carbon and oxygen isotope ratios are typical for the recognized sedimentary and diagenetic environment. A slightly meteoric influence is possible, although it has not been determined with certainty.
8. Lead isotope investigations on galena, gangue minerals, and host rock indicate that the lead is substantially more radiogenic than that of other strata-bound ore deposits in the Pucara basin (Gunnesch and Baumann 1986). Leaching of an old continental crust (Brazilian shield? underlying detrital sediments?) appears to be the probable source.
9. With advancing diagenetic evolution, slightly enrichment in radiogenic Sr and depletion in the proportion of heavier isotopes of O, C, and S is recognized.

10. A facies dependence on the $^{87}Sr/^{86}Sr$ ratios is recognized. Samples deposited in lagoonal or tidal flat facies display generally lower strontium ratios than those deposited in the oolitic barrier. This is due to a different diagenetic evolution.
11. This facies dependence is also reflected in the relative proportions of sphalerite in generations I and II. In samples deposited in the lagoon and in the tidal flat, sphalerite is very abundant in generation I, whereas in samples deposited in the barrier sphalerite occurs preferentially in generation II.

5.2 Presentation of the Model

A *genetic model* fulfilling these requirements is summarized as follows.

Detrital sediment-derived waters are discharged in early stages of diagenesis. The paleogeographic and facies reconstruction indicates that the ore-bearing sequence was deposited on a carbonate platform at the western margin of the Brazilian shield on tidal flats under preevaporitic conditions, and in lagoonal and internal barrier subenvironments. In this context, the existence of a discharge of waters which circulated through sediments rich in detrital material may be considered as very likely. A flow of these characteristics should display an increased radiogenic $^{87}Sr/^{86}Sr$ ratio compared to that of ocean water, and, perhaps, could show meteoric influence.

The facies dependence of the $^{87}Sr/^{86}Sr$ ratios support the existence of such a discharge in the early stages of diagenesis. In generation I of the samples deposited in lagoonal and tidal flat facies, with still recognized algal mat textures, evaporate pseudomorphs, etc., the dolomite (and sphalerite) formation is interpreted as a very early diagenetic process. On the other hand, generation I of the samples deposited in the barrier generally show an oolitic grainstone texture which is pervasively replaced by medium-crystalline dolomite. The dolomitization of the barrier calcarenites should take place, therefore, under moderate burial, but essentially contemporaneously with the early diagenetic dolomite formation in the lagoon.

Near the interface sediment-ocean water the influence of the more radiogenic, detrital sediment-derived fluid should be extremely low compared to that of seawater. Therefore, the strontium isotope composition of the early dolomite in the lagoon should be that of ocean water, or very slightly higher. The dolomitization of the barrier oolitic grainstones takes place in a deeper position in the sediment pile. The influence of the detrital sediment-derived fluid on the isotope composition of the intraformational fluid should be gradually larger compared to that of ocean water and, therefore, the $^{87}Sr/^{86}Sr$ ratios are increasingly radiogenic.

The clearly higher $^{87}Sr/^{86}Sr$ ratios of dolomites of generation I of the barrier compared to those of the tidal flat and lagoon are consistent with this interpretation.

The ore and dolomite depositions could occur at the interface between acid-reducing to neutral-reducing environments.

In a *progressing diagenetic stage*, with increasing burial in a progressively closed system, the influence of the intraformational fluid versus the marine ocean water increases gradually. The phases crystallizing now, specifically generations II and III of DCRs, display, therefore, also increasingly radiogenic $^{87}Sr/^{86}Sr$ ratios. They represent the crystallization of a residual phase in a system becoming progressively closed.

Such a partly closed system would also be consistent with the fractionation trends toward lighter C isotopes with an advancing diagenetic stage.

5.3 Conclusion

In conclusion, the present investigation allows one to recognize the influx of radiogenic continent-derived fluid in *early diagenetic stages.* With increasing cementation, the ratio seawater/continent-derived water in the intraformational fluid decreases, and, as pointed out above, the system will become progressively closed. New introduction of radiogenic brines during late diagenesis is therefore not necessary in order to explain increased $^{87}Sr/^{86}Sr$ ratios in the residual phase of diagenesis and in the precipitates therefrom.

Two sources for the acid-reducing waters which circulated through detrital sediments derived by erosion of old crust (as indicated by the Pb and Sr isotopes) are possible: (1) ascensional basinal brines, (2) surface-linked water movement (waters arriving from the east which circulated through still uncemented sediments rich in detritic material at the margin of the Brazilian shield). At the present stage of the investigation we favor the second possiblity because such a surface-linked hydrological system is known to act in early diagenetic dolomitization in similar environments.

Finally, it should be noted that the model for "early diagenetic ore formation" presented here constitutes an alternative to models also based on isotope studies and which propose late diagenetic or postdiagenetic basinal brine migration as the most important mechanism for the formation of Zn–Pb–(F–Ba) ore deposits in carbonate rocks (e.g., Kessen et al. 1981; Lange et al. 1983). In our opinion, late diagenetic or postdiagenetic processes do not explain many of the observations described above, for example, the clear association of the orebodies to certain facies and to early diagenetic dolomitization, or the isotope evolution and in particular the correlation between strontium isotope ratios and definite depositional environments. On the contrary, the present investigation underlines the role played by "surface-linked" ore formation processes, i.e., by ore-forming processes essentially linked to the environments in which early diagenesis takes place, in the genesis of some types of strata-bound Zn–Pb–(F–Ba) ore deposits in carbonate rocks characterized by a clear facies control.

Appendix: Some Exploration Guides for Similar Stratiform Carbonate-Hosted Zn–Pb–(F–Ba) Deposits

The paleogeography and the facies position of the ore deposits constitute the main empirical exploration guide at a regional scale. The ore deposits in the Alpujarrides described above and the San Vicente Mine are located in carbonate rocks deposited in an environment which can be pictured as the sum of the following elements: (1) a shallow-water carbonate platform at the margin of an emerged continent; (2) the shallow-water platform acts as a transition between detrital-evaporitic facies and open sea facies; (3) the orebodies occur in peritidal facies comprising tidal flat, lagoon, and barrier environments and do not occur in deeper facies; (4) evaporite molds in the tidal flat facies account for preevaporitic conditions; (5) evidence for early diagenetic dolomitization in the lagoon and tidal flat facies is recognized.

The available information indicates that the ores occur as flat lenses, of which the maximum length is arranged parallel to the facies belts occurring in the area.

In a similar way as in other carbonate-hosted Zn–Pb districts (e.g., Barbier 1979) the lithogeochemical investigations in the San Vicente Mine indicate that the ore-free samples of the ore-bearing units are not anomalous. In addition, the top and bottom contacts of the stratiform orebodies are very sharp. Several detailed profiles show that the Zn contents pass usually from the percent range to some tens ppm within a few centimeters. The same can be said for Pb. Since the orebodies are of lenticular shape, it is therefore possible to drill in the neighborhood of an orebody without detecting any Zn or Pb anomaly.

In several works the existence of "manganeses haloes" around Zn–Pb strata-bound ore deposits has been reported (e.g., Gwosdz and Krebs 1977; Clifford et al. 1986; Clifford et al., this Vol.). The lithogeochemical investigations performed in San Vicente have (1) demonstrated the existence of high Mn values in the dolomitic ore-bearing units, which (2) have no relation to volcanic activity, and which (3) are clearly linked to the dolomitization process. It appears that these Mn values, which are about six times higher than a world average of dolomitic rocks, can represent an indicator of the conditions favorable to Zn–Pb ore formation during diagenesis and, so far, constitute an important empirical exploration guide at the regional scale.

The stable and strontium isotope ratios do not show anomalous values compared to those of ore-free areas in the investigated ore districts. The ore formation is considered to be essentially a diagenetic process. In this process one important element appears to be the discharge of waters which circulated through sediments rich in detrital material. This discharge is specifically characterized by (slightly) more radiogenic strontium ratios by increasing the diagenetic stage. In this sense, in order to trace this definite type of diagenetic evolution, isotope investigations can be useful in exploration.

Acknowledgements. The present investigation was part of a research project supported by the European Communities (*Contract No. MSM-010-D*). We would like to thank the coordinator of this program J. Boissonnas and our colleagues from the B.R.G.M., Orléans and from the Service Géologique de Belgique, as well as Prof. G.C. Amstutz, for fruitful discussions in joint field trips and in several internal meetings. The rubidium content in carbonates was analyzed by AAS (Dr. R. Sobott, Preussag-Berkhöpen), and, for comparison, in a few samples, by neutron activation (Dr. E. Pernicka, Max-Planck-Institut für Kernphysik, Heidelberg).

References

Amstutz GC, Bubenicek L (1967) Diagenesis in sedimentary mineral deposits. In: G Larsen and GV Chillingar (eds) Diagenesis in sediments. Elsevier, Amsterdam, pp 417–475

Amstutz GC, Park WC (1971) The paragenetic position of sulfides in the diagenetic crystallization sequence. Soc Mining Geol Jpn, Spec Issue 3:280–282

Amstutz GC, Ramdohr P, El Baz F, Park WC (1964) Diagenetic behaviour of sulphides. In: Amstutz, G.C. (Ed.) Sedimentology and Ore Genesis. Elsevier, Amsterdam, pp 65–90

Barbier J (1979) Géochemie en roche autour de quelques gîtes ou indices de Pb–Zn en milieu carbonaté. Bull BRGM S–II:1–22

Bates RL, Jackson JA (eds) (1980) Glossary of geology. Am Geol Inst, Falls Church, Virginia, 749 pp

Burke WH Denison RE, Hetherington EA, Koepnick RB, Nelson HF, Otto JB (1982) Variations of seawater $^{87}Sr/^{86}Sr$ throughout Phanerozoic time. Geology (Boulde) 10:516–519

Clifford JA, Ryan P, Kucha H (1986) A review of the geological setting of the Tynagh Pb–Zn deposit, Ireland. In: Andrew CJ, Crowe RWA, Finlay S, Pennell WM, Pyne JF (eds) Geology and genesis of mineral deposits in Ireland. Irish Assoc for Econ Geol, Dublin, pp 419–439

Delgado F, Estévez A, Martín JM, Martín-Algarra A (1981) Observaciones sobre la estratigrafía de la formación carbonatada le los mantos alpujárrides (Cordillera Bética). Estud Geol (Madr) 37:45–57

Dickson JAD, Coleman ML (1980) Changes in carbon and oxigen isotope composition during limestone diagenesis. Sedimentology 27:107–118

Dunoyer de Segonzac G (1968) The birth and development of the concept of diagenesis (1866–1966). Earth Sci Rev 4:153–201

Fontboté JM, Fontboté L, Amstutz GC (1983) Observations on the setting of the strata-bound F–(Pb–Zn) deposits in the Triassic of the Alpujarrides (Betic Cordillera, southern Spain). In: Schneider HG (ed) Mineral deposits of the Alps and of the Alpine epoch in Europe. Springer, Berlin Heidelberg New York, pp 249–258

Fontboté L (1981) Strata-bound Zn–Pb–F–Ba deposits in carbonate rocks: new aspects of paleogeographic location, facies factors and diagenetic evolution. (With a comparison of occurrences from the Triassic of southern Spain, the Triassic/Liassic of central Peru and other localities). Diss Univ Heidelberg, 193 pp

Fontboté L, Amstutz GC (1983) Diagenetic Crystallization Rhythmites in Mississippi Valley Type Ore Deposits Int Conf on Mississippi Valley Type Lead-Zinc Deposits, Proc Vol Univ Missouri-Rolla, pp 328–337

Fontboté L, Gorzawski H (1987) Petrographic and geochemical indicators for the exploration of hidden ore deposits in sedimentary rocks. Final report Contract MSM-010-D, Commission of the European Communities, 221 pp (unpubl.)

Gonzalez E, Fontboté L (1986) Petrographic and lithogeochemical investigations on the Pucará Group in the carbonate-hosted Zn–Pb Deposit of San Vicente, Peru. Berliner Geowiss Abh, Reihe A, Sonderband (10. Geowiss. Lateinamerika-Kolloq, Berlin), pp 126–127

Gorzawski H, Fontboté L (1986) Geochemical (isotopic) tracing of diagenetic ore formation in the carbonate-hosted Zn–Pb deposit of San Vicente, Peru. Berliner Geowiss Abh, Reihe A, Sonderband (10 Geowiss Lateinamerika-Kolloq, Berlin), pp 128–129

Gunnesch KA, Baumann A (1986) Bleiisotopie in Zentralperu. Berliner Geowiss Abh, Reihe A, Sonderband (10. Geowiss, Lateinamerika-Kolloq, Berlin, Abstracts), pp 136–137

Gwosdz W, Krebs W (1977) Manganese halo surrounding Meggen ore deposit. Trans Inst Ming Metall 86 (5):73–77

Hannah JL, Stein HJ (1984) Evidence for changing ore fluid composition: stable isotope analyses of secondary carbonates, Bonneterre Formation, Missouri. Econ Geol 79:1930–1935

Kessen K, Woodruff MS, Grant NK (1981) Gangue mineral $^{87}Sr/^{86}Sr$ ratios and the origin of Mississippi Valley mineralizations. Econ Geol 76:913–920

Lange S, Chaudhuri S, Clauer N (1983) Strontium isotopic evidence for the origin of barites and sulfides from the Mississippi Valley-type ore deposits in southeast Missouri. Econ Geol 78:1255–1261

Lavado M (1980) Geologic aspects of the ore occurrences at the San Vicente Mine, San Ramon, central Peru. M Sci Thesis Univ of Texas, Dallas, 127 pp

Levin P (1975) Der petrographisch-geologische Rahmen des Chanchamayo-Gebietes in Ost-Peru und die Deutung seiner schichtgebundeneen Vererzungen. Entwurf einer Metallogenese der östlichen Zentral-Peru. Diss Univ Heidelberg, 242 pp

Levin P, Amstutz GC (1973) Neue Untersuchungen über schichtgebundene Lagerstätten im zentralen Ostperu. Münster Forsch Geol Paläontol 31/32:233–259

Martín JM, Torres-Ruiz J (1982) Algunas consideraciones sobre la convergencia de medios de depósito de las mineralizaciones de hierro y plomo-zinc-fluorita de origen sedimentario encajadas en rocas triásicas de los Complejos Nevado-Filábride y Alpujárride del sector central de la Cordillera Bética. Bol Geol Min Madrid, 93:314–329

Martín JM, Torres-Ruiz J, Velilla N, Fenoll Hach-Ali P (1984) Paleokarstic lead-(zinc)-fluorite deposits in shallowing upward sequences in the Triassic of the Alpujarrides (Betic Cordillera, southern Spain). In: Wauschkuhn A, Kluth C, Zimmermann R (eds) Syngenesis and epigenesis in the formation of mineral deposits. Springer, Berlin Heidelberg New York, pp 438–447

Martín JM, Torres-Ruiz J, Fontboté L (1987) Facies control of strata-bound ore deposits in carbonate rocks: the F–(Pb–Zn) ore deposits in the Alpine Triassic of the Alpujarrides, southern Spain. Miner Dep 22:216–226

Meyers WJ, Lohmann KC (1985) Isotope geochemistry of regionally extensive calcite cement zones and marine components in Mississippian limestones, New Mexico. In: Schneidermann N, Harris M (eds) Carbonate cements. Soc Econ Paleontol Mineral Spec Publ 36:223–239

Moore CH (1985) Upper Jurassic subsurface cements: a case history. In: Schneidermann N, Harris M (eds) Carbonate cements. Soc Econ Paleontol Mineral Spec Publ 36:291–308

Prinz P (1985) Zur Stratigraphie und Ammonitenfauna der Pucara-Gruppe bei San Vicente (Dpto. Junín, Peru). Newsl Stratigr 14:129–141

Schulz GG (1971) Die schichtgebundene Zinkblende Lagerstätte San Vicente in Ost-Peru und ihr geologischer Rahmen. Diss Univ Aachen, 165 pp

Veizer J, Hoefs J (1976) The nature of O^{18}/O^{16} and C^{13}/C^{12} secular trends in sedimentary carbonate rocks. Geochim Cosmochim Acta 40:1387–1395

Woronick RE, Land LS (1985) Late burial diagenesis, Lower Cretaceous Pearsall and Lower Glen Rose Formations, south Texas. In: Schneidermann N, Harris M (eds) Carbonate cements. Soc Econ Paleontol Mineral Spec Publ 36:265–275

Strata-Bound Mineralizations in the Carnic Alps/Italy

L. Brigo[1], P. Dulski[2], P. Möller[2], H.-J. Schneider[3], and R. Wolter[3]

Abstract

The strata-bound mineralizations of the Palaeocarnic Alps/Italy are characterized by a zonation of parageneses: predominating Hg–Cu sulphides in the west, ZnS in the centre, barite and fluorite in the east. They are bound to a palaeorelief of Devonian limestones transgressively overlain by Lower Carboniferous to Lower Permian clastic sediments. The mineralizations are largely caused by reworking protores (sulphides) or weathering of carbonates (barite and fluorite). The developing Devonian reef relief, karst cavities and faults acted as mechanical and geochemical traps. In situ remobilization and recrystallization did not affect the almost lithified limestones which do not exhibit significant dispersion halos of base metals or brines. This rather unusual type of mineralization cannot be prospected by lithogeochemical proximity indicators.

1 Introduction

Devonian carbonates crop out, more or less continuously, all along the 100 km of the Palaeocarnic chain/Italy (Fig. 1) and continue 70 km to the east (Caravanche region/Yugoslavia). The carbonates occur within a fossiliferous Palaeozoic sequence.

Geological models of the Palaeozoic chain have been summarized by Selli (1963), Flügel (1964), and Vai (1974). Recently, Spalletta et al. (1982a, b) proposed a model that is based on biostratigraphic analyses of some selected sections of the Carnic Alps. According to this model, the carbonate platforms were drowned and transgressively covered by Hercynian clastic sediments as the result of progressive synsedimentary extensional tectonics beginning in the Frasnian. They suggested that this model is valid for the Palaeocarnic chain as a whole. In contrast, Brigo

[1] Istituto di Mineralogia dell'Universitá di Ferrara, Corso Ercole 1° d'Este, 32, I-44100 Ferrara, Italy
[2] AG Geochemie, Hahn-Meitner-Institut für Kernforschung Berlin, Glienickerstr. 100, D-1000 Berlin 39
[3] Institut für Angewandte Geologie der Freien Universität Berlin, Wichernstr. 16, D-1000 Berlin 33

Mineral Deposits within the European Community
(ed. by J. Boissonnas and P. Omenetto)

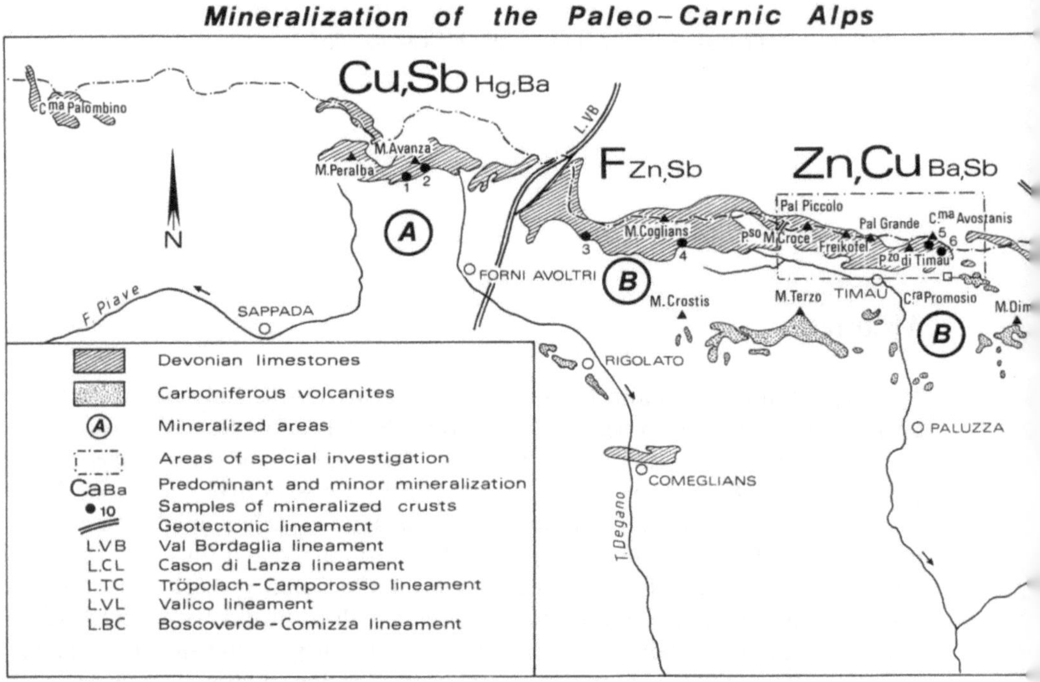

Fig. 1. Geological sketch map of the area of the Carnic Alps. Mineralized areas **A** through **E** are described in detail in the text. Typical parageneses are given; the size of the element symbols characterizes the relative frequency

and di Colbertaldo (1972) defined the mineralized horizon as an "important litho-geochemical guide horizon" for the paleaorelief of the Carnic Alps and Caravanches (Omenetto and Brigo 1974; Assereto et al. 1976; Brigo and Omenetto 1978; Venerandi-Pirri 1977). This paleaorelief is the result of an emersion phase which has been proved by the observed stratigraphic gaps between the Devonian-Carboniferous to Permian.

Besides the development of a reef-controlled relief (bioherms and biostromes), strong indications of synsedimentary fracturing are observable, particularly in the Timau area (Fig. 1B), causing horst and graben structures. Similar tectonic patterns can be tentatively recognized in the other areas, too, e.g. in M. Cavallo (Fig. 1C) and the Coccau area (Fig. 1E).

The Carnic Alps are cut by some regionally important lineaments such as the Val Bordaglia, Cason di Lanza, Tröpolach-Camporosso and Coccau-Thörl lines (Fig. 1). These lineaments separate zones of different mineralizations as well as areas displaying different stratigraphic gaps. Therefore, it has to be assumed that these zones already existed in the Palaeozoic. The orientation of these structural elements and their function as the limits of subsided basins or uplifted blocks are the consequence of large-scale transcurrent regional tectonics. Deep-rooting plate movements along the Gailtal line, as part of the regional Insubric lineament, acted as a pathway for Hercynian granites (Sassi et al. 1985). West of the Val

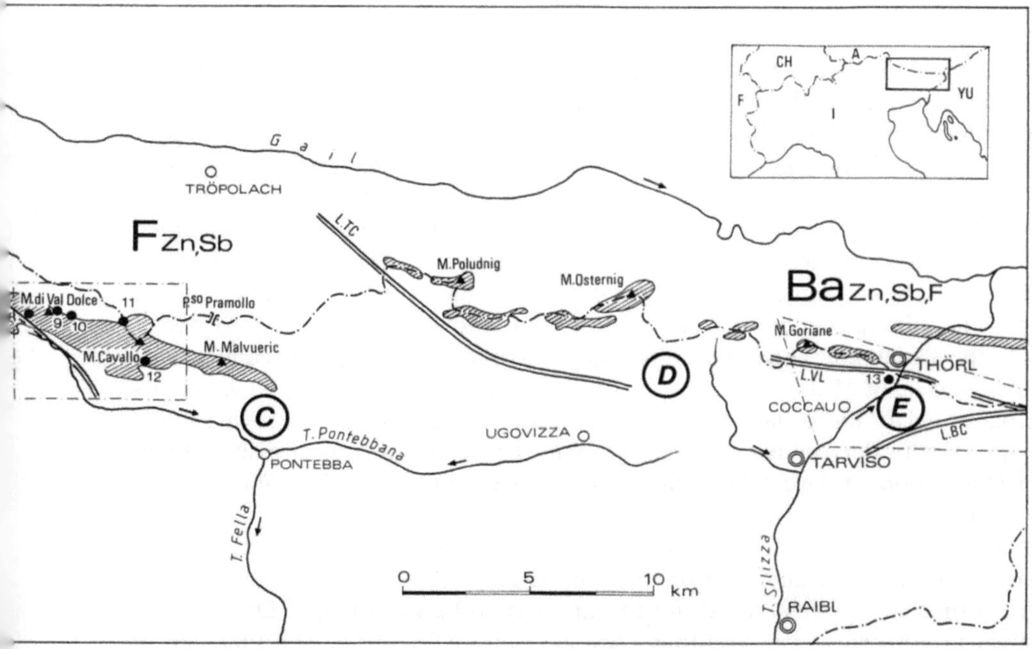

Bordaglia lineament and within the northern Austrian slope, the Palaeozoic sequence is weakly metamorphosed.

Base metal deposits associated with barite and/or fluorite extend over large, and palaeogeographically well-limited areas of the carbonates. The mineralizations are linked to a pronounced palaeorelief of Devonian limestones transgressively overlain by Lower Carboniferous to Lower Permian clastic sediments. Most of the mineralizations are of minor importance. The major ones were mined, e.g. in the areas of M. Avanza and, subordinately, in P. zo di Timau and Coccau/Thörl (Italy) and Rus-Stegovinik (Caravanche region, Yugoslavia).

The various terrains overlaying the palaeorelief such as Hochwipfel 'flysch' (Lower to Middle Carboniferous), Pontebba Supergroup (Upper Carboniferous) and Tarvisio breccia or Val Gardena sandstone (Lower to Middle Permian) indicate a complex palaeogeographic evolution strictly related to the synsedimentary Devonian tectonics and the polyphase Hercynian orogeny. The mainly compressive Alpine orogeny has only partially obliterated the old structures. Carboniferous magmatic volcanic events are known within the Hochwipfel formation to the south of the Carnic crest (Fig. 1).

With regard to the genetic problems of the mineralization only the time interval between Middle to Upper Devonian and Lower Carboniferous/Lower Permian is of particular interest.

2 Stratigraphic Position of the Mineralizations

The mineralizations are strata-bound to a carbonate sequence of Middle to Upper Devonian age. During this time interval, the Devonian reefs underwent an intensive (synsedimentary) tectonic fracturing forming local horsts and grabens. The upper tens of metres of this limestone relief host the mineralization, mostly on top (Fig. 2). The mineralized rocks are transgressively overlain by clastic series of considerably different ages. Within this time gap erosion and karstification of the emerged palaeorelief took place. Contemporaneously, the quantitatively most important part of the present mineralization was formed, consisting of silica crusts, karst fillings and mineralized breccias, the latter pointing to an unknown protore phase.

In the Timau area (Fig. 2) the mineralized host rock is stratigraphically proved by fossils to be of Middle to Upper Devonian. The transgressive clastic sediments of Hochwipfel formation lack biostratigraphic marks in this area. Francavilla (1966) palynologically determined the stratigraphic record of the transgressive Hochwipfel formation for the entire Carnic Alps to range from Namurian to Westphalian B. On the other hand, Flügel et al. (1959) proved an age of Tournaisian-Visean transition by conodonts from the lower part of a similar sequence 2 km north of the mineralized Pal Grande section (Fig. 1B, Angertal/Austria). With this information the time gap in this area shrinks to an interval between Upper Devonian to Lower Carboniferous. The resulting 5 Ma are the smallest range of a time gap as well as the most precise determination in the area of the Carnic Alps.

In the M. Cavallo-Val Dolce area (Fig. 1C) the host rock is believed to be of Middle Devonian age. It represents a karstic palaeorelief. Locally, pedogenetic reddish horizons and dolomitic crusts occur. The very complex palaeogeographic and tectonic Devonian cycle is transgressively overlain by the so-called Permo-Carboniferous molasse (base of the Pontebba Supergroup, mostly Meledis formation). This indicates a time gap of probably more than 20 Ma.

In the Coccau area the mineralized host rock is believed to be an undifferentiated and karstified Devonian limestone. The transgressive contact is formed by the Lower Val Gardena sandstone/Tarvisio breccia of Permian age, revealing a time gap of at least 60 Ma.

3 Characterization of Mineralizations

The mineralizations consist of a few sulphides associated with gangue minerals in regionally varying proportions. The sulphide minerals are arranged with reference to their abundance:

fahlore (different varieties), sphalerite, chalcopyrite, pyrite and marcasite, locally accompanied by galena, cinnabar, boulangerite and skutterudite.

Of probably economic interest are local enrichments of barite and fluorite. Calcite and quartz are common, siderite is present subordinately only.

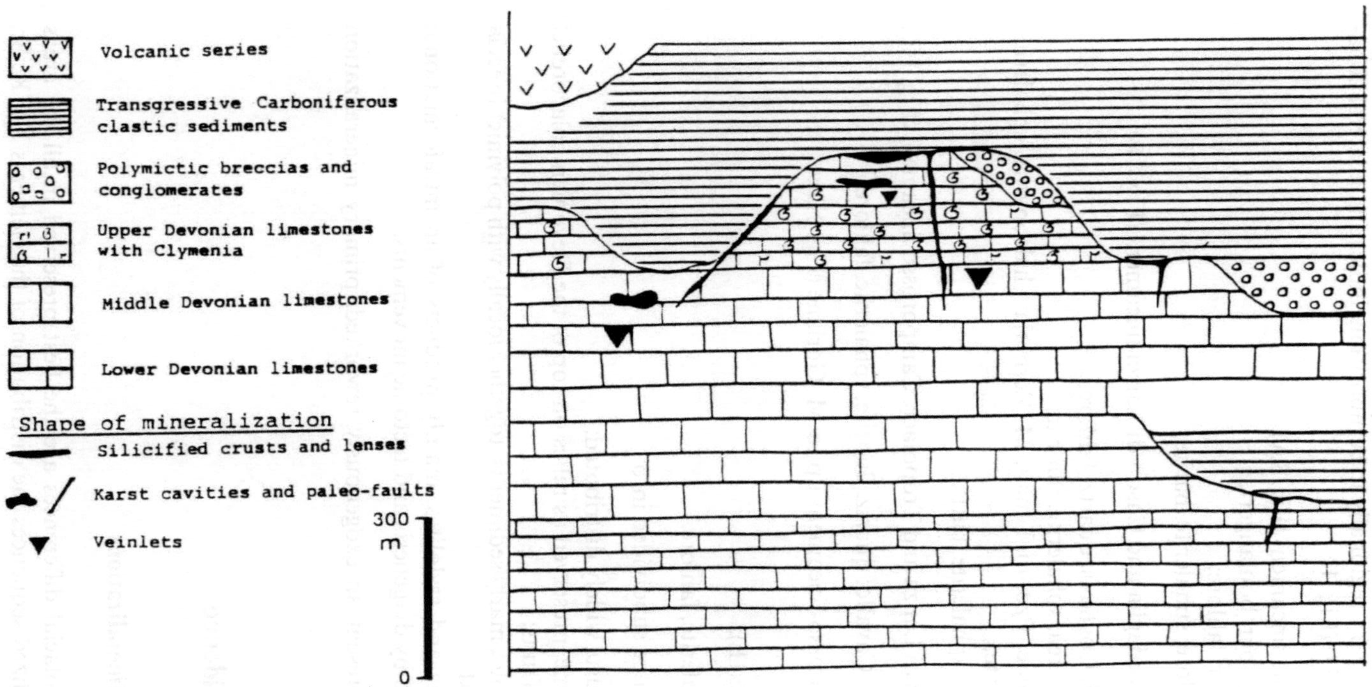

Fig. 2. Stratigraphic sketch profile of the Timau area (Brigo et al. 1986). The mineralization is stratabound to the Devonian limestone and occurs as crust on top of the paleorelief, in karst cavities and faults

The ore parageneses vary regionally in composition, indicating a lateral zonation (Venerandi-Pirri 1977) (Fig. 1):

West of the Val Bordaglia lineament (Fig. 1A):
M. Palombino: cinnabar, boulangerite;
M. Peralba: cinnabar, fahlore;
M. Avanza: Hg-fahlore, sphalerite, barite.

Area between the Val Bordaglia and Cason di Lanza lineament (Fig. 1B):

Monti di Volaia, M. Coglians: quartz, fahlore,
western part of this area: sphalerite, fluorite;
P. Piccolo–P. Grande, Pzo. di Timau, M. Zermula: chalcopyrite, sphalerite, fahlore, barite, siderite,
Comeglians: fahlore, sphalerite, barite.

Area between the Cason di Lanza and Tröpolach-Camporosso lineament (Fig. 1C):

C. ra Valbertad–M. Cavallo: quartz, fluorite, sphalerite, fahlore.

Area Tröpolach-Camporosso-lineament up to M. Goriane (Fig. 1D):
unmineralized.

Coccau-Thörl area (Fig. 1E):

barite, fluorite, sphalerite, fahlore.

The mineralizations can be subdivided into:

1. Patchy, silicated crusts widely distributed;
2. Polymict breccia type, preserved as relics on top of the Devonian palaeorelief and in open palaeofaults;
3. Rhythmic, bituminous, marly sediments, together locally with polymict breccia in karst cavities; and
4. Mineralized veinlets and stylolites, mainly products of an intraformational mobilization caused by diagenesis and tectonic movements.

Probably (1) to (3) represent an exogenously reworked, primary mineralization (protore).

4 Geochemical Evidence

4.1 Host Rocks and Mineralizations

In the Timau area lithofacial differences are the best preserved in all the areas examined. In unmineralized sequences the distribution of the elements Na, K, Sr, Mn, Fe, Co, Ni, Cu, Ag, Pb, As and F is strictly controlled by the lithofacies, corresponding with averages of limestone, sandstone and shale (Wedepohl 1969). This is particularly true for the clastic sediments of the overlaying Hochwipfel formation. Elements enriched in mineralizations of the Timau area are Ba, Cu, Zn, Sb (Ag, Pb, As). According to Kern (1985) and Moschik (1985), the background contents of most of these elements are again normal (Wedepohl 1969). In Devonian limestones sporadic enrichments of Ba and Zn are not part of a well-developed halo, but are erratic. They indicate minute barite and ZnS mineralizations in fissures and stylolites. In a few cases Ba and Zn enrichments may be first indications of a mineralization nearby. Generally, these elements cannot be used as 'proximity' indicators because a systematic increase towards known mineralization is not observed.

In the M. Cavallo-Val Dolce area (Fig. 1C) the transgressive Carboniferous sequence is nearly totally eroded. However, a few 100 m north of the mineralized Devonian, which is cut by major faults, the Upper Carboniferous outcrops. The geochemical investigations were restricted to the Devonian limestones and their mineralizations.

The unmineralized Devonian limestones show average element contents (Wedepohl 1969). Only Co, Zn and F are slightly enriched. Two types of mineralization may be distinguished in this area:

1. Predominant fluorite with accessory sphalerite; and
2. Predominant sphalerite with accessory Cu–Pb–Sb–Ag sulphides, mainly bound to dark, clay-bearing fillings of karst cavities.

The profile 'Val Dolce' (Fig. 3) cuts one of these mineralizations. Zn, Cu and Pb are enriched sporadically, usually accompanied by significant enrichments of fluorine. In this profile the distribution of fluorine itself seems to be important because all along the 15 km below a 'CaF_2 crust' the F content of the limestone exceeds the probable background value of 450 ppm by a factor of two. In this case F may serve as an important 'proximity indicator'.

The ZnS + (Cu, Pb, Sb, Ag) mineralization is very similar to the Timau paragenesis. Only barite is nearly absent. Nevertheless, these mineralizations are interpreted as having formed in karst cavities under sapropelic conditions. In the vicinity of this type of mineralization no sigificant fluorine enrichment could be traced.

In the Coccau area (Fig. 1E) the mineralization is covered by the Val Gardena sandstone/Tarvisio breccia (Lower Permian). Carboniferous units are largely missing. The mineralizations predominantly consist of a barite breccia body on top of a limestone (partly dolomitic limestone) sequence of Devonian age(?). In all the sampled cores a more or less intensive mineralization is present.

Val Gardena sandstone overlaying the mineralization is slightly enriched in fluorine and in Ba significantly. Microscope examinations reveal that these enrichments are due to barite or fluorite mineralizations in fissures or in the cement of the sandstone. The limestones and the dolostones are enriched in Ba, Co, Ni and Ag, the dolostones additionally in F. Co and Ni are generally bound to the amount of the $HCl–HNO_3$ insoluble residue.

The mineralization of Coccau is characterized by Ba and F (Zn, Ag, Pb). A mineralized breccia body can be distinguished from mineralizations in karst cavities. The breccia body, economically most important, consists predominantly of barite. The karst mineralization contains various sulphide minerals, besides the barite and

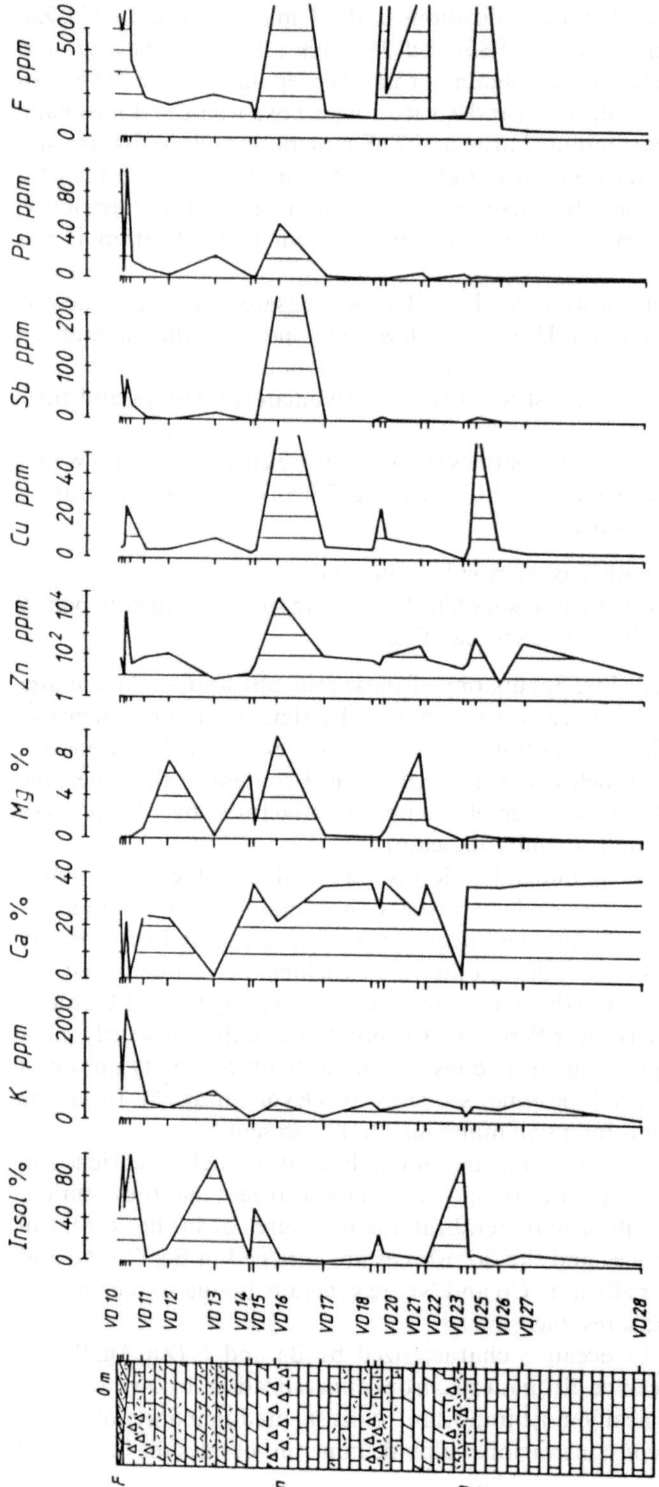

Fig. 3. A typical geochemical profile cutting the upper 22 m of the Devonian limestone. Three sulphide mineralizations are intercalated (*Cu–Zn, Zn, F*); on *top*, a fluorite crust

fluorite embedded in a bituminous clayish matrix. None of the profiles sampled are totally unmineralized. Below the barite breccia, isolated Ba, F or sulphide enrichments occur sporadically. The mineralizations are not surrounded by a well-developed halo. In the vicinity of the mineralizations, especially in the carbonate sequence, very small mineralizations occur in fissures disseminated in the host rock, pointing to economically interesting orebodies. Additionally, the Mn content of the host rock decreases on approach to these mineralizations at a distance of a few metres.

4.2 Palaeobrines

In previous investigations on Triassic Pb–Zn deposits (Wolter and Schneider 1983, 1985 and in press) a correlation between mineralization and an increasing salinity of brine relics was found. Therefore, leaching experiments have been performed on samples from mineralized and barren profiles of the Timau area. Ca^{2+}, Cl^- and $SO_4{}^{2-}$ are from the predominantly leached Devonian limestones; $SO_4{}^{2-}$, K^+ and Na^+ from the Carboniferous clastics. These components reflect the mineralogical composition of the matrix. Na^+ and Cl^- could be attributed to brines generated during the diagenesis of the limestones (Wolter and Schneider in press). The salinity of such brines is related to their Na/Cl ratio (Collins 1975).

From the Na^+/Cl^- ratio in leachates of Devonian limestones a salinity very similar to ocean water is derived, only slightly increased in a few samples. Na^+/Cl^- ratios cannot be used to estimate salinities of connate water trapped in the Carboniferous sediments because:

1. Chlorine contents, in general, are below the detection limit of conductometric titration;
2. The high Na^+ contents in the leachates must be caused by hydrolysis of clastic feldspar and clay minerals.

Na^+/Cl^- ratios in mineralized zones are extremely scattered. In contrast to the Triassic Pb–Zn deposits, Na^+/Cl^- ratios in this type of mineralization yield no information on fluids related to ore formation.

4.3 Microthermometry

Fluid inclusions in fluorite and barite are, in general, very small and mainly single phase. Their freezing temperatures are between $-13°$ to $-15\,°C$, indicating a salinity of 18 to 19 wt% NaCl equivalent.

The few two-phase inclusions observed homogenize at temperatures between 50° and 104 °C. Their freezing temperatures are in the range of $-10.8°$ to $-18.8\,°C$, indicating salinities of 14 to 23 wt% NaCl equivalent.

In one sample only, inclusions with high homogenization temperatures (170° to 230 °C) were detected with freezing temperatures, indicating 20 to 21 wt% NaCl equivalent. The frequent occurrence of very small ($<0.5\ \mu$) and single-phase inclusions indicates that the mineralization of fluorite and barite took place at low temperatures. This could be the result of mobilization during diagenesis.

4.4 Isotope Studies

Sulphur isotopes of barites fall into the range of $\delta^{34}S = +15.5$ to $+19.3‰$, with an average of $+17.7‰$ (Puchelt, priv. comm.; Drovenik, pers. comm.). These values are in good agreement with $\delta^{34}S$ of sulphate in ocean waters of Middle Devonian or Lower to Middle Carboniferous age (Hoefs 1980). It may be assumed that the source of barite-sulphur was ocean water or its brines.

Lead-isotope analyses of galena samples from Devonian mineralizations of the Carnic Alps display the same lead-isotope ratios as the important Triassic Pb–Zn deposits of the eastern Alps. These lead ratios indicate a Devonian to Carboniferous age (Köppel, pers. comm.), thus implying that the metal-bearing fluids were generated in a crustal source by a regional event.

4.5 REE in Fluorite and Barite

REE distribution patterns in Ca minerals can be used to characterize the physicochemical conditions during mineral formation (Möller 1983). The REE distribution patterns of fluorites from both areas are quite similar (Fig. 4). Compared with typical primary, hydrothermal REE distribution patterns, the fluorite samples show a significant depletion of light REE. This indicates severe remobilization. Evidence of an intensive fluorite mobilization is clearly observed in thin sections which show different fluorite generations such as fillings of fissures, cements of breccias and euhedral crystals.

Because barite is a common constituent of the mineralization, its geochemical characterization is of special interest for its genetic interpretation (Morgan and

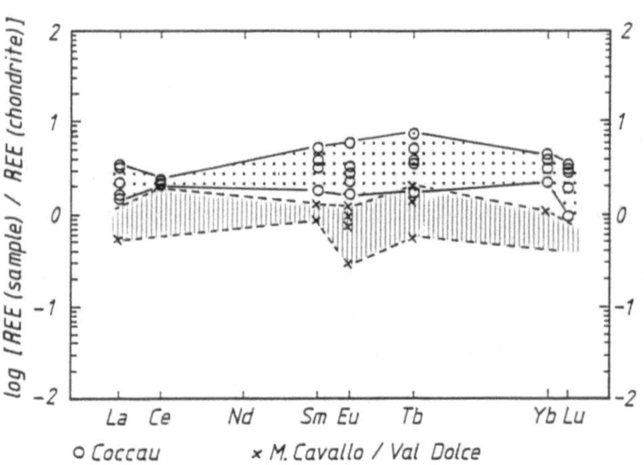

Fig. 4. REE distribution patterns of fluorites from two different areas of the Carnic Alps

Wandless 1980; Guichard et al. 1979). After dissolving barite by DOWEX ion exchanger (Luck 1985) REE contents were found to be below the detection limits. In detailed studies it was found that the REE contents in barite reported in literature are erroneous because in 20 samples from various deposits either REE were not found by ICP–AES or the REE were correlated with Ca^{2+}. Pure barite is nearly devoid of REE (except Eu).

4.6 Ga/Ge in Sphalerite

The Ga/Ge ratios of sphalerite yielded an estimate of the initial temperatures of metal-bearing solutions (Möller 1985). Ten sphalerite samples from different mineralizations and areas of the Carnic Alps were analyzed by spark-source, mass spectrometry. In all samples the log Ga/Ge ratio falls in the range of 0.21 to 0.59. Seven of ten samples have log Ga/Ge ratios between 0.44 and 0.59. These values indicate that the lowest initial model temperature of the metal ion-bearing solutions are backed up by very low Tl contents (<1 ppm) (Möller et al. in prep).

4.7 Carboniferous Volcanites

The distribution of the Carboniferous volcanites in time and space indicates a deep-rooted change of the geotectonic regime within the crustal segment studied. Previous works have described a basic volcanism of alkaline-olivine composition (Spalletta et al. 1982a). Recently, Flora et al. (1983) inferred from new petrological data a transitional composition of the Carboniferous spilites. Further studies are in progress (Coltorti et al., in prep.), probably indicating an anorogenic or continental origin for the basalts.

5 Discussion

The studied mineralizations obviously represent an extraordinary, nearly unknown type. Although the formation of the mineralization has been discussed in the past (Brigo and di Colbertaldo 1972; Omenetto and Brigo 1974; Assereto et al. 1976; Brigo and Omenetto 1978; Spalletta et al. 1982b), the offered interpretations do not solve all the genetic problems.

The horizontal extension of the mineralization over a nearly 100-km strike as well as its remarkable strata-bound occurrence, linked to the top of a Middle to Upper Devonian reef chain, points to a regional event. But this very event is documented solely by the mineralization which formed in a time interval that is not recorded stratigraphically. The mineralization cannot be derived from the overlaying series because the inconformable transgression of clastic sediments has proved to be of Lower to Middle Carboniferous age in the centre (Timau area, Fig. 1B) and of Lower Permian age (Coccau area, Fig. 1E) in the east. Additionally, the geochemical investigations on the transgressive clastics reveal a remarkable shortage of base metals. Thus, these formations cannot be considered as the source of the mineralizations.

On the other hand, the carbonate host rocks do not show any primary geochemical anomalies with reference to base metals. Therefore, the carbonate rocks also cannot be considered as the 'source bed' of mineralization. Only Ba (about 400 ppm) and F (about 1000 ppm) indicate obviously increased contents compared with the common average values of carbonate rocks (Wedepohl 1969). This may have been significant especially for the fluorite enrichments of the M. Cavallo area (Fig. 1C) and the barite deposit of Coccau (Fig. 1E). In this case the carbonate host rock could be regarded as the source for fluorite and barite. The enrichment of fluorite and barite took place under superficial conditions, which has been proven by REE distribution patterns, microthermometry of fluorite and sulphur-isotope ratios of barite.

Therefore, the formation of the widespread mineralization can be explained only by the occurrence of two different processes. The parageneses of the mineral assemblages show a remarkable regional variation of the base metal ratios: e.g. predominating Hg–Cu in the west (area A), Zn–Sb–Cu (area B) and Zn alone (area C) in the centre. Farther to the east (area E) the base metal content decreases to only accessory amounts. Overlooking this distinct paragenetic W–E zoning, a rather systematic variation is observed which may be explained either by different superficial sources (e.g. as erosional products) or by regionally differentiated hydrothermal springs. Because the Ga/Ge ratios of various sphalerites indicate an initial model temperature of the metal-bearing solutions of about 200 °C, a spotlike hydrothermal supply is rather plausible.

In the search for a corresponding magmatic event, the Middle Carboniferous volcanites at least represent the stratigraphically and regionally nearest activity located on a still continental crust. It may have been started by hydrothermal activities along the centre line of the Upper Devonian reef belt. In late Carboniferous the widely distributed granitic intrusions along the deeply rooted Insubric line mark the end of the Variscan orogeny, accompanied by strong phases of folding and faulting. This geotectonically radical change is introduced in Upper Devonian by a tectonic disintegration of the carbonate platforms with partial uplifting, intensive erosion and karstification. These processes were accompanied by local hydrothermal activities. This very period falls into the 'gap time' of the stratigraphic record.

The first stage of accumulation of base metal sulphides locally occurred in euxinic pools or patches. Contemporaneously, proceeding weathering of the Devonian carbonates supplied the required amounts of Ba and F for the formation of barite and fluorite. Superficial remobilization processes then concentrated the barite and fluorite together with the sulphide precipitates in karst cavities and open faults. A last stage of this complex development is represented by a widespread silification which 'seals off' the surface of the pronounced relief.

6 Conclusions

The mineralizations of the Palaeocarnic Alps are strata-bound, but neither 'syn-sedimentary' nor 'syngenetic' in a common sense. Rather, they should be specified as epigenetic. Moreover, the complex mineralization was formed by at least two different processes, one generated by ascending hydrotherms, the other by weathering. They met at various localities of the reef relief during the tectonically unsteady time, producing the combined ore lodes. The reef caverns and sinks acted as mechanical and geochemical traps. Since the carbonate host rock was already lithified when the karst cavities and open faults, or its surface were mineralized, it cannot exhibit distinct, far-reaching lithogeochemical proximity indicators.

The transgressive hanging clastic series are formed independently from the locations of mineralizations. Thus, these sedimentary sequences do not present any indications for the underlaying mineralization. Only Mn developed a secondary dispersion halo of a few metres.

This special type of mineralization can be explored only by intensive geological mapping and systematic geochemical soil prospecting.

Acknowledgements. This study was financially supported by the EEC under *contract No. MSM-008-D.* The authors wish to express their sincere thanks to Prof. Puchelt and Prof. Drovenik for arranging sulphur isotope measurements, and to Prof. Köppel for supplying lead isotope data.

References

Assereto R, Brigo L, Brusca C, Omenetto P, Zuffardi P (1976) Italian ore/mineral deposits related to emersion surfaces. A summary. Miner Dep 11:170–179

Brigo L, Colbertaldo D di (1972) Un nuovo orizzonte metallifero nel Paleozoico delle Alpi Orientali. In: Proceedings of the 2nd ISMIDA, Ljubljana, 4.-7. October, 1972, pp 109–124

Brigo L, Omenetto P (1978) Metallogenese der italienischen Ostalpen. In: Proceedings of the 3rd ISMIDA, Leoben, Oct. 7–10, 1977. Verh Geol Bundesanst 3:249–266

Brigo L, Möller P, Schneider H.J., Wolter R (1986), Mineralization of the Carnic Alps. In: Möller P, Brigo L, German, K, Schneider H.J. (eds), Litho-geochemical proximity indicators for strata-bound base metal deposits. EUR 10826 EN, 36–74

Collins AG (1975) Geochemistry of oilfield waters. Elsevier, Amsterdam, p 496

Coltorti M, Comin-Chiaramonte P, Princivalle F, Sinigoi S (in prep) Petrochemical characterization of Carboniferous volcanics of Paleo-carnic chain (Eastern Alps, Italy).

Flora O, Martino L, Comin-Chiaramonte P (1983) Some considerations on Paleozoic spilites of the Carnia (Italian Eastern Alps), Vol 5. Gortania, Atti del Museo di St Nat, pp 29–44

Flügel H (1964) Das Paläzoikum in Österreich. Mitt Geol Ges Wien 56:401–443

Flügel H, Gräf W, Ziegler W (1959) Bemerkungen zum Alter der "Hochwipfelschichten" (Karnische Alpen). Neues Jahrb Geol Paläontol Mh 1959:153–167

Francavilla F (1966) Spore nel Flysch Hochwipfel. Giorn Geol Bologna, Ser 2, 33:493–526

Guichard F, Church TM, Treuil M, Jaffrezic H (1979) Rare earth in barytes: distribution and effects on aqueous partitioning. Geochim Cosmochim Acta 43:983–997

Heger B (in press) Geologische Kartierung des Gebietes zwischen Pal Grande und Crete del Mezzodi (Karnische Alpen/Italien). Diploma Thesis, FB 24, Freie Universität Berlin, Berlin

Hoefs J (1980) Stable isotope geochemistry. Minerals, rocks and inorganic materials. Springer, Berlin Heidelberg New York, p 208. (Monograph Series of Theoretical and Experimental Studies, vol 9).

Kern M (1985) Vererzungen und Geochemie der Gesteine an der Grenze Devon/Karbon, im Gebiet des Pal Piccolo-Pal Grande und Pizzo di Timau (Karnische Alpen/Italien). Diploma Thesis, FB 24, Freie Universität Berlin, p 98, (Manuscript)

Luck J (1985) A new decomposition and separation method of insoluble geological samples for ICP. In: Extended abstracts of the Second International Symposium on Analytical Chemistry in Exploration, Mining and Processing of Materials, CSIR, Pretoria 15th–19th April, 1985, pp 72–73

Möller P (1983) Lanthanoids as a geochemical probe and problems in lanthanoid geochemistry. Distribution and behaviour of lanthanoids in non-magmatic phases. In: Sinha SP (ed) Systematics and the properties of the lanthanides. Reidel, Dordrecht, pp 561–616

Möller P (1985) Development and application of the Ga/Ge-geothermometer for sphalerite from sediment-hosted deposits. Monogr Ser Miner Dep 25: 15–30

Möller P, Dulski P, Szacki W (1987) Element distribution in sphalerite — a key to genesis of sediment-hosted sulfide mineralizations. Chem Geol (Submitted)

Morgan JW, Wandless GA (1980) Rare earth element distribution in some hydrothermal minerals: evidence for crystallographic control. Geochim Cosmochim Acta 44: 973–980

Moschik A (1985) Zur Paragenese und Geochemie des Barytvorkommens Coccau, Tarvisio (Italien). Diploma Thesis, FB 24, Freie Universität Berlin, Berlin, p 69

Omenetto P, Brigo L (1974) Metallogenesi nel quadro dell' orogene ercinico delle Alpi (con particolare riguardo al versante italiano). Mem Soc Geol Ital 13: 1–24

Sassi FP, Del Moro A, Kalvacheva R, Zanferrari A, Zirpoli G (1985) Chronological data and problems concerning the South Alpine basement in the Eastern Alps. Sassi e Julivert. IGCP no 5, Newslett 6: 111–115

Selli R (1963) Schema geologico delle Alpi Carniche e Giulie Occidentali. Giorn Geol 30: 1–136

Spalletta C, Vai GB, Venturini C (1982a) La catena Paleocarnica. In: Castel: Pariu A and Vai GB (ed) Guida alla geologia del Sudalpino centro-orientatle. Guida Geol Reg SGI, pp 281–292

Spalletta C, Vai GB, Venturini C (1982b) Controllo ambientale e stratigrafico delle mineralizzazioni in calcari devonodinantiani delle Alpi Carniche. Mem Soc Geol Ital 22: 101–110

Vai GB (1974) Stratigrafia e paleogeografia ercinica delle Alpi. Mem Soc Geol Ital 13: 7–37

Venerandi Pirri I (1977) Le paragenesi a Zn, Cu, Pb, Sn, Hg, Ni, As, fluorite, barite nel Devoniano della Catena Carnica. Rend Soc It Min Petr 33: 821–844

Wedepohl KH (ed) (1969) Handbook of geochemistry, vol 1 and 2. Springer, Berlin Heidelberg New York, p 1 and 5

Wolter R, Schneider HJ (1983) Saline relics of formation water in the Wettersteinkalk and their genetical connection with the Pb–Zn mineralization. In: Schneider HJ (ed) Mineral deposits of the Alps and of the Alpine epoch in Europe. Springer, Berlin Heidelberg New York, pp 223–230

Wolter R, Schneider HJ (1985) Solerelikte in Erz und Nebengestein der Blei-Zink-Lagerstätte Bleiberg-Kreuth. Arch Lagerstättenforsch Geol Bundesanst 6: 201–208

Wolter R, Schneider HJ (in press) Genetical significance of saline relics in carbonate host rocks of Alpine Pb–Zn deposits. In: Friedrich G (ed) Base metal sulfide deposits in volcanic and sedimentary environment, Springer, Berlin Heidelberg New York

The Geological Setting of Base Metal Mineralisation in the Rhodope Region, Northern Greece

R.W. Nesbitt[1], M.F. Billett[1], K.L. Ashworth[1], C. Deniel[1], D. Constantinides[2], A. Demetriades[2], C. Katirtzoglou[2], C. Michael[2], E. Mposkos[2], S. Zachos[2], and D. Sanderson[3]

Abstract

The Rhodope region of NE Greece is made up of a complex Lower Palaeozoic (?) basement sequence (metasediments, gneisses and amphibolites) unconformably overlain by Mesozoic and Tertiary formations. There is evidence for ophiolitic sequences in the structurally higher parts of the eastern basement and in the Mesozoic, Circum-Rhodope Belt. Metamorphic grades reach upper-amphibolite facies in the central Rhodope. Base metal mineralisation (Pb, Zn, Cu, Fe, Mn) occurs throughout the region in a variety of environments. Within the basement, sulphides are found associated with marbles (in the west) and amphibolite-serpentinites (in the east). Important base metal occurrences are found in the eastern Tertiary volcano-sedimentary basins, both as strata-bound, syngenetic sulphides and often as fault-controlled Pb–Zn deposits. All of the major deposits, both within the basement and cover sequences, appear to result from a major mineralisation event in the Tertiary. Evidence for this view comes from their homogeneous lead isotope values and the overall mineralisation style which appears to be controlled by large-scale fracture systems. Field and remotely sensed data provide the basis for a dynamic model in which extensional tectonics associated with arc magmatism, resulted in hydrothermal solutions migrating through well-defined fracture systems. Isotope data from the major sulphide occurrences support the model of large-scale homogenisation of crustal lead which was scavenged from the basement and cover sequences. The metals were subsequently dumped in chemically and physically favourable environments such as marble horizons or into actively depositing basins. Although the region is dominated by Pb–Zn occurrences, there is evidence from the Cu sulphides within the amphibolite-serpentinite sequences that local control did play an important role. Pb isotope data leave no doubt, however, that the mineralising solutions circulated through very large rock volumes outside of the mafic sequences. Overall, the study illustrates the potential of an exploration programme which uses remotely sensed data in combination with well-constrained ground studies.

[1] Department of Geology, University of Southampton, Southampton SO9 5NH, UK
[2] Institute of Geology and Mineral Exploration (IGME), 70 Mesoghion Ave., Athens 115 27, Greece
[3] Department of Geology, The Queen's University, Belfast BT7 1NN, Northern Ireland, UK

Mineral Deposits within the European Community
(ed. by J. Boissonnas and P. Omenetto)

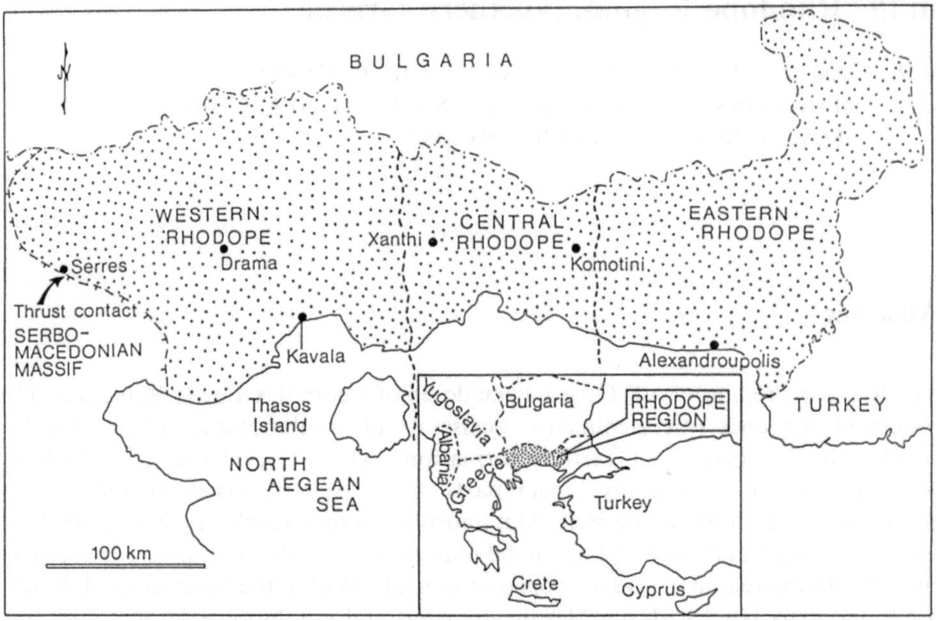

Fig. 1. Location of the Rhodope region and its sub-division into the western, central and eastern Rhodope

1 Introduction

The Rhodope region of Greece is situated in the northeastern part of that country and is bordered to the east by Turkey and to the north by Bulgaria (Fig. 1). Geologically, the region can be sub-divided into three units; the metamorphic basement (or basement massif), the Circum Rhodope Belt and the Tertiary volcano-sedimentary basins. The Massif itself extends northwards into Bulgaria and its western boundary is marked by the north-south fracture zone known as the Strimon Fracture Zone. To the west of this zone is the Serbo-Macedonian Massif and together the two metamorphic basement areas form an ancient continental fragment which was extensively reworked during the Alpine Orogeny. Within the tectonic framework of Greece, the massifs represent two of the 'internal' geotectonic zones of Marinos (1982).

Mineral exploration and small-scale mining activity for precious and base metals in the Rhodope region dates back to ancient Greece (Epitropou et al. 1983; Mack 1983). In recent years the economic potential of the area has increased by the exploration activity of the Greek Institute of Geology and Mineral Exploration (IGME). Further impetus has been provided by the EC-sponsored programme over the period 1983–1986. This chapter deals with aspects of the general geology with special emphasis on the economic geology of the eastern Rhodope region.

2 Regional Geology

2.1 Metamorphic Basement

The metamorphic basement which comprises greenschist to amphibolite facies metasedimentary and metaigneous rocks can be sub-divided into two major units:

1. The Amphibolite-Gneiss Series consisting of para- and ortho-gneiss, associated with mica schist, amphibolite and thin marble horizons. In the eastern Rhodope the upper part of this series comprises a 1- to 1.5-km-thick sequence of metamorphosed mafic and ultramafic rocks, known as the Amphibolite-Serpentinite Unit.

2. The Carbonate Series, consisting of marble intercalated with mica schist, chlorite schist and quartzite.

The geology of the western Rhodope is dominated by the Carbonate Series (Fig. 2) and estimates of thickness vary from about 1.5 km (Zachos and Dimadis 1983) down to 300 m (Papanikolaou 1986). The Series stratigraphically overlies the Amphibolite-Gneiss Series but tectonic repetition and folding makes detailed correlation difficult. The central and eastern Rhodope are dominated by the Amphibolite-Gneiss Series and in the central region this unit is associated with minor carbonate horizons and isolated occurrences of mafic and ultramafic rocks. In the eastern region, the upper part of the Amphibolite-Gneiss Series is made up

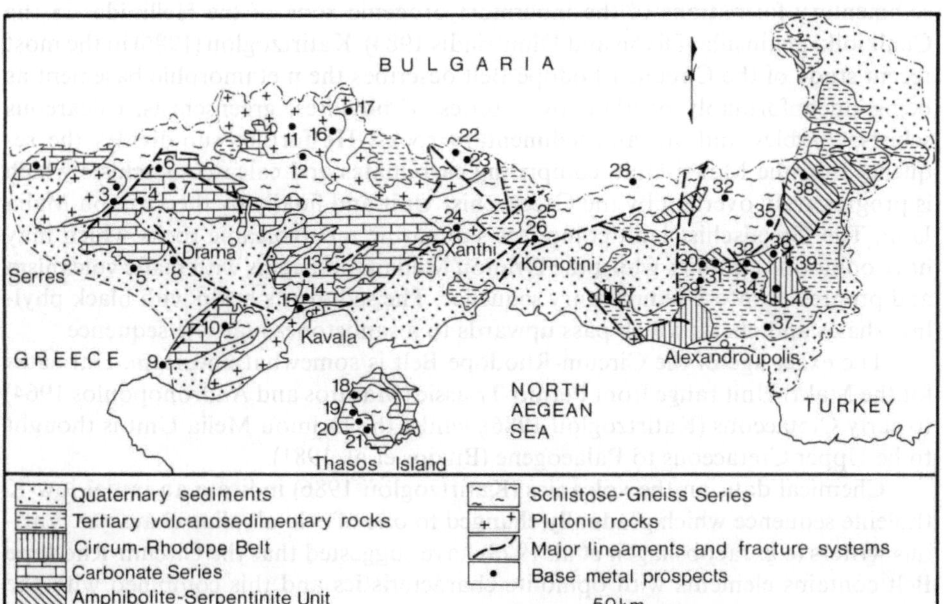

Fig. 2. Simplified geological map of the Rhodope region showing the major base metal prospects (after Bitzios et al. 1981). Each prospect is *numbered* and described in more detail in Table 1

of an Amphibolite-Serpentinite Unit which consists of metamorphised mafic and ultramafic rocks, which in part is extensively migmatised. This unit has been thrust from the southeast onto the gneissic basement and may represent part of a disrupted ophiolite complex. Other occurrences of serpentinised ultramafic rocks are found within the Amphibolite-Gneiss Series particularly in the northern part of the eastern Rhodope.

Overall, the geology of the metamorphic basement exhibits a number of strong contrasts from west to east. In particular, there is a significant decrease in the importance of marble and an increase in the volume of mafic rocks. The significance of this contrast is not fully understood and the lack of a firm geochronology further hampers a better understanding of tectonic and stratigraphic relations. The paucity of firm geochronological data has led Bulgarian geologists to use lithostratigraphic comparisons with other basement terrains and they have assigned a Proterozoic or an Archaean age to the Massif (Dimitrov and Zidarov 1969). Moorbath and Zagorcev (1983) reported a Rb/Sr age of 342 ± 27 Ma for a granitoid complex which intrudes the basement of southern Bulgaria. This implies that the Rhodope Massif must be of Lower Carboniferous age or older.

2.2 Circum-Rhodope Belt

The Mesozoic Circum-Rhodope Belt formations of the southeastern margin of the Rhodope Massif exhibit stratigraphic similarities with the Mesozoic volcano-sedimentary formations of the innermost orogenic zone of the Hellinides in the Chalkidiki peninsula (Dixon and Dimitriadis 1984). Katirtzoglou (1986) in the most recent study of the Circum-Rhodope Belt describes the metamorphic basement as being unconformably overlain by a series of phyllites, greenschists, calcareous schists, marbles and volcano-sedimentary rocks. He further sub-divides the sequence into the Makri Unit, comprising carbonates and calcareous schists which is progressively overlain by the Greenschist Unit and finally by the Drimou Melia Unit. The Greenschist Unit comprises submarine metavolcanic rocks which may have ophiolitic affinities whilst the Drimou Melia Unit begins with mafic volcanism and progresses into a sedimentary sequence. The sediments begin with black phyllitic shales and cherts which pass upwards to a sandstone-quartzite sequence.

The exact age of the Circum-Rhodope Belt is somewhat uncertain. Estimates for the Makri Unit range from Permo-Triassic (Maratos and Andronopoulos 1964) to early Cretaceous (Katirtzoglou 1986), whilst the Drimou Melia Unit is thought to be Upper Cretaceous to Palaeogene (Bitzios et al. 1981).

Chemical data on the volcanics (Katirtzoglou 1986) indicate an initial low-K tholeiite sequence which gradually changed to one of calc-alkaline character. Various writers (e.g. Jacobshagen et al. 1978) have suggested that the Circum-Rhodope Belt contains elements with ophiolite characteristics and this combined with the nature of the volcanics has led to models involving the development of island arcs and subduction-related volcanism. In these models, the Belt is folded, metamorphosed and thrust onto the Rhodope Massif during the late Cretaceous.

2.3 Tertiary Volcano-Sedimentary Basins

The Tertiary volcano-sedimentary basins occur along the southern and eastern margins of the Rhodope Massif (Fig. 2). The basins were initiated during mid-Eocene (Papadopoulos 1980) and rest unconformably and have tectonic contacts with both the metamorphic basement and the Circum-Rhodope Belt. Initial sedimentation (basal conglomerate, marl and limestones) was followed in the Upper Eocene and Oligocene by a mixed volcano-sedimentary sequence consisting of calcareous sandstones, shales, andesites and rhyolites. The volcanics which are dominantly of calc-alkaline affinity (Fytikas et al. 1984; Papavassiliou and Sideris 1984) form part of a major volcanic province in the north Aegean which is associated with a series of sub-volcanic intrusive rocks of intermediate composition. These form an important component in the development of base metal mineralisation, a subject which is discussed in a later section. Overlying the older formations throughout the Rhodope are a series of Neogene and Quaternary sediments which include both marine and terrestrial clastic sediments. These do not form part of the present study.

3 Metamorphism and Tectonism

The metamorphic rocks of the Rhodope region have been affected by various degrees of regional metamorphism and deformation. The complex tectonic history of the basement is manifest as a penetrative regional schistosity and a series of isoclinal to open fold structures. Papanikolaou and Panagopoulos (1981) and Papanikolaou (1986) describe three phases of deformation in the western Rhodope with the syntectonic second phase being the most important. The general model is one of a multiphase, recumbent isoclinal folding with associated thrusting — a model also supported by Bulgarian workers (Ivanov et al. 1979). The important conclusion by both Greek and Bulgarian geologists is that the major deformation is Tertiary in age, with post-Cretaceous nappes recorded by Ivanov (1985) on the northern margin of the Rhodope.

In contrast, the Lower Tertiary volcano-sedimentary sequence is relatively undeformed. Folding, however, does occur locally along the margins of the sedimentary basins (Billett and Nesbitt 1986) and this is thought to be related to a basin closure in the Oligocene.

The basement rocks of the Rhodope Massif exhibit a number of different metamorphic facies. Recent work on central Rhodope to the east of Xanthi by Mposkos (1986) has demonstrated the presence of several metamorphic zones ranging up to upper amphibolite facies. In the central zone (to the north of Komatini), the grade of metamorphism increases progressively to the NW and Mposkos (1986) has mapped the isograd separating the staurolite-kyanite zone from the kyanite-sillimanite zone. This upper amphibolite facies unit is confined to the migmatite zone which surrounds the Oraeo-Echinos granite. Further to the east, the eastern Rhodope is thrust onto the central Rhodope Block along the NE–SW trending Kimi-Pandrosos thrust (Mposkos 1986). To the east of the thrust, metamorphic grade is generally of lower to middle amphibolite facies with the highest metamorphic grade (kyanite) being found in the Smiyadi-Kimi Formation immediately east of the thrust. To the east of the thrust, the leucocratic biotite gneisses, which occupy much of the eastern zone of the block, are sub-divided by Mposkos (1986) into a staurolite-chloritoid zone (to the northwest) and a garnet zone (to the southeast). The eastern Rhodope therefore shows an increase in metamorphic grade from southeast to northwest. The distinctive Amphibolite-Serpentinite Unit in the extreme east of the Rhodope displays assemblages of actinolite-epidote-oligoclase within the amphibolitic members.

Retrograde metamorphism occurs throughout the Rhodope Massif and is characterised by greenschist facies conditions. This period of metamorphism is thought to be related to a regional metamorphism in the Mesozoic Circum-Rhodope Belt (Mposkos 1986). The major period of regional metamorphism in the crystalline basement must therefore be pre-Alpidic.

In several areas in the region, the major metamorphic formations are separated by thrusting. For example, the eastern Rhodope Massif is thrust onto the central Rhodope Massif to the ENE of Komotini (Fig. 2). Although major thrust movements in the Rhodope region are thought to be associated with the collision of the northern European plate with the African plate in the Upper Cretaceous (Robertson and Dixon 1984), it is most likely that they may represent reactivated basement structures (Zachos and Dimadis 1986).

This pre-existing structural control may also be true for many of the high angle normal and reverse faults which are found throughout the region. Recent work by Sanderson et al. (1986) has shown that these faults not only control the development of the Tertiary volcano-sedimentary basin but are also important structures for the localisation of the mineralisation. The thrust of Sanderson's argument is that the eastern Rhodope can be considered as lying at the dilational (extensional) termination of the right-lateral North Anatolian Fault system and that this influenced the development of the Eocene-Oligocene basins. Based on remotely sensed data and ground observations, Sanderson et al. (1986) conclude that the Tertiary basins are bounded by major faults even though this may be masked at the surface by the overlap of sedimentary fill. Thus, the Kirki-Essimi Basin formed as a consequence of a pull-apart at the southern edge of the basement massif as it was extended in a NE–SW direction along a right-lateral 060° trending fault. An interplay between these northeast fractures and the associated southeast extensional trending faults produced the zigzag outline of the southern margin of the basin with its general east-west trend. The significance of these fracture systems to controls of known mineralisation is discussed in the following sections.

4 Mineralisation

In this section we attempt to summarise the main features of the mineralisation in the Rhodope region, by considering in turn the metamorphic basement, the Circum-

Rhodope Belt and the Tertiary volcano-sedimentary basins. The main polymetallic prospects in the Rhodope region are located in Fig. 2 and listed and described in more detail in Table 1. Many of these prospects are located on the metallogenic map of Greece published by Zachos and Maratos (1965). However, this new listing contains additional evidence from several old prospects and also a number of new, recently discovered prospects.

4.1 Metamorphic Basement

Approximately 80% of the major base metal prospects in the Rhodope region occur in the metamorphic basement. Within this group three main types of mineralisation are recognized:

1. Vein-type mineralisation associated with faults;
2. Carbonate-hosted mineralisation; and
3. Mineralisation associated with serpentinite and amphibolite.

1. Vein-type mineralisation, related to fracture systems, occurs in metamorphic rocks of different composition. In the Tris Vrises area, for example (Prospect No. 35) base metal mineralisation occurs in 0.1–1-m-thick veins in leucocratic gneiss, whereas at Thermae (No. 23) mineralisation occurs in fractures cross-cutting amphibolite and marble. Metasomatic mineralisation occurs at Thermae where faults cut marbles. Wall rock alteration is generally present and pyritic, silicic, chloritic, carbonate and sericitic alteration zones can be recognised in the vicinity of the fractures. The Pb–Zn–Cu mineralisation of the Thermae area is similar to the economic ore deposits of the Madan area in southern Bulgaria (Bogdanov 1982). Ore concentrations at Madan are controlled mainly by NNW trending faults and mineralisation occurs where faults cross-cut marbles. Throughout the Greek Rhodope Massif mineralisation is generally confined to fault systems with a similar northwest-southeast orientation.

2. Carbonate-hosted base metal sulphide mineralisation is an important feature in the metamorphic basement. On Thasos Island (Fig. 2) four major Zn (Pb, Fe) sulphide prospects are situated at or close to the contact between the Amphibolite-Gneiss Series and the Kastro Marble Series (Fig. 3). Many of these base metal occurrences are lenticular in form with a limited vertical but a considerable lateral extent. The deposits appear to be both strata-bound and stratiform in character. The mineralisation is typically karstic, formed by groundwater circulation (Epitropou et al. 1983; Omenetto 1983). Similar mineralisation occurs in the Palia Kavala area (Prospect No. 12, 13, 14, 15) and in the western Rhodope, Mn (Fe) and Pb/Zn mineralisation occurs at a similar stratigraphic position (Prospect No. 6, 7, 11).

Marble is also associated with classic skarn-type mineralisation related to intrusive granitic rocks. A typical example of this type of mineralisation occurs at the Kimmeria deposit (No. 24), where magnetite-sphalerite-chalcopyrite-pyrite-gold is associated with molybdenite and scheelite. Several other base metal pro-

Table 1. Important base metal prospects in the Rhodope region

Number/name	Mineralization	Host rock[a]	Style/control
1 Angistro	Zn, Pb, As (Cu, Py, Ag, Au)	Marble/Gn-Sch	Strata-bound
2 Kato Vrontou	Py (Cu, Zn)	Granite	Vein-type
3 Vrontou	Py, As (Pb, Cu)	Marble	Skarn
4 Panorama	Fe, Py (Pb, Zn, W)	Marble	Skarn
5 Dafnoudi	Pb, Py (Cu, Au)	Marble/Gn-Sch	Strata-bound
6 Kato Nevrokopi	Fe (Pb, Zn, Cu)	Marble/Gn-Sch	Strata-bound
7 Granitis	Mn (Pb, Zn, Py)	Marble/Gn-Sch	Strata-bound
8 Kalithea	Fe (Pb)	Seds	Strata-bound
9 Ofrinio	Py, As (Pb, Zn)	Marble	Skarn
10 Asimo. Pangaeo	Py, As (Pb, Zn, Au, Ag)	Marble	Skarn
11 Finterna	Mn (Pb, Py, Zn)	Marble/Gn-Sch	Strata-bound
12 Radista	Pb, Zn, Mn (Py, Cu)	Marble/Gn-Sch	Strata-bound
13 Kouroutgou	Zn, Fe (Pb, Cu)	Marble/Gn-Sch	Strata-bound
14 Mandra Kari	Fe, Mn (Pb, Zn)	Marble/Gn-Sch	Strata-bound
15 Amigdaleon	Fe, Mn (Pb, Zn)	Marble/Gn-Sch	Strata-bound
16 Pefki	Pb, Zn, Py	Marble	Vein-type
17 Farasino	Pb, Py (Zn, Cu)	Marble	Vein-type
18 Sotiras	Zn, Pb (Fe)	Marble/Gn-Sch	Strata-bound
19 Marlou	Pb, Zn (Py)	Marble/Gn	Strata-bound
20 Koumaria	Zn (Pb, Fe)	Marble/Gn	Strata-bound
21 Vouves	Zn (Pb, Fe)	Marble/Gn	Strata-bound
22 Diasparto	Pb, Zn, Py (Cu)	Marble/Gn-Sch	Fracture
23 Thermae	Pb, Zn, Py (Cu, Ag)	Marble/Gn-Amp	Fracture
24 Kimmeria	Fe, Zn, Cu, Py (Pb, Mo, W)	Marble	Skarn
25 Kaloticho	Py, Pb (Cu, Zn)	Gn-Sch	Vein-type
26 Iasmos	Py, Pb (Cu, Zn)	Gn-Sch	Vein-type
27 Xilayani	Py (Cu, Zn, Au)	Greenschists	Strata-bound
28 Sarakini	Py, Cu (Pb, Zn)	Basic Gn	Vein-type
29 St. Philip	Zn, Pb, Py (Cu, Ag)	Seds	Fracture
30 King Arthur	Py, Zn, Cu, Pb	Volcs	Vein-type
31 Aberdeen	Py, Cu (Zn, Pb)	Amph/Serp	Fracture
32 Kechros	Zn, Cu, Py, Pb	Acid Gn	Vein-type
33 Baiko (1)	Fe (Cu, Py)	Marble/Amp	Strata-bound
(2)	Py, Cu (Zn, Pb)	Amp/Serp	Various
34 Essimi (1)	Py, Zn	Amp	Stratiform
(2)	Py, Zn, Pb (Cu, Ag)	Seds/Volcs	Stratiform
(3)	Py, Zn, Pb, Cu	Seds	Vein-type
(4)	Py, Cu	Intrusives	Porphyry
35 Tris Vrises	Zn, Pb, Py, Cu	Acid Gn	Fracture
36 Pessani	Py, Cu (Pb)	Amp/Serp	Stratiform
37 Pefka	Cu, Py, Pb (Zn, Sb, Ag, Au)	Volcs	Vein-type
38 Mikro Dherio	Py (Cu)	Greenschists	Strata-bound
39 Virini	Py, Zn, Pb, Cu	Seds/Volcs	Vein-type
40 Elva	Py, Cu	Greenschists	Strata-bound

[a] Gn = gneiss; Sch = schist; Seds = sediments; Amp = amphibolite; Serp = serpentinite; Volcs = volcanics)

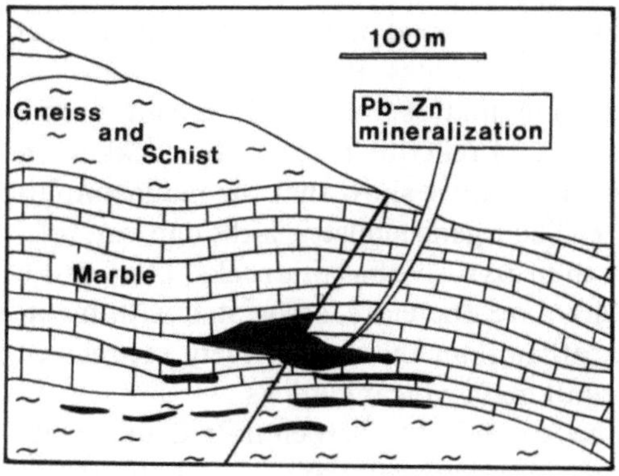

Fig. 3. Schematic cross-section of the Marlou Pb/Zn, prospect No. 19, on Thasos Island (Epitropou et al. 1983)

spects in the western Rhodope have a similar style of mineralisation, namely Vrontou (No. 3), Panorama (No. 4), Ofrinio (No. 9) and Asimotripes Pangaeo (No. 10).

3. Mineralisation related to serpentinites and amphibolites occurs in a sequence of metamorphosed mafic and ultramafic rocks in the eastern Rhodope, known as the Amphibolite-Serpentinite Unit. Recent research work in the eastern Rhodope has concentrated on investigating the geological controls of this type of mineralisation. Current ideas on its origin are discussed in Billett and Nesbitt (1986). Three major prospects occur in the Amphibolite-Serpentinite Unit; Aberdeen (No. 31), Baiko (No. 33) and Pessani (No. 36). A fourth prospect (Essimi, No. 34) has only been intersected by drilling. The mineralisation is predominantly Cu-rich in contrast to the Zn- and Pb-rich mineralisation of other parts of the Amphibolite-Gneiss Series and the Carbonate Series.

In the Baiko area (No. 33) of eastern Rhodope two types of mineralisation occur. Strata-bound magnetite-pyrite-chalcopyrite mineralisation occurs within a 40–50-m-thick sequence of intercalated amphibolite, marble and granitic bands. The mineralisation, which extends for at least 1 km along the strike, is concentrated in metasomatic, epidote-rich areas at the contact between mafic and carbonate bands (Billett and Nesbitt 1986). The second type of mineralisation is amphibolite-hosted and consists of pyrite-chalcopyrite mineralisation which is spatially related to the interplay of a major isoclinal fold with a Tertiary acid intrusive body. Within the tectono-stratigraphy of the Amphibolite-Serpentinite Unit, the pyrite-chalcopyrite mineralisation lies at the contact between a unit of serpentinite and a unit of mafic amphibolite. The second type of mineralisation at Baiko is therefore fundamentally different from the strata-bound magnetite-pyrite-chalcopyrite mineralisation.

The Aberdeen prospect lies on a north-south trending fracture zone close to an intersection with an east-west trending fracture. The mineralisation, which comprises chalcopyrite-pyrite ± magnetite, occurs in the lower part of a unit of serpentinites cross-cut by acid pegmatites, close to its contact with an underlying unit of mafic amphibolite. The mineralised zone varies in width from 3–7 m and extends for 300–350 m in a north-south direction.

Mineralisation at both Aberdeen and Baiko occurs at or close to the lithological boundary between serpentinites and mafic amphibolites. Petrochemical studies of both these units show that they comprise a comagmatic suite of mafic and ultramafic rocks (Billett and Nesbitt 1986). The presence of relict textures in the rocks, such as rhythmic layering, ophitic and pegmatitic textures, show that the mafic amphibolites represent a metagabbroic sequence. Their geochemistry, although variable, is comparable to intrusive gabbroic rather than extrusive basaltic rocks. The Cu–Fe sulphide occurrences at Aberdeen and Baiko are, therefore, associated with metamorphosed mafic and ultramafic rocks, rather than the more widely recognized association of base metal mineralisation with mafic volcanic rocks.

Pessani, the third major prospect in the Amphibolite-Serpentinite Unit, occurs in minor fractures along a similar contact between mafic and ultramafic rocks. Mineralisation comprises supergene and primary Cu and Fe oxides and sulphides with minor galena. The concentration of base metals along this contact is also shown by soil geochemical traverses across the area (Michael et al. 1984). The geochemical anomalies show that the amphibolite is preferentially enriched in base metals in the immediate vicinity of serpentinite bodies. Billett and Nesbitt (1986) suggest that fluids related to serpentinization have mobilized base metal sulphides in the surrounding amphibolite and produced metal concentrations at the margins of serpentinite bodies. Fluid flow and mineralisation may also be enhanced by the relatively ductile contact between serpentinite and amphibolite.

In the Amphibolite-Serpentinite Unit the occurrence of a Cu-dominated base metal mineralisation is directly related to the nature of the host rocks. The formation of sub-economic mineral prospects is a result of secondary processes, such as fracturing (Aberdeen and Pessani), folding (Baiko) and igneous intrusion (Baiko), which resulted in the concentration of base metals from low background levels in the mafic and ultramafic rocks. The fluids which effected this concentration of base metal sulphides, either originated from the serpentinites themselves or from the passage of hydrothermal fluids along the lithological contact between amphibolites and serpentinites. Economou and Naldrett (1984) have proposed a similar model for the formation of Cu–Fe sulphide halos which occur at the peripheries of podiform chromite bodies in serpentinites. They suggest that base metals precipitated from hydrothermal fluids which also caused serpentinization of the host rocks.

4.2 Circum-Rhodope Belt

The Circum-Rhodope Belt is found only in the eastern Rhodope region, where it is associated with a number of prospects. The Xilayani prospect (No. 27), which comprises disseminated and massive pyrite with minor chalcopyrite and rare sphalerite and galena, occurs in a unit of low grade metabasic rocks (Bitzios et al. 1981). The ore is lenticular in form and coplanar to the regional metamorphic fabric (Fig. 4). It is therefore considered to be a syngenetic, stratiform, volcanogenic deposit of Cyprus-type (Bitzios et al. 1981; Constantinides et al. 1983). A similar type of mineralisation occurs in the Mikro Dherio area (prospect No. 38), where dis-

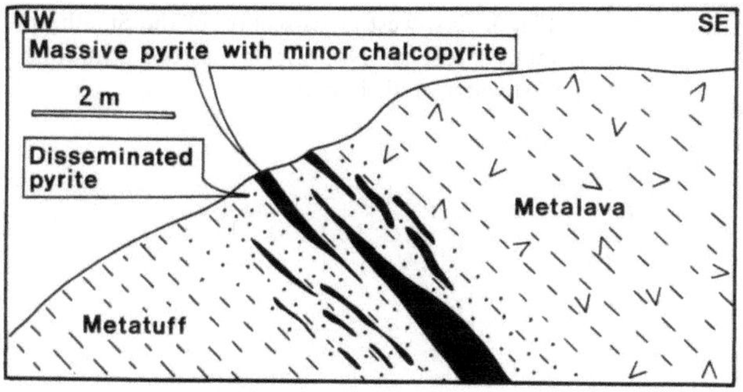

Fig. 4. Schematic cross-section illustrating the strata-bound mineralisation at the Xilayani prospect, No. 27 (Bitzios et al. 1981)

seminated pyrite and minor chalcopyrite are associated with sub-greenschist facies metatuff and metabasic lava.

4.3 Tertiary Volcano-Sedimentary Basins

Several important polymetallic mineral prospects occur within the Tertiary volcano-sedimentary basins of the central and eastern Rhodope. Three main types of mineralisation exist: (1) fault controlled, (2) strata-bound and (3) porphyry style. The St. Philip prospect (No. 29) represents the most significant area of base metal mineralisation in the eastern Rhodope region. It has known reserves of 1.2 million tonnes of ore at a grade of 9% combined Pb/Zn, with 50–150 ppm Ag. Sulphide mineralisation occurs as a fracture filling within a major NNW–SSE trending fault system (Fig. 5). According to Sanderson et al. (1986), this fault system is a left-lateral strike-slip fault which develops subsidiary (mineralised) 125° trending extensional fractures. This observation is an important one because the much larger Bulgarian deposits to the north are also developed along NNW trending fractures with many of the major lodes occurring in associated 125° fracture systems.

The King Arthur prospect (No. 30) lies 4 km north of the St. Philip mine. Mineralisation occurs in small veinlets and impregnations in brecciated zones along a NNW–SSE fracture system and comprises galena, sphalerite, chalcopyrite and pyrite (Govett and Galanos 1974). The King Arthur fault zone, which may lie on an extension of the St. Philip fracture system, separates Tertiary silicic volcanics from basement metamorphic rocks. This important observation suggests that mineralised fracture systems which cut the metamorphic basement are Tertiary in age (see Sect. 5).

Fracture controlled mineralisation also occurs in the Pefka area (prospect No. 37). Although base metals are related to silicified fracture zones, there is a close spatial relationship between mineralisation and Tertiary volcanic and sub-volcanic rocks.

Another important style of mineralisation associated with Tertiary rocks is the stratiform mineralisation of the Kirki-Essimi-Virini basin. This type of mineralisation is found in sub-economic levels in the Mili area of Essimi (No. 34). At Mili, pyrite, sphalerite, chalcopyrite and galena occur in discontinuous bands within a

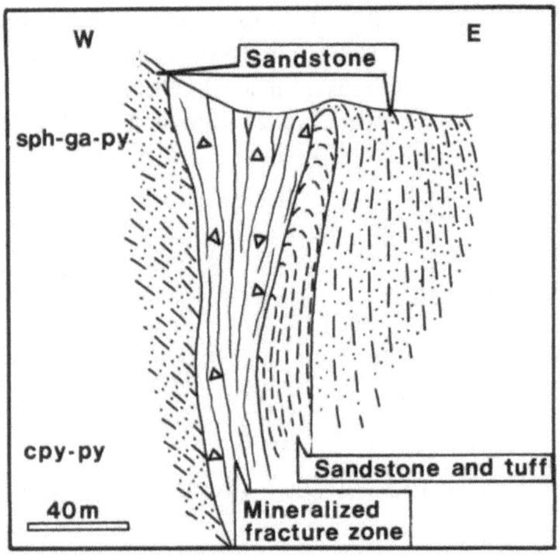

Fig. 5. East-west cross-section of the St. Philip mine, prospect No. 29, in eastern Rhodope (Ashworth et al., in press)

30-m-thick sequence of calcareous siltstone and fine-grained sandstone. The stratiform mineralisation at Mili appears to be spatially related to an extensive, possibly deep-seated north-south fracture system and to felsic intrusives which occur beneath the mineralised area (Ashworth et al. 1986; Arvanitides and Katirtzoglou 1985).

In addition to the stratiform mineralisation at Mili, small occurrences of porphyry-style pyrite-chalcopyrite mineralisation occur. These are associated with the alteration zones of high level intrusives which are present in the area (Kalogeropoulos and Katirtzoglou 1986) and their economic potential appears to be limited.

5 Discussion and Concluding Remarks on Mineral Potential

The preceding sections have attempted to give a brief overview of the varying styles of mineralisation in the whole of the Rhodope area. From the exploration viewpoint, we need to identify those types of mineralisation which have the most promise and arrive at an exploration strategy which will find extensions of known deposits and provide guidelines for a future search programme. Overall, it is clear that the style of mineralisation changes rather dramatically from west to east across the Rhodope. This is clearly a reflection of the changing rock types as the major marble horizons of the west give way to amphibolite-gneiss complexes and younger sedimentary basins of the east. However, despite this variation, the predominant style of mineralisation is base metal sulphides in vein systems. In our view, this broad pattern of mineralisation throughout the Rhodope suggests that irrespective of the age of the host rock, the vein systems are largely contemporaneous. It follows from the occurrence of mineralisation in the volcano-sedimentary basins that all of the vein-type mineralisation is lower Tertiary in age. In the following sections we examine the evidence for this view.

5.1 Age of Mineralisation

There are two principal lines of evidence supporting the Tertiary age of the mineralisation. These are structure, and lead isotopes. It is evident from the work of Sanderson et al. (1986) that the development of the volcano-sedimentary basins was in response to a major extensional tectonic event and that the same stress regime can be used as the basis for the development of the major fracture systems. Remote sensing and ground truth observations confirm that the fracture systems within the Tertiary basins are also found in the basement rocks. Hence, these data suggest that the development of the basins and the subsequent mineralisation of the associated fracture system are all part of the same tectonic package. In support of this model, we point to the similarity in structural control between the major Tertiary deposit of St. Philip (deposit 29) and the smaller fracture-controlled mineralisation within the basement rocks at Aberdeen (deposit 31). At the St. Philip mine, Sanderson et al. (1986) have shown that the fractures are produced by dilation or pull-apart at the termination of NNW left-lateral wrench faults. This results in a set of ESE extensional fractures which are commonly mineralised as are the NNW fractures. At Aberdeen the mineralisation occurs within basement amphibolites and is found at the intersection of NNW and ESE fractures. We take this to indicate that the mineralisation is Tertiary.

The second line of evidence concerning the age of mineralisation comes from lead isotope data. Deniel et al. (1986) present data from all of the major mineralisation localities of the eastern Rhodope including the basement veins (Aberdeen), basement stratiform (Baiko), Tertiary vein (Kirki etc.) and Tertiary stratiform style (Mili). They show that the Pb isotopes are largely homogeneous with high 208/204 ratios (indicating a continental rather than a mantle source) and that on the basis of Pb isotopes it is not possible to distinguish between mineralisation which occurs in the basement and that which is found in the Tertiary basins. It is important to point out that these data do not exclude the possibility of older mineralisation but the isotopes require that if this was the case then the Pb was remobilised and thoroughly mixed during the Tertiary to provide a large uniform reservoir.

5.2 Source of the Metals

It has been suggested in the preceding section that the dominant mineralisation styles are related to a sequence of Tertiary fracture systems which are themselves related to basin development. Within the Tertiary basins themselves, the dominant mineralisation is Pb and Zn with chalcopyrite being a minor component in all but the 'porphyry' style deposits. Ashworth et al. (1985) and Kalogeropoulos and Katirtzoglou (1986) have both proposed variants of a hydrothermal brine circulation model for this type of mineralisation. In these models, the altered nature of the sub-volcanic intrusives is taken to indicate that they were involved in the hydrothermal circulation and probably provided the heat source.

In contrast to the deposits hosted by Tertiary rocks, the vein and stratiform (Baiko) deposits within the basement Amphibolite-Serpentinite Unit are dominated by Cu. This suggests that the mafic nature of the host rock exerted an important influence on the bulk chemistry of the sulphide mineralisation. Billett and Nesbitt (1986) have suggested that the Cu-dominated mineralisation at Aberdeen is either the result of remobilisation of Cyprus-type Cu deposits originally present in the gabbros (= amphibolites) or more simply represents the operation of hydrothermal solutions scavenging Cu from the mafic rocks. These same authors point to the fact that the mineralisation within the Amphibolite-Serpentinite Unit is commonly found at the faulted contact of the two rock types. This suggests that the fault acted as a suitable pathway for either fluids derived from the serpentinite or for the more ubiquitous hydrothermal circulation system which was driven by the Tertiary magmatism. The homogeneous nature of the Pb isotopes support the latter model. Hence, we suggest that although the source of the Cu was local, the bulk of the evidence suggests that the fluids were circulating through a much large volume of rock.

5.3 Future Exploration

As a general exploration model we envisage a tectonic situation within the Greek Rhodope region in which a destructive plate margin existed in the N. Aegean during the early Tertiary (Papazachos and Papadopoulos 1977). Magmas resulting from this environment rose into extensional regions of the overlying crust and provided a heat source for circulating hydrothermal systems. The fluids not only scavenged metals from the igneous rocks themselves but also collected metals from the variety of rocks through which they passed. An important aspect of the model is the genetic linkage between magma genesis, basin development and the fracture system which provided pathways for the solutions. In the western Rhodope, the intersection of fractures with marble horizons allowed the hydrothermal solutions to react with the carbonates resulting in metal precipitation. In the eastern region, the solutions precipitated metals either as fracture fillings (e.g. at Kirki or Aberdeen) or dumped the metals into active sedimentation zones resulting in the Pb–Zn stratiform mineralisation (e.g. at Mili).

Irrespective of the model, it is clear that throughout the Rhodope region the bulk of the mineralisation is, in general, fracture controlled. In our view the immediate spin-off from the present EC-supported programme will be the application and extension of the fracture-pattern study of the type conducted by Sanderson et al. (1986). This work not only offers immediate target areas which should be investigated but also provides a kinematic model which can be tested by future ground and remote sensing observations.

Finally, it should be pointed out that although we have concentrated on one particularly important deposit type, we believe there is much room for work into other types. In particular, preliminary work carried out during this programme suggests that the potential for gold mineralisation is considerable.

Acknowledgements. Financial support for the project was supplied under EEC *Contracts No. MSM 133 GR and MSM 124 UK* and IGME. The authors would like to thank both the EEC and IGME for permission to publish this work. The interest shown by Dr L. Van Wambeke, Commission of the European Communities and Dr. C. Papavassiliou, IGME General Director, are much appreciated.

References

Arvanitides N, Katirtzoglou C (1985) A proposed genetic model of Tertiary polymetallic Pb–Zn–Cu–Ag sulfide mineral deposits in the Esimi area NE Greece. Fortschr Miner 63(1): 12

Ashworth KL, Billett MF, Constantinides D, Demetriades A, Katirtzoglou C, Michael C (1985) Base metal mineralization of the Evros region, N.E. Greece. Fortschr Miner 63(1): 13

Ashworth KL, Billett MF, Constantinides D, Demetriades A, Katirtzoglou C, Michael C (1986) A study of Tertiary polymetallic mineralisation and its geotectonic setting. Final Report to the EEC. Contract No MSM-124-UK

Billett MF, Nesbitt RW (1986) Base-metal mineralization associated with mafic and ultramafic rocks, eastern Rhodope massif, Greece. Trans Inst Miner Metall 95: B37–B45

Bitzios D, Constantinides D, Demades E, Demetriades A, Katirtzoglou C, Zachos S (1981) Mixed sulphide mineralization of the Greek Rhodope. Internal Report, Inst Geol Miner Expl, Athens, p 118

Bogdanov D (1982) Bulgaria. In: Dunning FW, Mykura W, Slater D (eds) Mineral deposits of Europe, vol 2: Southeast Europe. London Inst Miner Metal and Mineral Soc, pp 215–232

Constantinides D, Katirtzoglou C, Michael C, Demetriades A, Angelopoulos A, Constantinides E (1983) Metallogenic map of Evros country. Internal Report, Inst Geol Miner Expl (IGME) p 136

Deniel C, Nesbitt RW, Ashworth KL (1986) Lead isotopes and base metal mineralisation, eastern Rhodope Region, NE Greece. Final Report to the EEC (Contract MSM-124-UK)

Dimitrov DK, Zidarov N (1969) On the stratigraphy of the Archaic metamorphic complex in the Rhodope massif. Rev Bulg Geol Soc 30: Pt3

Dixon JE, Dimitriadis S (1984) Metamorphosed ophiolitic rocks from the Serbo-Macedonian Massif, near Lake Volvi, north-east Greece. In: Dixon JE, Robertson AHF (eds) The geological evolution of the eastern Mediterranean. Blackwell Scientific, Oxford, pp 603–619

Economou MI, Naldrett AJ (1984) Sulphides associated with podiform bodies of chromite at Tsangli, Eretria, Greece. Miner Dep 19: 289–297

Epitropou N, Constantinides D, Bitzios D (1983) The Marlou Pb–Zn mineralization of Thasos Island, Greece. In: Schneider HJ (ed) Mineral deposits of the Alps and of the Alpine epoch in Europe. Springer, Berlin Heidelberg New York, pp 366–374

Fytikas M, Innocenti F, Manetti P, Mazzouli R, Peccerillo A, Villari L (1984) Tertiary to Quaternary evolution of volcanism in the Aegean region. In: Dixon JE, Robertson AHF (eds) The geological evolution of the eastern Mediterranean. Blackwell Scientific, Oxford, pp 687–699

Govett GJ, Galanos DA (1974) Drainage and soil geochemical surveys in Greece: use of standardized data as an interpretative procedure. Trans Inst Miner Metall 83: B99–B111

Ivanov Z (1985) Position tectonique, structurale, géologique et évolution alpidique du Massif du Rhodope–Réunion extraord. Soc Géol Fr Bulg Guide, pp 1–31

Ivanov Z, Moskovski S, Kolceva K (1979) Basic features of the structure of the central parts of the Rhodope Massif. Geol Balc 9(1): 3–50

Jacobshagen V, Durr St, Kockel F, Kopp KO, Kowalczyk G (1978) Structure and geodynamic evolution of the Aegean region. Alps, Appennines, Hellenides, Stuttgart, pp 537–564

Kalogeropoulos S, Katirtzoglou C (1986) On the geochemical environment of sulphide ore deposition in the Esimi-Kirki Basin, Rhodope, N Greece. Final Report to the EEC. Contract MSM-133 GR

Katirtzoglou C (1986) Geology and mineralization of the area around Xylagani, Thrace, NE Greece. Final Report to the EEC. Contract No MSM-133 GR

Mack E (1983) Auriferous mineralization in northern Greece: history, exploration and evaluation. In: Scheider HJ (ed) Mineral deposits of the Alps and of the Alpine epoch in Europe. Springer, Berlin Heidelberg New York, pp 375–384

Maratos G, Andronopoulos V (1964) Strata of Melia, Alexandroupolis. Their age and position in the structure of Rhodope. Bull Geol Soc Greece 6/1

Marinos G (1982) Greece. In: Dunning FW, Mykura W, Slater D (eds) Mineral deposits of Europe, vol 2 southeast Europe, London. Inst Miner and Metal and the Mineral Soc, pp 233–253

Michael C, Demetriades A, Mastroyiannidou K, Angelopoulos A (1984) Mineral exploration in the Virini-Pessani area, Lefkimi, Evros county. Inst Geol Miner Expl, Athens, (internal report), p 68

Moorbath S, Zagorcev I (1983) Rubidium-strontium isotopic data on the age of the first granitoid complex (Smilovene and Hisar plutons) in Sastinska Sredna Gora. Geol Balc 13(3): 3–14

Mposkos E (1986) The metamorphic basement of central and eastern Rhodope. Final Report to the EEC. Contract No MSM-133 GR

Omenetto P (1983) Report on the visit to the Thassos ore deposits. Inst Geol Miner Expl, Athens, p 14

Papadopoulos P (1980) Geological map of Ferrae (1:50,000). Inst Geol Miner Expl (IGME) Athens

Papanikolaou DJ (1986) Contribution to the geology of central Rhodope: the N–S geotraverse Kavala-Sidironero. Final Report to the EEC. Contract MSM-133 GR

Papanikolaou DJ, Panagopoulos A (1981) On the structural style of southern Rhodope. Geol Balc 11: 13–22

Papavassiliou CT, Sideris C (1984) Geochemistry and mineralogy of Tertiary lavas of Sappai-Ferrai area (W Thrace) — Greece. Implications on their origin. Inst Geol Miner Expl Athens. Geochem Res Rep 4:21

Papazachos BC, Papadopoulos GA (1977) Deep tectonic and associated ore deposits in the Aegean area. In: Kallergis G (ed) Proc of the VIth Colloquium on the geology of the Aegean area, vol 3. IGME (Athens), pp 1071–1981

Robertson AHF, Dixon JE (1984) Introduction: aspects of the geological evolution on the eastern Mediterranean. In: Dixon JE, Robertson AHF (eds) The geological evolution of the eastern Mediterranean. Blackwell Scientific, Oxford, pp 1–74

Sanderson DJ, Benton M, Spence J (1986) Application of remote sensing to an investigation of the structural controls of mineralization in the eastern Rhodope Massif, Northern Greece. Final Report to the EEC. Contract No MSM-124-UK

Zachos K, Maratos G (1965) Carte metallogenique de la Grece, echelle 1:1,000,000. Athens: Institute for Geology and Subsurface Research

Zachos S, Dimadis E (1983) The geotectonic position of the Skaloti-Echinos granite and its relationship to the metamorphic formations of Greek Western and Central Rhodope. Geol Balc 13(5): 17–24

Zachos S, Dimadis E (1986) The geological evolution of the Tertiary basins of Central and Eastern Rhodope. Final Report to the EEC. Contract No MSM-12

Late Cretaceous Phosphate Stratiform Deposits of the Mons Basin (Belgium)

F. Robaszynski and M. Martin[1]

Abstract

The Late Cretaceous phosphate stratiform deposits of the Mons Basin are of economical relevance. Phosphate grains are found in the Maastrichtian "Ciply Phosphatic Chalk", which is a crumbly, brown to grey calcarenite with a P_2O_5 content of 5 to 10%. The stratified formation varies from several metres to more than 70 m in thickness and, on the southern border of the Ciply area, the overburden does not exceed 15 to 25 m. Recent drilling has provided new data on the thickness, geometrical setting, petrographical composition and reserves of the phosphate deposit.

1 Introduction: Resources and Reserves

From the end of the 19th century until World War II, the Ciply Phosphatic Chalk of the Mons area (SW Belgium) was a significant source of phosphate for Belgium. The "phosphatic sands", resulting from a natural atmospheric water leaching of phosphatic chalk as the grade attained 30–35% P_2O_5, were the first to be used. Then, phosphatic chalk with 8–10% P_2O_5 was extracted in underground galleries around Ciply and in open quarries around Saint Symphorien. From the entire phosphatic basin, more than 3 million tons were mined between 1880 and 1945. After the interruption of mining due to the economic competition from North African and US phosphate products, no quantitative information was available on the resources and reserves of the Ciply-St. Symphorien phosphatic basin.

A recent drilling campaign, undertaken by the Geological Survey of Belgium with support from the EC and the Belgian government, yielded unexpected results on the geometry and size of these deposits. The phosphatic basin has an area of about 23 km^2. The thickness of the phosphatic chalk generally exceeds 20 m and in one of the drill holes, it was found to reach 76 m; this is, to our knowledge, more or less without an equivalent in other post-Paleozoic phosphate deposits worldwide. *Resources* were evaluated at 960 million tons of phosphatic chalk at a grade

[1] Faculté Polytechnique, 9, rue de Houdain, 7000 Mons, Belgium

Mineral Deposits within the European Community
(ed. by J. Boissonnas and P. Omenetto)

comprised of between 5 and 10% P_2O_5. At the southern edge of the basin, a seemingly easy workable area, under an overburden of 10 to 25 m, has a surface estimated at about 1 km^2, which represents about 20 to 30 million tons with an average grade of 5–8% P_2O_5.

2 Geological Setting

The phosphate grains of the Mons Basin are included in a chalk matrix, chalk being the most common sediment deposited during the Late Cretaceous in the Boreal realm (Fig. 1). At that time, the Mons Basin was a dependence of the London-Paris Basin where three other areas of phosphate are preserved: Picardy (F), with more than ten mined localities, Yonne (F) with the outcrops of Saint-Martin-du-Tertre and Buckinghamshire (UK) with the locality of Taplow. In these three areas, phosphatic chalk was deposited during the Late Santonian to Early Campanian, whereas in the Mons Basin, phosphate sedimented during the Maastrichtian (Notholt and Highley 1979; Jarvis 1980).

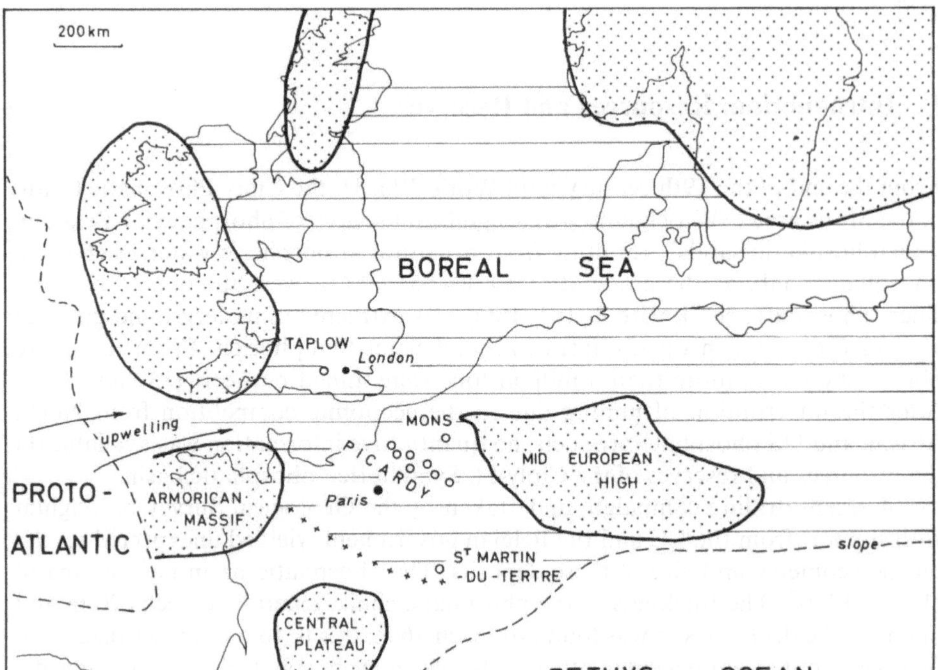

Fig. 1. Position of phosphatic chalks in the London-Paris Basin on a palaeogeographical map of the Late Cretaceous

TERTIARY	L.EOC.	YPRESIAN	"Paniselian" Sands Ieper Clays	
	PALEOCENE	LANDENIAN	Grandglise Sands	
		MONTIAN	Mons Limestone	
		DANIAN	Ciply Tuffeau *La Malogne Conglomerate*	
MESOZOIC	LATE CRETACEOUS	MAASTRICHTIAN	Saint-Symphorien Tuffeau	
			CIPLY PHOSPHATIC CHALK *Cuesmes Conglomerate*	WHITE CHALKS GROUP
			Spiennes Chalk	
		CAMPANIAN	Nouvelles Chalk Obourg Chalk Trivières Chalk	

Fig. 2. Position of the Ciply Phosphatic Chalk in the stratigraphical succession of the Mons Basin

The phosphatic formation of the Mons Basin is named "Ciply Phosphatic Chalk", grey chalk or brown chalk (cf. Fig. 2; *Craie phosphatée de Ciply* ou *Craie grise* ou *Craie brune*, Marlière in Sornay 1957, p. 96). The Ciply Chalk is preserved in two areas: the Ciply area and the Baudour area, where it generally occurs under a cover of Late Cretaceous and Tertiary-Recent sediments (Cornet and Briart 1878; Cornet 1905; Marlière 1947; Notholt and Highley 1979; Robaszynski 1984).

The Ciply Phosphatic Chalk represents a facies developed during the Maastrichtian times in a basin superimposed on the folded and eroded Westphalian Namur Basin. The two areas of Ciply and Baudour were probably linked before their separation by a very Late Cretaceous or post-Cretaceous relative uplift of the transverse Jemappes high (Fig. 3). Since there is very little literature pertaining to these areas (Cornet 1900) and recent data on the Baudour area are lacking, attention will be especially focussed on the Ciply area.

In the Ciply area, the Phosphatic Chalk occurs as a lens lying on the southern flank of a syncline and dipping to the north. Its thickness varies from several metres at the southern border to a little more than 70 m at the axis of the lens (Hyon boring) (Figs. 4 and 5). At the axis of the synclinal structure, the top of the Phosphatic Chalk is found about 100 m from the surface. The Ciply Chalk rests generally on the Spiennes white chalk and is covered by the Saint Symphorien Tuffeau (a calcarenite of Late Maastrichtian age) as shown in Fig. 2.

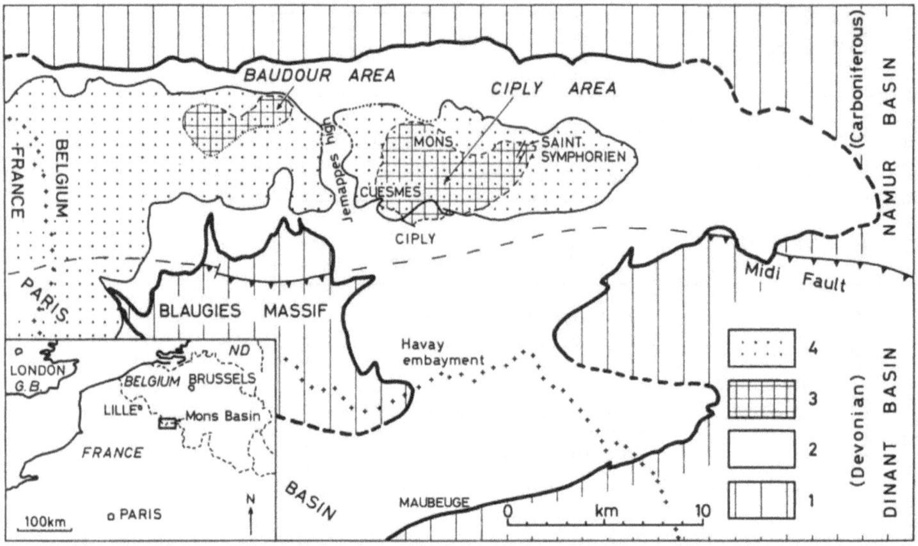

Fig. 3. Geological sketch-map of the Mons Basin and location of the Ciply and Baudour phosphatic chalk areas. *1* Paleozoic basement; *2* Cretaceous marls and white chalks; *3* Maastrichtian Ciply Phosphatic Chalk; *4* Tertiary calcarenites, sands and clays. Note: The largest part of the phosphatic strata in the Ciply and Baudour areas is overlain by Tertiary sediments. The extension of subsurface phosphate is shown under these Tertiary sediments

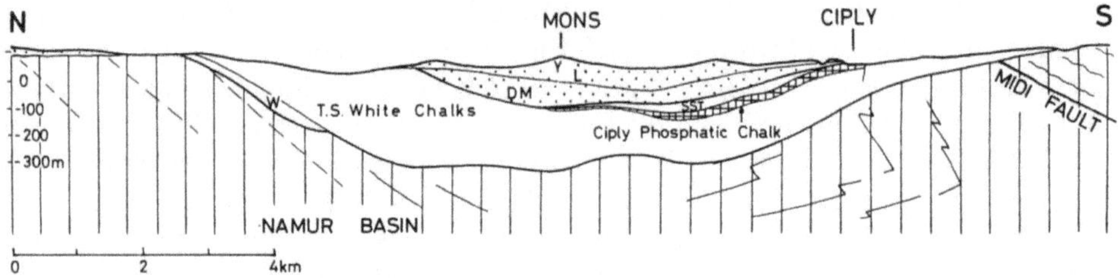

Fig. 4. North–south geological section of the Mons Basin with the general setting of the phosphatic layer (same patterns as in Fig. 3). *W* Wealden facies; *TS* Turonian-Senonian white chalks; *SST* Saint-Symphorien Tuffeau; *DM* Danian and Montian; *L, Y* Landenian, Ypresian

2.1 Sections Through the Ciply Phosphatic Chalk

Thirteen cored holes were drilled near the southern and eastern margins of the phosphatic basin and investigated by γ-ray logging using U-P_2O_5 correlations (cf. Quinif et al. 1981). The thickness of the phosphatic formation exceeds 20 m on the average but in some areas it reaches 70 m at grades of 5–10% P_2O_5, locally exceeding 14%. Details are shown in Fig. 6. The logs presented are those of seven old quarries and of the recent drill hole (Hyon) in which the maximum thickness was found. They reveal important variations in the thickness of the phosphatic formation: only a few metres to the south, with no cherts (Mortiau and Vandamme quarries), the

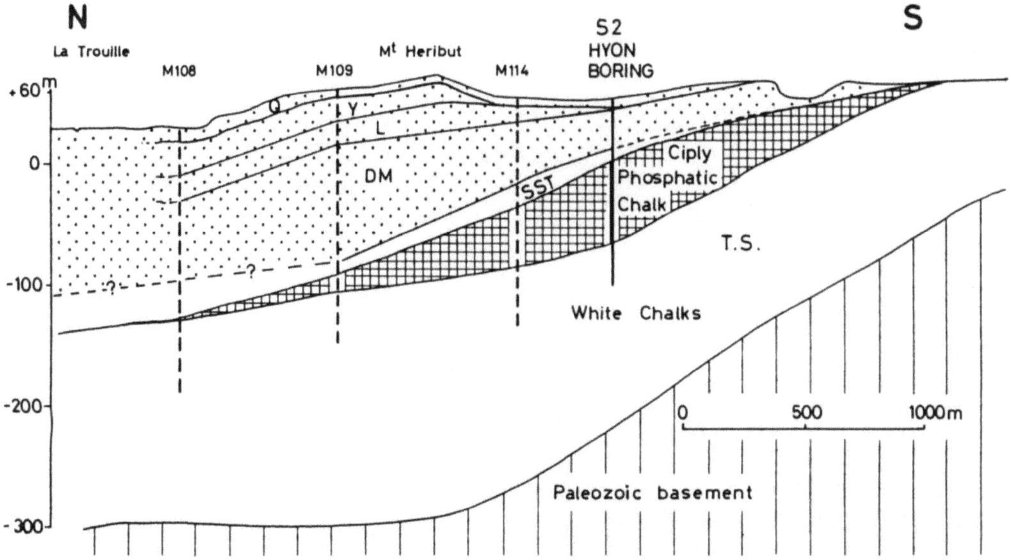

Fig. 5. Details of the southern part of the section presented in Fig. 4 based on data given by three old borings (M108 to 114) and the recent Hyon boring and showing the lenticular features of the phosphatic basin

thickness greatly increased northwards, with the presence of many chert beds near the base of the formation (Vienne and St. Symphorien quarries, Hyon borehole) (Robaszynski, in press). A detailed log of the Hyon borehole is shown in Fig. 7.

3 Characteristics of the Ciply Phosphatic Chalk

3.1 Lithology

The Ciply Chalk is a friable, light brown to light grey phosphatic calcarenite. On the border of the phosphatic basin, the base of the formation is marked by a conglomeratic bed containing phosphatic pebbles 0.5 to 1–2 cm in diameter: this is the Cuesmes Conglomerate. Still on the border, the formation is capped by a generally complex, burrowed hardground, 0.5- to 1-m-thick, which is strong enough to constitute a natural and resistant hanging wall when the crumbly phosphatic chalk was exploited by underground works (Rutot and Van Den Broeck 1886a and b). Several chert horizons are distributed in the lower part of the succession at the southern border of the phosphatic basin and are mostly represented towards the centre of the basin where the thickness of the phosphatic formation reaches several tens of metres. Abundant bioturbation traces and the presence of numerous benthic fossils such as Ostreids and Pectinids indicate a relatively shallow

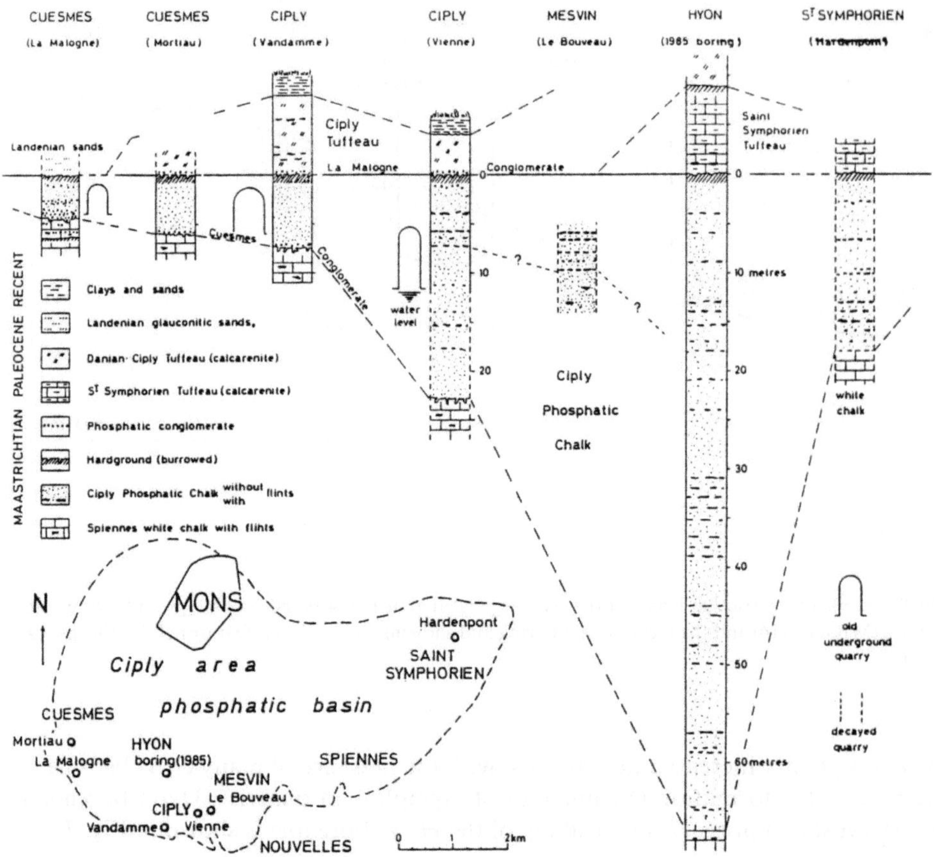

Fig. 6. Variations in thickness of the Ciply Phosphatic Chalk from six lithological outcropping sections and one boring from the southern to the eastern border of the Ciply area

marine environment which corresponds to the beginning of the large-scale Late Maastrichtian regression.

The fossil content is dominated by neritic and relatively shallow marine water species such as *Ostrea vesicularis*, *Aequipecten pulchellus*, *Catopygus fenestratus*, "*Pyrgopolon*" *mosae* and many rhynchonellids, terebratulids and benthic foraminifera such as *Osangularia*, *Rotalia*, *Bolivinoides*, *Gavelinella*, *Globorotalites*, *Stensioeina*, etc.

Some nektonic forms are represented by *Baculites vertebralis*, *Pachydiscus colligatus*, *Mosasaurus lemonnieri*, *Hainosaurus bernardi* and *Belemnella obtusa* which assign a Maastrichtian age to the formation.

3.2 Petrography

The Ciply Chalk is not truly a chalk but rather a phosphatic calcarenite with a P_2O_5 grade of around 10%. It consists mostly of particles of two size types. The first type

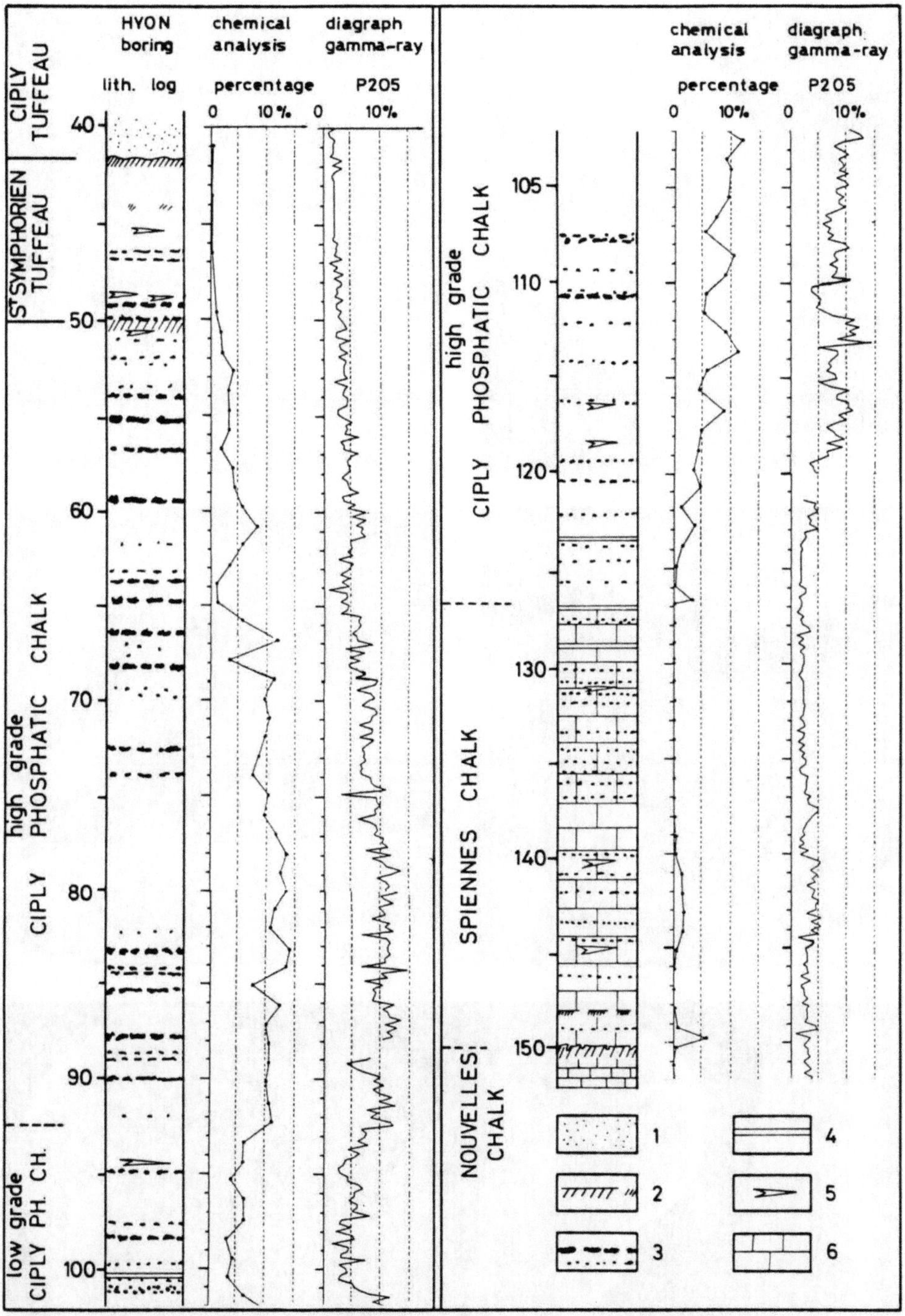

Fig. 7. Detailed lithological log of the Hyon boring between 40 to 152 m with locations of some sedimentological features. *1* Tertiary Ciply Tuffeau; *2* hardground; *3* flint band; *4* grey, slightly marly level; *5* belemnite level; *6* white chalk. The Hyon boring is located on figs. 5 and 6

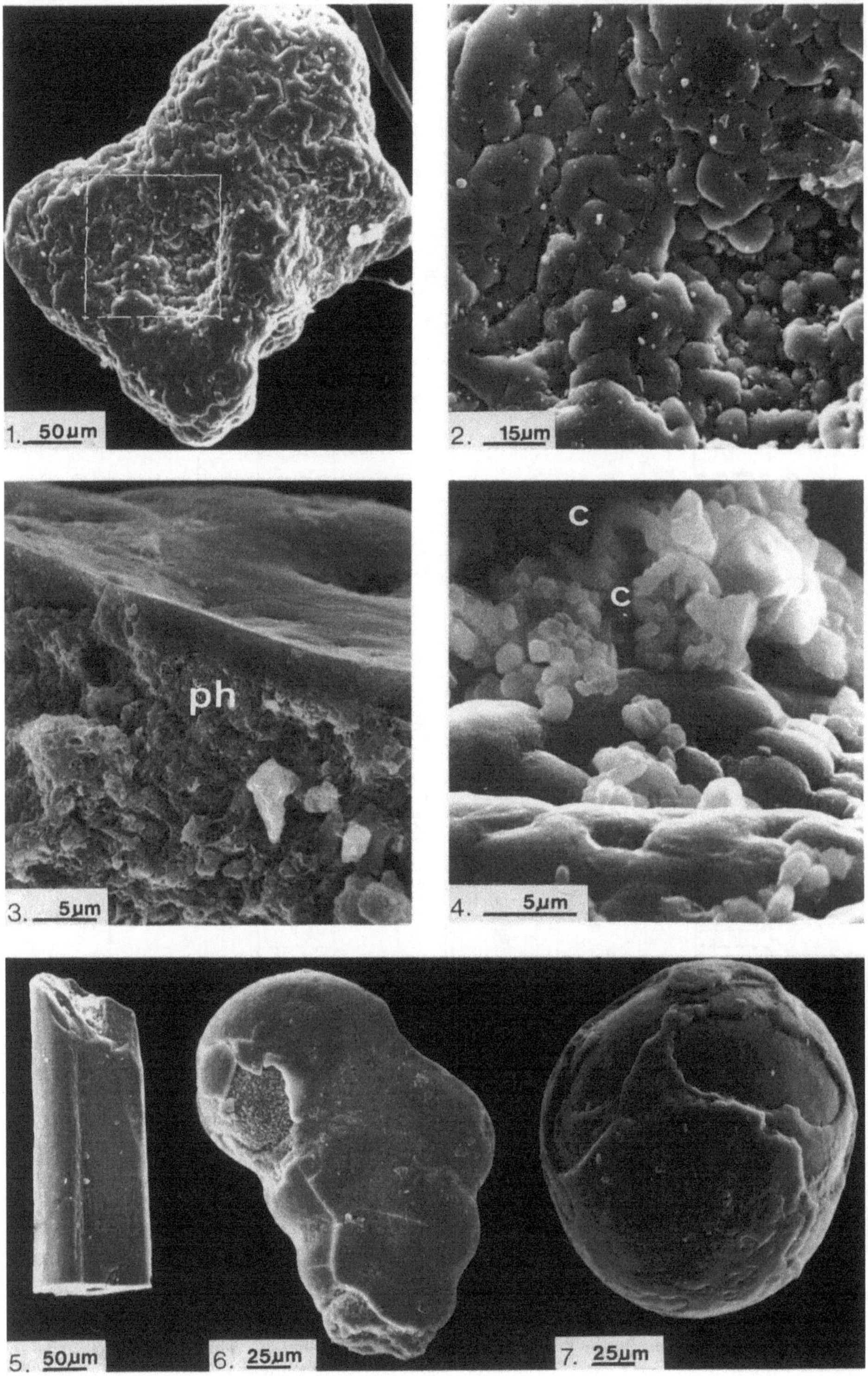

Plate 1

is represented by coarse carbonate and brown phosphate particles (about 55% in weight for the 0.100–0.270 mm size range, with a P_2O_5 content of about 14%). These particles consist mainly of microfossil fragments, rounded granules of more or less phosphatic white, yellow to brown grains, faecal pellets, bioclasts, intraclasts and small calcite crystals. The second type is represented by very fine carbonate particles such as intraclasts, bioclasts and coccoliths (about 20% in weight, size less than 0.05 mm, with a P_2O_5 content of about 1.5 to 2%).

In the first type of particles, four classes of phosphatic grains have been distinguished (Plate 1) after light microscope and scanning electron microscope (SEM) studies: brown to grey grains and white grains are the most abundant and are accompanied by some amber-coloured and spherical grains.

The *brown to grey* grains, the *white* grains and their intermediates have the porous internal structure of a chalky calcarenite, the first being the most affected by a centripetal phosphatization. Almost all are more or less coated with transparent phosphate which constitutes a brainlike cortex. Convolutions are a few micrometres in diameter, which seem to indicate a cyanobacterial activity which may have favoured precipitation of phosphate around peloid grains during the formation of

←

Plate 1. Phosphatic Grains of the Ciply Phosphatic Chalk (Mons Basin, Belgium)

1 Brown grain, from the brown to grey grain class, belonging to the 0.074–0.270 mm granulometric class. Typically a coated grain with a porous internal structure as demonstrated in Fig. 3. This kind of grain constitutes, together with the white ones and their intermediates, the most important part of the Ciply Chalk phosphatized elements. Average sample from the Vienne Quarry; 170.

2 Details from the brown grain of Fig. 1. The surface of the coated grain has brainlike features where the convolutions are a few micrometres in diameter. This aspect provides the grains with a resinous brightness under the binocular microscope; 570.

3 Grey grain, broken, from the brown to grey grain class. On the *upper part* of the picture, the cortex composed of transparent phosphate is about 3–5 μm thick. *Below*, the internal structure can be seen which is a more or less phosphatized, porous chalky calcarenite. On this grain, the brainlike surface is less marked as on other brown to grey grains; it appears smoother. In the special case of grey grains, a phyllitic structure (*ph*) with an iron content seems to be mixed with the porous chalky structure. Sample from Hardenpont old quarry; 1,800.

4 Contact between a brown grain (*below*) and chalky calcarenite matrix (*above*). The *upper part* of the picture represents the chalky calcarenite matrix composed of very fine carbonate particles such as coccoliths (*c*), small calcite cristals, etc. The *lower part* shows details of the brown grain cortex. Sample from the base of the Vienne Quarry; 2,600.

5 Amber-coloured grain. The grain is angular, transparent, amber-coloured and homogeneous and corresponds to a phosphatic skeletal element. Sample from the Vienne Quarry; 120.

6 Phosphatized microfossil. This is a very rare type of grain in the Ciply Phosphatic Chalk in comparison to Picardy Phosphatic Chalk (Paris Basin) where microfossils constitute the main part of the phosphatic elements. In Mons Basin, phosphatized microfossils are slightly phosphate-coated with smooth, cortex-like spherical grains (see Fig. 7). They belong to the white grain class because of their less phosphatic internal structure. Sample from Mesvin boring area; 260.

7 Spherical grain. This grain constitutes the lesser part of the phosphate elements. It seems to have an oolithic structure with a nucleus. Three of the numerous concentric smooth layers are well demonstrated near the *centre* of the picture. Generally, the concentric zones are strongly phosphatized, while the nucleus generally contains silica and iron. Sample from Vienne Quarry; 260.

phosphate in the Mons Basin (cf. Soudry and Champetier 1983). In the special case of grey grains, a phyllitic structure with an iron content seems to be mixed with the porous chalky structure. *The amber-coloured* grains correspond to phosphatic bone fragments: they are angular, transparent and homogeneous. The *spherical* grains seem to have an oolithic structure with a nucleus and concentric zones. In thin sections the four classes of grains can be well distinguished from the calcarenitic clasts as previously pointed out by Renard and Cornet (1891).

Phosphatized microfossils are very rare in the Ciply Phosphatic Chalk in comparison to the Picardy Phosphatic Chalk (Paris Basin) where they constitute the main part of phosphatic grains (Renard and Cornet 1891). They belong to the white grain class because of their less phosphatic internal structure.

3.3 Granulometry

As the Ciply Phosphatic Chalk is generally friable, it is fairly easy to study the distribution of grain sizes by sieving using under water techniques. Figure 8 shows a histogram of a sample collected at the base of old underground quarries at Mesvin at a stratigraphic level equivalent to the Vienne Quarry at Ciply. This sample represents significantly almost the entire Ciply Chalk. It should be noted that: (1) most of the phosphatic grains have a size between 0.073 and 0.177 mm; (2) thin particles less than 0.063 mm are abundant (about 25%) and contain very little

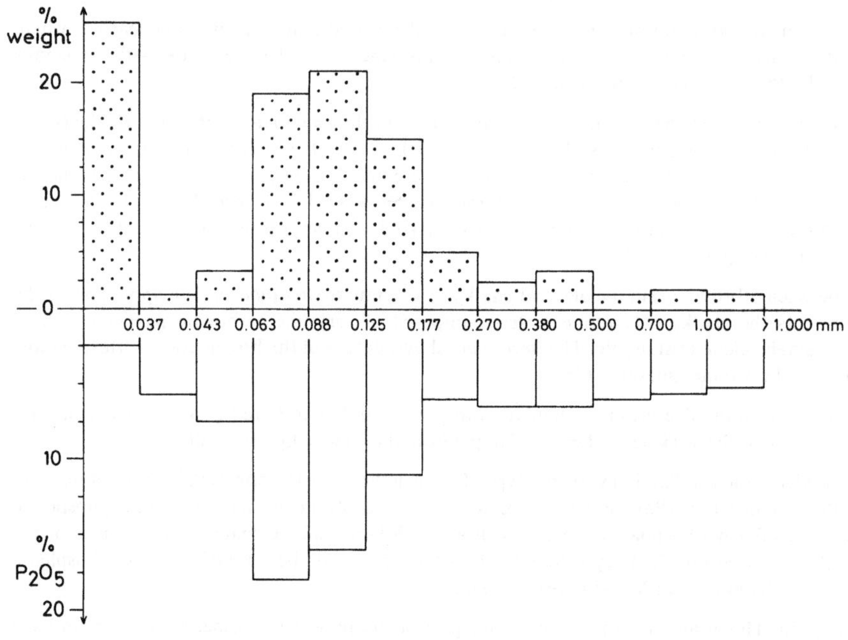

Fig. 8. Histogram of grain size classes of a Ciply Phosphatic Chalk sample from Mesvin (at the base of the section presented in Fig. 6). The P_2O_5 content of each grain size class is reported in the lower part of the diagram

phosphate. These characteristics have been verified by systematic granulometric analyses on samples from the Vienne Quarry and should be of great importance for the first phase of beneficiation. This phase consists first in a separation based on particle shape (by water sieving) and then further separation with a Wilfley table, which is a water thin film separator. This provides a product containing about 20% P_2O_5 with a recovery of 60–65%.

For example, 1000 kg bulk material at a grade of 10% P_2O_5 gives:

1. 500 kg carbonate-phosphate soil conditioner with a grade lower than 5% P_2O_5, this part can find commercial use without further preparation;
2. 200 to 300 kg mixed grains with a grade between 11 to 13% P_2O_5;
3. 200 to 300 kg preconcentrated phosphate with a grade between 20 to 23% P_2O_5.

Beneficiation. The preliminary phase for beneficiation is the separation of phosphatic grains from the calcitic ones. This is possible by using information obtained from lithology, mineralogy, petrography and grain relationships, followed by a second set of physical analyses such as density, granulometry and chemical characteristics. It was demonstrated that gravity methods are simple and useful for this first phase of "pre-beneficiation".

A second phase would assure a supplementary increase in P_2O_5 content. Since one relevant result of electron microscope and SEM petrographical studies showed that phosphatic grains are mostly calcarenite particles (more or less phosphatized) but all coated with phosphate, the aim of this phase would be to eliminate most of the calcite found in the grains. Several thermic tests such as thermal embrittlement (Champetier et al. 1980, 1984 a,b) would have to be made in the future.

3.4 Mineralogy and Chemical Composition

X-ray analysis of phosphatic particles from the Vienne Quarry and the Hyon borehole allowed the calculation of "a" and "c" lattice parameters. The a and c parameter values of 9,321 and 6,896 Å respectively, show that phosphate is mineralogically a member of the carbonate-fluorapatite or francolite group with an apatitic CO_2 content of 5.17%. Such an apatitic CO_2 content indicates a good phosphate solubility in acids (Slansky 1980).

Table 1 gives the chemical analyses of three Ciply Phosphatic Chalk samples with different P_2O_5 contents:

1. From Nouvelles, 7.84% P_2O_5;
2. From Vienne Quarry, 10.03% P_2O_5;
3. From Saint-Symphorien, in the north eastern part of the phosphatic basin, 13.45% P_2O_5.

4 Phosphatogenesis

The conditions of phosphatogenesis in the Mons Basin have not yet been completely elucidated, however, several points have been established. Some modes of the

Table 1. Chemical analysis of three samples representative of low-grade (Nouvelles), middle-grade (Ciply: Vienne Quarry) and high-grade (Saint-Symphorien) phosphatic chalks

sample from	CaO %	MgO %	MnO %	K_2O %	P_2O_5 %	SiO_2 %	Al_2O_3 %	Fe_2O_3 %	F %	CO_2 %	$H_2O + SO_3$ + O.M. %	U ppm	Σ %
Nouvelles	50.60	0.20	0.03	0.17	7.84	4.50	0.67	0.66	1.36		33.71	n.d.	99.74
Vienne Quarry	51.28	0.33	0.04	0.15	10.03	2.79	0.28	0.26	1.16	30.66	3.95	30	100.93
S^T-Symphorien	49.80	0.31	0.03	0.14	13.45	3.71	0.44	0.74	1.24		27.44	n.d.	97.30

O.M.: Organic Matter. n.d.: not determined.

phosphatic sedimentation in the Late Cretaceous chalks of the Paris Basin were previously discussed by Jarvis (1980) concerning the Santonian-Campanian phosphatic chalk of Picardy.

Morphology of Phosphate Deposits. In Picardy, phosphatic grains and their chalk matrix occur in units of 1- to 2-m thickness, forming cuvettes of one to several kilometres long including hard ground surfaces. In the Mons Basin, phosphatic chalk sedimented in basins of 4 to 7 km in diameter with a high rate of subsidence. This subsidence allowed a continuous accumulation of several tens of metres of phosphatic chalk interbedded with numerous chert layers.

Aspects of Phosphatic Grains. In Picardy, phosphatic particles are essentially formed of microfossils, generally benthic foraminifera, whereas in the Mons Basin they are mostly calcarenitic grains or peloids. In both cases, the matrix is composed of chalk.

Phosphatization is centripetal and represents a progressive impregnation of peloid grains by phosphate. A shining cortex often covers the grains. This coating results in the precipitation of phosphate, which is perhaps favoured by algal mucilage or bacteria.

Sedimentogenesis. The original grey colour of phosphatic grains (reduced iron) and the presence of pyrite, already demonstrated by Cornet (1905), suggest anoxic periods during sedimentation. Besides, the proportion of terrigenous material is very low. These conditions are close to those known for the formation of recent phosphates which require strong and persistent upwelling.

Origin of Phosphorus. As the phosphorus of recent phosphates originates from upwelling ocean waters, it is probable that the phosphorus of Cretaceous phosphates was produced in the same way. Several facts strengthen this hypothesis. The study of boreholes in the Mons Basin revealed transgressive periods during phosphate sedimentation, marked by several levels of pelagic organisms such as belemnites. Rises of sea level must have favoured the inflow of chemical elements such as phosphorus and silicium. Phosphorus was concentrated in phosphatic grains and silicium in cherts which are often abundant in the succesion. Such an hypothesis agrees with the existence of a Proto-Gulf Stream in the Cretaceous Atlantic ocean (Hart 1976) from which upwelling currents might have flowed onto a shelf extending from southern England to Picardy and to the Mons Basin either from the west or north.

5 Conclusion

The collation of previous data and new information provided by recent borings in the Mons Basin demonstrates the existence in the Ciply area of phosphate stratiform deposits of a possible economical importance as their thickness reaches several tens of metres. From a grade of about 10% P_2O_5, the physical and chemical characteristics of the Ciply Phosphatic Chalk show that phosphatic grains are concentrated mainly around 0.050 mm to 0.2 mm and this allows one to contemplate a physical technique of pre-beneficiation which can improve the grade from 20 to 23% P_2O_5 by granulometric (sieving) and gravimetric (Wilfley tables) techniques.

Acknowledgements. We would like to thank the Belgian Scientific Research Programming Office (SPPS, Service de Programmation de la Politique Scientifique), the European Economic Communities (*Contract No. MSM-079-B*) and the Belgian Geological Survey for financial support and for permission to publish some of the results obtained during the course of the program.

References

Champetier Y, Blazy P, Joussemet R, (1980) Enrichissement des phosphates carbonatés. Caractérisation gîtologique et pétrographique. Comportement au traitement thermique. Actes 2è Congr Int sur les Composés Phosphorés Boston, pp 283–312 15 fig 2 pl

Champetier Y, Gaballah I, Henin JP (1984a) Fragilisation thermique: un nouveau procédé de valorisation des facies phosphato-carbonatés indurés. Ind Min France, pp 150–156 10 fig

Champetier Y, Gaballah I, Henin JP (1984b) Thermal embrittlement: a new process in beneficiation of indurated, carbonated phosphatic facies. In: Won CP, Hausen DM, Hagni RD (eds) On applied Mineralogy in the minerals industry, Los Angeles, California, Febr. 22–25, 1984. Applied Mineralogy, pp 681–697 11 fig

Cornet FL, Briart A (1878) Sur la craie brune phosphatée de Ciply. Ann Soc géol Belg 5: M11–22 2 fig

Cornet J (1900) Etude géologique sur les gisements de phosphate de chaux de Baudour. Ann Soc Géol Belg 27: M3–32, 4 fig

Cornet J (1905) Sur les facies de la craie phosphatée de Ciply. Ann Soc Géol Belg 32: M137–146

Hart MB (1976) The mid-Cretaceous successions of Orphan Knoll (North-west Atlantic): micropalaeontology and palaeo-oceanographic implications. Can J Earth Sci 13: 1411–1421

Jarvis I (1980) The initiation of phosphatic chalk sedimentation. The Senonian (Cretaceous) of the Anglo-Paris Basin. SEPM Spec Publ 29: 167–192 18 fig

Marlière R (1947) Phosphates du Hainaut. Centenaire Ass Ing Liège Congrès, pp 330–334, figs. 3–4, p 337, fig 22

Notholt AJG, Highley DE (1979) Dossier on phosphate. Report for the commission of the European communities — DG XII — Research, science, education, raw materials R & D — XII, 1292, pp 78–234 28 fig, 32 tabl

Quinif Y, Charlet JM, Dupuis C, Robaszynski F (1981) Relations uranium–phosphate dans les craies phosphatées des Bassins de Mons et de Picardie. CR Acad Sci Paris 293(11): 913–916 1 fig, 2 tabl

Renard A, Cornet J (1891) Recherches micrographiques sur la nature et l'origine des roches phosphatées. Bull Acad R Belg 21(3, 2): 126–160, 2 pl

Robaszynski F (1984) Les gisements de phosphates. Hainaut. In: Bartholomé P et al. (eds) Métallogénie de la Belgique, des Pays-Bas et du Luxembourg. Mém Expl Carte Métallogénique Europe et Pays Limitr UNESCO Paris, pp 174–180, figs. 5–8

Robaszynski F (in press) The phosphatic Chalk of the Mons Basin. In: Sheldon RP, Notholt AJC (eds) Phosphate deposits of the world, vol 2. Cambridge University Press

Rutot A, Broeck E van den (1886a) La géologie des territoires de Spiennes, Saint-Symphorien et Havré. Ann Soc Géol Belg 13: M306–335, 7 fig

Rutot A, Broeck E van den (1886b) La géologie de Mesvin-Ciply. Ann Soc Géol Belg 13: 197–259, 16 fig

Slansky M (1980) Géologie des phosphates sédimentaires Mém BRGM 114: 90 33 fig, 19 tabl

Sornay J (ed) (1957) Lexique stratigraphique international, vol 1. Europe, Fasc 4a. VI. France, Belgique, Pays-Bas, Luxembourg — Crétacé. CNRs Paris, 403

Soudry D, Champetier Y (1983) Microbial process in the Negev phosphorites (Southern Israel). Sedimentology 30:421–423 14 fig

Geological maps of Belgium 1/25, 000e:
No. 151: Mons-Givry (1967) Explanatory text by R. Marlière, 71 p, 3 pl
No. 139: Beloeil-Baudour (1977) Explanatory text by R. Marlière, 63 p, 2 pl

Mineral Concentrations in the Recent Sediments Off Eastern Macedonia, Northern Greece: Geological and Geochemical Considerations

C. Perissoratis[1], S.A. Moorby[2], I. Angelopoulos[1], D.S. Cronan[2], C. Papavasiliou[3], N. Konispoliatis[4], F. Sakellariadou[2], and D. Mitropoulos[1]

Abstract

An extensive marine geological and geochemical research program has been carried out on the Strymonikos Plateau off eastern Macedonia, northern Greece. During the program the surface sediments were examined in order to locate possible placer deposits in the area. The plateau is covered mainly by fine-grained sediments such as silty clays and clayey silts, while sands are present along the shoreline, at the outer parts of the gulfs and landward of the shelf break. The fine-grained sediments were deposited by the rivers after the latest transgression, while the coarse sediments are the result of modern coastal erosion or represent relict and palimpsest sediments. The heavy minerals are represented mainly by amphiboles, pyroxenes, garnet, epidote and opaques (mainly pyrite), which occur along the coastal area, in the sandy sectors off the gulfs and near the shelf break. These occurrences are related either to present-day erosional products of the igneous and metamorphic formations on land, or to earlier fluvial action during lower sea level stages.

The geochemical data show that the major elements Si and Ca are related to the content of quartz and biogenic material in the sediments, while the distribution of Al and Fe show the influence of the river load in supplying aluminosilicate material to the area. The minor element distributions (Ti, Cr), however, do not always correlate with the heavy mineral content in the coarse fraction, probably because the latter is sometimes in the fine fraction of the sediment or is diluted by other phases in the sediments.

The evaluation of the data shows that there are no present-day, economically important heavy mineral occurrences in the surface sediments in this study area.

[1] Institute of Geology and Mineral Exploration, Athens, Greece
[2] Imperial College, London, UK
[3] University of Athens, Athens, Greece
[4] National Technical University of Athens, Athens, Greece

Mineral Deposits within the European Community
(ed. by J. Boissonnas and P. Omenetto)

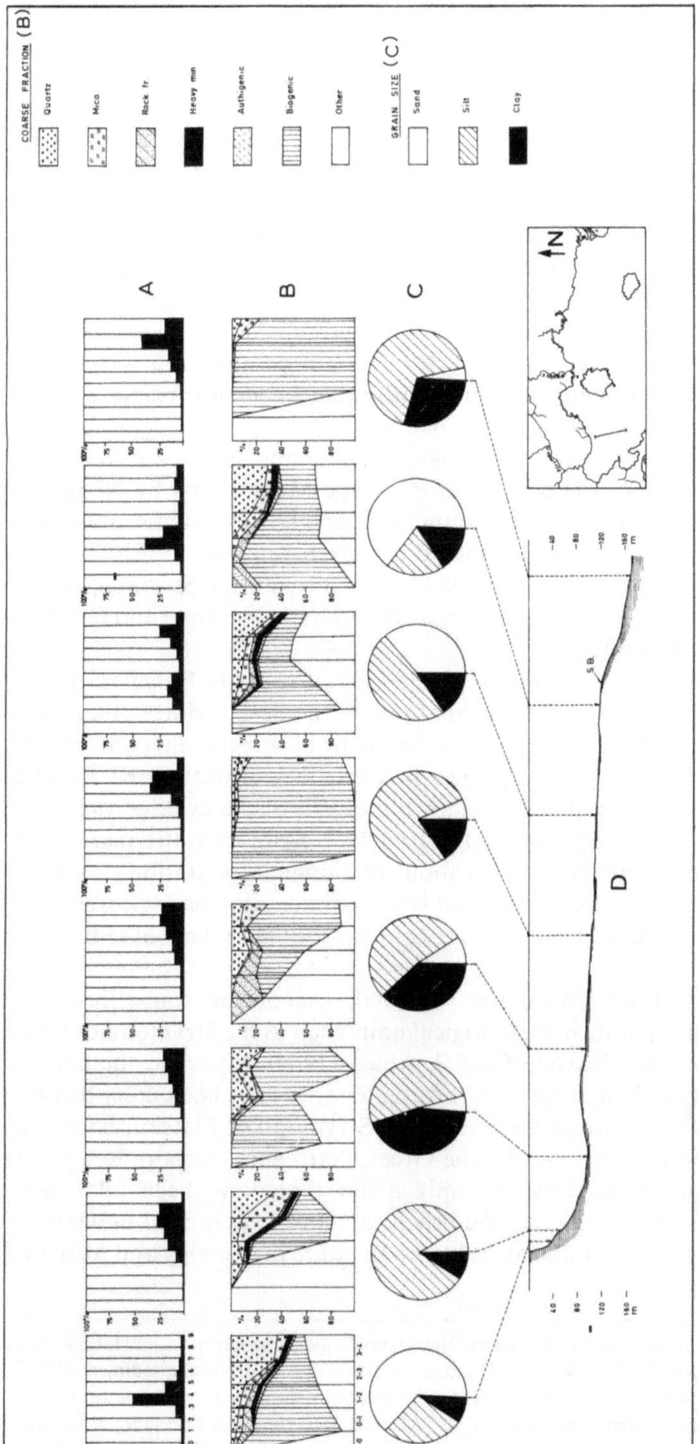

Fig. 1. Composite profile. **A** Histogram of surface sediments; **B** coarse-fraction analysis; **C** sand, silt, clay content; **D** bottom morphology (*S.B.* shelf break; *thin horizontal lines* indicate thickness of holocene sediments). Western part of index map shows Strymonikos Plateau (studied area) and location of profile. Eastern part is Samothraki Plateau

1 Introduction: Regional Geological Methods

The offshore sector of eastern Macedonia between the peninsula of Chalkidiki and Thassos Island and from the shoreline to the shelf break encompasses an area of 2200 km^2 with 240-km coastal length. This area, called Strymonikos Plateau (Figs. 1, 2A) along with the Samothraki Plateau, lying east of Thassos was surveyed by the Institute of Geology and Mineral Exploration of Greece between 1981 and 1984. The purpose of the research was to study the general geology and geochemistry of the seafloor sediments and to examine the possibility of the presence of heavy mineral concentrations. This chapter presents the results of this study in the Strymonikos Plateau area. Specifically, the texture and composition of the seafloor sediments are compared and combined with the geochemical results and an evaluation is made of the presence of placer deposits.

The surrounding land is covered by metamorphic and igneous formations, which belong to the Serbo-Macedonian and Rhodope Massifs, and by Neogene-Quaternary sediments (Fig. 2A). In the metamorphic and igneous (mafic and ultramafic) formations mineralization occurs including mixed sulphides of Pb, Zn, Cu, Fe, Au, etc. (Bornovas and Rondoyanni 1983). A large river, the Strymon, and a number of minor others drain the area running through the Neogene and Quaternary marls, clays and limestones.

The Strymon River drains an area of 10937 km^2. It starts in Bulgaria, where it drains the metamorphic and igneous formations of the Serbo-Macedonian and Rhodope massifs, while in Greece it drains the Neogene basin (graben) of Strymon. The basin narrows where the river reaches the sea, leaving little area for the formation of an extensive delta plain. As shown in Table 1, the average yield per year of the Strymon River is approximately 110 $m^3 s^{-1}$. Data from the years 1980 to 1985 indicate that the suspended load is about 18.5 ppm at the station closest to the sea. If we assume that the suspended load is 72% of the total load (Garrels and Mackenzie 1971), it is estimated that the river brings about 84000 m^3 of sediments per year to the sea.

The area has not been previously surveyed, except for some localized studies which examined separate morphological units such as the Strymonikos Gulf (Konispoliatis 1984) and the Kavalla Gulf (Lykousis 1984). However, the bathymetry, morphology and sediment types in the greater area have been described by Perissoratis et al. (1984). On the eastern part of the Strymonikos Plateau, extensive seismic surveys have been carried out by the Greek Petroleum Corporation, since this is an oil-producing area. However, only a few data have been published (Lalechos and Savoyat 1977). On land, the beach sands were examined previously with respect to their heavy mineral content by the Institute of Geology and Mineral

Fig. 2. A Map depicting land geology, bathymetry and thickness of holocene sediments. Land formations: [+ +] igneous; [VV] metamorphic; [ ] Neogene-Quaternary. Thickness of Holocene sediments: [ ] 0–5 m; [hatched] 5–10 m; [hatched] 10–15 m; [hatched] >15 m. *Lines with teeth* show faults and *line with small triangles*, position of shelf break (depth in m); PBG, Au, Mn etc: Mineralizations on land; 1: Ierissos Gulf, 2: Strymonikos Gulf, 3: Kavalla Gulf, 4: Strymonikos Plateau. B: Surface sediment distribution

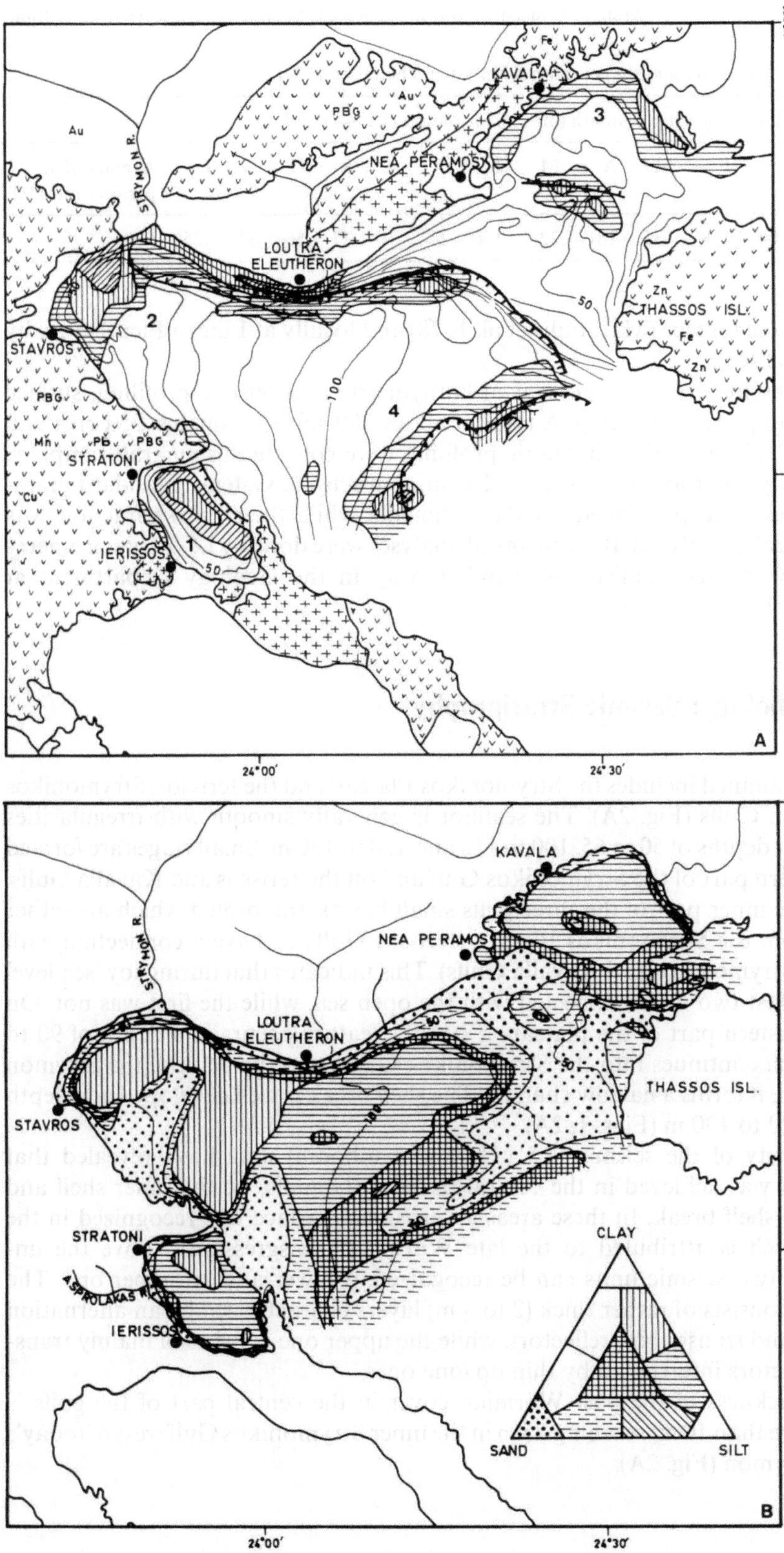
KAVALA
NEA PERAMOS
LOUTRA
ELEUTHERON
STRYMON R.
STAVROS
STRATONI
IERISSOS
THASSOS ISL.
PBG
Au
Fe
Zn
Mn
Pb
Cu
1
2
3
4
50
100
A
ASPROLAKAS R.
CLAY
SAND
SILT
B
24°00′
24°30′
41°00′
40°30′

Fig. 2A,B

Table 1. Hydrological data of the Strymon River (after Therianos 1974)

Drainage area (km^2)	Average yield per month (m^3s^{-1}) for the years 1944 to 1956												
	J	F	M	A	M	J	J	A	S	O	N	D	Average yield per year (m^3s^{-1})
10.937	104	118	135	181	228	156	65	31	27	56	83	135	110

Exploration of Greece (Markoulis et al. 1978) and locally at Loutra Eleutheron, by Papadakis (1975).

The present research consisted of bathymetric and seismic profiling, surface sediment sampling and coring. A total of about 380 surface samples, 25 cores and 2300 km of seismic and bathymetric profiling were collected using grab samplers, a 3-m-gravity corer and a 3.5 kHz and Uniboom seismic systems. Standard grain-size analyses were performed on the sediments while their coarse fraction was microscopically analyzed. Bulk chemical analyses were done by I.C.P. spectrometry in the Applied Geochemistry Research Group in the Geology Department at Imperial College, London.

2 Morphology: Seismic Stratigraphy

The area examined includes the Strymonikos Plateau and the Ierissos, Strymonikos and Kavalla Gulfs (Fig. 2A). The seafloor is generally smooth with irregularities occurring at depths of 50 to 65, 100 to 110 and 120 to 130 m. Small ridges are formed in the western part of the Strymonikos Gulf and off the Ierissos and Kavalla Gulfs. Thus, in the inner part of the three gulfs small basins are formed which are either isolated from the Strymonikos Plateau (Ierissos Gulf) or have a connection with the latter (Strymonikos and Kavalla Gulfs). This indicates that during low sea level stands the last two were connected with the open sea, while the first was not. On the northeastern part of the plateau, a steplike feature occurs at a depth of 90 to 110 m which continues into the Strymonikos Gulf up to the delta of the Strymon River, where it forms a narrow channel. The shelf break is well developed at a depth of about 120 to 130 m (Figs. 1, 2A).

The study of the seismic (3.5 kHz and Uniboom) data have revealed that penetration was achieved in the central parts of the gulfs, on the inner shelf and beyond the shelf break. In these areas an erosional surface was recognized in the profiles which is attributed to the late Würmian transgression. Above the unconformity two seismic units can be recognized, a lower and an upper one. The lower unit consists of rather thick (2 to 3 m) layers characterized by an alternation of opaque and transparent reflectors, while the upper one consists of mainly transparent reflectors interrupted by thin opaque ones.

The thickness of the post-Würmian cover in the central part of the gulfs is usually more than 10 m, reaching 25 m in the inner Strymonikos Gulf west of today's delta of Strymon (Fig. 2A).

Most of the morphological anomalies on the seafloor are of sedimentary origin. However, many normal faults were observed in the seismic profiles which have affected the uppermost sedimentary cover and seem to constitute the reactivation of older faults active during Neogene. The most prominent fault system is in the northeastern part of the Strymonikos Gulf. It has a NW-SE direction and forms a narrow channel (1 to 3 km wide). The fault continues to the SE, from south of Loutra Eleutheron to west of Thassos Island where it forms the steplike anomaly described earlier. This fault system probably constitutes the extension of the Nestos Neogene graben and is presumably the boundary of the Serbo-Macedonian and the Rhodope massifs. Other faults observed in the area are those off the Ierissos and Kavalla Gulfs which apparently uplifted parts of the Strymonikos Plateau and formed the small basins in the interior of the two gulfs. By contrast, the ridge at the Strymonikos Gulf is of sedimentary origin.

3 Sedimentation

3.1 Sediment Texture

Fine-grained sediments (silty clays, clayey silts and sand-silt-clay) predominate in the area studied and cover the central parts of the Ierissos, Strymonikos and Kavalla Gulfs as well as the central and western part of the Strymonikos Plateau (inner shelf). Coarser sediment types (sands, silty sands), on the other hand, are present in a narrow strip along the coast as well as on the outer shelf. The two areas are connected by a broad, sandy zone off the Kavalla Gulf between Nea Peramos and Thassos Island. Sand also covers the elevated areas off the Ierissos Gulf and at the western part of the Strymonikos Gulf. The histograms of the profiles (Fig. 1) show that the sands are usually medium- to fine-grained with the near-shore ones being well sorted and unimodal, whereas those on the outer shelf are poorly to very poorly sorted and/or polymodal. A better picture of the silt/clay relationship is given in Fig. 3, where particularly abundant silt occurs in the Ierissos Gulf, on the inner shelf and, to a minor extent off the northern and western coasts of Kavalla Gulf. On the other hand, clay is dominant over silt in the central Strymonikos Gulf, on the outer shelf and in the areas off Ierissos and Kavalla Gulfs.

3.2 Sediment Composition

Examination of the coarse fraction of the sediments showed that the constituents are mainly terrigenous (quartz, rock fragments, etc.) and biogenic (bioclasts and foraminiferal tests), with the occasional occurrence of authigenic material (glauconite, pyrite and iron oxides).

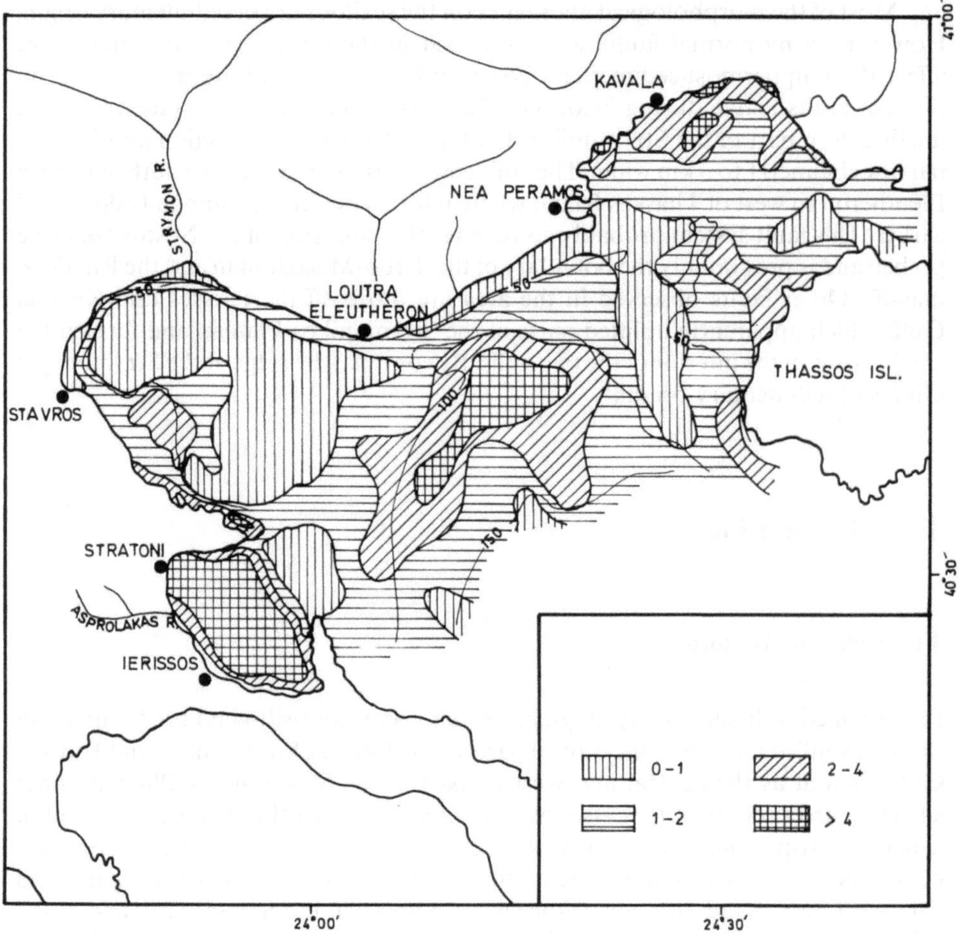

Fig. 3. Silt/clay ratio

Quartz is abundant in the nearshore sediments and on the outer shelf and diminishes in the central parts of the gulfs as well as on the inner shelf (Fig. 4). The highest concentration of quartz coincides with the areas where sand predominates, and is of angular shape in the outer shelf areas and more rounded in the nearshore ones. Rock fragment distribution shows high abundance in the same areas as quartz, but is restricted in the outer shelf where it occurs locally in its middle and eastern parts (Fig. 5). Finally, biogenic constituents, represented mainly by foraminiferal tests and to a minor degree by shell fragments, ostracodes, planktonic foraminifera etc. have high concentrations in the inner parts of the gulfs and the inner shelf, coinciding with areas where fine-grained sediments predominate. Abundant biogenic constituents also occur in sandy areas, but there the proportion of foraminifera tests diminishes, while there is an increase in the amount of worn and broken shell fragments. Carbonate distribution shows a high concentration in the central Strymonikos Gulf, the area off Kavalla Gulf and north of Thassos Island (Fig. 6).

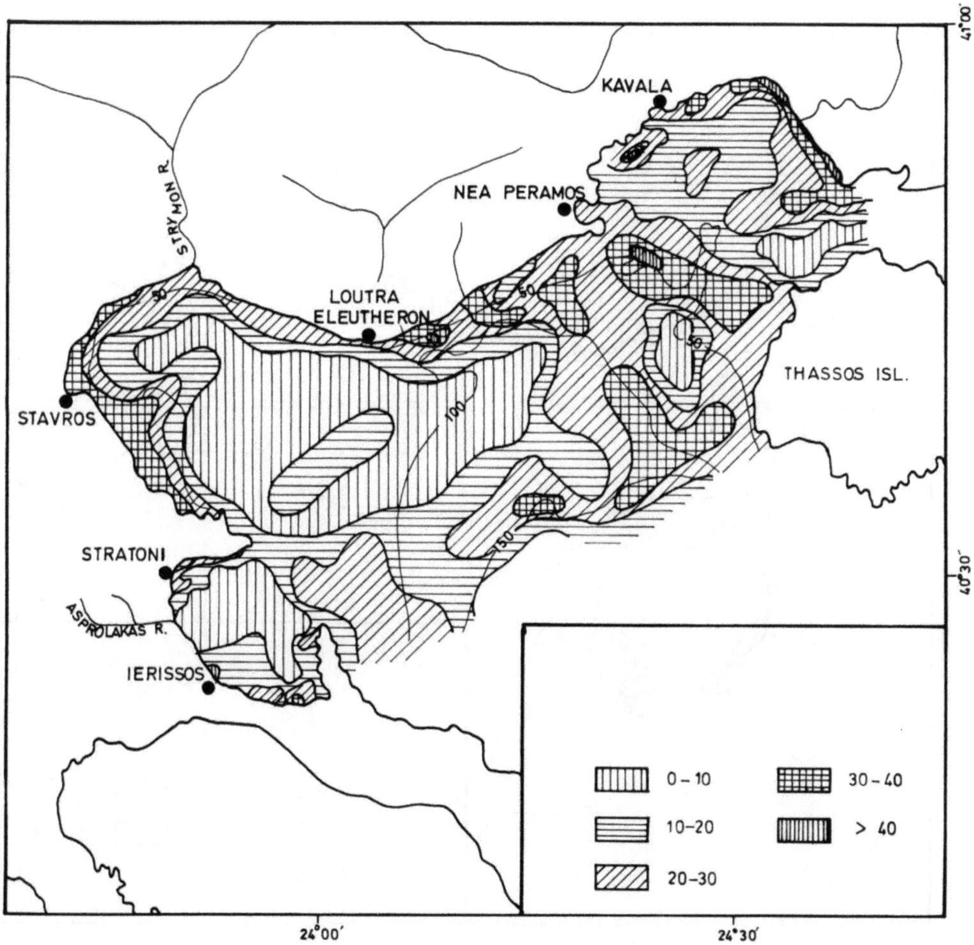

Fig. 4. Percentage of quartz (in coarse fraction)

4 Heavy Mineral Occurrences

As noted earlier, the land adjacent to the offshore area examined is heavily mineralized and, as a result, the possibility of recent placer deposits in the beach sands has been considered by other workers. Thus, Papadakis (1975) reported black sands from a short, sandy zone on the beach near Loutra Eleutheron. Although he cited no concentrations, he identified many transparent and opaque minerals such as hornblende, garnet, epidote, magnetite, ilmenite, goethite, pyrolusite, etc. Later, an extensive study aimed at locating concentrations of selected minerals (mainly zircon and rutile) was carried out in the coastal area of northern Greece by the Institute of Geology and Mineral Exploration of Greece (Markoulis et al. 1978; Mc Donald 1979). The study area also included the coasts of eastern Macedonia where, lo-

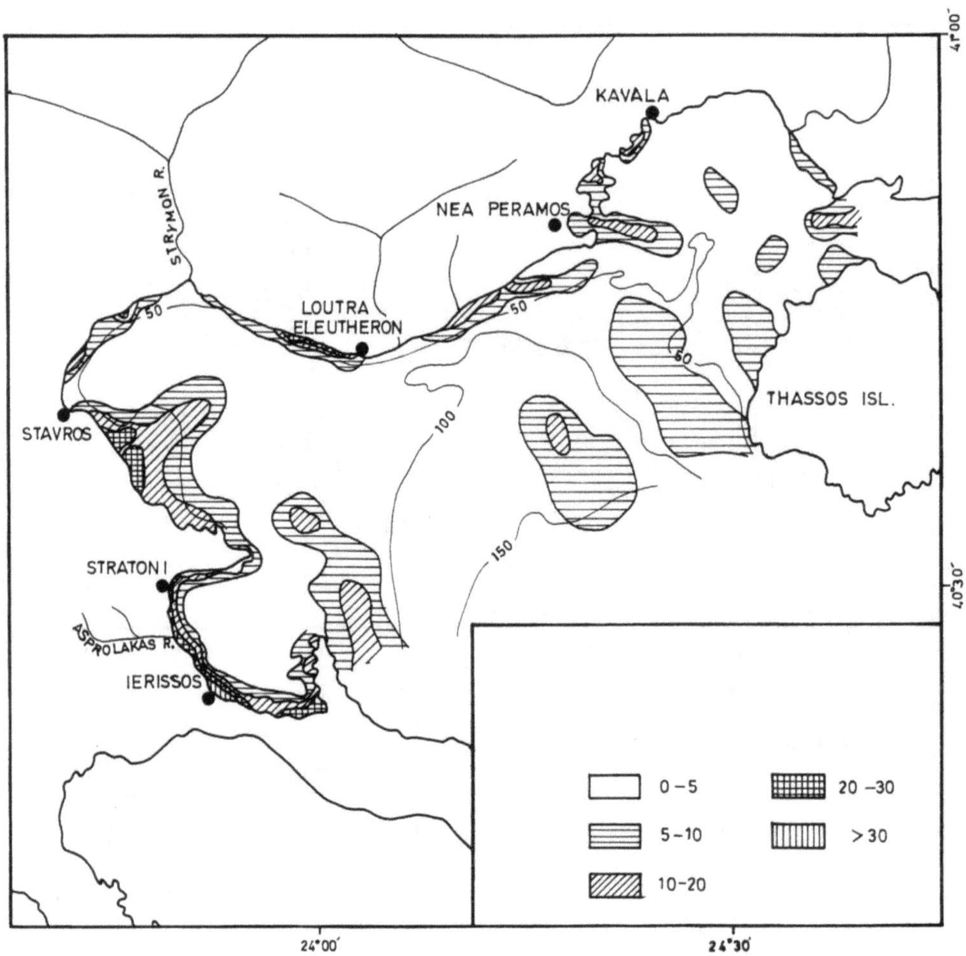

Fig. 5. Percentage of rock fragments (in coarse fraction)

cally, the heavy mineral content in the coarse fraction ranges from 3.06 to 12.71% (Table 2).

In the present study, the sand fraction of the marine sediments was examined for its heavy mineral content, as a continuation of the research on land. The main heavy minerals recognized were, according to their abundances, amphiboles, pyroxenes, epidotes, garnet, olivine, rutile, tourmaline, sphene and locally pyrite. The distribution of heavy minerals (Fig. 7) indicates that high concentrations exist off the northern and southern coasts of the Ierissos Gulf, near the shoreline around the Strymonikos Gulf, from Loutra Eleutheron to Nea Peramos and off the eastern coasts of the Kavalla Gulf. Offshore, high concentrations occur off the gulfs of Ierissos and Kavalla and at the western outer shelf near the shelf break.

A more detailed examination of the most abundant heavy minerals contained in the 2 to 3 phi sand fraction is presented in Figs. 8 and 9, which show that

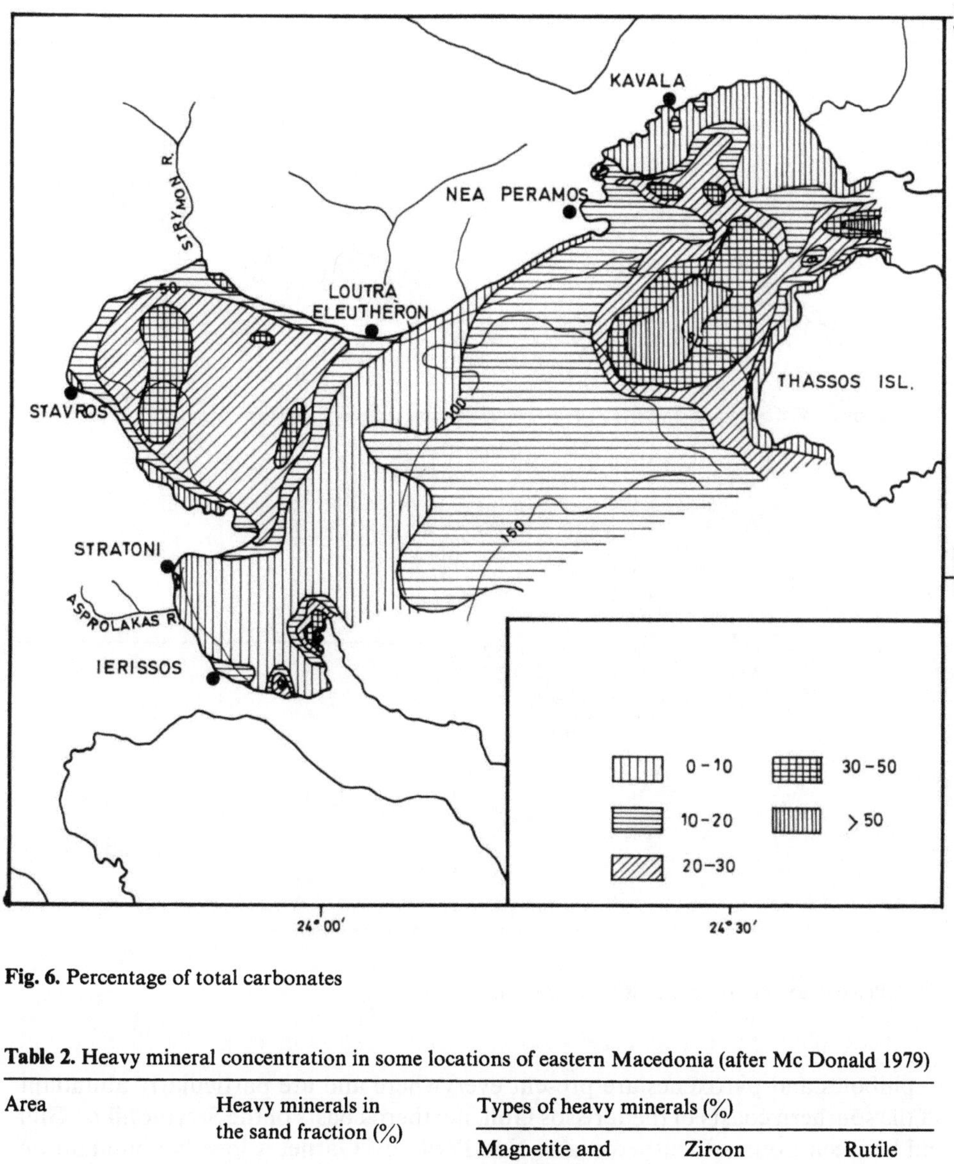

Fig. 6. Percentage of total carbonates

Table 2. Heavy mineral concentration in some locations of eastern Macedonia (after Mc Donald 1979)

Area	Heavy minerals in the sand fraction (%)	Types of heavy minerals (%)		
		Magnetite and ilmenite	Zircon	Rutile
Loutra Eleutheron	4.75–11.77	20.21–73.25	0.13–0.15	0.06–0.46
Strymon River	3.06–5.09	6.21–10.23	0.06–0.32	0.04–0.07
Stavros	7.98–12.71	2.99–6.39	0.09–0.65	0.09–0.66

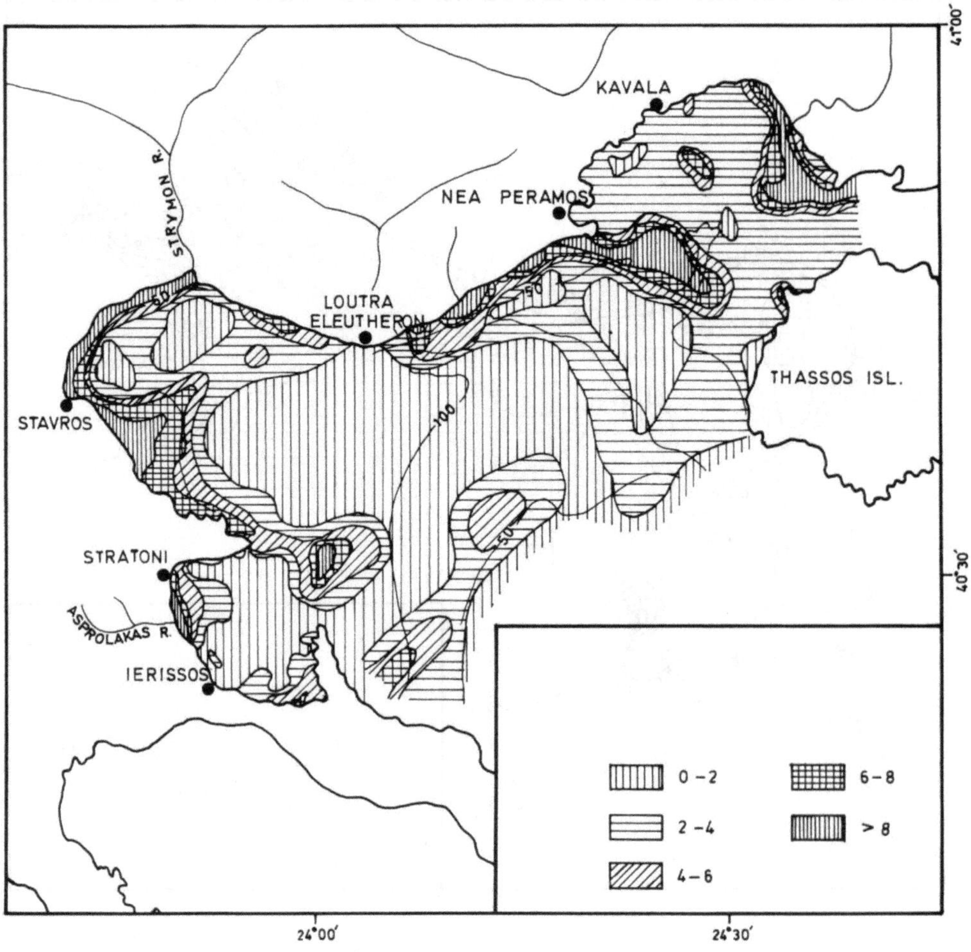

Fig. 7. Percentage of heavy minerals (in coarse fraction)

amphiboles and pyroxenes are present everywhere and are particularly abundant off the southern coasts of the Ierissos Gulf, northern coasts of the Strymonikos Gulf and between Loutra Eleutheron and Nea Peramos. Garnet is locally abundant off the southern coasts of Ierissos near the Strymon River off Nea Peramos, and off the eastern coast of the Kavalla Gulf.

Epidote is abundant between Nea Peramos and Thassos Island. Finally, pyrite is abundant locally in the sediments in the southern and northeastern parts of the Ierissos Gulf, while magnetite and ilmenite are present off the southwestern coast of the Strymonikos Gulf. The presence of all these minerals, except the opaques, was also noted in the offshore area. All sedimentary characteristics are shown in the composite profile of Fig. 1.

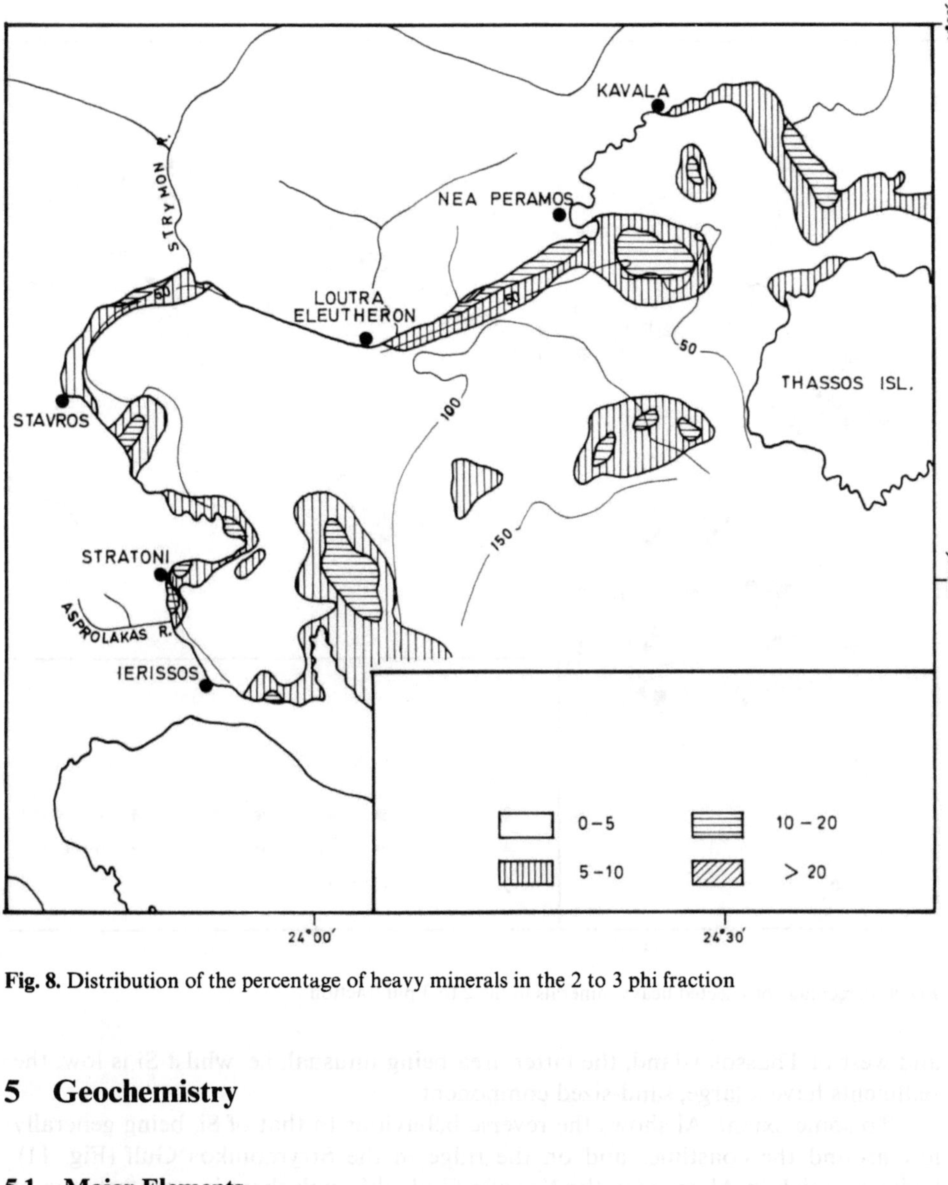

Fig. 8. Distribution of the percentage of heavy minerals in the 2 to 3 phi fraction

5 Geochemistry

5.1 Major Elements

The distribution of Si (Fig. 10) in the surface sediments reflects quite closely the percentage of sand-sized material in the sediment. The highest Si values occur in a very thin strip stretching along much of the mainland coastline, although values are generally low around the coastline of Thassos. The north-east trending ridge in the Strymonikos Gulf also shows elevated Si values but these sediments are also comparatively coarse-grained, as are the sediments in the SW part of the area near the Ierissos Gulf, which also display elevated Si values. Sediments near and to the south of Stratoni, however, do not show this trend. Si values are lowest to the north

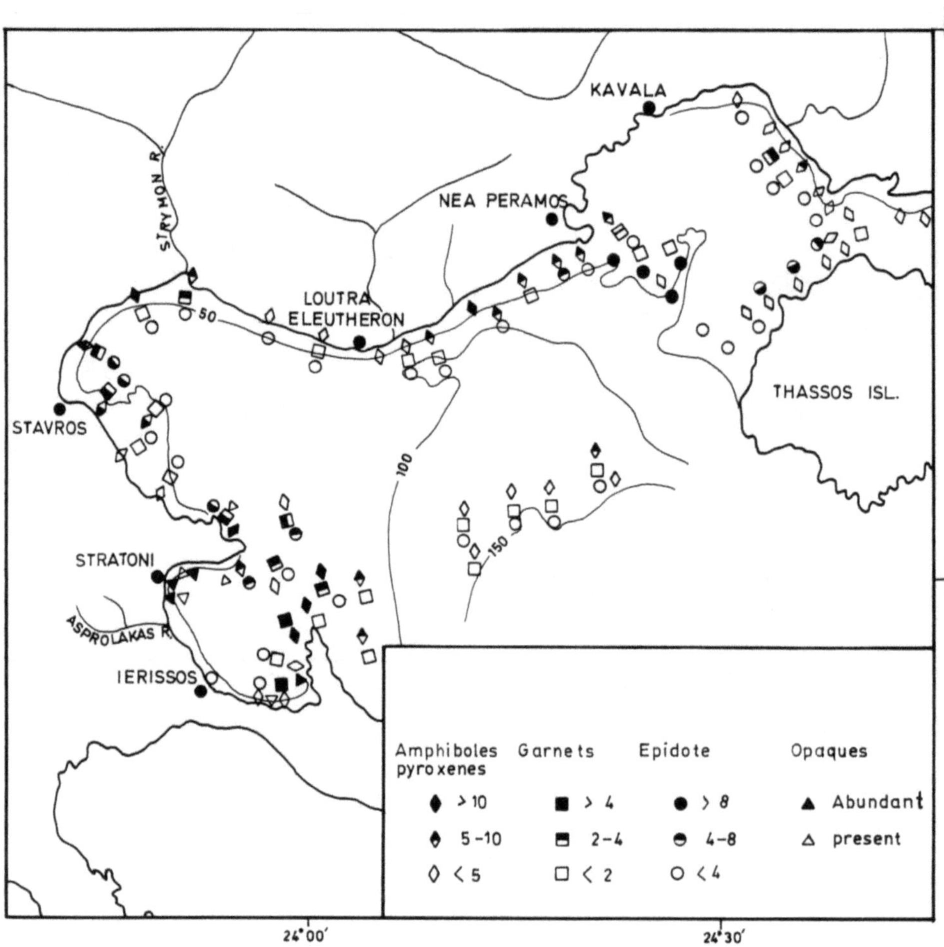

Fig. 9. Percentage of selected heavy minerals in the 2 to 3 phi fraction

and west of Thassos Island, the latter area being unusual, i.e. whilst Si is low, the sediments have a large, sand-sized component.

To some extent, Al shows the reverse behaviour to that of Si, being generally low around the coastlines and on the ridge in the Strymonikos Gulf (Fig. 11). Sediments rich in Al occur in the Kavalla Gulf, although there is a central area of lower Al content. Al-rich sediments cover much of the Strymonikos Plateau and this Al enrichment extends into the Strymonikos and Ierissos Gulfs which are largely floored by Al-rich sediment.

Calcium is generally low in surface sediments throughout much of the Ierissos Gulf, and the Strymonikos Gulf and Plateau (Fig. 12), although small patchy areas of comparatively Ca-rich sediment occur particularly on the ridge in the Strymonikos Gulf. There is a small area of Ca-enriched surface sediment just off the mainland east of Loutra Eleutheron, but the main area of Ca-enriched sediments is in a wide belt around Thassos Island. Particularly high values occur in the channel

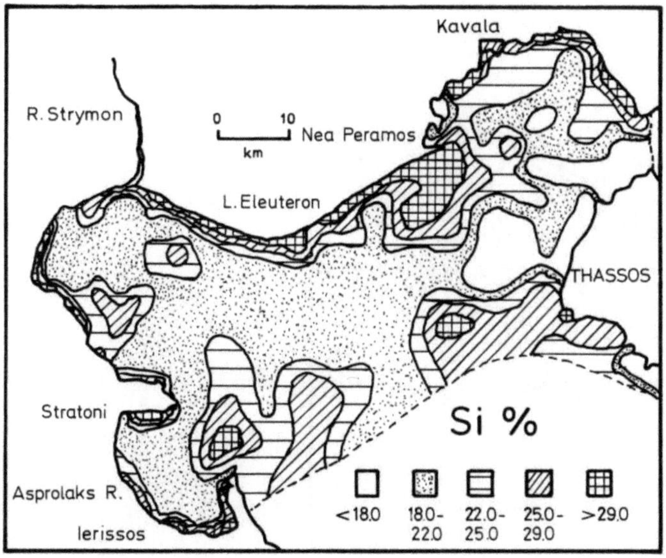

Fig. 10. Distribution percentage of Si

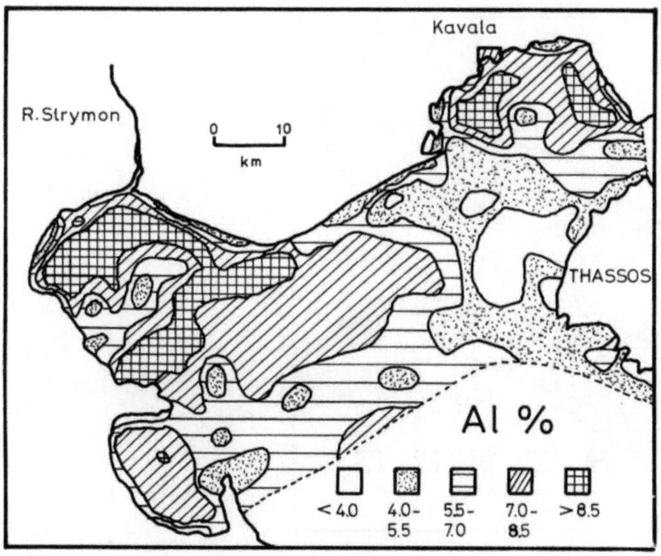

Fig. 11. Distribution percentage of Al

between Thassos and the mainland, off the southern coast and in a large area off the western coast. In the latter two cases these Ca-enriched sediments have an abundant, sand-sized component (Fig. 2B).

The distribution of Fe in the sediments (Fig. 13) is very similar to that of Al. Apart from a narrow coastal zone and the north-east trending ridge, the entire

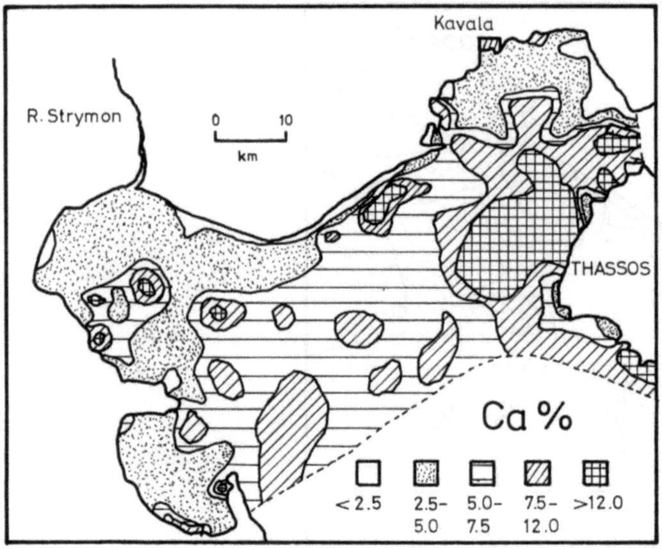

Fig. 12. Distribution percentage of Ca

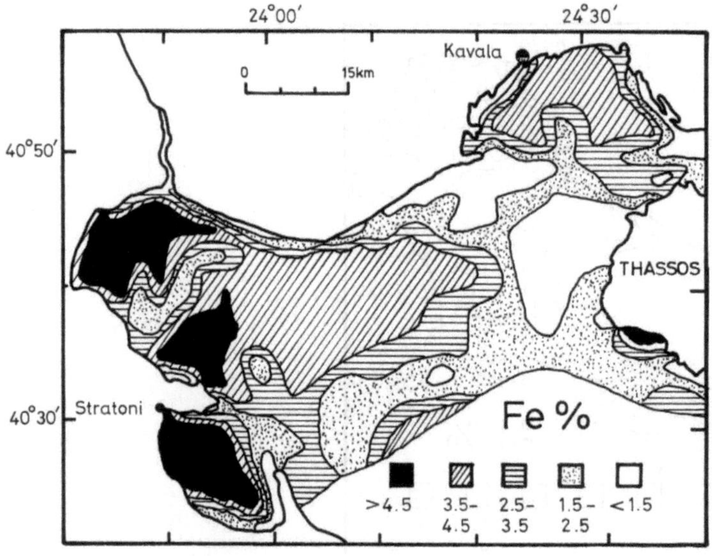

Fig. 13. Distribution percentage of Fe

Strymonikos Gulf comprises surface sediment rich in Fe, and this area of Fe enrichment extends eastwards over much of the Strymonikos Plateau. Surface sediments in the Kavalla Gulf also show Fe enrichment but not to quite the same extent as that in the Strymonikos Gulf. Iron, however (Fig. 13), shows unusual behaviour in the Ierissos Gulf, south of Ierissos town where it exhibits a pattern of distribution and concentrations similar to that seen in the Strymonikos and Kavalla Gulfs, i.e.

it closely follows Al and is highest in finer-grained sediment, reaching concentrations of about 5%. In the northern part of the gulf, however, Fe shows marked enrichment, several samples just off Stratoni containing 10–12% Fe. This area of Fe enrichment stretches southwards as far as Ierissos and southeastwards into the central part of the gulf.

5.2 Minor Elements

The distribution of Ti in the surface sediments (Fig. 14) shows many similarities to that of Fe, enrichment occurring over much of the Strymonikos Plateau and Gulf areas, with lower values on the north-east trending ridge. Ti enrichment also occurs in the Kavalla Gulf, but here, as in the Strymonikos Gulf, maximum values are less than 0.5%. A small area of Ti enrichment occurs south of the Strymonikos Gulf near the mouth of the Ierissos Gulf, where Ti reaches 0.55%, but an Fe-rich sample off the southern coast of Thassos contains the highest Ti value observed, 0.9%. However, within the Ierissos Gulf the distribution of Ti in the sediments does not closely follow that of Fe and Al, unlike its behaviour outside the gulf. The highest Ti concentrations, just over 0.5%, occur off the Asprolakkas River north of Ierissos town, and Ti concentrations decrease gradually to the north, south and east of this point.

Chromium (Fig. 15) shows many similarities to Ti in its distribution, being generally high in samples over much of the Strymonikos Plateau and Gulf, except in near-coastal samples. Chromium enrichment also occurs in the central northern part of the Kavalla Gulf, but maximum values in both areas do not exceed 220 ppm. One sample off the NW coast of Thassos contains 750 ppm but there is no evidence

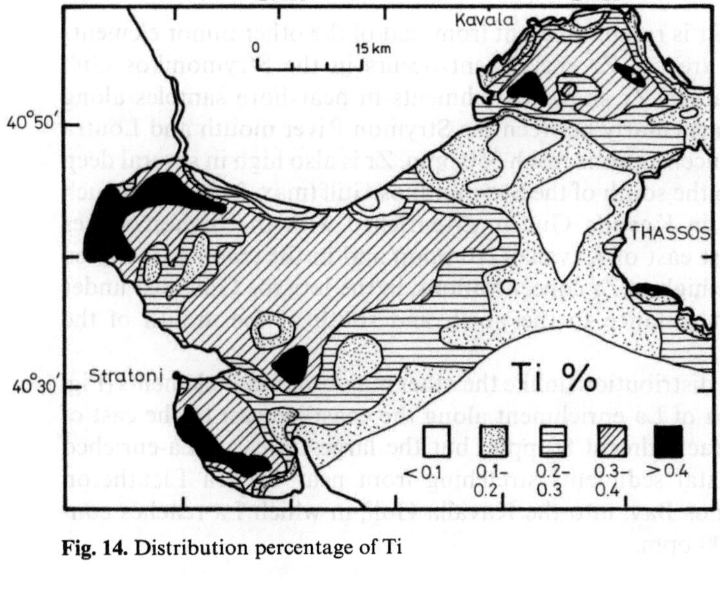

Fig. 14. Distribution percentage of Ti

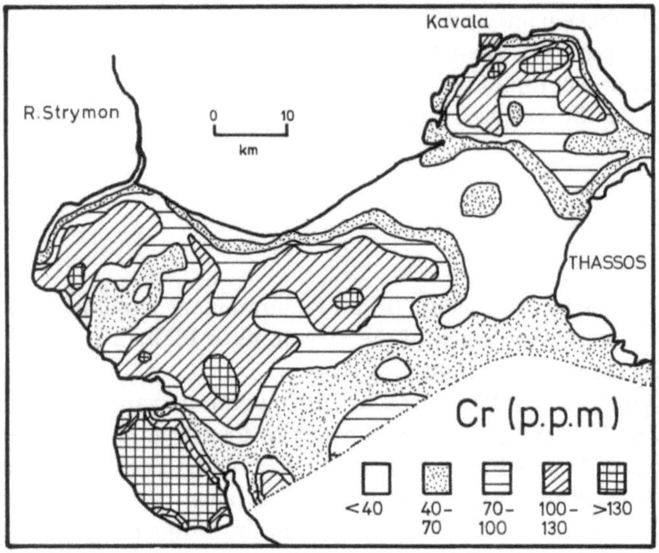

Fig. 15. Distribution percentage of Cr

of any extensive area of Cr-enriched sediment here. The behaviour of Cr in Ierissos Gulf sediments is different to the other elements considered. It is generally high throughout the area compared to concentrations seen, for example, in the Strymonikos and Kavalla Gulfs. In several nearshore samples south of Ierissos, Cr shows extreme enrichment, reaching a concentration of over 6000 ppm (0.6%) in one sample, but Cr values of over 200 ppm also occur off the Asprolakas River and in nearshore sediments to the north, indicating some river supply of Cr-rich material.

The distribution of Zr is rather different from that of the other minor elements (Fig. 16). No extensive area of Zr enrichment occurs in the Strymonikos Gulf or Plateau areas but, rather, Zr shows enrichments in nearshore samples along much of the coastline, particularly between the Strymon River mouth and Loutra Eleutheron, where Zr concentrations reach 350 ppm. Zr is also high in several deep water (100 m) samples to the south of the Strymonikos Gulf (max. 340 ppm). Much of the coastal sediment in Kavalla Gulf is Zr-enriched with the highest values occurring along the coast east of Kavalla (210 ppm) and in the small bay of Nea Peramos (250 ppm). The highest Zr concentrations, in the Ierissos Gulf, just under 200 ppm, occur a short distance to the north and south of the mouth of the Asprolakkas River.

Lanthanum shows a distribution unlike the other major or minor elements (Fig. 17). There is a small area of La enrichment along the coastline just to the east of Kavalla where values reach almost 80 ppm, but the largest area of La-enriched sediment occurs in coastal sediments stretching from near Loutra Eleutheron eastwards to Nea Peramos Bay, into the Kavalla Gulf, in which La reaches concentrations as high as 200 ppm.

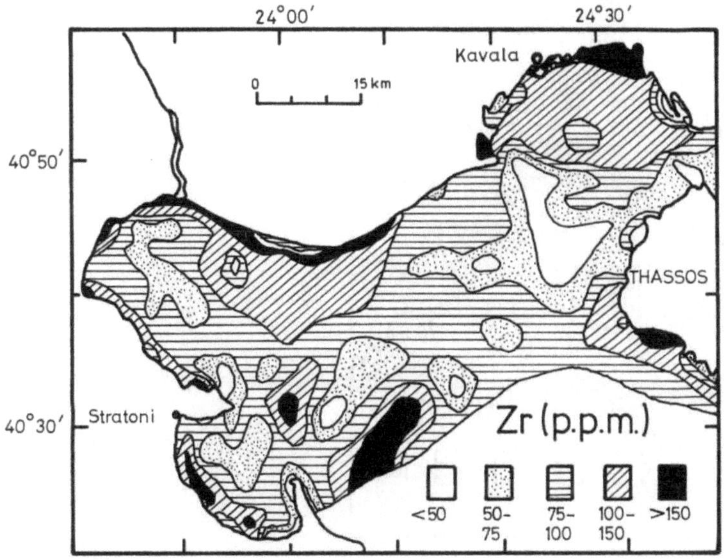

Fig. 16. Distribution percentage of Zr

6 Discussion and Conclusions

The recent sediments in the area are introduced by the rivers and by coastal erosion. The river bedload plays the most important role in sedimentation in the three gulfs, while coastal erosion seems to be the major factor in the area from Loutra Eleutheron to Nea Peramos. These latter sediments are redistributed by current action. Available data from the Hydrographic Service (pers. comm.) indicate that in the three gulfs the current movement forms a generally counterclockwise gyre. This pattern explains the accumulation of fine-grained surface sediments in the centre of the gulfs. On the Strymonikos Plateau the general direction of the water movement is from east to west close to the shore and from west to east further offshore.

Examination of the seismic data indicated that the recent sediments brought into the area after the Würmian transgression covered a considerable part of it. These sediments include the nearshore sands and the fine-grained sediments (silty clays, clayey silts, sand-silt-clay) further offshore. These sedimentary layers pinch out in the areas off the Ierissos and Kavalla Gulfs, in the western Strymonikos Gulf and on the outer shelf. There the sedimentary cover consists of coarse material (sands, silty sands), which have a bi- or polymodal character, contain rock fragments, have greater amounts of heavy minerals and contain worn and broken bioclasts. These sediments were apparently deposited under different sedimentation conditions than those of today, i.e. during earlier, lower sea level stands, although some of them have since been mixed with fine-grained material brought in by the currents. These sediments are characterized as relict or palimpsest according to Mc Manus (1975).

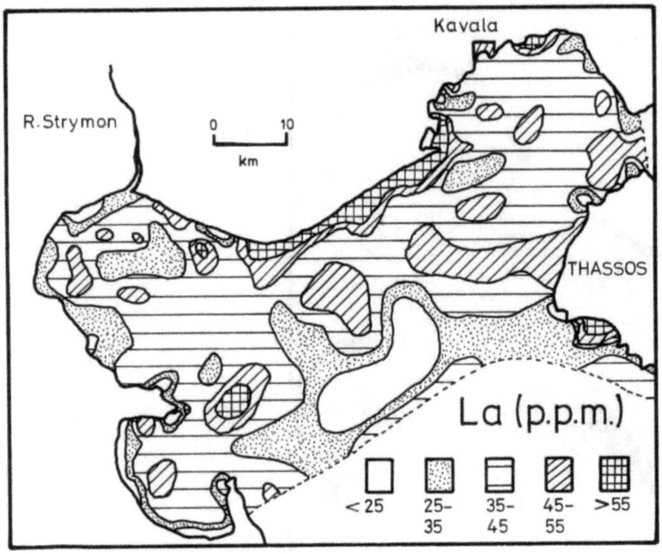

Fig. 17. Distribution percentage of La

Based on the above, the occurrences of heavy mineral concentrations in the area can be attributed either to the presence of recent erosional products of the mafic, ultramafic and metamorphic formations (nearshore sands) or to earlier depositional phases which deposited heavy minerals further offshore. Specifically, the occurrence of high amphibole and pyroxene concentrations in the nearshore sediments can be attributed to the erosion of the mafic and metamorphic formations which are rich in these minerals. The latter are supplied either by the rivers (Ierissos Gulf, Strymonikos Gulf) or are derived from coastal erosion (sector from Loutra Eleutheron to Nea Peramos).

Off the eastern coasts of the Kavalla Gulf, however, the concentrations of amphiboles, pyroxenes, garnets and epidotes have a different origin since no large river occurs there, nor do mafic or metamorphic formations outcrop on the nearby land. As shown by available seismic profiles, and data on land, the Nestos River was flowing in this area during earlier periods of the Holocene, depositing sediments rich in heavy minerals. Subsequent sorting by wave action disaggregated the abundant rock fragments into their mineral constituents, producing a heavy mineral-rich zone at a short distance from the shoreline (Fig. 7). The local concentrations of garnet off the southern coasts and pyrite off the northern coasts of the Ierissos Gulf can be attributed to local occurrences of ultramafic masses and known, mixed sulphide mineralization at these locations respectively.

In the offshore areas, heavy mineral concentrations occurs in sectors covered by relict sediments (off the Ierissos and Kavalla Gulfs, near the shelf break). This indicates that these concentrations were deposited during earlier periods when the sedimentological conditions were different from those of today. Specifically, during earlier, lower sea level stands (40 to 60 m lower than today), the uplifted areas just off the Ierissos and Kavalla Gulfs would have constituted high wave energy sectors, sorting the coarse sands deposited there. Similar conditions produced the concentrations of heavy minerals just north of the shelf break (Figs. 7, 8, 9) which was a coastal area during the latest Würmian regression when the sea level was about 120 m lower than today.

Regarding the geochemical data, the distribution of Si strongly reflects that of sand-sized material, indicating that this size fraction is probably composed largely of quartz and other silica-rich material. Its strong enrichment in the samples closest to land throughout much of the western part of the area and in the Kavalla Gulf indicates that coastal erosion may be an important supplier of material to nearshore sediments. This is reinforced by the observation that Si is low around the western coastline of Thassos which is formed largely of Si-poor marbles.

The main Ca-bearing component of the sediments is $CaCO_3$ and the distribution of Ca in the sediments, therefore, largely reflects the amounts of carbonate present, the major source of which is biogenic debris. Thus, Ca is highest in shallow-water areas where accumulation of sediment from other sources is low, such as on the ridge in the Strymonikos Gulf, on the shallow plateau area off the western coast of Thassos and in the channel to the north of Thassos where strong currents may sweep away fine-grained aluminosilicate material. Some of the high Ca values in sediments around Thassos may be due to a supply of detrital $CaCO_3$ from the marbles which form a major part of the western coastal areas of the island. The fact that most areas of Ca enrichment coincide with areas of coarse-grained sediment suggests that much of the carbonate material may consist of shell fragments.

The distributions of Al and Fe show clearly the influence of the Strymon River in supplying aluminosilicate-rich material to the Strymonikos Gulf and the importance of the west-east trending submarine canyon just off the coastline in distributing this material downslope to the deeper areas of the Strymonikos Plateau. Transport of this material further eastward is blocked by the shallow plateau area west of Thassos, but both Al and Fe distributions suggest that there is considerable southwards dispersal of aluminosilicate material, particularly in the shallower western part of the Strymonikos Plateau. No major river currently drains into the Kavalla Gulf and the Al- and Fe-rich fine-grained sediments in this gulf probably represent the results of reworking of older sediment.

In the southern and eastern parts of the Ierissos Gulf, Fe also behaves in the same way as in the Strymonikos and Kavalla Gulfs, following the distribution of Al quite closely. Just off Stratoni, however, severals samples show strong enrichment in Fe and these samples contain exceptionally high amounts of several other elements such as Mn (max. 9700 ppm) and Zn (max. 3600 ppm), suggesting that contamination occurs in sediments off Stratoni from the large mining operation based close to the sea just outside the town. That the contamination is man-made rather than natural is indicated by the fact that these metal enrichments are confined to the uppermost centimetres of sediment in cores taken close to Stratoni. The pattern of Fe distribution shows that this contaminant material is dispersed southeastwards throughout much of the gulf. A sill at the gulf's mouth, however, appears to prevent dispersion of this material beyond the gulf itself.

The distributions of Ti and Cr in the surface sediments suggest that, like Fe and Al, these elements occur predominantly in fine-grain-sized material. Their generally low concentrations and lack of coincidence with areas of heavy mineral enrichment (Fig. 7) indicate that discrete Ti- and Cr-bearing heavy minerals are either absent in the sediments of this area or are present in only comparatively small amounts and in the finer sediment fractions. There does not appear to be any clear reason for the high Cr content of the sample on the northwestern coast of Thassos, which may be due to the presence of significant amounts of Cr-bearing diopside. However, it is not known whether the marbles in this part of the island are diopside-bearing, so this possibility cannot be assessed at present. Apart from the samples discussed above, the highest concentration of Ti and Cr, and also of Zr, occurs in one sample just to the northeast of the mouth of the Ierissos Gulf. Several other samples in this area are also enriched in these elements, but to a lesser degree. These samples show some correspondence to the area of heavy mineral enrichment to the southwest (Fig. 7) which runs approximately along the 80-m isobath. Zirconium also shows enrichment in a distinct band which runs approximately along the 110-m isobath. These enrichments do not correspond to Fe or Al enrichments and, in view of their position, may be reflecting heavy mineral concentrations produced during periods of lowered sea levels. The enrichment pattern of Zr is different to that of the other elements discussed and indicates that some reworking and concentration of Zr-rich material are occurring in the immediate offshore environment throughout much of the area. Where heavy mineral data are available, however, areas of Zr enrichment do not correspond to areas of heavy mineral enrichment except in parts of the Kavalla Gulf, indicating that the Zr-bearing phases are not in the coarse fraction. The source of the Zr-rich material is not clear but may be from the reworking of beach sands.

Titanium, Cr and Zr do not appear to be present in significant amounts in the contaminant material at Ierissos Gulf, and their distribution indicates that Ti-, Cr- and Zr-rich material is supplied by the Asprolakkas River and is then dispersed both northwards and southwards fairly close to the coast, presumably by wave and current action. There is another source of Cr to the Ierissos Gulf sediments, however, which lies south of Ierissos. Here, the highly Cr-enriched nearshore sediments correspond exactly to the outcrop on land of a small ultrabasic igneous complex of ophiolitic character. Erosion of this material is the likely cause of the Cr enrichments. Dispersion from this source does not appear to be as widespread as that from the Asprolakkas River and this indicates that the material is probably derived from coastal erosion and is therefore much coarser grained than the river-supplied material.

The distribution of La is of interest. Although showing enrichment in the Ti-Cr-Zr-rich samples off the Ierissos Gulf and south of Thassos, the highest La concentrations all occur in a narrow strip of coastal sediment stretching eastwards from Loutra Eleutheron. This enrichment is interesting in that it is largely coincident with an area of heavy mineral enrichment (Fig. 7). It also appears to be closely related to the occurrence onshore of Tertiary acidic intrusive rocks (Fig. 2A) which may contain monazite, a mineral very rich in La.

The combined geological and geochemical data discussed above show that although in some areas significant amounts of heavy mineral occurrences exist, they are not considered to be of economic value, at least at present. In most cases the coarse-fraction analyses closely match the geochemical results and the observed amounts of heavy minerals mirror the respective element contents. However in at least two cases a discrepancy is observed between the two sets of data.

The first of these cases is in the northern part of Kavalla Gulf where in a low, heavy mineral content area (Fig. 7) there are high Ti, Cr and Fe contents. This area is characterized by a low sand and high silt content (Fig. 3). However, as indicated, during Upper Pleistocene-Lower Holocene lower sea level stands, the Nestos River was flowing out on the eastern coasts of the Kavalla Gulf and seismic data show that its route continued southwestward. The river mouth was at that time in the channel, formed at the middle of the Nea Peramos-Thassos ridge. Later, during the Holocene transgression, fine-grained sediments were deposited, while the Nestos River changed its flow into the Samothraki Plateau to the east of the present study area (Fig. 1 index map).

It is possible that this area, being a sector of coarse sediment deposition from the Nestos River before, had in its lower sedimentary horizons concentrations of heavy minerals which resulted in an enrichment of the overlying silty sediments in the elements Ti, Cr and Fe as a result of reworking. If this is true, then additional research is needed, by vibrocoring, in order to examine the possibility of placer deposits in the old channels and valleys of the Nestos River.

The other case, where there is a discrepancy between the geochemical and heavy mineral data, is on the northern part of the ridge from Nea Peramos to Thassos Island off the Kavalla Gulf. This is a sand-covered area where a high heavy mineral content of amphiboles, pyroxenes and epidote is correlated with low Fe, Ti and Cr values. No adequate explanation can be given with the data at hand. However, it is possible that since these sands contain large amounts of quartz, rock fragments and biogenic constituents, this resulted in a dilution of the elements concerned due to the overall increase in Ca, Si and Al.

Acknowledgements. Partial funding for this research was provided under EEC contract MSM-122-GR. We thank Dr. J. Boissonnas and Dr. L. van Wambeke, EEC scientific administrators, for their help and encouragement throughout the project.

References

Bornovas J, Rondoyanni (eds) (1983) Geological map of Greece, scale 1:500000. Institute of Geology and Mineral Exploration, Athens, Greece

Garrels MR, Mackenzie TF (1971) Evolution of sedimentary rocks. Norton, New York, 397 p

Konispoliatis N (1984) Study of recent sediments at Strymonikos bay. Doct Thesis, National Technical Univ of Athens, 109 p (in Greek)

Lalechos N, Savoyat ED (1977) La sédimentation dans le fossé Nord Egeén. VI Coll on the Geology of the Aegean Region, pp 591–609

Lykousis V (1984) Recent sedimentation at the gulf of Kavalla. Symp on Oceanography and Fisheries, pp 492–500 (in Greek)

Markoulis M, Orfanos B, Kaklamanis N, Takousis D (1978) Industrial minerals and rocks. Research for heavy minerals in sands of the coastal areas of eastern Macedonia and Thraki (Unpubl report). Inst Geol Min Exp, Athens, Greece, 75 p (in Greek)

Mc Donald EH (1979) The mineral sand deposits of Thrace and Macedonia unpubl. report on the project UN/TCD/GRE/77/007, pp 28
Mc Manus DA (1975) Modern versus relict sediments on the continental shelf. Geol Soc Am Bull 86: 1154–1160
Papadakis A (1975) The black sands from Loutra Eleutheron, near Kavalla, Greece, vol 15. Sci Ann Fac Phys Math Univ of Thessaloniki, pp 331–390.
Perissoratis C, Angelopoulos I, Mataragas D, Mitropoulos D, Konispoliatis N (1984) Bathymetry, morphology and characteristics of surface sediments from the area of Ierissos-Alexandroupolis (preliminary results). Symp on Oceanogr and Fisheries, Athens, pp 437–445 (in Greek)
Therianos AD (1974) Regime and geological distribution of the run-off of the Greek territory Bull Geol Soc Gr v.XI, p 1, pp 28–57 (in Greek).

Subject Index

MIX
Papier aus verantwortungsvollen Quellen
Paper from responsible sources
FSC® C105338

If you have any concerns about our products,
you can contact us on
ProductSafety@springernature.com

In case Publisher is established outside the EU,
the EU authorized representative is:
Springer Nature Customer Service Center GmbH
Europaplatz 3, 69115 Heidelberg, Germany

Printed by Libri Plureos GmbH
in Hamburg, Germany